The Flow of Complex Mixtures in Pipes

The Flow of Complex Mixtures in Pipes

George W. Govier, Sc. D.
Chairman, Energy Resources Conservation Board
Calgary, Alberta, Canada

Khalid Aziz, Ph. D.
Professor of Chemical Engineering
The University of Calgary
Calgary, Alberta, Canada

Second Edition

Society of Petroleum Engineers

Printed in the United States of America.

The paper used in this book meets the minimum requirements of ANSI/NSIO Z39.48-1992 (R1997). ♾

ISBN 978-1-55563-139-0

08 09 10 11 12 13 14 / 9 8 7 6 5 4 3 2 1

Society of Petroleum Engineers
222 Palisades Creek Drive
Richardson, TX 75080-2040 USA

http://store.spe.org/
service@spe.org
1.972.952.9393

To

Doris and Mussarrat

Preface

This book has been written to meet the need for an integrated presentation of the extensive and widely scattered information relating to the flow in pipes of non-Newtonian fluids and other complex mixtures. Recent years have produced many contributions of new data, new empirical correlations, and significant advances in our understanding of the mechanism and theory of the flow. The number of technical articles and review papers that have been written in the last two decades is all but staggering and well beyond what the average student or practicing engineer may hope to consult. Several excellent books, or sections in books, dealing with certain aspects of the overall topic have also appeared. They do not, however, even collectively, fulfill the need of the engineer who wants a theoretical background and an up-to-date practical understanding and working knowledge. The authors hope that this book will serve the need they see and be useful both for senior and graduate engineering students at universities and for the practicing engineer.

The basic idea of the book is to present the current state of knowledge of the flow in pipes of non-Newtonian fluids and a variety of multiphase mixtures in an orderly and unified way. The flow of these materials is related to the flow of single-phase fluids and the various complex systems are themselves interrelated.

The book aims at utility based on understanding. A real effort has been made to obtain a balanced blend of actuality and theory—of empiricism and understanding of mechanism. To the extent possible, the underlying principles and theory are developed to the point where they may be applied in the solution of practical problems. No apology is given, however, for the inclusion of empirical relationships where these are needed to meet the final objective of utility.

The book itself is an outgrowth and a considerable expansion of a graduate course of the same title given over a number of years by the first-named author at The University of Alberta and The University of Calgary. It reflects also the long-term research interests of the authors. The authors have drawn widely from the North American, European, and Asian technical literature. They have not avoided reference to their own research in the field, but where this is done it is, they believe, in proper perspective. The organization of the book and the material in the individual chapters is straightforward and we hope logical and convenient.

Two aspects of the treatment are somewhat novel. These are the systematic presentation of typical actual data and the section in most chapters on recommended design methods. The authors believe that the presentation of the actual data provides a real aid to a full understanding of the subject matter. Theory and empirical relations are important, but a feel for the subject is enhanced from a study of actual representative data. This is the reality. The discussion of recommended design methods which is included in most of the chapters is intended to serve two purposes. First, it is in a sense a summary of, or a listing of, references to the parts of the treatment of the chapter most important to the engineer in practice. Secondly, it sets forth the major steps and makes reference to the essential relations required in the normal type of design problem. When alternatives are available and there is a preferred choice among them, this also is indicated.

Chapter 1 presents an overall classification of complex mixtures, including an introductory practical treatment of the different types of fluids and a discussion of the measurement and interpretation of rheological properties. The complex mixtures considered are those of gas-liquid, immiscible liquids, gas-solid, and liquid-solid. These are classified into pseudo homogeneous and heterogeneous mixtures, the former permitting treatment as for single-phase fluids, and the latter requiring a recognition of the distinct behavior and the separating tendencies of the two phases.

The flow properties of fluids and their pressure and temperature dependence are dealt with in Chapter 2, and generalizations for their predictions are given. This treatment is included partly to serve as a refresher to an understanding of the physical properties that influence the flow behavior of fluids and mixtures and partly to aid the reader in estimating properties for which he has no measurements.

Chapter 3 reviews the fundamental concepts of the flow of single-phase Newtonian fluids, and Chapter 4 shows their application to pipe flow. These are review chapters containing a new form of presentation. They are important in setting the stage for the subsequent treatment of the complex mixtures. Chapters 5 and 6 deal with the pipe flow of non-Newtonian fluids or mixtures. In Chapter 5 the main categories of time-independent fluids are considered separately, and both the theoretical and selected empirical relationships are presented for the prediction of their pipe flow behavior. Chapter 6 extends the treatment to time-dependent fluids and presents methods applicable when the change in rheological behavior with time is important.

Chapter 7 treats the fundamental concepts applicable to the flow of all multiphase heterogeneous systems whether gas-liquid, liquid-liquid, gas-solid, or liquid-solid. Chapters 8, 9, 10, and 11 deal individually with the most important practical cases of multiphase flow in pipes. In Chapter 8 the vertical pipe flow of gas-liquid and liquid-liquid mixtures is considered in detail, the individual flow patterns each receiving separate treatment. Chapter 9 treats the related vertical flow of liquid-solid and gas-solid mixtures (vertical hydraulic and pneumatic transport). In Chapter 10 the authors consider the horizontal pipe flow of gas-liquid and liquid-liquid mixtures, again separately identifying and separately treating the various flow patterns. Chapter 11 concerns the horizontal flow of liquid-solid and gas-solid mixtures (horizontal hydraulic and pneumatic transport) and the various flow patterns encountered with them. Chapters 10 and 11 also deal briefly with the problem of flow in inclined pipes. Finally, Chapter 12 presents a comprehensive treatment of the present state of knowledge of the flow in pipes of solid capsules—a development of the last few years of The Research Council of Alberta.

The authors have drawn the line in their choice of subjects. First, the treatment is definitely restricted to steady-state, essentially isothermal, flow in circular pipes. Nonisothermal flow may be accounted for by combining the relations discussed with the appropriate heat balance and by integration or stepwise solution. Secondly, although related problems of heat and mass transfer to and in complex mixtures is not covered, the basic fluid mechanics discussed is of definite relevance. The text deals with all important aspects of the fluid mechanics of the flow of complex mixtures in circular pipes, but it is restricted to a consideration of such problems.

G. W. Govier
K. Aziz

Preface to the SPE Printing

The Flow of Complex Mixtures in Pipes was first published in 1972 by Van Nostrand Reinhold and reprinted by Krieger Publishing Company in 1977, with the last printing in 1987. For the past 10 years or so, the book has been largely unavailable except in libraries or occasionally from second-hand book sellers. Although the material in the book is more than 35 years old, the emphasis on fundamentals makes the book a useful reference even today. Many professionals in both industry and academia have used the book and continue to use it for applications ranging from flow in vertical, horizontal, and inclined pipes to the transport of solids to prop up hydraulic fractures in oil and gas reservoirs. It is at the request of such individuals that we have authorized the Society of Petroleum Engineers to reissue the book and make it generally available.

A book on a technical subject of current interest that is 35 years old would naturally be out of date in some respects. Every topic discussed in the book has been the subject of further research since the book was published, and a number of new books and many technical articles have been published on the subjects covered. We considered preparing a new edition of the book, but, with the excellent new publications on many of the topics and with the basic strength of the book being its presentation of fundamentals, we decided that a reprinting was more appropriate. One of us (K.A.) is currently active in the area of gas/liquid flow in pipes. Accordingly, we are adding a brief section summarizing the key developments in this area since the early 1970s.

As is usually the case, there were a number of printing errors in the original edition. Errata are included following "Recent Developments."

We are grateful to many friends, particularly Robert Mitchell of Halliburton, for spearheading this project with SPE. We thank Nicholas Petalas of Stanford University for help with the review of recent literature and Imraan Aziz for developing the new cover. Finally, we would like to thank SPE and particularly the SPE Books Committee for supporting this project.

We hope that the new availability of the book will be welcomed by academics and practicing engineers in their work dealing with the flow of complex mixtures in pipes.

George W. Govier
Retired

Khalid Aziz
Otto N. Miller Professor of Earth Sciences,
Stanford University

2008

Acknowledgments

The authors have had a great deal of institutional support as well as interest and assistance from a number of individuals in this undertaking. They express their appreciation and thanks to all.

A number of organizations and institutions have made contributions. These are The University of Calgary, The University of Alberta, the Research Council of Alberta, the Energy Resources Conservation Board (Alberta), the National Research Council (Canada), Imperial Oil Limited, and Marathon Oil Company. The University of Calgary has provided extensive library resources and, through the Faculty of Engineering and the University Stenographic Services, technical assistance both at the professional level and in drafting and typing. The University of Alberta, the Research Council of Alberta, the National Research Council, Imperial Oil Limited, and Marathon Oil Company all provided support for much of the early research of the authors in the flow of gas-liquid, liquid-liquid, and liquid-solid mixtures. The Energy Resources Conservation Board has suffered from some preoccupation of its Chairman with what, to the Board, is largely an extracurricular activity. It has also provided considerable drafting assistance and stenographic help.

The authors are indebted to a great many individuals. Dean R. A. Ritter of the Faculty of Engineering of The University of Calgary who, but for his appointment as Dean would have had the time to be a co-author, nonetheless gave the authors every assistance of his office. Dr. M. F. Mohtadi, Head of the Department of Chemical Engineering of The University of Calgary, also saw to it that the authors did not lack for assistance from his Department. Dr. G. A. Gregory, Assistant Professor of Chemical Engineering of The University of Calgary, gave special assistance in connection with Chapter 10. Dr. F. A. Seyer, Associate Professor of Chemical Engineering of The University of Alberta, reviewed the material in Chapters 1 and 5 on viscoelasticity and assisted the authors with these chapters. Dr. R. A. S. Brown of the Research Council of Alberta similarly reviewed the material of Chapter 8 and the authors are grateful for his useful suggestions. Mr. E. J. Wasp of Bechtel Incorporated, San Francisco, kindly reviewed Chapter 11 and offered a number of valuable comments. Mr. Jan Kruyer and Mr. H. Ellis of the Research Council of Alberta reviewed a draft of Chapter 12 and offered a number of helpful suggestions.

The authors also are grateful for the assistance provided with respect to Chapters 6 and 12, respectively, by Mr. Jimmy P. Batycky and Mr. Peter Holst, both Doctorate students in the Department of Chemical Engineering at The University of Calgary.

The authors feel a special indebtedness to the several persons who actually did the extensive drafting and typing work. They wish to thank Mr. N. Unterberger and the staff of The University of Calgary, Faculty of Engineering, Drafting and Photographic Services, and Mr. D. H. Hannah, Mr. Abe W. Sawatzky and others of the Energy Resources Conservation Board Drafting Department. They especially appreciate the meticulous and imaginative work done by Mr. Sawatzky in the preparation of the flow pattern drawings which make such a contribution in Chapters 8, 10 and 11.

The authors acknowledge gratefully the assistance and cooperation of Mrs. E. L. Wittig, Supervisor of the Stenographic Services Department of The University of Calgary. Renate Schomann (Mrs. M) of this Department handled the great burden of the typing of the difficult manuscript and did a masterful job of it. The exceptional interest which she took in the project and her dedication to it are most appreciated. Miss C. J. Esveld and Mrs. I. Constant, Secretaries of the Energy Resources Conservation Board, provided much assistance in the typing of early drafts of the material.

The authors also express their appreciation to Miss Kathy Govier who spent a good part of two summers preparing for the authors an extensive card file system and obtaining copies of more than 2000 technical papers. Lorna Patterson (Mrs. A. J.), Librarian for the Energy Resources Conservation Board, also assisted during the early stages of this endeavor.

Finally, the authors come to Maria Fogarasi (Mrs. J. C.), Research Associate at The University of Calgary Faculty of Engineering. Mrs. Fogarasi's services have been made available to the authors for a period of more than 18 months. For this the authors are indebted to the University, but for the dedication, competence, and great personal effort the authors are indebted to Mrs. Fogarasi herself. She has painstakingly reviewed every word, every equation, and every figure in the text. More than that she has contributed to all of these and especially to the preparation of the figures and the underlying calculations.

Recent Developments in the Flow of Gas/Liquid Mixtures in Pipes

The book before you presents a solid foundation for all kinds of complex flows in pipes. Two areas of particular interest to the authors that are also becoming extremely important for the oil and gas industry are the flow of two-phase (gas/oil, gas/water, or oil/water) and three-phase (oil/water/gas) mixtures in pipes. Multibillion-dollar projects involving the extraction of hydrocarbons from hostile environments like the North Sea, Gulf of Mexico, and the North Slope of Alaska are based on multiphase production through vertical, horizontal, and inclined pipes. In addition, drilling technology has advanced to where long horizontal wells several kilometers in length, with multiple laterals that result in complex tree-type networks, are becoming common. Pressure and liquid holdup profiles (steady and unsteady state) along wells are required for the optimum design and operation of such advanced wells that cost tens of millions of dollars to drill. As a result of this interest, a number of publications about the flow of complex mixtures in pipes particularly directed to the oil and gas industry has appeared over the last 35 years. It is clear from the review of the recent literature that the trend is clearly toward incorporating more and more mechanistic understanding in developing predictive models; however, no model is truly mechanistic, and some level of empiricism is necessary for closure. Here, we provide a brief introduction to key trends and relevant publications. This not a comprehensive review, but a starting point for engineers interested in the design of multiphase hydrocarbon-production systems.

Empirical and Semi-Mechanistic Approaches

A direct outcome of this book was a paper by Aziz et al. (1972), where they proposed an empirical flow-pattern map for vertical flow and mechanistic models for intermittent and bubble flows. For annular-mist flow, they used the Duns and Ros (1963) method, but with the new transition to this flow pattern as predicted by the Govier and Aziz (1972) flow-pattern map described in this book. Froth flow is treated by interpolation. To correct for the effect of fluid properties, the superficial phase velocity coordinates were modified in the flow-pattern map. Although the method was proposed for vertical flow, with simple changes to the hydrostatic head contribution, it has been also applied successfully to flow in inclined pipes.

Beggs and Brill (1973) defined three flow types (segregated, intermittent, distributed) to represent the observed flow patterns for horizontal pipes. Empirical correlations are used to determine the applicable flow pattern, and the pressure drop and holdup are evaluated accordingly. Although the calculations are performed based on a horizontal flow model, the method is said to be applicable to all pipe inclinations through the application of an inclination factor (which does not account for the flow patterns encountered in vertical/inclined flow). Pipe-roughness effects are not considered in the proposed model. The model correlations use nondimensional parameters, but the data on which the model is based are all for small-diameter pipes (1 and 1.5 in.) using air/water at standard conditions. This is one of the most widely used methods for pipe flows at all inclinations. Brown (1977) modified the Beggs and Brill (1973) method by introducing a transition region between the segregated and intermittent zones described in the original model.

In a series of papers, Mandhane et al. (1974, 1975, 1977) developed a composite approach based on a data bank compiled at the University of Calgary consisting of most of the data available in the early 1970s. Existing (empirical) correlations were tested for each flow pattern as determined from the Mandhane et al. (1974) map, and the best correlations were selected for the proposed model. This resulted in the following proposed correlations:

Flow Regime Predicted by Mandhane et al. (1974) Method	Liquid Holdup Correlation to Use	Pressure-Drop Correlation to Use
Elongated Bubble	Hughmark (1962)	Chenoweth and Martin (1956)
Stratified Smooth	Agarwal et al. (1973)	Agarwal et al. (1973)
Stratified Wavy	Chawla (1969)	Dukler et al. (1964)
Slug	Hughmark (1962)	Modified Lockhart and Martinelli (MGA 1977)
Annular, Annular-Mist	Lockhart and Martinelli (1949)	Chenoweth and Martin (1956)
Dispersed Bubble	Beggs and Brill (1973)	Modified Lockhart and Martinelli (MGA 1977)

Two modifications of the Lockhart and Martinelli (1949) correlation for pressure drop were proposed: one for slug flow, and one for dispersed bubble flow. These were based on an evaluation consisting of 4,000 data points. Although the authors proposed a correction to account for the effect of fluid properties, this effect is small, and the flow patterns are essentially functions of superficial velocities.

Mukherjee and Brill (1983, 1985a, 1985b) developed an empirical flow-pattern prediction-based model from experimental data using kerosene and lubricating oil as the liquid phases and compressed air as the gas phase over a wide range of pipe inclinations. The flow-pattern transitions defined in this model account for pipe inclination and correlations for pressure drop and holdup are included. A single empirical correlation is used to predict the holdup (with one set of coefficients for stratified flow, another set for uphill flow, and one other for downhill flow, the latter two being applied to all remaining flow patterns). Having determined the flow pattern and the holdup for the stratified region, the momentum balance equations for each phase are combined to eliminate the interfacial term and an iterative procedure is used to solve for the liquid height and, hence, the pressure drop. For the pressure drop under intermittent flow (bubble, slug), the mixture properties are used along with a "no-slip" friction factor. An entirely empirical approach is used for annular-mist flow.

Oliemans et al. (1986) proposed a mechanistic approach for annular-mist flow using the momentum balance between phases and incorporating the effects of droplet entrainment in the gas phase. Correlations were proposed for the interfacial velocity, interfacial roughness, and liquid entrainment.

Mechanistic Models

Taitel and Dukler (1976) presented the first truly mechanistic approach for horizontal flow-pattern determination. Although focus is on flow-pattern transitions, the mechanistic nature of the proposed method for the stratified flow regime directly leads to expressions for both the holdup and pressure gradient in this region. To simplify the solution of the momentum balance equations, certain assumptions are made (i.e., the interfacial friction factor in stratified flow is approximately equal to the gas/wall friction factor, and pipe-roughness effects are neglected).

Barnea (1987) extended the Taitel and Dukler (1976) approach to develop a comprehensive model for determining flow-pattern transitions over the entire range of pipe inclinations. The transition mechanisms are based on the works of Barnea et al. (1982a, 1982b), Barnea et al. (1985), Barnea and Brauner (1985), and Barnea (1986). The transition mechanisms proposed previously are summarized, and a systematic approach to determining the flow pattern is presented. Of note are the following assumptions made in this model:

- Pipe roughness is neglected in the calculation of friction factors.
- For stratified flow, the interfacial friction factor is assumed to be the same as the single-phase gas friction factors used in determining the wall/gas shear stress. The wall/liquid shear stress is also based on the single-phase-flow friction factor.
- For annular-mist flow, liquid entrainment in the gas core is neglected. The interfacial friction factor used is the one proposed by Wallis (1969).

Another model developed by Xiao et al. (1990) is similar to the Taitel and Dukler (1976) model, the main difference being the use of pipe roughness in the calculation of friction factors and the inclusion of pressure-drop and holdup calculations. The model is applicable to horizontal and near-horizontal pipes within a range of –15° to +15° pipe inclination. The flow patterns considered include stratified, intermittent, annular-mist, and dispersed bubble. The authors tested the model against 426 measurements from a variety of gas and oil pipelines with pipe diameters ranging from 2 to 26 in. One curious aspect of this model is the assumption of a flat interfacial surface and a uniform liquid height in the elongated bubble zone of the slug unit. In doing so, the authors propose using a stratified flow model in this region that can lead to significant discontinuities.

Petalas and Aziz (2000) presented a mechanistic model applicable to flow in horizontal and inclined pipes and a broad range of fluid properties. Empirical correlations are proposed for liquid/wall and liquid/gas interfacial friction in stratified flow, for the liquid fraction entrained and the interfacial friction in annular-mist flow, and for the distribution coefficient used in the determination of holdup in intermittent flow. Roughness effects (as well as liquid entrainment for annular-mist flow) are accounted for. The model was tested with data from 6,000 experiments representing a wide range of fluid properties and pipe inclinations. One of the key characteristics addressed by this model is the reduction of discontinuities at the flow-pattern transitions.

Drift-Flux Models

In modeling the interaction between wells and hydrocarbon reservoirs, it is necessary to use fully implicit coupled numerical models. For the practical application of such models, it is necessary that the flow in the wellbore be represented by a relatively simple model that is continuous and differentiable. Mechanistic and empirical models generally do not satisfy these conditions. Hence, the simple drift-flux model originally proposed by Zuber and Findlay (1965) for vertical bubble flow has been extended to handle two- and three-phase flows over all flow patterns and at all angles, including countercurrent flow. Oddie et al. (2003) present experimental data that form the basis of new models for two-phase flow developed by Shi et al. (2005a) and for three-phase flow by Shi et al. (2005b). The drift-flux model has two parameters: the profile parameter and the drift velocity. By correlating these parameters with the gas void fraction and by taking advantage of the limiting conditions when bubble flow exists at low in-situ gas fraction and annular-mist flow exists at high in-situ gas fraction, and by interpolating between these limits, the simple drift-flux model can be applied to all flow regimes. The flow of three phases is handled in two stages: first, the model is applied by considering the oil and water phases as a single liquid phase to calculate the in-situ gas fraction; second, the slip between two liquid phases is considered. In three-phase flow, gas entrained in the liquid has a significant influence on the slip between liquid phases. A model to handle this is also proposed.

Selected Recent Books

An SPE monograph by Brill and Mukherjee (1999) provided a comprehensive review of pressure-drop, flow-pattern, and holdup calculation procedures for flow through restrictions. The book also compares different calculation procedures, including many of the techniques developed by the authors, through examples.

Hetsroni (1982) edited a *Handbook of Multiphase Systems* that includes most of the topics discussed in this book. The *Handbook* is essentially a collection of papers by leading researchers with extensive references to previous work.

Cheremisinoff (1986) edited a six-volume *Encyclopedia of Fluid Mechanics* that includes volumes on gas/liquid, solid/gas, and slurry transport with contributions from many leading researchers in these fields.

Shoham (2006) provided a comprehensive treatment of mechanistic models for gas/liquid flow in pipes, including many of the developments by the author and his colleagues.

Hasan and Kabir (2002) presented the first comprehensive treatment of single and multiphase flow in wells with heat transfer. These two authors have made many significant contributions to the understanding and modeling of multiphase wellbore flows. References to their publications are contained in this book and in the papers on drift-flux modeling discussed here.

References

References mentioned above that are not in the individual chapter bibliographies are listed below.

Aziz, K., Govier, G.W., and Fogarasi, M. 1972. Pressure Drop in Wells Producing Oil and Gas. *J. Cdn. Pet. Tech.* **11** (3): 34-48.

Barnea, D., Shoham, O., and Taitel, Y. 1982a. Flow Pattern Transition for Downward Inclined Two Phase Flow; Horizontal to Vertical. *Chem. Eng. Sci.* **37** (5): 735–740.

Barnea, D., Shoham, O., and Taitel, Y. 1982b. Flow Pattern Transition for Vertical Downward Two Phase Flow. *Chem. Eng. Sci.* **37** (5): 741–744.

Barnea, D. and Brauner, N. 1985. Holdup of the Liquid Slug in Two Phase Intermittent Flow. *Intl. J. Multiphase Flow* **11** (1): 43–49.

Barnea, D., Shoham, O., Taitel, Y., and Dukler, A.E. 1985. Gas-Liquid Flow in Inclined Tubes: Flow Pattern Transitions for Upward Flow. *Chem. Eng. Sci.* **40** (1): 131–136.

Barnea, D. 1986. Transition from Annular Flow and from Dispersed Bubble Flow—Unified Models for the Whole Range of Pipe Inclinations. *Intl. J. Multiphase Flow* **12** (5): 733–744.

Barnea, D. 1987. A Unified Model for Predicting Flow-Pattern Transitions for the Whole Range of Pipe Inclinations. *Intl. J. Multiphase Flow* **13** (1): 1–12.

Beggs, H.D. and Brill, J.P. 1973. A Study of Two Phase Flow in Inclined Pipes. *JPT* **25** (5): 607–617; *Trans.,* AIME, **255.** SPE-4007-PA.

Brill, J.P. and Mukherjee, H. 1999. *Multiphase Flow in Wells*. Monograph Series **17,** SPE, Richardson, Texas.

Brown, K.E., 1977. *The Technology of Artificial Lift Methods* **1:** 134–138. Tulsa: Petroleum Publishing Co.

Cheremisinoff, N.P. (ed.). 1986. *Encyclopedia of Fluid Mechanics* (six volumes). Houston: Gulf Publishing Co.

Dukler, A.E., Wicks, M., and Cleveland, R.G. 1964. Frictional Pressure Drop in Two Phase Flow: A. A Comparison of Existing Correlations for Pressure Loss and Holdup. *AIChE J.* **10 (1):** 38–43.

Hasan, A.R. and Kabir, C.S. 2002. *Fluid Flow and Heat Transfer in Wellbores*. Richardson, Texas: SPE.

Hetsroni, G. (ed.). 1982. *Handbook of Multiphase Systems*. New York City: Hemisphere Publishing Corp.

Mandhane, J.M., Gregory, G.A., and Aziz, K. 1974. A Flow Pattern Map for Gas-Liquid Flow in Horizontal Pipes. *Intl. J. Multiphase Flow* **1:** 537–553.

Mandhane, J.M., Gregory, G.A., and Aziz, K. 1975. Critical Evaluation of Holdup Prediction Methods for Gas-Liquid Flow in Horizontal Pipes. *JPT* **27** (8): 1017–1026; *Trans.,* AIME, **259.** SPE-5140-PA.

Mandhane, J.M., Gregory, G.A., and Aziz, K. 1977. Critical Evaluation of Friction Pressure Drop Prediction Methods for Gas-Liquid Flow in Horizontal Pipes. *JPT* **29** (10): 1348–1358; *Trans.,* AIME, **263.** SPE-6036-PA.

Mukherjee, H. and Brill, J.P. 1983. Liquid Holdup Correlations for Inclined Two-Phase Flow. *JPT* **35** (5): 1003–1008. SPE-10923-PA.

Mukherjee, H. and Brill, J.P. 1985a. Empirical Equations to Predict Flow Patterns in Two-Phase Inclined Flow. *Intl. J. Multiphase Flow* **11** (3): 299–315.

Mukherjee, H. and Brill, J.P. 1985b. Pressure Drop Correlations for Inclined Two-Phase Flow. *Trans.,* ASME, *J. of Energy Res. Tech.* **107:** 549–554.

Oddie, G., Shi, H., Durlofsky, L.J., Aziz, K., Pffeffer, B., and Holmes, J.A. 2003. Experimental Study of Two and Three Phase Flows in Large Diameter Inclined Pipes. *Intl. J. Multiphase Flow* **29:** 527–558.

Oliemans, R.V.A., Pots, B.F.M., and Trompe, N. 1986. Modelling of Annular Dispersed Two-Phase Flow in Vertical Pipes. *Intl. J. Multiphase Flow* **12:** 711–732.

Petalas, N. and Aziz, K. 2000. A Mechanistic Model for Multiphase Flow in Pipes. *J. Cdn. Pet. Tech.* **39** (6): 43.

Shi, H., Holmes, J.A., Durlofsky, L.J. et al. 2005a. Drift-Flux Modeling of Two-Phase Flow in Wellbores. *SPEJ* **10** (1): 24–33. SPE-84228-PA.

Shi, H., Holmes, J.A., Diaz, L.R., Durlofsky, L.J., and Aziz, K. 2005b. Drift-Flux Parameters for Three-Phase Steady-State Flow in Wellbores. *SPEJ* **10** (2): 130–137. SPE-89836-PA.

Shoham, O. 2006. *Mechanistic Modeling of Gas-Liquid Two-Phase Flow in Pipes*. Richardson, Texas: SPE.

Taitel, Y. and Dukler, A.E. 1976. A Model for Predicting Flow Regime Transitions in Horizontal and Near Horizontal Gas-Liquid Flow. *AIChE J.* **22 (1):** 47–55.

Xiao, J.J., Shoham, O., and Brill, J.P. 1990. A Comprehensive Mechanistic Model for Two-Phase Flow in Pipelines. Paper SPE 20631 presented at the SPE Annual Technical Conference and Exhibition, New Orleans, 23–26 September.

Errata

Page	Location	Error
18	Page heading	Change Chraacteristics to Characteristics
42	Fig. 1.20, ordinate	Change 1.10, 1.15 to 0.10, 0.15
168	Fig. 4.17	Add to figure title: *k* in ft
173	Eq. 4.88	Limits of integral should be reversed
191	Fig. 5.4, abscissa	10^3 on top as on bottom
207	Fig. 5.12	Multiply ordinate values by 2
338	Following Eq. 8.3	Remove gram-cm-sec units and add cp and dynes/cm
339	Fig. 8.10	Figure is upside down (at the ordinate value of 0.2, X'=1.3)
344	Fig. 8.12, abscissa	Should go from 1 to 1,000
345	Fig. 8.13, abscissa	Should go from 1 to 1,000
350	Eq. 8.17	Replace G_G with G_M
350	Following Eq. 8.17	Replace $G_G = V_{SG}\,\rho_G$ with $G_M = V_{SG}\,\rho_G + V_{SL}\,\rho_L$
350	Following Eq. 8.17	Remove gram-cm-sec units and add cp and dynes/cm
358	Eq. 8.25	Add g_c in denominator on right-hand side
358	Eq. 8.27	Multiply *P* in denominator on right-hand side by g_c
358	Eq. 8.27	Add equal sign after $\Delta P / \Delta z$
358	Eq. 8.28	Replace *M* with M' and add g / g_c after $\overline{\rho}_M$
359	Eq. 8.30	Replace *M* with M'
360	Fig. 8.27, abscissa label	Remove N_{VL} /
387	Eq. 8.81	Terms after $(1-\overline{E}_G)$ should be in [..]
388	Eq. 8.83	Terms after $\pi D^2 / 4$ should be in [..]
404	Unnumbered equation after Eq. 8.126	Left-hand side should be V_{sf}
411	Last line above heading 5.12	Replace 8.163 with 8.143

434	First unnumbered equation	Right-hand side should be under $\sqrt{\ }$ sign
509	Table 10.2	Line M, column E_o should be 0.014
527	Before Eq. 10.8	Replace Hoogendoorn and Buitelaar (1961) with Hoogendoorn (1959) in two places
527	Eq. 10.8	Velocities in m/sec, D in m, g in m/sec^2
528	Fig. 10.19	Abscissa should start with 0.1, not 0
562	First line	Replace vave with wave
562	Eq. 10.86	Should be $Q_\beta = V_\beta A_\beta = V_{S\beta} A$
585	First line	Units of B are 1/ft
585	After Eq. 10.129	Add: where V_{SL} and V_{SG} are in m/sec, g_m in m/sec^2 and D in m
598	Last paragraph, line 2	Replace low with law
629	Row A, column heading Slip Velocity	Replace 0.48 with 0.47
637	Fig. 11.20, ordinate	Replace ΔL with ΔP
638	Fig. 11.22, ordinate	Replace ΔL with ΔP
648	Fig. 11.28, ordinate	Should be $F_L = \dfrac{V_{M_2}}{\left[2gD(s-1)\right]^{0.5}}$
666	Third line above Fig. 11.43	Replace E with E_S
674	Eq. 11.61	Second term on the right-hand side should be squared: $\left(\dfrac{C_{w2}}{Cw}\right)^2$
677	Fig. 11.49, abscissa label	$F_{r_M}\sqrt{\rho_T/(\rho_S-\rho_T)}$

Contents

List of Symbols

The symbols most commonly used throughout the text are defined. Specialized terms and constants used only within individual chapters are defined in the text itself and are not repeated here. In a few instances the same symbol is used with more than one meaning. In these cases the most common meanings are defined and the chapters or equations in which they are used are identified.

A	area normal to the velocity or cross-sectional area of pipe
A_b	area of the interphase boundary between the α and β phases (Ch. 7)
A_b	cross-sectional area of Taylor bubble (Ch. 8)
A_c	cross-sectional area of the gas core
A_e	external boundary area $= A_1 + A_2 + A_w$ (Ch. 7)
A_f	cross-sectional area of liquid film
A_G	cross section of the pipe occupied by the gas phase
A_L	cross section of the pipe occupied by the liquid phase
A_S	cross section of the pipe occupied by the solid phase
A_u	cross section of the pipe above a bed of solids (Ch. 11)
C	excess dimensionless pressure drop (Eq. 5.64)
C	volume fraction of a phase in the input (or output) mixture
C	compressibility of liquid defined by Equation 2.27
C_D	drag coefficient defined by Equation 1.2

C_L	lift coefficient (Eq. 12.70)
C_s	compressibility of saturated liquid
C_w	mass fraction of solid phase in the input (or output) mixture = M_S/M_M
C_{w1}, C_{w2}	mass fractions of the solids flowing in suspension and of the solids in the bed, respectively (see Section 3.22, Ch. 11)
C_0	distribution coefficient defined by Equation 7.102
c	concentration or volume fraction of particles (Ch. 1)
c	compressibility correction factor (Fig. 2.8)
c	the dimensionless clearance, ratio of the minimum distance between the pipe and capsule wall to the pipe diameter (Ch. 12)
c_v	volume fraction (Ch. 2)
$(c_v)_{\text{max}}$	maximum attainable volume concentration
c_w	weight or mass fraction (Ch. 2)
D	diameter of pipe
De	Deborah number (Eq. 1.63)
D_E	equivalent diameter = 4 times the cross-sectional area divided by the wetted perimeter
D_G	equivalent diameter D_E for the pipe section occupied by the gas phase
D_L	equivalent diameter D_E for the pipe section occupied by the liquid phase
D_o	orifice diameter
D_S	equivalent diameter D_E for the pipe section occupied by the solid phase
d	particle diameter
d	diameter of capsule (Ch. 12)
d_{app}	apparent particle diameter given by Equation 2.76
d_{av}	average particle diameter as determined by screen size or measurement

d_s	diameter of a sphere having the same volume as a particle or a bubble
d_{85}	particle diameter such that 85 percent by weight of particles are of $d < d_{85}$
E_α	cross-sectional average *in situ* volume fraction of the α phase
$\bar{E}_\alpha$	value of E_α averaged over a finite length of pipe
Eo	Eotvos number $= g\Delta\rho d_s^2/\sigma$ or $g\Delta\rho D^2/\sigma$ as appropriate
E_H	liquid head factor defined by Equation 10.64
e	*in situ* volume fraction of the liquid entrained in the gas core (Chs. 8, 10)
e	2.718, base of natural logarithms
e_m	momentum transfer coefficient defined by Equation 11.40
e_s	sediment transfer coefficient defined by Equation 11.38
exp	denotes 2.718 raised to a power
F'	Levy function defined by Equations 8.218 and 8.219
Fr	Froude number in the form $V/\sqrt{gD}$
$\hat{\text{Fr}}$	Froude number in the form $V^2/gD = \text{Fr}^2$
Fr_b	Bubble Froude number $V_b/\sqrt{gD}$
Fr_M	Froude number $= V_M/\sqrt{gD}$ as used by Gaessler (Ch. 11)
$\hat{\text{Fr}}_M$	Froude number $= V_M^2/gD$
Fr_0	Froude number $= V_0/\sqrt{gD}$ as used by Gaessler (Ch. 11)
$\hat{\text{Fr}}_{S\alpha}$	Froude number based upon superficial velocity of the α phase, $V_{S\alpha}^2/gD$
$\hat{\text{Fr}}_\alpha$	Froude number based upon average velocity of α phase, V_α^2/gD
F_e	fraction of flowing liquid entrained in the gas phase (Eq. 8.163)
F_w	wall effect factor (Eqs. 1.16 and 1.17)
f or f_α	single-phase friction factor defined by Equation 4.24 and determined at $\text{Re}_\alpha = DV_\alpha\rho_\alpha/\mu_\alpha$ or $D_E V_\alpha \rho_\alpha/\mu_\alpha$

f_{ann}	friction factor for annular space defined by Equation 12.62
f_{av}	average friction factor over the entrance length
f_B	Baxendell friction factor $= \frac{1}{2} f_{BTP}$ (Eq. 10.51)
f_{BT}	Tomita Bingham friction factor defined by Equation 5.107
f_{BTP}	Bertuzzi, Tek, and Poettmann friction factor defined by Equation 10.50
f_{eff}	effective friction factor defined by Equation 12.79
f_H	friction factor $= 4f$
f_{LM}	friction factor for single-phase liquid flow based upon $\text{Re}_{LM} = DV_M \rho_L/\mu_L$
f_{PC}	Poettmann-Carpenter friction factor defined by Equation 8.8
f_{PLT}	Tomita power law friction factor defined by Equation 5.86
f_R	Ros friction factor defined following Equation 8.22
f_S	solids friction factor defined by Equation 9.53, 9.58, or 11.65
f_{SN}	Shearer and Nedderman friction factor defined by Equation 8.184
$f_{S\alpha}$	single-phase friction factor determined at $\text{Re} = DV_{S\alpha}\rho_\alpha/\mu_\alpha$ and appropriate relative roughness
$f_{tp\alpha}, f_{tp\beta}, f_{tpM}$	two-phase friction factors defined by Equations 7.63, 7.64, and 7.65
G	mass velocity $= V\rho$
G	specific gravity of a gas relative to air (Ch. 4)
g	acceleration due to gravity
g_c	dimension conversion factor $= 32.2$ lb_m ft/lb_f sec^2
H	holdup ratio $= V_\alpha/V_\beta = C_\alpha E_\beta/C_\beta E_\alpha$ (Eq. 7.79)
He	Hedstrom number $= D^2 \tau_y \rho g_c/\eta^2$
h	maximum depth of lower stratified phase in horizontal pipe
h_G	height of gas phase above gas-liquid interface in horizontal pipe
i	pressure gradient, $\Delta P/\rho_W \Delta z$, feet of water per foot

i_W	pressure gradient for water flowing alone at velocity V_M, $\left(\frac{\Delta P}{\rho_W \Delta z}\right)_W$, feet of water per foot
	local superficial velocity of phase defined by Equation 7.89 or 7.90
j_M	$j_\alpha + j_\beta$ local volumetric flux density of the mixture (Eq. 7.91)
K	constant in the power law equation (consistency index) (Eq. 1.40 or 1.48)
K	Bankoff factor defined by Equation 7.83
K	Hanks' stability factor defined by Equation 4.7
K	Kozney constant (Eq. 1.22)
$\mathscr{K}$	von Kármán constant $= l/y$; a function of $\mathscr{R}$ for gas-solids flow
K'	Metzner-Reed modified power law coefficient defined by Equation 5.61
K_2'	wall correction factor given by Figure 8.37
k	measure of roughness, equals the height of relatively uniformly sized sand grains that would give the roughness effect observed in pipe flow (Ch. 4)
k	effective roughness of a liquid film
k	mass ratio of gas to liquid defined following Equation 8.15
k	ratio of capsule to pipe diameter (Ch. 12)
L	length of pipe
L	travel distance measured from the entrance of the pipe (Ch. 6)
ΔL	end-to-end spacing between capsules in a pipe (Ch. 12)
L_b	length of a Taylor bubble
L_c	value of L beyond which dP/dL for a time-dependent fluid is independent of L (Ch. 6)
L_c	length of a capsule (Ch. 12)
L_{cm}	value of L beyond which dP/dL for a time-dependent fluid is independent of both L and duration of flow

L_E	equivalent length of a pipe $= L + L_F$
L_F	equivalent length of any fittings valves or devices in a pipe
L_s	length of a liquid slug.
l	Prandtl mixing length defined following Equation 3.46
ln	logarithm to base e
log	logarithm to base 10
l_u	Prandtl mixing length for velocity (Ch. 8)
$l_{\rho u}$	Prandtl mixing length for momentum (Ch. 8)
M	mass rate of flow $= V\rho A$
$\mathcal{M}$	total mass in the system of volume $\mathcal{V}$
M'	mass of flowing mixture per barrel of oil defined by Equation 8.12
M_e	mass rate of flow of entrained liquid
m	molecular weight
N_D	Ros diameter number defined by Equation 10.11
N_L	Ros liquid viscosity number defined by Equation 10.13
N_{LB}	Brown liquid viscosity number defined by Equation 8.113
N_{LW}	Wallis liquid viscosity number defined by Equation 8.115
N_S	Ros slip velocity number defined by Equation 8.19
N_{VG}	Ros gas velocity number defined by Equation 10.10
N_{VL}	Ros liquid velocity number defined by Equation 10.9
N_1	flow pattern index defined preceding Equation 11.20
n	exponent in power law (flow behavior index) (Eq. 1.40 or 1.48)
n	ratio of the surface area per unit mass of particles to that of spheres of diameter d_{av} (Ch. 1)
n	the average number of bubbles in a horizontal plane (Ch. 8)
n	number of orifices (Eq. 8.31)

n'	Metzner-Reed modified power law exponent defined by Equation 1.68 or 5.61
P	pressure
P_B	Bagnold dispersive force (Eq. 11.46)
P_c	critical pressure
${}_pP_c$	pseudocritical pressure defined by Equation 2.36
P_r	P/P_c = reduced pressure
P_{11}	normal stress parallel to the direction of flow (Figure 1.24)
P_{22}, P_{33}	normal stresses perpendicular to the direction of flow (Figure 1.24)
ΔP	pressure drop over length L or Δz
ΔP^*, ΔP_G^*, ΔP_L^*	dimensionless pressure drops defined by Equations 8.195, 8.196, and 8.197
dP_f, ΔP_f	pressure drop due to friction defined by Equation 7.57 or 7.60
ΔP_f^*	pressure drop due to friction defined by Equation 7.73 or 7.74
$dP_{f\alpha}$, $\Delta P_{f\alpha}$	pressure drop due to friction for α phase flowing alone at velocity V_α
$dP_{fS\alpha}$, $\Delta P_{fS\alpha}$	pressure drop due to friction for α phase flowing alone at velocity $V_{S\alpha}$
ΔP_{GM}	pressure drop which would occur if the gas were flowing alone at the mass velocity G_M
ΔP_{HH}	hydrostatic head component of the pressure drop defined by Equation 7.72
ΔP_{KE}	pressure change due to kinetic energy or acceleration effects defined by Equation 7.60
ΔP_{LM}	pressure drop which would occur if the liquid phase were flowing alone at the mass velocity G_M
ΔP_{PE}	pressure change due to potential energy defined by Equation 7.60
ΔP_S	additional pressure drop due to presence of solids defined by Equation 9.52
$\Delta P_{S\alpha}$	total pressure drop for α phase flowing alone at velocity $V_{S\alpha}$
ΔP_T	total pressure drop over length L (Eq. 9.4)

p	perimeter of pipe
Q	volume rate of flow
Q_e	volume flow rate of the entrained liquid
Q_f	volume flow rate of the liquid in the film
R	radius of pipe
R	universal gas constant (Eq. 2.5)
$\mathcal{R}$	mass flow ratio defined by Equation 9.2
Re	Reynolds number $= DV\rho/\mu$
Re*	friction velocity Reynolds number $= RU^*\rho/\mu$ or $DU^*\rho/\mu$
$\mathrm{Re}_{\mathrm{ann}}$	Reynolds number for annulus defined by Equation 12.65
Re_B	Bingham Reynolds number $= DV\rho/\eta$
Re_{BT}	Tomita Bingham Reynolds number defined by Equation 5.106
Re_{BTP}	Bertuzzi, Tek, and Poettmann Reynolds number defined by Equation 8.15
Re_b	bubble Reynolds number defined following Equation 8.39
Re_{EA}	Eaton *et al.* Reynolds number defined by Equation 10.3
Re_f	film Reynolds number defined following Equation 8.126
Re_H	Hughmark Reynolds number defined by Equation 8.5
Re_{KG}	Kilbane-Greenkorn Reynolds number defined by Equation 5.125
Re_k	roughness Reynolds number defined by Equation 4.63
Re_{LM}	Reynolds number of the liquid phase based upon V_M, $= DV_M\,\rho_L/\mu_L$
Re_M	Reynolds number based upon the properties of the mixture
Re_{MR}	Metzner-Reed Reynolds number defined by Equation 5.63
R_P	pressure gradient ratio defined by Equation 12.39
R_P^*	pressure gradient ratio for a concentrically flowing capsule
Re_{PC}	Poettmann-Carpenter Reynolds number defined by Equation 8.14

Re_{PLC}	Clapp power law Reynolds number defined by Equation 5.97
Re_{PLT}	Tomita power law Reynolds number defined by Equation 5.87
Re_{PL1}	power law Reynolds number defined by Equation 5.29
Re_{PL2}	power law Reynolds number defined by Equation 5.30
Re_{SF}	superficial fluid phase Reynolds number, $DV_{SF}\rho_F/\mu_F$
Re_{SG}	superficial gas phase Reynolds number, $DV_{SG}\rho_G/\mu_G$
Re_{SL}	superficial liquid phase Reynolds number, $DV_{SL}\rho_L/\mu_L$
Re_α	Reynolds number based upon average velocity of α phase, $DV_\alpha\ \rho_\alpha/\mu_\alpha$
$\mathscr{R}_E$	Wicks and Dukler entrainment number defined in Section 7.32 of Chapter 8
R_L	Levy function of the density ratio ρ_L/ρ_c given by Figure 8.64
R_V	velocity ratio $= V_c/V_M$ (Chapter 12)
r	radial distance
r_i	radius to the gas-liquid interface in annular flow
r_P	radius of plug in laminar flow of Bingham or yield-pseudoplastic fluid
S	Wicks modified Reynolds number defined following Equation 11.101
S	slip velocity defined by Equation 7.76
S	entropy (Eq. 3.26)
S	hydrostatic head term defined following Equation 4.83
$\dot{S}$	absolute value of the rate of shear $= -du/dr$ for laminar flow in a pipe
S_b	center-to-center axial spacing of bubbles
s	density ratio, ρ_S/ρ_F or ρ_c/ρ
s	diameter ratio of rotational viscometer defined following Equation 1.71
s_P	specific surface of particles $=$ surface area per unit volume of actual particles (Chs. 1 and 2)

T	absolute temperature
T	torque of rotational viscometer (Sec. 5.2 of Ch. 1)
T_b	normal boiling point temperature
T_c	critical temperature
${}_pT_c$	pseudocritical temperature defined by Equation 2.35
T_r	reduced temperature $= T/T_c$
U^*	friction velocity defined by Equation 1.31, $= \sqrt{\tau_w g_c/\rho} = V\sqrt{f/2}$
U^*_{M1}, U^*_{M2}	friction velocity corresponding with V_{M1}, V_{M2}
u	local axial velocity
u^+	dimensionless local velocity $= u/U^*$ (Eq. 4.38)
V	cross-sectional average velocity
V	rise or fall velocity of particles (Ch. 1)
$\mathcal{V}$	total volume occupied by the mixture $= \mathcal{V}_\alpha + \mathcal{V}_\beta$
V_b	average bubble velocity in a flowing stream
V_{bs}	terminal rise velocity of bubble in stagnant liquid
V_c	capsule velocity defined by Equation 12.4
V_c	terminal settling velocity of solids in a mixture of concentration c
V_e	average velocity of the entrained liquid droplets
V_f	average velocity of the liquid film in annular-mist flow
V_L	average velocity of the liquid surrounding a capsule defined by Equation 12.5
V_{Lt}	threshold value of V_L (Ch. 12)
V_M	average velocity of the mixture $= V_{S\alpha} + V_{S\beta}$
V_{ML}	lift-off velocity (Ch. 12)
V_{Mt}	threshold value of V_M (Ch. 12)
V_{M1}	average mixture velocity above which particles are in symmetric suspension (Ch. 11)

V_0	terminal settling velocity of solids at low concentration
V_R	"slip" velocity (Sec. 3.22 of Ch. 12)
$V_{S\alpha}$, $V_{S\beta}$, V_{SF}, V_{SG}, V_{SL}, V_{SS}	superficial velocity of the phase indicated by the second subscript = the volume flow rate of the phase divided by the total cross-sectional area
V_{Sf}^*	dimensionless superficial film velocity defined by Equation 8.145
V_{SG}^*	dimensionless superficial gas velocity defined by Equation 8.146
V_{SG}'	superficial gas velocity at the bubble-slug transition (Eq. 8.85)
V_{sm}	maximum tolerable settling velocity (Ch. 1)
$\mathcal{V}_\alpha$	volume occupied by α phase
$V_{\alpha j}$	volume fraction weighted average drift velocity of α phase relative to mixture defined by Equation 7.103
$\mathcal{v}$	specific volume $= 1/\rho$
$\mathcal{v}_b$	molal liquid volume at T_b (Ch. 2)
v_b	volume of a bubble
v_{bo}	average bubble volume at generation site (Eq. 8.31)
v_{bo}^*	value of v_{bo} extrapolated to zero generation frequency
v_{boi}	ideal value of v_{bo} (Eq. 8.32)
$\mathcal{v}_c$	critical specific volume (Ch. 2)
v_f	volume of the liquid film surrounding a Taylor bubble
$\mathcal{v}_m$	volume per mole (Ch. 2)
v_P	volume of particle
v_r	local slip velocity defined by Equation 7.94
v_s	volume of a liquid slug
v_t	volume of tube of length L
$v_{\alpha j}$	local drift velocity of the α phase defined by Equation 7.95
W	work done by the system per unit mass of fluid

We	Weber number $= \rho V^2 D/\sigma g_c$
We_c	critical Weber number of gas phase defined in Section 7.32 of Chapter 8
We_{EA}	Eaton *et al.* Weber number defined by Equation 10.4
Ws	Weissenberg number $= (P_{11} - P_{22})/\tau_{21}$ (Eq. 1.57)
W^+	dimensionless liquid flow rate defined following Equation 8.183
W_B	buoyed weight of a capsule $= (\pi/4)d^2L(\rho_c - \rho)(g/g_c)$ (Ch. 12)
W_c	weight of a capsule $= (\pi/4)d^2L\rho_c(g/g_c)$ (Ch. 12)
W_L	weight of the fluid in an annulus $= (\pi/4)L(D^2 - d^2)\rho(g/g_c)$ (Ch. 12)
w_i	width of interface in stratified flow (Eq. 10.89)
X	Govier fluid-property parameter $= (\rho_G/\rho_A)^{1/3}\,Y$ (Sec. 3.1, Ch. 8)
X	Lockhart and Martinelli parameter defined by Equation 10.5
X'	Hughmark and Pressburg parameter defined by Equation 8.3
X_E	travel distance from the pipe entrance for the complete development of the velocity profile (Chs. 4, 5)
X_{E0}, $(X_E')_{\text{av}}$, $X_{E\infty}$	entrance lengths based on rheological properties at $\theta = 0$, θ'_{av}, and ∞ (Ch. 6)
x	center-to-center particle distance (Ch. 1)
x	$= \tau_y/\tau_w$ (Ch. 5)
Y	liquid pickup rate defined by Equation 10.114
Y	yield number $= D\tau_y g_c/V\eta$ (Ch. 5)
Y	Govier fluid-property parameter $= (\rho_L \sigma_{WA}/\rho_W \sigma)^{1/4}$ (Sec. 3.1, Ch. 8)
Y	White and Beardmore property number $= g\mu_L^4/\rho_L \sigma^3$ (Sec. 5.3, Ch. 8)
y	distance from pipe wall $= R - r$
y	height of "roughness" caused by interface waves or ripples
y^+	wall distance Reynolds number or dimensionless distance from the wall $= yU^*\rho/\mu$ (Eq. 4.39)

Z	compressibility factor defined by Equation 2.6
Z	Ryan and Johnson stability parameter defined by Equation 4.5
$\mathscr{Z}$	Hughmark correlation factor defined by Equation 8.4
Z_c	compressibility factor at the critical point
$Z^{(0)}, Z^{(1)}$	factors used in the Pitzer *et al.* compressibility factor correlation (Ch. 2)
Δz^+	"uphill" elevation change (Eq. 10.64)
Δz_b	axial center-to-center spacing of bubbles in vertical flow
Δz_c	critical value of Δz_b (Eq. 8.58)
α	velocity profile correction factor in kinetic energy term (following Eq. 3.20, and Eq. 5.67)
α	designates a phase, normally the less dense phase (Ch. 7)
α_c	Riedel factor (Eq. 2.86)
β	fractional friction factor defined by Equation 5.134
β	designates a phase, normally the more dense (Ch. 7)
β	Gaessler proportionality factor defined by Equation 11.50
β^*	Gaessler material constant defined by Equation 11.56
γ	Metzner-Reed consistency term $= g_c K' 8^{n'-1}$ defined following Equation 5.63
Δ	Zenz fluid-particle property function defined preceding Equation 11.31
δ	liquid film thickness, normally the mean value
δ^*	dimensionless film thickness defined following Equation 8.127
δ^+	dimensionless film thickness defined by Equation 8.181
δ_l	thickness of the laminar sublayer
ε	eccentricity, distance between the axes of the pipe and the capsule expressed as a fraction of the pipe radius (Ch. 12)
$\varepsilon_\alpha, \varepsilon_\beta$	local *in situ* volume fraction of the α or the β phase

ε_C	Chawla factor defined by Equation 10.16
ζ	dimensionless distance from the pipe wall $= y/R$
η	coefficient of rigidity for Bingham fluid (Eq. 1.47)
η	proportionality coefficient in Equation 12.54
η_a	apparent coefficient of rigidity
η_m	coefficient of sliding (moving) friction
η_s	coefficient of static friction
θ	angle between the pipe axis and a horizontal line (positive for uphill flow, negative for downhill flow)
θ	duration of shear (Chs. 1, 6)
θ	deflection of spring or torque measuring device (Sec. 5.2. of Ch. 1)
$\theta_b = \dfrac{\pi - \gamma}{2}$	angle defined in Figure 11.58
θ_E	rheological age of the entering fluid (Ch. 6)
θ_F	duration of flow (Ch. 6)
θ_p	time scale of process (Eq. 1.63)
θ_r	relaxation time of a material (Eq. 1.58)
λ	resistance coefficient (Eq. 1.20)
λ	Baker fluid-property parameter defined by Equation 10.1
μ	viscosity of fluid or of continuous fluid phase
μ_a	apparent viscosity defined by Equation 1.39
μ_c	viscosity of fluid at critical pressure and temperature
μ_M	viscosity of gas mixture
μ_0	viscosity of a pure gas at low pressures
μ_{0M}	viscosity of gas mixture at low pressure
μ_P	viscosity of fluid particle

μ_∞	limiting value of μ_a as $\dot{S} \rightarrow \infty$
ν	kinematic viscosity $= \mu/\rho$
ν_e	eddy kinematic viscosity defined by Equation 9.73
ξ	dimensionless film thickness, δ/R
ξ	Stiel and Thodos gas viscosity correlating factor defined following Equation 2.48
ξ_M	Dean and Stiel gas viscosity correlating factor defined following Equation 2.59
π	3.1416
ρ	density
$\Delta\rho$	density difference $= \rho_\beta - \rho_\alpha$
ρ_{mL}	molar density of the liquid phase
ρ_{mv}	molar density of the vapor phase
ρ_0	density at standard or reference pressure and temperature P_0 and T_0
ρ_{sL}	saturated liquid density
ρ_T	bulk density of a mixture of a fluid and suspended solid defined following Equation 11.57
σ	surface or interfacial tension
τ	shear stress
τ_y	yield value for Bingham or yield-pseudoplastic fluid (Eq. 1.47 or 1.48)
Φ	van Driest wall dampening term redefined by Gill and Scher by Equation 4.58
Φ	Graf and Acaroglu transport parameter defined by Equation 11.137
ϕ_S	Gaessler bed packing factor defined following Equation 11.57
ϕ_α, ϕ_β	Lockhart and Martinelli parameter $= \sqrt{dP_f/dP_{fS\alpha}}$, $\sqrt{dP_f/dP_{fS\beta}}$ (Eqs. 7.61, and 7.62)
Ψ'	Graf and Acaroglu shear intensity parameter defined by Equation 11.136

ψ	Baker fluid-property parameter defined by Equation 10.2
ψ	particle shape factor defined preceding Equation 1.15
ψ	Wicks correlating factor defined following Equation 11.101
ψ_1	particle shape factor defined by Equation 2.71
ψ_2	particle shape factor defined by Equation 2.73
ω	Zenz fluid-particle property function defined preceding Equation 11.31
ω	bubble generation frequency (Eq. 8.31)
ω	Pitzer *et al.* acentricity factor defined by Equation 2.10
ω_M	pseudoacentricity factor defined by Equation 2.41
∇	vector operator with components $\frac{\partial}{\partial x}, \frac{\partial}{\partial y}, \frac{\partial}{\partial z}$ in the Cartesian coordinate system

Overlines

$\overline{}$	averaged over a short interval of time

Superscripts

$*$	reduced to dimensionless form
$'$	turbulent fluctuating components (Eq. 3.34)
$\bullet$	refers to ΔP_f based upon Equation 7.73 or 7.74

Subscripts

ann	refers to the annulus
asym	refers to particles in asymmetric suspension flow
av	refers to average value
b	refers to a bed of solids
b	refers to bubble
c	refers to the gas core (Chs. 8, 10)
c	refers to transition between laminar and turbulent flow

c	refers to critical pressure and temperature conditions (Ch. 2)
c	refers to the capsule (Ch. 12)
ca	refers to average conditions at the capsule wall
cat	refers to average threshold conditions at the capsule wall (Ch. 12)
E	refers to entrance length
e	refers to entrained liquid
eff	means effective
F	refers to fluid (gas or liquid)
f	refers to irreversible or friction component
f	refers to liquid film
FRW	refers to fully rough wall turbulence (Ch. 4)
G	refers to gas phase
Gi	refers to the gas phase at the interface
GSL	refers to the gas in the liquid slug
HH	refers to hydrostatic head
i	refers to interfacial condition
i	refers to component i
L	refers to liquid
Li	refers to the liquid phase at the interface
l	refers to laminar flow conditions or the laminar sublayer
M	refers to mixture
$M1$, $M2$, $M3$, $M4$	refer to transition velocities for the horizontal flow of fluid-solid mixtures defined in Section 3.21 of Chapter 11
m	refers to sliding or moving conditions (Ch. 12)
m	refers to molar (Ch. 2)
m or max	refers to maximum value (Chs. 4, 5)

min	refers to minimum value
0	refers to center of pipe, i.e. at $r = 0$
0	refers to reference or standard pressure, temperature, or density
P	refers to the particle
PRW	refers to partially rough wall turbulence (Ch. 4)
r	refers to radial position
r	refers to relative velocity
r	refers to reduced conditions, i.e., the ratio of the actual to the critical condition (Ch. 2)
S	refers to solids
Sf	refers to superficial film flow rate based upon the full pipe cross section
SF, SL, SG, SS, Sc	refer to superficial flow rate based upon the full pipe cross section for fluid, liquid, gas, solid, or capsule, respectively
SW	refers to smooth wall turbulent flow (Ch. 4)
sym	refers to particles in symmetric suspension flow
t	refers to turbulent flow conditions
t	refers to threshold or startup conditions
tp	refers to two-phase
$tp\alpha$, tpG	refer to friction factor for two-phase flow, based upon the velocity and the properties of the light or the gas phase (Eq. 7.63)
$tp\beta$, tpL	refer to friction factor for two-phase flow, based upon the velocity and the properties of the heavy or the liquid phase (Eq. 7.64)
tpM	refers to friction factor for two-phase flow based upon the velocity and the properties of the mixture (Eq. 7.65)
u	refers to the part of a horizontal pipe above a stationary or moving bed of solids
W	refers to water
w	refers to the wall of pipe

wa	refers to average conditions at the pipe wall (Ch. 12)
wat	refers to average threshold conditions at the pipe wall (Ch. 12)
1, 2	refers to value at the inlet and outlet of the system, respectively
Boldface	denotes vectors
Boldface Greek	denotes tensors

Classification and Characteristics of Complex Mixtures

I

1 INTRODUCTION

The term "complex mixtures" as used in this text refers to all single-phase fluids that are non-Newtonian in their behavior and to all multiphase mixtures involving two or more fluids, or a fluid and a solid, which are capable of flowing in a pipe. The term is a convenience only and refers to fluid or fluid-solid systems for which the conventional fluid mechanics of single-phase Newtonian fluids do not apply.

Single-phase, homogeneous, fluids that do not follow the classical Newtonian behavior have become increasingly important to the engineer over the past decade. This is due partly to the fact that non-Newtonian behavior is now recognized in cases where previously it was not, and partly to the creation by the engineer of a great variety of synthetic compounds which, in their liquid state, exhibit non-Newtonian behavior. Among the natural systems, non-Newtonian behavior is encountered in the flow of many crude oils, especially at low temperatures, and with stable suspensions of fine solids in liquids. Many synthetic organic liquids are non-Newtonian and some result in non-Newtonian systems even in highly dilute solution in water. In the same way the last few years have shown a great increase in the understanding of the flow of gas-liquid, liquid-liquid, gas-solid, and liquid-solid mixtures, and the engineer now is aware of the large effects that may be caused by the presence of the second phase. Additionally, the technical and economic feasibility of pipeline transportation of mixed phases is increasing the number and type of applications. Gas-liquid flow is encountered in oil and gas wells, in oil and gas field gathering systems, in chemical processing plants, and in nuclear reaction systems. Liquid-liquid flow is encountered in oil wells, oil pipelining, and chemical processing. Gas-solid flow has long been encountered in pneumatic conveying and is important in dust collection systems and in fluidization and chemical processing. Liquid-solid mixtures have been transported by pipeline over short distances for many years. Recent improved understanding of their flow characteristics has led to a great expansion in the application of pipe transport in hydraulic dredging, mining, and long-distance movement of solids.

2 CLASSIFICATION OF COMPLEX MIXTURES

Figure 1.1 presents the broad classification of complex mixtures from the viewpoint of the effect of the phase condition on their flow behavior. Single-phase fluids are truly homogeneous and, although their flow behavior is not always simple, it is not complicated by variation in concentration of the phases. Multiphase mixtures, represented by the four types of two-phase mixtures previously discussed, may be qualitatively described in accordance with whether, under actual flow conditions, they result in

a. a fine dispersion—as small bubbles of gas, droplets of immiscible liquid or particles of solid, more or less uniformly dispersed in a continuous liquid phase, and small droplets of liquid or particles of solid more or less uniformly dispersed in a continuous gas phase;

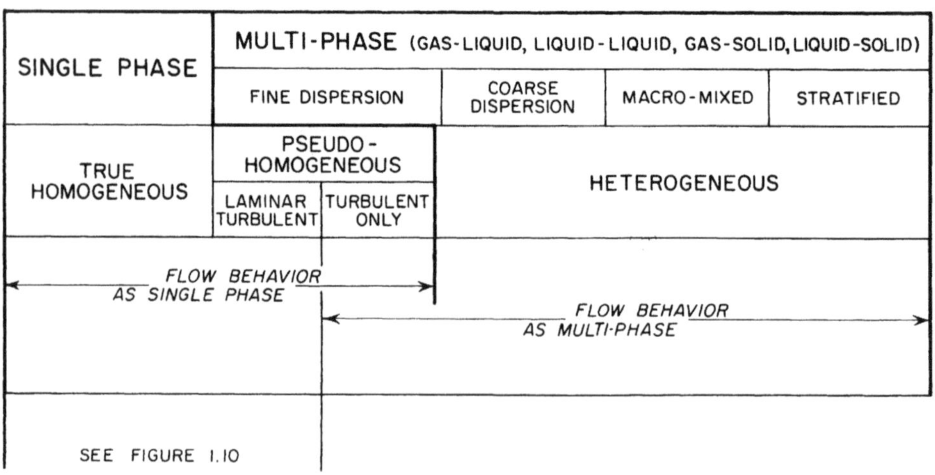

Figure 1.1. Classification of complex mixtures from the viewpoint of flow behavior.

b. a coarse dispersion—as large bubbles of gas, drops of an immiscible liquid, or particles of solid dispersed in a continuous liquid phase, and large drops of liquid or particles of solid dispersed in a continuous gas phase;
c. a macro-mixed flow pattern—as a froth or highly turbulent mixture of gas and liquid or of immiscible liquids under conditions where neither phase is continuous;
d. a stratified flow pattern—as in the case of a gas-liquid mixture or a mixture of two immiscible liquids under conditions where both phases are continuous.

Dispersions of sufficient "fineness" (submicron particle size) may be stable, supported by Brownian movement or electrical charge even in the absence of any turbulence, or supported as a result of high or special consistency properties of the continuous phase. These may be considered pseudohomogeneous and their flow behavior may be included with that of single-phase fluids. Additionally, some dispersions of moderate fineness may not be stable when static or under laminar flow conditions but may be maintained in a more or less uniformly dispersed state under conditions of sufficient turbulence. Such dispersions would not behave as single-phase fluids under laminar flow conditions but may be classified with single-phase fluids under conditions of sufficient turbulence.

3 PHASE SEPARATION AND SETTLING BEHAVIOR

3.1 Introduction

The multiphase mixtures of interest are those of gas-liquid, liquid-liquid, gas-solid, and liquid-solid which may be made to flow in a pipe. More complex systems, such as gas-liquid-liquid, or gas-liquid-solid, could also be considered, but the more complex behavior of these will be approximated by that of the two-phase systems. It is useful, in addition to the simple classification based on the type of phase present, to consider the range of behavior that may be encountered in the mixture under nonflowing conditions, i.e., to consider whether one phase will tend to rise or to settle in the other and, if so, at what rate. For this purpose we may treat the four types of two-phase mixtures as two: a mixture of two immiscible fluid phases and a mixture of a solid and a fluid phase. Both types will be considered as a mixture of "particles" in a fluid. (The cases of gas-liquid and liquid-liquid mixtures which are stratified are excluded as being completely "settled.") In both cases, if the mixture is not flowing and the phase densities are not identical, there will be a tendency for the mixture to settle or the phases to separate. Whether, in fact, settling (or rising) of a phase occurs, and, if so, at what rate depend upon the size, shape, viscosity, and density of the particles and the density and rheological properties of the continuous phase.

3.2 Terminal Settling Velocity

3.21 Solid Particles

The simplest case is that of a dispersion of solid particles in a Newtonian fluid where the particles are separated sufficiently so as not to collide or interact. Uniform, smooth, spherical particles of density ρ_P and diameter d are assumed in a fluid of density ρ and viscosity μ. It is further assumed that no electrostatic or external centrifugal forces are acting and that the system is under the action of gravity. Then, at equilibrium, the gravitational force causing the particle to rise ($\rho_P < \rho$) or fall ($\rho_P > \rho$) will be

$$F_g = \frac{\pi d^3}{6}(\rho_P - \rho)g \tag{1.1}$$

The rise or fall of the particles in the fluid results in a drag force which may be expressed as

$$F_D = C_D \frac{\rho V^2}{2} \frac{\pi d^2}{4} \tag{1.2}$$

where C_D is the drag coefficient, and V is the rise or fall velocity.

Dimensional analysis indicates that C_D should be a unique function of the particle Reynolds number, $dV\rho/\mu$, and this is confirmed by experience. The line labeled $\psi = 1.000$ in Figure 1.2 gives the relationship between C_D and $dV\rho/\mu$ as determined by actual measurements for solid spheres (Waddel, 1934a). The curve shows four regions blending into one another. At $dV\rho/\mu$ values less than about 1.0, the equation of the curve is

$$C_D = \frac{24\mu}{dV\rho} \tag{1.3}$$

and the relative motion between particle and fluid is laminar. This yields the relationship

$$F_D = 3\pi d\mu V \tag{1.4}$$

which is a form of Stokes' Law.

The second region of the curve of Figure 1.2 is one of transition and represents a gradually developing turbulence in the motion. Several entirely

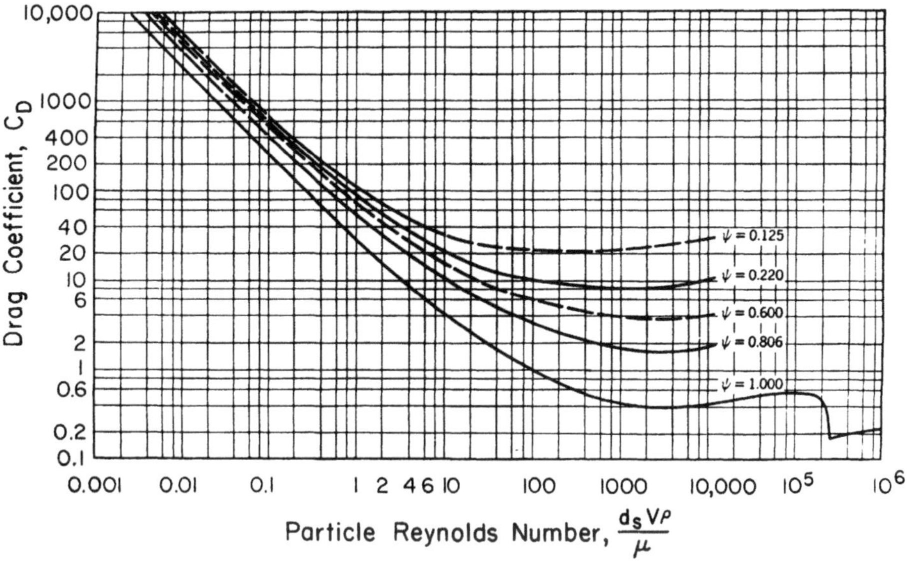

Figure 1.2. Dependence of drag coefficient on Reynolds number. (From G. G. Brown *et al., Unit Operations*, John Wiley & Sons, Inc., New York, 1950.)

empirical equations have been fitted to this part of the curve. That of Allen (1900) is convenient and gives a linear approximation to the curve for the approximate range $1 < dV\rho/\mu < 1000$

$$C_D = 30\left(\frac{dV\rho}{\mu}\right)^{-0.625} \tag{1.5}$$

The third region, representing fully developed turbulence except in the boundary layer, extends from $dV\rho/\mu$ of about 1000 to 2×10^5 and is described by the equation

$$C_D = 0.44 \tag{1.6}$$

which leads to one form of Newton's Law

$$F_D = 0.44\,\frac{\rho V^2}{2}\,\frac{\pi d^2}{4} \tag{1.7}$$

The final region extending from $dV\rho/\mu$ of 2×10^5 represents the situation where the boundary layer itself becomes turbulent and is described by the equation

$$C_D = 0.10 \tag{1.8}$$

It is not encountered in the free settling of particles.

After a relatively short initial period of acceleration, particles attain a uniform velocity under which the drag force just balances the gravitational force. This velocity is called the terminal velocity. Figure 1.2, or the equivalent equations for the drag coefficient, and Equations 1.1 and 1.2 enable the determination of the terminal settling velocity, V_0, of noninteracting, solid, spherical particles. In the three regions of direct interest the terminal settling velocities may be calculated as follows:

Laminar motion, $\dfrac{dV_0\rho}{\mu} < 1$

$$V_0 = \frac{g(\rho_P - \rho)d^2}{18\mu} \qquad \text{(Stokes' Law)} \tag{1.9}$$

This equation was derived from hydrodynamic principles by Stokes in 1851.

Transition region, (Allen, 1900) $1 < dV_0\rho/\mu < 1000$

$$V_0 = 0.20\left[g\,\frac{(\rho_P - \rho)}{\rho}\right]^{0.72} \frac{d^{1.18}}{(\mu/\rho)^{0.45}} \tag{1.10}$$

Turbulent motion, $dV_0\rho/\mu > 800$

$$V_0 = 1.74\left[g\,\frac{(\rho_P - \rho)}{\rho}\right]^{0.50} d^{0.50} \qquad \text{(Newton's Law)} \tag{1.11}$$

The equations for the laminar and turbulent settling of noninteracting, spherical particles may be represented graphically by the families of straight lines at the lower left and the upper right of Figure 1.3. The lines representing laminar flow conditions extend upward to conditions corresponding with a particle Reynolds number of unity. These conditions are indicated for various values of the ratio μ/ρ. Similarly the lines representing turbulent flow extend downward to conditions corresponding with a particle Reynolds number of 1000, again designated for various values of the ratio μ/ρ. Between the conditions corresponding with the Reynolds number of unity and of 1000, we have the transition situation. In Figure 1.3 the transition region is represented by smooth curves extending from the end of each laminar line to the beginning of the appropriate turbulent flow line. These smooth curves reflect the transition situation better than may be done through the use of Allen's equation which, as mentioned previously, fits a straight line to the drag coefficient curve in the region of Reynolds number 1–1000. For comparative purposes, the free settling line derived from Allen's equation for the condition of $(\rho_P - \rho)/\mu = 1000$, $\mu/\rho = 0.01$ and $(\rho_P - \rho)/\rho = 10$ is shown in Figure 1.3. The discontinuities at Reynolds number $= 1$ and Reynolds number $= 1000$ are apparent.

In addition to the improved representation of the transition conditions, Figure 1.3 has the advantage of eliminating the need for trial and error. Knowledge of the particle diameter and fluid properties immediately defines the region of flow and the applicable relations.

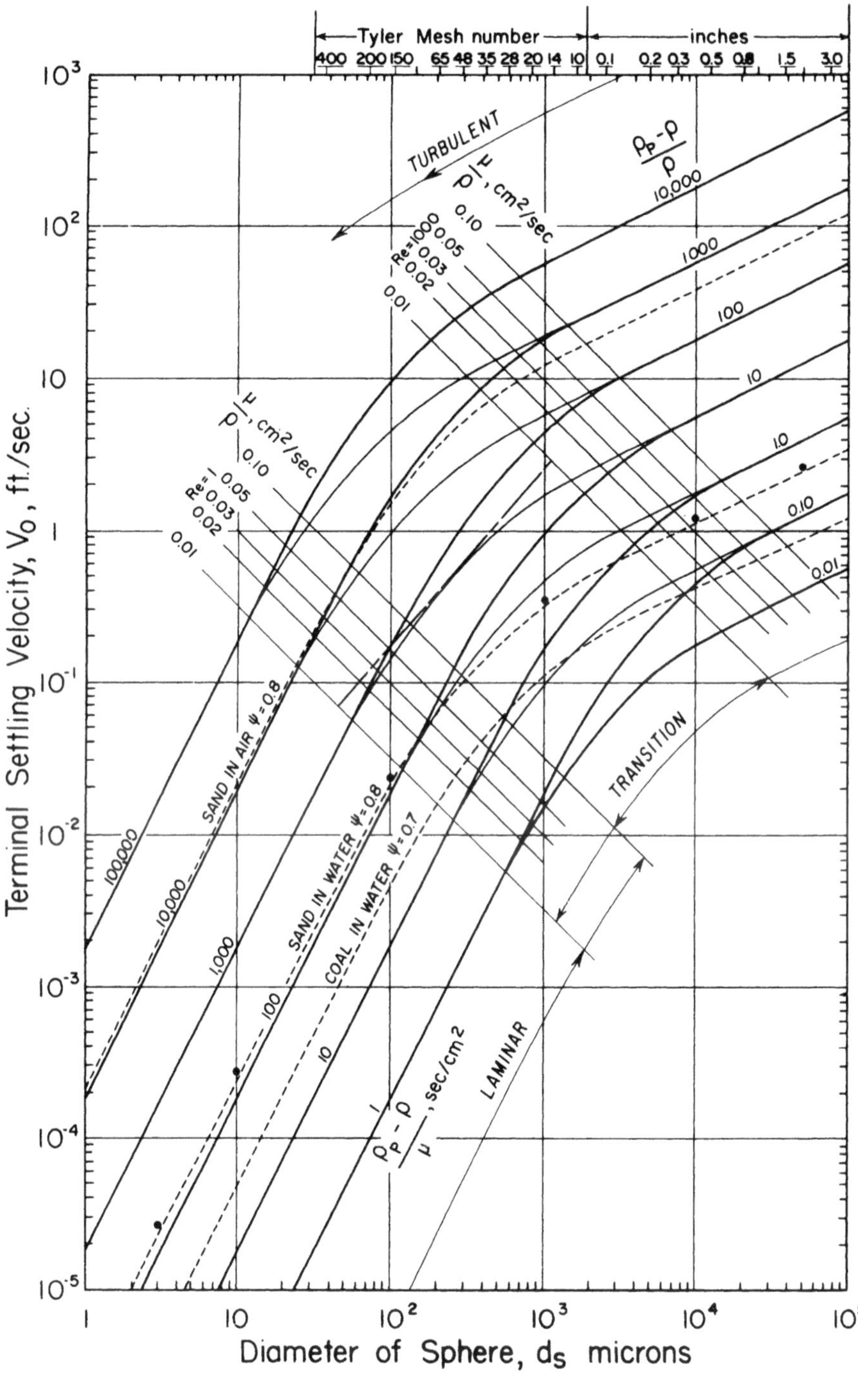

Figure 1.3. Terminal settling velocity of particles.
– – – – – – – Allen's Law, Equation 1.10;
· · calculated for sand in water from Equation 1.12, $\alpha = 1.277$, $\beta = 2.80$.

The above treatment leading to Stokes' and Newton's laws, and of the intermediate settling region, is adequate for our purpose, and Figures 1.3 and 1.5 (see Section 3.23) enable us to estimate the settling velocity of any particles in any fluid. We should mention, however, the interesting development of Swanson (1967), who has devised a single equation applicable to particles of any shape over a wide range of particle Reynolds numbers. The form of Swanson's equation is determined by considering the maximum thickness of the laminar boundary layer (see Chapter 3) on the particle and relating this to two coefficients which depend on the particle shape. The equation is

$$V = \frac{V_N}{\alpha}\left(\frac{1}{1 + \dfrac{\sqrt{48\beta\mu}}{d_{av}\,\rho V_N}}\right) \tag{1.12}$$

where $V_N = \sqrt{\dfrac{4gd_{av}(\rho_P - \rho)}{3\rho}}$,

d_{av} = average particle dimension as determined by screen size or measurement, cm,

α, β = boundary layer coefficients determined from actual settling data and Equation 1.12.

Table 1.1 gives Swanson's coefficients for typical shapes and particles. The coefficients are dimensionless. It is interesting to note that Equation 1.12 approaches Stokes' Law as the particle size decreases and Newton's Law as it increases.

TABLE 1.1

Boundary Layer Coefficients for the Swanson Equation[a]

Material	α	β	Shape
Quartz	1.277	2.80	irregular fragments
Galena	1.082	3.11	cubical cleavage
Kallodoc[b]	0.942	3.27	spheres
KCl	1.870	2.56	irregular fragments
Sphalerite	1.022	2.18	irregular fragments

[a] V. F. Swanson, *Trans. Soc. Mining Engrs.*, **238**, 160 (1967).
[b] A methyl methacrylate polymer.

Swanson's equation gives essentially the same results as may be obtained from Figures 1.3 and 1.5. This is illustrated for sand in water in Figure 1.3. Points calculated from Equation 1.12, assuming the particle shape to be characterized by Swanson's α of 1.277 and β of 2.80 for quartz particles, are shown. These points check almost exactly with the line calculated from Stokes' Law, Newton's Law, and the interpolated transition from Figure 1.3, all based upon a shape characterized by $\psi = 0.8$ (see Section 3.23).

3.22 Fluid Particles

All of the above assumes smooth, hard, spherical particles. Small bubbles of gas and droplets of liquid often approach a spherical shape, but, because of fluid circulation within the bubble or droplet, their terminal velocities may differ somewhat from those of solid particles. The laminar motion of small fluid spheres in an immiscible fluid has been studied theoretically by Hadamard (1911, 1912), Rybczynski (1911), and others, notably Levich (1962). Hadamard and Rybczynski each implicitly assumed highly purified fluids. The equation which they developed independently is

$$V_0 = \frac{gd^2(\rho_P - \rho)}{18\mu}\left(\frac{3\mu + 3\mu_P}{2\mu + 3\mu_P}\right) \tag{1.13}$$

where μ_P is the viscosity of the dispersed phase. (See Levich, 1962, for a clear and complete derivation of Equation 1.13.) We see that this is Stokes' equation multiplied by the factor $(3\mu + 3\mu_P)/(2\mu + 3\mu_P)$ which is always greater than unity. Circulation within a fluid drop lessens the velocity gradients adjacent the drop, decreases the energy dissipation, and results in a higher fall or rise velocity for a fluid as compared with a solid particle. For gas bubbles rising in a liquid $\mu_P \ll \mu$, and Equation 1.13 reduces to

$$V_0 = \frac{gd^2(\rho_P - \rho)}{12\mu} \tag{1.14}$$

For liquid droplets falling in a gas $\mu_P \gg \mu$ and Equation 1.13 reduces essentially to Stokes' equation (Equation 1.9).

Levich (1962) and others have found that experimental data only rarely conforms with the predictions of Equation 1.13 or 1.14 and that actual rise or fall velocities of small fluid particles in laminar motion are normally close to the prediction of Stokes or between that of the Stokes and the Hadamard–Rybczynski equations. Levich discusses the reasons why the latter equation applies well to the fall or rise velocity of large drops or bubbles in laminar motion and fails when applied to small fluid particles unless the fluids are exceptionally pure. With small particles in ordinary systems, trace impurities may have surfactant properties which inhibit internal circulation within the particle and cause it to behave essentially as a solid particle, i.e., in accordance with Stokes' equation. Larger particles are not so greatly affected, and even when surfactants are present the behavior is close to that given by the Hadamard–Rybczynski equation. Levich develops and presents supporting evidence for a theory relating the retardation of the fall or rise velocity (from the Hadamard–Rybczynski value in the direction of the Stokes value) to the concentration of surfactants. The theory confirms the increase in the retardation effect with decrease in particle size.

Marrucci, Apuzzo, and Astarita (1970) have reviewed the predictive equations for the terminal velocities of liquid drops in Newtonian fluids and shown how they may be applied to non-Newtonian systems. With the latter they find that the interface rigidity is normally such that extensions of Stokes' law are preferable to those of the Hadamard–Rybczynski, or similar equations.

We defer to Chapter 8 a discussion of the rise or fall velocity of fluid particles under conditions of transition or turbulent flow.

3.23 Irregularly Shaped Particles

The drag coefficient of irregularly shaped particles would be expected to depend upon a Reynolds number based on a characteristic dimension, one or more ratios to describe the shape of the particle, and a particle orientation factor. Notwithstanding the complexities involved, considerable success has been had with the use of a "shape factor" in conjunction with a characteristic particle dimension. Among the several shape factors in use, the sphericity, ψ, employed by Waddel (1934b) is convenient. This is defined (Brown and Associates, 1950) as follows:

$$\psi = \frac{\text{surface area of a sphere of volume equal to that of the particle}}{\text{surface area of the particle}}$$

For geometrically regular particles of known dimensions ψ may be calculated directly. In the case of irregularly shaped particles for which screen analysis results are available, Brown and Associates have shown that

$$\psi = \frac{d_{av}}{nd_s} \tag{1.15}$$

where d_{av} is the average screen size of the particle, d_s is the diameter of a sphere having the same volume as the particle, and n is the ratio of the surface area per unit mass of the particles to that of spheres of diameter d_{av}. Brown and Associates have computed ψ and n values for a number of mineral and regularly shaped particles. Their results are reproduced in Figure 1.4 and Table 1.2.

In Figure 1.2, the lines for ψ values other than 1 reflect Waddel's correlation of drag coefficient in terms of the particle Reynolds number based upon d_s and

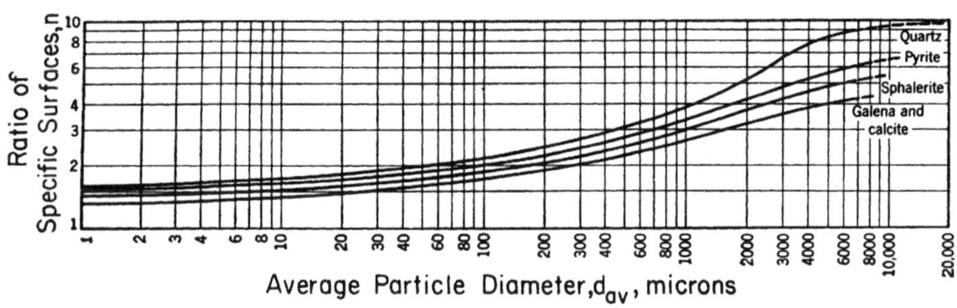

Figure 1.4. Dependence of ratio of specific surface on average particle size. (From G. G. Brown *et al., Unit Operations*, John Wiley & Sons, Inc., New York, 1950.)

TABLE 1.2

Sphericity and the Value of d_s Related to Screen Size[a]

Shape	Sphericity, ψ	$\frac{d_s}{d_{av}}$
Sphere	1.00	1.00
Octahedron	0.847	0.965
Cube	0.806	1.24
Prisms		
$a \times a \times 2a$	0.767	1.564
$a \times 2a \times 2a$	0.761	0 985
$a \times 2a \times 3a$	0 725	1.127
Cylinders		
$h = 3r$	0.860	1.31
$h = 10r$	0.691	1.96
$h = 20r$	0.580	2.592
Disks		
$h = r$	0.827	0.909
$h = r/3$	0.594	0.630
$h = r/10$	0.323	0.422
$h = r/15$	0.254	0.368

[a] G. G. Brown *et al.*, *Unit Operations*, John Wiley & Sons, Inc., New York, 1950.

the sphericity, ψ. As is to be expected, irregularly shaped particles have higher drag coefficients than spheres of the same volume.

The effect of particle shape on the settling velocity, as reflected through the effect of ψ and Re on the drag coefficient, may be expressed directly as the ratio $V_{0,\psi}/V_{0,\psi=1}$ as shown in Figure 1.5 which was derived from Figure 1.2 and Equations 1.1 and 1.2.

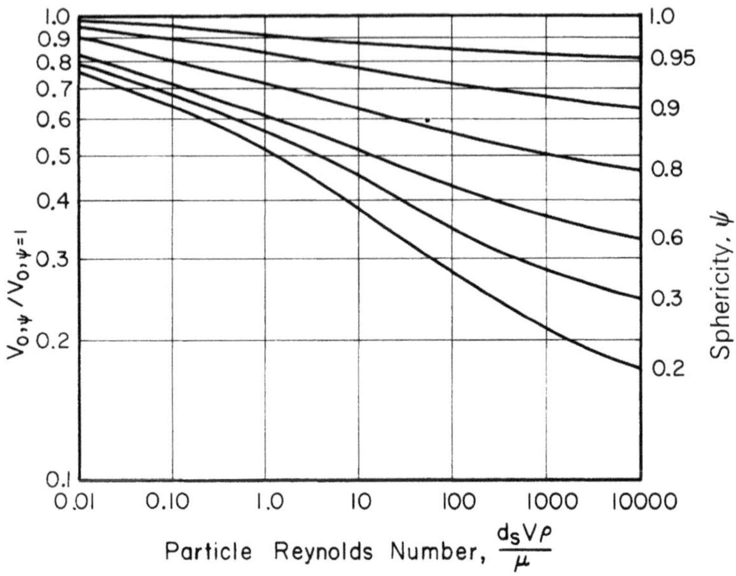

Figure 1.5. Effect of particle shape on terminal settling velocity.

3.24 Effect of Tube Wall

Another complication in the rise or fall of single or widely separated particles in a fluid is the interference to the motion caused by the presence of a vessel wall. The wall effect for vertical tubes has been the subject of extensive experimental and theoretical study. Happel and Brenner (1965) present a detailed review of this work. Empirical correction factors by which the wall-free settling velocity should be multiplied to obtain the actual velocity as given by Brown and Associates (1950) are:

For laminar flow (Francis, 1933)

$$F_w = 1 - \left(\frac{d}{D}\right)^{2.25} \tag{1.16}$$

and for turbulent flow (Munroe, 1888)

$$F_w = 1 - \left(\frac{d}{D}\right)^{1.5} \tag{1.17}$$

where D is the diameter of the tube. These equations give reasonable order of magnitude results for solid spheres traveling on the tube axis. More sophisticated relations based upon theoretical analysis and experiment are available for both solid and fluid spheres. We discuss the most relevant of these in Section 4.4 of Chapter 8.

3.3 Effect of Concentration and Hindered Settling Velocity

With increase in the concentration of the particles in a fluid the questions of hydrodynamic interference and particle collision arise. Figure 1.6, based on purely geometrical considerations, indicates the center-to-center interparticle distance, x/d, for spherical or cubic particles as a function of the volume fraction of the dispersed phase. For volume concentrations of even 1 percent the interparticle distance is only four diameters; at about 5 percent the distance is only about 2.5 diameters, and at 10 percent only about two diameters. From this one might expect hydrodynamic interference, interparticle collision, and interaction to be the rule rather than the exception at volume concentrations in excess of 2 or 3 percent. Uniformly sized particles rising or falling in laminar motion would have uniform velocities, and collisions would not be expected even as the concentration is increased. Nonetheless, hydrodynamic interference does exist at volume concentrations as low as 1 or 2 percent, and where the particles do not agglomerate the settling velocity is reduced.

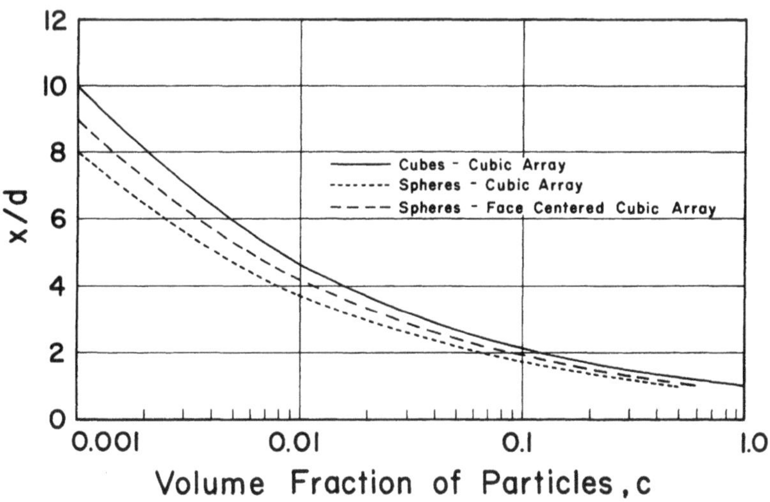

Figure 1.6. Center-to-center interparticle distances in dispersed mixtures.

The problem in the laminar flow region has been treated theoretically by several investigators. Burgers (1941, 1942) derived the following simple equation for random suspensions:

$$\frac{V_c}{V_0} = \frac{1}{1 + 6.88c} \tag{1.18}$$

where V_c is the settling velocity of particles in a concentration or volume fraction of particles, c. Famularo and Happel (1965) show that for randomly dispersed spherical particles the hydrodynamic interference results in

$$\frac{V_c}{V_0} = \frac{1}{1 + \gamma c^{1/3}} \tag{1.19}$$

where γ is a constant in the range of 1.30 ± 0.24.

Happel and Brenner (1965) give a detailed discussion of the laminar flow of a fluid through an assemblage of solid particles (equivalent to the rise or fall of swarms of particles in a fluid). They develop fundamental relationships for the drag force exerted on spherical particles for a number of different idealized cases spanning the particle concentration range from very dilute to concentrated. Their analysis supports the approach of Famularo and Happel and an equation of the type of Equation 1.19 for dilute concentrations.

Famularo and Happel also derive an expression for the ratio of the laminar settling velocity of vertically oriented doublet spheres to that of a single sphere and show that

$$\frac{V_c}{V_0} = \frac{1}{\lambda(1 + 2.8\lambda\gamma c^{1/3})} \tag{1.20}$$

where λ is a resistance coefficient dependent on the center-to-center distance of the spheres in the doublet and given as follows:

$\frac{d}{x} = \frac{\text{sphere diameter}}{\text{center-to-center distance}}$	Coefficient λ
0	1.0
0.266	0.836
0.648	0.702
1.0	0.645

The great significance of Equation 1.20 is that it shows how the terminal settling velocity of particles is increased by agglomeration even to doublets. Equations 1.19 and 1.20 are depicted graphically in Figure 1.7 along with the earlier relation of Burgers.

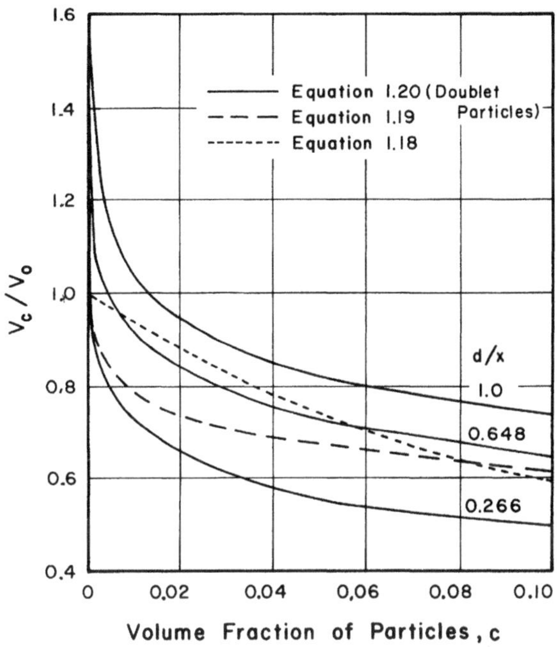

Figure 1.7. Effect of concentration on terminal settling velocity.

In turbulent motion random collisions are to be expected resulting in the speeding up of some particles, and the slowing down of others. The number and effect of these collisions increase with increasing particle concentration. Furthermore, in mixed-size particle suspensions, the larger, faster moving particles tend to overtake the smaller ones and collisions occur; in the case of particles that have no tendency to coalesce or agglomerate, the collisions tend to slow down the larger, and speed up the smaller, particles.

In the case of particles that tend to agglomerate, the collisions result in clusters or flocs that may contain many particles and, in addition, occluded fluid within the particle cluster. These clusters, whether of gas bubbles, liquid droplets, or solid particles, have settling velocities much different from the velocity of single particles and appropriate for the overall size, shape, and effective density of the cluster. Clusters that are stable under the relatively quiet conditions of settling ordinarily tend to break down under flow conditions; the degree of breakdown is dependent on the shear stresses to which they are subjected.

When the concentration of particles of mixed size is such that collisions and interaction are common, and as a result the particles tend to settle *en masse*, the process is known as hindered settling. The settling rate in hindered settling is much lower than that during free settling of the same particles. The concept of particles settling in an effective fluid of the bulk density and consistency of the mixture has been offered to explain this (see Brown and Associates, 1950). On this basis, Stokes' Law becomes

$$V_c = \frac{gd^2(\rho_P - \rho_M)}{18\mu_M} \tag{1.21}$$

where ρ_M is the bulk density of the suspension, and μ_M is the "effective viscosity" of the suspension.

The effective density and viscosity explanation fails if the particle size range is small and if the particles are settling together, or if the particles are flocculated and settling through the clear fluid. An alternate and better explanation lies in the recognition that as particles settle they displace fluid upward so that their velocity relative to the fluid exceeds their absolute velocity. The relative velocity effect is small at low concentrations but, as settling progresses and the concentration increases, the effect can become significant. Under these conditions the settling rate may be determined as the negative of the superficial velocity of the clear fluid displaced upward through the bed of particles under the action of the difference in the hydrostatic heads of the bulk mixture and the clear fluid.

Following this approach, Kearsey and Gill (1963) have shown how the Carman–Kozeny equation describing the laminar flow of a fluid through a porous medium may be applied to hindered settling. The Carman–Kozeny equation, developed by Kozeny (1927) and Carman (1937), is

$$V = \frac{g_c}{K} \frac{(1-c)^3}{c^2} \frac{1}{\mu s_P^2} \frac{\Delta P}{L} \tag{1.22}$$

where K is the Kozeny constant, really a function of porosity, particle shape, orientation, and size range. The range of values of K is 3 to 6, and 5 is the value commonly used. This equation and several variations are discussed in detail by Scheidegger (1960). The pressure gradient causing the flow is denoted by $\Delta P/L$, and s_P is the specific surface of the particles.

For the sedimentation process the difference in the hydrostatic gradients is

$$[(1-c)\rho + c\rho_P]\frac{g}{g_c} - \rho\frac{g}{g_c}$$

so that

$$\frac{\Delta P}{L} = c(\rho_P - \rho)\frac{g}{g_c} \tag{1.23}$$

and therefore

$$V_c = \frac{g}{K} \frac{(1-c)^3(\rho_P - \rho)}{\mu s_P^2 c} \tag{1.24}$$

For a sphere whose specific surface is expressed as

$$s_p = \pi d^2/(\pi d^3/6) = 6/d$$

the equation becomes

$$V_c = \frac{gd^2}{36K}\frac{(1-c)^3(\rho_P - \rho)}{c\mu} \tag{1.25}$$

illustrating that particle size and density, and fluid density and viscosity, play the same roles as in laminar free settling. Consideration of the term $(1 - c)^3/c$ indicates the sensitivity of the hindered settling velocity to concentration. As particle concentration is increased tenfold from a volume fraction of 0.03 to 0.30, the term decreases in the ratio of about 1 to 27. Taking the value of K as 5.0, one may determine the ratio of the velocity in hindered settling where Equation 1.25 applies, to that in free settling where Equation 1.9 applies as

$$\frac{V_c}{V_0} = \frac{(1-c)^3}{10c} \tag{1.26}$$

This ratio is plotted versus c in Figure 1.8. The ratio becomes unity at a volume concentration of about 0.078 when the separation distance between the particles (Figure 1.6) is about one diameter. This degree of separation seems

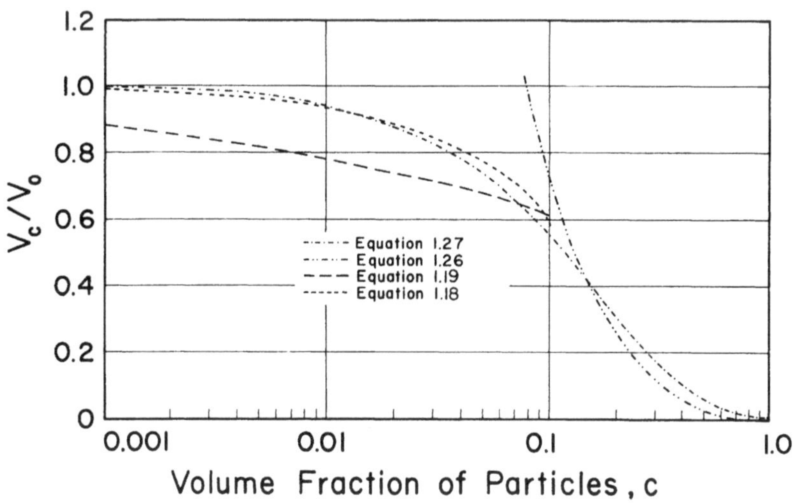

Figure 1.8. Effect of concentration on terminal settling velocity.

reasonable but is of no special significance. For nonspherical particles both Stokes' Law and Equation 1.25 are modified, but similar considerations apply.

Happel and Brenner (1965) compare their theoretical developments for several idealized concentrated particle assemblages with the Carman–Kozeny equation. They find a surprisingly good agreement.

Thomas (1963) gives the following wholly empirical expression for the reduction of settling velocity with increase in concentration and indicates that it is valid in the range $1 > V_c/V_0 > 0.08$,

$$2.303 \log \frac{V_c}{V_0} = -5.9c \tag{1.27}$$

This relationship is also shown in Figure 1.8. This figure is useful in depicting the approximate regions of free and hindered settling and the rapid decrease in settling velocity once particle interference becomes significant.

The Kozeny-based relationship may be expected to apply, at least in principle, up to concentrations where the particles are in contact and offer support to one another. From this point on the pressure gradient causing upward flow of the fluid will become rapidly less than that given by Equation 1.23. In case of hard spheres, the settlement process is complete once the particles come into mutual contact, and rearrangement occurs caused by particles sliding over one another. In case of flocculated particles or clusters containing occluded fluid which are subjected to deformation under stress, further settlement may occur as a result of the weight of particles on those below. This is said to be caused by compaction.

As irregularly shaped particles, and particularly flocs, come into direct contact with one another, the subsequent settlement is resisted by the resulting "structure." From a rheological viewpoint the system develops a yield stress and deformation occurs only if the net forces acting upon the concentrated suspension exceed its yield stress. When they do, and compaction results, the particles move closer together and the yield stress of the structure increases. This process continues until an equilibrium is established, at which point the weight of the "overburden" is supported by the structure of the fluid-saturated, compacted sediment.

Gal-Or and Waslo (1968) have extended work of Levich (1962) on single fluid particles in the presence of surfactants and of Happel (1958, 1959) on swarms of solid particles to develop a theoretical relationship for the effect of both particle and surfactant concentration on the fall or rise velocity of particle swarms. This analysis is based on a "free surface" cell model as used by Happel. Gal-Or (1970) gives an alternate solution to the same problem based upon a zero vorticity boundary condition first used by Kuwabara

(1959). The solutions of Gal-Or and Waslo and of Gal-Or are of special interest since they include the Stokes, Hadamard–Rybczynski, Levich, and Kuwabara solutions as limiting cases. The indicated references should be consulted for details.

3.4 Pseudohomogeneous Fluids

Whether a fluid-fluid or fluid-solid dispersion will behave as a pseudohomogeneous fluid or not depends on the degree of turbulence as well as on the phase separation velocity which we have seen to depend on the particle size, shape, density, and concentration, and the fluid density and viscosity. We should note that the volume fraction of particles, which may affect the settling velocity, is itself influenced by the settling velocity in flowing systems. We deal with this more fully later.

Laminar Flow For laminar horizontal flow, unless the settling velocity is very low and the pipe through which flow is to occur is short, even a relatively fine dispersion must be treated as a two-phase system. An indication of the settling velocity which may be tolerated without any significant settling may be obtained from the relationship

$$V_{sm} = \frac{4Vx^2}{LD} \tag{1.28}$$

where V_{sm} = the maximum tolerable settling velocity,
V = the average pipe flow velocity,
x = the average interparticle distance,
L and D = the pipe length and diameter, respectively.

The relationship is derived on the assumption that a particle within a distance of x from the wall could be permitted to settle on the wall by the time the fluid carrying it, and moving at its local velocity, reached the end of the pipe. An alternate practical approach is to say that a certain amount of settling may be tolerated and the mixture still treated as pseudohomogeneous. For example, if settlement at a distance of $0.05D$ is thought to be permissible, the maximum settling velocity, from Equation 1.28, would be

$$V_{sm} = \frac{DV}{100L} \tag{1.29}$$

The use of these two equations, together with Figures 1.3, 1.5, and 1.8, will enable one to determine whether a significant amount of settling would be expected in a horizontal line and, if not, that single-phase methods could be used.

Settlement in laminar flow in a vertical line is much less serious partly because of the normally shorter lengths and partly because the settlement is directly opposed by the fluid motion. If one assumes that settlement is not serious so long as the particles within a distance of y from the wall have settling velocities no greater than the local velocity at the point, the relationship

$$V_{sm} = \frac{8yV}{D} \tag{1.30}$$

may be derived.

Turbulent Flow In the case of turbulent flow in a horizontal pipe, one would expect some mixtures, which under static or laminar flow conditions exhibited significant phase separation velocities, to behave as pseudohomogeneous systems because of the mixing action of turbulence.

Experimental work and limited theoretical studies indicate that the turbulent driving force overcoming the tendency of particles to settle in horizontal pipe flow is a function of the friction velocity defined as

$$U^* = V\sqrt{\frac{f}{2}} = \sqrt{\frac{g_c D\Delta P}{4\rho L}} = \sqrt{\frac{\tau_w g_c}{\rho}} \tag{1.31}$$

where f, ΔP, and τ_w are the friction factor, pressure drop, and wall shear stress, respectively, as discussed in detail in Chapter 4. Based upon the foregoing, dimensional considerations suggest (Thomas, 1961)

$$\frac{V_0}{U^*_{M2}} = \phi\left(\frac{dU^*_{M2}\rho}{\mu}\right) \tag{1.32}$$

where U^*_{M2} is the friction velocity corresponding with an average mixture velocity V_{M2} above which the particles will be suspended and, in the case of very fine particles, the system approximates pseudohomogeneity.

Thomas (1961, 1964) has shown that, for fine particles, a dilute suspension becomes substantially homogeneous as soon as lift forces develop which prevent cumulative deposition of the particles on the wall of the pipe. He has correlated his own and other data regarding V_{M2}, for a variety of small particles (22-66 microns) suspended in air and in water flowing in pipes ranging from 1 to 12 inches in diameter by the equation

$$\frac{V_0}{U^*_{M2}} = 0.01\left(\frac{dU^*_{M2}\rho}{\mu}\right)^{2.71} \tag{1.33}$$

This equation is considered valid for particles sufficiently small to be totally immersed in the laminar sublayer. We will show in Chapter 4 that this means that $dU^*_{M2}\rho/\mu < 5$. Equation 1.33 is valid for fine particles at volume fraction solids up to 0.06 or 0.08.

For larger particles Thomas (1964) proposes the empirical equation

$$\frac{V_0}{U_{M2}^*} = 4.90\left(\frac{dU_{M2}^*\rho}{\mu}\right)\left(\frac{\mu}{DU_{M2}^*\rho}\right)^{0.60}\left(\frac{\rho_P - \rho}{\rho}\right)^{0.23} \tag{1.34}$$

Here U_{M2}^* corresponds, as in Equation 1.33, to V_{M2}, to the minimum velocity for full suspension of the particles, but the particles of this larger size range would not be uniformly suspended and the assumption of pseudohomogeneity would be a crude one.

Spells (1955) also has developed a relationship based on dimensional considerations for the velocity required to produce pseudohomogeneity. His equation tested for particles from 50 to 500 microns is

$$V_{M1} = 0.1196\left[gd\left(\frac{\rho_P - \rho}{\rho}\right)\right]^{0.816}\left(\frac{D\rho_M}{\mu}\right)^{0.633} \tag{1.35}$$

where V_{M1} is the minimum velocity for full symmetric suspension of the particles.

Neither Equation 1.34 nor 1.35 is fully confirmed, and about all we may expect is order of magnitude accuracy in predictions from them.

Other aspects of the minimum velocity in a horizontal pipe for pseudohomogeneity are discussed in Chapter 11.

For vertical upward flow the situation in turbulent flow is essentially similar to that in laminar flow, the variance being due to the difference in the turbulent and the laminar velocity profiles. The maximum tolerable settling velocity corresponding with any radial position in the pipe may be taken as the local turbulent velocity at that point and determined with the aid of Figure 4.9, Chapter 4.

The foregoing gives ways of predicting the pipe velocities above which suspensions of particles of given terminal settling velocities may be expected to behave as pseudohomogeneous fluids. Under these circumstances the pipe-flow design methods appropriate for single-phase fluids may be applied. If the suspension is dilute and if it behaves as a Newtonian fluid, the normal methods suitable for Newtonians are applicable, with the density and viscosity being taken as those of the mixture. Many mixtures, and especially the more concentrated suspensions, exhibit non-Newtonian behavior, and, although they may be pseudohomogeneous, the methods appropriate for their non-Newtonian characteristics must be used.

3.5 Recommended Design Procedure

On the basis of the foregoing, it is possible to set out a procedure for determining whether or not in a particular case a fluid-fluid or a fluid-solid system may be treated as pseudohomogeneous for pipeline flow calculation purposes.

If a mixture may be so treated, design methods for homogeneous fluids are applicable; if not, the design methods must take account of the presence of the two or more phases. It is assumed that the volume fractions and the separate properties of the two phases, the size or size range and the shape of the dispersed phase, and the approximate velocity at which the mixture is to be transported are all known. The procedure is as follows:

1. If the mixture is truly homogeneous or if the dispersed particles are ultrafine and the proposed transport velocities are high, provided electrostatic effects are not present, it may be apparent without further consideration that design methods for homogeneous fluids are suitable.
2. If the mixture is clearly one of separate phases, which will tend to separate rapidly (as in the case of coarse coal particles in water), it may be apparent without further consideration that multiphase design methods must be used.
3. If neither 1 nor 2 is applicable, it is necessary to estimate the phase separation velocity of the mixture and determine, having regard for the proposed mixture transport velocity and the length of the pipeline, whether the system will behave as one which is pseudohomogeneous or not.
4. Ideally, laboratory tests should be conducted with the system at the concentrations of interest to determine the actual phase separation velocity.
5. In the absence of laboratory tests the low concentration phase separation velocity in an infinite medium may be estimated for spherical particles from Equation 1.9, 1.10, or 1.11 or directly from Figure 1.3.
6. If the particles are not spherical or nearly so, the separation velocity may be corrected for particle shape from Figure 1.5.
7. If the proposed pipeline is to be vertical, the wall effect on the phase separation velocity may be estimated from Equation 1.16 or 1.17. (See also Figure 8.30, Chapter 8.)
8. If the concentration of the dispersed phases exceeds 2-3 volume percent, the effect of particle interference or hindered settling may be estimated from Equation 1.27 or Figure 1.8.
9. Given the estimated phase separation velocity corrected for particle shape, wall effect, and particle concentration, its significance in determining the pipeline flow behavior of the system is given
   (a) for horizontal laminar flow, by Equation 1.28 or 1.29,
   (b) for vertical laminar flow, by Equation 1.30,
   (c) for horizontal turbulent flow of fine particles, by Equation 1.33, and of coarse particles, by Equation 1.34,
   (d) for vertical turbulent flow, by a relationship similar to Equation 1.30 for the turbulent flow profile expected; and the applicability or otherwise of pipeline design methods for homogeneous systems is determined.

4 CLASSIFICATION OF SINGLE-PHASE AND PSEUDOHOMOGENEOUS MULTIPHASE MIXTURES

4.1 Introduction

Purely viscous, single-phase fluids and pseudohomogeneous multiphase fluid mixtures that are stable even in the absence of turbulence may be classified in accordance with the nature of their response to shearing stresses under conditions resulting in unidirectional laminar flow. (In viscoelastic fluids, i.e., fluids exhibiting elastic as well as viscous properties, the normal stresses developed during steady shear and the relaxation phenomenon that occurs during non-steady processes are also important. These matters are discussed later.)

Consider the system of two parallel moving plates separated by the fluid of interest as shown in Figure 1.9. In the figure x and z are, respectively, the Cartesian coordinates in the direction of flow and perpendicular to the plates. The shearing stress imposed on the fluid is $F/A = \tau_{zx}$, or simply, τ. The fluid is subjected to strain at the rate du_x/dz, which is the velocity gradient or the rate of shear, to the absolute value of which we assign the symbol, $\dot{S}$. One would expect a relationship between the imposed shearing stress, τ, and the resulting rate of shear, $\dot{S}$; and experience confirms that these quantities are related, although differently for different fluids, and differently for the same fluid under different temperature and/or pressure conditions.

The relationship

$$\tau = f(\dot{S}) \tag{1.36}$$

which is characteristic of a given fluid or fluid system at any given pressure and temperature, is known as the constitutive or rheological equation for the fluid. For purely viscous fluids this equation describes the rheological behavior of the fluid. The graphical representation of the equation is known as the rheogram for the fluid.

In the completely general case the functional relationship between the velocity gradient and the imposed shear stress is not a simple one. This is

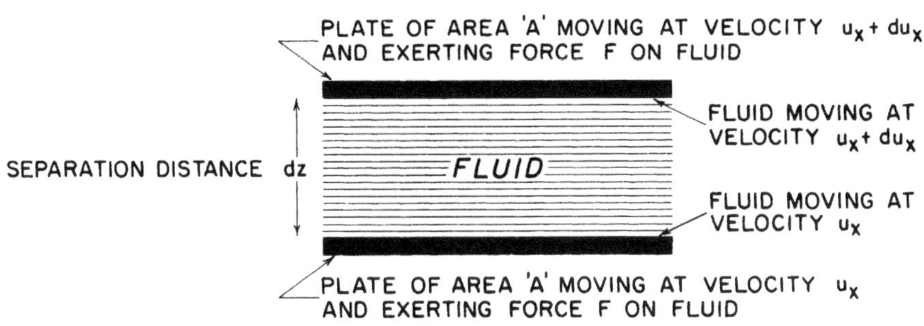

Figure 1.9. Schematic diagram of basis for Equation 1.36.

particularly evident with disperse multiphase mixtures, or solutions, or even pure liquids containing extremely long molecules. Here the velocity gradient generated by a shear stress is influenced by changes in the internal physical structure which, in turn, are induced by the shear action. In some systems these changes in structure merely involve changes in orientation and alignment of the particles or molecules in the direction of the shear stress. In other systems the physical and chemical interactions between the dispersed particles produce particle agglomerates or structures which are fragile and sensitive to shear. In even more complex cases the particles or long-chain molecules may possess elastic properties, and these may influence the response of the fluid to the shear stress.

In general terms, then, we may expect the shearing stress within a fluid to be related not only to the rate of shear but to all other factors which determine the concentration and resistance of macromolecules, particles, or particle agglomerates. This may be expressed as

$$\tau = f(\dot{S}, c_P, v_P, o_P, \psi_P, k_{PF}, k_{PD}, k_{PA}, k_{PM}, \theta, E) \tag{1.37}$$

where c_P, v_P, o_P, ψ_P are, respectively, the concentration, average volume, orientation, and the shape of the macromolecules, particles, or agglomerates at time θ after the start of shearing,

k_{PF} and k_{PD} are coefficients descriptive of the rate of formation or destruction of macromolecules, particles, or agglomerates,

k_{PA} and k_{PM} are coefficients descriptive of the rate of alignment or misalignment of the macromolecules, particles, or agglomerates in the direction of the shear stress,

E is a measure of the elastic properties of the composite system.

Although considerable effort has been expended in the search for exact relationships among the variables which characterize the rheological behavior of fluids and fluid systems, thoroughly satisfactory constitutive equations have so far been devised only for the simplest case, the so-called Newtonian fluid. One of the major problems is the difficulty of determining, either theoretically or experimentally, the independent effect of each of the many influencing factors. On the other hand, it is possible, through reference to Equation 1.37, to explain qualitatively certain limiting forms of rheological behavior, and this serves as a basis of classification.

Fluids are classified into various rheological types on the basis of the form of their rheological equations or rheograms. The scheme of classification is shown in Figure 1.10.

The major division is between fluids that are described as purely viscous and those that exhibit both viscous and elastic properties. Purely viscous fluids are those which, on removal of the shearing force, do not recover from any

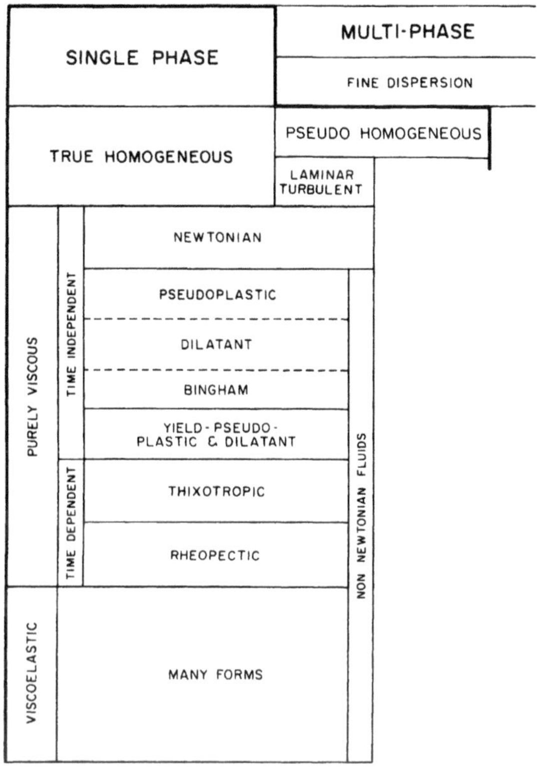

Figure 1.10. Rheological classification of complex mixtures that behave as single-phase systems.

deformation they may have undergone under its action. The coefficients of any terms involving E in Equation 1.37 are zero. Viscoelastic fluids are those which, on removal of the shearing force, do, in part, recover from the deformation they have undergone during the shearing action. Viscoelastic fluids may be considered as substances intermediate between purely viscous fluids, on the one hand, and purely elastic solids, which recover fully from deformation up to their yield stresses, on the other hand. The coefficients of terms involving E in Equation 1.37 are not all zero.

The purely viscous fluids may be divided first into the categories of time-independent and time-dependent. The first category includes those fluids which, if they exhibit development of structure or orientation of particles, have rates k_{PF}, k_{PD}, k_{PA}, and k_{PM} which are sufficiently high that, for all intents and purposes, the fluids attain an equilibrium condition immediately. Such high values of the rate constants are equivalent to a situation where the coefficients of all terms involving θ in Equation 1.37 are zero. The time-dependent category includes those other fluids that have lower rates of development or decay of structure or orientation, take a measurable time to attain their equilibrium conditions, and whose behavior therefore depends both upon the rate constants and duration of shear measured from a time of known condition. The coefficients of terms involving θ in Equation 1.37 are not all zero.

The time-independent group of purely viscous fluids may be subdivided into several different types. The first category is that for which the rheological behavior is such that the rheogram passes through the origin of the τ and $\dot{S}$ coordinates. This means that the fluid responds to the smallest applied shearing stress or that there is no "yield" strength required to be overcome before flow commences. In the language of Equation 1.37, this means that the rate constants are all high, that the coefficients of terms involving E and θ are all zero, and that $c_P = 0$, or if $c_P \neq 0$, v_P, ψ_P, and o_P are such that the macromolecules, particles, or agglomerates offer no resistance to the initiation of movement of the fluid. Within this "no yield" category are the Newtonian, the pseudoplastic, and the dilatant fluids. These will be discussed individually later.

The second category of purely viscous fluids is that for which a finite shearing stress is required before flow commences. The rheogram for these fluids does not pass through the origin of coordinates but intercepts the τ axis at a finite value known as the yield stress and characteristic of the fluid. In terms of Equation 1.37, this means that the rate constants are all high, that the coefficients of all terms involving E and θ are zero, but that $c_P \neq 0$ and v_P, ψ_P, and o_P are such that the macromolecules, particles, or agglomerates offer a finite resistance to the initiation of movement of the fluid. In this "yield stress" category of fluids are the Bingham fluids and the yield-pseudoplastics. These also are discussed individually later.

The time-dependent group of purely viscous fluids includes the thixotropic and the rheopectic fluids. These are similar to the pseudoplastic or yield-pseudoplastic fluids except that their behavior is dependent upon the duration of shear. In Equation 1.37 the rate constants, or some of them, are not high, the coefficients of all terms involving E are zero, the coefficients of all terms involving θ are not zero, and c_P, v_P, ψ_P and o_P, or one or more of them, change with time. Details are given later.

The less well understood viscoelastic fluids exhibit a complex blending of viscous and elastic behavior. They are not readily classified into distinct types and include many combinations of influence of the rate phenomena, orientation, and structural strength. In general, none of the rate constants, or the coefficients of terms involving θ, c_P, v_P, o_P, or ψ_P need be zero and the coefficients of all terms involving E are not zero. Viscoelastic fluids are discussed further in a later section.

4.2 *Time-Independent Viscous Fluids*

4.21 Newtonian Fluids

Newtonian fluids are so named because their rheological behavior follows the postulation of Newton that fluids might be expected to respond to an applied shearing stress by flowing in a manner such that the velocity gradient is strictly proportional to the applied stress. Newton postulated this for all fluids subject only to the restriction that the stress not be so high as to cause turbulence within the fluid. The Newton hypothesis is represented by the constitutive or rheological equation

$$\tau = \frac{\mu}{g_c}\dot{S} \tag{1.38}$$

where μ is the constant of proportionality in the relationship characteristic, at any given pressure and temperature, of the Newtonian fluid. The constant of proportionality is known as the viscosity of the fluid and this single property defines the rheological behavior of fluids which follow Equation 1.38.

In fact, it has been found that a vast number of fluids do, under actual test, behave this way and are Newtonian. These include all gases; many liquids, and especially those having relatively simple, nonelongated molecules; and pseudohomogeneous suspensions of spherical particles in gases or simple liquids.

The rheograms for three typical Newtonian fluids—air, water, and glycerine—are presented in Figures 1.11 and 1.12. These rheograms are based on measured data and thus reflect the actual behavior of these three fluids. Glycerine is a high viscosity fluid, water a medium viscosity fluid, and air a low viscosity fluid. Their actual viscosities at 1 atmosphere pressure and 15°C are:

	Viscosity, centipoises
glycerine	2330
water	1.1405
air	0.0182

The rheograms for these fluids exhibit the characteristics of all Newtonians. The arithmetic rheograms of Figure 1.11 show straight lines passing through the origin. The slope of the lines is the viscosity of the fluid. The logarithmically plotted rheograms also show straight lines but these are of unity slope. They have intercepts at $\dot{S} = 1$ equal to their viscosities.

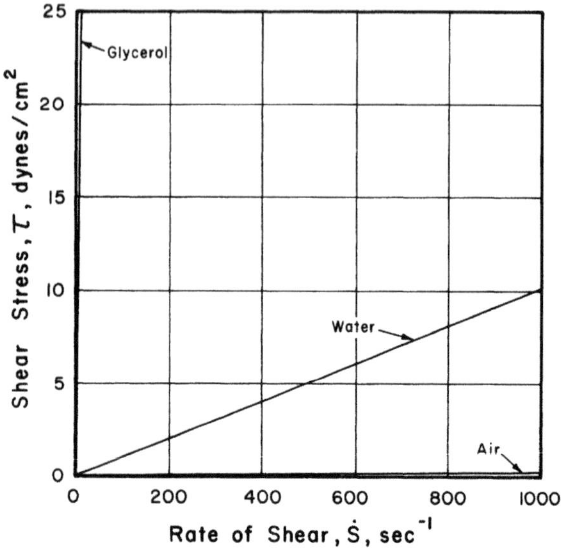

Figure 1.11. Rheograms for typical Newtonian fluids—arithmetic coordinates.

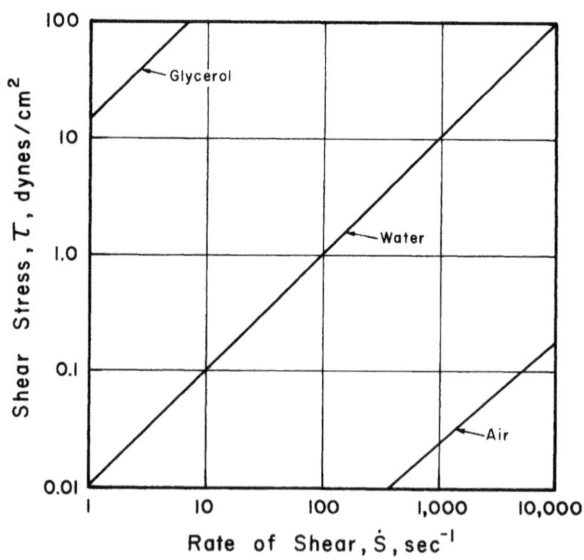

Figure 1.12. Rheograms for typical Newtonian fluids—logarithmic coordinates.

4.22 Pseudoplastic Fluids

Pseudoplastic fluids are those fluids whose behavior is time-independent, for which an infinitesimal shear stress will initiate motion, and for which the rate of increase in shear stress with velocity gradient decreases with increasing velocity gradient. This type of behavior is widely encountered in solutions or suspensions in which large molecules or fine particles form loosely bounded aggregates or alignment groupings that are stable and reproducible at any given shear rate, but which rapidly and reversibly break down or reform with increase or decrease in shear rate. There is no single or simple form of constitutive equation that accurately describes the rheological behavior of pseudoplastics, although several empirical equations are useful over limited ranges of velocity gradient. Even the simplest but most limited of these involves two constants to characterize the fluid as compared with the single property for Newtonians. Other equations, applicable over wider ranges of velocity gradient, incorporate three or more such constants. Rodriguez (1966) reviews many of these.

Apparent Viscosity Although it is clear that only Newtonian fluids possess a property which strictly may be called viscosity and all non-Newtonians require two or more parameters to describe their consistency behavior, nonetheless it is often convenient to refer to an "apparent viscosity" of a non-Newtonian. The apparent viscosity, μ_a, is defined by

$$\mu_a = g_c \frac{\tau}{\dot{S}} \tag{1.39}$$

and varies with $\dot{S}$.

For pseudoplastics, μ_a declines from μ_0 at $\dot{S} = 0$ asymptotically to a limiting value μ_∞ as $\dot{S} \to \infty$.

The Power Law Widely used in engineering calculations is the so-called power law

$$\tau = \frac{K}{g_c} \dot{S}^n \tag{1.40}$$

In this equation K is the consistency index or power law coefficient, and n is the flow behavior index or power law exponent. The flow behavior index, n, is readily determined as the slope of a plot of τ versus $\dot{S}$ on logarithmic coordinates. The value of n is less than unity for pseudoplastics. We may note that if $n = 1$, Equation 1.40 reduces to Equation 1.38, that for a Newtonian fluid. Also, from Equation 1.39 we see that the apparent viscosity for a pseudoplastic described by the power law is given by

$$\mu_a = K\dot{S}^{n-1}$$

and decreases with increase in shear rate.

The power law fits experimental data for many pseudoplastics at intermediate velocity gradients and over a range of 10 to 100 fold, but fails both at very low and very high velocity gradients. As the velocity gradient for a pseudoplastic approaches zero, the shear stress approaches zero and actual data conform with the equation for a Newtonian fluid. Similarly, at sufficiently high velocity gradients, the rate of increase of shear stress with the velocity gradient becomes constant and data again conform to the Newtonian equation. Notwithstanding these shortcomings, we will see that, for engineering problems involving flow in a pipe, the power law, if fitted to data at and immediately below a value of τ corresponding with the shear stress at the pipe wall, gives excellent accuracy.

The Prandtl–Eyring Equation A second two-constant equation (Eyring, 1936; Prandtl, 1928) describing the behavior of pseudoplastics is an equation based upon Eyring's kinetic theory of liquids. This is

$$\tau = \frac{A}{g_c} \sinh^{-1}\left(\frac{\dot{S}}{B}\right) \tag{1.41}$$

where A and B are rheological coefficients characteristic of the material. The apparent viscosity of a fluid obeying the Prandtl–Eyring equation is given by

$$\mu_a = \frac{A \sinh^{-1}\left(\dfrac{\dot{S}}{B}\right)}{\dot{S}}$$

Like the power law, this equation fits experimental data over limited ranges of velocity gradient. Unlike the power law, it does reduce asymptotically to the Newtonian law, with $\mu = A/B$, at low velocity gradients.

The so-called Powell–Eyring equation (Christiansen, Ryan, and Stevens, 1955; and Powell and Eyring, 1944) is formed from the Prandtl–Eyring relationship by the addition of a term linear in the velocity gradient, in effect to give an equation which indicates the shearing stress to be composed of a Newtonian and a Prandtl–Eyring component. This equation is discussed in Skelland (1967).

The Ellis Equation A more flexible but entirely empirical equation is that by Ellis (see Skelland, 1967). The Ellis equation, which contains three rheological constants, is explicit in the velocity gradient rather than the shear stress

$$\dot{S} = (\phi_0 + \phi_1 \tau^{\alpha-1})\tau \tag{1.42}$$

where ϕ_0, ϕ_1, and α are rheological coefficients characteristic of the material. We may note that for a fluid described by the Ellis equation

$$\mu_a = \frac{g_c}{\phi_0 + \phi_1 \tau^{\alpha-1}}$$

and also that, if $\phi_1 = 0$, the equation reduces to the Newtonian equation with $\phi_0 = g_c/\mu$; if $\phi_0 = 0$, it reduces to the power law with $\alpha = 1/n$ and $\phi_1 = (g_c/K)^{1/n}$. When $\alpha > 1$, the equation approaches the Newtonian equation at low shear stresses, whereas with $\alpha < 1$ it tends toward the Newtonian equation at high shear stresses.

The Reiner–Philippoff Equation Another three-parameter equation for pseudoplastics is the Reiner–Philippoff relation (Philippoff, 1935)

$$\dot{S} = \frac{g_c}{\mu_\infty + \dfrac{\mu_0 - \mu_\infty}{1 + (\tau/\tau_s)^2}}\,\tau \tag{1.43}$$

This equation reduces to Newton's Law both at very low shear rates with $\mu = \mu_0$ and at very high shear rates with $\mu = \mu_\infty$. The apparent viscosity is given by

$$\mu_a = \mu_\infty + \frac{\mu_0 - \mu_\infty}{1 + (\tau/\tau_s)^2}$$

The Sisko Equation A simple and versatile three-constant equation has been proposed by Sisko (1958) based on the rudimentary concept of additive Newtonian and non-Newtonian (power law) stresses, as follows:

$$\tau = a\dot{S} + b\dot{S}^c \tag{1.44}$$

where a, b, and c are the three characterizing properties. The equation may also be expressed as

$$\mu_a = (a + b\dot{S}^{c-1})g_c$$

Sisko found the equation to give an excellent match over a very wide range of shear rate with measured data for a series of hydrocarbon greases of varying composition and properties.

The Cross Equation Cross (1965) has developed from mechanistic considerations a three-coefficient constitutive equation for pseudoplastics which appears superior to any others of comparable simplicity. His equation is expressed in terms of apparent viscosity

$$\mu_a = \mu_\infty + \frac{\mu_0 - \mu_\infty}{1 + \alpha \dot{S}^{2/3}} \tag{1.45}$$

where μ_0, μ_∞ = the apparent viscosities at zero and infinity rate of shear,
α = a coefficient, related to the shear stability of the structure.

Cross' equation has been tested for a wide variety of pseudoplastics, including aqueous dispersions of polyvinyl acetate and of an acrylic copolymer; aqueous solutions of sodium carboxymethyl cellulose, and of ammonium polymethacrylate; an aqueous suspension of limestone; a nonaqueous pigment paste; and a nonaqueous polymer solution. The 2/3 exponent on the shear rate has some theoretical justification and is confirmed by these data. With the constants μ_0, μ_∞, and α, properly selected from actual data on each system, the equation reproduces the data extremely well.

The Meter Equation A versatile four-constant equation is proposed by Meter (1963) and used by Meter and Bird (1964) in their analysis of the laminar pipe flow of pseudoplastics. Meter's equation is

$$\mu_a = \mu_\infty + \frac{\mu_0 - \mu_\infty}{1 + \left(\dfrac{\tau}{\tau_m}\right)^{\alpha - 1}} \tag{1.46}$$

where μ_0, μ_∞ = the apparent viscosities at zero and infinity (or very high) rates of shear,
τ_m = the shear stress at which $\mu_a = \frac{1}{2}(\mu_0 + \mu_\infty)$.

When $\mu_\infty \ll \mu_0$, as is often the case, τ_m is the shear stress when $\mu_a \simeq \frac{1}{2}\mu_0$ and is designated $\tau_{1/2}$.

4.23 Dilatant Fluids

Dilatant fluids are those whose behavior is time-independent, for which an infinitesimal shear stress will start motion, but for which the rate of increase in shear stress with velocity gradient increases as the velocity gradient is increased. Mathematically they are similar to the pseudoplastics and indeed the empirical equations for pseudoplastics apply to the dilatants with appropriately different values of certain rheological constants. With reference to the power law, the flow behavior index, n, is greater than unity for dilatant fluids and less than unity for pseudoplastics.

Dilatant fluids are much less common than pseudoplastics and dilatancy is observed only in certain ranges of concentration in suspensions of irregularly shaped solids in liquids. It is due to the shift, under shearing action, of a close-packed particle system to a more open arrangement with the consequent entrapment or immobilization of some of the liquid. This "drying out" effect results in an increase in resistance to displacement and consequently an increasing slope in the $\tau - \dot{S}$ curve. The effect is generally reversible. The limited concentration range in which dilatancy is observed depends upon the size distribution and shape of the particles; at lower or higher concentrations the behavior is pseudoplastic, yield-pseudoplastic, or approximately Bingham.

Metzner and Whitlock (1958) studied the dilatancy phenomena with a variety of suspensions over a range of concentrations and shear rates. Their data clearly indicate that dilatancy is restricted to the relatively high concentration and high rate of shear ranges. The concentrations in which they observed dilatant behavior were those above 27-30 percent and dilatancy became apparent only at shear rates of 100 sec^{-1} for 40–47 percent suspensions and above about 500 sec^{-1} for 27–30 percent suspensions. The development of dilatancy appears to be independent of the viscosity of the suspending medium, at least in the range of 1–42 centipoises. Typical data of Metzner and Whitlock, showing the development of dilatancy with increase in concentration and shear rate for suspensions of 0.2–1.0 micron titanium dioxide particles in water, are given in Figure 1.13.

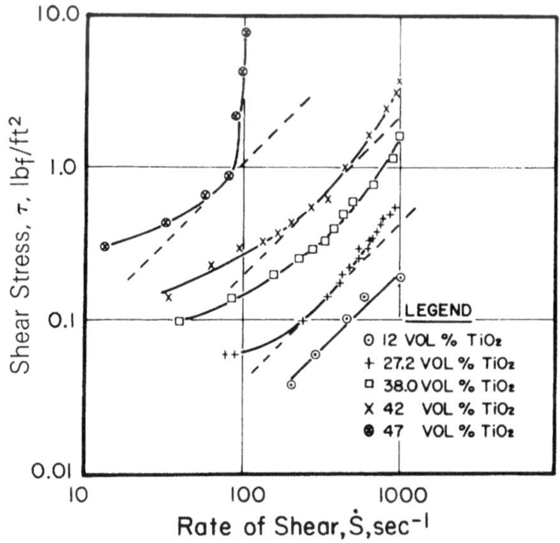

Figure 1.13. Effect of concentration on the development of dilatancy. Dashed lines represent Newtonian behavior—for reference only. (Modified after Metzner, A. B., and M. Whitlock, *Trans. Soc. Rheol.*, **2**, 239, 1958.)

Although Metzner and Whitlock, and others before them, have suggested that dilatancy is observed only with ultrafine particles of the order of 5 microns or less in diameter, there would appear to be no rational reason for this. In fact, we would expect that, in addition to concentration, only particle shape and surface condition would be relevant to whether or not dilatancy would occur. Data of Govier, Shook, and Lilge (1957) on water suspensions of magnetite, galena, and ferrosilicon tend to confirm this and, in any case, demonstrate the development of dilatancy in galena and ferrosilicon suspensions with mean particle sizes of 20–30 microns (size ranges from about 5 to 70 microns). The data for the galena suspensions are shown in Figure 1.14.

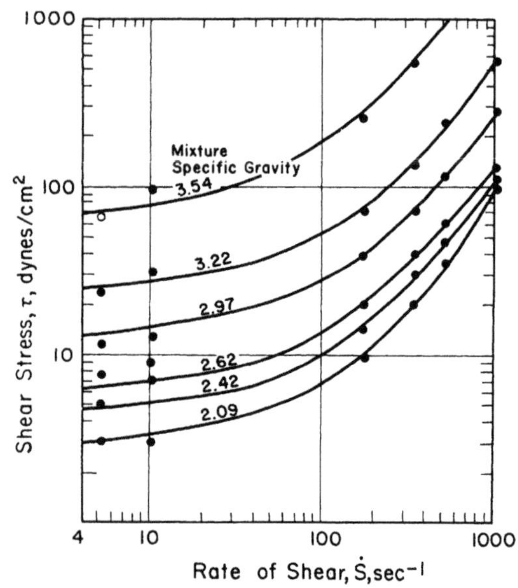

Figure 1.14. Rheograms for galena suspensions. (From Govier, G. W., C. A. Shook, and E. O. Lilge, *Trans. Can. Inst. Mining Met.*, **60**, 147, 1957.)

Metzner and Whitlock's explanation of the phenomenon of dilatancy is quoted as follows:

> Two mechanisms may account for the inflection and subsequent increases in slope of the curve. In the first place, as the shearing stresses reach levels at which the size of the aggregates or flowing particles has been reduced to a minimum, the fluid system should approach Newtonian behavior or constant viscosity. (This factor by itself should account for the inflections noted in the flow curves of other suspensions which approach Newtonian behavior at high shearing rates.) Second, and of much greater importance in suspensions which may become rheologically dilatant, at increasing stresses the system begins to expand volumetrically, as indicated by the data of Table 11 [see Table 1.3].

TABLE 1.3

Comparison of the Minimum Shearing Stresses at which Volumetric Dilation, Rheological Dilatancy, and Fracture of the Suspension First Occur[a]

System	Concentration of Solids, % by Volume	Shear Stress, lbf/ft², Corresponding to: Volumetric Dilation	Rheological Dilatancy	Visual and Audible Fracture in Viscometer
Titania in 42-cp sucrose solution	25.0	0.9	Above 2.2	—
	30.3	1.3	2.0	6.8
	38.5	—	3.5	5.8[b]
	40.2	0.9	2.0	4.4
Titania in 18-cp sucrose solution	30.0	0.8	1.0	2.9
	42.5	0.9	1.8	4.4[c]
Titania in water	27.2	—	0.14	0.43
	38.0	—	0.4	0.80
	42.5	0.4	0.6	1.0
	47.0	0.4	0.8	2.2

[a] A. B. Metzner and M. Whitlock, *Trans. Soc. Rheol.*, **2**, 239 (1958).
[b] Observed stress corresponding to audible cracking and fracture in the capillary tube was 5–6 psf.
[c] Observed stress corresponding to audible cracking and fracture in capillary tube was 5–6 psf for this suspension.

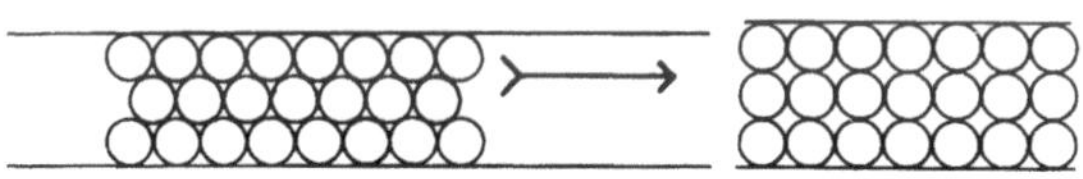

(a) (b)

Figure 1.15. Dilation (volumetric) of a closely packed suspension under shear. (a) Suspension at rest; particles fit into voids in next adjacent layer. (b) Suspension under rapid shear; particles have no time to fall into voids in adjacent layers and each layer glides over the adjacent one. (From Metzner, A. B., and M. Whitlock, *Trans. Soc. Rheol.*, **2**, 239, 1958.)

Accordingly, the flow mechanism begins to change to one in which entire layers of particles begin to glide over adjacent layers, as pictured by Reiner's diagrams depicting flowing dilatant suspensions. Figure 5 [see Figure 1.15] shows schematically why such expansion must occur in concentrated (closely packed) suspensions at high shear rates and indicates how this expansion may result in entire layers of particles gliding over the adjacent ones. However, to the present author's knowledge, no one has yet ventured to explain why such gliding of entire layers should lead to rheological dilatancy, i.e., increases in viscosity with increases in shear rate. The above picture of momentum transfer between adjacent layers or streamlines offers a possible explanation, however. At the lower shear rates discussed earlier, momentum transfer occurred between individual groups of particles in adjacent laminae. Now, however, with many particles in a given lamina or streamline moving past an adjacent one, momentum interchange, i.e., a change of velocity, of one particle in a lamina must concomitantly bring all the others following closely behind it to experience the same change in velocity. (If the particle in question moves from a region of high velocity to one of lower velocity. If the particle were to be accelerated, on the other hand, it would simultaneously need to accelerate all of the others in front of it.) Since, as one moves to progressively increasing shear rates within this high shear rate region the particles become progressively more perfectly aligned into laminae, one progressively increases the effective mass of the particles which experience momentum transfer with adjacent laminae whenever only a single particle moves laterally. It is suggested, that this mechanism becomes of importance as soon as some volumetric dilation occurs. However, since this dilation occurs while the flow is still pseudoplastic, it is not until one reaches considerably higher shear rates that this mechanism is able to predominate over those leading to pseudoplasticity, hence one does not experience rheological dilatancy until such higher shear stresses are reached.

4.24 Bingham Fluids

The Bingham fluid is to some extent a limiting or idealized case. It is a fluid for which a finite shearing stress is required to initiate motion and for which there is a linear relationship between the shearing stress in excess of the initiating stress and the resulting velocity gradient.

Materials that behave as, or nearly as, Bingham plastics include thickened hydrocarbon greases, certain asphalts and bitumens, some emulsions, and pseudohomogeneous suspensions of ultrafine or fine particles in liquids at intermediate concentrations (such suspensions may be Newtonian or pseudoplastic at low concentrations, and dilatant at very high concentrations). Included among the suspensions which, within a range of concentration and especially at low shear rates, behave as Bingham fluids are water suspensions of clay, fly ash, finely divided minerals, quartz, metallic oxides, sewage sludge, and paint systems. Aqueous solutions of certain long-chain polymers (i.e. carboxymethylene) exhibit yield points above a certain low concentration and may be Bingham-like, but they are usually better represented by the yield-pseudoplastic relations.

The constitutive equation for a Bingham fluid is

$$\tau - \tau_y = \frac{\eta}{g_c}\dot{S} \qquad (1.47)$$

where τ_y = the yield stress,
η = the coefficient of rigidity.

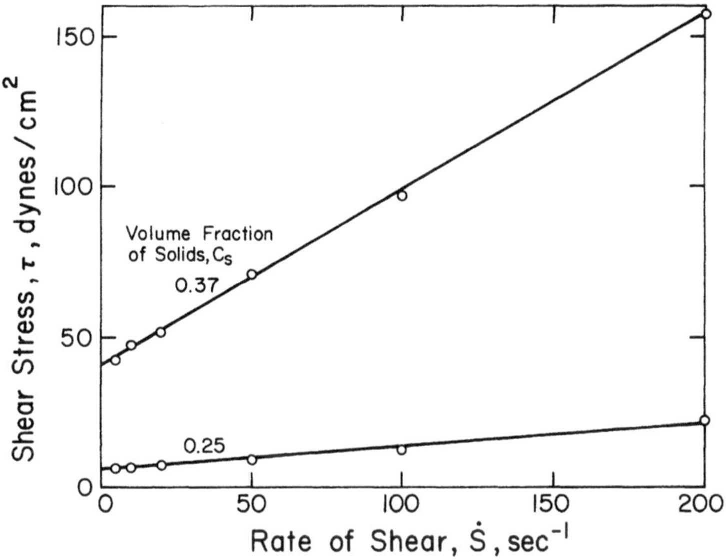

Figure 1.16. Rheograms for water suspensions of finely divided galena; typical Bingham plastics—arithmetic coordinates. (From data of Govier, G. W., C. A. Shook, and E. O. Lilge, *Trans. Can. Inst. Mining Met.*, **60**, 147, 1957.)

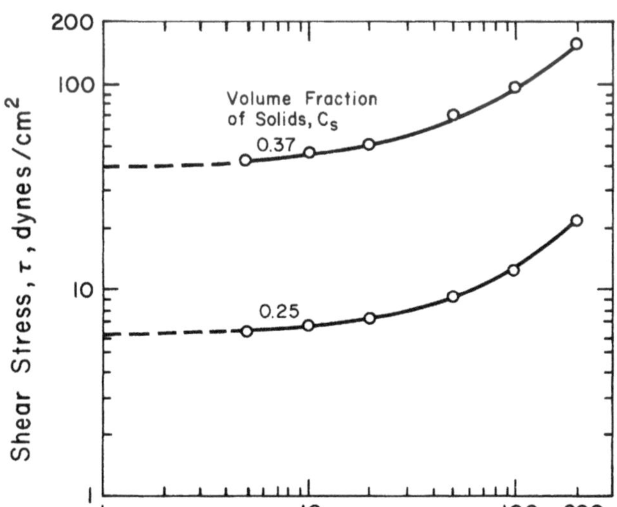

Figure 1.17. Rheograms for water suspensions of finely divided galena; typical Bingham plastics—logarithmic coordinates. (From data of Govier, G. W., C. A. Shook, and E. O. Lilge, *Trans. Can. Inst. Mining Met.*, **60**, 147, 1957.)

Yield values for Binghams vary from very small, less than 0.10 dynes/cm^2 for some sewage sludges (Dick and Ewing, 1967), to very large, more than 10^{10} dynes/cm^2 for some asphalts and bitumens (Pilpel, 1966). Coefficients of rigidity vary from values not much greater than the viscosity of water to nearly 1000 poises for some paints and to much higher values for asphalts and bitumens.

Experimental data for a typical suspension exhibiting near Bingham behavior is given in Figures 1.16 and 1.17. These data are for water suspensions of finely divided galena (100 percent less than 74 microns, 65 percent less than 26 microns) at shear rates below 200 sec^{-1}. The yield stress, τ_y, is obtained by extrapolation to zero velocity gradient. The coefficient of rigidity corresponds with the slope of the lines in Figure 1.16.

4.25 Yield-Pseudoplastics

A number of materials exhibit a yield stress, as in the case of Bingham plastics, but the relationship between the shearing stress in excess of that initiating flow and the resulting velocity gradient is not linear. Commonly the relationship exhibits convexity to the shear stress axis, and fluids showing this behavior may be called yield-pseudoplastics. (In the less common case when the curve is concave to the shear stress axis the fluid could be described as yield-dilatant.) Many clay-water and similar suspensions behave as yield-pseudoplastics, particularly at intermediate levels of concentration.

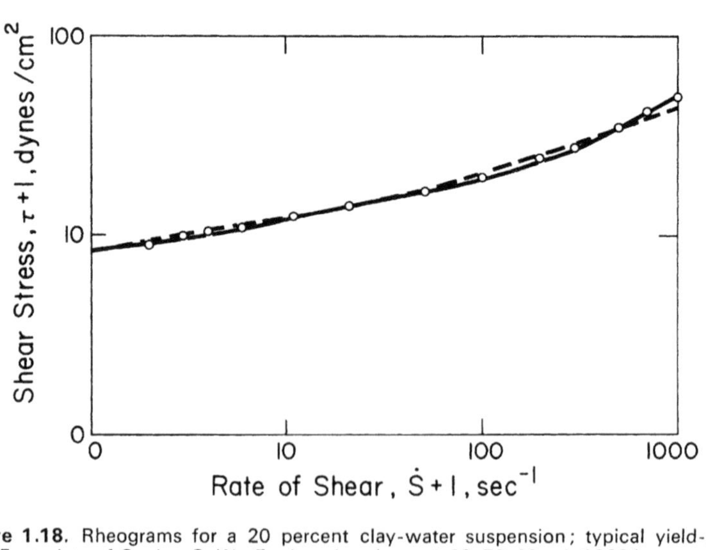

Figure 1.18. Rheograms for a 20 percent clay-water suspension; typical yield-pseudoplastic. (From data of Govier, G. W., *Engineering Journal*, **43**, 79, March 1960.)

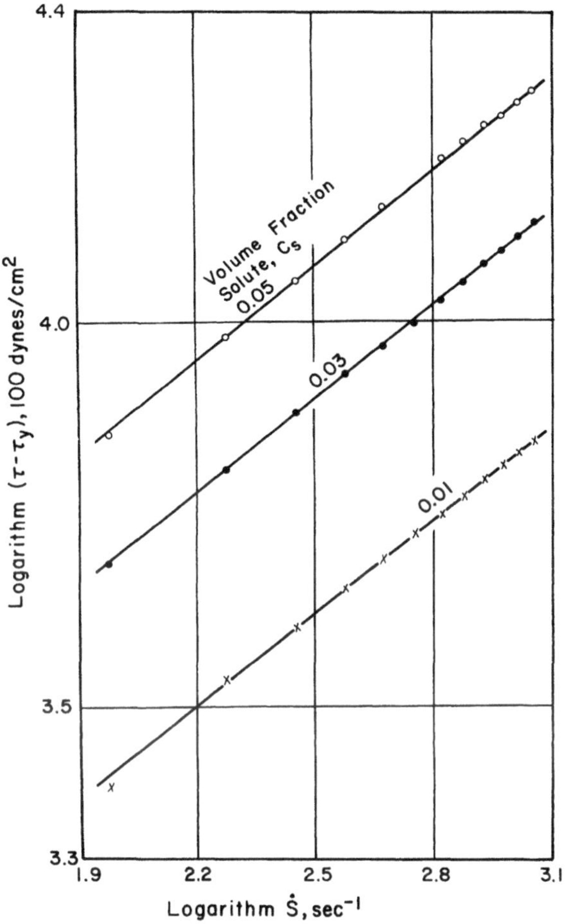

Figure 1.19. Rheograms for carbopol-water dispersions that follow the yield-power law equation. (Modified after Fischer, W. H., W. H. Bauer, and S. E. Wiberley, *Trans. Soc. Rheol.,* **5,** 221, 1961.)

As in the case of pseudoplastics, there is no single theoretically based constitutive equation for yield-pseudoplastics, but any of the empirical equations for pseudoplastics may be modified for yield-pseudoplastics. Illustrating this for the power law equation we have

$$\tau - \tau_y = \frac{K}{g_c} \dot{S}^n \tag{1.48}$$

where τ_y, K, and n are the characterizing coefficients. This three-parameter equation, first proposed by Herschel and Bulkley (1926), describes the behavior of yield-pseudoplastics reasonably well except at high shear rates.

Figure 1.18 gives typical data for a 20 percent clay-water suspension which could be described as a yield-pseudoplastic. (Actually, depending on how the data are extrapolated to zero velocity gradient, the system could also be described as a simple pseudoplastic.) The solid curve is drawn through the data points, and the dashed line represents Equation 1.48 with values of $\tau_y = 7.3$ dyne/cm^2, $K = 1.296$ dyne (sec)^n/cm^2, and $n = 0.483$.

Data of Fischer, Bauer, and Wiberley (1961) on 0.5 to 5.0 percent aqueous solutions of carboxypolymethylene (carbopol) illustrate the applicability of the yield-power law equation. These are shown in Figure 1.19. (Data of Dodge and Metzner, 1959, show that lower concentration solutions of carbopol, 0.2 to 0.5 percent, have no yield stress and follow the simple power law.)

4.3 *Time-Dependent Viscous Fluids*

4.31 Introduction

So far the types of purely viscous, time-independent fluids have been considered, where, in terms of Equation 1.37, the fluid behavior adjusts instantaneously to newly applied shear stresses and elastic effects are negligible.

A further consideration is to extend this analysis to systems in which the structural configuration and particle orientation are sensitive to shear, but the rate of structural rearrangement, which is described with the terms k_{PF}, k_{PD}, k_{PA}, and k_{PM} in Equation 1.37, is relatively slow. Then, since the rheological behavior is influenced by the structural changes in the system, the shear stress observed at any specified shear rate will vary with time until an equilibrium structure is achieved. As in the case of the pseudoplastic fluids, the nature of the equilibrium structure will depend on the shear rate, and will normally tend toward improved alignment of the particles and reduced resistance to flow as the shear rate is increased. In the event that the rate of structural reformation is of the same order as the rate of decay, the entire process may be considered as reversible. Under conditions of varying shear rate, the shear stress will adjust to a value which is appropriate to the prevailing shear rate. Furthermore, termination of shear will result in the reformation of the original structure. For those fluids in which the reconstruction process is extremely slow, changes in shear stress with shear rate will appear to be irreversible.

All fluids that display slow changes in rheological properties are broadly classified as time-dependent. If the shear stress is observed to decrease with the duration of shear, the fluid is classified as thixotropic. Strictly speaking, the classical definition of thixotropy pertains only to those fluids that exhibit reversible structural changes. The term "false-body" behavior is then frequently applied to those fluids in which the changes are, or appear to be, irreversible. However, since the difference between the two types of behavior is more a matter of degree than of kind, this proliferation of the terminology is somewhat unnecessary. In the following discussion, all fluids that experience structural decay with time under constant sustained shear will be considered as thixotropic.

A rather uncommon form of time-dependent behavior, which is known as rheopexy, occurs when, under conditions of constant sustained shear, the rate of structural reformation, though still low, exceeds the rate of decay. In this case the shear stress required to maintain any shear rate will gradually increase with time until an equilibrium structure is again achieved.

4.32 Thixotropic Fluids

Thixotropic behavior, as we have defined it, is rather widely encountered. Water suspensions of bentonitic clays are thixotropic, a property exploited in their use as drilling fluids in the petroleum industry. Many crude oils, especially at temperatures below normal atmospheric temperatures, exhibit thixotropic behavior. A typical thixotropic crude oil is that produced from the Pembina Field in Alberta. Figures 1.20 and 1.21 describe its behavior. Figure 1.20 shows the decay of structure (decrease of shear stress) with time under three approximately constant shear rates. Figure 1.21 shows the actual rheograms for the crude oil with separate curves for each of several shear durations. It should be noted that each curve has the characteristics of a yield-pseudoplastic or a pseudoplastic. In fact, a thixotropic fluid may be considered to be a yield-pseudoplastic whose properties change with the time of application of shear, ultimately approaching equilibrium or limiting values.

The behavior of thixotropic fluids is frequently studied in terms of so-called hysteresis curves resulting when a fluid has been subjected to shear, first, at successively increasing rates of shear, and then to successively decreasing rates of shear. Although such studies are useful to identify the presence of thixotropic behavior, they provide data of little direct value with respect to the pipe flow of the fluid and are not discussed here.

The description of the behavior of a thixotropic liquid at each of several different durations of shear, as in Figures 1.20 and 1.21, is useful and, for most practical purposes, quite adequate. To describe a thixotropic liquid as a

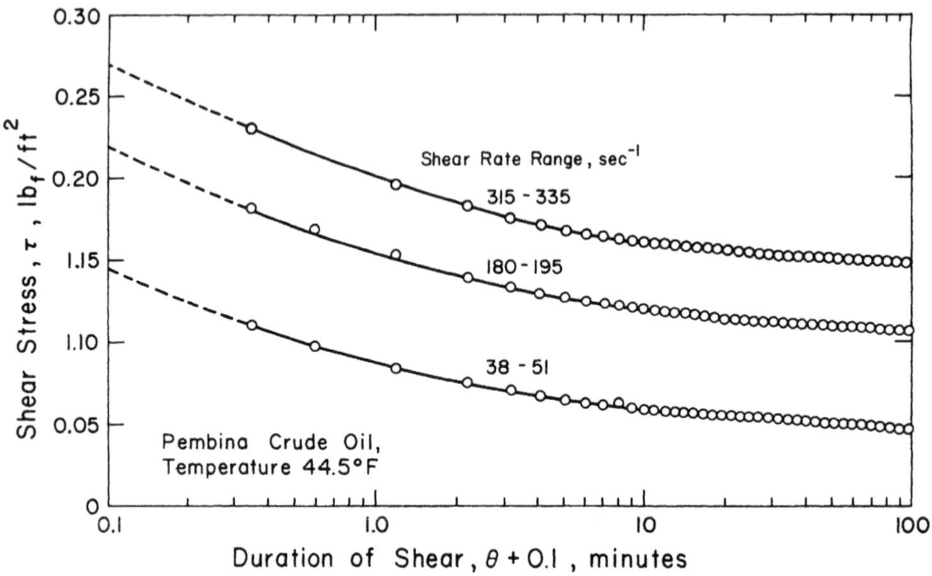

Figure 1.20. Structural decay of a typical thixotropic crude oil. (From data of Ritter, R. A., Doctoral Dissertation, Univ. Alberta, Edmonton, Alberta, 1961.)

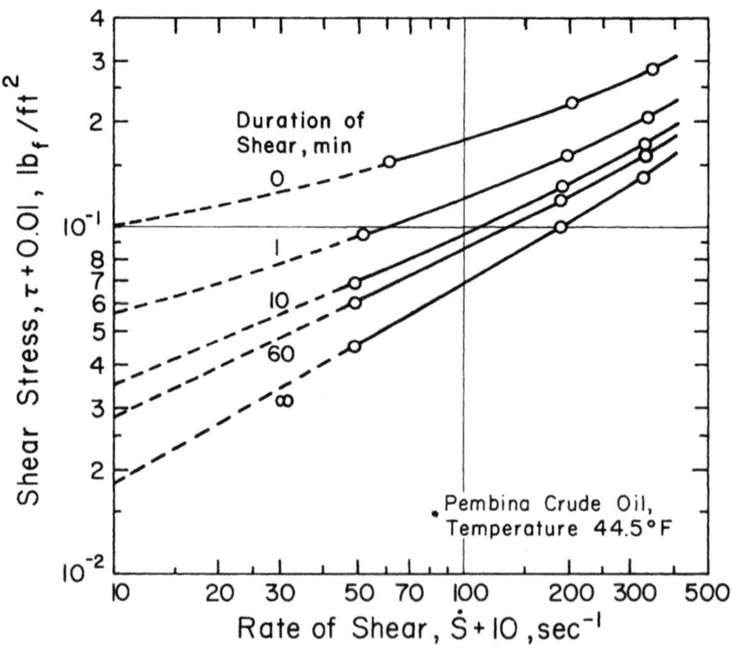

Figure 1.21. Rheograms for a typical thixotropic crude oil at constant duration of shear. (From data of Ritter, R. A., and G. W. Govier, *Can. J. Chem. Eng.*, **48**, 505, 1970.)

changing pseudoplastic, however, does little to reveal the cause of the behavior and is incomplete in describing the rate of change of properties with time. Many investigators have attempted the difficult task of formulating quantitative relationships between the rheological behavior and time. The more useful of these studies have depicted the thixotropic systems as containing particles or macromolecules which may undergo agglomeration or dispersion, and alignment or entanglement. This is the qualitative explanation we have already discussed. In the following we highlight a few of the many proposed relationships and discuss one of the more directly useful theories in some detail.

Goodeve (1939), Storey and Merrill (1958), and Billington (1960 a,b) propose relatively simple theories based upon kinetic considerations. The characteristics of these theories have been discussed by Ritter (1961). Many of their features have been incorporated in the Ritter and Govier theory treated later.

Hahn, Ree, and Eyring (1959) propose equations for describing the thixotropic behavior, based on the Powell–Eyring (1944) equation for pseudoplastic fluids, with time-dependent multipliers on the coefficients of the terms on the right-hand side of the Powell–Eyring equation. Their equations involve seven material constants which may be obtained, although through an involved process, from fundamental rheological data. With careful evaluation of the material constants, the Hahn *et al.* relationships provide a good empirical fit to actual data.

Moore (1959) proposed two relatively simple relations involving five material constants to describe the thixotropic behavior of a fluid not possessing a yield value. His relations are

$$\tau = \frac{1}{g_c}(\mu_0 + c\lambda)\dot{S}$$

$$\frac{d\lambda}{d\theta} = a - (a + b\dot{S})\lambda$$

where μ_0, a, b, c are material constants, and λ is a structural parameter having values 0 and 1 in the limits where the structure is fully broken down and fully developed. The evaluation of the five material constants from simple rheological tests is straightforward. For a constant structure these equations predict Newtonian behavior not always observed in practice.

Denny and Brodky (1962) develop another kinetic-based theory of thixotropic behavior involving six material constants which they indicate to be easily determined and readily interpreted. Their equations, however, do not permit the construction of the basic τ–θ–$\dot{S}$ relations from the material constants.

Slibar and Paslay (1962) and Paslay and Slibar (1963) proposed and illustrated the application of a set of constitutive equations to describe the behavior of Bingham-like time-dependent fluids. They express the yield value of the fluid as a function of time and shear rate while assuming the coefficient of rigidity to be a constant. Their relations involve five material constants, which may be evaluated from rheological tests. The values predicted do not appear to be as reliable as those of Hahn *et al.* for similar materials.

Pinder (1964) developed and tested a simple relationship describing the change in apparent viscosity with time for a thixotropic suspension of a crystalline gas hydrate in water. His equation is based upon assumed second-order and zero-order reaction rates for the formation and the destruction of structure. The equation involves three readily determinable material constants to describe the time dependency at any given shear rate. It is probably applicable only to systems closely similar to the system tested by Pinder.

Cheng and Evans (1965) describe the general form and the necessary overall characteristics of a set of constitutive equations for a thixotropic fluid which may possess a yield value. Their basic relations may be considered generalizations of those proposed by Moore (1959), Hahn *et al.* (1959) and Peter (1964). They take the form

$$\tau = f_1(\lambda, \dot{S})\dot{S}$$

$$\frac{d\lambda}{d\theta} = f_2(\lambda, \dot{S})$$

where f_1 and f_2 are functions to be determined by rheological tests. Using these relations, and a series of postulates concerning their nature, Cheng and Evans offer descriptions of the behavior of thixotropic fluids under a variety of shear treatments. They illustrate their description by reference to Moore's (1959) equations.

Fredrickson (1970) has presented a discussion of the necessary characteristics of a successful model for the thixotropy of purely viscous suspensions of rigid solids in liquids. He develops a simple four constant model, shows that it has the proper qualitative characteristics, and indicates how the material constants could be evaluated by rheological testing. His approach would appear to have real merit for solid-liquid suspensions.

The Ritter and Govier model (Ritter, 1961, Ritter and Govier, 1970) is more readily applied than others of comparable general suitability since it is expressed directly in terms of τ, $\dot{S}$, and θ. The model assumes that the formation of structures, networks, or agglomerates of particles is analogous to a second-order chemical reaction, and the breakdown of the structure is analogous to a series of consecutive first-order reactions. Ritter and Govier express the development and breakdown of structure in terms of a structural stress defined as

$$\tau_s = \tau - \tau_\mu \tag{1.49}$$

where τ = the shearing stress on the fluid,
$\tau_\mu = (\mu/g_c)\dot{S}$ is the shearing stress caused by the so-called Newtonian component of the fluid,
μ = the Newtonian viscosity of the fluid under conditions where all structure may be assumed broken down (independent of both duration and rate of shear)

Their equation is

$$\log\left(\frac{\tau_s - \tau_{s\infty}}{\dfrac{\tau_{s0}^2}{\tau_{s\infty}} - \tau_s}\right) = -K_D\left(\frac{\tau_{s0} + \tau_{s\infty}}{\tau_{s0} - \tau_{s\infty}}\right)\log\theta - \log K_{DR} \tag{1.50}$$

where τ_{s0}, $\tau_{s\infty}$ = the structural stresses at a given shear rate after zero and infinite duration of shear, given by $\tau_0 - (\mu/g_c)\dot{S}$, and $\tau_\infty - (\mu/g_c)\dot{S}$, respectively,
θ = the duration of shear, in minutes,
K_D = a constant independent of shear rate, but characteristic of the liquid, which is related to the first-order structural decay process, minute^{-1},
K_{DR} = a dimensionless measure of the interaction between the network or structure decay and the reestablishment processes.

The coefficient K_{DR} may be evaluated as

$$K_{DR} = \frac{\tau_{s0}^2 - \tau_{s1}\tau_{s\infty}}{\tau_{s1}\tau_{s\infty} - \tau_{s\infty}^2} \tag{1.51}$$

where τ_{s1} is the structural stress at time = 1 minute.

Ritter and Govier tested their equation with data on the thixotropic Pembina crude oil referred to earlier. The data confirmed the general form of Equation 1.50 and that K_D and K_{DR} are independent of, or insensitive to, the rate of shear.

Equation 1.50, with experimentally determined values of μ, K_{DR}, and K_D (independent of shear rate) and of τ_{s0} and $\tau_{s\infty}$ (dependent upon shear rate), defines the shear stress–time relationship for any shear rate. If we express the $\tau - \dot{S}$ relationships at zero and infinity rate of shear as

$$\tau_0 - \tau_{y0} = \frac{K_0}{g_c} \dot{S}^{n_0} \quad (1.52)$$

$$\tau_\infty - \tau_{y\infty} = \frac{K_\infty}{g_c} \dot{S}^{n_\infty} \quad (1.53)$$

we see that Equations 1.50, 1.52, and 1.53, with the values of μ, K_{DR}, K_D, τ_{y0}, $\tau_{y\infty}$, n_0, n_∞, K_0, and K_∞ determined from measurements, give a complete description of the variation of shear stress with shear rate and time. The description is accurate only if the coefficients are determined by the testing of a sample of liquid which has been exposed to both a thermal and a shearing history identical to that of the fluid to be described. This important point is discussed further in Chapter 6.

4.33 Rheopectic Fluids

Rheopectic or negative thixotropic behavior is less commonly encountered than thixotropic behavior. It is characterized by an increase with time of the stress required to maintain a constant rate of shear. At any given rate of shear the stress increases to approach an asymptotic maximum value. The time required for this maximum or equilibrium stress depends upon the particular system but is often in the range of 10–200 minutes. As with thixotropic fluids, the rheological data of interest are shear stress–time curves at constant shear rate, and the complete rheogram is a family of shear stress–shear rate curves for a range of shear durations from zero to the time of equilibrium.

Data of Steg and Katz (1965) on a 2000 molecular weight saturated polyester, given in Figure 1.22, are of interest. This system is time-independent at rates of shear below 1000 $\sec^{-1}$ but becomes rheopectic at shear rates of 1377 $\sec^{-1}$ and higher, and is strikingly rheopectic at a shear rate of 8267 $\sec^{-1}$.

The relationship between shearing stress and rate of shear at each of several durations of shear is given in Figure 1.23. From this it appears that this particular fluid, after adequate shear duration, behaves as a power law pseudoplastic over rates of shear extending to below that at which rheopexy develops. Steg and Katz explain the phenomenon in these words:

> At relatively low shear rates the systems disperse the shearing forces without undergoing any changes, and no time effects are observed. With the increase of the rate of shear above a certain critical value, the energy introduced into the system is not just dispersed in the form of heat but also effects the equilibrium in

intermolecular or perhaps also intramolecular bonding between the molecules. The distances between some of the polyester molecules aligned in the field of shear decrease, and the linear molecules build anisodiametric agglomerates, mostly due to the formation of strong polar and hydrogen bonds between them. These agglomerates can be seen as quasicrystallites in solution. As the shielding

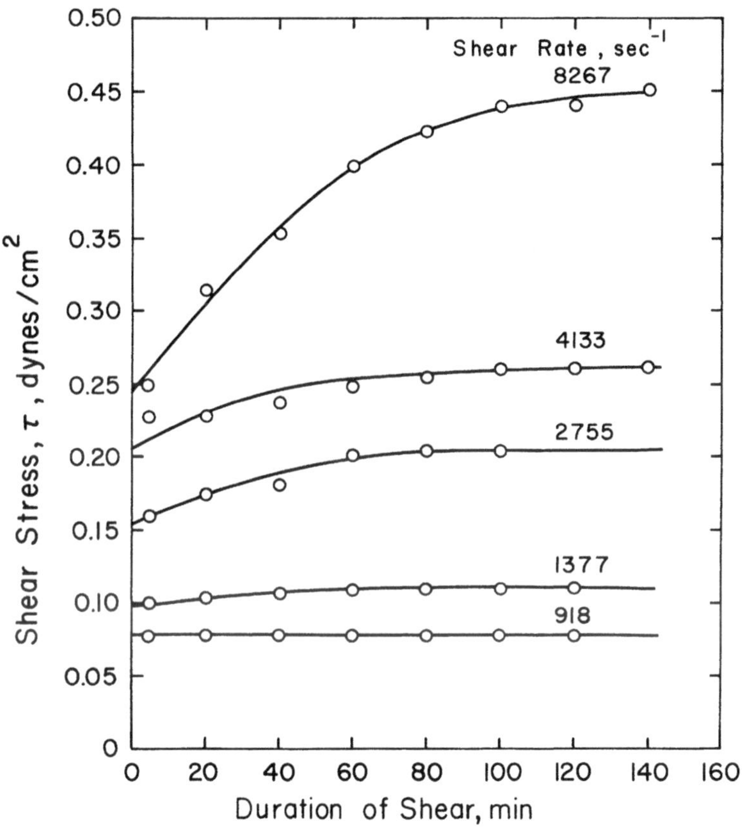

Figure 1.22. Structural buildup for a 2000 molecular weight polyester; typical rheopectic fluid. (Modified from Steg, I., and D. Katz, *J. Appl. Polym. Sci.*, **9**, 3177, 1965.)

of the polar polyester groups by the mediocre solvent is only slight, due to weak interaction, solvent molecules are probably "squeezed out" in the field of shear from the space between the approaching and interacting polyester molecules. A certain amount of energy is introduced by shear so as to affect the thermodynamic equilibrium in the fluid and to build the new structures with a higher free energy content. The quasicrystallites, oriented in the field of shear, can be held together by more intermolecular forces, probably of weaker nature than the ones building the quasicrystallites, thus forming a larger network structure. A somewhat similar explanation for negative thixotropy in a different system was offered by Eliassaf, Silberberg, and Katchalsky, and the model corresponds to the general definition given by Weltmann that in rheopectic materials an almost perfect orientation of anisodiametric particles in the field of shear is necessary. The relative ease of straightening of the macromolecules in the field of shear can be attributed to their limited length and to the internal stiffness of the polyester chains.

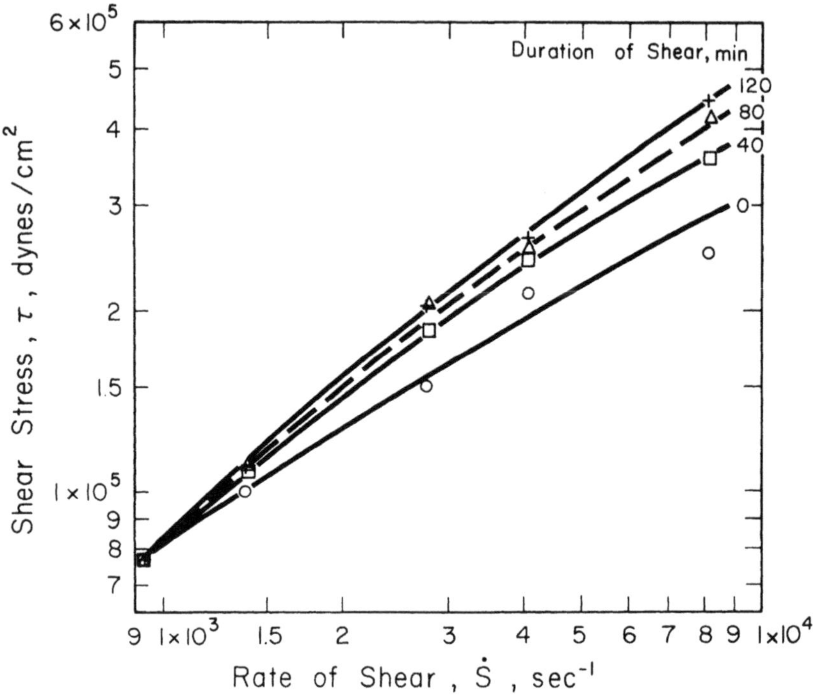

Figure 1.23. Rheograms for a 2000 molecular weight polyester; typical rheopectic fluid (From data of Steg, I., and D. Katz, *J. Appl. Polym. Sci.*, **9**, 3177, 1965.)

4.4 Viscoelastic Fluids

4.41 Introduction

As we mentioned earlier, the term "viscoelastic fluid" is applied to the most general of fluids, those which exhibit the effect of partial elastic recovery and whose viscous properties may be non-Newtonian and time-dependent. In one sense, all liquids and pseudohomogeneous liquid-liquid or liquid-solid mixtures could be considered viscoelastic with the coefficients of certain of the variables indicated in Equation 1.37 approaching zero in limiting cases.

Viscoelastic effects may be important in sudden changes in flow rate (start-up and stop), in rapid oscillatory flows, in flows at high shear rates as encountered in extrusion processes, and in flows where changes of cross section are encountered. On the other hand, in most normal steady-state flow, and in unsteady-state flow involving large process times, most materials behave essentially as purely viscous matrials. An important exception to this is in the turbulent pipe flow of viscoelastic fluids where a number of workers (e.g., Metzner and Park, 1964; Lummus, Fox, and Anderson, 1961; Crawford, 1962; Fabula, 1963; and Savins, 1964) have found very large effects of viscoelasticity on the pressure gradient. In Chapter 5 this will be explored further. In the following discussion only those characteristics of viscoelastic fluids that govern the major effects observed in pipe flow are considered. In particular, we will consider features that influence the important phenomenon of drag reduction observed in tubulent flow.

Where elastic effects are present, they manifest themselves through the development of normal stresses perpendicular to the direction of shear which are of a different magnitude than the normal stress parallel to the direction of flow. Additionally and important in the present context, the phenomenon of elastic recoil, or stress relaxation, occurs whenever a viscoelastic material is subjected to a rapid change in deformation. Qualitatively this phenomenon may be observed as the decay of the shearing stress with respect to time when a motion, such as that depicted by Figure 1.9, is suddenly stopped.

Figure 1.24, depicting an infinitesimal element of fluid subjected to a shearing motion, illustrates the situation where, in general, for a viscoelastic

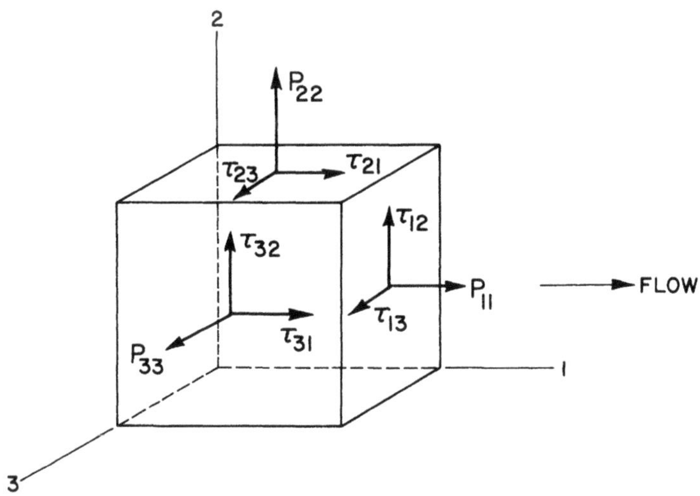

Figure 1.24. Element of fluid subject to shear stress. (Stresses on the hidden faces are not shown.)

material, the three total normal stresses are unequal: $P_{11} \neq P_{22} \neq P_{33}$.* Each of the normal stresses is composed of the sum of an arbitrary isotropic stress (or pressure), P, and a normal stress, τ_{ij}, $i = j$, i.e.,

$$P_{11} = -P + \tau_{11} \tag{1.54a}$$

$$P_{22} = -P + \tau_{22} \tag{1.54b}$$

$$P_{33} = -P + \tau_{33} \tag{1.54c}$$

The other stresses, i.e., τ_{ij}, $i \neq j$ are the tangential stresses. (It should be noted that in much of the literature of rheology the symbol p is used where we use P to maintain consistency with the bulk of the literature related to flow in pipes.)

If the isotropic pressure is defined such that

$$P = -\tfrac{1}{3}(P_{11} + P_{22} + P_{33}) \tag{1.55}$$

(see Fredrickson, 1964, for discussion of the limitations of this definition), then from Equations 1.54a, 1.54b, and 1.54c we have

$$\tau_{11} + \tau_{22} + \tau_{33} = 0 \tag{1.56}$$

and therefore only two of the normal stresses are independent.

Usually, for incompressible fluids, the arbitrary isotropic pressure is eliminated from consideration by treating the difference between the normal stresses, such as $(P_{11} - P_{22})$ and $(P_{22} - P_{33})$. Thus, at least in simple laminar shearing flow, a complete characterization of a viscoelastic material must involve measurement of the shearing stress plus two of the normal stress differences as functions of the deformation rate.

Data such as those of Huppler (1965), Dean and Roisman (1969), and Ginn and Metzner (1969) indicate that the normal stress difference $P_{11} - P_{22}$ is substantially greater than the difference $P_{22} - P_{33}$; therefore $P_{11} - P_{22}$ is called the principal normal stress difference and $P_{22} - P_{33}$ the secondary normal stress difference. In most analyses the secondary normal stress difference has been neglected because of insufficient data for its proper definition, or it has been assumed zero in accordance with the hypothesis of Weissenberg (1947, 1948). The relative significance of the elastic and the viscous effects was discussed by Weissenberg (1948) and is indicated by what is now known as the Weissenberg number, Ws, defined as the ratio of the principal normal stress difference to the shear stress parallel to flow

$$\mathrm{Ws} = \frac{P_{11} - P_{22}}{\tau_{21}} \tag{1.57}$$

(For simplicity we drop the subscript on τ in the following.)

*The numerical subscripts of P in this section are not standard throughout the book. They are employed only when the use of alphabetical subscripts (xx, xy, etc.) is not convenient.

For purely viscous liquids Ws = 0, and for elastic solids Ws = ∞. Limited data (Metzner and Park, 1964) indicate that, for turbulent pipe flow, elastic effects may be neglected when Ws ≤ 0.10.

In practice a wide variety of materials exhibit viscoelastic behavior under appropriate flow conditions. Essentially all molten polymers are highly elastic, as are solutions of long-chain molecules, such as polyethylene oxide, polyacrylamides, sodium carboxymethyl cellulose, and a number of others. Significantly many of these materials find large-scale use in such processes as coagulation, oil-well fracturing operations, and potentially in high-capacity pipe lines where small amounts of polymeric additives are effective in reducing the pressure drop.

Data on a typical viscoelastic fluid, a 5 percent solution of polyisobutylene in decalin, are shown in Figure 1.25 taken from Ginn and Metzner (1963). It

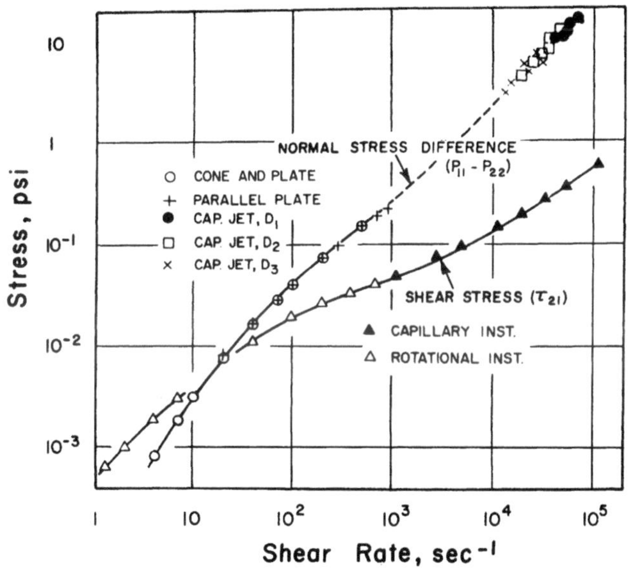

Figure 1.25. Normal and shear stress relationships for 5 percent polyisobutylene in decalin. (From Ginn, R. F., and A. B. Metzner, Fouth Int. Congress on Rheology, Brown University, 1963, *Proc.* Part 2, p. 583. Interscience Publishers, New York, 1965.)

is to be noted that the normal stress difference is comparable to the shear stress at low rates of shear but becomes an order of magnitude greater at high rates of shear.

4.42 Constitutive Equations

Although a large number of complex constitutive equations have been developed for viscoelastic materials, no simple practical equations are available that are capable of quantitatively describing the behavior of real viscoelastic materials over the range of flow conditions of practical significance.

(The reader is referred to Fredrickson, 1964; Middleman, 1969; Spriggs, Huppler, and Bird, 1966; and Meister and Biggs, 1969, for discussions of the various types and characteristics of general constitutive equations that have been proposed.) On the other hand, there are a number of relatively simple constitutive equations which, although restricted in applicability, are useful to us in our consideration of pipe flow problems. The White–Metzner (1963) constitutive equation, which appears to be the simplest and most practical of these, is

$$\tau_{ij} = -2\,\mu_a d_{ij} + \theta_r \frac{\delta \tau_{ij}}{\delta t} \tag{1.58}$$

where d_{ij} = components of deformation rate tensor, the absolute value of which is equal to $\dot{S}$ for simple shear,

$\delta/\delta t$ = the Oldroyd convected derivative, the time rate of change of a quantity relative to a coordinate system moving and deforming with the fluid (Oldroyd, 1950),

μ_a = apparent viscosity,

θ_r = relaxation time of the material.

The apparent viscosity is not expressed in terms of property parameters and rate of shear in this equation, but this could be done through the power law or other empirical equation for a time-independent viscous fluid.

It will be noted that for materials with $\theta_r = 0$, the equation is simply that of a purely viscous fluid and that the normal stresses, as well as the relaxation phenomenon, arise from the term involving the relaxation time and the convected derivative. Of primary interest from an engineering viewpoint is that only one parameter in addition to the apparent viscosity is necessary to portray the effects of elasticity in any flow situation.

For a steady laminar shearing motion, Equation 1.58 predicts that

$$\tau = \frac{\mu_a \dot{S}}{g_c} \tag{1.59}$$

$$P_{11} - P_{22} = 2\theta_r \tau \dot{S} \tag{1.60}$$

and

$$P_{22} - P_{33} = 0 \tag{1.61}$$

From Equation 1.60 we may observe the relationship between the relaxation time and the Weissenberg number

$$\theta_r = \frac{\mathrm{Ws}}{2\dot{S}}$$

Additionally we see that Equation 1.58 implicitly introduces the assumption of equality between P_{22} and P_{33} which, in view of our earlier comments on the small magnitude of the secondary normal stress relative to the first, is not seriously restrictive.

The White and Metzner equation (Equation 1.58) may be tested by observing that Equations 1.59 and 1.60 may be combined to yield

$$P_{11} - P_{22} = 2\theta_r \frac{g_c}{\mu_a} \tau^2 \tag{1.62}$$

indicating that $P_{11} - P_{22}$ should be proportional to τ^2 as long as the ratio θ_r/μ_a is constant. Figure 1.26, taken from Ginn and Metzner (1963), demonstrates this to be the case for the previously discussed polyisobutylene-decalin system. Further support at high deformation rates has been published by Meister and Biggs (1969).

In nonsteady flows the fluid relaxation time, θ_r, is of special importance in determining whether or not the material will exhibit elastic effects or effects of a purely viscous nature. Nonsteady flows considered here include those which are nonsteady from a Lagrangian viewpoint, for it is the changes that a given element of fluid experiences with respect to time, rather than changes at a point in the flow field, that are important in determining the response of

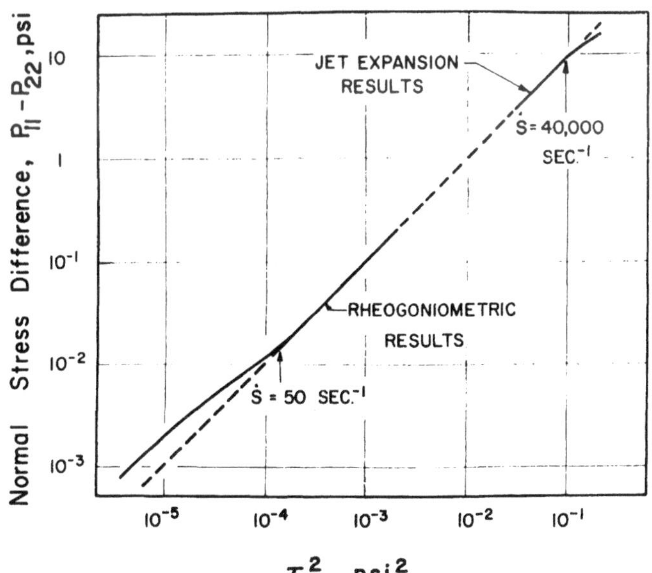

Figure 1.26. Test of White-Metzner constitutive equation. (From Ginn, R. F., and A. B. Metzner, Fourth Int. Congress on Rheology, Brown University, 1963, *Proc.* Part 2, p. 583, Interscience Publishers, New York, 1965.)

the fluid. Turbulent flow, flow in the entry of a duct, and elongational flow, such as accompanies the stretching of a sheet or rod of material, are examples of nonsteady flow in a Lagrangian sense. The relaxation time of a fluid may be defined in a number of ways (Metzner, White, and Denn, 1966), depending on the details of the flow situation and/or the particular constitutive equation chosen to represent the behavior of the fluid. For example, if the steady shearing motion depicted by Figure 1.9 is suddenly stopped and the decay of stresses observed, the relaxation time may be taken as the time for the stresses to decay to a certain percentage of their original value. In the particular case of Equation 1.58, the stresses are predicted to relax exponentially and θ_r is the time necessary for the stresses to reach $1/e$ of their original value.

Table 1.4 presents relaxation-time data for a variety of polymer solutions.

TABLE 1.4

Typical Relaxation times for Polymer Solutions

(The Rate of Shear Is Approximately 15,000 sec^{-1}).

Solution	Concentration, Weight %	Relaxation Time, 10^{-4} sec	References
CMC[a]	2.0	2.6	1
PIB[b]	3.0	7.5	1
PIB	5.0	5.9	1
Polyox	2.0	35	1
Polyacrylamide	0.4	10	2

REFERENCES:

1. A. B. Metzner, J. L. White, and M. M. Denn, *Chem. Eng. Progress*, **62**, 81 (1966).
2. F. A. Seyer and A. B. Metzner, *Can. J. Chem. Eng.*, **45**, 121 (1967).

[a] Carboxymethylcellulose.
[b] Polyisobutylene.

Relaxation times nearly three orders of magnitude larger than those of Table 1.4 have been observed for some biological fluids (King, 1966).

In the general case of nonsteady flows it has been shown by Metzner, White, and Denn (1966), Astarita (1965, 1967), and others that the parameter which describes whether or not elastic effects are important is the Deborah number defined as

$$\mathrm{De} = \frac{\text{relaxation time of fluid, } \theta_r}{\text{time scale of process or residence time of fluid, } \theta_p} \qquad (1.63)$$

In view of Equation 1.62 we see that for a fluid described by the White-Metzner equation,

$$\mathrm{De} = \frac{\mathrm{Ws}}{2\dot{S}\theta_p}$$

Qualitatively, for De much less than unity, the relaxational processes are sufficiently rapid, or the time scale of the process sufficiently long, to enable the material to respond to any changes in deformation that may occur. The material, therefore, behaves essentially as though it were purely viscous. An important example is the steady laminar flow through a tube for which the time scale is infinity in a mathematical sense. Conversely, for Deborah numbers in the order of unity, the rapidly changing deformations do not allow the material to relax significantly; and in the limit for very large Deborah number, the behavior is predicted to be that of an elastic solid.

Although the predictions outlined in the preceding discussion are dependent on the assumed constitutive equation, they appear generally valid for materials and flow situations of interest in pipelining applications.

5 RHEOLOGICAL MEASUREMENTS AND INTERPRETATION OF DATA

Most engineers are familiar, at least in a general way, with the techniques of obtaining and interpreting laboratory data to determine the viscosity of a Newtonian fluid, but the measurement and interpretation of the rheological properties of non-Newtonian fluids and mixtures are less well understood. Skelland (1967) discusses these matters clearly and concisely. Here only a brief review will be given with emphasis on data interpretation rather than details of apparatus.

5.1 Viscometric Equipment

Three different types of apparatus are in common use for measuring viscosity or other rheological properties. Each of these is designed to create conditions of laminar shear, as indicated schematically in Figure 1.9, and to permit the measurement of quantities from which the shear stress and the rate of shear may be determined. The types are the following:

a. The capillary viscometer is a device that causes a sample of fluid to flow at a measured rate in laminar motion under a measured pressure gradient through a precision bore capillary tube of known diameter and length. Precautions are taken to maintain constant temperature conditions, and corrections are applied for entrance and kinetic energy effects. This viscometer has the advantage of mechanical simplicity and of permitting the attainment of high rates of shear. On the other hand, as will be seen more fully in Chapter 5 where the laminar flow of non-Newtonian fluids in pipes will be discussed, the sample is subjected to a rate of shear that varies from zero at the axis of the capillary to a maximum at the wall. Also ordinary capillary viscometers do not permit the same sample of fluid to be subjected to sustained flow and thus are not suitable for measuring the behavior of time-dependent fluids.

b. The concentric cylinder or rotational viscometer permits a sample of fluid placed in the annular space between a stationary and a rotating cylinder to be subjected to shear, and the torque acting upon the stationary cylinder to be measured. The rate of shear is determinable from the geometry of the system and the speed of the rotating cylinder. The shear stress is obtained from the measured torque. A variant of the design involves the use of a single rotating cylinder, which may be immersed in a sample of the fluid in any suitable container. Again precautions must be taken to ensure laminar flow and constant temperature, and end-effect corrections may be required.

The rotational viscometer is suitable for use over a wide range of rates of shear (although not the highest). With a small annular gap the shear rate is nearly constant through the fluid and this, coupled with the fact that a particular sample can be subjected to sustained shear, makes it a most useful instrument for the study of the viscous properties of non-Newtonian fluids.

c. The cone-and-plate viscometer or the similar parallel-plate viscometer is designed to subject a sample of fluid maintained in the narrow space between a rotating, flat, circular plate and an inverted cone or a parallel plate to laminar shear. Generally, in the absence of strong secondary flows, these instruments are useful over a reasonable range of shear rates and are ideal for the study of viscous properties of non-Newtonian fluids at low and moderate shear rates. For the cone-and-plate geometry, the angle between the cone face and the flat plate is normally less than 3 degrees. Measurements of the rotational speed and the torque permit the calculation of the rate of shear and the shear stress. Precautions are required to ensure isothermal conditions and the absence of secondary flows, but end effect corrections are usually negligible. One of the chief differences resulting from the cone-and-plate and the parallel-plate arrangements is the near constancy of the rate of shear in the cone-and-plate geometry. This allows a direct calculation of shear rate and shear stress without the necessity of differentiating the torque–rotational speed data, as must be done with the parallel-plate geometry. It should be noted, however, that the analysis for the cone and plate is only approximate for situations where elastic effects are present, and that secondary flows may cause errors in the interpretation of the data. The secondary flows are minimized by using small cone angles and low rotational speeds. A detailed discussion of these effects and indications of the range of utility of the cone-and-plate viscometer has been presented by Ginn and Metzner (1969). Generally, for an unknown fluid it is not possible to predict the magnitude of the secondary flows in the cone-and-plate geometry and the parallel plate device is preferable. Notwithstanding this, the cone-and-plate data of Ginn and

Metzner and of Meister and Biggs (1969) show excellent agreement with capillary tube data and suggest that, with proper corrections for inertial effects, a cone-and-plate instrument with a cone angle of less than 2 degrees will give results suitable for most applications.

Each of the preceding instruments with minor modifications permits the measurement of viscoelastic effects. Savins (1966) discusses this in some detail, and Meister and Biggs (1969) discuss the use of the capillary instrument and the cone-and-plate viscometer.

With the cone-and-plate viscometer, the normal forces are easily related to the rise of the fluid in capillary tubes attached to the upper cone or to measurement of the total force on the cone. As noted earlier, however, the utility of this device is limited to modest shear rates. With appropriate experimentally determined corrections for inertial effects, Meister and Biggs found the cone-and-plate instrument suitable for shear rates from about 1 to 3×10^3 sec^{-1}. This corresponds to the lower range of utility of the capillary tube device.

The concentric cylinder may also be used to determine normal forces by exploitation of the Weissenberg effect (Weissenberg, 1947, 1948) wherein the difference in pressure on the inner and outer cylinders is measured. Generally this device, which is also limited to modest shear rates, gives only a crude estimate of the normal forces owing to the small differences in pressure that must be measured while maintaining negligible inertial effects. Additionally it appears that large unknown aberrations owing to "pressure-hole" effects may complicate the measurements (Ginn and Metzner, 1969; Tanner and Pipkin, 1969).

The low shear rate limitations of the cone-and-plate and concentric cylinder instruments are largely overcome with a capillary tube device. In this case the normal forces may be related to the thrust on the capillary tube or, alternately, to the diameter of the jet of fluid discharged from the capillary. Meister and Biggs indicate that measurements of the thrust and diameter of a jet of fluid from a capillary yields reliable results for the normal stress difference for shear rates in the range 4×10^3 to 10^5 sec^{-1}. Powell and Middleman (1969) note, however, that when jet diameter measurements are used, there may be a systematic discrepancy in the data dependent upon tube diameter.

The relaxation of the jet to its final diameter has also been used to provide a direct estimate of the fluid relaxation time (Metzner *et al.*, 1966).

5.2 Interpretation of Rheological Measurements

The interpretation of the laboratory measurements for engineering use involves the calculation of points to define the rheogram or flow curve for the fluid and, in the case of viscoelastic fluids, the calculation of the normal stresses. This means the calculation of the rates of shear from the measured data and the corresponding shear stresses, and the normal stress difference in the case of viscoelastic fluids, and requires a complete understanding of the details of the flow of the fluids in the instrument.

With the capillary instrument, the measured data related to the viscous properties, after application of any appropriate corrections, are:

Q = volumetric flow rate,
D = inside diameter of capillary,
L = effective length of capillary, and
ΔP = pressure drop due to laminar friction.

The average fluid velocity is

$$V = \frac{4Q}{\pi D^2} \tag{1.64}$$

A simple steady-state force balance yields

$$\frac{\pi D^2}{4} \Delta P = \pi D L \tau_w \tag{1.65}$$

or

$$\tau_w = \frac{D \Delta P}{4L} \tag{1.66}$$

where τ_w is the shearing stress at the wall of the capillary. The rate of shear at the capillary wall may be determined from the general equation of Rabinowitsch (1929) and Mooney (1931), which is

$$\dot{S}_w = \left(\frac{1 + 3n'}{4n'}\right) \frac{8V}{D} \tag{1.67}$$

where

$$n' = \frac{d \ln \dfrac{D \Delta P}{4L}}{d \ln \dfrac{8V}{D}} \tag{1.68}$$

is the slope of the logarithmic plot of $D\Delta P/4L$ versus $8V/D$. The slope, n', may vary and the applicable value should be used for each value of $8V/D$. For Newtonian fluids $n' = 1$.

The interpretive procedure is, therefore,

a. to calculate τ_w from the data after any appropriate corrections, and Equation 1.66;

b. to plot τ_w versus $8V/D$ on logarithmic coordinates and determine n' as a function of $8V/D$;
c. to calculate $\dot{S}_w$ from Equation 1.67;
d. to plot τ_w versus $\dot{S}_w$ on arithmetic or logarithmic coordinates as preferred, and
e. if desired, to fit an appropriate one of the constitutive equations to the data.

The viscous property data obtained with a concentric cylinder instrument are:

D_1 = diameter of the inner cylinder assumed to be stationary,
D_2 = diameter of the outer cylinder,
θ = spring deflection being a measure of the torque on the stationary cylinder,
h = height of stationary cylinder immersed in fluid, corrected for end effect,
N = speed of rotating cylinder, sec^{-1}.

From a simple force balance, the following equation is obtained

$$\tau_{D1}(\pi D_1 h)\frac{D_1}{2} = T = K\theta \tag{1.69}$$

or

$$\tau_{D1} = \frac{2K\theta}{\pi D_1^2 h} \tag{1.70}$$

where T is the torque on the stationary cylinder, and K is a suitable spring constant.

The relationship between the rate of shear and the geometry of the system is given by an equation due to Krieger and Maron (1954):

$$\dot{S}_{D1} = \frac{4\pi N}{1 - s^{-2}} F_{KM} \tag{1.71}$$

where $s = D_2/D_1$
F_{KM} = the Krieger–Maron correction factor,

$$F_{KM} = 1 + \frac{s^2 - 1}{2s^2}\left(1 + \tfrac{2}{3}\ln s\right)\left(\frac{1}{n''} - 1\right) + \frac{s^2 - 1}{6s^2}\ln s\left(\left(\frac{1}{n''} - 1\right)^2 + \frac{d\left(\frac{1}{n''} - 1\right)}{d\log T}\right) \tag{1.72}$$

$n'' = \mathrm{d}\ln T/d\ln N$ is the slope of a logarithmic plot of T versus N.

When n'' is constant, or nearly so, the equation simplifies somewhat and may be approximated by a series expansion.

Calderbank and Moo-Young (1959) have calculated values of F_{KM} from the approximate equation for a series of values of s and n''. These are given by Skelland (1967). For Newtonian fluids, $n'' = 1$ and $F_{KM} = 1$.

The procedure for interpreting data from the concentric cylinder viscometer is, therefore,

a. to calculate τ_{D1} from θ, Equation 1.70 and the appropriate value of K;
b. to plot θ versus N on logarithmic coordinates and determine n'' as a function of N;
c. to calculate F_{KM} for the values of s and N of interest from Equation 1.72;
d. to calculate $\dot{S}_{D1}$ from Equation 1.71 with the appropriate values of N, s, and F_{KM};
e. to plot τ_{D1} versus $\dot{S}_{D1}$ on arithmetic or logarithmic coordinates as preferred, and
f. if desired, to fit an appropriate one of the constitutive equations to the data.

For the rotating cylinder viscometer, the procedure is the same as for the concentric cylinder instrument except that the shear rate at the surface of the rotating cylinder is given by

$$\dot{S} = \frac{4\pi N}{n''} \tag{1.73}$$

The interpretation of the viscous behavior data from the cone-and-plate instrument is easiest of all. The data consist of

$D = 2R$ = diameter of the stationary plate,
ϕ = angle between the cone and plate,
N = rotational speed of cone, sec^{-1},
θ = spring deflection being a measure of the torque on the plate.

The rate of shear at a radius r is given by

$$\dot{S}_r = \frac{2\pi r N}{r \tan\phi} \simeq \frac{2\pi N}{\phi} \tag{1.74}$$

since for small angles $\tan\phi \simeq \phi$. As mentioned earlier, for the range of cone angles employed, the rate of shear is essentially independent of r, i.e., is constant through the fluid.

If $T = K\theta$ is the torque (where K is a suitable spring constant), then from a force balance, and since τ is constant under the constant rate of shear, we have

$$T = 2\pi\,\tau \int_0^R r^2 dr \tag{1.75}$$

from which

$$\tau = \frac{3K\theta}{2\pi R^3} \tag{1.76}$$

From these relations the interpretation procedure is

a. to calculate τ from the deflection, and spring constant using Equation 1.75;
b. to calculate $\dot{S}$ from the cone speed and Equation 1.76;
c. to plot τ versus $\dot{S}$ on arithmetic or logarithmic coordinates as preferred. and
d. if desired, to fit an appropriate one of the constitutive equations to the data.

In the foregoing we have dealt with the determination of the shear stress–rate of shear relationship for a fluid from viscometric measurements. A special feature of this for non-Newtonian fluids is the determination of whether or not the fluid exhibits a yield value and, if so, its magnitude. Either or preferably both of two approaches may be used here. First, and regardless of the type of instrument used, the shearing stress required to initiate fluid motion may actually be measured; secondly the shear stress data obtained at several low rates of shear may be extrapolated to zero rate of shear. The latter procedure is somewhat dangerous unless reliable data at very low shear rates are available. Gill and Russell (1954) and Russell (1960) describe the use of the capillary instrument for yield value determination.

As mentioned previously, the normal stress difference characterizing viscoelastic behavior may be obtained by proper interpretation of suitable measurements made with modified cone-and-plate, parallel-plate, concentric cylinder, or capillary instruments. We will deal here only with the interpretation of the cone-and-plate and the capillary data.

The interpretation of the speed and torque data for a cone-and-plate rheogoniometer is the same as for a simple cone-and-plate viscometer. White and Metzner (1962) show how the rheogoniometer capillary rise data may be converted to the corresponding principal normal stress difference.

White and Metzner assume that the centrifugal forces are negligible and derive the following expression for the equilibrium of forces

$$\frac{g}{g_c}\rho\left(-\frac{dh}{d\ln r}\right) = P_{11} + P_{22} - P_{33} \tag{1.77}$$

where h is the capillary rise in a tube at a radial position r. This indicates that the capillary rise, when plotted versus the logarithm of radial position, should give a straight line with a slope of $-(P_{11} + P_{22} - 2P_{33})g_c/\rho g$. Using the approximation $P_{22} = P_{33}$ (Weissenberg hypothesis), the slope of the line is

$-(P_{11} - P_{22})g_c/\rho g$. The interpretive procedure for obtaining the principal normal stress difference is, therefore,

a. at each value of $\dot{S}$ to plot the capillary rise height, h, versus the logarithm of the radial position, r, and determine the slope of the best fit straight line;
b. to multiply the slope by the quantity $-\rho(g/g_c)$ to obtain $P_{11} - P_{22}$;
c. to repeat a and b at other values of $\dot{S}$ and to plot $P_{11} - P_{22}$ versus $\dot{S}$.

Where the total force on the cone is measured, the principal normal stress difference is given by the simple relation

$$P_{11} - P_{22} = \frac{2F''}{\pi R^2} \tag{1.78}$$

where F'' is the normal force exerted by the fluid corrected for momentum and surface tension effects (Ginn and Metzner, 1969), and R is the radius of the fluid-filled region between the cone and the plate. Interpretation of total force data merely requires application of Equation 1.78 to empirically corrected force measurements at each of several rates of shear.

With the capillary tube instrument the final diameter of the jet is readily obtained from photographs which may be taken simultaneously with the viscous measurements. Denoting the axial coordinate with the subscript 1 and the radial coordinate with the subscript 2, the principal normal stress difference is given by (Metzner, Houghton, Sailor, and White, 1961; Savins, 1966)

$$\frac{(P_{11} - P_{22})_w}{\rho V^2} = \left(\frac{1 + n'}{n'}\right)\left(\frac{1 + 3n'}{1 + 2n'}\right) - \frac{(D/D_j)^2}{n'}\left(1 + n' + \frac{d \ln D/D_j}{d \ln \dfrac{8V}{D}}\right) \tag{1.79}$$

where the subscript w indicates that the normal stress difference is determined at the wall shear rate, D_j is the equilibrium diameter of the fluid jet, and the remaining variables have their usual meaning.

Equation 1.79 is valid provided the flow is laminar. At very low flow rates, surface tension effects may become significant and it is recommended that Reynolds numbers in excess of 150 be used (Savins, 1966). For purely viscous materials the normal stress difference is zero, $n' = 1.0$, and the constant diameter ratio of 1.15 prevails. For highly elastic systems, diameter ratios as low as 0.47 have been observed (Metzner *et al.*, 1961).

The procedure for determining the normal stress at a given shear rate from data at several values of $8V/D$ is therefore as follows:

a. determine n' from Equation 1.68;
b. determine $\dot{S}_w$ at the chosen value of $8V/D$ from Equation 1.67;

c. prepare a logarithmic plot of D/D_j versus $8V/D$ and determine the slope at the chosen value of $8V/D$;
d. calculate the principal normal stress difference from Equation 1.79.

As an alternative to measurement of the jet diameter, measurements of the thrust created by the fluid jet on the capillary tube may be used to calculate the normal stress. The total thrust, T_c, may be determined by mounting the capillary tube on a calibrated spring system and measuring the deflection of the spring. Using the same assumptions as for Equation 1.79, the normal stress difference is given by

$$\frac{(P_{11} - P_{22})_w}{\rho V^2} = \left(\frac{1 + n'}{n'}\right)\left(\frac{1 + 3n'}{1 + 2n'}\right) - \frac{4T_c}{\pi D^2 \rho V^2}\left(1 + \frac{1}{2n'}\frac{d \ln T_c}{d \ln \frac{8V}{D}}\right) \tag{1.80}$$

Interpretation of data using this equation is exactly the same as for Equation 1.79 except that the jet diameter measurements are replaced by measurements of the jet thrust, T_c.

Bibliography

Allen, H. S., *Phil. Mag.*, **50**, 323 (1900).

Astarita, G., *Ind. Eng. Chem. Fundam.*, **4**, 354 (1965); **6**, 257 (1967).

Billington, E. W., *Proc. Phys. Soc.*:
a. **75**, 40 (1960);
b. **76**, 127 (1960).

Brown, G. G., and Associates, *Unit Operations*, John Wiley & Sons, Inc., New York, 1950.

Burgers, J. H., *Proc. Koninkl. Akad. Wetenschap.*, Amsterdam, **42**, 1045 (1941); **45**, 9 (1942).

Calderbank, P. H., and M. B. Moo-Young, *Trans. Inst. Chem. Engrs.*, **37**, 26 (1959).

Carman, P. C., *Trans. Inst. Chem. Engrs*, **15**, 150 (1937).

Cheng, D. C.-H., and F. Evans, *Brit. J. Appl. Phys.*, **16**, 1599 (1965).

Christiansen, E. B., N. W. Ryan, and W. E. Stevens, *A.I.Ch.E. Journal*, **1**, 544 (1955).

Crawford, H. R., Report to Bureau of Naval Weapons. U. S. Navy (1962).

Cross, M. M., *J. Colloid Sci.*, **20**, 417 (1965).

Denn, M. M., and J. J. Roisman, *A.I.Ch.E. Journal*, **15**, 454 (1969).

Denny, D. A., and R. S. Brodky, *J. Appl. Phys.*, **33**, 2269 (1962).

Dick, R. I., and B. B. Ewing, *Journal W.P.C.F.*, **39**, 543 (1967).

Dodge, D. W., and A. B. Metzner, *A.I.Ch.E. Journal*, **5**, 189 (1959).

Eyring, H. J., *J. Chem. Phys.*, **4**, 283 (1936).

Fabula, A. G., Fourth Int. Congress on Rheology, Brown University (1963), *Proc.* Part 3, p. 455, Interscience Publishers, New York, 1965.

Famularo, J., and J. Happel, *A.I.Ch.E. Journal*, **11**, 981 (1965).

Fischer, W. H., W. H. Bauer, and S. E. Wiberley, *Trans. Soc. Rheol.*, **5**, 221 (1961).

Francis, A. W., *Physics*, **4**, 403 (1933).

Fredrickson, A. G., *Principles of Continuum Mechanics*, Prentice-Hall, Englewood Cliffs, N. J., 1964.

Fredrickson, A. G., *A.I.Ch.E. Journal*, **16**, 436 (1970).

Gal-Or, B., *Can. J. Chem. Eng.*, **48**, 526 (1970).

Gal-Or, B., and S. Waslo, *Chem. Eng. Sci.*, **23**, 1431 (1968).

Gill, F., and R. J. Russell, *Ind. Eng. Chem.*, **46**, 1264 (1954).

Ginn, R. F., and A. B. Metzner, Fourth Int. Congress on Rheology, Brown University (1963), *Proc.* Part 2, p. 583, Interscience Publishers, New York, 1965.

Ginn, R. F., and A. B. Metzner, *Trans. Soc. Rheol.*, **13**, 429 (1969).

Goodeve, C. F., *Trans Faraday Soc.*, **35**, 342 (1939).

Govier, G. W., *Engineering Journal*, **43**, 79, March (1960).

Govier, G. W., C. A. Shook, and E. O. Lilge, *Trans. Can. Inst. Mining Met.*, **60**, 147 (1957).

Hadamard, M. J., *Compt. Rend. hebd. Séances Acad. Sci.*, Paris, **152**, 1735 (1911); **154**, 109 (1912).

Hahn, S. J., T. Ree, and H. Eyring, *Ind. Eng. Chem.*, **51**, 856 (1959).

Happel, J., *A.I.Ch.E. Journal*, **4**, 197 (1958); **5**, 174 (1959).

Happel J., and Brenner, H., *Low Reynolds Number Hydrodynamics*, Prentice-Hall Inc., Englewood Cliffs, N. J., 1965.

Herschel, W. H., and R. Bulkley, *A.S.T.M. Proc.*, **26**, 621 (1926); *Kolloid-Z.*, **39**, 291 (1926).

Huppler, J. D., *Trans. Soc. Rheol.*, **9** *Pt.* **2**, 273 (1965).

Kearsey, H. A., and L. E. Gill, *Trans. Inst. Chem. Engrs.*, **41**, 296 (1963).

King, R. G., *Rheol. Acta*, **5**, 41 (1966).

Kozeny, J., *Sitz.-Ber. Wiener Akad.*, Abt. 11a, **136**, 271 (1927).

Krieger, I. M., and Maron, S. H., *J. Appl. Phys.*, **25**, 72 (1954).

Kuwabara, S., *J. Phys. Soc.* (Japan), **14**, 527 (1959).

Levich, V. G., *Physicochemical Hydrodynamics*, Prentice-Hall Inc., Englewood Cliffs, N. J., 1962.

Lummus, J. L., J. E. Fox, and D. B. Anderson, *Oil and Gas J.*, **59**, 87, No. 50 (1961).

Marrucci, G., G. Apuzzo, and G. Astarita, *A.I.Ch.E. Journal*, **16**, 538 (1970).

Meister, B. J., and R. D. Biggs, *A.I.Ch.E. Journal*, **15**, 643 (1969).

Meter, D. M., Doctoral Dissertation, University of Wisconsin, Madison, Wisconsin (1963).

Meter, D. M., and R. B. Bird, *A.I.Ch.E. Journal*, **10**, 878 (1964).

Metzner, A. B., W. T. Houghton, R. A. Sailor, and J. L. White, *Trans. Soc. Rheol.*, **5**, 133 (1961).

Metzner, A. B., and M. G. Park, *J. Fluid Mech.*, **20** *Pt.* **2**, 291 (1964).

Metzner, A. B., J. L. White, and M. M. Denn, *Chem. Eng. Progress*, **62**, 81 (1966); *A.I.Ch.E. Journal*, **12**, 863 (1966).

Metzner, A. B., and M. Whitlock, *Trans. Soc. Rheol.*, **2**, 239 (1958).

Middleman, S., *The Flow of High Polymers*, Interscience Publishers, New York, 1969.

Mooney, M., *J. Rheol.*, **2**, 210 (1931).

Moore, F., *Brit. Ceram. Soc. Trans.*, **58**, 470 (1959).

Munroe, H. S., *Trans. Amer. Inst. Mining Met. Engrs.*, **17**, 637 (1888-1889).

Oldroyd, J. G., *Proc. Roy. Soc., Ser. A*, **200**, 523 (1950).

Paslay, P. R., and A. Slibar, *Ingenieur-Archiv.*, **32**, 66 (1963).

Peter, S., *Rheol. Acta*, **3**, 178 (1964).

Philippoff, W., *Kolloid-Z.*, **71**, 1 (1935).

Pilpel, N., *Chem. Process Eng.*, **47**, 57 (1966).

Pinder, K. L., *Can. J. Chem. Eng.*, **42**, 132 (1964).

Powell, R. E., and H. Eyring, *Nature*, **154** (3909), 427 (1944).

Powell, R. L., and S. Middleman, *Trans. Soc. Rheol.* **13**, 111 (1969).

Prandtl, L., *Z. Angew. Math. Mech.*, **8**, 85 (1928).

Rabinowitsch, B., *Z. physik. Chem., Ser. A.*, **145**, 1 (1929).

Ritter, R. A., Doctoral Dissertation, Univ. Alberta, Edmonton, Alberta (1961).

Ritter, R. A., and G. W. Govier, *Can. J. Chem. Eng.*, **48**, 505 (1970).

Rodriguez, F., *Trans. Soc. Rheol.*, **10**, 169 (1966).

Russell, R. J., *J. Inst. Petrol.*, **46**, 199 (1960).

Rybczynski, W., *Bull. Acad. Sci., Cracow, Ser. A*, **40** (1911).

Savins, J. G., *Soc. Petrol. Engrs. J.*, **4**, 203 (1964).

Savins, J. G., in *Encyclopedia of Industrial and Chemical Analysis*, Vol. 3, John Wiley & Sons, Inc., New York, 1966.

Scheidegger, A. E., *The Physics of Flow Through Porous Media*, University of Toronto Press, Second Ed., 1960.

Seyer, F. A., and A. B. Metzner, *Can. J. Chem. Eng.*, **45**, 121 (1967).

Sisko, A. W., *Ind. Eng. Chem.*, **50**, 1789 (1958).

Skelland, A. H. P., *Non-Newtonian Flow and Heat Transfer*, John Wiley & Sons, Inc., New York, 1967.

Slibar, A., and Paslay, P. R. in Second-Order Effects in Elasticity Plasticity and Fluid Dynamics, Int. Symp., Haifa, Israel (1962) *Proc.* p. 314, edited by M. Reiner and D. Abir, Pergamon Press, New York, 1964.

Spells, K. E., *Trans. Inst. Chem. Engrs.*, **33**, 79 (1955).

Spriggs, T. W., J. D. Huppler, and R. B. Bird, *Trans. Soc. Rheol.*, **10**, 191 (1966).

Steg, I., and D. Katz, *J. Appl. Polym. Sci.*, **9**, 3177 (1965).

Storey, B. T., and E. W. Merrill, *J. Polym. Sci.*, **33**, 361 (1958).

Swanson, V. F., *Trans. Soc. Mining Engrs.*, **238**, 160 (1967).

Tanner, R. I., and A. C. Pipkin, *Trans. Soc. Rheol.* **13**, 471 (1969).

Thomas, D. G., *A.I.Ch.E. Journal*, **7**, 423 (1961); **9**, 310 (1963); **10**, 303 (1964).

Waddel, H.:
a. *J. Appl. Phys.*, **5**, 281 (1934).
b. *J. Franklin Inst.*, **217**, 459 (1934).

Weissenberg, K., *Nature*, **159**, 310 (1947).

Weissenberg, K., *First Int. Rheological Congress* (1948), *Proc.* Part 1, p. 29, North-Holland Publishing Co., Amsterdam, 1949.

White, J. L., and A. B. Metzner, in *Progress in International Research on Thermodynamic and Transport Properties*, pp. 748–762, Academic Press, New York, 1962.

White, J. L., and A. B. Metzner, *J. Appl. Polym. Sci.*, **7**, 1867 (1963).

The Flow Properties of Fluids

2

1 INTRODUCTION

The flow behavior of single-phase fluids in pipes and elsewhere is largely dependent on the density and the consistency of the fluid. For simple Newtonian systems, the consistency is measured by the single property, viscosity. We have seen that more than one quantity is necessary to express the consistency of non-Newtonian fluids as follows: for the Bingham plastic, yield value and coefficient of rigidity; for a pseudoplastic described by the power law, consistency index and flow behavior index; for a pseudoplastic described by the Prandtl–Eyring equation, the coefficients A and B; for a pseudoplastic described by the Ellis equation, the coefficients ϕ_1, ϕ_2 and α; etc., as described in Chapter 1.

The pipe and other flow behavior of multiphase fluids is dependent not only on the density and consistency of the individual phases, but, as we will see later, on their interfacial tension, the state of their dispersion, and their relative quantities.

Wherever possible, in any flow calculation we should endeavor to use accurately measured values of the density, consistency, and interfacial tension corresponding with the temperature and pressure of the flow system. For the most common, simple, single-phase fluids, such values are available as functions of pressure and temperature. Useful references are: J. H. Keenan, and F. G. Keyes, *Thermodynamic Properties of Steam*, John Wiley and Sons, New York (1961); J. H. Perry, *Chemical Engineers' Handbook*, McGraw-Hill, Toronto (1963); *Engineering Data Book*, Natural Gas Processors Suppliers

Association, Tulsa, Oklahoma (1966); *Gas Engineers' Handbook*, Industrial Press, New York (1965); *Technical Data Book—Petroleum Refining*, American Petroleum Institute, New York (1966); *International Critical Tables of Numerical Data*, National Research Council, U.S.A., McGraw-Hill, New York (1926–1933); *Handbook of Chemistry and Physics*, Chemical Rubber Co., Cleveland, Ohio (1967); R. C. Reid, and T. K. Sherwood, *The Properties of Gases and Liquids*, McGraw-Hill, New York (1966).

Table A-1 of the Appendix gives the more important physical properties significant in flow calculations for the most common fluids. More extensive tabulations are available in Reid and Sherwood and the *Engineering Data Book*.

More often than not the engineer is concerned with the flow of a fluid or mixture whose density and consistency properties are unknown or imperfectly known, perhaps at only one or two temperatures or at one pressure. In such cases broad generalizations regarding the variations of these properties with the nature of the fluid or mixture and with pressure and temperature are invaluable. Such generalizations have been developed over the years. Some, such as the equation of state for a gas, relate a property (density) to pressure and temperature with considerable precision for all fluids of a type. Others are both more restrictive in their application and less precise.

In this chapter we will discuss the most important generalizations available for the prediction and pressure-temperature variation of fluid density, Newtonian fluid viscosity, measures of non-Newtonian fluid consistency, and interfacial tension. The treatment of density and of viscosity will be reasonably satisfying since much is known about these two properties and their pressure-temperature dependence. The treatment of measures of non-Newtonian consistency and of interfacial tension is less satisfying, reflecting the greater complexity and less complete understanding of these properties. The excellent treatment of methods of estimating properties of liquids and gases by Reid and Sherwood (1966) will be referred to frequently and may be consulted for a more comprehensive treatment of a number of the methods here discussed.

2 DENSITY

2.1 Single-Component Fluids

2.11 General

The density of pure single-component gases and liquids has been the subject of scientific study since the days of Boyle and Gay-Lussac, and the general form of its pressure-temperature dependence, including the transition from the gaseous to the liquid state, is well understood. Figure 2.1 presents typical data for a pure single-component substance in its two fluid states. The data happen to be for methane, but all pure fluids behave similarly. Following tradition in the presentation of phase diagrams, specific volume, the reciprocal of density, is shown as a function of pressure and temperature. The critical point, the single-phase gas region, the single-phase liquid region, and the two-phase region are all shown. Data such as those for Figure 2.1 are available for many pure fluids in the references previously quoted. These may be in graphical form, in the form of empirical equations closely reproducing the data, or in the form of extensive tabulations such as the familiar steam tables. When such data for a fluid of interest are available, they should normally be used in place of the generalizations discussed later.

An equation relating the variation of density (or its reciprocal, the specific volume) to pressure and temperature is known as an equation of state. Many attempts have been made to develop theoretical and mixed theoretical-empirical equations of state covering the entire range of fluid behavior. The famous van der Waals (1873) equation is the simplest and best known of these. This equation is

$$\left(P + \frac{a}{v_m^2}\right)(v_m - b) = RT \tag{2.1}$$

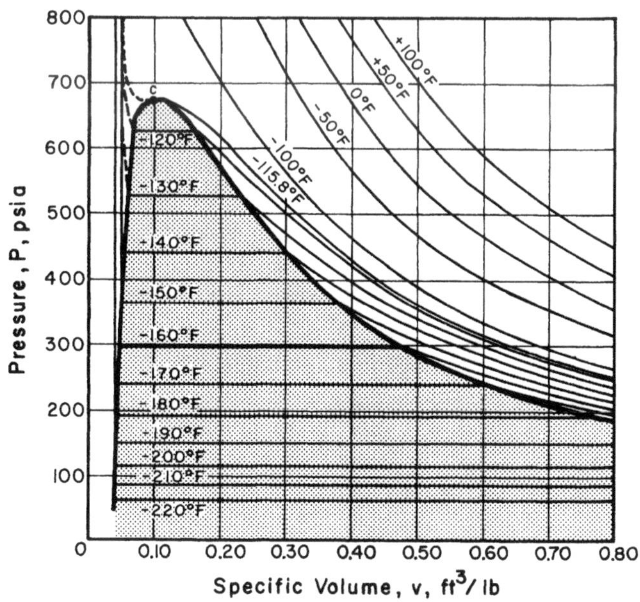

Figure 2.1. Pressure-temperature dependence of specific volume of methane. (From D. L. Katz *et al.*, *Handbook of Natural Gas Engineering*, McGraw-Hill, New York, 1959. Used with permission of McGraw-Hill Book Co.)

where P = pressure,
v_m = molal volume,
T = absolute temperature,
R = gas constant,
$a = 27R^2T_c^2/64P_c$,
$b = RT_c/8P_c$,
T_c = critical temperature,
P_c = critical pressure.

With the appropriate values of P_c and T_c, this equation broadly reproduces the form of Figure 2.1, although it does not apply in the two-phase region and its quantitative accuracy is not good in the liquid region or near the two-phase region.

Among the other generally applicable, relatively simple (two parameter) equations of state, that of Redlich and Kwong (1949) is probably the most useful. It is

$$P + \frac{a'}{T^{0.5}v_m(v_m + b')}(v_m - b') = RT \tag{2.2}$$

where $a' = 0.4278 \dfrac{R^2T_c^{2.5}}{P_c}$, and $b' = 0.0867 \dfrac{RT_c}{P_c}$.

Like the van der Waals equation, the Redlich–Kwong relation is not applicable in or near the two-phase region and gives only rough approximations in the liquid region. Kennedy and Bhagia (1969) have expressed the Redlich–Kwong constants for individual substances as empirical functions of temperature. With the appropriate constants and coefficients for the functions, densities of pure substances may be predicted to within about 0.25 percent.

Equations with more than two parameters to characterize the fluid are inherently more accurate, but their utility is restricted because the characterizing constants are not known for many fluids. Two examples of such equations are the well known equations of Beattie and Bridgeman (1927),

$$P = RT\rho_m + \left(B_0RT - A_0 - \frac{Rc}{T^2}\right)\rho_m^2 + \left(-B_0bRT + A_0a - \frac{RB_0c}{T^2}\right)\rho_m^3 + \frac{RB_0bc\rho_m^4}{T^2} \tag{2.3}$$

and Benedict, Webb, and Rubin (1940),

$$P = RT\rho_m + \left(B_0RT - A_0 - \frac{C_0}{T^2}\right)\rho_m^2 + (bRT - a)\rho_m^3 + a\alpha\rho_m^6 + \frac{c\rho_m^3}{T^2}(1 + \gamma\rho_m^2)\exp(-\gamma\rho_m^2) \tag{2.4}$$

where $\rho_m = 1/v_m$ = molal density. In the Beattie–Bridgeman equation, five constants, a, A_0, b, B_0, and c are used to characterize the fluid; whereas in the Benedict, Webb, and Rubin (BWR) equation, eight constants, A_0, B_0, C_0, a, b, c, γ, and α are used. Determination of these constants is a major problem essentially restricting the use of the equations to the smoothing interpolation and extrapolation of measured data. The constants for the Beattie–Bridgeman and Benedict–Webb–Rubin equations of state are given for a number of fluids in the original references and, for the latter, by Cooper and Goldfrank (1967).

The Beattie–Bridgeman equation is suitable for all gases but not for liquids. Benedict, Webb, and Rubin indicate that their equation is applicable to the estimation of the density of saturated liquids as well as of all gases. Cooper and Goldfrank indicate that the equation predicts pressures to within 0.5 percent for densities to 1.8 times the critical. Inasmuch as neither equation is explicit in density, solution for density is cumbersome. The nature of the equations requires that calculation be carried through to five or six significant figures and that the precise values of the atomic masses incorporated in the development of the equations be used.

2.12 Gases

In the gas region at low pressures and remote from the two-phase boundary, the simple ideal gas law

$$Pv_m = RT \tag{2.5}$$

is reasonably accurate and adequate for many engineering calculations. More generally applicable to the gas region and appropriate for engineering calculations at pressures above a few atmospheres or under conditions near the two-phase boundary is the engineering gas law

$$Pv_m = ZRT \tag{2.6}$$

where Z = compressibility factor.

The compressibility factor is defined by Equation 2.6 and is the factor which validates the equation. Many attempts have been made to correlate values of Z, calculated from measured data through Equation 2.6. Those methods based on the hypothesis that all fluids behave similarly at the same ratios of their critical pressures and temperatures (corresponding states) have been extremely successful. Leland and Chappelear (1968) give a good review of the corresponding states principle and its application.

The simplest correlation based on the concept of corresponding states is

$$Z = \phi(P_r, T_r) \tag{2.7}$$

where $P_r = P/P_c$ = reduced pressure,
$T_r = T/T_c$ = reduced temperature.

This correlation was proposed in graphical form for a series of "permanent" gases by Standing and Katz (1942). Since then new and better data have permitted some refinements. The compressibility factor chart of Viswanath and Su (1965) is probably the best generally applicable correlation of this simple type available today for pure gases. This is reproduced in Figure 2.2. The Viswanath and Su chart predicts the compressibility of most common pure gases other than hydrogen, sulfur dioxide, and hydrogen sulfide to within an average deviation of less than 2 percent and a maximum error of less than about 10 percent.

Sommerfeld and Perry (1968) have expressed the Beattie–Bridgeman equation in reduced form, evaluated the reduced constants using values for nitrogen, and rearranged it to give the following:

$$Z = \frac{P v_m}{RT} = \left[1 + \frac{B_0^*}{v_r}\left(1 - \frac{b^*}{v_r}\right)\right]\left(1 - \frac{c^*}{v_r T_r^3}\right) - \frac{A_0^*}{T_r v_r}\left(1 - \frac{a^*}{v_r}\right) \tag{2.8}$$

where $B_0^* = \dfrac{B_0}{v_{cN}} = 0.56015$; $\qquad b^* = \dfrac{b}{v_{cN}} = -0.076707$;

$c^* = \dfrac{c}{v_{cN} T_{cN}^3} = 0.23274$; $\qquad A_0^* = \dfrac{A_0}{R v_{cN} T_{cN}} = 1.4428$;

$a^* = \dfrac{a}{v_{cN}} = 0.290508$;

v_{cN} = critical molal volume of nitrogen;

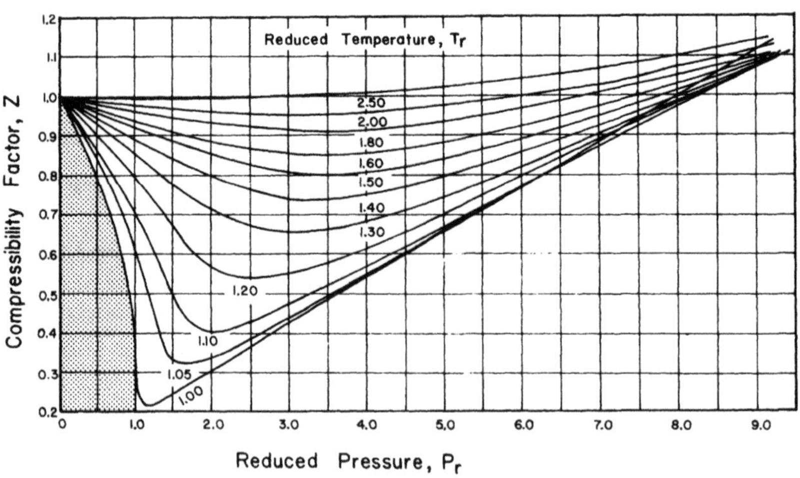

Figure 2.2. Compressibility factor, *Z*, for pure gases. (From Viswanath, D. S., and G.-J. Su, *A. I. Ch. E. Journal*, **11**, 202, 1965.)

$v_r = v_m/v_{mc}$ = reduced molal volume; P is in atmospheres, T in degrees Kelvin, and volumes in liters per mole.

For calculation purposes, the equation is further rearranged to

$$\phi(Z) = Z^4 - Z^3 - a_2 Z^2 - a_3 Z - a_4 = 0 \tag{2.9}$$

where $a_2 = \alpha\beta$; $\quad a_3 = \alpha^2\gamma$; $\quad a_4 = \alpha^3\delta$; $\quad \alpha = \dfrac{Z_c P_r}{T_r}$;

$$\beta = B_o^* - \frac{A_o^*}{T_r} - \frac{c^*}{T_r^3};$$

$$\gamma = B_o^* b^* + \frac{b^* c^*}{T_r^3} + \frac{A_o^* a^*}{T_r};$$

$$\delta = \frac{B_o^* b^* c^*}{T_r^3}.$$

In this form the equation may be solved for Z using the Newton–Raphson iteration procedure (see, for example, Lapidus, 1962). This method gives results comparable in accuracy to Figure 2.2 and is readily adapted to computer application.

Although compressibility factor correlations equivalent to Equation 2.7 are extremely useful and, for nonpolar gases of simple molecular structure, quite reliable, for completely general application and for maximum accuracy the effect of variables other than the reduced pressure and temperature must be considered.

The clear need for this is the known fact that the compressibility factor at the critical pressure and temperature is not the same for all pure substances but varies from 0.230 for steam to 0.304 for hydrogen. Table 2.1 lists the values of Z_c, i.e., Z at $T_r = P_r = 1$, for a number of pure compounds, showing the type of compounds that deviate most from the average. (The Viswanath and Su chart gives a value of $Z_c = 0.28$.)

Lydersen, Greenkorn, and Hougen (1955) have used the compressibility factor at the critical point, Z_c, in addition to the critical pressure and the critical temperature to characterize the fluid more accurately. They have prepared charts for the compressibility factor for several values of the critical compressibility factor, Z_c. This gives an improved correlation near the critical point but not generally.

Another approach more directly related to the fundamental behavior of fluid molecules has been presented by Pitzer *et al.*, (1955). They introduce an "acentricity" factor which is a measure of the deviation of intermolecular forces from the simple idealized case. The acentricity factor is defined as

$$\omega = -\log P_{vr,\,0.7} - 1.000 \tag{2.10}$$

TABLE 2.1

Values[a] of Z_c for Typical Pure Compounds

Compound	Z_c
Methane	0.290
Ethane	0.285
Propane	0.277
i-Butane	0.283
n-Butane	0.274
n-Pentane	0.269
n-Hexane	0.264
n-Heptane	0.259
Benzene	0.274
Toluene	0.271
Ammonia	0.242
Carbon dioxide	0.274
Hydrogen (normal)	0.304
Nitrogen	0.291
Oxygen	0.292
Steam	0.230

[a] From R. C. Reid and T. K. Sherwood, *The Properties of Gases and Liquids*, 2nd. Ed., McGraw-Hill, New York, 1966. Used by permission of McGraw-Hill Book Co.

where $P_{vr,\,0.7}$ is the reduced vapor pressure at a reduced temperature of 0.7. This definition is based on the observation that for simple fluids, such as argon, neon, krypton, and methane, $P_{vr,\,0.7}$ is close to 0.1 and Equation 2.10 indicates $\omega = 0$. For most other fluids ω lies between 0 and 0.4. When vapor pressure data are available, ω should be calculated from Equation 2.10. Reed and Sherwood discuss several alternate methods of estimating ω in the absence of vapor pressure data. One of these is based upon the approximate linearity between Z_c and ω, i.e.,

$$Z_c = 0.291 - 0.080\omega \tag{2.11}$$

For simple fluids where $\omega = 0$, the compressibility factor is a function only of the reduced pressure and the reduced temperature. For other and more complex gases the compressibility factor may be expressed as

$$Z = Z^{(0)} + \omega Z^{(1)} \tag{2.12}$$

where $Z^{(0)}$ is the factor for simple fluids and $\omega Z^{(1)}$ is a correction for the acentricity of other than simple fluids. Pitzer *et al.* present tables of $Z^{(0)}$ and $Z^{(1)}$ covering the range of T_r from 0.8 to 4, and P_r up to 9. Satter and Campbell (1963) have independently checked these and extended the P_r range to 14. These extended values and Pitzer's original data are reproduced in Tables A-2 to A-5 (see Appendix), and are presented graphically in Figures 2.3 and

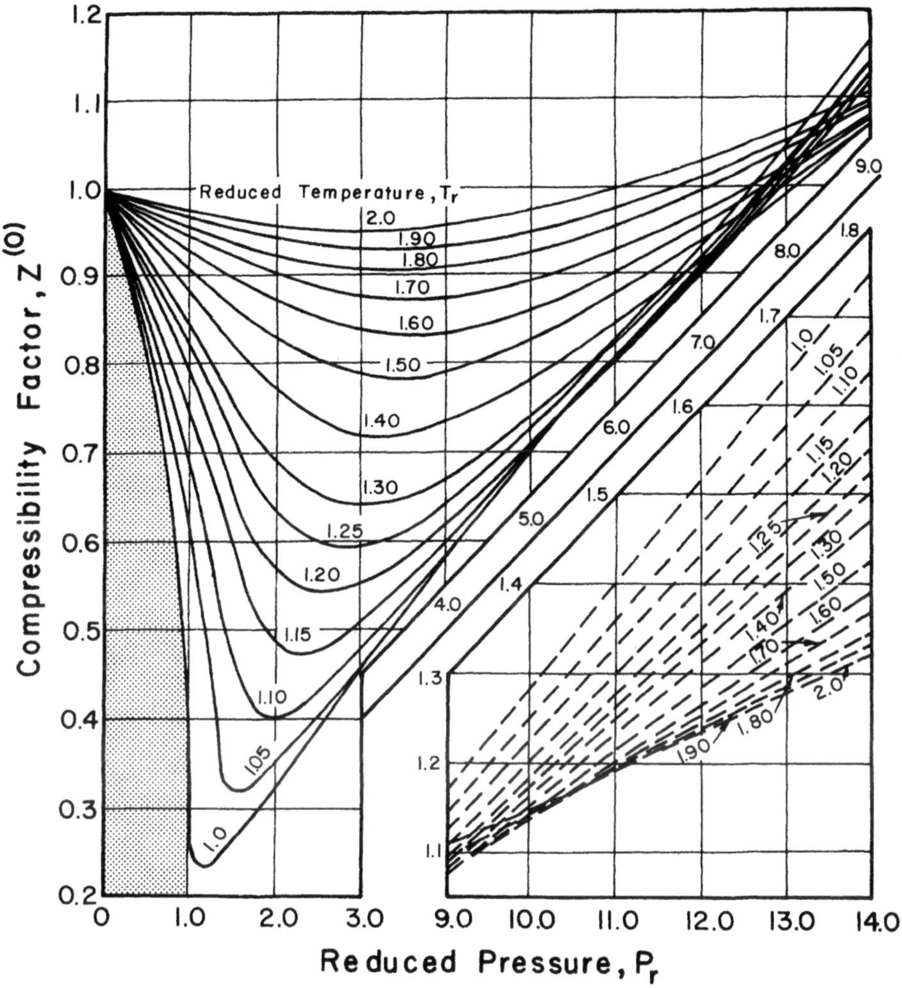

Figure 2.3. Compressibility factor, $Z^{(0)}$, for pure gases. (From Satter, A., and J. M. Campbell, *Soc. Petrol. Engrs. J.*, **3**, 333, 1963.)

2.4. The accuracy with which compressibility factors of pure gases (and hence densities of pure gases) may be predicted by the Pitzer method (as extended by Satter and Campbell) is considerably greater than by the method of Viswanath and Su. Root-mean-square deviations are of the order of 1 percent for all common pure gases except carbon dioxide.

The Pitzer method discussed here is one of the procedures recommended in the API Technical Data Book (1966). In addition this data book recommends the method of Hirschfelder, Buehler, McGee, and Sutton (1958, 1962) for

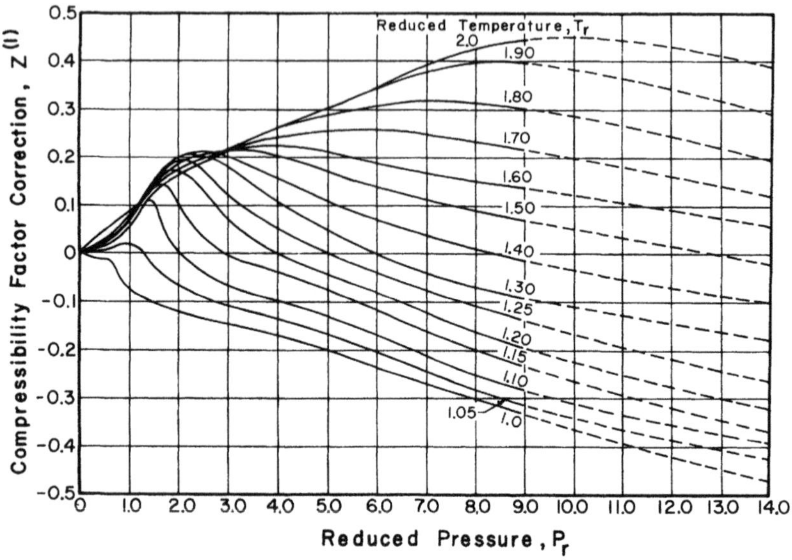

Figure 2.4. Compressibility factor correction, $Z^{(1)}$, for pure gases. (From Satter, A., and J. M. Campbell, *Soc. Petrol. Engrs. J.*, **3**, 333, 1963.)

computer calculations. This latter procedure is claimed to be as accurate as the Pitzer method.

In pipeline flow problems it is often necessary to determine the saturated vapor density. For hydrocarbons this may be accomplished directly through the use of the Bradford and Thodos (1968) correlation, which is

$$\rho_{sv} = \rho_c[1 + \beta(1 - T_r) + \gamma(1 - T_r)^2 - \delta(1 - T_r)^n] \tag{2.13}$$

The constants in this equation are related to the critical compressibility factor by

$$n = 0.160 + 0.586Z_c \tag{2.14}$$

$$\delta = 2.785 - 3.544Z_c \tag{2.15}$$

$$\beta = 2.924 - 7.34Z_c \tag{2.16}$$

$$\gamma = \delta - \beta - 1 \tag{2.17}$$

Equation 2.13 applies in the range $0.9 \leq T_r \leq 1.0$ with and average deviation of 2.2 percent. If the reduced temperature is less than 0.9, the saturated vapor density depends upon the temperature and may be calculated from the relation

$$\rho_{sv} = \rho_c[k \exp(-m/T_r)] \tag{2.18}$$

Empirical equations for k and m are

$$k = 688 \times 10^5 \exp(-46.06)Z_c \tag{2.19}$$

$$m = 55.96 \exp(-7.99Z_c) \tag{2.20}$$

Equation 2.18 for $T_r < 0.9$ predicts densities with an average deviation of about 3.0 percent.

2.13 Liquids

The density behavior of pure, single-component liquids is not so readily predicted as that for gases. Although much less sensitive to pressure, the density of liquids and their temperature dependence are both considerably dependent upon their molecular structure. The most useful generalizations relate to homologous series of compounds or compounds otherwise rather similar. For example, Alani and Kennedy (1960) have modified the van der Waals equation of state so that it may be used for the accurate prediction of densities of liquid hydrocarbons. They assume the van der Waals constants a and b to be functions of temperature and obtain the following relationships.

$$a = K \exp(n/T) \tag{2.21}$$

$$b = mT + C \tag{2.22}$$

In these equations, K, n, m, and C are constants for each hydrocarbon and must be determined experimentally. Table A-6 (see Appendix) gives values of these constants for several hydrocarbons. The authors claim that the average absolute deviation in the molal volume calculated by this method does not exceed 0.33 percent.

For saturated liquids there are several methods for the estimation of density. Lyckman, Eckert, and Prausnitz (1965) have developed the following quadratic equation

$$\frac{\rho_c}{\rho_{sL}} = V_r^{(0)} + \omega V_r^{(1)} + \omega^2 V_r^{(2)} \tag{2.23}$$

where ρ_c is the critical density, ρ_{sL} is the saturated liquid density, ω is Pitzer's acentricity factor, and $V_r^{(0)}$, $V_r^{(1)}$, $V_r^{(2)}$ are generalized functions that depend on temperature and are given in Table A-7. Lyckman *et al.* suggest the use of the linear relationship (Equation 2.11) between Z_c and the acentricity factor where more reliable information is not available. Application of Equation 2.23 to liquids having acentricity factors up to 0.6 results in average deviations of less than 1 percent in most cases, but up to 7 percent in some cases. The density of saturated hydrocarbon liquids has been correlated by Bradford and Thodos (1968) by an equation similar to that which they developed for saturated vapors. Their equation is

$$\rho_{sL} = \rho_c[1 + \beta(1 - T_r) + \gamma(1 - T_r)^2 + \delta(1 - T_r)^n] \tag{2.24}$$

where n, β, γ, and δ are empirical constants expressed by Equations 2.14 to 2.17. Equation 2.24 approximates experimental liquid densities for several hydrocarbons with an overall average deviation of 1.1 percent in the entire temperature range.

Other equations of state, e.g., Equations 2.2, 2.3, and 2.4, presented for the estimation of gas densities may be applied for liquids. The accuracy of prediction, however, does not usually justify the effort required in the use of these equations.

If an accurate specific volume or density measurement is available at one temperature, the density of saturated liquids may be estimated at any other temperature using Benson's (1948) method.

$$\rho_{sL} = 2\rho_c + \left(\frac{T_c - T}{T_c - T_b}\right)(\rho_{Lb} - 2\rho_c) - \rho_{sv} \tag{2.25}$$

where T_b and ρ_{Lb} are the boiling point temperature and the density at the boiling point, respectively. If ρ_{Lb} is not available, it may be replaced by the density at any other temperature, and T_b is replaced by the corresponding temperature. The saturated vapor phase density ρ_{sv} may be calculated by methods already discussed.

A similar method attributed to Goldhammer (1910) is given by the equation

$$\frac{(\rho_{sL} - \rho_{sv})_2}{(\rho_{sL} - \rho_{sv})_1} = \left(\frac{T_c - T_2}{T_c - T_1}\right)^n \tag{2.26}$$

where the exponent n is temperature sensitive above $T_r = 0.85$ and ranges in value from 0.25 to 0.33. Values of n for some common liquids reported by Fishtine (1963) are shown in Table A-8 (see Appendix).

At constant temperature the variation of density of liquids with pressure is usually expressed in terms of a compressibility defined as

$$C = -\frac{1}{v}\left(\frac{\partial v}{\partial P}\right)_T \tag{2.27}$$

(where $v = 1/\rho$ = specific volume) or its reciprocal, the bulk modulus, K. In general, C is a point function varying for a given liquid with both pressure and temperature. Over pressure regions where the variation of specific volume with pressure is small and nearly linear, either of two essentially equivalent equations may be used

$$\rho = \rho_0[1 + C(P - P_0)] \tag{2.28}$$

or

$$\rho = \frac{\rho_0}{1 - C(P - P_0)} \tag{2.29}$$

where ρ_0 = density at pressure P_0 (usually one atmosphere),
ρ = density at pressure P.

Equations 2.28 and 2.29 are often used even where the variation of specific volume (or density) is not linear, C being interpreted as an average value effective over the range of pressure P_0 to P.

An alternative approach applicable with greater accuracy over a wider range of pressure involves the integration of Equation 2.27, defining C to give

$$\rho = \rho_0 \exp C(P - P_0) \tag{2.30}$$

This still assumes that C is constant over the pressure range. Liquid compressibilities suitable for use with any of Equations 2.28, 2.29, and 2.30 are given in Table A-9 (see Appendix).

Hayward (1967) gives an excellent discussion of various equations proposed to account for the variation of compressibility with pressure and recommends a form originally proposed by Tait (1888). This is

$$\rho = \frac{\rho_0}{1 - \dfrac{P}{K_0 + mP}} \tag{2.31}$$

where K_0 is the bulk modulus, or the reciprocal of C, at zero pressure, and m is a constant equal to the average slope of the variation of K or $1/C$ with P.

Equation 2.31 reproduces the data of Amagat (1893) on the variation of the density of water with pressure to within about six parts in a million to pressures of 3000 atmospheres.

Chueh and Prausnitz (1969) have developed a generalized correlation for the density of liquids that involves the use of the saturated liquid density. The correlation is

$$\rho_L = \rho_{sL}[1 + 9C_s(P - P_s)]^{1/9} \tag{2.32}$$

where P_s is the saturation pressure at the temperature of interest. The compressibility of saturated liquids, C_s, is correlated by

$$C_s = \frac{m}{RT_c\,\rho_c}(1 - 0.89\omega^{1/2})\,[\exp(6.9547 - 76.2853T_r + 191.3060T_r^2 - 203.5472T_r^3 + 82.7631T_r^4)] \tag{2.33}$$

where m is the molecular weight. This correlation is valid in the range

$$0.4 \leq T_r \leq 0.98.$$

An empirical method for predicting liquid densities near the critical point has been proposed by Yen and Woods (1966). This method is particularly suitable for use with digital computers.

An alternate to the simplified concept of a compressibility coefficient for a liquid is the concept of an equation of state involving two coefficients, each a function of temperature. Chaudhuri, Mathur, and Stager (1968) reviewed these two-coefficient equations and proposed a new one which appears superior. Their equation is

$$P - P_0 = K_c\left[\left(\frac{\rho}{\rho_0}\right)^4 - \left(\frac{\rho}{\rho_0}\right)^2\right]^q \tag{2.34}$$

where K_c and q are coefficients characteristic of the liquid and dependent upon temperature.

This equation has been tested for aldehydes, alcohols, water, and alkanes (hexane to dodecane) and has been found to give excellent accuracy. Values

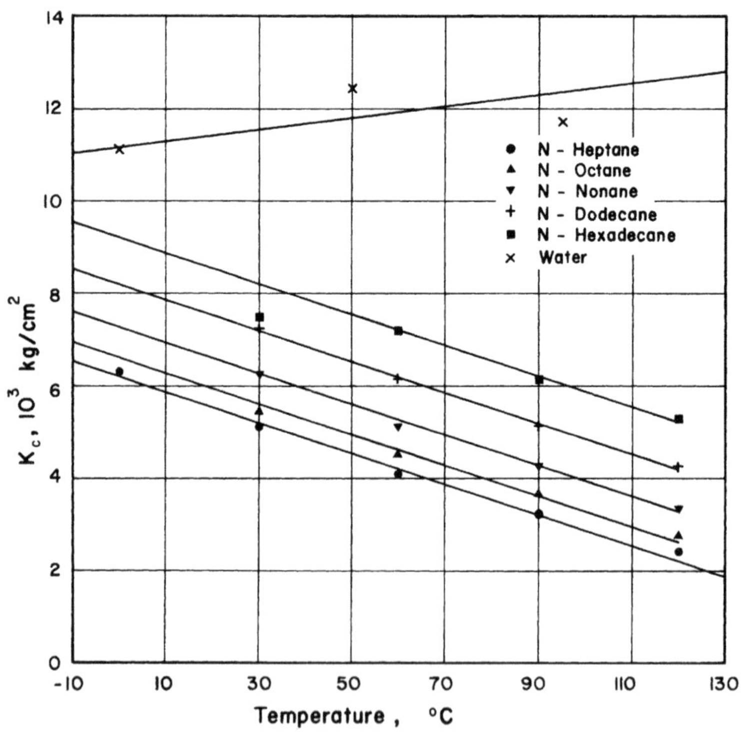

Figure 2.5. Temperature dependence of the constant, K_c, in Equation 2.34. (Modified after Chaudhuri, P. M., G. P. Mathur, and R. A. Stager, *Can. J. Chem. Eng.*, **46**, 132, 1968.)

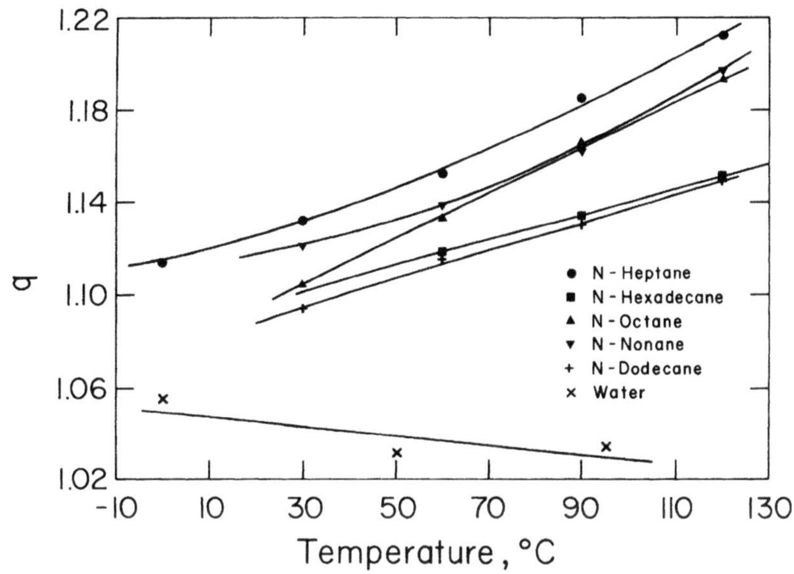

Figure 2.6. Temperature dependence of the constant, q, in Equation 2.34. (Modified after Chaudhuri, P.M., G. P. Mathur, and R. A. Stager, *Can. J. Chem. Eng.*, **46**, 132, 1968.)

of the constants K_c and q for water and the alkanes are given as functions of temperature in Figures 2.5 and 2.6 modified after Chaudhuri *et al.* Coefficients for the aldehydes and the alcohols are available in the original paper.

Other methods for the prediction of liquid densities are available in the API Technical Data Book (1966).

2.2 Multicomponent Fluids

2.21 General

There are few general correlations for multicomponent fluids which may be applied to both the liquid and the gaseous state and, as in the case of single-component fluids, they are not so reliable as the correlations applicable to the individual phases. Kennedy and Bhagia (1969) have extended their work with the Redlich–Kwong equation (see Section 2.11) to describe gas-condensate fluids encountered in petroleum reservoirs. The correlation requires the use of a number of empirical constants and figures. The reader is referred to the original work for details.

2.22 Gases

The density behavior of multicomponent gases is similar to, but more complex than, that of a pure single-component gas. As in the case of the pure compounds, when actual density data are available at the pressures and temperatures of interest, these should be used. When these are lacking or incomplete, modifications of the generalizations discussed earlier may be used.

Kay (1936) proposed that for gas mixtures the compressibility factor for use in the engineering gas law could be estimated by the methods used for pure components provided the critical properties required were replaced by pseudocritical properties. The pseudocritical properties were defined as the molal average critical pressure and temperature as

$$_pT_c = \sum y_i T_{ci} \tag{2.35}$$

$$_pP_c = \sum y_i P_{ci} \tag{2.36}$$

where y_i is the mole fraction of component i in the gas. With substitution of the pseudocritical properties in the calculation of the reduced pressure and temperature, Figure 2.2 for pure gases or its equivalent gives a reasonable approximation of the correct compressibility factor for the mixture.

Brown (1940) and others have used Kay's definitions of pseudocritical properties and found that for mixtures of natural gases containing paraffinic hydrocarbons in a homologous series, but free of carbon dioxide and hydrogen sulfide, much improved accuracy resulted from a $Z - {}_pP_r - {}_pT_r$ plot constructed from actual data on such gas mixtures. This correlation, widely used in the petroleum industry, is reproduced in Figure 2.7.

The presence of significant quantities of carbon dioxide or hydrogen sulfide in a natural gas may affect the compressibility factor by 10 percent or more. Robinson, Macrygeorgos, and Govier (1960) have developed a simple, reasonably accurate method, based upon actual data, of adjusting the compressibility factor determined by Brown's correlation for the presence of hydrogen sulfide or carbon dioxide. They defined a correction factor as

$$c = \frac{Z_a}{Z}$$

where Z = compressibility factor determined from Figure 2.7,
Z_a = compressibility factor corrected for the presence of carbon dioxide and hydrogen sulfide.

Values of c are given in Figure 2.8.

Robinson and Jacoby (1965) present a modification of the Redlich–Kwong equation, which gives good precision with natural-gas mixtures containing any proportions of nitrogen, carbon dioxide, or hydrogen sulfide. Standard

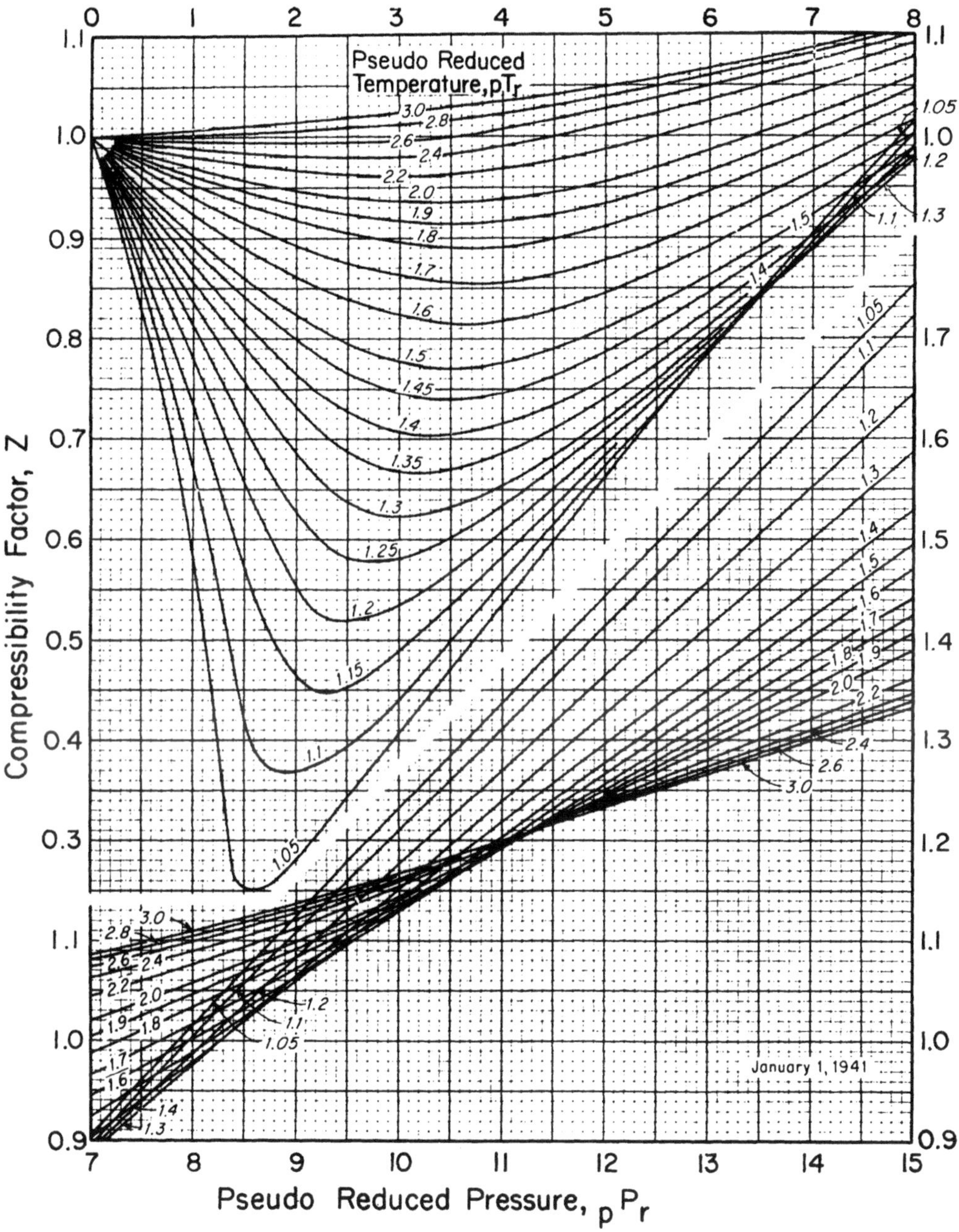

Figure 2.7. Compressibility factor, *Z*, for natural gases. (From Standing, M.B., and D. L. Katz, *Trans. A.I.M.E.*, **146**, 140, 1942.)

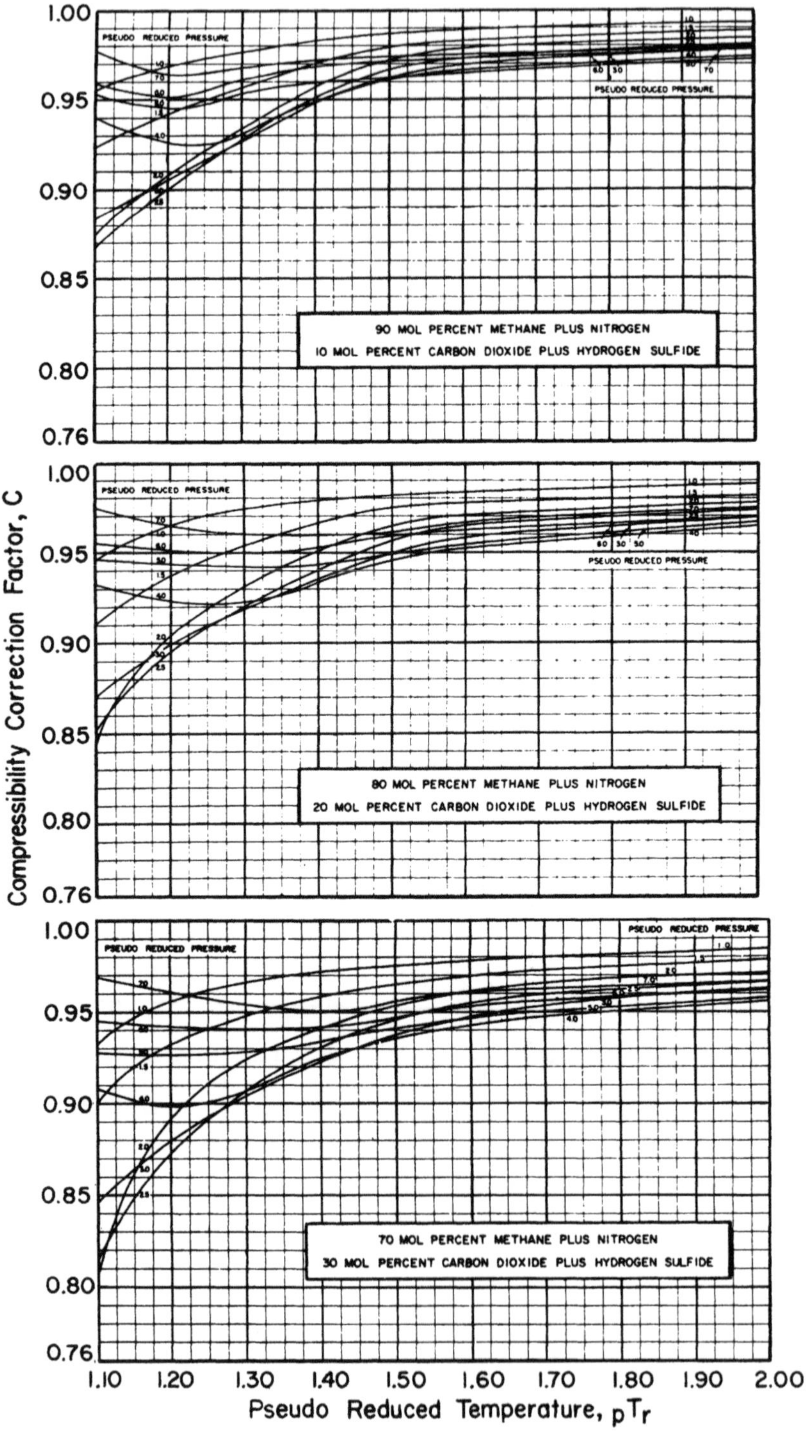

Figure 2.8. Compressibility correction factor, *c*, for sour natural gases. (From Robinson D. B., C. A. Macrygeorgos, and G. W. Govier, *Petrol. Trans.*, **219**, 54, 1960.)

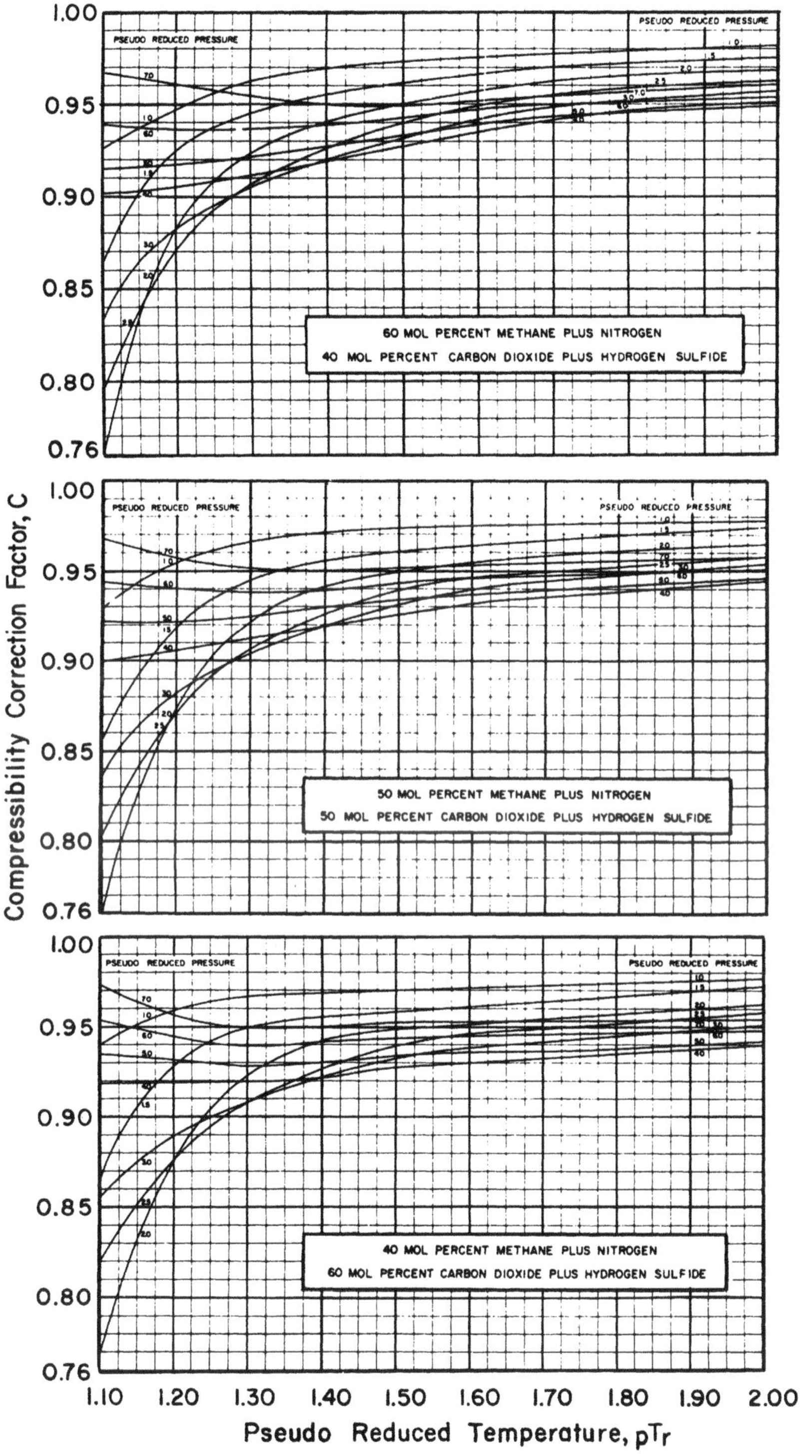

Figure 2.8. *Continued*

deviations and maximum errors between predicted and measured values are within 0.3 and 1.1 percent, respectively. The original paper should be consulted for details.

Satter and Campbell (1963) have reviewed a number of alternate ways of calculating pseudocritical properties to permit the use of Pitzer's correlation for gas mixtures. They recommend the method of Stewart, Burkhardt, and Voo (1959), where a modified pseudocritical temperature and a modified pseudocritical pressure are defined by the equations

$$ {}_pT_c' = \frac{K^2}{J} \tag{2.37} $$

$$ {}_pP_c' = \frac{{}_pT_c'}{J} \tag{2.38} $$

where

$$ J = \frac{1}{3}\sum y_i\left(\frac{T_c}{P_c}\right)_i + \frac{2}{3}\left[\sum y_i\left(\frac{T_c}{P_c}\right)_i^{1/2}\right]^2 \tag{2.39} $$

$$ K = \sum y_i\left(\frac{T_c}{P_c^{1/2}}\right)_i \tag{2.40} $$

and a pseudoacentricity factor defined by

$$ \omega_M = \sum y_i\omega_i \tag{2.41} $$

is employed.

Pitzer's compressibility factor relation is modified to

$$ Z = Z^{(0)} + \omega_M Z^{(1)} \tag{2.42} $$

and $Z^{(0)}$ and $Z^{(1)}$ are obtained from Figure 2.3 and Figure 2.4 using the modified pseudocritical temperature and pressure calculated from Equations 2.37 and 2.38.

The API Technical Data Book (1966) recommends the use of Pitzer's method with ${}_pT_c$, ${}_pP_c$ and ω_M calculated from Equations 2.35, 2.36, and 2.41, respectively. In addition, it presents an extension of the Hirschfelder *et al.* (1958, 1962) method for gas mixtures. This latter procedure is suitable for computer calculations and is comparable in accuracy to the Pitzer method.

Wichert and Aziz (1971) have tested most of the available methods for the prediction of compressibility factors for sour natural gases. Predictions were compared with 1085 experimental data points for sour natural gases covering the following range of concentrations:

	Mole Fraction
Hydrogen sulfide	0–0.74
Carbon dioxide	0–0.54
Hydrogen sulfide + carbon dioxide	0–0.82

They found the method of Robinson *et al.* (1960), within its range of applicability, to be the most reliable of the available methods and to reproduce the data to average absolute and maximum absolute deviations of 1.0 and 12.6 percent. The Robinson and Jacoby (1965) modification of the Redlich–Kwong equation was found to be the best existing correlation for the entire range of composition, pressure, and temperature tested by Wichert and Aziz.

Following Buxton and Campbell (1967), Wichert and Aziz have developed a modification of the Pitzer method applicable to natural gases containing either or both hydrogen sulfide and carbon dioxide. The method involves the use of empirically adjusted values of ${}_pT_c'$ and ${}_pP_c'$ based upon the hydrogen sulfide and carbon dioxide contents of the gas. For gases containing carbon dioxide and no hydrogen sulfide, it reduces to the equivalent of the Buxton and Campbell method. Wichert and Aziz found this method to reproduce the 1085 experimental data points referred to earlier to within an average absolute and a maximum absolute deviation of 0.9 and 7.1 percent.

Wichert and Aziz also propose a modification to the Redlich–Kwong equation of state for application to sour natural gases. This modification also involves empirical adjustments based upon the hydrogen sulfide and carbon dioxide contents of the gas. They compared the predictions of this method with the previously mentioned 1085 data points and found average absolute and maximum absolute deviations of 0.94 and 5.8 percent. This method is superior to that of Robinson *et al.* (1960) with respect to the maximum deviation, is comparable to the Wichert–Aziz modification of the Pitzer method, and is recommended for use where the frequency of calculations warrant computer methods.

2.23 Liquids

As in the case of gases, the relationship between density and molecular structure, and the pressure and temperature dependence of density, is more complex for multicomponent liquids than for pure liquids. With few exceptions, data are insufficient to make generalizations other than that the methods suitable for pure liquids may be applied with approximate results.

The equations of state may be used for liquid mixtures as they were for gas mixtures. The constants appearing in the equations must be calculated from the available values of these constants for pure fluids. The method of Alani and Kennedy is applicable for hydrocarbon mixtures if the constants a and b in the van der Waals equation are replaced by molal average values of these constants for the mixture. For liquefied natural gases, treatment of the C_{7+} components requires special attention and is discussed in some detail by Alani and Kennedy (1960).

The methods described for estimating the effect of pressure and temperature on the density of single-component liquids may be used to approximate the effect on density of multicomponent liquids.

2.3 Pseudohomogeneous Multiphase Systems

Frequently the mixture density of a pseudohomogeneous multiphase system is of interest. Normally this may be calculated simply be assuming additive volumes. Thus

$$\rho_M = \frac{1}{\sum \frac{c_{wi}}{\rho_i}} \tag{2.43}$$

where c_{wi} = weight fraction of the individual phases, or

$$\rho_M = \sum c_{vi} \rho_i \tag{2.44}$$

where c_{vi} = volume fraction of the individual phases.

3 NEWTONIAN VISCOSITY

3.1 Single-Component Fluids

3.11 General

The pressure-temperature dependence of the viscosity of pure liquids and gases has been studied theoretically and experimentally for many years. Figure 2.9 presents data, covering both the gas and liquid phase, for propane. These are typical for most pure fluid substances. The well known decrease of viscosity with temperature in the liquid phase and its increase in the gas phase are apparent. Data of this type are available for many pure compounds. References consolidating such data include: Comings, Mayland, and Egly (1944); Katz *et al.* (1959); Huang, Swift, and Kurata (1966); Rao and Doraiswamy (1966); Giddings, Kao, and Kobayashi (1966); Gonzalez and Lee (1966); and Gonzalez *et al.* (1967).

Correlations based upon the concept of a residual viscosity, discussed in Section 3.12, have applicability to both gases and liquids.

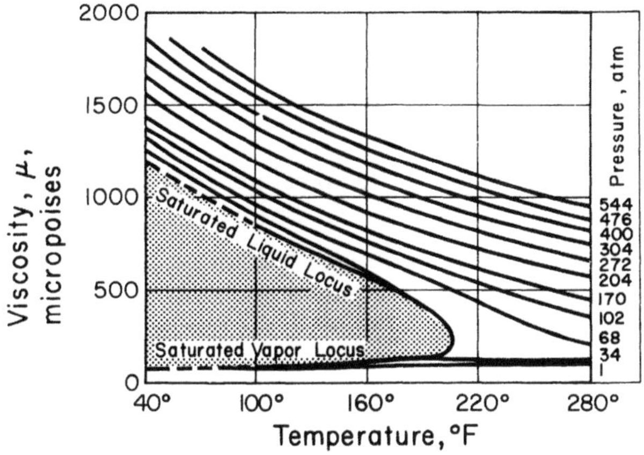

Figure 2.9. Pressure-temperature dependence of the viscosity of propane. (From Giddings J. G., J. T. F. Kao, and R. Kobayashi, *J. Chem. Phys.*, **45**, 578, 1966.)

3.12 Gases

For pure gases, equations based upon kinetic theory are of use in the prediction of viscosity at low pressures (referred to as the dilute-gas viscosity) and these and semitheoretical methods based upon them give good results. Also useful are empirical methods based on the same concept of corresponding states, which led to the excellent compressibility factor correlations.

Bromley and Wilke (1951) have modified the basic Chapman-Enskog equation (see Chapman and Cowling, 1939) and their relationship

$$\mu_0 = \frac{0.00333(mT_c)^{1/2}}{v_c^{2/3}}[f(1.33T_r)] \tag{2.45}$$

where $f(1.33T_r) = 1.058T_r^{0.645} - \dfrac{0.261}{(1.9T_r)^{0.9 \log(1.9T_r)}}$,

μ_0 = dilute-gas viscosity, centipoise,
m = molecular weight,
v_c = critical volume, cm^3/g mole,
T_c = critical temperature, °K,

gives results for the viscosity of pure gases at low pressure usually within about 3 percent accuracy.

Table A-10 (see Appendix) taken from Perry gives values of $f(1.33T_r)$ for a range of values of $(1.33T_r)$.

An alternate, somewhat less accurate correlation is that of Arnold (1933):

$$\mu_0 = \frac{27.0 m^{1/2} T^{3/2}}{v_b^{2/3}(T + 1.47T_b)} \tag{2.46}$$

where μ_0 is in micropoises,

T_b = normal boiling point temperature, °K,

v_b = molal liquid volume at T_b, cm^3/g mole.

Reid and Sherwood (1966) give a discussion on the different methods of estimating the molal volume of liquid at the normal boiling point. Experimental values of v_b for a few substances are shown in Table A-11 (see Appendix).

Arnold's relationship is particularly convenient for estimating the effect of temperature on the viscosity of gases at low pressure.

Stiel and Thodos (1961) have developed the following relationships for correlating the dilute-gas viscosity of pure nonpolar gases at low pressure.

$$\mu_0 = \frac{34.0 \times 10^{-5}}{\xi} T_r^{0.94} \qquad \text{for } T_r \le 1.5 \tag{2.47}$$

$$\mu_0 = \frac{17.78 \times 10^{-5}}{\xi} (4.58T_r - 1.67)^{0.625} \qquad \text{for } T_r > 1.5 \tag{2.48}$$

where $\xi = T_c^{0.167} m^{-0.5} P_c^{-0.667}$.

For polar gases an additional molecular structure parameter is found necessary and Stiel and Thodos (1962) select the compressibility factor at the critical pressure and temperature. Their relationships are

$$\mu_0 = \frac{1 \times 10^{-5}}{\xi Z_c^{1.25}} (7.55T_r - 0.55) \tag{2.49}$$

for substances exhibiting hydrogen bonding, $T_r < 2.00$, and

$$\mu_0 = \frac{1 \times 10^{-4}}{\xi Z_c^{0.667}} (1.90T_r - 0.29)^{0.80} \tag{2.50}$$

for substances not exhibiting hydrogen bonding, $T_r < 2.5$. These relationships may be used to predict the dilute-gas viscosity and its temperature dependence with good engineering accuracy. Reid and Sherwood (1966) show an average error of 1.8 percent for nonpolar gases and 3.2 percent for polar gases. (In Equations 2.47 to 2.50 the units are as defined following Equation 2.45 and P_c is in atmospheres.)

For the normal paraffinic hydrocarbons, a number of specific correlations of dilute-gas viscosity have been developed. The widely used correlation of Carr, Kobayashi, and Burrows (1954) takes the simple form of

$$\mu_0 = \phi(m, T) \tag{2.51}$$

Giddings and Kobayashi (1964) have modified the original Carr *et al.* correlation by the consideration of more recent data. Their correlation reproduced as Figure 2.10 may be considered the best for normal paraffinic hydrocarbons (and their mixtures, as discussed later). The accuracy which may be expected is of the order of ± 2 percent.

The correlations so far discussed are for the low-pressure or dilute-gas viscosity and apply reasonably well up to a few atmospheres. Beyond 5–10 atmospheres the effect of pressure becomes significant. Again the concept of corresponding states has proven useful and correlations of the form

$$\mu = \mu_0 \, \phi(P_r, T_r) \tag{2.52}$$

or

$$\mu = \mu_c \, \phi(P_r, T_r) \tag{2.53}$$

where μ_c = viscosity of the gas at the critical point, have been developed. Carr *et al.* (1954) based their correlation for the effect of pressure on the viscosity of normal paraffinic hydrocarbons on Equation 2.52. Others, e.g.,

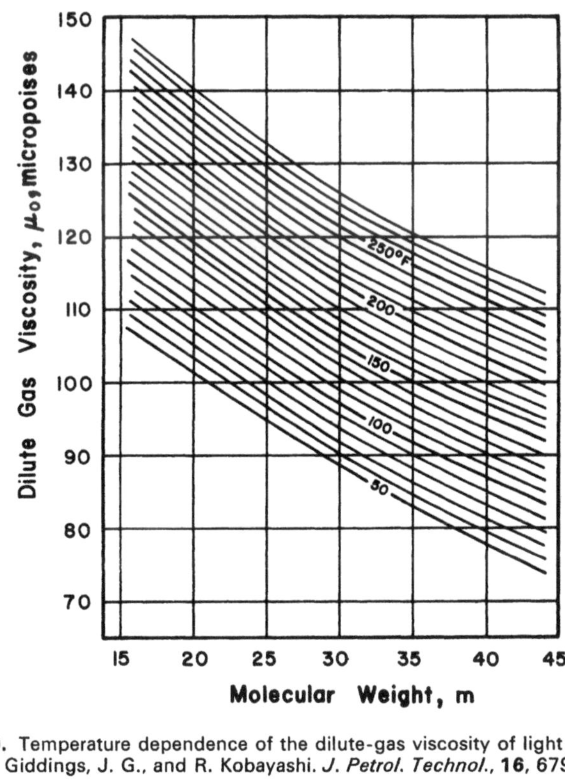

Figure 2.10. Temperature dependence of the dilute-gas viscosity of light paraffinic hydrocarbons. (From Giddings, J. G., and R. Kobayashi. *J. Petrol. Technol.*, **16**, 679, 1964.)

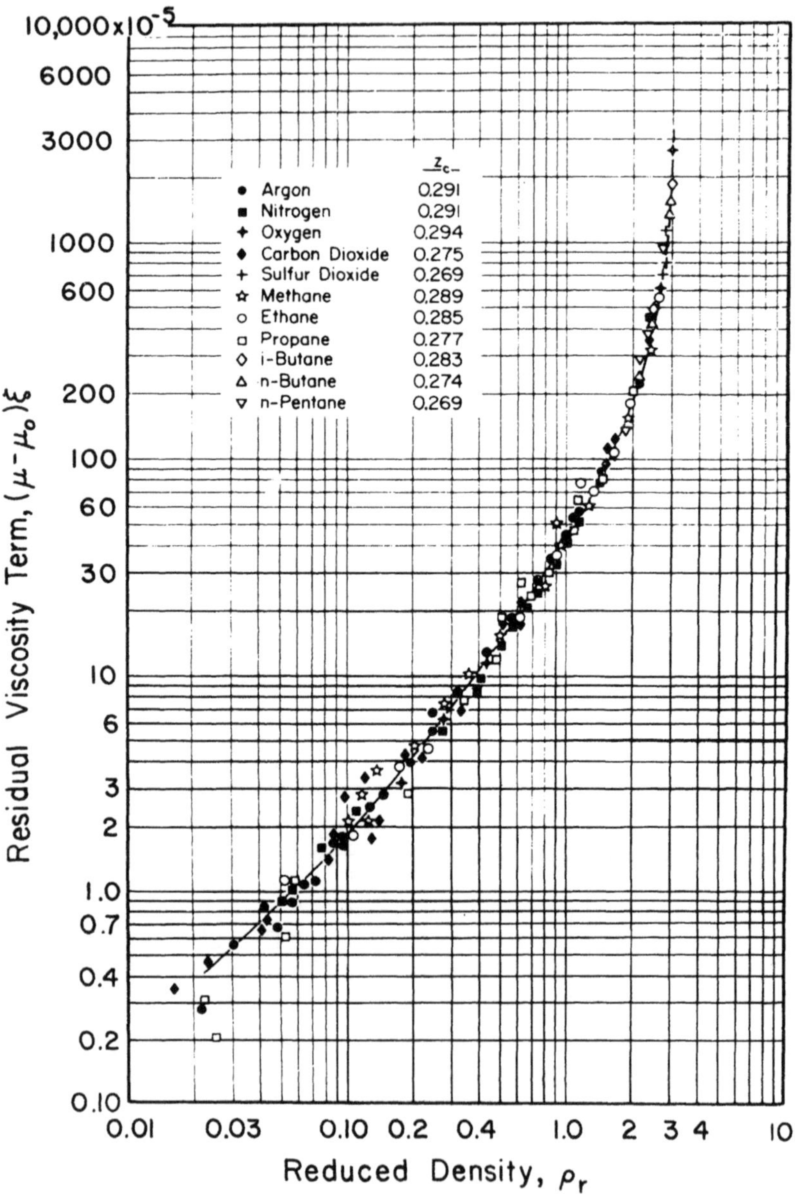

Figure 2.11. Density dependence of the residual viscosity term for normally behaving substances. (From Jossi, J. A., L. I. Stiel, and G. Thodos, *A.I.Ch.E. Journal*, **8**, 59, 1962.)

TABLE 2.2

Summary of the Relationships for Residual Viscosity of Pure Fluids (Centipoise)

Equation	Fluids Represented	Restrictions	Refs.
$[(\mu-\mu_0)\xi+10^{-4}]^{1/4}=0.10230$ $+0.023364\rho_r+0.058533\rho_r^2$ $-0.040758\rho_r^3+0.0093324\rho_r^4$	nonpolar[a]	$0.02\leq\rho_r\leq 3.0$	1
$(\mu-\mu_0)\xi=[23.12e^{1.079\rho_r}$ $-25]\times 10^{-5}$	nonpolar[a]	$0.30\leq\rho_r\leq 2.0$	1
$(\mu-\mu_0)\xi=11.0\times 10^{-5}$ $\times[e^{1.584\rho_r}-1]$	nonpolar[a]	$\rho_r<0.30$	1
$[(\mu-\mu_0)\xi+10^{-4}]^{1/4}=0.10616$ $-0.042426\rho_r+0.17553\rho_r^2$ $-0.12295\rho_r^3+0.028149\rho_r^4$	hydrogen	$0.08\leq\rho_r\leq 2.5$	1
$[(\mu-\mu_0)\xi+10^{-4}]^{1/4}=0.10970$ $+0.022655\rho_r+0.035749\rho_r^2$ $-0.032153\rho_r^3+0.0089998\rho_r^4$	ammonia	$0.03\leq\rho_r\leq 3.0$	1
$[(\mu-\mu_0)\xi+10^{-4}]^{1/4}=0.10721$ $+0.040646\rho_r+0.0026282\rho_r^2$ $-0.0054430\rho_r^3+0.0017979\rho_r^4$	water	$0.01\leq\rho_r\leq 3.0$	1
$(\mu-\mu_0)\xi=15.00\times 10^{-5}$ $\times[e^{1.28\rho_r}-1]$	methane-carbon dioxide mixture	$\rho_r\leq 2.0$	2
$(\mu-\mu_0)\xi=16.56\times 10^{-5}\rho_r^{1.111}$	polar[b]	$\rho_r\leq 0.10$	3
$(\mu-\mu_0)\xi=0.607\times 10^{-5}$ $\times[9.045\rho_r+0.63]^{1.739}$	polar[b]	$0.10<\rho_r\leq 0.90$	3
$\log\{-\log[(\mu-\mu_0)\xi]\}$ $=0.6439-0.1005\rho_r-\Delta$ where $\Delta=0$ for $0.9<\rho_r\leq 2.2$ and $\Delta=4.75\times 10^{-4}[\rho_r^3-10.65]^2$ for $2.2<\rho_r<2.6$.	polar[b]	$0.9<\rho_r<2.6$	3

[a] Normal fluids, including argon, nitrogen, oxygen, carbon dioxide, sulfur dioxide, and paraffinic hydrocarbons.

[b] Polar substances, including chloroform, *i*-propanol, methyl chloride, sulfur dioxide, diethyl ether, hydrazine, ethyl alcohol, methyl alcohol, ammonia, and freon.

REFERENCES

1. J. A. Jossi, L. I. Stiel, and G. Thodos, *A.I.Ch.E. Journal*, **8**, 59 (1962).
2. K. J. DeWitt and G. Thodos, *Can. J. Chem. Eng.*, **44**, 148 (1966).
3. L. I. Stiel and G. Thodos, *A.I.Ch.E. Journal*, **10**, 275 (1964).

Michels and Gibson (1931); Michels, Botzen, and Schuurman (1954); and Michels, Schipper, and Rintoul (1953) have shown that the increment in viscosity due to pressure (termed the residual viscosity, $\mu - \mu_0$) is not dependent upon temperature but is a function of molecular size and structure and of fluid density. A number of workers, e.g., Brebach and Thodos (1958); Shimotake and Thodos (1958); Starling, Eakin, and Ellington (1960); Eakin *et al.* (1962), have applied this concept to pure gases.

Jossi, Stiel, and Thodos (1962) have correlated the residual viscosities of nonpolar gases and liquids in terms of their ξ factor and the reduced density, ρ/ρ_c. Their correlation is reproduced in Figure 2.11. It does not apply to such polar substances as ammonia, water or hydrogen, which are represented by separate lines on a figure such as 2.11. Stiel and Thodos (1964) have extended this study to polar compounds and found that the same type of relations apply. The equivalent in equation form of Figure 2.11 is given in Table 2.2 along with special equations for polar compounds. These equations are particularly useful for computer calculations.

3.13 Liquids

The viscosity of pure liquids is highly sensitive to molecular structure and temperature, but relatively insensitive to moderate pressure changes. At extremely high pressures (1000 atmospheres and more), liquid viscosity is increased significantly and the pressure effect is approximately linear.

With the present state of the theory of liquid viscosity no theoretically based predictions are available and, in the absence of proper data, recourse must be made to empirical methods, which give approximate values only. When no viscosity data are available, the method of Thomas (1946) is suitable for estimating the viscosity at temperatures below the normal boiling point. Thomas' relationship is

$$\mu = 0.1167\rho_L^{1/2}10^{\gamma} \tag{2.54}$$

where μ is in centipoises, ρ is in g/cm^3,

$$\gamma = \frac{B(1 - T_r)}{T_r}$$

and B is a constant determined by summing atomic and group contributions from Table A-12 (see Appendix). Reid and Sherwood indicate that errors from this method are usually less than 20 percent except for such compounds as alcohols, acids, naphthenes, amines, aldehydes, and multihalogenated compounds, where the method is inapplicable.

In the temperature range between the normal boiling point and the critical temperature, the method of Stiel and Thodos discussed earlier with respect to pure gases may be applied satisfactorily to pure liquids. Errors less than 10 to 15 percent may be expected if experimental values for the liquid density are used.

Of special practical interest is that the temperature dependence of the viscosity of pure liquids is well represented by the Andrade equation (Andrade, 1930),

$$\mu = A \exp B/T \tag{2.55}$$

where A and B are constants and T is temperature in absolute degrees, permitting reliable interpolation of data at two or more temperatures.

3.2 Multicomponent Gases

The viscosity of a multicomponent gas mixture is best determined by actual measurement, and measured values are available for many important mixtures, including air and various mixtures of normal paraffinic hydrocarbons with and without carbon dioxide, hydrogen sulfide, and nitrogen. Some references containing such data are: Wilke, (1950); Carr, Kobayashi, and Burrows (1954); Dean and Stiel (1965); Mathur and Saxena (1965); Giddings, Kao, and Kobayashi (1966); Lee, Gonzalez, and Eakin (1966); and Brokaw (1968).

When such data are not available, the mixture viscosity may be calculated with good accuracy by recently developed methods. As with pure gases, the approach is to determine the dilute-gas viscosity and, for higher pressures, to add the residual viscosity.

The dilute-gas viscosity of a gas mixture may be determined from that of its components through the empirical equation of Herning and Zipperer (1936)

$$\mu_{0M} = \frac{\sum y_i \mu_{0i} (m_i)^{1/2}}{\sum y_i (m_i)^{1/2}} \tag{2.56}$$

This relation represents the data for nonpolar compounds to within 2 or 3 percent accuracy. Where more than a few percent of polar compounds are present, the more accurate, but more complex, theoretical expression of Wilke (1950) may be used,

$$\mu_{0M} = \frac{\mu_{01}}{1 + \left(\frac{y_2}{y_1}\right)\phi_{12} + \left(\frac{y_3}{y_1}\right)\phi_{13} + \cdots} + \frac{\mu_{02}}{1 + \left(\frac{y_1}{y_2}\right)\phi_{21} + \left(\frac{y_3}{y_2}\right)\phi_{23} + \cdots} + \cdots$$

where

$$\phi_{ij} = \frac{[1 + (\mu_{0i}/\mu_{0j})^{1/2}(m_j/m_i)^{1/4}]^2}{2\sqrt{2}[1 + (m_i/m_j)]^{1/2}}$$

y_i, μ_{0i} and m_i are, respectively, the mole fraction, the dilute-gas viscosity and the mole weight of component i.

Dean and Stiel (1965) have developed an alternate to the Herning and Zipperer and the Wilke methods that does not require that the dilute-gas viscosities of the pure components be separately calculated. Their approach is to use equations of the Stiel and Thodos type for pure argon with suitably determined values of pseudocritical temperature, volume, and compressibility factors. The equations are (units same as in Eqs. 2.47–2.50)

$$\mu_{0M}\xi_M = 34.0 \times 10^{-5} T_r^{0.889} \qquad \text{for } T_r \leq 1.5 \quad (2.58)$$

$$\mu_{0M}\xi_M = 166.8 \times 10^{-5}(0.1338T_r - 0.0932)^{0.556} \qquad \text{for } T_r > 1.5 \quad (2.59)$$

where
$$\xi_M = \frac{{}_pT_c^{1/6}}{(\sum y_i m_i)^{1/2}\,{}_pP_c^{2/3}}$$

$${}_pT_c = \sum y_i T_{ci}$$

$$v_{cM} = \sum y_i v_{ci}$$

$$Z_{cM} = \sum y_i Z_{ci}$$

$${}_pP_c = \frac{Z_{cM} R\,{}_pT_c}{v_{cM}}$$

For both binary and multicomponent mixtures they found an average deviation of only 1.44 percent for more than 984 experimental data points.

For light paraffinic hydrocarbons, Giddings and Kobayashi's correlation with molecular weight and temperature (Figure 2.10) may be used directly.

Residual viscosities for gas mixtures appear to be well correlated with the relations discussed earlier for pure gases with pseudocritical properties and the average molecular weight being used as appropriate. For light paraffinic hydrocarbon mixtures in both the gas and liquid phases, Giddings and Kobayashi have developed a specific correlation of residual viscosity in terms of average molecular weight and reduced density. This correlation, reproduced in Figure 2.12, gives excellent accuracy for hydrocarbon mixtures.

3.3 Multicomponent Miscible Liquids

Even given the viscosities of the components, it is not generally possible at the present state of knowledge to predict the viscosity of a miscible liquid mixture of several components with any reliability. For binary mixtures of components of known viscosity, and where the viscosities of at least two mixtures are known, the viscosities of other mixtures may be predicted reasonable well by a semitheoretical method due to McAllister (1960) and discussed by Reid and Sherwood.

In the absence of mixture viscosity data the Stiel and Thodos method for

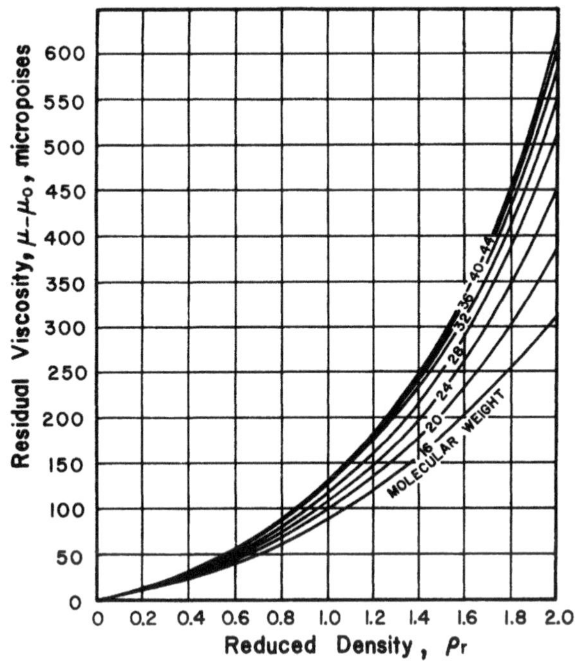

Figure 2.12. Density dependence of residual viscosity for light hydrocarbons and their mixtures. (From Giddings, J. G., and R. Kobayashi, *J. Petrol. Technol.*, **16**, 679, 1964.)

gas mixtures may be used. Also additive partial viscosities, reciprocal viscosities, or logarithm of viscosities may be assumed as

$$f(\mu_M) = \sum x_i f(\mu_i) \tag{2.60}$$

where $f(\mu) = \mu$, $1/\mu$, or $\ln \mu$, and x_i is the volume, weight, or mole fraction. This latter method is useful for mixtures of homologous compounds where previous experience indicates the most suitable viscosity function and concentration expression.

3.4 Pseudohomogeneous Suspensions of Spherical Particles in Liquids

Pseudohomogeneous suspensions of spherical noninteracting particles in Newtonian liquids exhibit Newtonian behavior and their viscosity is determined by the viscosity of the carrier liquid and the concentration and size distribution of the suspended particles. Although more of academic than practical interest, the concentration dependence of the viscosity of such suspensions forms a background for comparison of the more complex rheological behavior of suspensions of irregularly shaped particles.

Einstein (1905, 1906, 1911) first analyzed the problem for dilute systems of solid spherical particles, finding that

$$\mu_M = \mu(1 + 2.5c_v) \tag{2.61}$$

where μ_M is the viscosity of the suspension, μ is the viscosity of the carrier liquid, and c_v is the volume fraction of particles. This equation has been shown to be valid only to concentrations to the order of 1 to 2 percent.

Guth and Simha (1936) have extended the Einstein analysis to make the following expression valid within about 10 percent to concentrations of approximately 20 percent.

$$\mu_M = \mu(1 + 2.5c_v + 14.1c_v^2) \tag{2.62}$$

Many expressions have been proposed for the viscosity of more concentrated suspensions of solid spheres, but none are wholly successful. Thomas (1965) proposes the empirical equation

$$\mu_M = \mu[1 + 2.5c_v + 10.05c_v^2 + 0.00273 \exp(16.6c_v)] \tag{2.63}$$

and indicates that it fits the available data reasonably well over the entire range.

The most comprehensive theoretical analysis of highly concentrated suspensions is that of Frankel and Acrivos (1967) who show that, for concentrations exceeding about 0.80 times the maximum, the relative viscosity depends upon the ratio between the concentration and the maximum concentration rather than the concentration itself. This is a reflection of the effect of maximum packing density, which could be particularly significant with mixed size particles. The Frankel–Acrivos equation is

$$\mu_M = 1.125\mu\left[\frac{[c_v/(c_v)_{max}]^{1/3}}{1 - [c_v/(c_v)_{max}]^{1/3}}\right] \tag{2.64}$$

where $(c_v)_{max}$ is the maximum attainable concentration. This equation checks very well with the average of many data considered reliable and brings into line data that cannot be correlated in terms of c_v alone.

Data of Landel, Moser, and Bauman (1963) on water suspensions of glass beads, aluminum powder, copper powder, and ammonium perchlorate with mean particle diameters ranging from less than 10 to more than 100 microns are correlated in terms of experimentally determined values of $(c_v)_{max}$, the maximum obtainable volume concentration of particles. Although some of these suspensions exhibited non-Newtonian behavior at the lowest shear rates, they all behaved as Newtonians at the high shear rates at which the viscosity was calculated. The correlation of Landel *et al.* is

$$\frac{\mu_M}{\mu} = \left(1 - \frac{c_v}{(c_v)_{max}}\right)^{-2.5} \tag{2.65}$$

and is illustrated in Figure 2.13. Although there is considerable scattering of the data, the correlation is rather remarkable since it suggests that μ_M/μ is a unique function of $c_v/(c_v)_{max}$ independent of the polarity of the medium, the type of particle, the surface area of the particle, the particle size, and, to a degree, the shape and size distribution of the particle.

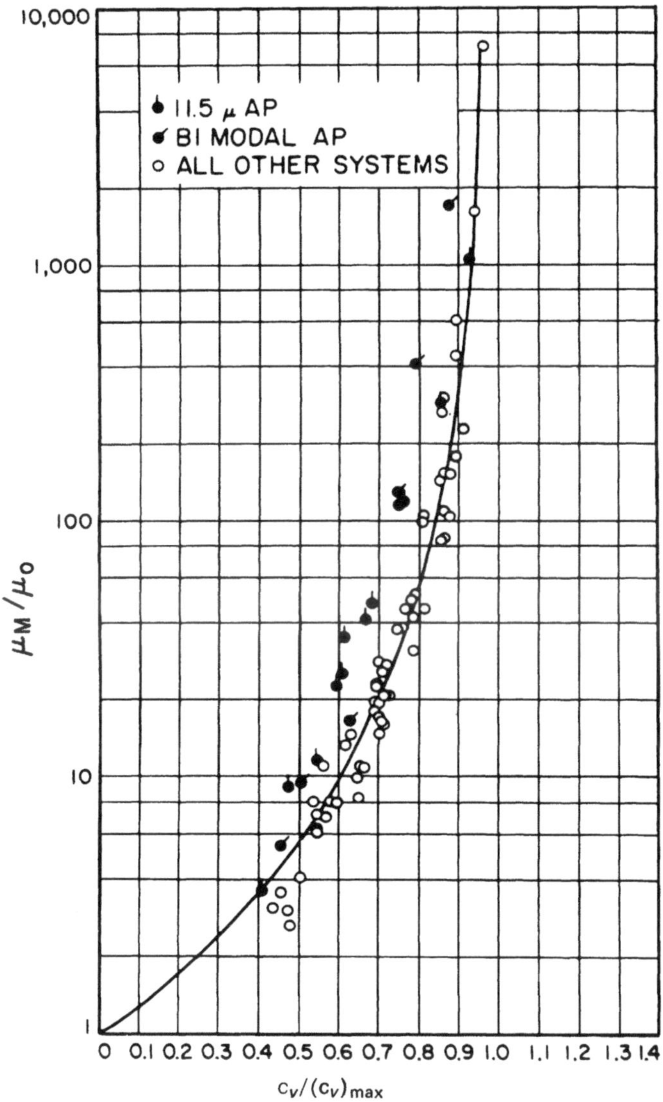

Figure 2.13. Comparison of experimental data with prediction of Equation 2.65. AP refers to ammonium perchlorate particles. (From Landel, R. F., B. G. Moser, and A. J. Bauman, Fourth Int. Congress on Rheology, Brown Univ. 1963, *Proc.* Part 2, p. 663, Interscience Publishers New York, 1965.)

Taylor (1932) extended the Einstein analysis to suspensions of spherical liquid droplets obtaining

$$\mu_M = \mu\left[1 + 2.5c_v\left(\frac{\mu_P + \frac{2}{5}\mu}{\mu_P + \mu}\right)\right] \tag{2.66}$$

where μ_P is the viscosity of the suspended liquid. This equation is useful for obtaining order of magnitude viscosities for emulsions containing less than 2–3 volume percent of the dispersed phase.

4 CONSISTENCY MEASURES FOR COMMON NON-NEWTONIAN MATERIALS

4.1 General

Generally, detailed and careful rheological measurements, as discussed in Chapter 1, are necessary to identify a material as a particular type of non-Newtonian and to characterize its consistency in a quantitative way. The present state of our theoretical knowledge is far short of permitting predictions of the rheological behavior from molecular structure or system composition. Nonetheless, in a few instances, limited generalizations may be made. These are largely based on empirical observations. They may give order of magnitude values of rheological constants (as in the case of suspensions of fine and ultrafine particles) or they may be useful in extrapolations to other temperatures or other concentrations.

4.2 Bingham Plastics

4.21 Homogeneous Materials

Thickened hydrocarbon greases, asphalt, and bitumen are among the few homogeneous (nonsettling) materials which behave as, or nearly as, Bingham fluids. No generalizations for the prediction of the Bingham constants for materials of these types are known. Table 2.3 gives typical values of the yield values and coefficient of rigidity for a few greases and other substances.

An equation of the form of Equation 2.55

$$\eta = A \exp\left(\frac{B}{T}\right) \tag{2.67}$$

where A, B are constants, and T is temperature in absolute degrees, may be used to correlate the coefficient of rigidity with temperature and has also been found useful for describing the effect of temperature on the yield value.

TABLE 2.3

Typical Bingham (or Bingham-like) Plastics and Their Properties

Substance	Density, g/cm^3	Yield Value, τ_y, dynes/cm^2	Coefficient of Rigidity, η, centipoises	Refs.
Luting (dispersion of synthetic wax in polyisobutene)		15,700–19,600	1,000,000	1
Calcium soap grease (at 120°C)		7,100	6,500	2
Aluminum soap grease (at 120°C)		5,100	5,000	2
Lithium stearate grease, 14G8		6,500	100	3
Flocculated aqueous china-clay Suspensions (80% under 1 micron)				4
No. 6	1.149	78	4.0	
No. 4	1.207	250	6.7	
No. 1	1.280	590	13.1	
Aqueous suspensions of Milicz clay ($d_{av} = 70$ microns)				5
Weight percent solids = 13.9		23	8.7	
= 16.8		53	13.6	
= 19.6		130	25.0	
Aqueous clay suspensions				6
No. V.	1.36	66.5	19.4	
No. III.	1.44	200	32.8	
No. I.	1.49	345	44.7	
Aqueous suspensions of cement rock (92% under 74 microns) Weight percent solids = 54.3	1.52	38	6.86	7
Sewage sludges				6
Weight percent solids = 10				
Chicago plant		6.7	52	
Stuttgart plant		48	115	
Sludge No. 5 (Imhoff tank) Weight percent solids = 14	1.06	31	24.5	6
Aqueous suspensions of coal char				8
Weight percent: total solids / −325-mesh solids				
45.4 / 28.6	1.23	26.2	31	
50.4 / 47.5	1.22	20	28	

REFERENCES

1. W. A. Dukes, in *Rheology of Dispersed Systems*, p. 139, Pergamon Press, London (1959).
2. N. Pilpel, *Chem. Process Eng.*, **45**, 395 (1964).
3. C. R. Singleterry, and E. E. Stone, *J. Colloid, Interface Sci.*, **6**, 171 (1951).
4. L. Valentik, and R. L. Whitmore, *Brit. J. Appl. Phys.*, **16**, 1197 (1965).
5. W. Parzonka, *Compt. Rend. Acad. Sci. Paris.*, **259**, (No. 13), 2073 (1964).
6. D. H. Caldwell, and H. E. Babbitt, *Ind. Eng. Chem.*, **33**, 249 (1941).
7. B. O. A. Hedstrom, *Ind. Eng. Chem.*, **44**, 651 (1952).
8. M. E. Sacks, M. J. Romney, and J. F. Jones, *Ind. Eng. Chem. Process Des. Develop.*, **9**, 148 (1970).

4.22 Pseudohomogeneous Materials

The commonest pseudohomogeneous materials that exhibit (or approximate) Bingham plastic behavior are relatively concentrated suspensions of ultrafine or fine particles in a liquid. Typical such mixtures and their rheological properties are given in Table 2.3. The data on clay-water suspension of Govier and Winning (1948) are also illustrative and show the typical effect of concentration on the Bingham properties. These are shown in Figures 2.14 and 2.15.

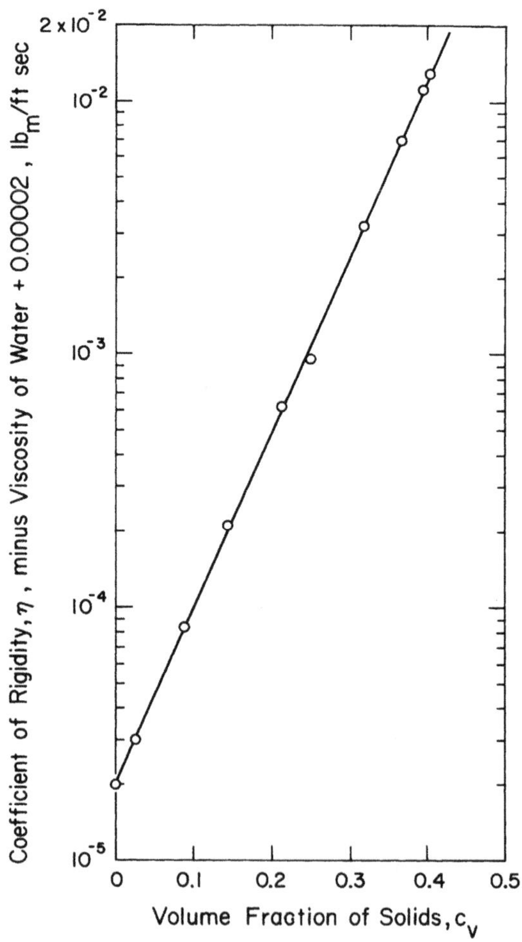

Figure 2.14. Coefficient of rigidity of clay suspension in water. (From data of Govier, G. W., and M. D. Winning, The Flow of Non-Newtonian Fluids, Paper Presented at the Meeting of the American Institute of Chemical Engineers, Montreal, Que., 1948.)

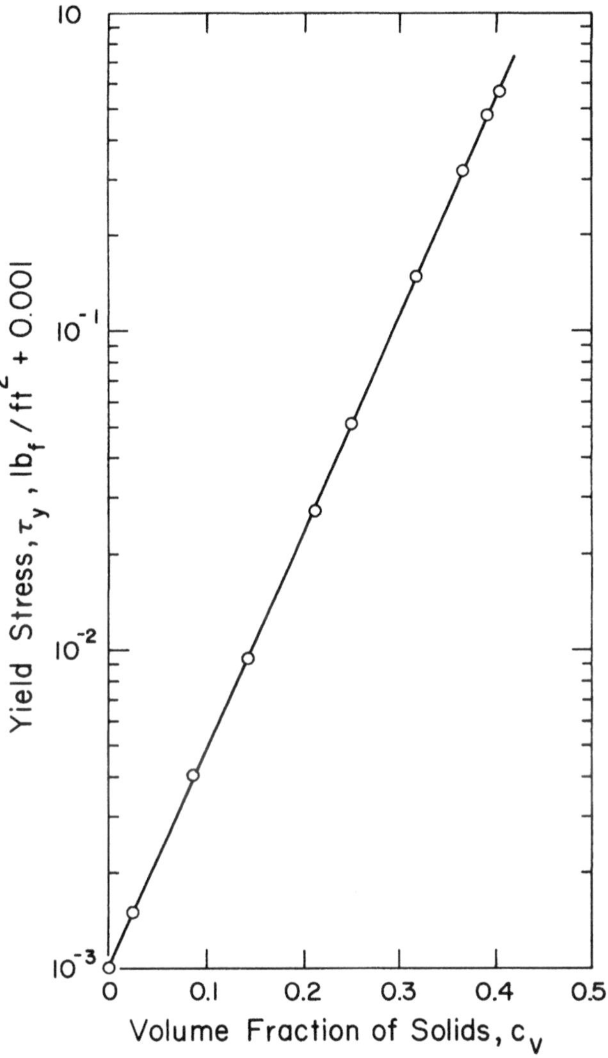

Figure 2.15. Yield stress of clay suspensions in water. (From data of Govier, G. W., and M. D. Winning, The Flow of Non-Newtonian Fluids, Paper Presented at the Meeting of the American Institute of Chemical Engineers, Montreal, Que., 1948.)

Figure 2.16 gives data of Dick and Ewing (1966) for an activated sewage sludge replotted in the form of Figure 2.15, which is useful for smoothing, interpolating, and extrapolating.

Measurements on a wide variety of such suspensions confirm the dependence of the two Bingham parameters on the size, shape, and concentration of the particles (or their agglomerates) and the viscosity of the carrier liquid.

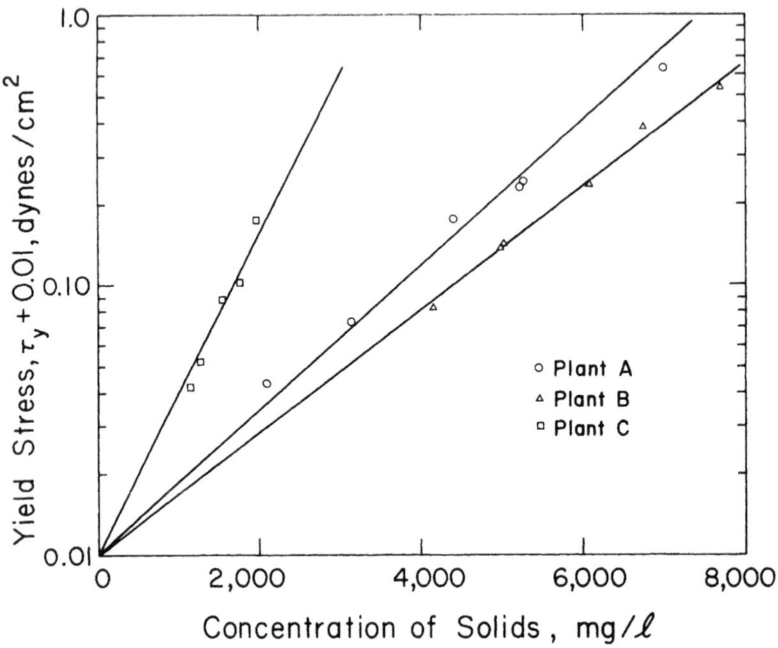

Figure 2.16. Yield stress of activated sewage sludges. (From the data of Dick, R. I., and B. B. Ewing, *Journal W.P.C.F.*, **39**, 543, 1967.)

Arrhenius (1887) was probably the first to propose specific relationships between both the yield stress and the coefficient of rigidity and the concentration of solids, but Thomas (1961) has shown that relationships of certain forms may be applied to a broad range of materials. The equations, as he proposes them, are

$$\tau_y = k_1 c_v^3 \tag{2.68}$$

$$\frac{\eta}{\mu} = \exp\,(k_2 c_v) \tag{2.69}$$

where τ_y = yield stress,
η = coefficient of rigidity,
μ = viscosity of suspending medium,
c_v = volume fraction solids.

Both k_1 and k_2 are constants characterizing the suspension and dependent on particle size and shape and electrolyte concentration. These equations closely fit Thomas' data on suspensions of titanium dioxide, graphite, thorium oxide, and kaolin in the particle-size range 0.35 to 13 microns and in concentrations of from 0.02 to 0.23 volume fraction solids.

Thomas has also shown the dependence of k_1 and k_2 on particle diameter and shape. For k_1 he finds

$$k_1 = \frac{210}{(d')^2} \psi_1 \tag{2.70}$$

where k_1 is in units of lb_f/ft^2,
d' is the particle diameter in microns,
ψ_1 is a shape factor given by

$$\psi_1 = \exp\left[0.7\left(\frac{s_P}{s_0} - 1\right)\right] \tag{2.71}$$

with s_P = surface area per unit volume of the actual particles,
s_0 = surface area per unit volume of a sphere or cube of equivalent dimension = $6/d$.

These equations appear suitable for particles of irregular crystalline shape but not for spheroidal particles for which ψ_1 is approximately 0.1. For k_2 he gives

$$k_2 = 2.5 + \frac{14}{\sqrt{d'}} \psi_2 \tag{2.72}$$

where ψ_2 is a second shape factor defined as

$$\psi_2 = \sqrt{\frac{s_P}{s_0}} \tag{2.73}$$

Thomas' equations are supported by his and other data for particles in the size range of about 0.4 to 20 microns.

Thomas (1963) has extended the above work on dispersed ultrafine and fine particles to cover flocs of such particles of overall dimensions up to 115 microns. He shows that

$$\tau_y = k_1 \alpha^4 c_v^3 \tag{2.74}$$

where $k_1 = 1.55 \times 10^{-2}\ \text{lb}_f/\text{ft}^2$,
α = ratio of the volume of immobilized dispersing fluid to the volume of solids related approximately to the particle and floc apparent diameters through the equation

$$\frac{d_f}{d_{\text{app}}} = (1 + \alpha^2) \tag{2.75}$$

where d_f = apparent floc diameter,
d_{app} = apparent particle diameter given by

$$d_{\text{app}} = d(s_0/s_P) \exp\left(-\tfrac{1}{2}\ln^2 \sigma\right) \tag{2.76}$$

where σ = logarithmic standard deviation.

Additionally his work indicates that

$$\frac{\eta}{\mu} = \exp\,[(2.5 + \alpha)c_v] \tag{2.77}$$

where μ is the viscosity of the suspending medium. With these relations, and an estimate of α, or d_f/d_{app}, it is possible to make good order of magnitude estimates of τ_y and η for flocculated suspensions, as functions of the volume fraction and shape of the primary particles.

The coefficients of Bingham suspensions may also be correlated by exponential equations of the general form

$$\frac{\eta}{\mu} = A \exp\,(Bc_v) \tag{2.78}$$

and

$$\tau_y = M \exp\,(Nc_v) \tag{2.79}$$

where A, B, M, and N are constants related to the particle size, its shape, and the nature of its surface.

The effect of temperature on the rheological properties of Bingham suspensions has been observed by a number of workers including Govier and Winning (1948), Michiyoshi *et al.* (1966), and Parzonka (1964). Generally temperature influences η much in the way it influences μ for a Newtonian liquid. In the case of τ_y also, there is an effect but of much lower order. As in the case of homogeneous Bingham fluids, Equation 2.67 may be used to correlate values of η over a rather wide temperature range. Parzonka (1964) shows that a simple power law expression

$$\eta = a(T)^{-n} \tag{2.80}$$

where T is the Fahrenheit or centigrade temperature, is applicable over moderate ranges (say 50 Fahrenheit degrees). Equations of these forms may also be employed to describe the modest influence of temperature on the yield value.

4.3 Pseudoplastics

4.31 Homogeneous Materials

No broad generalizations are available for predicting the behavior of homogeneous pseudoplastics. For any given family of pseudoplastics, however, as in the case of water solutions of different concentrations of the same solute, there is a systematic trend in the value of the coefficients for any rheological equation that suits the data. Illustrative of this are data of Shaver and Merrill (1959) on water solutions of ammonium alginate. The rheograms over the rate of shear range 200 to 20,000 sec^{-1} are given in Figure 2.17 from which it is apparent that the power law may be applied. The power law constants determined by Shaver and Merrill are shown as functions of concentration in Figures 2.18 and 2.19. The smooth trend is apparent. Trends of this type may be anticipated with other similar solutes, but measurements on solutions of at least a few different concentrations are needed to define the properties.

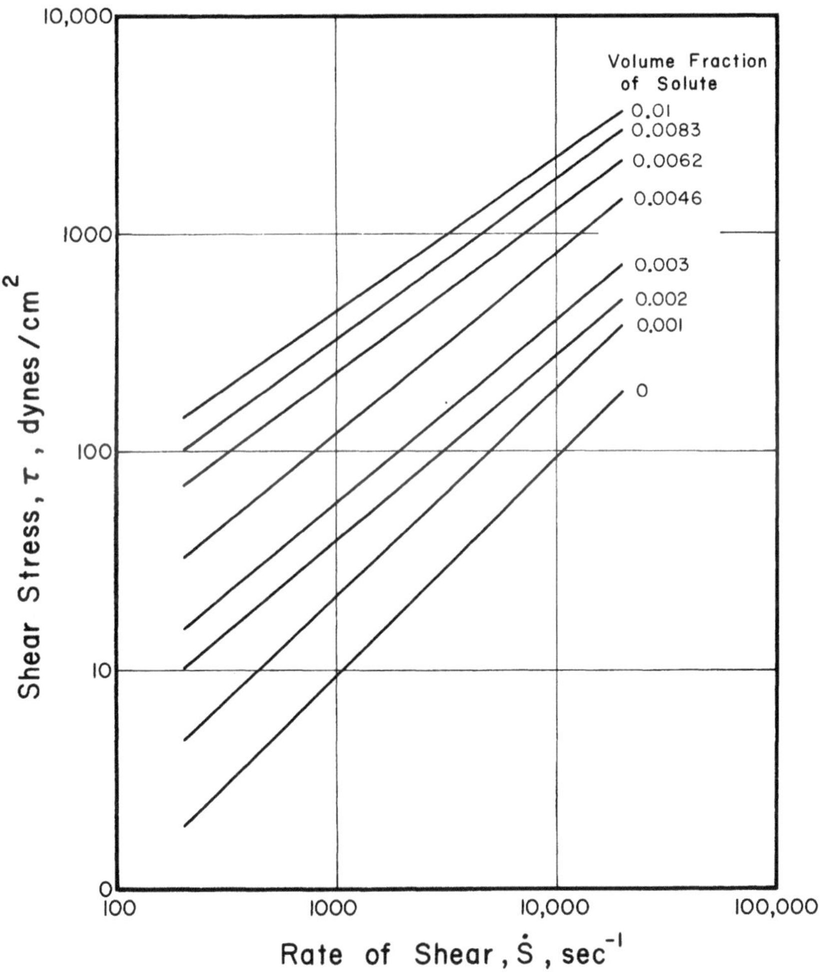

Figure 2.17. Rheograms for aqueous solutions of ammonium alginate. (From data of Shaver, R. G., and E. W. Merrill, *A.I.Ch.E. Journal*, **5**, 181, 1959.)

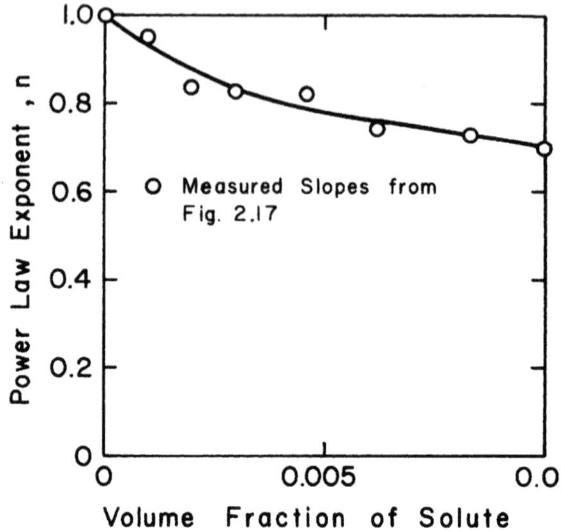

Figure 2.18. Power law exponent, *n*, for aqueous solutions of ammonium alginate. (From data of Shaver, R. G., and E. W. Merrill, *A.I.Ch.E. Journal*, **5**, 181, 1959.)

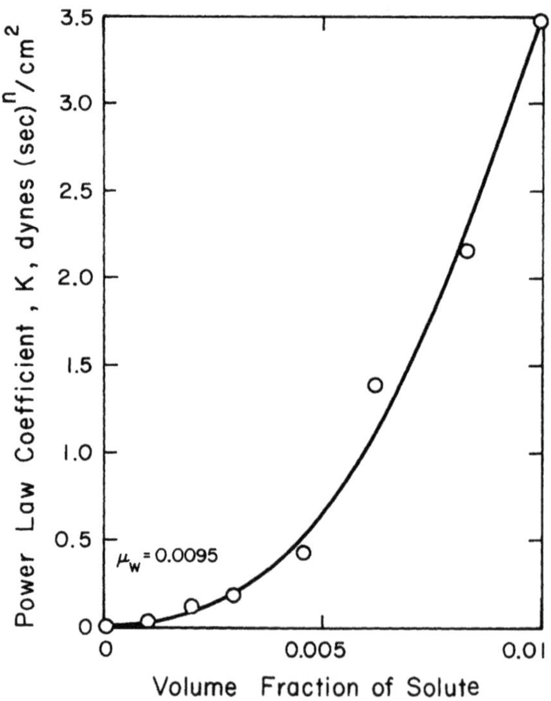

Figure 2.19. Power law coefficient, *K*, for aqueous solutions of ammonium alginate (From data of Shaver, R. G. and E. W. Merrill, *A.I.Ch.E. Journal*, **5**, 181, 1959.)

4.32 Pseudohomogeneous Materials

The situation with respect to pseudohomogeneous materials is much the same as with the homogeneous pseudoplastics. There are no broad generalizations, and all we may do is illustrate the trends in the rheological coefficients for typical families of pseudohomogeneous suspensions.

Data of Govier (1960) on a series of clay-water suspensions and of Govier, Shook, and Lilge (1957) on magnetite-water suspensions have all been interpreted over an intermediate shear range (10 to 1000 sec^{-1}) in terms of the power law model. The variation of the power law coefficients with concentration in these three systems is shown in Figures 2.20 and 2.21. An analysis of the type carried out by Thomas (1961, 1963) for pseudohomogeneous Bingham plastics should be possible for pseudoplastic suspensions. This would permit the prediction of K and n values in terms of the concentration, size, shape, etc., of the particles.

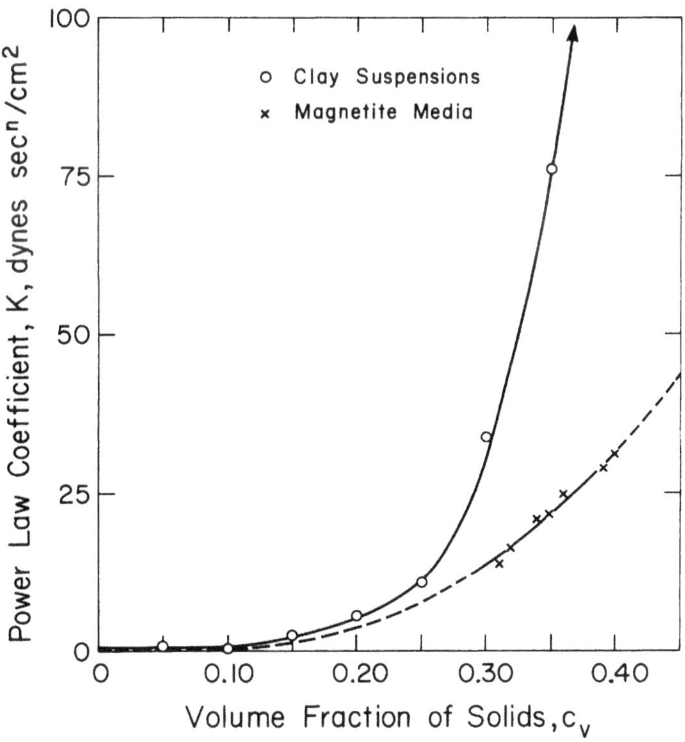

Figure 2.20. Power law coefficient, K, for water suspensions of clay and of magnetite. (From data of Govier, G. W., C. A. Shook and E. O. Lilge, *Trans. Can. Inst. Mining Met.*, **60**, 147, 1957 and of Govier, G. W., *Engineering Journal*, **43**, 79, March, 1960.)

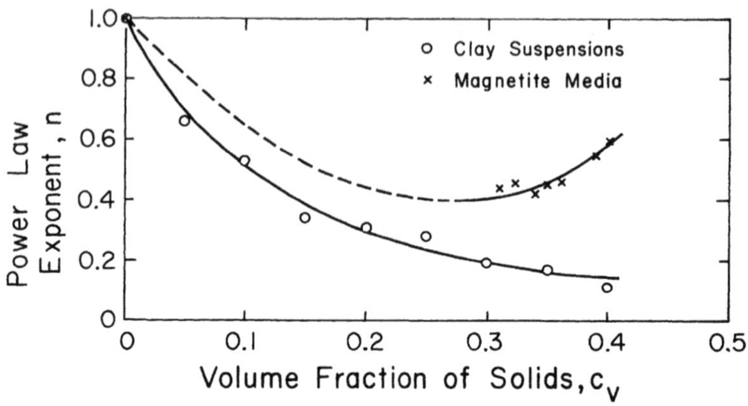

Figure 2.21. Power law exponent, *n*, for water suspensions of clay and of magnetite. (From data of Govier, G. W., C. A. Shook, and E. O. Lilge, *Trans. Can. Inst. Mining Met.*, **60**, 147, 1957 and of Govier, G. W., *Engineering Journal*, **43**, 79, March 1960.)

5 SURFACE TENSION AND INTERFACIAL TENSION

5.1 Relationship Between Surface Tension and Interfacial Tension

5.11 Gas-Liquid

Surface tension occurs at the boundary between a liquid and a gas and results from unequal forces exerted on the surface molecules. The surface molecules are attracted sideways and toward the bulk liquid but experience considerably less attraction in the direction of the gas phase, resulting in the surface layer being in tension and tending to contract to the smallest area compatible with the mass of liquid and the container walls. The surface tension for a liquid below its saturated vapor will be greater than the surface tension of the same liquid below a different gas, since minute amounts of the gas will be dissolved in the surface layer, thus reducing the intermolecular attraction.

5.12 Liquid-Liquid

At an interface between two or more immiscible or partially miscible liquid phases, an interfacial tension exists. This interfacial tension is analogous to the surface tension between a liquid and a vapor, although it is numerically different. Antonoff's (1907) rule (Equation 2.81) gives a useful estimation of interfacial tension,

$$\sigma_i = |\sigma_1 - \sigma_2| \tag{2.81}$$

where σ_i = interfacial tension,
σ_1, σ_2 = surface tension of mutually saturated phases measured against a common gas.

In Equation 2.81 it is important to use the surface tension values of liquid saturated with respect to the other component since, as seen before, small concentrations of dissolved, dissimilar molecules can affect the surface tension.

5.2 Surface Tension of Single-Component Liquids

5.21 Estimation from Density

A relationship between the molal densities of the liquid and vapor and the surface tension was postulated by Macleod (1923),

$$\sigma = \{[P](\rho_{mL} - \rho_{mv})\}^4 \tag{2.82}$$

where σ = surface tension, dynes/cm
ρ_{mL}, ρ_{mv} = molal densities, g mole/cm^3
$[P]$ = constant, characteristic of each substance.

The constant, $[P]$, called the parachor by Sugden (1924), has been shown to be an additive function of the atoms and groups of atoms in the molecule and to be almost independent of temperature. A group contribution tabulation for the parachor is given in Table A-13 (see Appendix) taken from Reid and Sherwood. It should be noted that Equation 2.82 is not applicable to highly polar compounds such as water.

5.22 Estimation from Refractive Index

Surface tension can be calculated quite accurately (2 to 3 percent) using a combination of the Lorentz–Lorenz (1880) relation between refractive index and the molar refraction and the Macleod correlation. The relationship obtained for σ in dynes/cm is

$$\sigma = \left\{\frac{[P](n^2 - 1)}{[R_D](n^2 + 2)}\right\}^4 \tag{2.83}$$

where $[R_D]$ = molar refraction based on the sodium D-line,
n = refractive index for the wavelength of the sodium D-line.

In this relation, the simplifying assumption that $\rho_L \gg \rho_v$ has been made. $[R_D]$ is similar to $[P]$ in that it can be evaluated from group contributions. Values of the group contributions are given in Table A-14 (see Appendix) taken from Reid and Sherwood.

For most liquids, the empirical Eykman (1895) equation for refractive index applies and gives quite accurate results.

$$\rho_L = \frac{n^2 - 1}{C(n + 0.4)} \tag{2.84}$$

where C is a constant, independent of the temperature. It may be determined from known values of n and ρ_L at the same temperature. Table A-15 (see Appendix) taken from Dreisbach (1948) shows values of C for a few substances.

5.23 Estimation from Viscosity

Several relations between surface tension of a liquid in air and viscosity have been postulated. The most useful of these (it can be applied over the entire liquid-phase region) was put forward by Pelofsky (1966),

$$\ln \sigma = \frac{B}{\mu} + \ln A \tag{2.85}$$

where A and B are constants. Pelofsky (1966) tabulated values for A and B. Table A-16 (see Appendix) reproduces his values for selected liquids. Pelofsky has shown the relation also to hold true for solutions, regardless of whether temperature or concentration were kept constant.

5.24 Estimation from Corresponding States

Brock and Bird (1955) advocate the estimation of surface tension by an empirical method based upon the concept of corresponding states. Their equation is

$$\sigma = P_c^{2/3} T_c^{1/3} (0.133\alpha_c - 0.281)(1 - T_r)^{11/9} \tag{2.86}$$

where α_c is a characterizing parameter known as the Riedel factor. The Riedel factor is related to the acentricity factor ω and may be calculated from it through the equation

$$\alpha_c = 7.00 + (\omega - 0.242)/0.203 \tag{2.87}$$

Alternately and for most normal liquids, Miller (1963) has shown that α_c may be closely estimated from the relation

$$\alpha_c = 0.9076[1 + T_{br} \ln P_c/(1 - T_{br})] \tag{2.88}$$

where $T_{br} = T_b/T_c$.

The Brock and Bird method is said to be capable of accuracies within about 3 percent and has the advantage that it does not require a knowledge of density, refractive index, or viscosity.

5.25 Effect of Temperature

Surface tensions at temperatures 1 and 2 are related by Lövgren (1941) by the equation

$$\frac{\sigma_1}{\sigma_2} = \left(\frac{1 - T_{r1}}{1 - T_{r2}}\right)^{1.2} \tag{2.89}$$

Equation 2.89 applies only for nonpolar compounds, the value of the exponent being different for polar compounds. Reid and Sherwood (1966) give a method for estimating the value of the exponent for polar compounds.

5.26 Effect of Pressure

No accurate relation for the effect of pressure on surface tension has been obtained since, at higher pressures, a pressurizing gas must be used and this can affect the surface tension (see Section 5.1). Slowinski *et al.* (1957) have obtained a rough approximation of the effect of pressure.

$$\sigma - \sigma_P = -KPRT \tag{2.90}$$

where K is a constant, independent of the liquid but dependent on the temperature and the pressurizing gas.

Approximate values of K for selected pressurizing gases at 25°C are given in Table 2.4, calculated from data of Slowinski *et al.*

TABLE 2.4

Range of the Constant, K, in the Slowinski Equation[a]

Pressurizing Gas	$K \times 10^6$, dynes mole/atm^{2}cm^4, at 25°C			
	Over water		Over *n*-hexane	
	20 atm	100 atm	20 atm	100 atm
CO_2	28.6225		8.2726	
C_2H_6	20.8857		16.3116	
CH_4	8.3557		6.7675	
N_2	3.5913	2.7975	3.5126	2.8577
H_2	0.9428	0.7175	1.6284	1.0529

[a] From the data of E. J. Slowinski, Jr., E. E. Gates, and C. E. Waring, *J. Phys. Chem.*, **61**, 808 (1957).

5.3 Surface Tension of Multicomponent Liquids

For nonaqueous mixtures the previously discussed relations for single-component mixtures have been modified and can be applied with surprising accuracy.

The Macleod-Sugden correlation with the assumption that $\rho_{LM} \gg \rho_{vM}$ becomes

$$\sigma_M = \{[P_M](\rho_{LM})\}^4 \tag{2.91}$$

where

$$[P_M] = \sum [P_i]x_i$$

and ρ_{LM} is the molal density of the mixture, P_i and x_i are, respectively, the parachor and the mole fraction for component i.

The relationship between refractive index and surface tension for multicomponent nonaqueous mixtures is

$$\sigma_M = \left\{\frac{[P_M]}{[R_{DM}]}\left(\frac{n_M^2 - 1}{n_M^2 + 2}\right)\right\}^4 \tag{2.92}$$

where $[P_M]$, $[R_{DM}]$, and n_M are all of the form

$$[P_M] = \sum [P_i]x_i$$

For dilute solutions of organic materials in water, Meissner and Michaels (1949) extended Szyszkowski's (1908) work and suggested the relation

$$\sigma = \sigma_W\left[1 - 0.411 \log\left(1 + \frac{x}{a}\right)\right] \tag{2.93}$$

where σ_W = surface tension of pure water,
x = mole fraction of organic material,
a = constant, characteristic of the organic material.

Table A-17 (see Appendix), taken from Meissner and Michaels, lists the constant a for several organic compounds. It should be noted that Equation 2.93 is valid only for $x < 0.01$.

A complex set of equations which apply over a wide range of concentrations was obtained by Tamura, *et al.* (1955).

Bibliography

Alani, G. H., and H. T. Kennedy, *Petrol. Trans.*, **219**, 288 (1960).

Amagat, E. H., *Ann. Chim. Phys.*, **29**, 68 (1893).

American Petroleum Institute, *Technical Data Book—Petroleum Refining*, New York, 1966.

Andrade, E. N. da C., *Nature*, **125**, 309 (1930).

Antonoff, J. N., *J. Russ. Phys.-Chem. Soc.*, **39**, 342 (1907).

Arnold, J. H., *J. Chem. Phys.*, **1**, 170 (1933).

Beattie, J. A., and O. C. Bridgeman, *J. Amer. Chem. Soc.*, **49**, 1665 (1927).

Benedict, M., G. B. Webb, and L. C. Rubin, *J. Chem. Phys.*, **8**, 334 (1940).

Benson, S. W., *J. Phys. Colloid Chem.*, **52**, 1060 (1948).

Bradford, M. L., and G. Thodos, *Can. J. Chem. Eng.*, **46**, 277 (1968).

Brebach, W. J., and G. Thodos, *Ind. Eng. Chem.*, **50**, 1095 (1958).

Brock, J. R., and R. B. Bird, *A.I.Ch.E. Journal*, **1**, 174 (1955).

Brokaw, R. S., Viscosity of Gas Mixtures, Report N.A.S.A. T.N. D-4496, Lewis Res. Center, Cleveland, Ohio, April 1968.

Bromley, L. A., and C. R. Wilke, *Ind. Eng. Chem.*, **50**, 1095 (1951).

Brown, G. G., Deviations of Natural Gas from Ideal Gas Laws, Univ. Michigan, Ann Arbor, Mich., 1940.

Buxton, T. S., and J. M. Campbell, *Soc. Petrol. Engrs. J.*, **7**, 80 (1967).

Caldwell, D. H., and H. E. Babbitt, *Ind. Eng. Chem.*, **33**, 249 (1941).

Carr, N. L., R. Kobayashi, and D. B. Burrows, *Trans. A.I.M.E.*, **201**, 264 (1954).

Chapman, S., and T. G. Cowling, *The Mathematical Theory of Non-uniform Gases*, Cambridge University Press, New York, 1939.

Chaudhuri, P. M., G. P. Mathur, and R. A. Stager, *Can. J. Chem. Eng.*, **46**, 132 (1968).

Chueh, P. L., and J. M. Prausnitz, *A.I.Ch.E. Journal*, **15**, 471 (1969).

Comings, E. W., B. J. Mayland, and R. S. Egly, *Bull.* 354, Eng. Exp. Sta., Univ. Illinois, Urbana, Ill., 1944.

Cooper, H. W., and J. C. Goldfrank, *Hydrocarbon Processing*, **46**, 141 (1967).

Dean, D. E., and L. I. Stiel, *A.I.Ch.E. Journal*, **11**, 526 (1965).

DeWitt, K. J., and G. Thodos, *Can. J. Chem. Eng.*, **44**, 148 (1966).

Dick, R. I., and B. B. Ewing, *Journal WPCF*, **39**, 543 (1967).

Dreisbach, R. R., *Ind. Eng. Chem.*, **40**, 2269 (1948).

Dukes, W. A. in *Rheology of Dispersed Systems*, edited by C. C. Mill, pp. 139–166, Pergamon Press, London 1959.

Eakin, B. E., K. E. Starling, J. P. Dolan, R. T. Ellington, *J. Chem. Eng. Data*, **7**, 33 (1962).

Einstein, A., *Ann. Physik*, **17**, 459 (1905); **19**, 289 (1906); **34**, 591 (1911).

Eykman, J. F., *Rec. Trav. Chim.*, **14**, 185 (1895).

Fishtine, S. H., *Ind. Eng. Chem. Fundam.*, **2**, 149 (1963).

Frankel, N. A., and A. Acrivos, *Chem. Eng. Sci.*, **22**, 847 (1967).

Giddings, J. G., J. T. F. Kao, and R. Kobayashi, *J. Chem. Phys.*, **45**, 578 (1966).

Giddings, J. G., and R. Kobayashi, *J. Petrol. Technol.*, **16**, 679 (1964).

Goldhammer, D. A., *Z. Physik Chem.*, **71**, 577 (1910).

Gonzalez, M. H., R. F. Buckacek, and A. L. Lee, *Soc. Petrol. Engrs. J.*, **7**, 75 (1967).

Gonzalez, M. H., and A. L. Lee, *J. Chem. Eng. Data*, **11**, 357 (1966).

Govier, G. W., *Eng. Journal*, **43**, 79 (1960).

Govier, G. W., C. A. Shook, and E. O. Lilge, *Trans. Can. Inst. Mining Met.*, **60**, 147 (1957).

Govier, G. W., and M. D. Winning, The Flow of Non-Newtonian Fluids, Paper Presented at the Meeting of the American Institute of Chemical Engineers, Montreal, Que., 1948.

Guth, E., and R. Simha, *Kolloid-Z.*, **74**, 266 (1936).

Hayward, A. T. J., *Brit. J. Appl. Phys.*, **18**, 965 (1967).

Hedstrom, B.O.A., *Ind. Eng. Chem.*, **44**, 651 (1952).

Herning, F., and L. Zipperer, *Gas u. Wasserfach*, **79**, 49 (1936).

Hirschfelder, J. O., R. J. Buehler, H. A. McGee, Jr., and J. R. Sutton, *Ind. Eng. Chem.*, **50**, 375 (1958). *Ind. Eng. Chem. Fundam.*, **1**, 224 (1962).

Huang, E. T. S., G. W. Swift, and F. Kurata, *A.I.Ch.E. Journal*, **12**, 932 (1966).

Jossi, J. A., L. I. Stiel, and G. Thodos, *A.I.Ch.E. Journal*, **8**, 59 (1962).

Katz, D. L. and Associates, *Handbook of Natural Gas Engineering*, McGraw-Hill Book Co., New York, 1959.

Kay, W. B., *Ind. Eng. Chem.*, **28**, 1014 (1936).

Kennedy, H. T., and N. S. Bhagia, *Soc. Petrol. Engrs. J.*, **9**, 279 (1969)

Landel, R. F., B. G. Moser, and A. J. Bauman, Fourth Int. Congress on Rheology, Brown Univ. (1963), *Proc.* Part 2, p. 663, Interscience Publishers, New York, 1965.

Lapidus, L., *Digital Computation for Chemical Engineers*, McGraw-Hill Book Co., New York, 1962.

Lee, A. L., M. H. Gonzalez, and B. E. Eakin, *J. Petrol. Technol.*, **18**, 997 (1966).

Leland, T. W., Jr., and P. S. Chappelear, *Ind. Eng. Chem.*, **60**, 15 (1968).

Lövgren, N., *Svensk. Kem. Tidskr.*, **53**, 359 (1941).

Lyckman, E. W., C. A. Eckert, and J. M. Prausnitz, *Chem. Eng. Sci.*, **20**, 703 (1965).

Lydersen, A. L., R. A. Greenkorn, and O. A. Hougen, Report No. 4 Eng. Expt. Sta., Univ. Wisconsin, Madison, Wis., Oct. 1955.

Macleod, D. B., *Trans Faraday Soc.*, **19**, 38 (1923).

Mathur, S., and S. C. Saxena, *Appl. Sci. Res., Ser. A*, **15**, 203 (1965).

McAllister, R. A., *A.I.Ch.E. Journal*, **6**, 427 (1960).

Meissner, H. P., and A. S. Michaels, *Ind. Eng. Chem.*, **41**, 2782 (1949).

Michels, A., A. Botzen, and W. Schuurman, *Physica*, **20**, 1141 (1954).

Michels, A., and R. O. Gibson, *Proc. Roy. Soc., Ser. A*, **134**, 288 (1931).

Michels, A., A. C. J. Schipper, and W. H. Rintoul, *Physica*, **19**, 1011 (1953).

Michiyoshi, I., R. Matsumoto, K. Mizuno, and M. Nakai, *Int. Chem. Eng.*, **6**, 382 (1966).

Miller, D. G., *Ind. Eng. Chem. Fundam.*, **2**, 78 (1963).

Natural Gas Processors Suppliers Association, *Engineering Data Book*, 8th Edition, Tulsa, Oklahoma, 1966.

Parzonka, W., *Compt. Rend. Acad. Sci. Paris*, **259**(13), 2073 (1964).

Pelofsky, A. H., *J. Chem. Eng. Data*, **11**, 394 (1966).

Perry, J. H., *Chemical Engineers' Handbook*, McGraw-Hill Book Co., Toronto, 1963.

Pilpel, N., *Chem. Process Eng.*, **45**, 395 (1964); **47**, 57 (1966).

Pitzer, K. S., D. Z. Lippmann, R. F. Curl, Jr., C. M. Huggins, and D. E. Peterson, *J. Amer. Chem. Soc.*, **77**, 3433 (1955).

Quayle, O. R., *Chem. Rev.*, **53**, 439 (1953)

Rao, K. B., and L. K. Doraiswamy, *Indian J. Technol.*, **4**, 141 (1966).

Redlich, O., and J. N. S. Kwong, *Chem. Rev.*, **44**, 233 (1949).

Reid, R. C., and T. K. Sherwood, *The Properties of Gases and Liquids*, McGraw-Hill Book Co., New York, 1966.

Robinson, D. B., C. A. Macrygeorgos, and G. W. Govier, *Petrol. Trans.*, **219**, 54 (1960).

Robinson, R. L., Jr., and R. H. Jacoby, *Hydrocarbon Processing*, **44**, 141 (1965).

Satter, A., and J. M. Campbell, *Soc. Petrol. Engrs. J.*, **3**, 333 (1963).

Shaver, R. G., and E. W. Merrill, *A.I.Ch.E. Journal*, **5**, 181 (1959).

Shimotake, H., and G. Thodos, *A.I.Ch.E. Journal*, **4**, 257 (1958).

Singleterry, C. R., and E. E. Stone, *J. Colloid, Interface Sci.*, **6**, 171 (1951).

Slowinski, E. J., Jr., E. E. Gates, and C. E. Waring, *J. Phys. Chem.*, **61**, 808 (1957).

Smithonian Institution, *Smithonian Physical Tables*, Washington, D.C., 1933.

Sommerfeld, J. T., and G. L. Perry, *Hydrocarbon Processing*, **47**, 109 (1968).

Standing, M. B., and D. L. Katz, *Trans. A.I.M.E.*, **146**, 140 (1942).

Starling, K. E., B. E. Eakin, and R. T. Ellington, *A.I.Ch.E. Journal*, **6**, 438 (1960).

Stewart, W. F., S. F. Burkhardt, and D. Voo, Prediction of Pseudo-critical Parameters for Mixtures. Paper presented at the A.I.Ch.E. Meeting, Kansas City, Mo., 1959.

Stiel, L. I., and G. Thodos, *A.I.Ch.E. Journal*, **7**, 611 (1961); **8**, 229 (1962); **10**, 275 (1964).

Sugden, S., *J. Chem. Soc.*, **125**, 27 (1924).

Szyszkowski, B. von, *Z. physik. Chem.*, **64**, 385 (1908).

Tait, P. G., *The Voyage of H.M.S. Challenger*, Vol. 2, Pt. 4, H.M.S.O., London, 1888.

Tamura, M., M. Kurata, and H. Odani, *Bull. Chem. Soc. Japan*, **28**, 83 (1955).

Taylor, G. I., *Proc. Roy. Soc., Ser. A*, **138**, 41 (1932).

Thomas, D. G., *A.I.Ch.E. Journal*, **7**, 431 (1961); **9**, 310 (1963).

Thomas, D. G., *J. Colloid Sci.*, **20**, 267 (1965).

Thomas L. H., *J. Chem. Soc., Part 2*, 573 (1946).

Valentik, L., and R. L. Whitmore, *Brit, J. Appl. Phys.*, **16**, 1197 (1965).

Viswanath, D. S., and G.-J. Su, *A.I.Ch.E. Journal*, **11**, 202 (1965).

Wichert, E., and K. Aziz, *Can. J. Chem. Eng.*, **49**, 267 (1971).

Wilke, C. R., *J. Chem. Phys.*, **18**, 517 (1950).

Yen, L. C., and S. S. Woods, *A.I.Ch.E. Journal*, **12**, 95 (1966).

Basic Concepts of the Flow of Newtonian Fluids

3

1 INTRODUCTION

It is assumed that the reader has a general knowledge of elementary fluid mechanics. Nonetheless, it seems desirable, partly for purposes of review, but more especially to create the appropriate perspective, that we highlight in this chapter the fundamental concepts of the flow, in general, of single-phase Newtonian fluids. This sets the stage for the review of the pipe flow of these fluids preparatory finally to a discussion of the treatment of the pipe flow of complex mixtures.

General problems in the flow of single-phase Newtonian fluids may, in principle at least, be solved by the simultaneous solution of five basic equations:

1. the equation of state of the fluid, giving the dependence of the fluid density on pressure and temperature;
2. the constitutive equation of the fluid, characterizing its behavior when subjected to shear stresses resulting in laminar flow;
3. the continuity equation, or the law of conservation of mass;
4. the momentum equation, or the application of Newton's second law of motion to a fluid system;
5. the energy equation, or the law of conservation of energy.

We have dealt with the equation of state and the constitutive equations in detail in Chapters 1 and 2 and here we will only refer to them. Together they describe the fluid under consideration. The other three equations, which will be discussed in detail here, condense empirical observations concerning the conservation of mass, momentum, and energy into relations applicable to fluids in motion.

Although the set of the five relations, together with a description of the boundary conditions, theoretically suffices for the solution of even complex problems of fluid flow, in actual practice the application of the momentum and the energy equations may pose insurmountable difficulties. An alternate both to the momentum equation and to the rigorously applied energy equation is

6. the so-called mechanical energy equation introducing an energy degradation or friction term.

This equation finds wide application.

The first three of the basic equations will be found useful and indeed essential in the solution of all flow problems; the fourth is useful in many problems of laminar flow; the fifth, in simplified form, is of value in problems of nonisothermal flow; and the sixth, the mechanical energy equation, is useful in a host of problems involving either laminar or turbulent flow.

2 THE EQUATION OF STATE

Equations showing the effect of pressure and temperature on the density of both gases and liquids were discussed in Chapter 2. It suffices here to say that, whether directly or implicitly, some relationship between density and pressure is involved in every flow problem, and frequently the effect of temperature is also of importance. For the present purposes it may be assumed that an equation of state exists for any single-phase fluid under consideration

$$\rho = \phi(P, T, \text{fluid}) \tag{3.1}$$

where ρ = density of fluid,
P = pressure,
T = absolute temperature,
ϕ designates "some function of,"

and that it may be approximated to a satisfactory degree of accuracy. Commonly in engineering calculations the engineering gas law (Equation 2.6), discussed in Chapter 2, is adequate in dealing with the flow of gases, and the assumption of incompressibility suffices in the case of liquids. We should nonetheless not lose sight of the fact that the complete equation of state, for the particular fluid, is required for a precise understanding of the flow of the fluid.

3 THE CONSTITUTIVE EQUATION

The concept of the constitutive equation was introduced in Chapter 1 and it was shown that for one-dimensional laminar flow of a Newtonian fluid it is

$$\tau = \frac{\mu}{g_c} \dot{S} \tag{1.38}$$

where τ = shear stress,
$\dot{S}$ = absolute value of the rate of shear,
μ = viscosity,
$\mu = \phi(P, T, \text{fluid})$

It is important to note that, although Equation 1.38 defining viscosity applies only to conditions of laminar flow, viscosity plays a role in turbulent flow. It should also be noted that there is no alternative to the use or assumption of some constitutive equation, since the relationship between shear stress and velocity gradient arises in every flow problem. One can assume a viscosity of zero (inviscid fluid) and obtain useful results in certain types of high-speed flow problems (aerodynamics). In other problems, including pipe flow, a knowledge of the actual viscosity of the fluid may be critical.

4 THE CONTINUITY EQUATION

The application of the law of conservation of mass may be made on a microscopic basis, referring to a differential element of fluid, or on a macroscopic basis, referring to the volume of a finite system.

Microscopic The microscopic continuity equation may be derived readily (see Knudsen and Katz, 1958; Bird, Stewart, and Lightfoot, 1960, or other references in fluid mechanics) by equating the input of mass to a small elemental volume to the accumulation of mass in and the output of mass from the volume. The equation in its most general form is

$$-\frac{\partial \rho}{\partial t} = \nabla \cdot (\rho \mathbf{v}) \tag{3.2}$$

where $\mathbf{v}$ is the velocity vector of the fluid at a point, and t is time. In Cartesian coordinates Equation 3.2 may be written in full as

$$-\frac{\partial \rho}{\partial t} = \frac{\partial}{\partial x}(\rho u) + \frac{\partial}{\partial y}(\rho v) + \frac{\partial}{\partial z}(\rho w) \tag{3.3}$$

where u, v, and w are velocity components in the coordinate directions x, y, and z. In the case of one-dimensional pipe flow, the equation simplifies to

$$-\frac{\partial \rho}{\partial t} = \frac{\partial}{\partial x}(\rho u) \tag{3.4}$$

where the x-coordinate direction is oriented along the pipe axis.

The equations may also be expressed in cylindrical coordinates as

$$-\frac{\partial \rho}{\partial t} = \frac{1}{r}\frac{\partial}{\partial r}(\rho r v_r) + \frac{1}{r}\frac{\partial}{\partial \theta}(\rho v_\theta) + \frac{\partial}{\partial z}(\rho w) \tag{3.5}$$

where r and θ are the radial and angular coordinates, and v_r, v_θ, and w are the components of velocity in the three coordinate directions. Equation 3.3 or 3.4 is perfectly general applying to any flow of any single-phase fluid. For specific application the equation may often be simplified, i.e., the density ρ may be constant, or one or more of the velocities may be constant, or zero.

Macroscopic Integration of the microscopic equation over a finite volume, $\mathscr{V}$, yields the overall material balance or the macroscopic equation of continuity,

$$\frac{d\mathscr{M}}{dt} = M_1 - M_2 \tag{3.6}$$

In the case of one-dimensional pipe flow, Equation 3.6 becomes

$$\frac{d\mathscr{M}}{dt} = \rho_1 V_1 A_1 - \rho_2 V_2 A_2 = -\Delta(\rho V A) \tag{3.7}$$

where V = average velocity,

A = area normal to the velocity,

$\mathscr{M} = \int_{\mathscr{V}} \rho d\mathscr{V}$ = total mass contained in the system of volume, $\mathscr{V}$,

M = mass rate of flow,

subscripts 1 and 2 refer to the inlet and outlet of the system, respectively.

The steady-state form of Equations 3.6 and 3.7 is obtained by equating the left-hand side to zero.

5 THE MOMENTUM EQUATION

As an expression of Newton's second law of motion applied to a fluid system, the momentum equation is a statement of the concept that (mass) (acceleration) = net force acting. The forces acting upon a mass of fluid, causing acceleration unless they neutralize one another, are the force of gravity, normal or hydrostatic pressure forces, and shear or tangential forces (caused by the resistance of the fluid to deformation).

As in the case of the continuity equation, the momentum equation may be expressed either in the microscopic or the macroscopic form.

Microscopic A detailed derivation of the microscopic equation is given by Knudsen and Katz (1958), Bird, Stewart, and Lightfoot (1960) and others. The equation in its most general form is

$$\rho \frac{D\mathbf{v}}{Dt} = \rho\mathbf{g} - g_c\nabla P - g_c[\nabla \cdot \boldsymbol{\tau}] \tag{3.8}$$

where $\mathbf{g}$ = gravitational force,

$\boldsymbol{\tau}$ = shear stress tensor,

D/Dt = the substantial derivative, in Cartesian coordinates given by

$$\frac{D}{Dt} = \frac{\partial}{\partial t} + u\frac{\partial}{\partial x} + v\frac{\partial}{\partial y} + w\frac{\partial}{\partial z}$$

The term on the left-hand side represents the product of mass per unit volume and acceleration. The terms on the right-hand side represent various forces acting on a unit volume of the system. This is a vector equation and will yield three partial differential equations for the three components of velocity. The expanded form of the equation will depend upon the selected coordinate system. In the form of Equation 3.8 the momentum equation involves no assumptions, but before it may be applied in any useful way the components of the shear stress tensor, $\boldsymbol{\tau}$, must be related to the velocity gradients causing them. This involves the constitutive equation for the fluid.

In the simple case of a Newtonian fluid the normal and the tangential stresses may be replaced by terms involving the viscosity and the velocity gradients resulting in the momentum equation or the equation of motion for the Newtonian fluid. The equation in its general form, allowing for variation in density and viscosity, is given by Bird *et al.*

For the special case of a Newtonian fluid of constant viscosity and density, the equation reduces to

$$\rho \frac{D\mathbf{v}}{Dt} = \rho\mathbf{g} - g_c\nabla P + \mu\nabla^2\mathbf{v} \tag{3.9}$$

Equation 3.9 is the well known equation of motion for Newtonian fluids (Navier–Stokes equation). Equation 3.9 is expressed in Cartesian and cylindrical coordinates in Table 3.1. For the special case of constant density and zero viscosity, the equation reduces to Euler's equation

$$\rho \frac{D\mathbf{v}}{Dt} = \rho\mathbf{g} - g_c\nabla P \tag{3.10}$$

TABLE 3.1

The Equation of Motion[a]

In the Cartesian Coordinate System:

x-component

$$\rho\left(\frac{\partial u}{\partial t}+u\frac{\partial u}{\partial x}+v\frac{\partial u}{\partial y}+w\frac{\partial u}{\partial z}\right)=-g_c\frac{\partial P}{\partial x}+\mu\left(\frac{\partial^2 u}{\partial x^2}+\frac{\partial^2 u}{\partial y^2}+\frac{\partial^2 u}{\partial z^2}\right)+\rho g_x$$

y-component

$$\rho\left(\frac{\partial v}{\partial t}+u\frac{\partial v}{\partial x}+v\frac{\partial v}{\partial y}+w\frac{\partial v}{\partial z}\right)=-g_c\frac{\partial P}{\partial y}+\mu\left(\frac{\partial^2 v}{\partial x^2}+\frac{\partial^2 v}{\partial y^2}+\frac{\partial^2 v}{\partial z^2}\right)+\rho g_y$$

z-component

$$\rho\left(\frac{\partial w}{\partial t}+u\frac{\partial w}{\partial x}+v\frac{\partial w}{\partial y}+w\frac{\partial w}{\partial z}\right)=-g_c\frac{\partial P}{\partial z}+\mu\left(\frac{\partial^2 w}{\partial x^2}+\frac{\partial^2 w}{\partial y^2}+\frac{\partial^2 w}{\partial z^2}\right)+\rho g_z$$

In the Cylindrical Coordinate System:

r-component

$$\rho\left(\frac{\partial v_r}{\partial t}+v_r\frac{\partial v_r}{\partial r}+\frac{v_\theta}{r}\frac{\partial v_r}{\partial \theta}-\frac{v_\theta^2}{r}+w\frac{\partial v_r}{\partial z}\right)$$

$$=-g_c\frac{\partial P}{\partial r}+\mu\left[\frac{\partial}{\partial r}\left(\frac{1}{r}\frac{\partial}{\partial r}(rv_r)\right)+\frac{1}{r^2}\frac{\partial^2 v_r}{\partial \theta^2}-\frac{2}{r^2}\frac{\partial v_\theta}{\partial \theta}+\frac{\partial^2 v_r}{\partial z^2}\right]+\rho g_r$$

θ-component

$$\rho\left(\frac{\partial v_\theta}{\partial t}+v_r\frac{\partial v_\theta}{\partial r}+\frac{v_\theta}{r}\frac{\partial v_\theta}{\partial \theta}+\frac{v_r v_\theta}{r}+w\frac{\partial v_\theta}{\partial z}\right)$$

$$=-g_c\frac{1}{r}\frac{\partial P}{\partial \theta}+\mu\left[\frac{\partial}{\partial r}\left(\frac{1}{r}\frac{\partial}{\partial r}(rv_\theta)\right)+\frac{1}{r^2}\frac{\partial^2 v_\theta}{\partial \theta^2}+\frac{2}{r^2}\frac{\partial v_r}{\partial \theta}+\frac{\partial^2 v_\theta}{\partial z^2}\right]+\rho g_\theta$$

z-component

$$\rho\left(\frac{\partial w}{\partial t}+v_r\frac{\partial w}{\partial r}+\frac{v_\theta}{r}\frac{\partial w}{\partial \theta}+w\frac{\partial w}{\partial z}\right)=-g_c\frac{\partial P}{\partial z}+\mu\left[\frac{1}{r}\frac{\partial}{\partial r}\left(r\frac{\partial w}{\partial r}\right)+\frac{1}{r^2}\frac{\partial^2 w}{\partial \theta^2}+\frac{\partial^2 w}{\partial z^2}\right]+\rho g_z$$

[a] R. B. Bird, W. E. Stewart, and E. N. Lightfoot, *Transport Phenomena*, John Wiley & Sons, New York, 1960.

These equations find extensive application in all areas of fluid mechanics. Equation 3.9 is restricted in its practical use to laminar flow where the term involving the viscous shear may be evaluated. Equation 3.10 and equations based on it, although not restricted to laminar flow, apply only to fluids of low viscosity (gases) and to situations where the viscous forces are small relative to the others.

For one-dimensional flow in a pipe along the x-axis we have the following simplified form of the Navier–Stokes equation:

$$\rho\left(\frac{\partial u}{\partial t} + u\frac{\partial u}{\partial x}\right) = -g_c\frac{\partial P}{\partial x} + \mu\frac{\partial^2 u}{\partial x^2} + \rho g_x \tag{3.11}$$

where g_x is the component of the gravitational force along the x-axis.

Macroscopic Integration of the microscopic momentum equation over the volume of a system yields

$$\frac{d}{dt}\int_{\mathscr{V}} \rho \mathbf{v}\, d\mathscr{V} = \rho_1 V_1^2 \mathbf{A}_1 - \rho_2 V_2^2 \mathbf{A}_2 + g_c P_1 \mathbf{A}_1 - g_c P_2 \mathbf{A}_2 - g_c \mathbf{F} + \mathscr{M}\mathbf{g} \tag{3.12}$$

where the vector $\mathbf{A}$ has the magnitude A and the direction of the time-smoothed velocity, V, at A and the vector $\mathbf{F}$ is the resultant shear force acting on the system.

A particularly useful form of the macroscopic momentum equation is that applicable to steady-state situations when the left hand side of Equation 3.12 becomes zero and the equation reduces to a simple balance of forces:

$$\mathbf{F} g_c = \rho_1 V_1^2 \mathbf{A}_1 - \rho_2 V_2^2 \mathbf{A}_2 + g_c P_1 \mathbf{A}_1 - g_c P_2 \mathbf{A}_2 + \mathscr{M}\mathbf{g} \tag{3.13}$$

This is the steady-state form of the so-called Bernoulli equation. In the case of a one-dimensional pipe flow at an angle θ to the horizontal, the equation simplifies to

$$g_c(P_1 A_1 - P_2 A_2 - F) = \mathscr{M} g \sin\theta \tag{3.14}$$

6 THE ENERGY EQUATION

The law of conservation of energy, as applied to a fluid system in motion, is of particular use in connection with problems on nonisothermal flow where the changing temperature of the fluid is of interest.

Again the equation may be expressed either in microscopic or macroscopic form.

Microscopic In the microscopic form the equation is a statement of the balance between the initial energy content of a small elemental volume, the external energy which may be added to or removed from it, and its final energy content. The forms of energy, ordinarily important and included in the equation, are the internal energy, the kinetic energy, pressure-volume or compression energy, and the potential energy or energy of position in the gravitational field of the earth. If we neglect surface, chemical and electrical effects we obtain the energy equation in the following form:

$$\frac{\rho}{g_c}\frac{D}{Dt}\left(U + \Phi + \frac{\mathbf{v}\cdot\mathbf{v}}{2}\right) = -(\nabla\cdot\mathbf{q}) - (\nabla\cdot P\mathbf{v}) - (\nabla\cdot[\boldsymbol{\tau}\cdot\mathbf{v}]) \qquad (3.15)$$

where U = internal energy per unit mass,
$\mathbf{q}$ = heat flux vector,
and the term Φ is defined by $\mathbf{g} = -\nabla\Phi$

It is convenient to subtract from this equation an equation representing the change of kinetic energy, which may be derived from Equation 3.8 as follows:

$$\frac{\rho}{g_c}\frac{D}{Dt}\frac{\mathbf{v}\cdot\mathbf{v}}{2} = P(\nabla\cdot\mathbf{v}) - (\nabla\cdot P\mathbf{v}) + \frac{\rho}{g_c}(\mathbf{v}\cdot\mathbf{g}) - (\nabla\cdot[\boldsymbol{\tau}\cdot\mathbf{v}]) + (\boldsymbol{\tau}:\nabla\mathbf{v}) \quad (3.16)$$

The subtraction of Equation 3.16 from Equation 3.15 yields (see Bird *et al.*, 1960):

$$\frac{\rho}{g_c}\frac{DU}{Dt} = -(\nabla\cdot\mathbf{q}) - P(\nabla\cdot\mathbf{v}) - (\boldsymbol{\tau}:\nabla\mathbf{v}) \qquad (3.17)$$

where in Cartesian coordinates

$$\boldsymbol{\tau}:\nabla\mathbf{v} = \tau_{xx}\left(\frac{\partial u}{\partial x}\right) + \tau_{yy}\left(\frac{\partial v}{\partial y}\right) + \tau_{zz}\left(\frac{\partial w}{\partial z}\right) + \tau_{xy}\left(\frac{\partial u}{\partial y} + \frac{\partial v}{\partial x}\right) + \tau_{yz}\left(\frac{\partial v}{\partial z} + \frac{\partial w}{\partial y}\right) + \tau_{zx}\left(\frac{\partial w}{\partial x} + \frac{\partial u}{\partial z}\right)$$

This latter is a form of the energy equation which is suitable when the thermal effects are of greatest interest. Equation 3.17 may be further modified by expressing dU as $C_v dT$, where C_v is the specific heat at constant volume:

$$\rho C_v \frac{DT}{Dt} = -(\nabla . \mathbf{q}) - T\left(\frac{\partial P}{\partial T}\right)_\rho (\nabla . \mathbf{v}) - (\boldsymbol{\tau}:\nabla\mathbf{v}) \qquad (3.18)$$

Equations 3.15, 3.17, and 3.18 are all general forms of the energy equation, equally free of assumptions. Unfortunately the difficulties in their application to practical problems, particularly with regard to the stress term, make simplifying assumptions necessary.

If a Newtonian fluid of constant thermal conductivity is assumed, and $\mathbf{q}$ is expressed in terms of temperature gradients and $\boldsymbol{\tau}$ in terms of velocity gradient, the following equation is obtained:

$$\rho C_v \frac{DT}{Dt} = k\nabla^2 T - T\left(\frac{\partial P}{\partial T}\right)_\rho (\nabla\cdot\mathbf{v}) + \mu\phi \qquad (3.19)$$

where k = thermal conductivity, and ϕ = the so-called dissipation function. The product $\mu\phi$ represents viscous heating of the fluid; this term is usually negligible in pipe flow calculations. Even Equation 3.19 is difficult to apply in any but the simplest cases, and further assumptions are often made, including ideal gas behavior, constant pressure, or constant density.

Macroscopic The unsteady state form of the macroscopic energy equation may be developed in a manner similar to that used for the macroscopic momentum equation. We are concerned mainly with the steady-state case and will not discuss the general case of this equation. Under steady-state conditions the energy equation is

$$\Delta\left(U + \frac{P}{\rho} + \frac{\overline{(u^3)}}{2Vg_c} + \frac{g}{g_c}z\right) = \mathcal{Q} - W \tag{3.20}$$

where $\mathcal{Q}$ = heat added to the system per unit mass of fluid,
W = work done by the system per unit mass of fluid.

The kinetic energy term $\overline{(u^3)}/2Vg_c$ is usually written as $V^2/2\alpha g_c$, where α depends upon the velocity profile.

7 THE MECHANICAL ENERGY EQUATION

Microscopic By taking the scalar product of $\mathbf{v}$ with the microscopic momentum balance, an equation relating various forms of mechanical energy may be derived. Again, the derivation is discussed in more detail by Bird *et al.* The equation is

$$\frac{\partial}{\partial t}\frac{\rho}{g_c}\frac{\mathbf{v}\cdot\mathbf{v}}{2} = -\left(\nabla\cdot\frac{\rho}{2g_c}(\mathbf{v}\cdot\mathbf{v})\mathbf{v}\right) - (\nabla\cdot P\mathbf{v}) - P(-\nabla\cdot\mathbf{v}) - (\nabla\cdot[\boldsymbol{\tau}\cdot\mathbf{v}])$$
$$-(-\boldsymbol{\tau}:\nabla\mathbf{v}) + \frac{\rho}{g_c}(\mathbf{v}\cdot\mathbf{g}) \tag{3.21}$$

The terms from left to right are: rate of increase in kinetic energy per unit volume, net rate of input of kinetic energy by bulk flow, rate of work done by the pressure of the surrounding fluid on the volume element, rate of reversible conversion to internal energy, rate of work done by viscous forces on the volume element, rate of irreversible conversion to internal energy, and rate of work done by the gravity force on the volume element. Equation 3.21 is derived from the momentum equation and hence does not constitute an independent equation. This equation may be used in place of one of the momentum equations.

Macroscopic A more useful form of the mechanical energy balance may be derived by integrating Equation 3.21 over the entire system volume to give the macroscopic mechanical energy balance. The equation for the steady-state case is

$$\frac{1}{2g_c}\Delta\left(\frac{V^2}{\alpha}\right)+\frac{g}{g_c}\Delta z+\int_{P_1}^{P_2}\frac{dP}{\rho}+W+\mathscr{F}=0 \tag{3.22}$$

where $\mathscr{F}$ is the "friction loss" per unit mass. The term $\mathscr{F}$ accounts for all energy degraded to internal energy as a result of irreversibilities. It may be considered as the term required to balance the equation. The integral appearing in Equation 3.22 for a gas having a constant compressibility factor is

$$\int_{P_1}^{P_2}\frac{1}{\rho}\,dP=\frac{ZRT}{m}\ln\frac{P_2}{P_1} \tag{3.23}$$

and for an incompressible fluid is

$$\int_{P_1}^{P_2}\frac{1}{\rho}\,dP=\frac{1}{\rho}(P_2-P_1) \tag{3.24}$$

The unsteady-state form of Equation 3.22 finds little use and has been omitted here.

The macroscopic mechanical energy balance may also be derived from the macroscopic energy equation. For this derivation we first write Equation 3.20 in the form

$$\int dU+\int d(Pv)+\frac{1}{\alpha g_c}\int V\,dV+\frac{g}{g_c}\int dz=\mathscr{Q}-W \tag{3.25}$$

where the integration is performed over the volume of the system. The the quantity $U+(P/\rho)$ is the enthalpy per unit mass, H, and according to the well-known thermodynamic relationship may be expressed as

$$dU+d(Pv)=dH=TdS+v\,dP \tag{3.26}$$

where $v=1/\rho=$ specific volume, and

$$\int TdS=\mathscr{Q}+\mathscr{F} \tag{3.27}$$

Substituting the definition of enthalpy from Equation 3.26 into Equation 3.25, we obtain

$$\int TdS+\int v\,dP+\frac{1}{\alpha g_c}\int V\,dV+\frac{g}{g_c}\int dz=\mathscr{Q}-W \tag{3.28}$$

Substituting for the $\int TdS$ from Equation 3.27, we obtain an equation equivalent to Equation 3.22.

The mechanical energy equation in the form of Equation 3.22 is of great practical utility, but the term $\mathscr{F}$ must be separately evaluated. In the simple cases of laminar flow this may be done from the momentum equation; otherwise we must rely on the results of actual measurements of generalizations from them. It is often convenient to apply Equation 3.22 in differential form, i.e.,

$$\frac{dP}{\rho} + \frac{V\,dV}{\alpha g_c} + \frac{g}{g_c}\,dz - \frac{dP_f}{\rho} = 0 \tag{3.29}$$

where $\int -dP_f/\rho = \mathscr{F}$

8 MODES OF FLOW

Quite apart from the theoretical relationships that have been described, there is a most extensive background of empirical and experimental knowledge of the flow of fluids. Most important, and not predictable from the relationships discussed, is that fluids may flow in either laminar or turbulent motion. The distinction between these two modes of flow was first made clear by Osborne Reynolds in 1883. His classical experiments showed that at sufficiently low velocities a fluid flowed in laminae and without eddy currents, whereas at greater velocities eddy currents developed and the flow was characterized by an apparently random motion. Reynolds also showed that the criterion for the breakdown of laminar motion and the transition to turbulent flow was a dimensionless quantity, which now bears his name, the Reynolds number:

$$\mathrm{Re} = \frac{XV\rho}{\mu}$$

where X is some appropriate linear dimension characterizing the flow boundary.

The Reynolds concept, originally determined empirically, is now given greater importance through the recognition that the Reynolds number is a measure of the ratio of the inertial to the viscous forces which are developed in a flowing fluid. We may see this directly from the Navier–Stokes equation (Equation 3.9), where the inertial forces may be represented by terms of the form $\rho V^2/X$ and the viscous forces by terms of the form $\mu V/X^2$. Also if we cast the Navier–Stokes equation in dimensionless form by substituting

$$\mathbf{v}^* = \frac{\mathbf{v}}{V}$$

$$P^* = g_c \frac{P}{\rho V^2}$$

$$t^* = \frac{tV}{X}$$

$$\nabla^* = X\nabla$$

we have

$$\frac{D\mathbf{v}^*}{Dt^*} = \frac{gX}{V^2} \cdot \frac{\mathbf{g}}{g} - \nabla^* P^* + \frac{\mu}{XV\rho} \nabla^{*2}\mathbf{v}^* \tag{3.30}$$

In this equation the scale factors are the Reynolds number, $Re = XV\rho/\mu$, and the Froude number, $\hat{\text{Fr}} = V^2/gX$. Systems having the same values of Re and $\hat{\text{Fr}}$ would therefore behave similarly, provided the assumptions involved in the derivation of the equation for the system in question are justified, and provided also that no new dimensionless parameters are introduced through the boundary and initial conditions.

Related to the existence of the two modes of flow is the existence of the so-called boundary layer encountered where a fluid flows over a solid surface (or a gas phase flows relative to a liquid phase). It may be predicted from theoretical considerations, and it is known from extensive observations, that a steep velocity gradient occurs adjacent the interface. The region in which this gradient is observed is called the boundary layer. At sufficiently low values of the Reynolds number, the flow in the boundary layer is laminar; at higher values, the boundary layer may be composed of a laminar sublayer, a transition region, and a turbulent region.

In flow over a flat plate oriented parallel to the flow, the thickness of the boundary layer is zero at the leading edge of the plate, and from there it increases with increasing flow distance. At first the boundary layer is wholly laminar, but depending upon the flow conditions the laminar layer becomes unstable after the boundary layer has attained a certain critical thickness. The laminar boundary layer will then break up and be replaced by a much thinner laminar sublayer, a transition zone, and a turbulent zone extending indefinitely into the fluid.

Analogous phenomena are encountered in the flow of a fluid in a pipe where the boundary layer that grows from the pipe wall eventually extends to the center of the pipe.

9 LAMINAR FLOW

In ideal laminar flow in a duct of constant cross section there are no components of velocity (apart from random molecular and Brownian movement) normal to the direction of flow. This is also true in practice provided the viscosity of the fluid remains constant. Under nonisothermal conditions variations in viscosity normal to the flow give rise to transverse components of velocity, but these are usually small. Also we will see that in the case of certain non-Newtonian fluids, the consistency of the fluid may change with travel distance and this too leads to small transverse components of velocity. Notwithstanding the slight deviations from ideal laminar flow caused by variations in consistency, the flow is still considered to be laminar.

In the case of Newtonian fluids where the flow is wholly laminar and where the geometry of the flow is simple as in the flow in pipes, annular spaces, and other regular ducts, or over spheres, cylinders, or other regularly shaped objects, analytical solutions of the momentum equation enable us to predict the growth of the boundary layer, the velocity profile, and the energy dissipation. With more complex geometries, solutions based upon the application of the momentum equation may still be obtained with the aid of numerical methods. Even where the momentum equation may be solved, however, we find many applications where the mechanical energy equation, with the appropriate value for the friction term, is more convenient to use.

10 TURBULENT FLOW

Turbulent flow is characterized by fluctuating components of velocity in all directions—both parallel and transverse to the direction of flow. For flow in a duct of constant cross section the time-average values of the velocity components normal to the flow are zero. The intensity of the turbulence may be measured in terms of the root-mean-square of the fluctuating velocities, or the ratio of this term to the mean velocity of flow. Intensities so measured may have values as great as 10 percent of the mean velocity. Normally the turbulence intensity is greater in the direction of flow than transverse to it. Remote from a flow boundary turbulence tends to be truly random, but near a boundary wall turbulence is suppressed, and anisotropy develops with the intensity of the velocity fluctuations normal to the wall, being decreased more than those parallel to the main flow.

When the flow is turbulent, almost insurmountable complications prevent the solution of the fundamental equations. These are associated with the fact that the fluid velocities referred to in the Navier–Stokes equations are the instantaneous fluctuating point values. Although there are suggestions in the literature that the Navier–Stokes equations do not apply when the fluid is in turbulent motion (Ladyzhenskaya, 1963), all theoretical indications are to the effect that the Navier–Stokes equations are the proper relationships for both laminar and turbulent flows. The problem concerns their solution, not their applicability. The full description of the fluctuating velocities requires an enormous number of Fourier components. The solution of the equations depends strongly on minute details of the initial and boundary data, and these details are never available in practice. For this reason we must content ourselves with a statistical description of both the initial and the boundary data, and the resulting velocity field.

The objective of all statistical theories of turbulence is the determination of probability distribution functions which would yield the likelihood of occurrence of any combination of velocities at points in space and time. The work of Batchelor (1953), Townsend (1956), Hinze (1959), Harlow and Nakayama (1967), and many others is aimed in this direction. This is, undoubtedly, the proper approach to the complete understanding of the problem. At the present time, however, practical application of these detailed statistical theories to the engineering problems is extremely difficult. In order to keep our discussion simple and brief we will not discuss the statistical approaches, but instead we will present a more directly useful approach based upon the so-called mixing length concept.

The Navier-Stokes equation for the x-component of velocity in the case of an incompressible fluid with no gravitational forces may be expressed as

$$\frac{\partial u}{\partial t} + u\frac{\partial u}{\partial x} + v\frac{\partial u}{\partial y} + w\frac{\partial u}{\partial z} = -\frac{g_c}{\rho}\frac{\partial P}{\partial x} + \frac{\mu}{\rho}\nabla^2 u \tag{3.31}$$

and the equation of continuity is given by

$$\frac{\partial u}{\partial x} + \frac{\partial v}{\partial y} + \frac{\partial w}{\partial z} = 0 \tag{3.32}$$

Multiplying Equation 3.32 by u and adding the resulting equation to 3.31 yields, after rearrangement,

$$\frac{\partial u}{\partial t} + \frac{\partial(u^2)}{\partial x} + \frac{\partial(uv)}{\partial y} + \frac{\partial(uw)}{\partial z} = -\frac{g_c}{\rho}\frac{\partial P}{\partial x} + \frac{\mu}{\rho}\nabla^2 u \tag{3.33}$$

It is convenient at this stage to write the velocity and the pressure as terms composed of two parts,

$$\left.\begin{aligned} u &= \bar{u} + u' \\ v &= \bar{v} + v' \\ w &= \bar{w} + w' \\ P &= \bar{P} + P' \end{aligned}\right\} \tag{3.34}$$

where $\bar{u}$, $\bar{v}$, $\bar{w}$, and $\bar{P}$ are values of the three components of velocity and of pressure averaged over a short interval of time at a particular point in space, and u', v', w', and P' are the turbulent fluctuating components. Substituting

Equation 3.34 into Equation 3.33 and following the usual rules for averaging, we obtain

$$\rho\left[\frac{\partial \bar{u}}{\partial t} + \frac{\partial(\overline{u^2})}{\partial x} + \frac{\partial}{\partial y}(\bar{u}\bar{v}) + \frac{\partial}{\partial z}(\bar{u}\bar{w})\right] = -g_c\frac{\partial P}{\partial x} + \mu\nabla^2\bar{u}$$

$$- \frac{\partial}{\partial x}(\rho\overline{u'^2}) - \frac{\partial}{\partial y}(\rho\overline{u'v'}) - \frac{\partial}{\partial z}(\rho\overline{u'w'}) \quad (3.35)$$

Equation 3.35 is identical to Equation 3.33 with the mean quantities replacing the instantaneous values, except in the last three terms involving the turbulent fluctuating velocities. Similar equations may also be derived for the y- and z-components of velocity introducing six more terms containing the turbulent fluctuating velocities. The terms $\rho\overline{u'^2}$, $\rho\overline{u'v}$, etc., are known as Reynolds stresses, and they express the physical fact that, in turbulent flow, momentum is transferred by the movement of the fluid, associated with the fluctuating velocity components, in addition to the ordinary transfer of momentum by molecular motion. In terms of the shear stress, τ, Equation 3.35 is expressed as

$$\rho\left[\frac{\partial \bar{u}}{\partial t} + \frac{\partial(\overline{u^2})}{\partial x} + \frac{\partial}{\partial y}(\bar{u}\bar{v}) + \frac{\partial}{\partial z}(\bar{u}\bar{w})\right] = g_c\left[-\frac{\partial \bar{P}}{\partial x}\right.$$

$$\left. - \frac{\partial}{\partial x}(\tau_{xx}) - \frac{\partial}{\partial y}(\tau_{yx}) - \frac{\partial}{\partial z}(\tau_{zx})\right] \quad (3.36)$$

where

$$\tau_{xx} = -\frac{1}{g_c}\left(\mu\frac{\partial \bar{u}}{\partial x} - \rho\overline{u'^2}\right) \quad (3.37)$$

$$\tau_{yx} = -\frac{1}{g_c}\left(\mu\frac{\partial \bar{u}}{\partial y} - \rho\overline{u'v'}\right) \quad (3.38)$$

$$\tau_{zx} = -\frac{1}{g_c}\left(\mu\frac{\partial \bar{u}}{\partial z} - \rho\overline{u'w'}\right) \quad (3.39)$$

The shear stress may be looked upon as made up of two parts: a viscous stress component and a turbulent stress component.

In the same way, the mean and fluctuating velocity components may be substituted for the velocities in the continuity equation. In this case, however, the fluctuating components cancel and the equation of continuity, when averaged over a short interval of time, becomes

$$\frac{\partial \bar{u}}{\partial x} + \frac{\partial \bar{v}}{\partial y} + \frac{\partial \bar{w}}{\partial z} = 0 \quad (3.40)$$

It is not possible, even in principle, to solve the equations of motion for turbulent flow, resulting from the above analysis, without additional information. In three-dimensional flow we have, as unknowns, the three mean and the three fluctuating components of velocity and the pressure; and only four equations. This is why this type of analysis of turbulent flow problems must rely upon some experimental information for its solution. In particular, we must relate the Reynolds stress terms to the mean velocity components.

The Prandtl Mixing Length Theory One of the simplest methods of estimating the Reynolds stress terms is derived from the Prandtl (1925) momentum transfer or mixing length theory. Most commonly the theory is applied to two-dimensional turbulence with mean flow in one direction. The theory is based upon the assumption that a "lump" of fluid moving from some position, y_1, to a new position y_2, carries with it the mean x-momentum of the lump at level y_1. A simple equation for the instantaneous excess of x-momentum at y_2 for unidirectional flow in the x-direction over a flat plate may be written as

$$\rho u' = -(y_2 - y_1)\frac{d}{dy}(\rho\bar{u}) \tag{3.41}$$

Here y is the direction normal to the flow boundary. Multiplying Equation 3.41 by v', and averaging over a short period of time we can derive the expression

$$-\overline{\rho u'v'} = \overline{v'(y_2 - y_1)}\frac{d}{dy}(\rho\bar{u}) \tag{3.42}$$

for one of the components of the Reynolds stress. At this stage an additional assumption is made that there is complete correlation between the components u' and v', i.e.,

$$\overline{u'^2} = \overline{v'^2} = -\overline{u'v'} \tag{3.43}$$

Then from Equation 3.41,

$$\overline{u'^2} = \overline{(y_2 - y_1)^2}\left(\frac{d\bar{u}}{dy}\right)^2 \tag{3.44}$$

Equating Equations 3.43 and 3.44 and multiplying by ρ, we obtain

$$-\overline{\rho u'v'} = \rho\overline{(y_2 - y_1)^2}\left(\frac{d\bar{u}}{dy}\right)^2 \tag{3.45}$$

Substitution of Equation 3.45 into Equation 3.38 yields

$$\tau_{yx} = -\frac{1}{g_c}\left[\mu\frac{d\bar{u}}{dy} + \rho l^2\left(\frac{d\bar{u}}{dy}\right)^2\right] \tag{3.46}$$

where l is the Prandtl mixing length defined by

$$l^2 = \overline{(y_2 - y_1)^2}$$

The term l may be visualized as an effective distance over which a "lump" of fluid loses its identity and becomes mixed with the surrounding fluid—hence the term *mixing length*. In Equation 3.45 the sign of $\overline{u'v'}$ depends upon the sign of $d\bar{u}/dy$ and Equation 3.46 is usually written as

$$\tau_{yx} = -\frac{1}{g_c}\left[\mu\frac{d\bar{u}}{dy} + \rho l^2 \left|\frac{d\bar{u}}{dy}\right|\left(\frac{d\bar{u}}{dy}\right)\right] \tag{3.47}$$

The quantity $l^2 \mid d\bar{u}/dy \mid$ has the units of kinematic viscosity and is often referred to as the eddy kinematic viscosity, ν_e.

The foregoing analysis still does not give us a method of evaluating the turbulent stress terms. In order to do this, we must have a method of evaluating the mixing length, l.

The simplest procedure for this is based upon the assumption that l is directly proportional to the distance, y, from the wall, i.e.,

$$l = \mathscr{K} y \tag{3.48}$$

where $\mathscr{K}$ is a universal constant with a value of approximately 0.4. Other empirical relations (von Kármán, 1930; Van Driest, 1956; Gill and Scher, 1961) have also been proposed for relating l to the geometric, velocity, or velocity gradient factors characterizing the flow, but surprising success has been obtained using the simple expression of Equation 3.48.

With the aid of Equations 3.47 and 3.48 we may, in principle, integrate Equation 3.36 to obtain the time-average velocity profile and the energy dissipation associated with turbulent flow. We will apply this to the case of pipe flow of Newtonian fluids, in Chapter 4, and of non-Newtonian fluids, in Chapter 5. We will also find this approach useful in dealing with certain two-phase flow problems.

Bibliography

Batchelor, G. K., *The Theory of Homogeneous Turbulence*, Cambridge University Press, New York, 1953.

Bird, R. B., W. E. Stewart, and E. N. Lightfoot, *Transport Phenomena*, John Wiley & Sons, Inc., New York, 1960.

Gill, W. N., and M. Scher, *A.I.Ch.E. Journal*, 7, 61 (1961).

Harlow, F. H., and P. I. Nakayama, *Physics of Fluids*, **10**, 2323 (1967).

Hinze, J. O., *Turbulence*, McGraw-Hill Book Co., New York, 1959.

Knudsen, J. G., and D. L. Katz, *Fluid Dynamics and Heat Transfer*, McGraw-Hill Book Co., New York, 1958.

Ladyzhenskaya, O. A., *The Mathematical Theory of Viscous Incompressible Flow*, Gordon and Breach Science Publishers, New York, 1963.

Prandtl, L., *Z. angew. Math. u. Mech.*, **5**, 136 (1925).

Townsend, A. A., *The Structure of Turbulent Shear Flow*, Cambridge University Press, New York, 1956.

Van Driest, E. R., *J. Aeronaut Sci.*, **23**, 1007 (1956).

von Kármán, T., *Nachr. Ges. Wiss. Göttingen, Math.-physik. Klasse* (1930); also *NACA* TM 611 (1931).

The Flow of Newtonian Fluids in Pipes

4

1 INTRODUCTION

In this chapter we review the application of the fundamental concepts to the pipe flow of Newtonian fluids. We will see that the basic equations discussed earlier lead to useful results in certain cases, notably laminar, steady-state flow, but we will find that generalizations which have resulted from the correlation of much experimental data are necessary in the solution of many problems.

The equations we develop are directly applicable to the pipe flow of Newtonian fluids, and they form the necessary background for our later study of non-Newtonian fluids. We deal only with the case of isothermal, steady-state flow. Flow under conditions of varying temperature may be handled either by assuming appropriate average fluid properties, or, in principle, by expressing the density and viscosity of the fluid as functions of temperature, and integrating the relations. Unsteady or transient flow is beyond the scope of this treatment.

2 CRITERIA FOR LAMINAR, TRANSITIONAL, OR TURBULENT FLOW

We have already referred to the important observation that fluids may flow in laminar or turbulent motion and that the Reynolds number is the general criterion of the mode of flow. In the case of flow in a pipe, as one would anticipate from boundary layer consideration, the flow pattern is not fully established until some distance downstream from the entrance. Moreover under certain circumstances the stability of laminar flow is dependent upon the flow conditions upstream of the pipe entrance and the nature of the entrance (whether square or rounded). We are, therefore, concerned with the following questions:

a. Under what conditions, sufficiently remote from any kind of entrance, and regardless of the type of flow upstream of the entrance, will the flow be laminar or turbulent?
b. What is the distance from the entrance required for the development of stabilized laminar or turbulent flow? (Presumably this will depend upon the kind of entrance and the type of flow upstream of the entrance.)

Few of these matters are predictable from theory and we rely mainly on the results of observations.

2.1 Criteria for Laminar and Turbulent Flow Remote from the Entrance

The generally accepted criterion for the end of stable laminar flow and the beginning of turbulent flow in a pipe is when the bulk Reynolds number, $\mathrm{Re} = DV\rho/\mu$, equals 2100. Notwithstanding this, laminar flow may be observed at Reynolds numbers as high as 50,000 under special conditions where the pipe entrance is rounded and the system is free of vibrations (Schlichting, 1968).

On the other hand, deviations from true laminar flow (even under stabilized conditions) are observed at Re values as low as 1200 (Sencal and Rothfus, 1953). These take the form of the development of a sinuous or wavy core which is first evident, threadlike, around $\mathrm{Re} = 1225$ and which increases in thickness and waviness with increasing Reynolds numbers. As the core grows within the range $1500 < \mathrm{Re} < 2100$, a stable spiral motion develops within it and this motion eventually results in the generation of eddies. At $\mathrm{Re} > 2100$ the eddies are generated with increasing frequency until at $\mathrm{Re} >$ about 3000 the motion is entirely turbulent except in the region immediately adjacent the wall. Therefore, the overall transition from laminar to turbulent flow in a pipe, well downstream from the entrance, occurs over the bulk Reynolds number range from 1225 to 3000, the major change occurring at $\mathrm{Re} = 2100$. In the bulk Reynolds number range above 1225, the velocity of the still truly laminar film immediately adjacent the wavy or spiraling core is given by the "wave point" Reynolds number, $Du_f\rho/\mu = 2450$, where u_f is the local velocity in the truly laminar layer adjacent the wavy core at radius r_f.

We will see later (Section 3.1) that the local velocity at any point within the truly laminar region between the wall and the core is given by

$$u = -\frac{g_c}{4\mu}(R^2 - r^2)\frac{dP_f}{dx} \tag{4.1}$$

where $-dP_f/dx$ = pressure loss caused by friction,
u = point velocity in the x-direction,
$R = D/2$ = radius of the pipe.

By definition of the friction factor, f,

$$-\frac{dP_f}{dx} = \frac{2f\rho V^2}{g_c D} \tag{4.2}$$

where V is the average velocity in the x-direction. This relationship is discussed in more detail in Section 3.2 of this chapter. Combining Equations 4.1 and 4.2, we obtain

$$u = \frac{f}{8}\,\text{Re}\,V\left[1 - \left(\frac{r}{R}\right)^2\right] \tag{4.3}$$

Multiplying Equation 4.3 by $D\rho/\mu$ and evaluating it at radius $r = r_f$, where $u_f = 2450\mu/D\rho$, it is found that

$$1 - \left(\frac{r_f}{R}\right)^2 = \frac{19{,}600}{(\text{Re}\sqrt{f})^2} \tag{4.4}$$

This relation has been found (Prengle and Rothfus, 1955) to be applicable to the prediction of the thickness of the truly laminar layer adjacent the wall to values of Re as high as 25,000.

The values of bulk and wave-point Reynolds number just discussed prevail where no special precautions are taken to preserve the laminar flow. They represent the stable conditions in a pipe, well downstream of its entrance when the entering fluid is initially in a high level of turbulence, and normal (ordinarily rather high) levels of vibration.

Ryan and Johnson (1959) have made a stability analysis of laminar flow based upon the assumption of small perturbations being applied to the equation of motion. Their analysis is not restricted to Newtonian fluids and, in fact, is most useful when applied to non-Newtonian fluids, as we will see in Chapter 5. The Ryan and Johnson stability parameter is

$$Z = -\frac{Ru\rho}{\tau_w}\left(\frac{\partial u}{\partial r}\right) \tag{4.5}$$

The quantity Z has a value of zero at the pipe wall and at the pipe center. Its maximum value occurs at a radial position dependent upon the laminar velocity profile. For a Newtonian fluid, utilizing Equation 4.19, it may be shown that

$$Z_{max} = \mathrm{Re}\sqrt{\frac{4}{27}} \qquad \text{at } \frac{r}{R} = \frac{1}{\sqrt{3}} \tag{4.6}$$

where Z_{max} is the maximum value of Z. From this, and since the critical value of the Reynolds number, Re_c, is 2100 for Newtonian fluids, Ryan and Johnson deduce that

$$(Z_{max})_c = 2100\sqrt{\frac{4}{27}} = 808$$

is a general criterion for the onset of turbulence in pipe flow. Furthermore, Ryan and Johnson's analysis indicates that turbulence originates at a radius $R/\sqrt{3}$ rather than at the axis.

Hanks (1963, 1969) has taken a different approach to the laminar stability problem and proposes a generalized stability parameter applicable to laminar flow in any geometry. He defines a factor K as

$$K = \frac{|\rho \mathbf{v} \times \xi|}{g_c |\nabla \cdot \tau|} \tag{4.7}$$

being the ratio of the magnitude of the acceleration force term to the viscous force term in the fundamental equation of motion. The vector ξ is the vorticity ($\xi = \nabla \times \mathbf{v}$). The factor K, like Ryan and Johnson's Z, varies from zero at the wall, through a maximum, to zero at a line of symmetry in the velocity field. For steady-state rectilinear flow, the expression for K simplifies to

$$K = \frac{1}{2}\rho \frac{|\nabla(\mathbf{v} \cdot \mathbf{v})|}{|\rho \mathbf{g} - g_c \nabla P)|} \tag{4.8}$$

Hanks shows that for pipe flow of a Newtonian fluid

$$K_{max} = \mathrm{Re}\sqrt{\frac{1}{27}} \tag{4.9}$$

and therefore reduces to the Ryan and Johnson criterion with

$$(K_{max})_c = \frac{(Z_{max})_c}{2} = 404 \tag{4.10}$$

where K_{max} is the maximum value of K, and $(K_{max})_c$ is the value of K_{max} at the transition from laminar to turbulent flow.

2.2 Distance from the Entrance Required for the Development of Stabilized Flow

As we would expect, the distance from an entrance required for the development of stabilized flow has been found to depend upon the nature of the motion upstream of the entrance, the nature of the entrance, and whether the ultimate stabilized flow is laminar or turbulent. We will consider first the situation where the ultimate flow is laminar. Here the boundary layer, which grows from the wall with the distance from the entrance, remains entirely laminar. The final velocity profile is established when the boundary layer extends to the center of the pipe. Several investigators have attempted, by theoretical studies (Schiller, 1922; Prandtl and Tietjens, 1931; Langhaar, 1942; McComas, 1967), and by some experimental work (McComas and Eckert, 1965), to determine the travel distance, X_E, required for the full establishment of the laminar velocity profile following the entrance of a fluid to a pipe. Collins and Schowalter (1963) and Astarita, Greco, and Peluso (1968) have studied the development of the velocity profile for non-Newtonian fluids, and their results may also be applied to Newtonian fluids. These are discussed in Section 2.7 of Chapter 5.

The results of all these investigations may be expressed in the form

$$\frac{X_E}{D} = \phi\,(\text{Re}) \tag{4.11}$$

The theoretical analysis of McComas (1967) involves a minimum of serious assumptions and is most nearly consistent with the experimental observations. Assuming viscosity effects negligible in the axial fluid filament from the entrance to the point of full development of the laminar velocity profile, McComas derives the relation

$$\frac{X_E}{D\,\text{Re}} = \frac{3 - K}{C} \tag{4.12}$$

where $C = 4\,\text{Re} f$, and $K = 1.3333$ = overall velocity development coefficient, related to the ratio of the fully developed local velocity to the average velocity. K is determined by integration, over the area of the pipe, of the difference between the cube and the square of such ratios from values of the local velocity given by Equation 4.1. Equation 4.12 reduces to

$$\frac{X_E}{D} = 0.0260\ \text{Re} \tag{4.13}$$

Experimental data (McComas and Eckert, 1965, cf. McComas, 1967) suggest the value of the coefficient ranging from 0.030 to 0.035. For practical purposes a constant of 0.028 is recommended. It is interesting to note that this equation indicates full development of the velocity profile at $X_E = 2.6\ D$ for Re = 100, $X_E = 52D$ for Re = 2000.

When the ultimate flow pattern is turbulent, no theoretical relations are available and the travel distance required for stabilization is particularly sensitive to the nature of the entrance. Knudsen and Katz (1958) quote experimental work which indicates that the distance required to achieve a fully developed profile is

$$\frac{X_E}{D} \geq 50 \tag{4.14}$$

Limited experimental data reported by Schlichting (1968) indicate that very large entrance lengths are required for the development of stabilized flow in the region 2100 < Re < approximately 3500. This is indicated qualitatively in Figure 4.1. Probably both the Reynolds number range over which the effect is observed and the entrance length itself are affected by the upstream flow details.

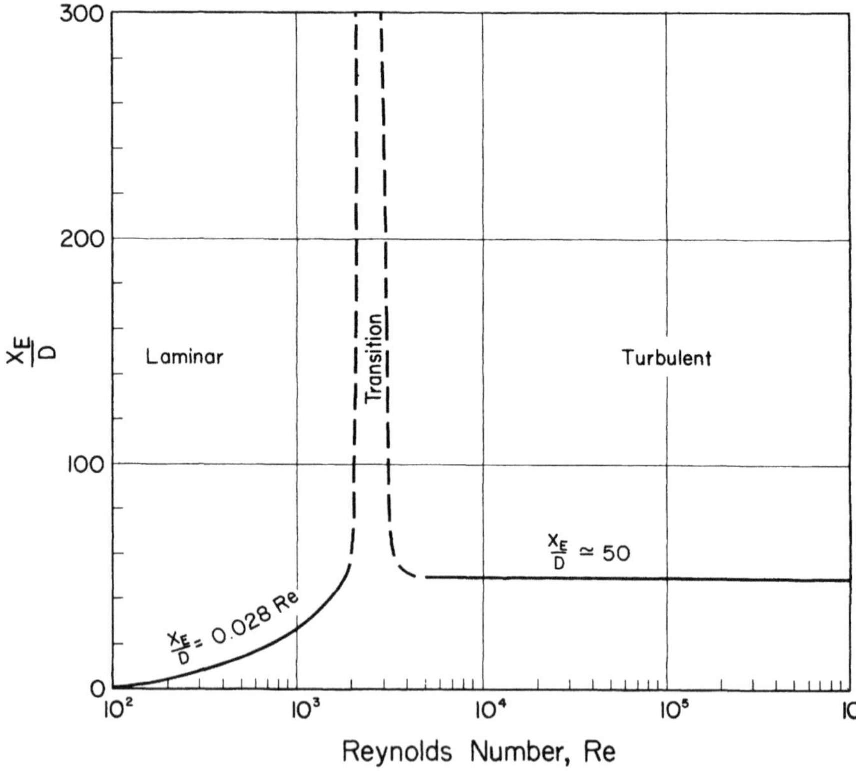

Figure 4.1. Laminar, turbulent, and transition entry lengths for Newtonian fluids. (From Seyer, F. A., Private Communication, 1970.)

3 STEADY-STATE LAMINAR FLOW

We turn now to a detailed consideration of laminar flow in a pipe. We restrict ourselves to isothermal steady flow and deal first with the fully developed flow which occurs beyond X_E. The applicable fundamental relations are those discussed in Chapter 3 and specifically the equation of state, the constitutive equation, the continuity equation, the momentum equation, and the mechanical energy equation.

3.1 The Hagen–Poiseuille Relationships

It is convenient to consider first the flow through a horizontal pipe of constant cross section and subsequently to generalize from this. We assume an incompressible Newtonian fluid in steady-state flow. The equation of state, the continuity equation, and the momentum equation are

$$\rho = \text{constant} \tag{4.15}$$

$$\frac{\partial u}{\partial x} = 0 \tag{4.16}$$

$$\frac{dP_f}{dx} = \frac{\mu}{g_c}\left(\frac{1}{r}\frac{\partial u}{\partial r} + \frac{\partial^2 u}{\partial r^2} + \frac{1}{r^2}\frac{\partial^2 u}{\partial \theta^2}\right) \tag{4.17}$$

The constitutive equation has been incorporated into Equation 4.17. In Equation 4.17, dP_f/dx is the pressure gradient caused by viscous shear within the fluid, and between the fluid and the wall, i.e., by fluid friction.

The flow is symmetrical about the axis of the pipe so that $\partial^2 u/\partial\theta^2 = 0$ and Equation 4.17 becomes

$$\frac{g_c}{\rho}\frac{dP_f}{dx} = \frac{\mu}{\rho r}\frac{\partial}{\partial r}\left(r\frac{\partial u}{\partial r}\right) \tag{4.18}$$

Under steady flow conditions, dP_f/dx is constant. The boundary conditions are

$$r = 0, \quad \frac{\partial u}{\partial r} = 0; \qquad r = R, \quad u = 0$$

The first integration gives the relationship for the velocity gradient

$$\frac{\partial u}{\partial r} = -\frac{g_c r}{2\mu}\frac{dP_f}{dx} \tag{4.19}$$

A second integration yields Equation 4.1, and a third integration results in

$$Q = -\frac{\pi R^4 g_c dP_f}{8\mu \; dx} \tag{4.20}$$

or

$$V = -\frac{R^2 g_c}{8\mu}\frac{dP_f}{dx} = -\frac{D^2 g_c}{32\mu}\frac{dP_f}{dx} \tag{4.21}$$

Equations 4.20 and 4.21 are alternate forms of the Hagen-Poiseuille equation for laminar flow in a pipe.

Equations 4.19, 4.1, and 4.20 give the effect of pressure gradient due to friction, fluid properties and geometry upon the velocity gradient, the local velocity, and the discharge. From these relations by rearrangement and/or solution for values of $r = 0$ and $r = R$, we may obtain all of the various relationships of practical interest.

These theoretical relationships have been extensively verified by experimental measurements and may be considered entirely valid for the conditions assumed, i.e., for isothermal, steady-state, stabilized flow of an incompressible Newtonian fluid. Of special interest is the confirmed observation that the laminar velocity profiles and the pressure gradient are unaffected by wall roughness. Precisely the same equations are applicable to both smooth and rough pipes.

For compressible fluids, analogous, but somewhat more complex equations, may be derived by incorporation of the appropriate equation of state. We will not deal specifically with the laminar flow of compressible fluids here since only rarely is the pipe flow of gases or vapors in laminar motion. However, the friction factor approach discussed in Section 4 of this chapter in connection with turbulent flow is equally applicable to laminar flow, provided the appropriate friction factor is employed.

3.2 The Friction Factor Concept

The equations discussed so far in this section are sufficient for the determination of all the quantities of interest in the laminar pipe flow of a Newtonian fluid. However, it is instructive to express the relation between V and dP_f/dx in another way—a way we are obliged to use when dealing with turbulent flow. First, we may observe that the ratio of the wall shear stress, τ_w, to the kinetic energy per unit volume of fluid, $\rho V^2/2g_c$, is a dimensionless quantity indicative of the relative significance of the wall shear stress. We call this dimensionless ratio the *friction factor* i.e.,

$$f = \frac{2\tau_w g_c}{\rho V^2} \tag{4.22}$$

or, since from a simple force balance

$$\tau_w = -\frac{D}{4}\frac{dP_f}{dx} \tag{4.23}$$

f may be defined by

$$-\frac{dP_f}{dx} = \frac{2f\rho V^2}{g_c D} \tag{4.24}$$

This is the well known Fanning equation. Other similar relations have been used to define other friction factors. Commonly used in civil engineering work, and often confused with f defined by Equation 4.24, is f_H defined by the Darcy–Weisbach equation:

$$-\frac{dP_f}{dx} = \frac{f_H \rho V^2}{2g_c D} \tag{4.25}$$

We will use the friction factor f (Equation 4.24) throughout this work. It is important to note that f is defined by these equations so that there can be no question about their validity. The question is: What is the value of the friction factor, so defined?

Clearly, in laminar flow the value of f must be such that Equation 4.21 and 4.24 yield the same answer for the pressure gradient. Equating these, and solving for f, we find that

$$f = \frac{16}{\text{Re}} \tag{4.26}$$

This expression, together with Equation 4.24 defining f, is the mathematical equivalent of Equation 4.20 or 4.21.

3.3 Entrance Effects

Equations 4.1, 4.3, 4.19 and 4.20 do not apply in the entrance region of the pipe, since the velocity profile is not stabilized in this region. Work of Langhaar (1942), Sparrow, Lin, and Lundgren (1964), and others leads to the counterpart of Equation 4.3 applicable in the entrance region, $X < X_E$. Langhaar's equation is

$$\frac{u}{V} = \frac{I_0[\phi(X)] - I_0[\phi(X)r/R]}{I_2[\phi(X)]} \tag{4.27}$$

where X is the distance downstream from the entrance, and I_0 and I_2 are modified Bessel functions of the first kind. The quantity $\phi(X)$ is given as a function of X/D Re in Figure 4.2 plotted from the results of Langhaar (1942).

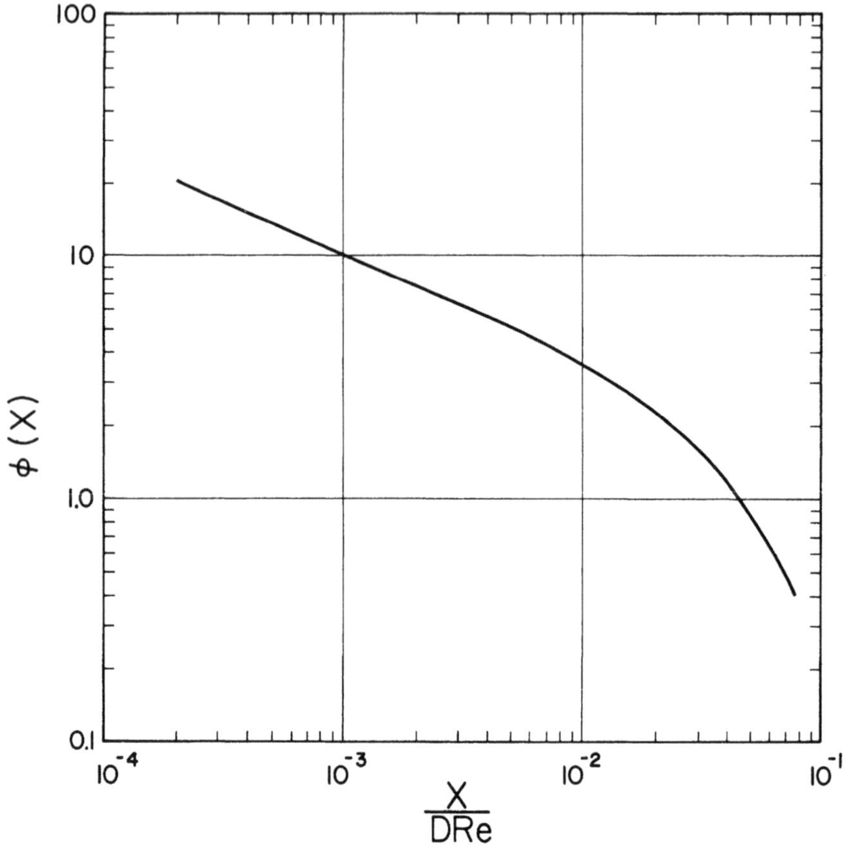

Figure 4.2. The function $\phi(X)$. (From Knudsen, J. G., and D. L. Katz, *Fluid Dynamics and Heat Transfer*, McGraw-Hill, New York, 1958. Used with permission of McGraw-Hill Book Co.)

The alternate relationship to Equation 4.20 for calculating the flow rate from the pressure gradient, or vice versa, is given by Langhaar in the friction factor form, analogous to Equation 4.26. Langhaar defines f_{av} as the average friction factor applicable over the length X. The result of his analysis is presented in Figure 4.3 along with results of experiments by Kline and Shapiro (1953) correlated empirically by the equation

$$f_{av}\frac{X}{D} = 3.435\sqrt{\frac{X}{D\,\mathrm{Re}}} \tag{4.28}$$

over the range

$$10^{-5} < \frac{X}{D\,\mathrm{Re}} < 10^{-3}$$

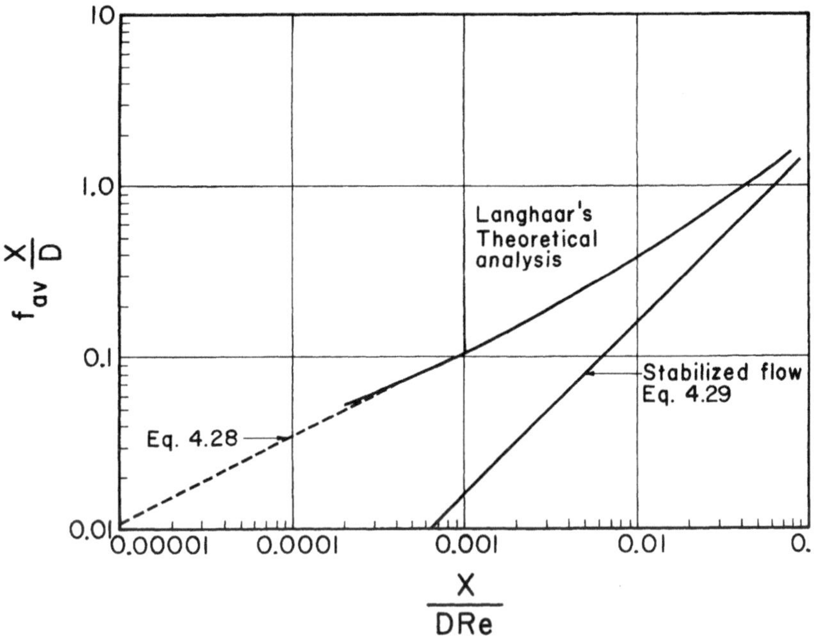

Figure 4.3. Average friction factors for laminar flow in the entrance region of a circular pipe. (From Knudsen, J. G., and D. L. Katz, *Fluid Dynamics* and *Heat Transfer*, McGraw-Hill, New York, 1958. Used with permission of McGraw-Hill Book Co.)

Langhaar's theoretical analysis is reasonably well confirmed, and its range extended, by Kline and Shapiro's data. For comparison purposes the equation for f for stabilized flow is also shown in Figure 4.3 in the form

$$f\frac{X}{D}=\frac{16}{\mathrm{Re}}\frac{X}{D} \tag{4.29}$$

Both the higher friction factor in the entrance region and its asymptotic approach with increasing X/D to the limiting value given by Equation 4.29 are apparent.

Sparrow *et al.* (1964), making somewhat different assumptions in their theoretical analysis, obtain a solution that appears to match the experimental data of Pfenninger (1952) and Reshotko (1958) even better than Langhaar's solution. Their solution is in the form

$$\frac{u}{V}=\phi\left(\frac{r}{R},\frac{4X^*}{D\,\mathrm{Re}}\right)$$

where X^* is a "stretched" axial coordinate related to X, the true physical distance from the entrance. Figures 4.4 and 4.5 modified after Sparrow *et al.*

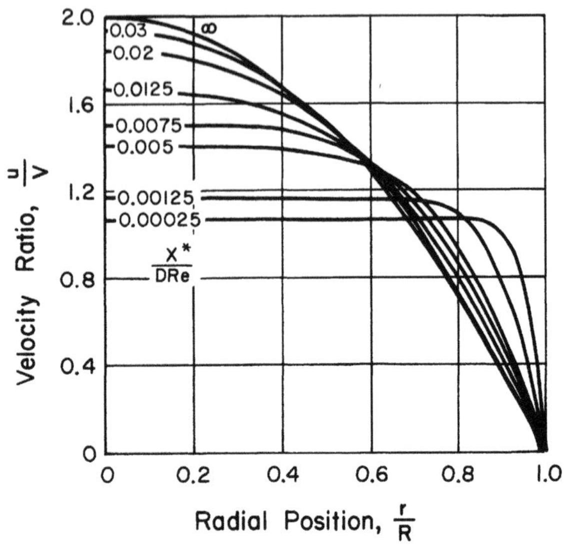

Figure 4.4. Laminar velocity profiles during stabilization. (From Sparrow, E. M., S. H. Lin, and T. S. Lundgren, *Physics of Fluids*, **7**, 338, 1964.)

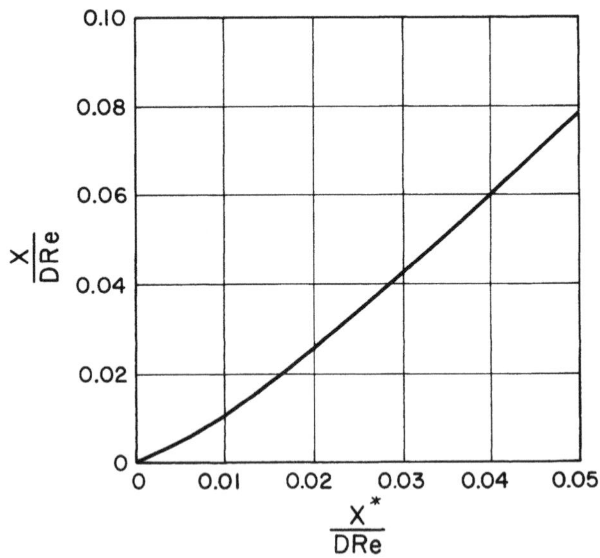

Figure 4.5. Relationship between X and the stretched coordinate, X^*. (From Sparrow, E. M., S. H. Lin, and T. S. Lundgren, *Physics of Fluids*, **7**, 338, 1964.)

give their solution. Figure 4.4 also gives a good pictorial representation of the transformation of the initial blunt profile to the parabolic profile.

It is interesting to compare the prediction of Figure 4.4 with that of McComas (1967) for the length of the entrance section. If we define the entrance length as that required for the attainment of an axial velocity u_0 of some arbitrary high fraction of the fully stabilized axial velocity u_{max}, we may develop Figure 4.6 from Figures 4.4 and 4.5. This shows how sensitive the ratio $X_E/D\,\text{Re}$ is to the stabilization criterion employed. McComas' value of $X_E/D\,\text{Re} = 0.026$, and the experimental evidence in the range of $X_E/D\,\text{Re} = 0.03$ to 0.035 correspond with 92 to 95 percent axial stabilization in accordance with Sparrow *et al.*

Sparrow *et al.* extend their analysis to include the pressure drop in the entrance section. Their approach is to express the total pressure drop over the length $X < X_E$ as

$$\frac{2\Delta P g_c}{\rho V^2} = 16\left(\frac{4X}{D\,\text{Re}}\right) + K(X) \tag{4.30}$$

where $K(X)$ is a correction term to account for the effects of flow development, and given as a function of $X/D\,\text{Re}$ in Figure 4.7. The prediction of Langhaar may also be expressed in terms of $K(X)$ and compared with the analysis of Sparrow *et al.* This has been done in Figure 4.7, which shows that the two analyses are in quite good argeement. (This is not surprising since the analysis of Sparrow *et al.* is essentially a refinement of that of Langhaar.)

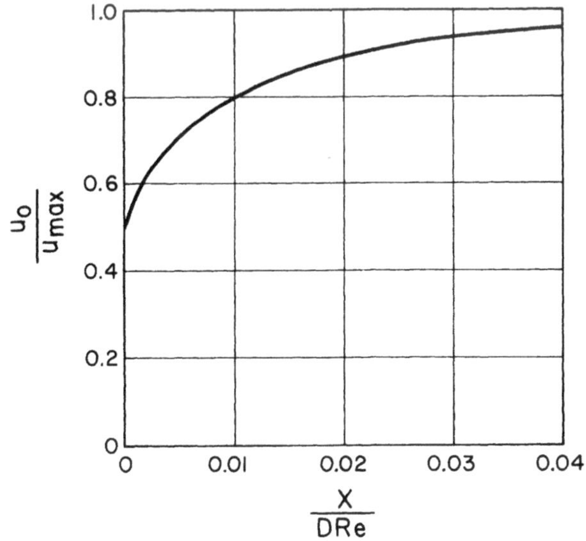

Figure 4.6. Criterion of stabilization.

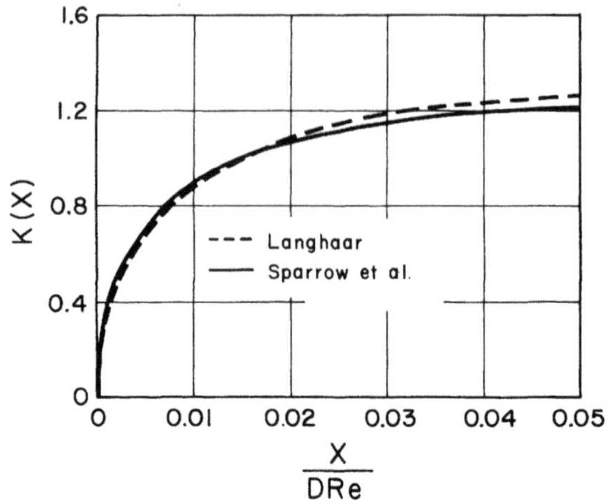

Figure 4.7. Correction term, $K(X)$, to account for the effects of flow development. (From Sparrow, E. M., S. H. Lin, and T. S. Lundgren, *Physics of Fluids*, **7**, 338, 1964.)

The entrance region analyses of Collins and Schowalter (1963) for power law pseudoplastic fluids and of Michiyoshi *et al.* (1966) for Bingham fluids, discussed in Section 2.7 of Chapter 5, are also applicable to Newtonian fluids and should be consulted.

Schmidt and Zeldin (1969) summarize the various techniques that have been used for the prediction of the excess pressure drop in the entrance section and present the results of their own analysis based upon a finite difference solution of the Navier-Stokes equation. Comparing the results of the several analyses, they show that the theoretical predictions of the asymptotic values of K range from 1.08 to 1.41. Their own analysis indicates K to range from 1.314 at Re = 10,000 to 1.397 at Re = 100. They report experimentally determined values from several sources ranging from 1.00 to 1.45.

We conclude that the analyses of Langhaar and of Sparrow *et al.*, yielding K values in the range of 1.24 to 1.28, are suitable for practical application.

3.4 General Form of the Pressure Drop Equation

We may now remove the restriction of a horizontal, constant diameter pipe and consider the more general case of a pipe system in which there may be changes in elevation, sections of different diameters, pipe fittings, and devices for change of direction and flow regulation. The mechanical energy balance (Equation 3.29) is applicable with

$$-\frac{dP_f}{\rho} = \frac{2fV^2}{g_c D}\,dL = \frac{32\mu V}{\rho g_c D^2}\,dL$$

For the normal case, in laminar flow, of an incompressible fluid we may integrate over the length of pipe, L, in any constant diameter section to obtain

$$\Delta P = \frac{g}{g_c}\Delta z\rho + \frac{\rho}{2g_c\alpha_l}(V_2^2 - V_1^2) + \frac{2f\rho V^2 L_E}{g_c D} \tag{4.31}$$

in terms of the friction factor, or

$$\Delta P = \frac{g}{g_c}\Delta z\rho + \frac{\rho}{2g_c\alpha_l}(V_2^2 - V_1^2) + \frac{32\mu V L_E}{g_c D^2} \tag{4.31a}$$

in terms of the Hagen–Poiseuille equation. In these equations $\alpha_l = 0.50$ is the velocity profile correction factor in the kinetic energy terms, and $L_E = L + L_f$, where L_f is the equivalent length under laminar flow conditions of any fittings, valves, or devices in the pipe section.

Equivalent lengths of fittings, etc., for laminar flow are not well known but may be approximated from the equivalent lengths for turbulent flow (see Section 4.6, Table 4.1) through the relationship (Crane Company, 1957)

$$\left(\frac{L_f}{D}\right)_l = \frac{\text{Re}}{1000}\left(\frac{L_f}{D}\right)_t \tag{4.32}$$

applicable to conditions where Re < 1000; where Re > 1000 the equivalent lengths are assumed the same as for turbulent flow.

Often with incompressible fluids the kinetic energy term is zero or negligible.

Laminar flow of compressible fluids is seldom encountered in practice but where it is, pipe flow problems may be handled by the methods discussed in Section 4.6 substituting α_l where appropriate and using a friction factor appropriate for laminar flow.

4 STEADY-STATE TURBULENT FLOW

Our knowledge of turbulent flow in a pipe, even under stabilized steady-state conditions, is based largely upon the accumulated results of experimental observations of the time-average velocity profile and, particularly, of the pressure gradient. It is well established that both the velocity profile and the pressure gradient are extremely sensitive to the condition of the pipe wall, whether smooth or rough, and if rough, to the nature of the roughness.

In our treatment of turbulent flow we will deal first with flow in a smooth wall pipe, then consider the effect of roughness, and finally deal with fully rough wall turbulence and partially rough wall turbulence.

Data on the pressure gradient, and to a lesser extent, on the time-average velocity profile for turbulent flow in smooth tubes have been accumulated from the time of Reynolds (1883). The most noteworthy contributors of such data were probably Stanton (1911), Stanton and Pannell (1914), Nikuradse (1930, 1932), Deissler (1951), Senecal and Rothfus (1953), and Laufer (1953).

4.1 Smooth Wall (SW) Turbulent Flow

4.11 Empirical Friction Factor Relations

From the time of the earliest suggestion by Blasius in 1913, the pressure gradient data have been expressed in terms of a friction factor, as defined by Equation 4.24 (or, in some work, a factor four times as large) and correlated with the bulk Reynolds number. Over the years the evidence supporting the existence of a single line correlation has mounted; and empirical, graphical representations of data covering a limited range of variables have given way to well confirmed relations applicable to a wide or the full range of the variables. Here we will mention a few of the best such relations. The Blasius equation (Blasius, 1913)

$$f = 0.079\ \mathrm{Re}^{-0.25} \tag{4.33}$$

is simple and entirely empirical, but it fits the reliable data to a good degree of accuracy in the range $3000 < \mathrm{Re} < 100{,}000$. A number of equations of this type with different values of the constants are in use, such as the following (Knudsen and Katz, 1958):

$$f = 0.046\ \mathrm{Re}^{-0.2} \tag{4.34}$$

These equations are often referred to as Blasius-type equations.

For the full practical range of turbulent flow, say $3000 < \mathrm{Re} < 3{,}000{,}000$, the more complex equation due to Nikuradse (1932) provides an excellent fit to essentially all reliable data

$$\frac{1}{\sqrt{f}} = 4.0 \log (\mathrm{Re}\sqrt{f}) - 0.40 \tag{4.35}$$

This equation is based on a form suggested by von Kármán (1930), and discussed later, with the constants determined from Nikuradse's data.

An empirical equation which closely approximates the Nikuradse equation but has a simpler form and is explicit in f, is that of Drew, Koo, and McAdams (1932):

$$f = 0.00140 + 0.125\ \mathrm{Re}^{-0.32} \tag{4.36}$$

for a range of Reynolds number from 3000 to 3,000,000.

4.12 Velocity Profile Relations

Experimental data on the time-average velocity profile have also been correlated in a number of different ways. One of the earliest of these is the so-called power law velocity distribution proposed by Prandtl (1927a) and developed from the Blasius friction factor equation. The power law distribution equation is

$$\frac{u}{u_{\max}} = c\left(\frac{y}{R}\right)^{1/7} \tag{4.37}$$

where c is a proportionality factor close to unity, and y is the distance from the wall. This equation correlates velocity data reasonably well for $0.1 < y/R < 1.0$ and for $3000 < \mathrm{Re} < 100{,}000$. Actually, for a best fit of experimental data, the exponent on the right-hand side of the equation varies from 1/6 at $\mathrm{Re} = 4 \times 10^3$ to 1/10 at $\mathrm{Re} = 3.2 \times 10^6$ (Schlichting, 1968). The power law equation does not hold near the wall. Furthermore, the equation does not yield the required zero value of du/dr at the center line.

A number of semiempirical equations have been developed which extend the range of application and improve the accuracy of expression of the profile. In these expressions usually a dimensionless velocity,

$$u^+ = \frac{u}{U^*} \tag{4.38}$$

where $U^* = \sqrt{\tau_w g_c/\rho}$ is the friction velocity, and a dimensionless distance from the wall, or Reynolds number

$$y^+ = \frac{U^* \rho y}{\mu} \tag{4.39}$$

are related by an expression which is independent of the Reynolds number. In order to develop a complete profile three different regimes are considered separately. Experimental observations indicate that in the immediate neighborhood of the wall the laminar shearing stress predominates, i.e.,

$$\tau_w = -\frac{\mu}{g_c}\left(\frac{du}{dy}\right)_w = -\frac{\mu}{g_c}\frac{u}{y} \tag{4.40}$$

Substituting this value of τ_w into Equation 4.38, we obtain

$$\left(\frac{u}{U^*}\right)^2 = \frac{uy\rho}{\mu} = \frac{u}{U^*}\left(\frac{U^* y\rho}{\mu}\right)$$

or

$$u^+ = y^+ \tag{4.41}$$

Experimental data of Nikuradse (1932), Reichardt (1940), and Lindgren and Chao (1969) conform to Equation 4.41 for values of y^+ less than about 5. The data carefully taken and interpreted by Lindgren and Chao give especially convincing support. For this reason the so-called laminar sublayer on smooth pipe walls is taken to extend to $y^+ = 5$.

From the limit of the laminar sublayer to a value of y^+ of approximately 30, the relative importance of the laminar shearing stress diminishes and the role of the turbulent stress terms increases. The region extending from $y^+ = 5$ to $y^+ = 30$ is called the buffer or transition zone. Beyond this zone ($y^+ > 30$) the viscous shear terms play a relatively minor role which may be neglected. It is convenient to consider the velocity profile for the turbulent core before we deal with the transition zone.

One of the earliest and in many ways the most useful development of the velocity profile in the turbulent core is due to Prandtl (1921, 1925, 1927b, 1933) based on his concept of a mixing length. In this analysis the effect of fluid viscosity is neglected: the shear stress is assumed constant and equal to the value at the wall. Introducing these assumptions in Equation 3.47, we have

$$\sqrt{\frac{\tau_w g_c}{\rho}} = -\mathscr{K} y \frac{du}{dy} \tag{4.42}$$

Introducing U^* on the left-hand side of Equation 4.42 and rearranging, we obtain

$$du = \frac{U^*}{\mathscr{K}} \frac{dy}{y} \tag{4.43}$$

which yields, upon integration,

$$u = \frac{U^*}{\mathscr{K}} \ln y + C \tag{4.44}$$

This equation may also be expressed in dimensionless form as

$$u^+ = \frac{1}{\mathscr{K}} \ln y^+ + C_1 \tag{4.45}$$

Equation 4.45 provides a remarkably good fit to available data, even well away from the wall. Although based on some theoretical reasoning, it must be recognized as largely empirical because of the many assumptions, some of which are quite questionable. Analysis of Nikuradse's smooth pipe data shows the constant $\mathscr{K}$ to have a value of 0.40 and C_1 to be in the range of 5.5 to 5.6 (Ross, 1953). Taking the commonly used value of 5.5 for C_1, we have

$$u^+ = 5.5 + 2.5 \ln y^+ \tag{4.46}$$

or

$$u^+ = 5.5 + 5.756 \log y^+$$

This is a relationship which correlates turbulent velocity profile data satisfactorily over a range of $30 < y^+ < 10^5$. At values of $y^+ < 30$, i.e., near the wall, the data deviate substantially from the equation, reflecting the existence of a buffer zone and the laminar sublayer.

Vasilescu (1967) has approached the turbulent shear problem differently, avoiding the concept of a mixing length, by expressing τ_t, the contribution to shear due to turbulence, as

$$\tau_t = c\rho y u\left(\frac{du}{dy}\right) \tag{4.47}$$

where c is a constant evaluated from Nikuradse's data as 0.0219. From this relationship Vasilescu developed the velocity equation

$$u^+ = \sqrt{210 \log y^+ - 90} \tag{4.48}$$

for smooth wall turbulence. Vasilescu shows that this relation gives results closely similar to the Nikuradse equation (Equation 4.46).

Bogue and Metzner (1963) make reference to some of the inadequacies of Equation 4.45 or its equivalent, and refer to empirical corrections to it, proposed by Millikan (1939), Reichardt (1951), and Hinze (1959). On the basis of a comprehensive analysis of all available data they propose an empirically corrected equation in the form

$$u^+ - c(\zeta, f) = 5.57 \log y^+ + 5.57 \tag{4.49}$$

where

$$c(\zeta, f) = 0.05\sqrt{\frac{2}{f}}\exp\left[\frac{-(\zeta - 0.8)^2}{0.15}\right] \tag{4.50}$$

and

$$\zeta = y/R \tag{4.51}$$

The function, $c(\zeta, f)$, is given in graphical form in Figure 4.8, taken from Bogue and Metzner (1963).

The velocity profile correlations, Equations 4.46, 4.48, and 4.49, all apply in the turbulent core, whereas Equation 4.41 applies in the laminar sublayer. The correlation of velocity profile in the transition zone is more difficult since both the laminar and turbulent shear stress terms must be included in the analysis. The equation for the transition zone, in addition to representing the data in this zone, must also provide a smooth transition to the laminar

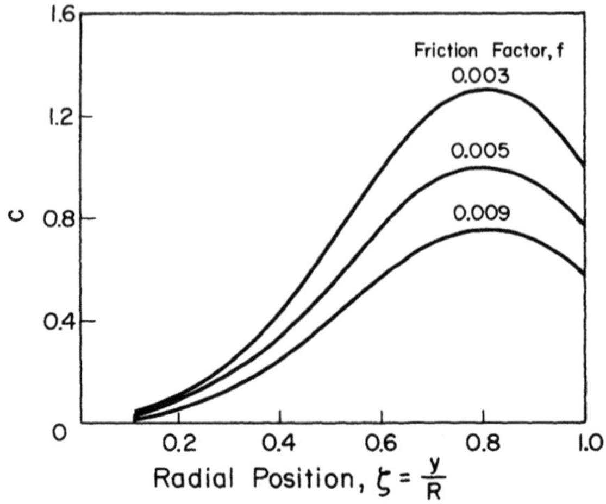

Figure 4.8. Empirical correction functions. (From Bogue, D. C., and A. B. Metzner, *Ind. Eng. Chem. Fundam*, **2**, 143, 1963.)

sublayer equation and the turbulent core. An empirical equation that gives a reasonable fit to experimental data is

$$u^+ = -3.05 + 11.513 \log y^+ \qquad (4.52)$$

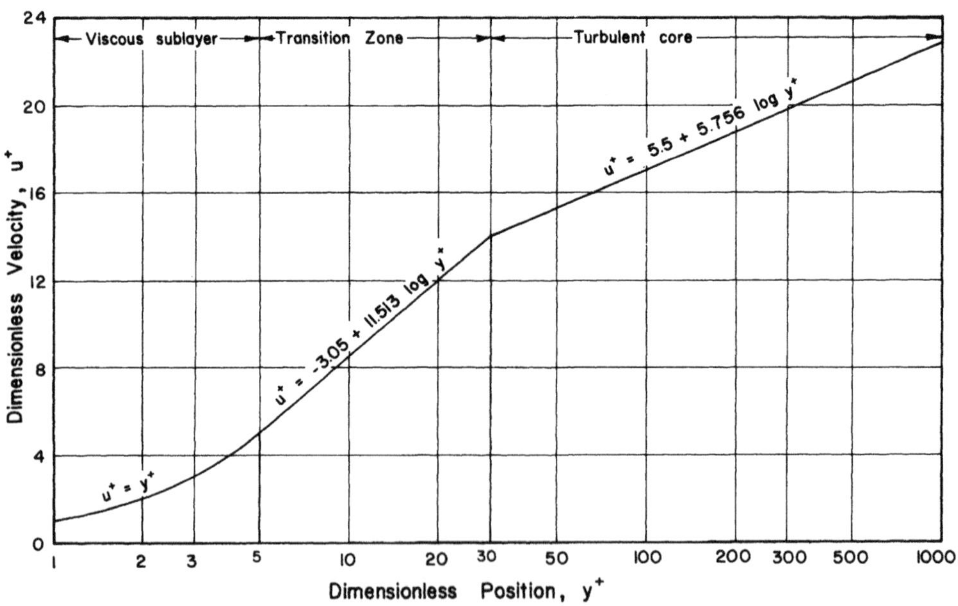

Figure 4.9. Universal Velocity Profile.

This equation represents a straight line on semilogarithmic coordinates joining the points $y^+ = 5$ and $y^+ = 30$ taken to represent the end of laminar sublayer and the start of the turbulent core, respectively.

Equations 4.41, 4.46, and 4.52 together constitute the "universal velocity profile." They are depicted graphically in Figure 4.9.

4.13 Friction Factor Relations Based on Velocity Profile Analysis

Von Kármán (1930), adopting the mixing length concept and the resulting velocity profile relationship of Prandtl, developed an equation for the friction factor in terms of a Reynolds number based upon the maximum velocity and the tube radius. He did not evaluate the numerical constants of the equation, but this was done by Nikuradse (to yield Equation 4.35) based on his own and other friction factor—average velocity data. Nikuradse changed von Kármán's form of the Reynolds number to the conventional form.

Rouse (1946) integrated the Prandtl velocity profile in a manner somewhat different from von Kármán. He substituted the point velocity, u as given by Equation 4.46, into the equation for the average velocity to obtain

$$V = \frac{Q}{A} = \frac{2\pi \int_0^R urdr}{\pi R^2} \tag{4.53}$$

This results in the following relationship for smooth pipes:

$$\frac{V}{U^*} = 5.75 \log \frac{U^* D\rho}{\mu} + 0.02 \tag{4.54}$$

where

$$U^* = \sqrt{\tau_w g_c/\rho} = V\sqrt{f/2} \tag{4.55}$$

Equation 4.54 may be rewritten in terms of f as

$$\frac{1}{\sqrt{f}} = 4.06 \log (\text{Re}\sqrt{f}) - 0.6 \tag{4.56}$$

The constants of Equation 4.56 are almost identical with those of the Nikuradse equation (Equation 4.35). This is rather remarkable in view of the many assumptions involved in Equation 4.46 and the fact that the constants in Equation 4.56 were determined from velocity profile data only. It suggests that the friction factor is not sensitive to minor variations in the velocity profile.

Van Driest (1956), Gill and Scher (1961), Hanks (1968) and Quarmby (1969) have made useful improvements on the universal velocity profile as developed by Prandtl. Van Driest introduced a "wall dampening term" into Prandtl's relationship between l and y such that

$$l = \mathscr{K} y\left[1 - \exp\left(-\frac{\Phi y}{R}\right)\right] \tag{4.57}$$

In this relation Φ is a constant related to the Reynolds number. He also retained the viscous shear term dropped by Prandtl. Gill and Scher (1961) redefined Van Driest's damping term and made allowance for the variations of shear stress with the radius. They expressed Φ as

$$\Phi = \frac{\mathrm{Re}^* - a}{b} \tag{4.58}$$

where $\mathrm{Re}^* = \dfrac{RU^*}{\mu}\rho$

a = constant = 60, selected so that Φ becomes zero at the laminar-turbulent transition, taken by Gill and Scher as Re = 1800 for which $\mathrm{Re}^* = 60$,

b = constant = 22, determined from experimental data of Nikuradse (1932) and Deissler (1950).

From Equations 3.35, 3.46, and 4.57 and substituting r^* for r/R, we may write

$$r^* - \frac{1}{\mathrm{Re}^*}\left(-\frac{du^+}{dr^*}\right) - \mathscr{K}^2(1 - r^*)^2\{1 - \exp[-\Phi(1 - r^*)]\}^2\left(-\frac{du^+}{dr^*}\right)^2 = 0 \tag{4.59}$$

Assuming that the constants $\mathscr{K}$, a, and b are determinable, the velocity profile is established solely through Re^*.

Hanks' treatment introduces modifications to that of Gill and Scher and provides a theoretical basis for one of the two arbitrary constants involved in the definition of Φ. We shall follow his approach. It may be shown through the definitions of U^* and f that

$$\mathrm{Re}^* = \frac{\sqrt{2}}{4} B, \qquad \text{where} \quad B = \mathrm{Re}\sqrt{f}$$

With this definitions of B, Φ may be expressed as $(B - B_c)/2b\sqrt{2}$, where $B_c = \mathrm{Re}_c\sqrt{f_c} = 2100\sqrt{16/2100} = 183.303$. Substituting these relations into Equations 4.58 and 4.59, we obtain

$$r^* - \frac{2\sqrt{2}}{B}\left(-\frac{du^+}{dr^*}\right) - \mathscr{K}^2(1 - r^*)^2\{1 - \exp[-\Phi(1 - r^*)]\}^2\left(-\frac{du^+}{dr^*}\right)^2 = 0 \tag{4.60}$$

Equation 4.60 may be solved for du^+/dr^* and integrated to obtain

$$u^+ = \frac{\sqrt{2}}{\mathscr{K}^2 B} \int_{r^*}^{1} \frac{\left(1 + \frac{\mathscr{K}^2}{2} B^2 r^* A\right)^{1/2} - 1}{A^2} dr^* \tag{4.61}$$

where

$$A = (1 - r^*)^2\{1 - \exp[-\Phi(1 - r^*)]\}^2$$

The average velocity is obtained as

$$V = 2U^* \int_0^1 r^* u^+ dr^* \tag{4.62}$$

From the solution of these equations and the definition of U^*, f, and B, Hanks has compared both the predicted velocity profile and the predicted friction factor with experimental data from the following sources: Stanton and Pannell (1914), Nikuradse (1932), Rothfus (1948), Senecal and Rothfus (1953), Shaver (1957), Walker (1957), Dodge (1958), Hanks (1960), and Wells *et al.* (1966). Hanks chose the same value as Gill and Scher for both constants, i.e., $\mathscr{K} = 0.36$ and $b = 22$ (which provided a good fit to the high Reynolds number friction factor data). His test of the velocity profile prediction is shown in Figures 4.10 and 4.11. From Figure 4.10 we note that the prediction of V/u_{max} is excellent through the region, $2100 < Re < 3000$, but somewhat high at $Re > 3000$, being about 7 percent high at $Re = 10^5$. These observations are confirmed in Figure 4.11, although even at $Re = 1.05 \times 10^5$ the predicted profile compares very favorably with the experimental observations except in the central core.

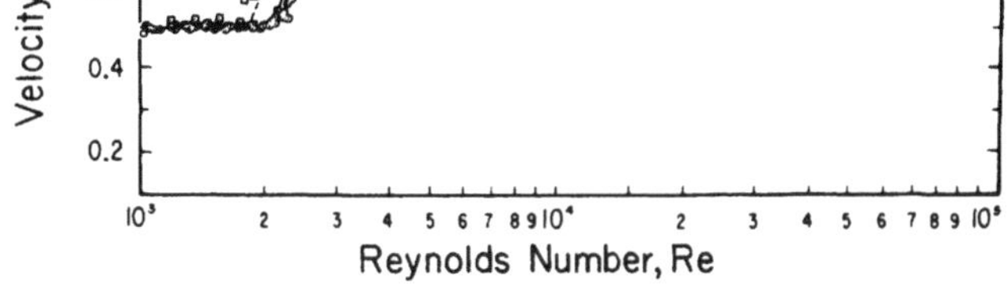

Figure 4.10. Velocity profile predictions. (From Hanks, R. W., *A.I.Ch.E. Journal*, **14**, 691, 1968.)

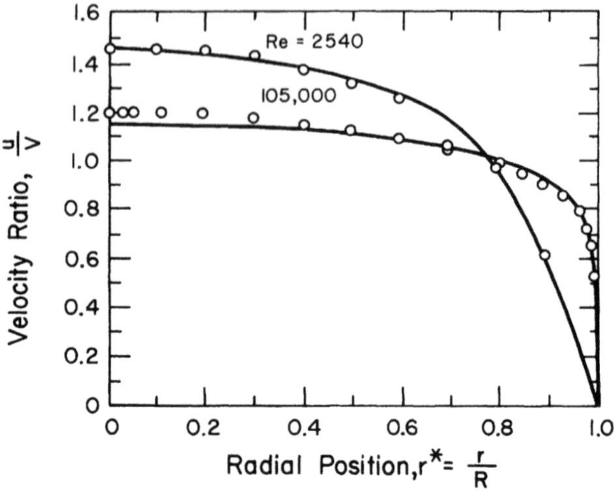

Figure 4.11. Velocity profiles. (From Hanks, R. W., *A.I.Ch.E. Journal*, **14**, 691, 1968.)

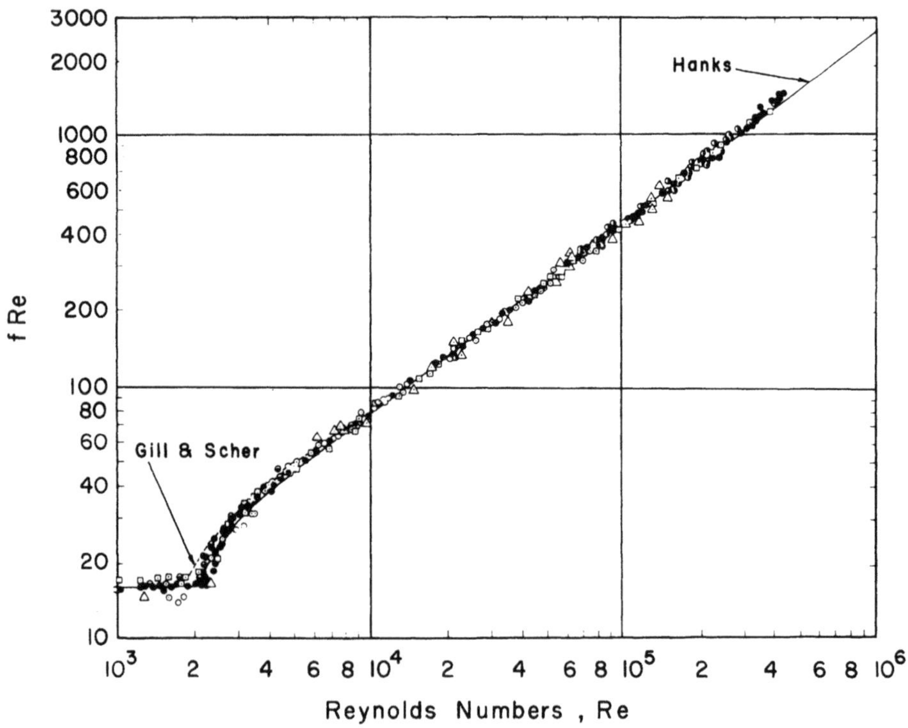

Figure 4.12. Friction factor relationship. (From Hanks, R. W., *A.I.Ch.E. Journal*, **14**, 691, 1968.)

Hanks' friction factor prediction is compared with the experimental data in Figure 4.12. Here the agreement is excellent for all values of $Re > 2100$, and is especially good within the transition range $2100 < Re < 3000$. The Nikuradse equation (Equation 4.35) essentially overlies the data and Hanks' equation for $Re > 4000$ but deviates significantly from the data at $Re < 4000$.

4.2 Wall Roughness and Its Effectiveness

In most practical applications the inside wall of a pipe is not physically smooth, certainly not in a microscopic sense; and in certain regions of turbulent flow the roughness of the wall may have a marked effect on the velocity profile and the pressure gradient due to friction.

Roughness may be attributable to the nature of the wall substance and the method of its manufacture as in the case of new steel, cast iron, and cement-asbestos pipe. Further, it may be influenced by erosion and corrosion. Such roughness is thought of as natural roughness. In addition, roughness may be artificially created, usually for experimental purposes, as by the attachment of sand grains or small channels to the wall surface.

A full description of roughness would require a complete definition of its geometry, including the height, length, width, and shape of any protrusions or indentations and their distribution. This is seldom possible, and it has become customary to think of roughness either as natural or artificial; to measure artificial roughness in terms of the mean height of the sand grain, or other protuberance; and finally to relate natural roughness empirically to the artificial roughness.

Additionally it is recognized from dimensional analysis considerations that the effect of roughness is not due to its absolute dimensions but rather to its dimensions relative to that of the pipe. Thus a relative roughness is defined as k/D, where k is the mean protruding height of relatively uniformly sized, uniformly distributed, tightly packed sand grains in the case of pipe artificially roughened in this manner. In case of naturally rough walls, k is the height of the uniform sand grains which would give the roughness effect observed.

The effectiveness of wall roughness in turbulent flow has been found to depend not only on the relative roughness, k/D, but also on the Reynolds number. This is attributed to the laminar sublayer which exists in contact with the wall. If the sublayer is sufficiently thick to cover the wall roughness (as at low Reynolds numbers), the roughness will not be effective; however, if the sublayer is thin compared with the roughness of the wall (as at high Reynolds numbers) the roughness will be effective.

The thickness of the laminar sublayer is determinable from the experimentally observed fact that Equation 4.41 is applicable to a value of $y^+ = 5$. We may now define a roughness Reynolds number by substituting in the expression for y^+ the value $y = k$ to obtain

$$\mathrm{Re}_k = \frac{kU^*\rho}{\mu} \tag{4.63}$$

which may be rearranged to

$$\mathrm{Re}_k = \frac{k}{D}\,\mathrm{Re}\sqrt{\frac{f}{2}} \tag{4.64}$$

We would expect that as long as the roughness was "buried" in the laminar sublayer, i.e., y^+ or $\mathrm{Re}_k < 5$, it would be ineffective. In fact it is found (Schlichting, 1968) that roughness makes itself felt at $\mathrm{Re}_k > 3$, and to a value of Re_k of about 70 the friction factor depends upon both the relative rough-

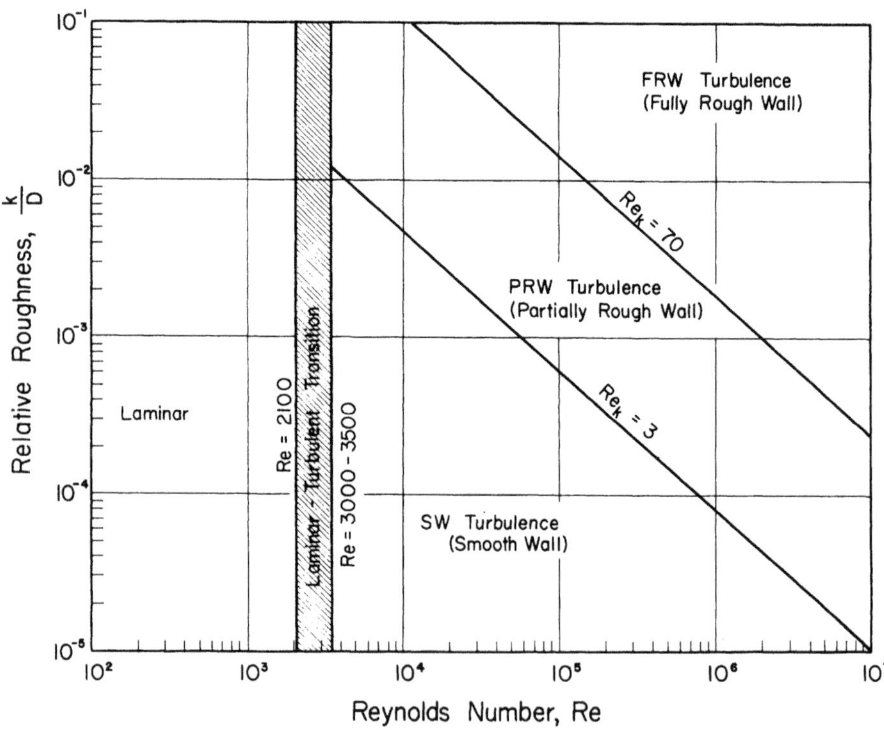

Figure 4.13. Regions of laminar and turbulent flow.

ness, k/D, and the Reynolds number. Above $\mathrm{Re}_k = 70$ the Nikuradse data, and other recent data, show that the friction factor is dependent upon the relative roughness alone. Thus, we have three regions of turbulent flow in a pipe:

a. smooth wall (SW) turbulence
b. partially rough wall (PRW) turbulence
c. fully rough wall (FRW) turbulence

with the boundaries between SW and PRW turbulence given by $\mathrm{Re}_k = 3$, and the boundary between PRW and FRW turbulence given by $\mathrm{Re}_k = 70$.

These regions of turbulent flow are shown in Figure 4.13 along with the region of laminar flow and the transition from laminar to turbulent flow.

The condition of SW turbulence is identical with that of turbulent flow in a smooth pipe which we have already dealt with. We will now consider the case of FRW turbulence and subsequently treat PRW turbulence by combining the treatments for smooth pipe and FRW turbulence.

4.3 Fully Rough Wall (FRW) Turbulent Flow

Nikuradse (1933) is responsible for the most comprehensive studies of turbulent flow in pipes of well defined roughness, prepared by cementing sand grains to the inside of the walls. Nikuradse obtained both velocity profile and pressure gradient data. His velocity profile data are correlated according to Prandtl's (1927a) modification of Equation 4.44 in which the distance from the wall, y, is made dimensionless by use of k rather than $\mu/U^*\rho g_c$, i.e.,

$$u^+ = \frac{1}{k}\ln\frac{y}{k} - \frac{1}{k}\ln\frac{\mu/U^*\rho g_c}{k} + C_1$$

or

$$u^+ = \frac{1}{k}\ln\frac{y}{k} + C_2 \qquad (4.65)$$

Actual data covering D/k values from 30 to 1014 are reasonably well correlated by a value of $C_2 = 8.5$ in Equation 4.65. The resulting equation for velocity profile in the FRW region is

$$u^+ = 5.756 \log\frac{y}{k} + 8.5 \qquad (4.66)$$

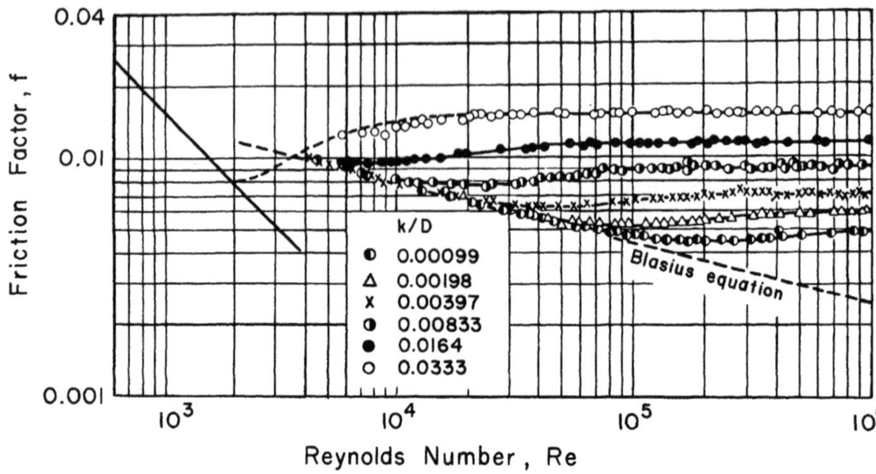

Figure 4.14. Friction factor relationship—rough pipe data of Nikuradse. (From Nikuradse, J. *V.D.I.-Forschungsheft* No. 361, 1933.)

Nikuradse's friction factor data for rough pipes, reproduced in Figure 4.14, reflect all three regions of turbulent flow. He correlated the data in the FRW regime empirically through the equation

$$\frac{1}{\sqrt{f}} = 4 \log \left(\frac{D}{2k}\right) + 3.48 \tag{4.67}$$

The von Kármán equation is essentially the same

$$\frac{1}{\sqrt{f}} = 4.06 \log \left(\frac{D}{2k}\right) + 3.36 \tag{4.68}$$

He obtained this in a manner similar to that used in the derivation of Equation 4.55.

4.4 Partially Rough Wall (PRW) Turbulent Flow

In the region of PRW turbulence the velocity profile is affected by both k/D and Re. Clauser (1954) and others have shown that Nikuradse's velocity profile data in the PRW turbulent region may be expressed by the smooth pipe relation (Equation 4.46) adjusted by a "velocity shift," Δu^+

$$u^+ = 5.5 + 5.756 \log y^+ - \Delta u^+ \tag{4.69}$$

The velocity shift is dependent both on Re_k and on the nature of the roughness (whether artificial or natural). Robertson, Martin, and Burkhart

(1968) discuss this in detail and show the relationship between Δu^+ and Φ_1 (a quantity defined by Equation 4.71):

$$\Delta u^+ = 5.6\left[\frac{\Phi_1 - 3.038}{2} + \log(\mathrm{Re}_k) + 1\right] \tag{4.70}$$

The velocity shift for naturally rough pipes may be calculated from Equation 4.70 using a value of Φ_1 given by Equation 4.73 discussed later. Then the velocity profile may be evaluated from Equation 4.69.

In the region of PRW turbulence the pressure gradient is also sensitive to both Re and k/D and our only approach is empirical.

Following Robertson *et al.* (1968) we may define

$$\Phi_1 = \frac{1}{\sqrt{f_{\mathrm{FRW}}}} - \frac{1}{\sqrt{f_{\mathrm{PRW}}}} = 4\log\frac{D}{2k} + 3.48 - \frac{1}{\sqrt{f_{\mathrm{PRW}}}} \tag{4.71}$$

where f_{FRW} and f_{PRW} denote the friction factor in FRW and PRW turbulence, respectively. The function Φ_1 would be expected to vary with Re_k in a manner dependent upon the nature of the roughness only. A relationship between Φ_1 and Re_k is implicit in the PRW turbulent data of Nikuradse, and in similar data on artificially roughned pipe by Colebrook (1939), Colebrook and White (1937), and Hama (1954). The relationships are illustrated in Figure 4.15 taken from Robertson *et al.* (1968). This figure also shows the Φ_1 values consistent with actual data of Robertson *et al.* in naturally rough, old,

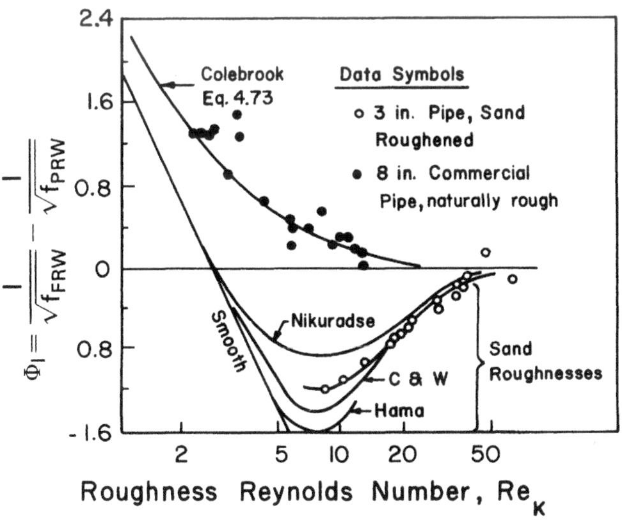

Figure 4.15. Dependence of Φ_1 on type of roughness, and roughness Reynolds number. (From Robertson, J. M., J. D. Martin, and T. H. Burkhart, *Ind. Eng. Chem. Fundam.*, **7** 253, 1968). Note : C & W refers to Colebrook and White (1937).

corroded, 8-inch steel pipe. These values deviate considerably, as expected, from the values for sand-roughened pipe, indicating a different relationship between f_{FRW} and f_{PRW} for natural roughness.

Colebrook (1939) has proposed the following empirical equation for the PRW turbulent region in naturally rough pipe

$$\frac{1}{\sqrt{f}} = 4 \log \frac{D}{2k} + 3.48 - 4 \log \left(1 + 9.35 \frac{D}{2k \, \mathrm{Re}\sqrt{f}}\right) \tag{4.72}$$

This equation is the equivalent of the expression

$$\Phi_1 = 4 \log\left(1 + \frac{3.3}{\mathrm{Re}_k}\right) \tag{4.73}$$

Colebrook's PRW turbulent equation has been widely used, and it is the basis of modern friction factor charts. Its suitability in the PRW turbulent region is confirmed by the work of Robertson *et al.*, as indicated in Figure 4.15, where the data points for the 8-inch, naturally rough pipe follow fairly well the solid line depicting Equation 4.73.

Wood (1966) has developed an empirical equation, explicit in f, which is the close equivalent of the Colebrook relationship and which is more suitable for computer evaluation of f from k/D and Re. Wood's equation is

$$f = a + b \, \mathrm{Re}^{-c} \tag{4.74}$$

where

$$a = 0.026\left(\frac{k}{D}\right)^{0.225} + 0.133\left(\frac{k}{D}\right) \tag{4.75}$$

$$b = 22\left(\frac{k}{D}\right)^{0.44} \tag{4.76}$$

$$c = 1.62\left(\frac{k}{D}\right)^{0.134} \tag{4.77}$$

Although Wood's equation deviates as much as 4 percent in extreme cases from the Colebrook equation, the uncertainties in our ability to predict k/D are such that the values of f calculated from the Wood's equation may be considered equally as good as those calculated from the Colebrook equation.

4.5 The Friction Factor Chart

The turbulent flow friction factor relations, i.e., Equation 4.35 for a smooth pipe or SW turbulence, Equation 4.72 for PRW turbulence, and Equation 4.67 for FRW turbulence, are presented graphically in Figure 4.16 modified

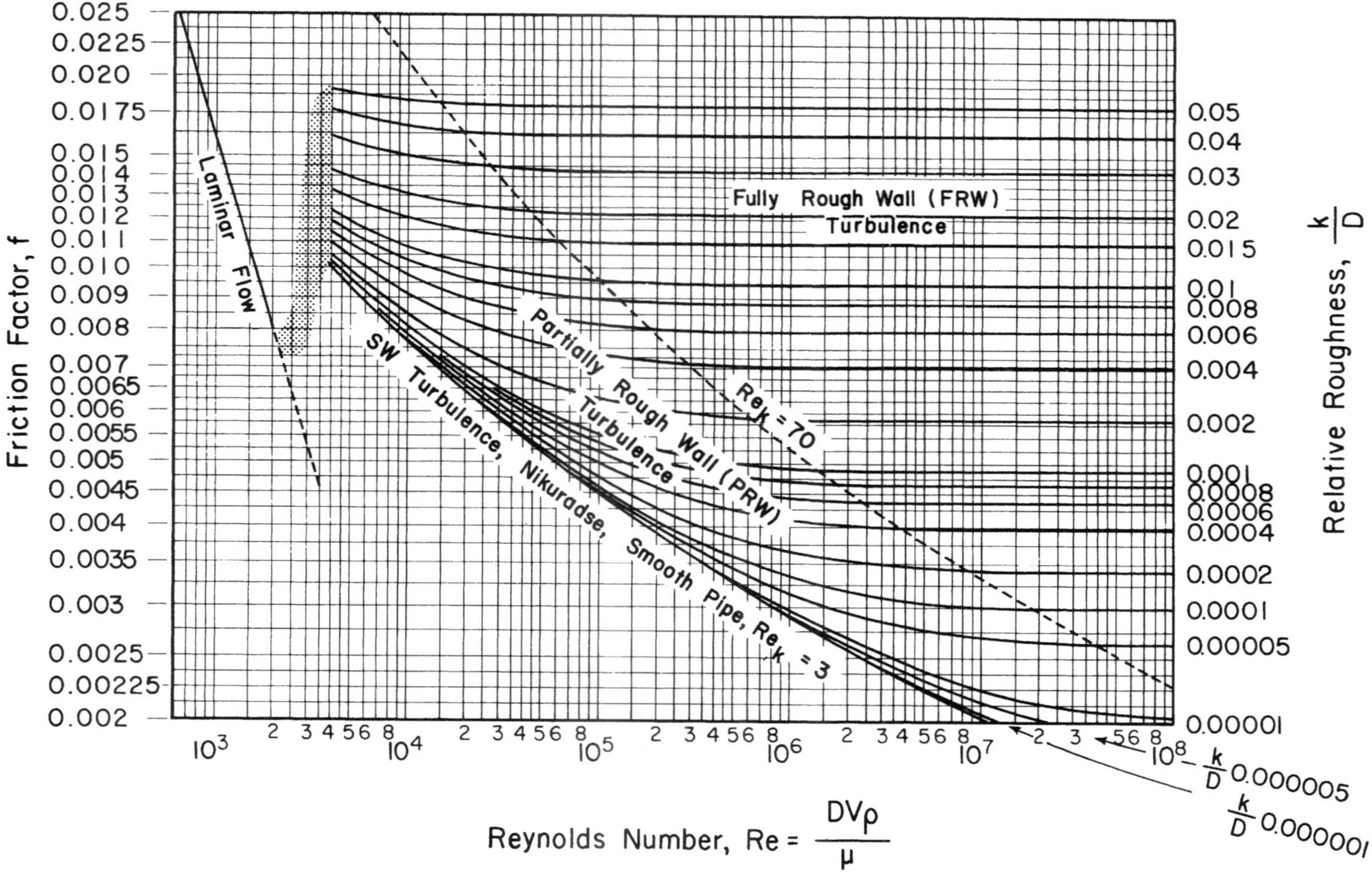

Figure 4.16. Pipe friction factors for turbulent flow. (Modified after Moody, L. F., *Trans. ASME*, **66**, 671, 1944.)

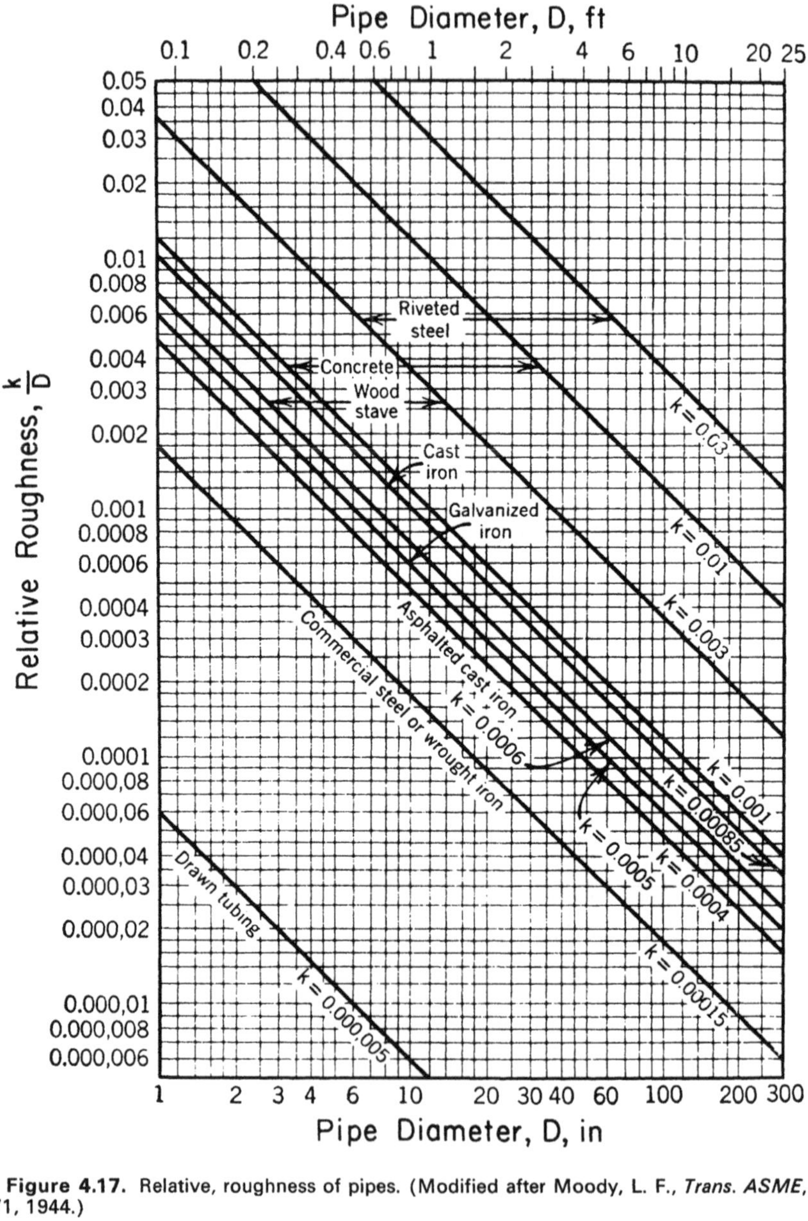

Figure 4.17. Relative, roughness of pipes. (Modified after Moody, L. F., *Trans. ASME*, **66**, 671, 1944.)

after Moody (1944). These relations and their graphical counterpart represent our best present knowledge of the effect of Re and k/D on the turbulent friction factor. We must remember, however, that k is the equivalent sand grain roughness and that natural roughness must be expressed in terms of the sand grain roughness, which would result in the same friction factor. This is not easily achieved; in fact, the only way it can be done is by comparison of the behavior of a naturally rough pipe with a sand-roughened pipe. Moody (1944) has made such comparisons, and his widely used chart giving the absolute and relative sand grain roughness of a variety of pipe-wall materials is reproduced as Figure 4.17. Moody's roughnesses are typical of the materials indicated, but we cannot expect them to be precise for any given material. When it is possible to obtain the results of a flow test through a pipe of identical or closely similar material to that to be used, one may calculate the friction factor, and hence the effective k/D and use it in further calculations.

4.6 General Form of the Pressure Drop Equation

As we did in the case of laminar flow, we may now apply the mechanical energy balance to obtain the expression for the overall pressure gradient in a pipeline. We start with Equation 3.29 and with dP_f/ρ written in terms of the friction factor

$$\frac{dP}{\rho} + \frac{VdV}{\alpha g_c} + \frac{g}{g_c}dz + \frac{2fV^2dL}{g_cD} = 0 \tag{4.78}$$

For an incompressible fluid, integration over the length of the pipe yields

$$\Delta P = \frac{g\rho}{g_c}\Delta z + \frac{\rho}{2g_c}\left(\frac{V_2^2}{\alpha_2} - \frac{V_1^2}{\alpha_1}\right) + \frac{2f\rho V^2L_E}{g_cD} \tag{4.79}$$

where α_1 and α_2 are the velocity profile correction factors in the kinetic energy terms and are in the range of 0.80 to 0.90, depending upon the Reynolds number, and $L_E = L + L_f$, where L_f is the equivalent length under turbulent flow conditions of any fittings, valves, or other devices in the line. Equivalent lengths applicable to turbulent flow are given for selected fittings in Table 4.1. Further details are available in the original paper (Crane Company, 1957).

Often, as with laminar flow, the kinetic energy term is negligible when the flowing fluid is incompressible.

For compressible fluids the integration of Equation 4.78 is somewhat more involved because the density and the velocity vary with pressure.

TABLE 4.1

Representative Equivalent Length[a] in Pipe Diameters (*L*/*D*) of Various Valves and Fittings[b]

		Description of Product		Equivalent Length in Pipe Diameters, L/D
Globe valves	Stem perpendicular to run	With no obstruction in flat, bevel, or plug type seat	Fully open	340
		With wing or pin guided disc	Fully open	450
	Y-pattern	(No obstruction in flat, bevel, or plug type seat) —With stem 60 degrees from run of pipe line	Fully open	175
		—With stem 45 degrees from run of pipe line	Fully open	145
Angle valves		With no obstruction in flat, bevel, or plug type seat	Fully open	145
		With wing or pin guided disc	Fully open	200
Gate valves	Wedge, disc Double disc or plug disc		Fully open	13
			Three-quarters open	35
			One-half open	160
			One-quarter open	900
	Pulp stock		Fully open	17
			Three-quarters open	50
			One-half open	260
			One-quarter open	1200
Conduit pipe line gate, ball, and plug valves			Fully open	3[c]
Check valves	Conventional swing		0.5[d]... Fully open	135
	Clearway swing		0.5[d]... Fully open	50
	Globe lift or stop; stem perpendicular to run or Y-pattern		2.0[d]... Fully open	Same as globe
	Angle lift or stop		2.0[d]... Fully open	Same as angle
	In-line ball	2.5 vertical and 0.25 horizontal[d]	... Fully open	150
Foot valves with strainer		With poppet lift-type disc	0.3[d]... Fully open	420
		With leather-hinged disc	0.4[d]... Fully open	75
Butterfly valves (8 inches and larger)			Fully open	40

[a] Based on clean commercial steel pipe.
[b] Reproduced from Technical Paper No. 410, "Flow of Fluids," Courtesy Crane Co., U.S.A.
[c] Exact equivalent length is equal to the length between flange faces or welding ends.
[d] Minimum calculated pressure drop (psi) across valve to provide sufficient flow to lift disc fully.

TABLE 4.1 (cont.)

	Description of Product			Equivalent length in Pipe Diameters, L/D
Cocks	Straight-through	Rectangular plug port area equal to 100% of pipe area	Fully open	18
	Three-way	Rectangular plug port area equal to 80% of pipe area (fully open)	Flow straight through	44
			Flow through branch	140
Fittings	90 degree standard elbow			30
	45 degree standard elbow			16
	90 degrees long radius elbow			20
	90 degree street elbow			50
	45 degree street elbow			26
	Square corner elbow			57
	Standard tee	With flow through run		20
		With flow through branch		60
	Close pattern return bend			50

Katz *et al.* (1959) review the development of the practical forms of the equation. We will present here various integrated forms of the equation applicable to the pipe flow of gases and vapors.

The most convenient form of the equation of state is

$$\rho = \frac{mP}{ZRT} = \frac{29GP}{ZRT} \tag{4.80}$$

where m = molecular weight of the gas and G = specific gravity of the gas with respect to air. Substituting for ρ in Equation 4.78 from Equation 4.80, we obtain

$$\frac{ZRT}{mP}\,dP + \frac{V\,dV}{\alpha g_c} + \frac{g}{g_c}\,dz + \frac{2fG_m^2 Z^2 R^2 T^2}{g_c D m^2 P^2}\,dL = 0 \tag{4.81}$$

where $G_m = V\rho$, $\mathrm{lb}_m/\mathrm{ft}^2$ sec. Now the following quantities are either constant over the length L or may be assumed so at average values:

- G_m, constant for steady-state flow;
- T, constant for isothermal flow or equal to T_{av} for cases of modest temperature variation;
- f, constant, provided k/D is constant, since $DV\rho/\mu = DG_m/\mu$ is ordinarily constant or nearly so;

Z, variable with P and T but normally taken as constant at an average value, Z_{av}.

For simplicity in handling we now neglect the term $VdV/\alpha g_c$ (we may consider its effect on the overall pressure drop later).

Denoting the constant terms by C_1 and C_2, i.e.,

$$C_1 = \frac{Z_{av} R T_{av}}{m} \quad \text{and} \quad C_2 = \frac{2fG_m^2 Z_{av}^2 R^2 T_{av}^2}{g_c D m^2}$$

we have

$$C_1 \frac{dP}{P} + \frac{g}{g_c} dz + C_2 \frac{dL}{P^2} = 0 \tag{4.82}$$

Replacing dz by $(\Delta z/L)dL$ and solving, we obtain

$$P_2^2 = P_1^2 \left[\exp\left(\frac{-2\Delta z g}{C_1 g_c} \right) \right] + \frac{C_2 L}{\Delta z} \left[\exp\left(\frac{-2\Delta z}{C_1} \frac{g}{g_c} \right) - 1 \right] \tag{4.83}$$

Replacing the values of C_1 and C_2, defining a quantity

$$S = \frac{2\Delta z m g}{R Z_{av} T_{av} g_c} = \frac{2\Delta z G g}{53.33 Z_{av} T_{av} g_c}$$

and substituting for G_m its equivalent in terms of cubic feet per hour at standard conditions of P_0 and T_0

$$G_m = \frac{1}{3600} \frac{4Q_0}{\pi D^2} \frac{29 G P_0}{Z_0 R T_0}$$

we obtain our final equation in commonly used units

$$Q_0 = 1.6156 \frac{T_0}{P_0} \left[\frac{(P_1^2 - e^S P_2^2) D^5 S}{L f Z_{av} G T_{av} (e^S - 1)} \right]^{1/2} \tag{4.84}$$

where Q_0 = flow rate, standard cubic feet per hour,
P = pressure, psia,
D = diameter, inches,
L = length, miles,
Δz = elevation change in feet,
T = temperature, degrees Rankine.

For a horizontal line $\Delta z = 0$, $S = 0$, and the equation simplifies to

$$Q_0 = 1.6156 \frac{T_0}{P_0} \left[\frac{(P_1^2 - P_2^2) D^5}{L f Z_{av} G T_{av}} \right]^{1/2} \tag{4.85}$$

In the case of a vertical pipe, such as a gas well, the equation becomes

$$Q_0 = 117.4\,\frac{T_0}{P_0}\left[\frac{(P_1^2 - e^S P_2^2)D^5 S}{\Delta z f Z_{av} G T_{av}(e^S - 1)}\right]^{1/2} \tag{4.86}$$

The friction factor, f, required in Equation 4.84, 4.85 or 4.86 may be determined for the appropriate pipe relative roughness, k/D, and Reynolds number from Figure 4.16. It is convenient to express the Reynolds number for a gas in terms of the volumetric flow rate at standard conditions. From the definition of Re and the expression given for G_m we obtain

$$\mathrm{Re} \simeq \frac{0.48\,Q_0\,G}{\mu D} \tag{4.87}$$

for $P_0 = 14.65$ psia,
$T_0 = 60°\mathrm{F}$,

where μ = viscosity, centipoise, and the other terms are defined following Equation 4.84.

We should note that in the integration leading to Equations 4.84, 4.85 and 4.86 no account is taken of the effect of any fittings, valves or other devices in the pipe. Account may be taken of any fittings by replacing L in Equation 4.84 or 4.85 with $L_E = L + L_f$. This may not be done with Equation 4.86. For vertical pipes containing fittings, Equation 4.84 should be used with L replaced by L_E, but using the true change of elevation in the S term.

Sukkar and Cornell (1955) have shown that, if G_m, T, and f are taken as constants and if kinetic energy effects are neglected, Equation 4.81 may be expressed for flow in a vertical pipe as

$$\int_{P_{r1}}^{P_{r2}} \frac{Z/P_r}{1 + B(Z/P_r)^2}\,dP_r = \frac{0.01877 G \Delta z}{T_{av}} \tag{4.88}$$

where

$$B = \frac{667(4)(24)^2(520)}{10^{12}(14.65)}\,\frac{P_0}{T_0}\,\frac{f Q_0^2 T_{av}^2}{D^5 P_c^2},$$

P_c = critical pressure, psia
P_r = reduced pressure,

and the other terms are defined following Equation 4.84. For natural gases, using the relationship between Z, P_r and T_r given in Figure 2.7, Sukkar and Cornell integrated the left-hand side of Equation 4.88 for a series of values of B and T_r, and for ranges of P_r. Values of their integrals are given in Katz *et al.* (1959). Equation 4.88, with the Sukkar and Cornell integrals, is an alternate to Equation 4.86 for flow in natural gas wells.

The Sukkar and Cornell method does not lend itself to the direct inclusion of the effect of fittings but the pressure drop due to fittings may be separately estimated by the method previously discussed for incompressible fluids.

In any long pipeline, or even in a gas well as deep as 15,000 feet, the kinetic energy effect is small; where it is desirable to account for this, however, P_2, as calculated from the foregoing equations, may be reduced by an amount

$$\Delta P_{KE} = \frac{\rho}{2\alpha g_c}(V_2^2 - V_1^2) \tag{4.89}$$

There are many alternate, and several approximate forms of Equations 4.84 through 4.86 in common use (Katz *et al.*, 1959). The better of these are based upon the equations given here with various substitutions made for the friction factor. Having regard to the complex relationship between f, Re and k/D, such equations which incorporate approximations for f can hardly be expected to be as accurate as the fundamental equations used with a properly chosen f, based upon the correct Re, and the best information concerning the relative roughness.

The Panhandle friction formula

$$\frac{1}{f} = 208\left(\frac{GQ}{D}\right)^{0.1461} \tag{4.90}$$

provides a reasonable approximation to the smooth pipe Nikuradse equation over a wide range of Reynolds numbers. There are two forms of the Panhandle flow equation, the original (see Katz *et al.*, 1959), and the revised or Panhandle-Eastern equation (see May and June, 1956, issues of *Pipeline Engineer*). The Panhandle-Eastern equation is

$$Q_0 = 737\left(\frac{T_0}{P_0}\right)^{1.02} D^{2.530}\left[\frac{P_1^2 - P_2^2}{G^{0.961}T_{av}Z_{av}L}\right]^{0.510} \mathscr{E} \tag{4.91}$$

where Q_0 is in standard cubic feet per day. Allowance for roughness in the Panhandle and the Panhandle-Eastern equation is made through a so-called efficiency factor, $\mathscr{E}$, defined as the ratio of the actual flow to the flow predicted through the use of the friction factor incorporated in the equation. The pipeline efficiency factor $\mathscr{E}$ is merely an alternate device to account for pipe roughness. It will vary with the age, cleanness, and diameter of the pipe. Its value in a particular case is usually based upon experience in the operation of a pipeline similar in diameter, material, and method of construction to that for which the calculation is being made. For preliminary designs $\mathscr{E}$ is usually taken as 0.85 to 0.95 for diameters ranging, respectively, from less than 12 inches to more than 30 inches.

Many other equations, such as that of Weymouth (1912) based on relatively crude approximation of the friction factor, were used in the period prior to 1950 or so but their use today is not recommended.

5 RECOMMENDED DESIGN METHODS

5.1 Introduction

The choice and use of the relations that we have discussed for the solution of actual design problems depend upon whether the flowing fluid may be assumed incompressible and of constant properties and, if not, upon the integration of the appropriate differential relations to account for variation in density and viscosity. Liquids may almost always be considered incompressible, and for gases an often suitable criterion is that incompressibility may be assumed if the anticipated pressure drop is less than about 10 percent of the upstream pressure. Frequently the change with pressure of density and viscosity may be expressed in such a way that direct integration of the differential equations is possible. In other instances, step calculations may be made with average flow properties over pipe sections such that the pressure drop is of the order of 10 percent of the upstream pressure over each section.

We will treat in detail only the incompressible case, expecting the reader to apply it in stepwise calculations or to use suitably integrated forms of the equations where the fluid is compressible or the viscosity changes.

In later chapters we will present the design method only for the assumed problem of determining the pressure drop (and the flow pattern and holdup where applicable) from specified input rates, the fluid or mixture properties and the pipe size, orientation, and roughness. Here, however, we will give rather full details for alternative design problems.

5.2 Entrance Effects

Entrance effects are normally rather modest and in their calculation we may assume constancy of density and viscosity (at average values if convenient). Also we assume for convenience that D, k/D, Q (or V) and the fluid properties are known. The effects may be estimated as follows:

1. Calculate Re for the known D, Q (or V) and the average fluid properties.

2. If Re is less than 2100,
   a. the entrance length may be estimated from Equation 4.13;
   b. if the entrance length is less than 10 percent of the total length, its effect may be neglected;
   c. if otherwise, ΔP_f for the entrance length may be calculated through Equations 4.28 and 4.30.
3. If Re is greater than 2100, the entrance length may be of the order of $50D$, as indicated by Equation 4.14 or Figure 4.1. There is no method available to us for the accurate prediction of the entrance length and the pressure drop over this length. Except in the case of very short pipelines, it is usually acceptable to assume $X_E = 0$.

5.3 Effect of Bends and Fittings

The effect of bends, fittings, valves, enlargements, and contractions in the pipe system is estimated using the equivalent length concept discussed in Section 3.4 for laminar flow and Section 4.6 for turbulent flow. The pipe length for purposes of the friction pressure drop calculation is increased by the sum of the equivalent lengths of the fittings in the system. When the flow is laminar the individual L_E values are determined from Equation 4.32 and Table 4.1 when the flow is turbulent Table 4.1 is used directly.

Note that the sum of the L_E terms is not added to the pipe length for purposes of calculating a hydrostatic head or potential energy effect.

5.4 Potential Energy or Hydrostatic Head Effects

For incompressible fluids we may calculate the potential energy effect from the first term of Equation 4.79, i.e.,

$$\Delta P_{PE} = \frac{g}{g_c} \rho \Delta z$$

For compressible fluids the term requires integration to account for the variation in density,

$$\Delta P_{PE} = \frac{g}{g_c} \int_1^2 \rho \, dz$$

For gases ΔP_{PE} is not usually calculated separately, but, if desired, it may be determined from Equation 4.84 with $Q_0 = 0$, i.e.,

$$P_1^2 = e^S P_2^2$$

For gas flow at low pressure, ΔP_{PE} is small and may often be neglected.

5.5 Acceleration or Kinetic Energy Effects

The kinetic energy effect may be estimated from Equation 4.89. This effect may be neglected for liquids and is usually small for gases.

5.6 Pressure Gradient for the Steady Flow of Incompressible Fluids (Liquids; and Gases and Vapors when $\Delta P < 0.10P_1$)

5.61 Steps in Estimating ΔP when D and Q (or V) Are Known

(A) Re < 2100

1. The pressure drop due to friction, ΔP_f, may be calculated through application of Equation 4.20 to the entire length of the pipe if the entrance effects are negligible. Otherwise it may be applied to the entire length less the entrance length, in which case the entrance ΔP_f is added to the ΔP_f calculated from Equation 4.20.
2. Calculate the total pressure drop, ΔP, by summing the values of ΔP_{KE}, ΔP_{PE}, and ΔP_f (Equation 4.31a).

(B) Re > 2100

1. Estimate k/D for the pipe with the aid of Figure 4.17.
2. From the known Re and k/D estimate f from Figure 4.16.
3. Calculate ΔP_f for the entire pipe length from the last term of Equation 4.79.
4. Calculate the total pressure drop, ΔP, from Equation 4.79.

5.62 Steps in Estimating D when Q (or V) and ΔP (or ΔP_f) are Known

1. If ΔP is the known quantity, calculate ΔP_f from Equation 4.79 by subtracting the potential and kinetic energy contributions from the total pressure drop.
2. Assuming laminar flow, calculate D from the integrated form of Equation 4.20.
3. Calculate Re; if Re < 2100 the calculation of step *2* is correct.
4. Otherwise, determine by trial and error the correct D, and from it the correct Re, k/D, and f so that

$$\Delta P_f = \frac{2f\rho V^2 L_E}{g_c D}$$

The term, k/D, and f may be estimated from Figures 4.17 and 4.16.
5. If appropriate, make adjustments for the effect of the entrance.

5.63 Steps in Estimating Q (or V) when D and ΔP Are Known

1. Calculate ΔP_f from Equation 4.79 as in step *1* of Section 5.62.
2. Assume laminar flow, and calculate Q (or V) from Equation 4.20 (or 4.21).
3. Calculate Re; if Re < 2100, the calculation of step *2* is correct.
4. Otherwise, determine by trial and error the correct V, and from it the correct Re, and obtain the correct f by using the appropriate k/D so that

$$\Delta P_f = \frac{2f\rho V^2 L_E}{g_c D}$$

5. If appropriate, make adjustments for the effect of the entrance.

5.7 Pressure Gradient for the Steady Flow of Compressible Fluids (Gases and Vapors when $\Delta P > 0.10 P_1$)

The procedures for calculating ΔP, Q_0, or D for compressible fluids are the same in principle as for incompressible fluids. Normally, however, the flow is turbulent and this assumption should be made initially and checked. Accordingly, the starting point of the calculation is one of Equations 4.84, 4.85, 4.86 or the equivalent. When ΔP or $(P_1^2 - P_2^2)$ is the quantity sought, and D and Q_0 are known, f may be determined directly from Re (Equation 4.87) and k/D. Otherwise f must be assumed, and later checked against the values of Re and k/D. The iteration involved is not extensive since, even though D may not be known, k/D is probably known or may be estimated independently of the calculations.

Bibliography

Astarita, G., G. Greco, and L. Peluso, *Ind. Eng. Chem. Fundam.*, **7**, 595 (1968).

Blasius, H., *Mitt. Forschungsarb.* No. 131, (1913).

Bogue, D. C., and A. B. Metzner, *Ind. Eng. Chem. Fundam.*, **2**, 143 (1963).

Clauser, F. R., *J. Aeron. Sci.*, **21**, 91 (1954).

Colebrook, C. F., *J. Inst. Civil Engrs.*, **11**, 133 (1939).

Colebrook, C. F., and C. M. White, *Proc. Royal Soc. London, Ser. A.*, **161**, 367 (1937).

Collins, M., and W. R. Schowalter, *A.I.Ch.E. Journal*, **9**, 804 (1963).

Crane Co., Flow of Fluids Through Valves, Fittings and Pipe, Technical Paper No. M-409, Montreal, Que., May 1942; Technical Paper No. 410, Chicago, Ill., 1957.

Deissler, R. G., Analytical and Experimental Investigation of Adiabatic Turbulent Flow in Smooth Tubes, Nat. Advisory Committee Aeronautics. Tech. Note No. 2138, July 1950, 41 pp.

Deissler, R. G., *Trans. A.S.M.E.*, **73**, 101 (1951).

Dodge, D. W., Doctoral Dissertation, University of Delaware, Newark, Del., (1958).

Drew, T. B., E. C. Koo, and W. H. McAdams, *Trans., A.I.Ch.E.* **28**, 56 (1932).

Gill, W. N., and M. Scher, *A.I.Ch.E. Journal*, **7**, 61 (1961).

Hama, F. R., *Trans. Soc. Naval Arch. Marine Engrs.*, **62**, 333 (1954).

Hanks, R. W., Doctoral Dissertation, University of Utah, Salt Lake City, Utah, (1960).

Hanks, R. W., *A.I.Ch.E. Journal*, **9**, 45 (1963); **14**, 691 (1968); **15**, 25 (1969).

Hinze, J. O., *Turbulence*, McGraw-Hill Book Co., New York, (1959).

Katz, D. L., and Associates, *Handbook of Natural Gas Engineering*, McGraw-Hill Book Co., New York, 1959.

Kline, S. J., and A. H. Shapiro, National Advisory Committee for Aeronautics, Technical Note 3048, (1953).

Knudsen, J. G., and D. L. Katz, *Fluid Dynamics and Heat Transfer*, McGraw-Hill Book Co., New York, 1958.

Langhaar, H. L., *J. Appl. Mech., Trans. A.S.M.E.*, **9**, A55 (1942).

Laufer, J., The Structure of Turbulence in Fully Developed Pipe Flow, NACA Technical Note 2954, June 1953, see also NACA Report 1175, 1954.

Lindgren, E. R., and J. Chao, *Physics of Fluids*, **12**, 1364 (1969).

McComas, S. T., *J. Basic Eng., Trans. A.S.M.E., Ser. D*, **89**, 847 (1967).

McComas, S. T., and E. R. G. Eckert, *J. Appl. Mech., Trans, A.S.M.E., Ser. E*, **30**, 765 (1965).

Michiyoshi, L., K. Mizuno, and Y. Hoshiai, *Int. Chem. Eng.*, **6**, 373 (1966).

Millikan, C. B., Fifth Int. Congress on Applied Mechanics, *Proc.* p. 386, John Wiley and Sons Inc., New York, 1939.

Moody, L. F., *Trans. A.S.M.E.*, **66**, 671 (1944).

Nikuradse, J., Third International Congress of Applied Mechanics (1930) *Proc.* Vol. 1, p. 239, Sveriges Litografiska Tryckerier, Stockholm, 1931.

Nikuradse, J., Gesetzmässigkeiten der turbulenten Strömung in glatten Rohren, Forschungsheft No. 356, V. D. I. Verlag, Berlin, 1932, 36 pp.

Nikuradse, J., Strömungsgesetze in rauhen Rohren, Forschungsheft No. 361. V. D. I. Verlag, Berlin, 1933, 22 pp.

Pfenninger, W., Experiments with Laminar Flow in the Inlet of a Tube at High Reynolds Numbers with and without Boundary Layer Suction, Northrop Aircraft Inc., Hawthorne, California, 1952.

Prandtl, L., *Z. angew. Math. u. Mech.*, **1**, 431 (1921); **5**, 136 (1925).

Prandtl, L.:
a. *Ergebnisse der Aerodynamischen Versuchsanstalt zu Göttingen* **3**, 1 (1927).
b. Second Int. Congress on Applied Mechanics, *Proc.* p. 62, Zurich, (1927).

Prandtl, L., *Z. Ver. deut. Ing.*, **77**, 105 (1933); also *NACA* TM 720 (1933).

Prandtl, L., and O. Tietjens, *Hydro- und Aeromechanik*, Vol. 2, p. 28, Springer-Verlag, Berlin, 1931.

Prengle, R. S., and R. R. Rothfus, *Ind. Eng. Chem.*, **47**, 379 (1955).

Quarmby, A., *J. Mech. Eng. Sci.*, **11**, 14 (1969).

Reichardt, H., *Z. angew. Math. u. Mech.*, **20**, 297 (1940); **31**, 208 (1951).

Reshotko, E., Report No. 20–364, Jet Propulsion Laboratory, Pasadena, California, 1958.

Reynolds, O., *Roy. Soc. Phil. Trans., Ser. A*, **174**, 935 (1883).

Robertson, J. M., J. D. Martin, and T. H. Burkhart, *Ind. Eng. Chem. Fundam.*, **7**, 253 (1968).

Ross, D., Third Midwestern Conference on Fluid Mechanics, Univ. of Minnesota, 1953, *Proc.* p. 651.

Rothfus, R. R., Doctoral Dissertation, Carnegie Institute of Technology, Pittsburgh, Pa., 1948.

Rouse, H., *Elementary Mechanics of Fluids*, John Wiley & Sons, Inc., New York, 1946.

Ryan, N. W., and M. M. Johnson, *A.I.Ch.E. Journal*, **5**, 433 (1959).

Schiller, L., *Z. angew. Math. u. Mech.*, **2**, 96 (1922).

Schlichting, H., *Boundary-Layer Theory*, McGraw-Hill Book Co., New York, 1968.

Schmidt, F. W., and B. Zeldin, *A.I.Ch.E. Journal*, **15**, 612 (1969).

Senecal, V. E, and R. R. Rothfus, *Chem. Eng. Progress*, **49**, 533 (1953).

Shaver, R. G., Sc. D. Thesis, Massachusetts Institute of Technology, Cambridge, Mass., 1957.

Sparrow, E. M., S. H. Lin, and T. S. Lundgren, *Physics of Fluids*, **7**, 338 (1964).

Stanton, T. E., *Proc. Royal Soc., Ser. A*, **85**, 366 (1911).

Stanton, T. E., and J. R. Pannell, *Roy. Soc. Phil. Trans., Ser. A*, **214**, 199 (1914).

Sukkar, Y. K., and D. Cornell, *J. Petrol. Technol.*, **7**, 43 (1955).

Van Driest, E. R., *J. Aeronaut. Sci.*, **23**, 1007 (1956).

Vasilescu, A. A., *Rev. Roum. Sci. Techn., Méc. Appl.* (Bucarest) **12**, 115 (1967).

von Kármán, T., *Nachr. Ges. Wiss. Göttingen, Math.-physik, Klasse* (1930) also *NACA* TM 611 (1931).

Walker, J. E., Doctoral Dissertation, Carnegie Institute of Technology, Pittsburgh, Pa., 1957.

Wells, C. S., Jr., J. Harkness, and W. A. Meyer, Report No. 0-71000/6R-22, L.T.V. Research Center, Dallas, Texas, Dec. 1966.

Weymouth, T. R., *Trans. A.S.M.E.*, **34**, 203 (1912).

Wood, D. J., *Civil Eng.-A.S.C.E.*, **36** (12), 60 (1966).

The Flow of Time-Independent Non-Newtonian Fluids in Pipes

5

1 INTRODUCTION

We now turn to a treatment of the pipe flow of time-independent non-Newtonian materials. Like Newtonian fluids, they may flow in laminar, transitional, or turbulent motion. Also, a finite distance of travel from the entrance of a pipe is required for the development of stabilized flow. In our discussion of Newtonian fluids in Chapter 4 we dealt with the criteria for laminar, transitional, and turbulent flow remote from the entrance, and also with the travel distance required for stabilization, before we discussed laminar and turbulent flow. In the case of the non-Newtonian fluids it is more convenient to deal first with the stabilized flow conditions and then to turn to the question of transition and entrance effect.

When the constitutive equation of the fluid is known or assumed, and in stabilized laminar flow, the pipe flow relations may be derived by simple theoretical analysis paralleling the treatment leading to the Hagen-Poiseuille relationships for Newtonian fluids. In turbulent flow, as with Newtonian fluids, theoretical analysis leads us a certain distance but it must be supplemented by experimental observations.

We start our discussion with a development of the Rabinowitsch-Mooney equation, which is important in the interpretation of rheological data taken with capillary instruments, and generally applicable to any time-independent, non-Newtonian material in laminar flow. The results of the same type of analysis of pseudoplastics (described by various constitutive equations), the ideal Bingham plastic, and yield-pseudoplastics are then discussed. We deal next with the transition from laminar to turbulent flow. Finally we present the more complex case of turbulent flow, and trace recent, semitheoretical and empirical developments.

The treatment in this chapter is restricted to steady state isothermal flow in plain pipes. Skelland (1967) gives a more extensive discussion covering nonisothermal flow and flow in annuli and other shapes.

2 STEADY-STATE LAMINAR FLOW

2.1 The Rabinowitsch-Mooney Relations

A perfectly general relationship for the steady, stabilized, laminar flow of any time-independent purely viscous fluid may be developed from a generalized constitutive equation. The resulting equations, although usually referred to as the Rabinowitsch-Mooney relations, according to Savins, Wallick, and Foster (1962) were first developed by Herzog and Weissenberg (1928). Rabinowitsch (1929) and Mooney (1931) recognized the full significance of the relationships.

We start from a constitutive equation which, however complex, may be expressed as

$$f(\tau) = \dot{S} \tag{5.1}$$

where τ = shear stress,
$\dot{S}$ = absolute value of the rate of shear = $-du/dr$ in a circular pipe.

For steady flow in a pipe

$$Q = 2\pi \int_0^R urdr \tag{5.2}$$

where Q = total volumetric flow rate,
u = velocity at radius r,
R = radius of the pipe.

Equation 5.2 may be expressed as

$$Q = \pi \left[ur^2 - \int r^2 du \right]_0^{R^2} \tag{5.3}$$

Since $u = 0$ when $r = R$ (no slip at the wall), this simplifies to

$$Q = -\pi \int_0^{R^2} r^2 du \tag{5.4}$$

A simple force balance on a cylindrical element of radius r and length dL yields

$$-\pi r^2 dP = 2\pi r \tau dL \tag{5.5}$$

from which

$$\tau = -\frac{r}{2}\frac{dP}{dL} \tag{5.6}$$

and when $r = R$

$$-\pi R^2 dP = 2\pi R \tau_w dL \tag{5.7}$$

where τ and τ_w denote the shear stress at radius r and R, respectively. Equations 5.5 and 5.7 show that

$$r^2 = \frac{R^2}{\tau_w^2}\tau^2 \qquad \text{and} \qquad dr = \frac{R}{\tau_w}\,d\tau$$

Combining Equations 5.1 and 5.4, and substituting the foregoing for r^2 and dr, we obtain

$$Q = \pi \int_0^{\tau_w} \frac{R^3\tau^2}{\tau_w^3} f(\tau)d\tau \tag{5.8}$$

If the value of τ_w from Equation 5.7 (with $-dP/dL$ replaced by $\Delta P/L$) is inserted and R is replaced by $D/2$, the following equation results

$$\frac{8Q}{\pi D^3} = \frac{2V}{D} = \frac{1}{\left(\frac{D\Delta P}{4L}\right)^3}\int_0^{D\Delta P/4L} \tau^2 f(\tau)d\tau \tag{5.9}$$

where L denotes the length of the pipe. This is a perfectly general relationship of the discharge Q, the pressure gradient $\Delta P/L$, the diameter D, and the constitutive equation $f(\tau) = -du/dr$.

Equation 5.9 may be rearranged and differentiated with respect to $D\Delta P/4L$ to yield

$$f(\tau_w) = -\left(\frac{du}{dr}\right)_w = 3\left(\frac{8Q}{\pi D^3}\right) + \frac{D\Delta P}{4L}\frac{d\left(\frac{8Q}{\pi D^3}\right)}{d\left(\frac{D\Delta P}{4L}\right)} \tag{5.10}$$

a general relationship for the rate of shear at the wall. Equation 5.10 is more simply written as in Chapter 1

$$-\left(\frac{du}{dr}\right)_w = \frac{8V}{D}\left(\frac{1+3n'}{4n'}\right) \tag{1.67}$$

where

$$n' = \frac{d \ln (D\Delta P/4L)}{d \ln (8V/D)} \tag{1.68}$$

The importance of Equation 1.67 in the interpretation of rheological data taken in a capillary viscometer was shown in Chapter 1. Since Equations 5.9 and 1.67 are perfectly general, they may be used as the starting point for deriving the relations for the laminar flow of fluids whose rheological behavior is described by various constitutive equations. Alternatively, if rheological data are available, Equation 5.9 may be used directly to find the flow rate–pressure drop relationship by numerical evaluation of the integral. Savins, Burdyn, and Wallick (1963) and Savins and Cox (1963) describe such a method of estimating the pressure gradient in laminar flow of time-independent non-Newtonian fluids based upon an experimentally determined τ–$\dot{S}$ curve. They carry out a numerical integration of the Rabinowitsch-Mooney equation (Equation 5.9) using actual rheological data, taken at successively increasing rates of shear, in place of an assumed rheological equation.

Before proceeding with the application to non-Newtonian materials it is of interest to see whether Equation 5.9 gives the expected result for the laminar flow of a Newtonian fluid. In this case Equation 5.1 becomes

$$\frac{g_c}{\mu}\tau = -\frac{du}{dr} \tag{5.11}$$

Substituting for $f(\tau)$ in Equation 5.9, we obtain

$$\frac{2V}{D} = \frac{1}{\left(\dfrac{D\Delta P}{4L}\right)^3}\frac{g_c}{\mu}\int_0^{D\Delta P/4L} \tau^3 d\tau = \left(\frac{4L}{D\Delta P}\right)^3 \frac{g_c}{\mu}\frac{1}{4}\left(\frac{D\Delta P}{4L}\right)^4 \tag{5.12}$$

that is,

$$\frac{8V}{D} = \frac{g_c}{\mu}\left(\frac{D\Delta P}{4L}\right) \tag{5.13}$$

From Equation 5.13 we know that n', as defined by Equation 1.68, is unity for a Newtonian fluid. Inserting $n' = 1$ in Equation 1.67 we have

$$-\left(\frac{du}{dr}\right)_w = \frac{8V}{D} \tag{5.14}$$

which is the same result obtained from Equation 4.19 and 4.21 of Chapter 4.

2.2 The Relations for Pseudoplastic Fluids

2.21 The Power Law Fluid

The relation describing the stabilized laminar pipe flow of a power law pseudoplastic may be derived, in a manner analogous to the Hagen–Poiseuille equation, from the fundamental momentum equation and the power law rheological equation.

Combining the power law

$$\tau = \frac{K}{g_c} \dot{S}^n \tag{1.40}$$

and

$$\tau = -\frac{r}{2} \frac{dP}{dL} \tag{5.6}$$

we obtain

$$-\frac{du}{dr} = \left(-\frac{g_c}{2K} \frac{dP}{dL}\right)^{1/n} r^{1/n} \tag{5.15}$$

Integrating this equation and substituting $u = 0$ at $r = R$,

$$u = \left[\left(-\frac{g_c}{2K} \frac{dP}{dL}\right)^{1/n} \frac{n}{1+n}\right] (R^{(1+n)/n} - r^{(1+n)/n}) \tag{5.16}$$

$$Q = 2\pi \int_0^R urdr \tag{5.2}$$

from which the equation for the total flow rate is

$$Q = \pi \left(-\frac{g_c}{2K} \frac{dP}{dL}\right)^{1/n} \left(\frac{n}{1+3n}\right) R^{(1+3n)/n} \tag{5.17}$$

and the average velocity is given by

$$V = \left(-\frac{g_c}{2K} \frac{dP}{dL}\right)^{1/n} \left(\frac{n}{1+3n}\right) R^{(1+n)/n} \tag{5.18}$$

Equations 5.15 and 5.16 allow us to calculate the rate of shear and the velocity at any radial position.

The relations of Equations 5.15 and 5.16 for various values of n, and for $R = 1$ and $g_c \Delta P / 2K \Delta L = 1$, are shown in Figures 5.1 and 5.2. These figures illustrate the effect of n on the velocity gradient and the velocity profiles for a given pipe size and a fixed pressure gradient and value of K.

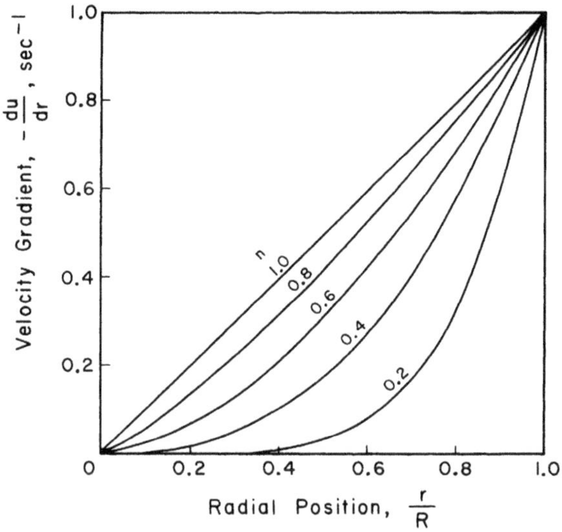

Figure 5.1. Effect of the power law exponent on the velocity gradient in laminar flow.

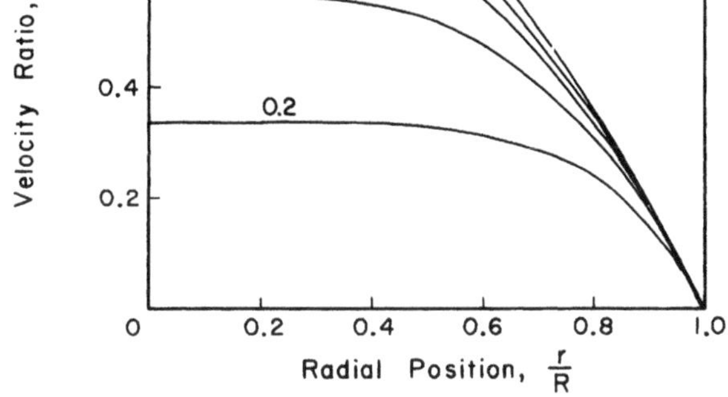

Figure 5.2. Effect of the power law exponent on the velocity profile in laminar flow.

The rate of shear at the wall and the velocity at the center line are of special interest. These are

$$-\left(\frac{du}{dr}\right)_w = \left(-\frac{g_c}{2K}\frac{dP}{dL}\right)^{1/n} R^{1/n} \tag{5.19}$$

and

$$u_m = \left[\left(-\frac{g_c}{2K}\frac{dP}{dL}\right)^{1/n} \frac{n}{1+n}\right] R^{(1+n)/n} \tag{5.20}$$

Equation 5.20 can only be an approximation with respect to any real pseudoplastic, since at the center line the shear stress is zero, and we know that the power law equation does not apply as the shear stress approaches zero. From Equations 5.18 and 5.20 we see that

$$\frac{V}{u_m} = \frac{1+n}{1+3n} \tag{5.21}$$

An alternate approach to obtain Equation 5.17 or 5.18 directly is from the generalized Equations 5.9 and 1.67.

If we substitute

$$f(\tau) = \tau^{1/n}\left(\frac{g_c}{K}\right)^{1/n} \tag{5.22}$$

in Equation 5.9 and perform the integration we obtain

$$\frac{2V}{D} = \frac{8Q}{\pi D^3} = \frac{n}{1+3n}\left(\frac{D\Delta P}{4L}\right)^{1/n}\left(\frac{g_c}{K}\right)^{1/n} \tag{5.23}$$

which, on rearrangement, yields Equations 5.17 and 5.18. We may also note that for the power law fluid $n' = n$ and from Equation 1.67

$$-\left(\frac{du}{dr}\right)_w = \frac{8V}{D}\left(\frac{1+3n}{4n}\right) \tag{5.24}$$

Equation 5.18 may be rearranged and put in friction factor form. Solving the equation for $\Delta P/L$, we obtain

$$\frac{\Delta P}{L} = \frac{4V^n K}{g_c}\left(\frac{2+6n}{n}\right)^n \frac{1}{D^{1+n}} \tag{5.25}$$

Now by definition

$$f = \frac{g_c D\Delta P}{2\rho V^2 L} \tag{4.24}$$

from which

$$f = \left(\frac{2V^{n-2}K}{D^n\rho}\right)\left(\frac{2+6n}{n}\right)^n = \left(\frac{16}{\dfrac{D^nV^{2-n}\rho}{K}}\right)\left(\frac{1}{8}\right)\left(\frac{2+6n}{n}\right)^n \tag{5.26}$$

or

$$f = \left(\frac{16}{\mathrm{Re}_{PL1}}\right)\left(\frac{1}{8}\right)\left(\frac{2+6n}{n}\right)^n \tag{5.27}$$

and

$$f = \frac{16}{\mathrm{Re}_{PL2}} \tag{5.28}$$

where Re_{PL1} and Re_{PL2} are alternate forms of what may be called the "power law Reynolds number," defined as

$$\mathrm{Re}_{PL1} = \frac{D^nV^{2-n}\rho}{K} \tag{5.29}$$

and

$$\mathrm{Re}_{PL2} = \frac{D^nV^{2-n}\rho}{K}\,8\left(\frac{n}{2+6n}\right)^n \tag{5.30}$$

both of which reduce to the conventional Newtonian Reynolds number when $n = 1$. We will wish to refer again to Re_{PL1} and Re_{PL2} and the relationship between them. From the definitions we see that

$$\frac{\mathrm{Re}_{PL2}}{\mathrm{Re}_{PL1}} = 8\left(\frac{n}{2+6n}\right)^n \tag{5.31}$$

This relationship is shown graphically in Figure 5.3.

Clearly in terms of Re_{PL2}, the friction factor–Reynolds number relation for a power law fluid in laminar flow is identical with that for a Newtonian fluid. In terms of Re_{PL1} the relationship varies with the value of n.

Since it would be shear coincidence if the transition from laminar to turbulent flow in a power law fluid occurred at the same Reynolds number as for a Newtonian fluid, the Re_{PL1} form of the power law Reynolds number seems preferable. This form allows for ready display of the variation in the critical Reynolds number with n. Figure 5.4 shows the variation of f with Re_{PL1} and Re_{PL2}. The figure suggests either that the Reynolds number for transition from laminar to turbulent flow is lower for pseudoplastics than for

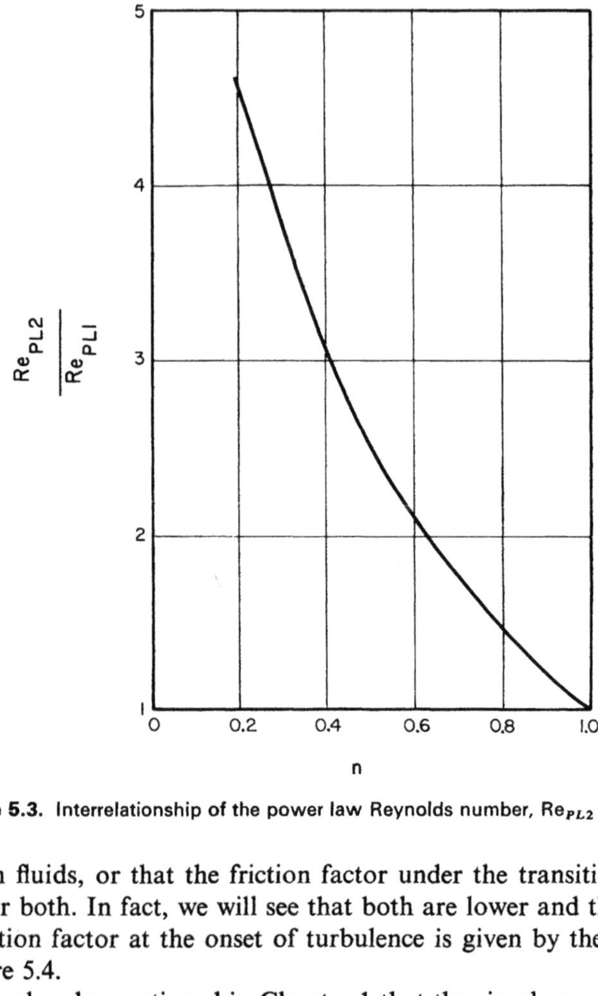

Figure 5.3. Interrelationship of the power law Reynolds number, Re_{PL2} and Re_{PL1}.

Newtonian fluids, or that the friction factor under the transition condition is lower, or both. In fact, we will see that both are lower and that the locus of the friction factor at the onset of turbulence is given by the line labeled f_c in Figure 5.4.

We have already mentioned in Chapter 1 that the simple power law equation, although not applicable as shear rates approach zero, gives a good fit to actual data for most pseudoplastics over at least one and often two decades of shear rate. This means that if the power law is fitted to data at and immediately below a value equivalent to the wall shear rate (or stress) anticipated in a particular application of laminar pipe flow, it will accurately represent the rheological behavior of the fluid from the wall to the inner region where the shear rate is, say, one tenth of the wall shear rate. From Equation 5.15 we see that, with a fluid having an n value of 0.5, the power law would be accurate over the range $r = R$ to $r = 0.316R$; with $n = 0.75$ the range would be extended to $r = 0.178R$. Metzner (1968) has shown that even if the velocity

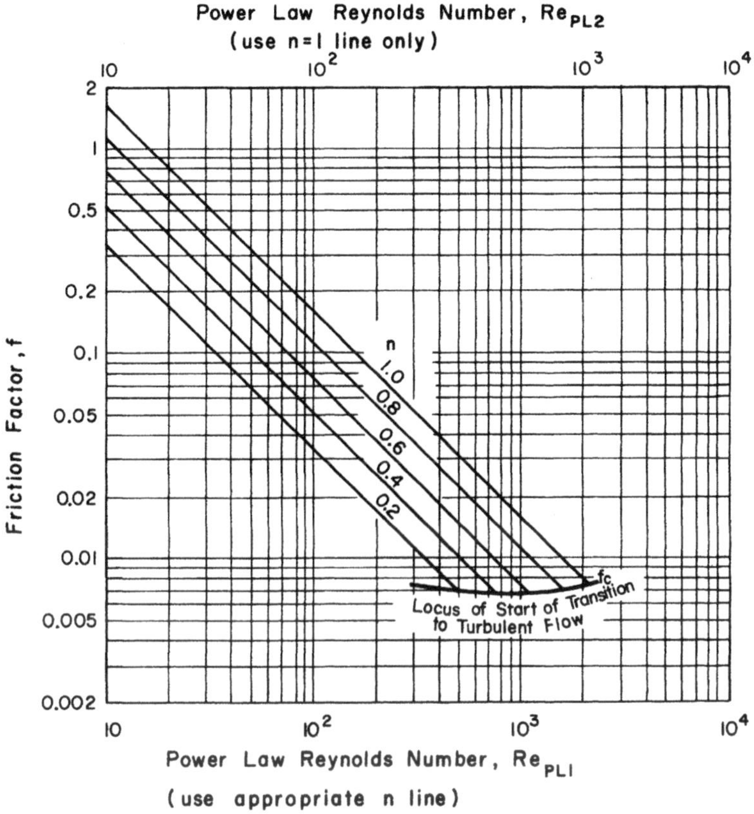

Figure 5.4. Friction factors for laminar flow of power law pseudoplastics. Transition locus line calculated from Equation 5.70.

profile were truncated and taken as pluglike in the region $r = 0$ to $r = 0.5R$, the error in the volumetric discharge calculated from the profile would be only 6 percent for n close to unity and 3 percent for $n = 0.5$. Accordingly, if the power law is properly fitted to the range $\dot{S} \leq (du/dr)_w$ or $\tau \leq \tau_w$, this simple rheological equation will permit calculations of pressure drop and flow rate in laminar flow in circular pipes of accuracy entirely satisfactory for normal engineering purposes. We should note that this conclusion does not apply to flow in annuli, where the local velocity remains finite in the region in which the shear stress becomes zero, and the fit of the power law equation becomes poor.

Notwithstanding the eminent suitability of a properly fitted power law equation for ordinary laminar pipe flow problems involving the pressure drop–flow rate relationship, it is not suitable for determining the complete velocity or rate of shear profile. For this reason, and also to give greater breadth to our discussion, we continue the analysis by considering the laminar flow of pseudoplastics described by a few of the other rheological equations discussed in Chapter 1.

2.22 The Ellis Fluid

The approach of Section 2.21 may be followed for a fluid described by the Ellis equation (see Skelland, 1967)

$$\dot{S} = (\phi_0 + \phi_1 \tau^{\alpha-1})\tau \tag{1.42}$$

to yield the following relations, derivable directly from Equations 1.42 and 5.6:

$$\frac{du}{dr} = \left(\frac{r}{2}\frac{dP}{dL}\right)\left[\phi_0 + \phi_1\left(-\frac{r}{2}\frac{dP}{dL}\right)^{\alpha-1}\right] \tag{5.32}$$

$$u = \phi_0\left(-\frac{dP}{4dL}\right)(R^2 - r^2) + \frac{\phi_1}{\alpha+1}\left(-\frac{dP}{2dL}\right)^{\alpha}(R^{\alpha+1} - r^{\alpha+1}) \tag{5.33}$$

$$Q = \frac{\pi D^3}{8}\left[\frac{\phi_0}{4}\left(\frac{D\Delta P}{4L}\right) + \frac{\phi_1}{\alpha+3}\left(\frac{D\Delta P}{4L}\right)^{\alpha}\right] \tag{5.34}$$

$$V = \frac{D}{2}\left[\frac{\phi_0}{4}\left(\frac{D\Delta P}{4L}\right) + \frac{\phi_1}{\alpha+3}\left(\frac{D\Delta P}{4L}\right)^{\alpha}\right] \tag{5.35}$$

$$\left(\frac{du}{dr}\right)_w = \left(\frac{R}{2}\frac{dP}{dL}\right)\left[\phi_0 + \phi_1\left(-\frac{R}{2}\frac{dP}{dL}\right)^{\alpha-1}\right] \tag{5.36}$$

$$u_m = \phi_0\left(-\frac{dP}{4dL}\right)R^2 + \frac{\phi_1}{\alpha+1}\left(-\frac{dP}{2dL}\right)^{\alpha}R^{\alpha+1} \tag{5.37}$$

$$\frac{V}{u_m} = \frac{\alpha+1}{2(\alpha+3)}\left[\frac{(\alpha+3)\phi_0\tau_w^{\alpha-1} + 4\phi_1}{(\alpha+1)\phi_0\tau_w^{\alpha-1} + 2\phi_1}\right] \tag{5.38}$$

Because the shear rate and velocity profiles given by Equation 5.32 and 5.33 depend upon all the three rheological constants of the Ellis equation, it is not possible to illustrate the profiles as in the case of the power law fluid.

Equations 5.34 and 5.36 may also be obtained directly from Equations 5.9 and 5.10 by substituting

$$f(\tau) = (\phi_0 + \phi_1\tau^{\alpha-1})\tau$$

In principle, Equation 5.35 could be put in friction factor form analogous to Equation 5.26, but the complexity of the relationship would be such that the result would be less convenient to use than Equation 5.35 itself.

2.23 The Powell-Eyring Fluid

The methods of Section 2.21 may be extended to pseudoplastics described by the Powell–Eyring equation (Christiansen, Ryan, and Stevens, 1955, and Powell and Eyring, 1944)

$$\tau = \left(\frac{\mu_\infty}{g_c}\right)\dot{S} + \frac{1}{B'}\sinh^{-1}\left(\frac{1}{A'}\dot{S}\right) \tag{5.39}$$

where the units of the constants are as follows:

$$\mu_\infty = \text{lb}_\text{m}/\text{ft sec}$$
$$B' = \text{ft}^2/\text{lb}_\text{f}$$
$$A' = \text{sec}^{-1}$$

Following Hanks (1963) and Sisko (1958), we may rewrite Equation 5.39 as

$$r^* T_w = \omega\Gamma + \sinh^{-1}\Gamma \tag{5.40}$$

where $r^* = \dfrac{r}{R}$,

$$T_w = B'\tau_w,$$

$$\Gamma = \frac{1}{A'}\dot{S},$$

$$\omega = A'B'(\mu_\infty/g_c).$$

Substituting Equation 5.40 into Sisko's (1958) equation for the volumetric flow rate, which is in the form

$$\frac{8Q}{\pi D^3 A'} = \frac{1}{3}\Gamma_w - \frac{1}{3}\int_0^{\Gamma_w}(r^*)^3 d\Gamma \tag{5.41}$$

and integrating, we obtain

$$\begin{aligned}\frac{8Q}{\pi D^3 A'} = \frac{1}{3}\Gamma_w - \frac{1}{3T_w^3}\Bigg\{&\frac{1}{4}\omega^3\Gamma_w^4 + \omega^2\Bigg[\Gamma_w^3\sinh^{-1}\Gamma_w - \Gamma_w^2\sqrt{1+\Gamma_w^2} \\ &+ \frac{2}{3}(1+\Gamma_w^2)^{3/2} - \frac{2}{3}\Bigg] + \frac{3}{4}\omega[(2\Gamma_w^2+1)(\sinh^{-1}\Gamma_w)^2 - 2\Gamma_w \\ &\times\sqrt{1+\Gamma_w^2}\sinh^{-1}\Gamma_w + \Gamma_w^2] + \Gamma_w(\sinh^{-1}\Gamma_w)^3 \\ &- 3\sqrt{1+\Gamma_w^2}(\sinh^{-1}\Gamma_w)^2 + 6[\Gamma_w\sinh^{-1}\Gamma_w - \sqrt{1+\Gamma_w^2}+1]\Bigg\}\end{aligned} \tag{5.42}$$

where $T_w = \omega\Gamma_w + \sinh^{-1}\Gamma_w$.

As in the case of the Ellis fluid, the equations describing the pipe flow of a pseudoplastic following the Powell–Eyring equation are too complex to represent graphically or to put in friction factor form. Nonetheless the Powell–Eyring relations are useful where a close representation of the behavior of a pseudoplastic is desired over wide ranges of shear rates.

2.24 The Meter Fluid

The four-constant Meter equation (Meter, 1963; Meter and Bird, 1964) described in Chapter 1 may also be used to develop an equation for the laminar flow of a pseudoplastic. Following the procedure we have used in arriving at the other relations and starting from the relation

$$\tau = \frac{\mu_a}{g_c}\dot{S}$$

where

$$\mu_a = \mu_\infty + \frac{\mu_0 - \mu_\infty}{1 + (\tau/\tau_m)^{\alpha-1}} \tag{1.46}$$

we obtain (Meter and Bird, 1964),

$$u = \frac{\tau_w R g_c}{2\mu_0}\left\{\left[1 - \left(\frac{r}{R}\right)^2\right] + \frac{2}{\alpha+1}\left(\frac{\tau_w}{\tau_m}\right)^{\alpha-1}\left[1 - \left(\frac{r}{R}\right)^{\alpha+1}\right]\right\}$$
$$- \left(\frac{\mu_\infty}{\mu_0}\right)\left(\frac{\tau_w}{\tau_m}\right)^{\alpha-1}\left[\left(\frac{2}{\alpha+1}\right)(1 - r^{*\alpha+1}) + \frac{1}{\alpha}\left(\frac{\tau_w}{\tau_m}\right)^{\alpha-1}(1 - r^{*2\alpha})\right] + \mathcal{O}\left[\left(\frac{\mu_\infty}{\mu_0}\right)^2\right] \tag{5.43}$$

$$Q = \frac{\pi R^3 \tau_w g_c}{4\mu_0}\left[1 + \frac{4}{\alpha+3}\left(\frac{\tau_w}{\tau_m}\right)^{\alpha-1}\right]$$
$$- \left(\frac{\mu_\infty}{\mu_0}\right)\left(\frac{\tau_w}{\tau_m}\right)^{\alpha-1}\left[\frac{4}{\alpha+3} + \frac{2}{\alpha+1}\left(\frac{\tau_w}{\tau_m}\right)^{\alpha-1}\right] + \mathcal{O}\left[\left(\frac{\mu_\infty}{\mu_0}\right)^2\right] \tag{5.44}$$

$$V = \frac{\tau_w R g_c}{4\mu_0}\left[1 + \frac{4}{\alpha+3}\left(\frac{\tau_w}{\tau_m}\right)^{\alpha-1}\right]$$
$$- \left(\frac{\mu_\infty}{\mu_0}\right)\left(\frac{\tau_w}{\tau_m}\right)^{\alpha-1}\left[\frac{4}{\alpha+3} + \frac{2}{\alpha+1}\left(\frac{\tau_w}{\tau_m}\right)^{\alpha-1}\right] + \mathcal{O}\left[\left(\frac{\mu_\infty}{\mu_0}\right)^2\right] \tag{5.45}$$

$$f = \frac{128}{\left[\left(\frac{D^2\tau_m\rho}{\mu_0^2}\right)\left(\frac{\tau_w}{\tau_m}\right)\left\{\left[1 + \frac{4}{\alpha+3}\left(\frac{\tau_w}{\tau_m}\right)^{\alpha-1}\right] - \left(\frac{\mu_\infty}{\mu_0}\right)\left(\frac{\tau_w}{\tau_m}\right)^{\alpha-1}\left[\frac{4}{\alpha+3} + \frac{2}{\alpha+1}\left(\frac{\tau_w}{\tau_m}\right)^{\alpha-1}\right] + \cdots\right\}^2\right]} \tag{5.46}$$

where $\mathcal{O}$ means "a quantity of the order of." We may note from Equation 5.46 that in laminar flow the friction factor for a Meter fluid depends upon four dimensionless ratios, corresponding to the four parameters of the Meter equation.

2.25 The Sisko Relations

Sisko's constitutive equation based on simple additive Newtonian and power law non-Newtonian stresses, as discussed in Chapter 1, provides the basis for his relationships for laminar flow in a circular tube. In terms of true wall shear rate or wall shear stress, the relationships are complex and Sisko (1958) finds it convenient to express them in terms of apparent wall shear rate, $8V/D$. The equations will not be given here, but we may note that Sisko reports excellent agreement between his calculated velocity profiles and pressure gradients and actual data on a series of hydrocarbon greases. Figure 5.5

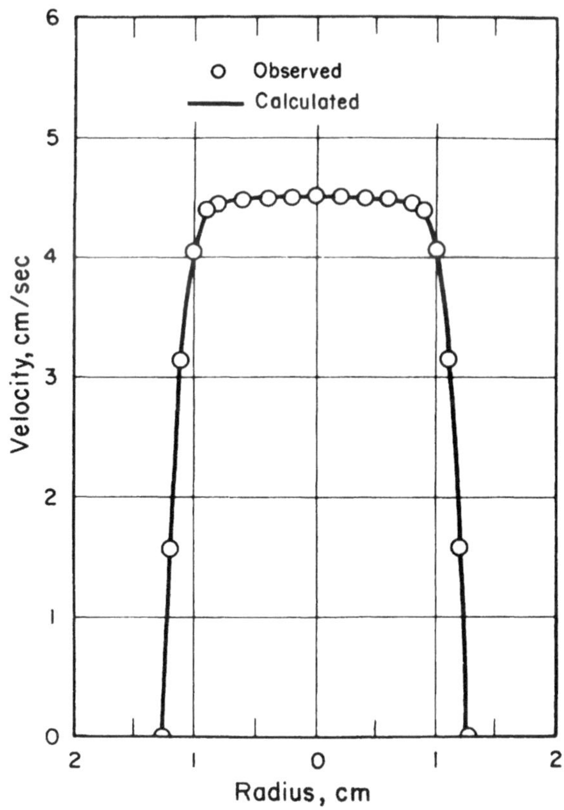

Figure 5.5. Comparison of observed and calculated velocity profiles for flow of grease (pipe radius 1.27 cm). (From Sisko, A. W., *Ind. Eng. Chem.*, **50**, 1789, 1958.)

illustrates the agreement between the predicted and measured velocity profile for a flowing hydrocarbon grease.

2.3 The Relations for a Bingham Fluid

Again, the same approach as that taken in Section 2.21 for the power law fluid may be applied to the Bingham fluid. Combining Equation 5.6 and the equation of the Bingham fluid (Equation 1.47), we obtain

$$\dot{S} = -\frac{g_c}{\eta}\left(\frac{r}{2}\frac{dP}{dL} + \tau_y\right) \tag{5.47}$$

In this case we observe that, when $\tau = -rdP/2dL \leq \tau_y$, the fluid does not shear and adjacent laminae are stationary relative to one another. This occurs for values of $r \leq r_P$, where

$$r_P = \frac{2\tau_y}{\left(-\dfrac{dP}{dL}\right)}$$

For $R > r > r_P$ the laminae of the fluid are in relative motion and we may integrate Equation 5.47. Inserting $u = 0$ when $r = R$, we obtain

$$u = \frac{g_c}{\eta}\left[-\frac{dP}{4dL}(R^2 - r^2) - \tau_y(R - r)\right] \tag{5.48}$$

When $0 < r < r_P$ the fluid moves as a plug at a uniform velocity given by

$$u_P = \frac{g_c}{\eta}\left\{-\frac{dP}{4dL}\left[R^2\left(\frac{2\tau_y}{-\dfrac{dP}{dL}}\right)^2\right] - \tau_y\left[R - \left(\frac{2\tau_y}{-\dfrac{dP}{dL}}\right)\right]\right\} \tag{5.49}$$

We obtain the discharge as the sum of the flow through the sheared region $(R > r > r_P)$ and the plug $(r < r_P)$ as follows:

$$Q = \int_{r_P}^{R} 2\pi r u\, dr + \pi r_P^2 u_P \tag{5.50}$$

This simplifies to

$$\frac{2V}{D} = \frac{8Q}{\pi D^3} = \frac{g_c}{4\eta}\left(\frac{D\Delta P}{4L}\right)\left[1 - \frac{4\tau_y}{3}\left(\frac{4L}{D\Delta P}\right) + \frac{\tau_y^4}{3}\left(\frac{4L}{D\Delta P}\right)^4\right] \tag{5.51}$$

Equation 5.51 is known as the Buckingham equation after Buckingham (1921) who first developed it.

Equations 5.47 and 5.48 provide the means for calculating the shear rate and velocity profiles in terms of the pressure gradient and the yield stress. Illustrative profiles for Bingham fluids for a range of values of the ratio $\tau_y g_c/\eta$ are given in Figures 5.6 and 5.7.

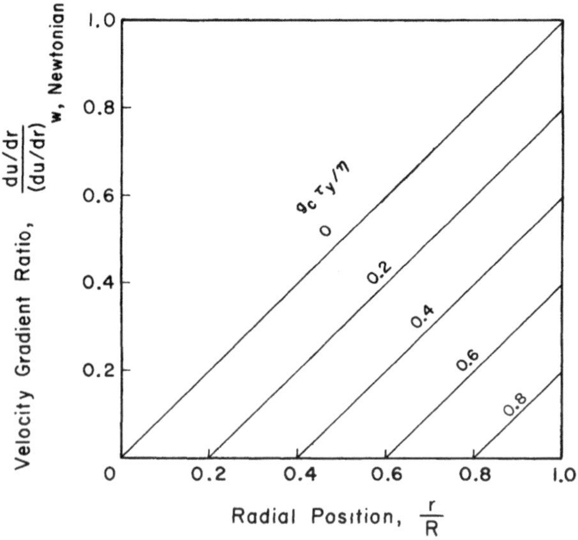

Figure 5.6. Effect of yield value on the velocity gradient in laminar flow.

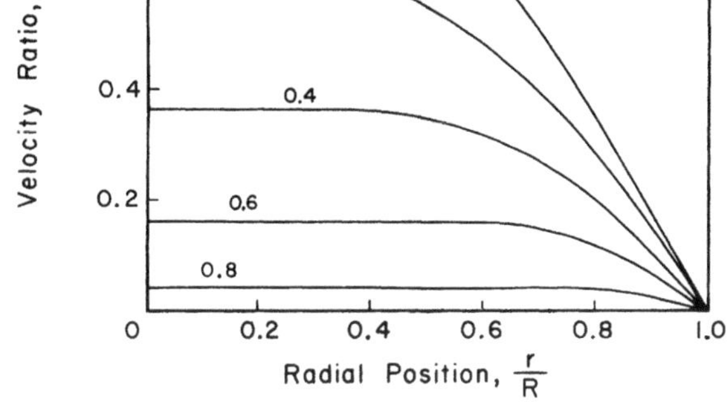

Figure 5.7. Effect of yield value on the velocity profile in laminar flow.

Equation 5.51 may be rearranged to the following friction factor form

$$f = \frac{16\eta}{DV\rho}\left(\frac{1}{1 - 4x/3 + x^4/3}\right) \tag{5.52}$$

where $x = \dfrac{\tau_y}{\tau_w} = \dfrac{4L\tau_y}{D\Delta P}$

By dimensional analysis Govier and Winning (1948) showed that for a Bingham plastic

$$f = \phi_1\left(\frac{DV\rho}{\eta}, \frac{D\tau_y g_c}{V\eta}\right) \tag{5.53}$$

where $DV\rho/\eta = \mathrm{Re}_B$ is a Bingham modified Reynolds number and $D\tau_y g_c/V\eta = Y$ was termed the yield number. They correlated experimental data in terms of these groups and in agreement with the Buckingham equation. Hedstrom (1952), also by dimensional analysis, showed that

$$f = \phi_2\left(\frac{DV\rho}{\eta}, \frac{D^2\tau_y \rho g_c}{\eta^2}\right) \tag{5.54}$$

The second of these groups has been termed the Hedstrom number, He, and is simply the product of the Bingham Reynolds number and the yield number.

Hedstrom (1952), and others (Perkins and Glick, 1954; Metzner, 1956, and Govier, 1959) have shown that Equation 5.51 or 5.52 may also be rearranged without further assumption to either a friction factor–Bingham Reynolds number–yield number, or a friction factor–Bingham Reynolds number–Hedstrom number form. The results are

$$\frac{1}{\mathrm{Re}_B} = \frac{f}{16} - \frac{Y}{6\,\mathrm{Re}_B} + \frac{Y^4}{3f^3\,\mathrm{Re}_B^4} \tag{5.55}$$

and

$$\frac{1}{\mathrm{Re}_B} = \frac{f}{16} - \frac{\mathrm{He}}{6\,\mathrm{Re}_B^2} + \frac{\mathrm{He}^4}{3f^3\,\mathrm{Re}_B^8} \tag{5.56}$$

Equation 5.55 is the simpler but it has the disadvantage of containing the velocity in both the yield number and the Bingham Reynolds number, whereas Equation 5.56 contains the velocity only in the Reynolds term.

Rearrangement of Equation 5.55 indicates that the product ($f\,\mathrm{Re}_B$) is a simple function of Y having a value ranging upward from 16 when $Y = 0$ (Newtonian fluid). Govier's (1959) plots of $f\,\mathrm{Re}_B$ versus Y and f versus Re_B at constant values of Y are reproduced in Figures 5.8 and 5.9. Hedstrom's similar figure of f versus Re_B at constant values of the Hedstrom number is shown in Figure 5.10. Each of these figures is the equivalent of the Buckingham equation and gives the equivalent of the relationship of Equation 5.51 between average velocity and pressure gradient.

Figures 5.9 and 5.10 suggest that the transition from laminar to turbulent flow for Bingham plastics occurs at a substantially higher Reynolds number than for Newtonian fluids. We will see that this in fact is the case. The transition line shown in Figure 5.9 is derived from Figure 5.21 discussed later.

McMillen (1948) has rearranged the Buckingham equation in terms of dimensionless ratios and presented generalized solutions of the equation in numerical and graphical forms. For normal pipeline calculations, it is doubtful if these solutions have any advantage over the direct application of Equations 5.55 and 5.56 or Figures 5.9 or 5.10.

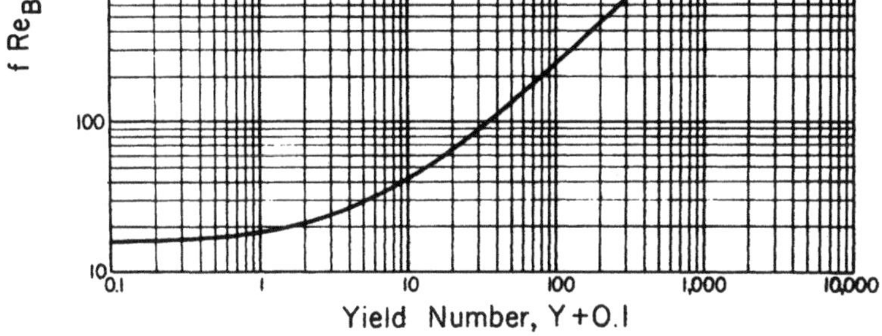

Solutions for the Buckingham Equation (Eq.5.55)

Yield Number, Y	Product, $f\,\mathrm{Re}_B$
0	16.00
1	18.67
2	21.32
3	23.97
5	29.20
10	41.94
20	66.42
30	90.09
50	136.1
100	247.5
200	463.8
300	676.4
500	1,096
1,000	2,133
2,000	4,186
3,000	6,226
5,000	10,290
10,000	20,407

Figure 5.8. Friction factor–Bingham Reynolds number products for laminar flow of Bingham plastics. (From Govier, G. W., *Chem. Eng.*, **66**, 138, 1959, Reprinted by special permission of McGraw-Hill, Inc., New York.)

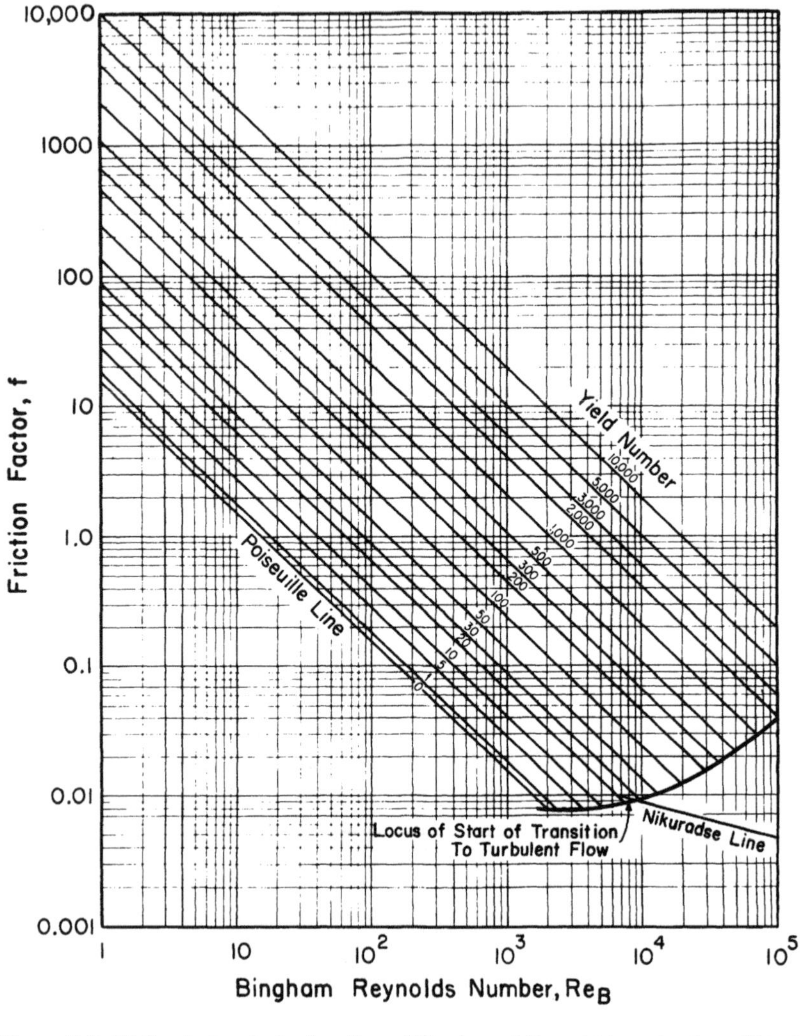

Figure 5.9. Friction factors for laminar flow of Bingham platics–yield number form. Transition locus line derived from Figure 5.21. (Modified after Govier, G. W., *Chem. Eng.*, **66**, 138, 1959. Reprinted by special permission of McGraw Hill, Inc., New York.)

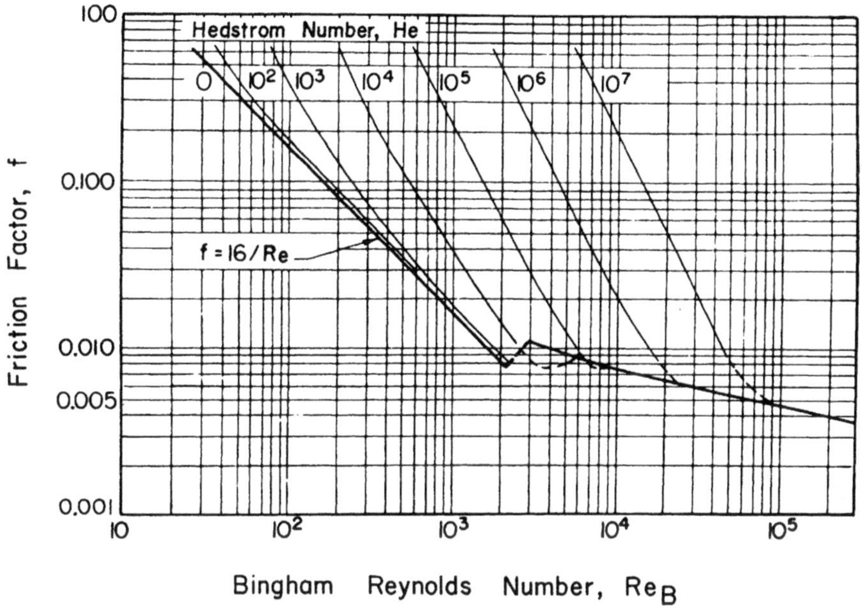

Figure 5.10. Friction factors for laminar flow of Bingham plastics–Hedstrom number form. (From A. B. Metzner in *Advances in Chemical Engineering*, Academic Press Inc., New York, 1956, Vol. 1, p. 77; original from Hedstrom, B.O.A., *Ind. Eng. Chem.*, **44**, 651, 1952.)

2.4 The Relations for a Yield-Pseudoplastic Fluid

We will consider only one example of a yield-pseudoplastic, the fluid for which shear stresses in excess of the yield stress are related to the rate of shear through the power law as expressed by Equation 1.48

$$\tau - \tau_y = \frac{K}{g_c} \dot{S}^n \tag{1.48}$$

If we follow the procedure of Section 2.21 for such a fluid, we obtain

$$-\frac{du}{dr} = \left(\frac{g_c}{K}\right)^{1/n} \left(-\frac{r}{2}\frac{dP}{dL} - \tau_y\right)^{1/n} \tag{5.57}$$

which may be integrated with $u = 0$ when $r = R$ to give

$$u = \frac{\left(\frac{g_c}{K}\right)^{1/n}}{\left(-\frac{dP}{2dL}\right)} \frac{n}{1+n} \left[\left(-\frac{R}{2}\frac{dP}{dL} - \tau_y\right)^{(1+n)/n} - \left(-\frac{r}{2}\frac{dP}{dL} - \tau_y\right)^{(1+n)/n}\right] \tag{5.58}$$

Integrating once again to obtain the volumetric discharge, we have

$$\frac{2V}{D} = \frac{8Q}{\pi D^3} = \frac{\left(\frac{g_c}{K}\right)^{1/n}}{\left(\frac{D\Delta P}{4L}\right)^3} \left(\frac{D\Delta P}{4L} - \tau_y\right)^{(1+n)/n}$$

$$\times \left[\frac{\left(\frac{D\Delta P}{4L} - \tau_y\right)^2}{\frac{1+3n}{n}} + \frac{2\tau_y\left(\frac{D\Delta P}{4L} - \tau_y\right)}{\frac{1+2n}{n}} + \frac{\tau_y^2}{\frac{1+n}{n}} \right] \tag{5.59}$$

This equation reduces to Equation 5.51 when $n = 1$ (Bingham fluid) and to Equation 5.23 when $\tau_y = 0$. The complexity of Equation 5.59 is such that it cannot readily be converted to friction factor form, but we may see that

$$f = \phi\left(\frac{D^n V^{2-n}\rho}{K}, \frac{D\tau_y g_c}{V\eta}, \text{terms involving } n\right) \tag{5.60}$$

In practice, Equation 5.59, if being employed to estimate pressure gradient, is best solved by trial and error. For a first trial the Buckingham equation (n taken equal to unity) may be used. The actual pressure gradient for the yield-pseudoplastic will be somewhat less than for the assumed Bingham plastic.

2.5 Viscoelastic Fluids

In steady-state laminar flow of a viscoelastic fluid in a pipe of constant diameter, none of the elastic forces enter the microscopic momentum or equivalent force balance. This fact may be rigorously demonstrated on simplification of the general equation of motion given in Chapter 3 (Equation 3.8). Sylvester and Rosen (1970) present experimental data confirming that in steady stabilized laminar flow the pressure gradient is unaffected by the elastic properties of a fluid. Thus the flow is dependent only on the viscous properties of the material, and we may treat the laminar steady-state pipe flow of viscoelastic fluids by applying to them the most appropriate of the rheological equations applicable to purely viscous fluids, usually those of pseudoplastics.

2.6 The Metzner and Reed Generalized Approach

Metzner and Reed (1955) have developed a generalized approach applicable to the laminar pipe flow of any time-independent fluid, based upon the Rabinowitsch equation in the form of Equation 1.67

$$-\left(\frac{du}{dr}\right)_w = \frac{8V}{D}\left(\frac{1+3n'}{4n'}\right) \tag{1.67}$$

where $n' = \dfrac{d\ln(D\Delta P/4L)}{d\ln(8V/D)}$

The coefficient n' is determinable from the slope of a log-log plot of $D\Delta P/4L$ versus $8V/D$ where $\Delta P/L$ is the pressure gradient at a flow velocity V in a pipe or capillary tube of diameter D under laminar flow conditions. The log-log plot itself characterizes the rheological properties of the non-Newtonian fluid and may be used directly for pipe scale up or design purposes.

Metzner and Reed carry the approach further by writing the equation

$$\frac{D\Delta P}{4L} = K'\left(\frac{8V}{D}\right)^{n'} \tag{5.61}$$

which represents the tangent to the logarithmic plot of $D\Delta P/4L$ versus $8V/D$ at any particular value of $D\Delta P/4L$ or $8V/D$. (If the plot is a straight line, then Equation 5.61 is the equation of the line.) Now if we recall that

$$f = \frac{g_c D\Delta P}{2\rho V^2 L} \tag{4.24}$$

and define a generalized Reynolds number, Re_{MR} such that

$$f = 16/\mathrm{Re}_{MR} \tag{5.62}$$

we may combine Equations 4.24 and 5.62 and substitute $D\Delta P/4L$ from Equation 5.61 to obtain the following solution for Re_{MR} :

$$\mathrm{Re}_{MR} = \frac{D^{n'}V^{2-n'}\rho}{\gamma} \tag{5.63}$$

where

$$\gamma = g_c K' 8^{n'-1}$$

Equation 5.62 with Re_{MR} defined by Equation 5.63 then represents a generalized relationship applicable to the laminar pipe flow of any time-independent fluid. Figure 5.11 taken from Metzner and Reed (1955), illustrates the conformity of data from a variety of pseudoplastic, yield-pseudoplastic, and

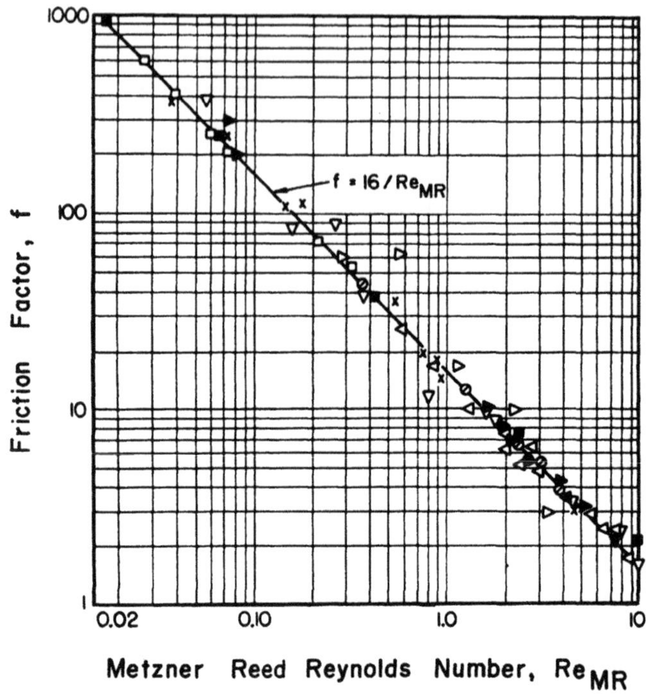

Figure 5.11. Friction factors for laminar flow of non-Newtonian fluids in general. (From Metzner, A. B. and J. C. Reed, *A.I.Ch.E. Journal*, **1**, 434, 1955.)

Bingham-like fluids over an intermediate range of Re_{MR}. The source of the data shown in the figure is given in Table 5.1. Metzner and Reed indicate that stable laminar flow of time-independent non-Newtonian fluids extends to a value of Re_{MR} of 2000 to 2500.

The Metzner and Reed generalized approach has been verified for dilatant materials by Griskey and Green (1971), who conducted tests with suspensions of corn starch in ethylene glycol and in ethylene glycol-glycerine with small amounts of water in a $1\frac{1}{4}$ inch pipe. These dilatant mixtures are described by the power law (Equation 5.61) with n' in the range of 1.15 to 2.50. In the range of $Re_{MR} = 12$ to 400 (velocity from 0.37 to 1.88 ft/sec) the measured pressure drops were within 5% of the values predicted by the friction factor—Reynolds number relationship of Metzner and Reed. Griskey and Green draw attention to the fact that when $n' > 2$ the friction factor increases with increasing velocity, contrary to the behavior of Newtonian and pseudoplastic fluids, but in accordance with the Metzner-Reed correlation for $n' > 2$

TABLE 5.1

Rheological Properties[a] for Fluids Shown in Figure 5.11

Symbol used in Figure	Nominal Pipe Size, in.	Composition of Fluid	Rheological Properties n'	γ	Source of Data
⊘	$\frac{7}{8}$ and $1\frac{1}{2}$	3.0% CMC in water	0.566	2.80	1
◁	$\frac{7}{8}$ and $1\frac{1}{2}$	10% napalm in kerosene	0.520	1.18	1
▼	8, 10, and 12	4% paper pulp in water	0.575	6.13	2
●	$\frac{3}{4}$ and $1\frac{1}{4}$	14.3% clay in water	0.350	0.0344	3
▷	$\frac{3}{4}$ and $1\frac{1}{4}$	21.2% clay in water	0.335	0.0855	3
×	$\frac{3}{4}$ and $1\frac{1}{4}$	25.0% clay in water	0.185	0.204	3
▽	$\frac{3}{4}$ and $1\frac{1}{4}$	31.9% clay in water	0.251	0.414	3
□	$\frac{3}{4}$ and $1\frac{1}{4}$	36.8% clay in water	0.176	1.07	3
■	$\frac{3}{4}$ and $1\frac{1}{4}$	40.4% clay in water	0.132	2.30	3
▶	$\frac{1}{8}$, $\frac{1}{4}$, $\frac{1}{2}$ and 2	23% lime in water	0.178	1.04	4

[a] A. B. Metzner and J. C. Reed, *A.I. Ch. E. Journal*, **1**, 434 (1955).

SOURCE OF DATA:
1. W. E. Stevens, Doctoral Dissertation, Univ. of Utah, Salt Lake City, 1953.
2. Allis-Chalmers Mfg. Co. Bulletin No. 1649.
3. G. W. Govier and M. D. Winning, A.I.Ch.E. Meeting, Montreal, Que., 1948.
4. G. E. Alves *et al.*, *Chem. Eng. Progr.*, 48, 385 (1952).

resulting in the velocity term of the Reynolds number being raised to a negative power.

Metzner (1957) also demonstrates the relationship between n', K', and γ and the actual rheological properties of the different types of fluids. For a Newtonian fluid,

$$n' = 1; \qquad \gamma = \mu; \qquad K' = \frac{\mu}{g_c}$$

For a power law pseudoplastic,

$$n' = n; \qquad K' = \frac{K}{g_c}\left(\frac{1+3n}{4n}\right)^n; \qquad \gamma = K\left(\frac{1+3n}{4n}\right)^n 8^{n-1}$$

For a Bingham plastic,

$$n' = \frac{1 - 4x/3 + x^4/3}{1 - x^4}$$

$$K' = \tau_w \left[\frac{\eta}{g_c \tau_w (1 - 4x/3 + x^4/3)} \right]^{n'}$$

$$\gamma = g_c 8^{n'-1} \tau_w \left[\frac{\eta}{g_c \tau_w (1 - 4x/3 + x^4/3)} \right]^{n'}$$

The generalizations of Metzner and Reed are of great value in instances where the rheological behavior of the fluid is not adequately described by one of the simpler constitutive equations, or when one is involved in the direct scale-up from data taken in a small diameter pipe. For the simple cases of the power law pseudoplastic and the Bingham plastic, however, the previously derived relations are at least equally convenient and a better appreciation of the transition from laminar to turbulent flow may be obtained from Figures 5.4 and 5.9 than from the Metzner and Reed generalization.

2.7 *Entrance Effects*

As in the case of Newtonian fluids, the velocity profile for pseudoplastic, Bingham, and yield-pseudoplastic fluids does not develop immediately upon the entrance of the fluid to a pipe but a boundary layer is generated and gradually extends toward the pipe center line to produce the ultimate velocity profile.

The entrance effect for a power law pseudoplastic has been studied by Bogue (1959), Tomita (1961), and Collins and Schowalter (1963). Each of these workers based his analysis on the fundamental equation of motion, the continuity equation, and the power law constitutive equation, but each introduced different assumptions and approximation techniques. Collins and Schowalter's work is probably the most accurate and is somewhat more conservative than that of Bogue, predicting slightly higher pressure losses in the range $0.4 < n < 1.0$. Boger and Ramamurthy (1970) give data for a number of purely viscous fluids that fall between the predictions of Bogue and of Collins and Schowalter.

Collins and Schowalter's results are expressed in terms of the power law Reynolds number, $\text{Re}_{PL1} = D^n V^{2-n} \rho / K$ and the ratio X/D, where X is the travel distance from the entrance. Figure 5.12 taken from their work gives values of $2X_D/D \, \text{Re}_{PL1}$ as a function of n, where X_D is the travel distance for the development of a center line velocity equal to 98 percent of the stabilized value. Figures 5.13a, 5.13b, and 5.13c give Collins and Schowalter's velocity profile results at travel distances of $0.25X_D$, $0.50X_D$, and X_D.

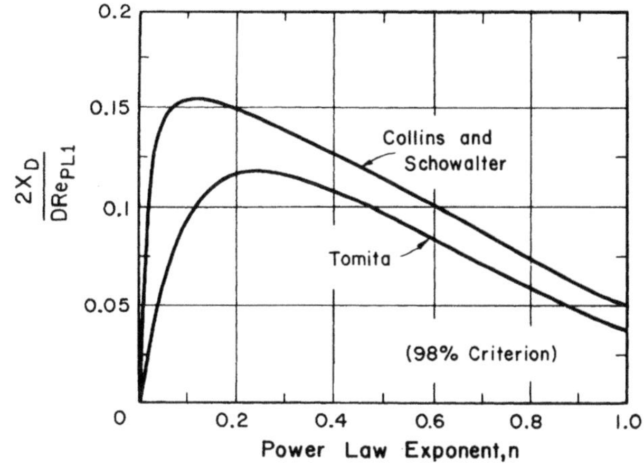

Figure 5.12. Entrance length for laminar flow of power law pseudoplastics. (From Collins M. and W. R. Schowalter, *A.I.Ch.E. Journal*, **9**, 804, 1963.)

For the power law fluid the pressure drop in the entrance length X_D may be expressed as

$$\frac{2\Delta P g_c}{\rho V^2} = \frac{2^{n+2}\left(\frac{1+3n}{n}\right)^n}{\mathrm{Re}_{PL1}}\left(\frac{2X_D}{D}\right) + C \tag{5.64}$$

where C is an excess dimensionless pressure drop which includes the kinetic energy effect due to the acceleration of the fluid from zero velocity. Collins and Schowalter's results are given in Figure 5.14. These results are confirmed for Newtonian fluids ($n = 1$) by data of Kreith and Eisenstadt (1957) and for carbopol and CMC (power law fluids) by data of Dodge quoted by Bogue (1959).

Hornbeck (1965) has performed a similar analysis for the Powell–Eyring fluid. His results,which include a solution for the fully developed pressure gradient, are expressed in terms of dimensionless ratios which incorporate the three rheological constants of the Powell–Eyring equation. Hornbeck's solutions should be referred to in cases where the rate of shear varies over a wide range and where a rheological model, which properly reflects the approach to Newtonian behavior at low and high shear rates, is warranted.

For the Bingham fluid, Michiyoshi, Mizuno, and Hoshiai (1966) have carried out a theoretical analysis of the entrance region of laminar pipe flow. They express their entrance length results in terms of a relationship between $(X_E/D)(1/\mathrm{Re}_B)$ and the dimensionless ratio $x = \tau_y/\tau_w$ as shown in Figure 5.15. For the case of $x = 0$ (Newtonian fluid) their relationship gives $X_E/D\ \mathrm{Re}_B =$

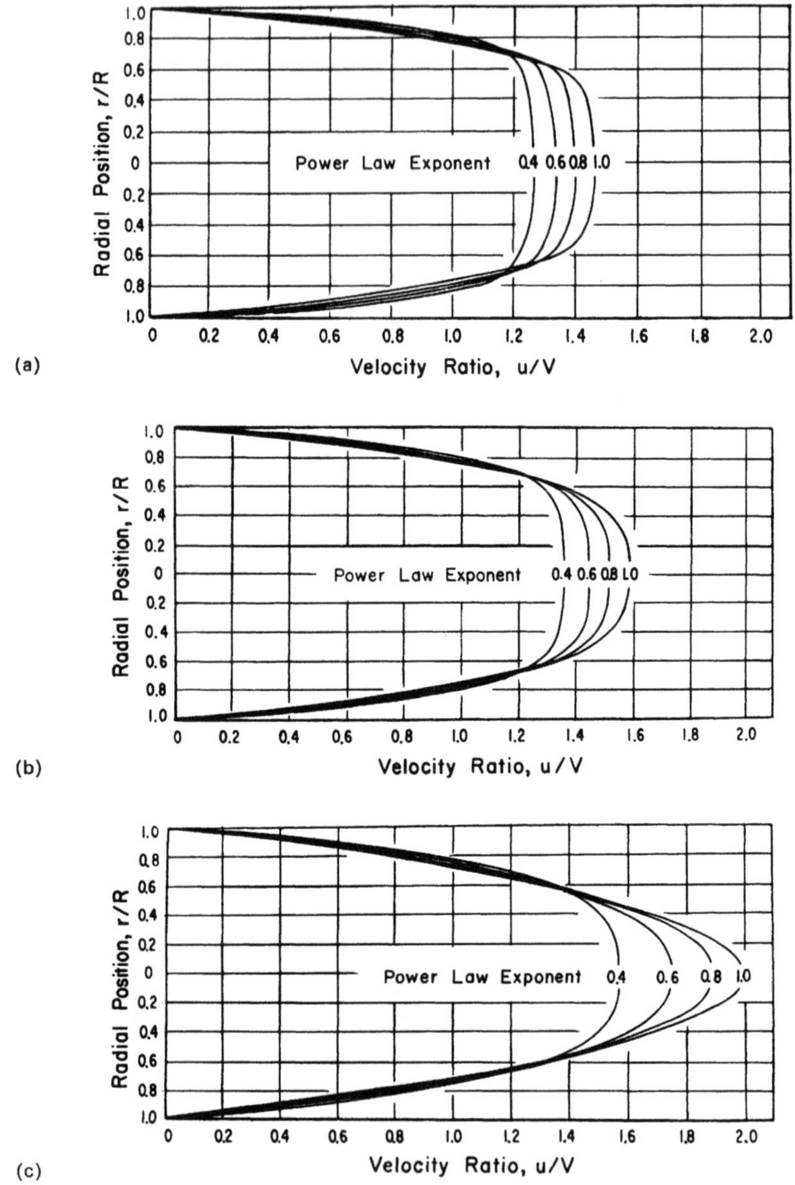

Figure 5.13. Velocity profiles for the laminar flow of power law pseudoplastics
a. Downstream distance, $X = 0.25X_D$
b. Downstream distance, $X = 0.5X_D$
c. Downstream distance, $X = X_D = 0.98X_E$
(From Collins M., and W. R. Schowalter, *A.I.Ch.E. Journal*, **9**, 804, 1963.)

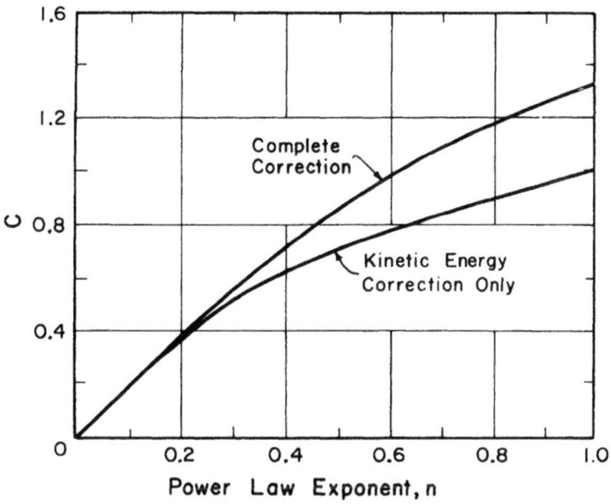

Figure 5.14. Excess dimensionless entrance length pressure drop, *C*, for the laminar flow of power law pseudoplastics, (From Collins, M., and W. R. Schowalter, *A.I.Ch.E. Journal*, **9** 804, 1963.)

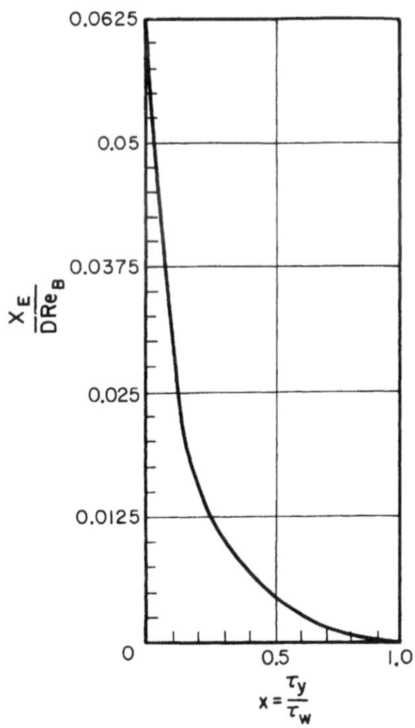

Figure 5.15. Entrance length for the laminar flow of a Bingham fluid. (From Michiyosi, L., K. Mizuno, and Y. Hoshiai, *Int. Chem. Eng.*, **6**, 373, 1966.)

0.061 compared with Nikuradse's experimental value, as reported by Michiyoshi *et al.*, of 0.0625 and a value of approximately 0.04 for u_0/u_{max} of about 0.95, as indicated in Figure 4.6. The relationship developed by Michiyoshi *et al.* for the pressure gradient in the entrance region is

$$\frac{2\Delta P g_c}{\rho V^2} = \frac{96}{\mathrm{Re}_B(x^4 - 4x + 3)}\left(\frac{2X_E}{D}\right) + C \tag{5.65}$$

where the first term is the pressure drop under stabilized flow conditions and C is the excess pressure drop. Figure 5.16 gives their solution for C as a function of x. It is to be noted that the excess dimensionless pressure drop in the entrance region decreases, from 1.25 for $x = 0$ (Newtonian), as x increases and becomes 0 for $x = 1$ (full diameter plug flow). The analysis of Michiyoshi *et al.* is supported by data for water and water suspensions of finely divided alumina.

Chen, Fan, and Hwang (1970) criticize the analysis of Michiyoshi *et al.* on the grounds that certain of the assumptions are unsupported. Chen *et al.* develop relationships comparable to those of Figures 5.15 and 5.16 but reflecting both a higher entrance length and a higher pressure drop correction. Their paper should be consulted for details.

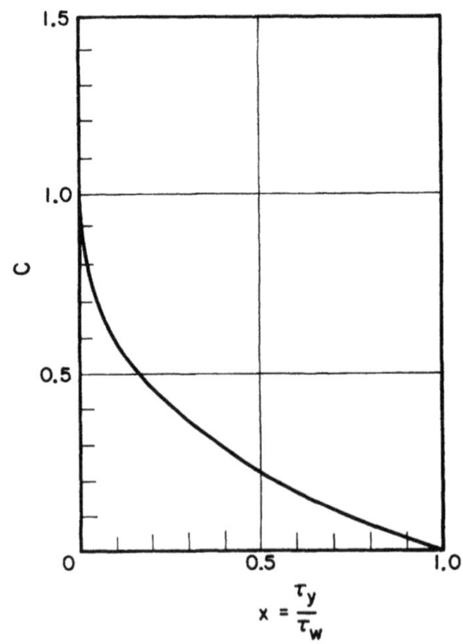

Figure 5.16. Excess dimensionless entrance length pressure drop, C, for the laminar flow of a Bingham fluid. (From Michiyosi, L., K. Mizuno, and Y. Hoshiai, *Int. Chem. Eng.*, **6**, 373, 1966.)

Astarita, Greco, and Peluso (1968) have obtained data on entrance losses for both viscoelastic and purely viscous fluids and correlated these using an equation similar to Equation 5.64. They found that the viscoelastic fluids did not behave much differently from the purely viscous fluids, but that for both the entrance pressure loss was substantially greater than predicted by theory. The fact that Astarita *et al.* obtained only modest shear rates in their experiments suggests the possibility that the Deborah number (see Chapter 1, Section 4.42) for the flow may not have been sufficiently high for elastic effects to become important. Boger and Ramamurthy (1970), on the other hand, found a marked decrease in the excess pressure gradient (Collins and Schowalter's C) with increase in the elastic properties of a series of methocel solutions. Further data and analyses are required to clarify the entrance behavior of viscoelastic fluids.

2.8 General Form of the Pressure Drop Equation

Expressing the mechanical energy balance in its integrated form for a pipe system in which changes of direction, elevation, and diameter may occur, we have, for any constant diameter section, as in Chapter 4,

$$\Delta P = \frac{g}{g_c}\,\Delta z\rho + \frac{\rho}{2g_c}\left(\frac{V_2^2}{\alpha_2} - \frac{V_1^2}{\alpha_1}\right) + \frac{2f\rho V^2 L_E}{g_c D} \tag{5.66}$$

Here the quantity α is no longer identically 0.5, as in the case of laminar flow of Newtonians, but, strictly speaking, should be evaluated specifically for the non-Newtonian fluid and the velocity profiles in question. The procedure involves carrying out the integration indicated by

$$\alpha = \frac{V^3}{\dfrac{2}{R^2}\displaystyle\int_0^R ru^3 dr} \tag{5.67}$$

In practice, unless the kinetic energy term is of particular significance (one of V_1 or V_2 large with respect to the other), the integration is not warranted and α may be taken to be approximately 0.5. For the simple case of the power law fluid, the integration results in (Skelland, 1967)

$$\alpha = \frac{(1 + 2n)(3 + 5n)}{3(1 + 3n)^2} \tag{5.68}$$

a relation given graphically in Figure 5.17. Skelland discusses the integration for a Bingham fluid and gives the result in terms of the dimensionless ratio

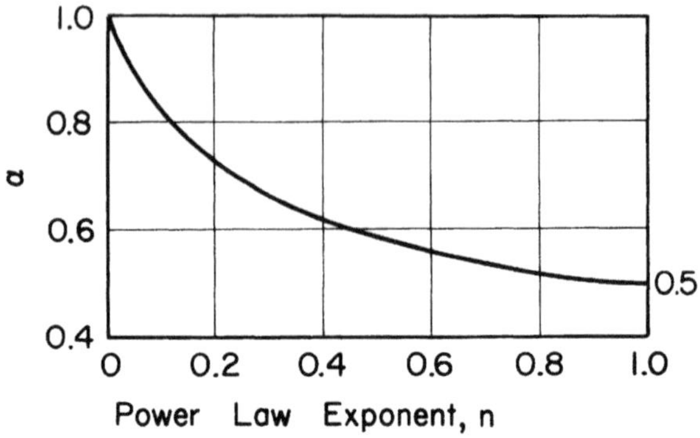

Figure 5.17. Velocity profile correction factor, α, for kinetic energy terms in laminar flow of power law pseudoplastic.

$x = \tau_y/\tau_w$. The equation involves x to powers ranging to 8 and is cumbersome to use. Metzner (1956) reports that

$$\alpha = \frac{1}{2 - x} \tag{5.69}$$

is an approximation, valid to 2.5 percent. This approximation may also be used for yield-pseudoplastic fluids.

The friction factor in the last term of Equation 5.66 is the one appropriate for the pipe diameter, velocity, and rheological behavior, as discussed in Sections 2.2 to 2.5, and corrected where appropriate for entrance effects as discussed in Section 2.6. The equivalent length under laminar flow conditions of any pipe fitting is not strictly the same as for laminar flow of Newtonians. Relatively few data are available on equivalent lengths of fittings for laminar flow of non-Newtonians, and often we are forced to use data based upon Newtonian fluids. The limited work of Weltmann and Keller (1957) indicates that the procedure is reasonably sound. Griskey and Green (1971) find a significant effect of the flow behavior index of dilatant materials on the equivalent length of fittings.

3 TRANSITION FROM LAMINAR TO TURBULENT FLOW

Regardless of the type of non-Newtonian fluid, laminar flow gives way to turbulent flow at sufficiently high values of the Reynolds number. Experience confirms our intuition that the transition from laminar to turbulent flow occurs at different Reynolds number values for different fluids. The Ryan and Johnson (1959) general analysis of the situation was described in Chapter 4. As applied to purely viscous pseudoplastics described by the power law, their analysis predicts

$$f_c = \frac{1}{404}\frac{(1+3n)^2}{n}\left(\frac{1}{2+n}\right)^{(2+n)/(1+n)} \tag{5.70}$$

In terms of the modified power law Reynolds number, this may be expressed as

$$(\mathrm{Re}_{PL2})_c = \frac{6464\,n}{(1+3n)^2\left(\dfrac{1}{2+n}\right)^{(2+n)/(1+n)}} \tag{5.71}$$

A comparison of this prediction with actual data for pseudoplastics is given in Figure 5.18 taken from Hanks and Christiansen (1962).

The approach of Ryan and Johnson or Hanks may be applied to determine the transition condition for Ellis or Powell–Eyring fluids or for any time-independent purely viscous fluid for which a constitutive equation is known or assumed.

In the case of Bingham plastics, Hanks (1963) shows that the transition from laminar to turbulent flow occurs at a critical value of Re_B given by

$$(\mathrm{Re}_B)_c = \frac{\mathrm{He}}{8x_c}\left(1 - \frac{4}{3}x_c + \frac{1}{3}x_c^4\right) \tag{5.72}$$

where x_c is the critical value of x defined by the relation

$$\frac{x_c}{(1-x_c)^3} = \frac{\mathrm{He}}{16{,}800} \tag{5.73}$$

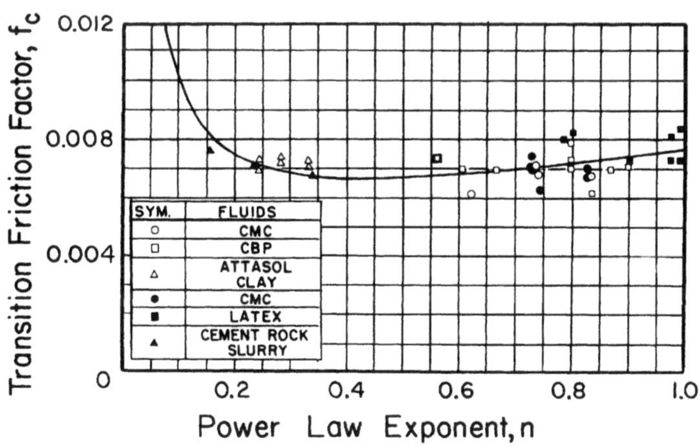

Figure 5.18. Comparison of Johnson's predicted f_c with data for pseudoplastic fluids. (From Hanks, R. W., and E. B. Christiansen, *A.I.Ch.E. Journal*, **8**, 467, 1962.)

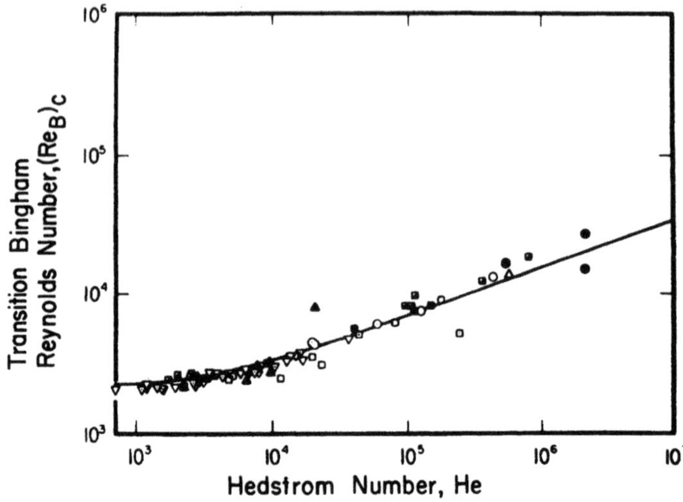

Figure 5.19. Variation of transition Reynolds number with Hedstrom number for flow of Bingham plastics. Solid curve calculated from Eqs. 5.72 and 5.73; data from selected references. (From Hanks, R. W., and D. R. Pratt, *Soc. Petrol. Engrs. J.*, **7**, 342, 1967.)

Hanks and Pratt (1967) solve these equations for $(Re_B)_c$ and x_c in terms of He and obtain the results shown in Figures 5.19 and 5.20 taken from their work. The predictions of Figures 5.19 and 5.20 are extensively confirmed by data of Caldwell and Babbitt (1941), Wilhelm, Wroughton, and Loeffel (1939), Alves, Boucher, and Pigford (1952), and others.

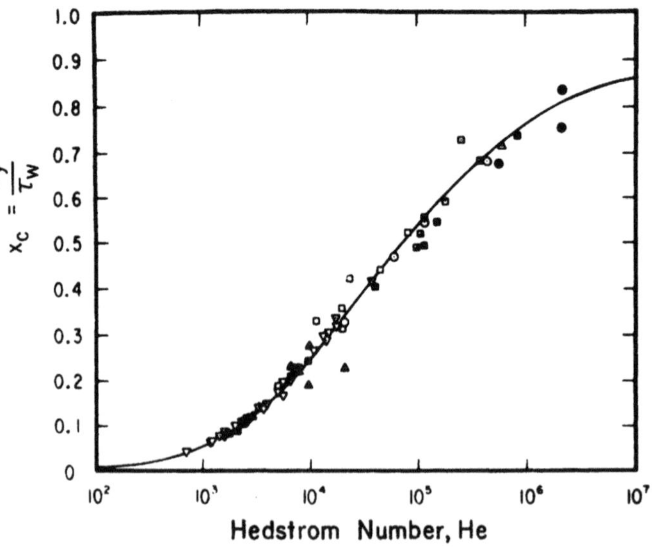

Figure 5.20. Variation of $x_c = (\tau_y/\tau_w)_c$ with Hedstrom number for laminar flow of Bingham plastics. Solid curve calculated from Eq. 5.73; data from selected references. (From Hanks, R. W., and D. R. Pratt, *Soc. Petrol. Engrs. J.*, **7**, 342, 1967.)

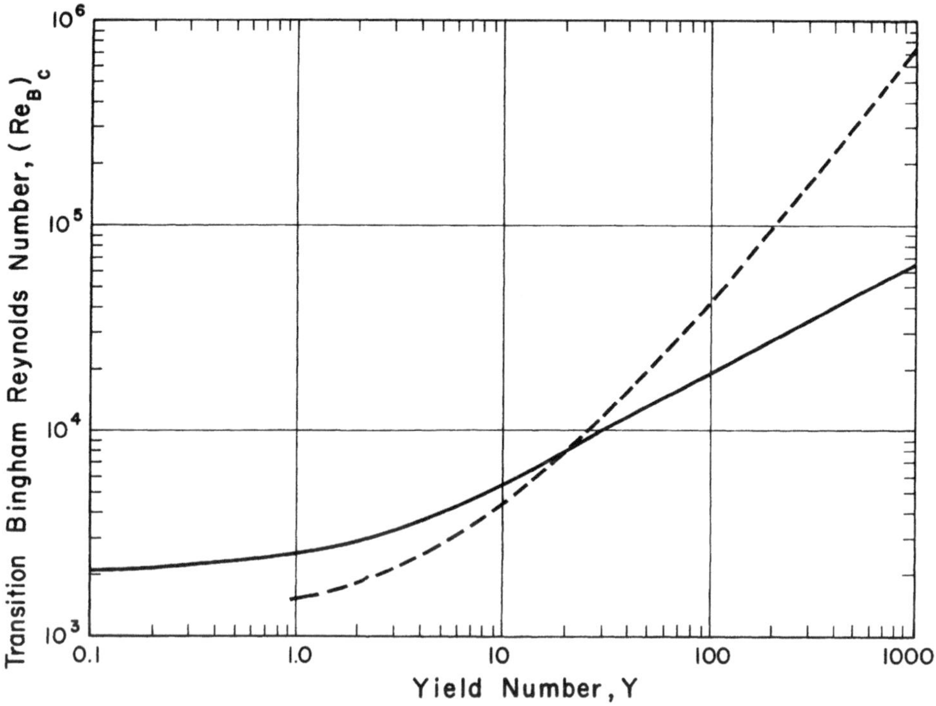

Figure 5.21. Transition Bingham Reynolds number.

——————$Y = \frac{He}{Re_B}$, Hanks and Pratt (1967)

- - - - - - Extrapolated from f–Re–Y lines of Fig. 5.9.

The relationship of Figure 5.19 may also be expressed in terms of the yield number, Y, since $Y = He/Re_B$. This is shown as the solid line in Figure 5.21, with a dashed line representing the abscissa values of the points of intersection of extensions of the lines of constant Y with the Nikuradse line of Figure 5.9. Thus, the intersections of the straight lines in Figure 5.9, representing the Buckingham equation, with the Nikuradse line do not represent the transition points. The transition to turbulent flow occurs at higher Reynolds numbers than indicated by these intersections for $Y < 20$ and at lower Reynolds numbers for $Y > 20$.

If we neglect the small last term in the denominator of Equation 5.52, i.e., $x^4/3$, and substitute for τ_w in terms of the friction factor f, we obtain

$$f \simeq \frac{16\eta}{DV\rho}\left(1 + \frac{\tau_y g_c D}{6\eta V}\right) \tag{5.74}$$

Since this equation approximately describes the laminar flow of Bingham plastics and it is of the same form as the equation for Newtonians, with μ

replaced by $\eta[1 + (\tau_y g_c D/6\eta V)]$, it is commonly assumed that the transition from laminar to turbulent flow will occur when

$$\frac{DV_c \rho}{\eta\left(1 + \dfrac{\tau_y g_c D}{6\eta V_c}\right)} \simeq 2100 \tag{5.75}$$

For tubes larger than about 1 inch in diameter, the term $\tau_y g_c D/6\eta V$ is large compared to unity and we have

$$V_c \simeq 19\sqrt{\frac{\tau_y g_c}{\rho}} \tag{5.76}$$

This equation, although approximate, is important, since it indicates (Thomas, 1963) that where τ_y is much greater than about 0.1 lb_f/ft^2, the critical velocity exceeds the normal economic transport velocity. This means that turbulent flow of Bingham plastics having a yield stress above about 0.1 lb_f/ft^2 is rarely encountered.

Metzner and Park (1964) and others have shown that the transition from laminar to turbulent flow is substantially influenced by elastic properties and that the critical modified Reynolds number, Re_{MR}, may be increased to beyond 10,000 for visoelastic materials. The effect is well indicated in Figure 5.22 taken from Metzner and Park, in which the data points are seen to follow the laminar relation to Reynolds numbers much in excess of those for transition of purely viscous systems. This is discussed further in connection with the turbulent flow of viscoelastic fluids.

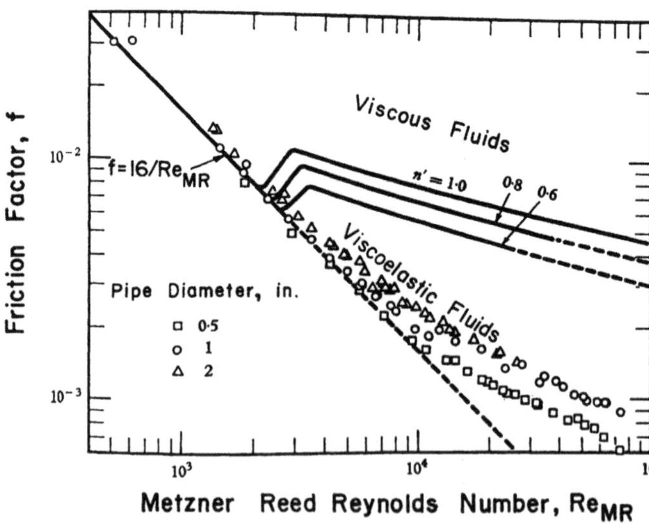

Figure 5.22. Friction factor relationships for viscous and typical viscoelastic fluids (0.3 per cent solution of J–100 in water). (From Metzner, A. B., and M. G. Park, *J. Fluid Mech.*, **20**, 291, 1964.)

4 STEADY-STATE TURBULENT FLOW

4.1 Introduction

In the last decade a number of advances have been made in our understanding of the turbulent pipe flow of power law pseudoplastics, and some of this work is applicable also to the turbulent flow of other time-independent viscous non-Newtonians. Here we will discuss the significant contributions of Dodge and Metzner (1959), Tomita (1959), and Clapp (1961), all of which are specifically applicable to the power law fluid. The analysis of Torrance applicable to power law, Bingham, and yield-pseudoplastic fluids is of special interest because it extends to the case of fully developed rough wall turbulence. Meter's analysis for a Meter fluid is also treated.

A scale-up correlation due to Lord, Hulsey, and Melton (1967) will also be discussed. This is applicable to power law fluids, i.e., to any fluid for which pressure drop data taken in a capillary tube or pipe display a linear relation between $\log(D\Delta P/4L)$ and $\log(8V/D)$. Our present understanding of the turbulent flow of viscoelastic fluids is limited, although the very substantial "drag reduction" attributable to elastic properties is well recognized. In this connection we will consider the work of a number of investigators who have correlated the drag reduction effect in several different ways.

4.2 The Relations for Pseudoplastic Fluids

4.21 The Dodge and Metzner Relations

Dodge and Metzner (1959) carried out a semitheoretical analysis of the turbulent flow in smooth tubes of time-independent viscous non-Newtonian fluids described by the power law. Applying the techniques of dimensional analysis for the power law fluid, as Millikan (1939) has done for Newtonian fluids, they derived the expression

$$\frac{1}{\sqrt{f}} = A_{1n} \log\left[\mathrm{Re}_{MR} f^{(1-n'/2)}\right] + C_n' \tag{5.77}$$

where $C_n' = A_{1n} \log\left[\frac{1}{8}\left(\frac{2+6n'}{n'}\right)^{n'}\right] + C_n$. The terms A_{1n} and C_n' are functions of n' to be determined experimentally. If n' is taken as unity, Equation 5.77 reduces to

$$\frac{1}{\sqrt{f}} = A\left(\log \frac{DV\rho}{\mu}\sqrt{f}\right) + C \tag{5.78}$$

which is of the same form as the Nikuradse equation (Equation 4.35), with $A = 4.0$ and $C = -0.40$. For the power law fluids which they tested, Dodge and Metzner found empirically that

$$A_{1n} = \frac{4.0}{n'^{0.75}} \tag{5.79}$$

and

$$C'_n = -\frac{0.4}{n'^{1.2}} \tag{5.80}$$

Incorporating these values into Equation 5.77, we obtain

$$\frac{1}{\sqrt{f}} = \frac{4.0}{n'^{0.75}} \log\left[\mathrm{Re}_{MR} f^{(1-n'/2)}\right] - \frac{0.40}{n'^{1.2}} \tag{5.81}$$

This relationship is reflected in graphical form in Figure 5.23 taken directly from Dodge and Metzner. These authors indicate excellent agreement between calculated and experimental friction factors over values of n' from 0.36 to 1.0 and Re_{MR} from 2900 to 36,000 with solid-liquid suspensions and polymeric solutions. They show that non-power law clay suspensions also conform to

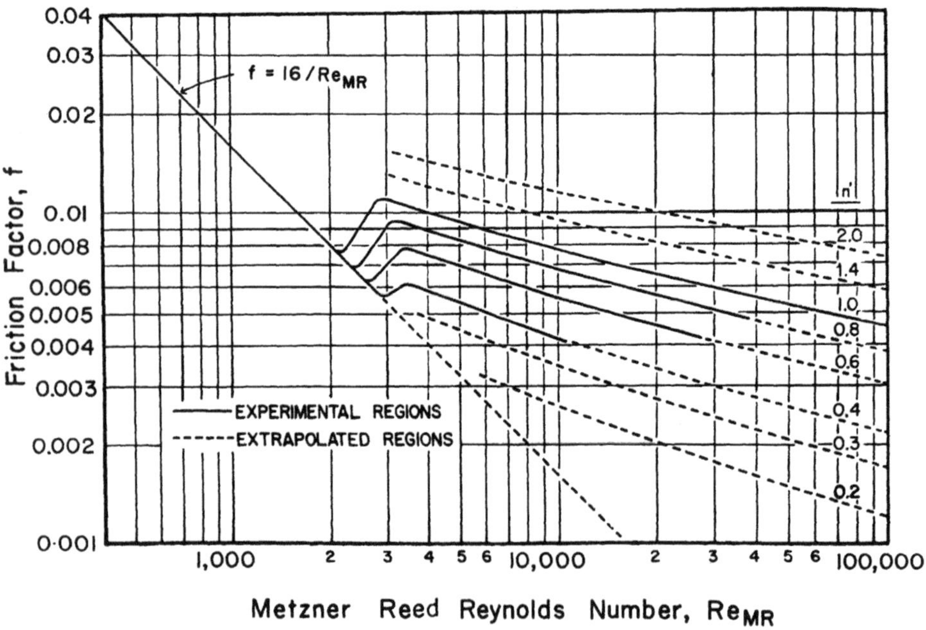

Figure 5.23. Friction factor relationships for Newtonian and non-Newtonian fluids. (From Dodge, D. W., and A. B. Metzner, *A.I.Ch.E. Journal*, **5**, 189, 1959.)

the correlation provided that the exponent n' is evaluated at a stress corresponding with the wall shear stress.

Savins, Burdyn, and Wallick (1963) and Savins and Cox (1963) describe a computer program for the application of the Dodge and Metzner analysis and illustrate it for the prediction of the pressure gradient for the flow in pipes of "rheologically complex" oils. The method involves an iterative calculation to determine K' and n' at the wall shear stress.

Dodge and Metzner also reported anomalous behavior for a solution of 0.3 percent sodium carboxymethyl cellulose (CMC). For this solution the position of the f-Re_{MR} line was influenced by diameter, and for all diameters it fell below the line for the other power law fluids of the same n' value. The effect, now attributed to viscoelasticity, was not observed in the other systems tested by Dodge and Metzner, and Figure 5.23 should be free of it.

Shaver and Merrill (1959) published the results of their experimental study of the turbulent flow of a series of pseudoplastics in 3/8- and 3/4-inch smooth pipes simultaneously with Dodge and Metzner. They correlated their own data in terms of f, Re_{PL2}, and n. Over most values of Re_{PL2} and n, however, their friction factors are lower than the general results of Dodge and Metzner and more in line with Dodge and Metzner's anomalous results for the CMC solutions. In fact, many of Shaver and Merrill's data were taken with CMC or other solutions now known to be viscoelastic. For these reasons, despite the good correlation of their own data, the Shaver and Merrill correlation cannot be considered of general applicability.

Dodge and Metzner's analysis results in the following velocity profile through the turbulent core

$$u^+ = \frac{5.66}{n'^{0.75}} \log y^+ - \frac{0.40}{n'^{1.20}} + \frac{2.458}{n'^{0.75}} \left[1.960 + 1.255n' - 1.628n' \log\left(3 + \frac{1}{n'}\right)\right] \qquad (5.82)$$

where $u^+ = u/U^*$,

$$y^+ = \frac{y^n (U^*)^{2-n} \rho}{K},$$

$$U^* = \sqrt{\tau_w g_c/\rho} = V\sqrt{f/2}.$$

Bogue and Metzner (1963) have reviewed the earlier work of Dodge and Metzner and propose as an improvement on Equation 5.82 the following relation

$$u^+ - c(\zeta, f) = 5.57 \log (y^+)^{1/n} + I(n, \mathrm{Re}_{PL2}) \qquad (5.83)$$

where the function $I(n, \mathrm{Re}_{PL2})$ is given in Table 5.2 and $\zeta = y/R$. They also point out that, if the velocity profile is normalized in the form of either u/V or $(u_{max} - u)/U^*$, there is no significant difference between the velocity profile for Newtonians and power law pseudoplastics, i.e., for both

$$\frac{u}{V} = \sqrt{\frac{f}{2}}\,[5.57 \log \zeta + c(\zeta, f) + 3.63 + 0.984] \tag{5.84}$$

$$\frac{u_{max} - u}{U^*} = 5.57 \log \zeta + [c(1, f) - c(\zeta, f)] \tag{5.85}$$

TABLE 5.2

Values[a] of the function $I(n, \mathrm{Re}_{PL2})$

n	Re_{PL2}			
	5,000	10,000	50,000	100,000
1.0	5.57	5.57	5.57	5.57
0.8	6.01	5.92	5.69	5.58
0.6	6.78	6.51	5.89	5.60
0.4	8.39	7.70	6.27	5.60

[a] D. C. Bogue and A. B. Metzner, *Ind. Eng. Chem. Fundam.*, **2**, 143, 1963.

4.22 The Tomita Relations

Tomita (1959), on the basis of similarity considerations, has defined a modified Fanning friction factor and a modified Reynolds number, applicable to power law fluids only, as follows:

$$f_{PLT} = \frac{2D\Delta P g_c(1 + 2n)}{3L\rho V^2(1 + 3n)} \tag{5.86}$$

$$\mathrm{Re}_{PLT} = \frac{6[(1 + 3n)/n]^{1-n}}{2^n[(1 + 2n)/n]}\,\frac{D^n V^{2-n}\rho}{K} \tag{5.87}$$

We may note that Tomita's friction factor is related to the conventional Fanning friction factor by

$$f_{PLT} = \frac{1 + 2n}{1 + 3n}\left(\frac{4}{3}\right) f \tag{5.88}$$

and his modified Reynolds number to Re_{PL2} by

$$\mathrm{Re}_{PLT} = \frac{3}{4}\left(\frac{1 + 3n}{1 + 2n}\right)\mathrm{Re}_{PL2} \tag{5.89}$$

With these definitions the laminar flow of power law pseudoplastics is correlated by

$$f_{PLT} = \frac{16}{\mathrm{Re}_{PLT}} \tag{5.90}$$

Applying Prandtl's mixing length concept, Tomita then showed that in turbulent flow in smooth tubes

$$\sqrt{\frac{1}{f_{PLT}}} = 4 \log\left(\mathrm{Re}_{PLT}\sqrt{f_{PLT}}\right) - 0.40 \tag{5.91}$$

The approximate validity of this equation is confirmed by some 40 data points taken with starch pastes and lime slurries.

4.23 The Clapp Relations

Clapp (1961) has followed the Prandtl and von Kármán approach, which we discussed in Chapter 4 in connection with Newtonian fluids, to derive for turbulent flow in smooth tubes a universal velocity profile for the power law fluid and, from this, a friction factor correlation. He obtains the following three equations for the velocity profile:

Laminar sublayer: $0 < y^+ < 5^n$

$$u^+ = (y^+)^{1/n} \tag{5.92}$$

Buffer layer: $5^n < y^+ < y_2^+$

$$u^+ = \frac{5.0}{n} 2.303 \log y^+ - 3.05 \tag{5.93}$$

Turbulent core: $y^+ > y_2^+$

$$u^+ = \frac{G}{n} 2.303 \log y^+ + H \tag{5.94}$$

where $u^+ = (y^+)^{1/n}$

$$y^+ = \frac{(U^*)^{2-n}\rho y^n}{K}.$$

In these relations y_2^+, G, and H are constants to be evaluated from experimental data. The close relationship between this set of equations and those we derive in Chapter 4 for Newtonian fluids is to be noted. From Equation

5.94 for the turbulent core, we may obtain, by integration over the pipe cross section, the equation for the average velocity

$$V = 2U^*\left[\frac{H}{2} - \frac{3G}{4} + \frac{G}{2n} 2.303 \log \frac{R^n(U^*)^{2-n}\rho}{K}\right] \tag{5.95}$$

Then by substituting

$$U^* = V\sqrt{\frac{f}{2}}$$

and rearranging, the following expression is obtained for the friction factor

$$\frac{1}{\sqrt{f}} = \sqrt{2}\left[\frac{H}{2} - \frac{3G}{4} + \frac{G}{2n} 2.303 \log \text{Re}_{PLC}(\sqrt{f})^{2-n} + \frac{0.174(5n-8)G}{n}\right] \tag{5.96}$$

where

$$\text{Re}_{PLC} = \frac{8\rho V^2}{K\left(\frac{8V}{D}\right)^n} = \frac{D^n V^{2-n}\rho}{K8^{n-1}} = \frac{\text{Re}_{PL1}}{8^{n-1}} \tag{5.97}$$

In order to evaluate the necessary constants, Clapp made detailed velocity traverses with fluids having n values ranging from 0.698 to 0.813 under Reynolds number conditions varying from 5480 to 42,800. It was found that the turbulent core data could not be completely correlated by Equation 5.94 but an empirical modification of it

$$u^+ = \frac{G}{n} 2.303 \log y^+ + \frac{H}{n} \tag{5.98}$$

with values of $G = 2.78$ and $H = 3.8$, as proposed by Deissler (1951) for Newtonian fluids, correlated the non-Newtonian data to within ± 5 percent. Figure 5.24 compares the predictions of Clapp for n values of 0.700 and 0.765 with the actual data and with the Dodge and Metzner prediction for $n = 0.700$. Clapp's equation suits his data better than does the Dodge and Metzner relationship.

Clapp has also taken Deissler's values of G and H, as confirmed by the velocity data, for use with Equation 5.98 in the equivalent of Equation 5.96 to yield

$$\frac{1}{\sqrt{f}} = \frac{2.69}{n} - 2.95 + \frac{4.53}{n} \log\left(\text{Re}_{PLC}(\sqrt{f})^{2-n}\right) + \frac{0.68}{n}(5n-8) \tag{5.99}$$

with which he was able to correlate his friction data to within 5 percent. For a Newtonian fluid ($n = 1$), Equation 5.99 reduces to

$$\frac{1}{\sqrt{f}} = 4.53 \log(\text{Re}\sqrt{f}) - 2.3 \tag{5.100}$$

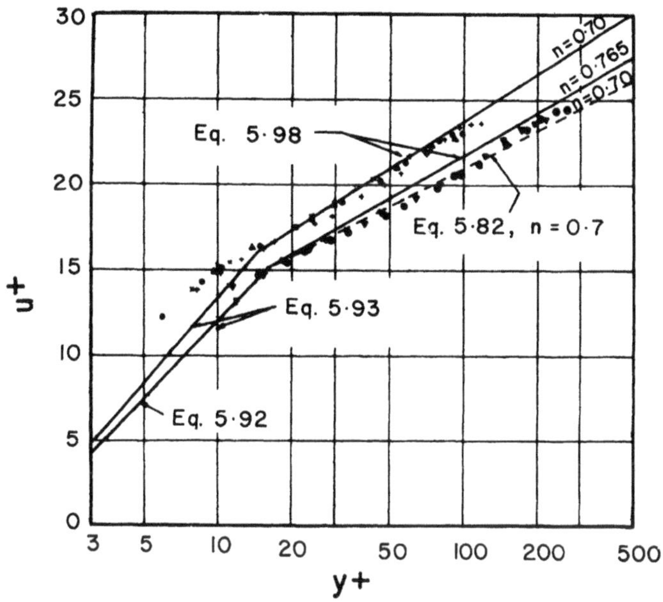

Figure 5.24. Comparison of velocity distritution equations with Experimental Results. (From Clapp, R. M., Conf. on Int. Development in Heat Transfer, University of Colorado, 1961, *Proc.* p. 652.)

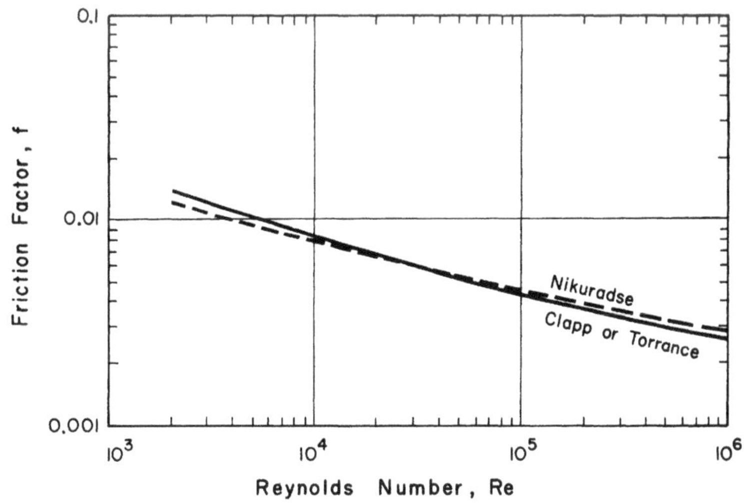

Figure 5.25. Comparison of the Clapp or Torrence equation (Eq. 5.100) and the Nikuradse equation (Eq. 4.35) for friction factors for Newtonian fluids.

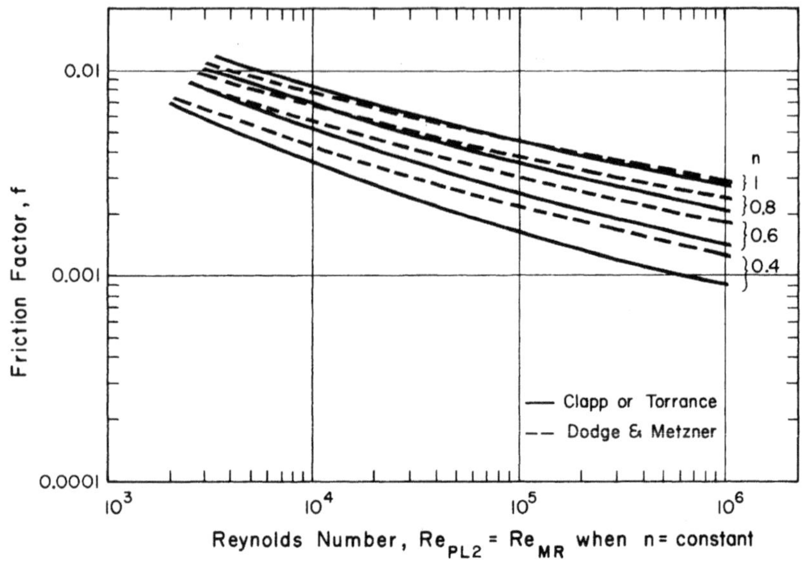

Figure 5.26. Comparison of the Dodge and Metzner equation (Eq. 5.81) and the Clapp or Torrance Equation (Eq. 5.99) for friction factors for power law pseudoplastics.

an equation similar to Nikuradse's Equation 4.35. Figure 5.25 shows the comparison of Clapp's equation with Nikuradse's equation. Figure 5.26 compares the f–Re correlation of Clapp with that of Dodge and Metzner. These correlations are closely similar at $\mathrm{Re} < 15{,}000$ but significant deviations occur at $\mathrm{Re} > 20{,}000$. Clapp's correlation, although based on a limited range of n, is probably superior in the higher Re region, since it is confirmed to $\mathrm{Re} = 42{,}800$, whereas Dodge and Metzner had few data above $\mathrm{Re} = 20{,}000$.

4.24 The Torrance Relations

The analysis of Torrance (1963) for the turbulent flow in smooth tubes of yield-pseudoplastic fluids discussed in Section 4.4 is also applicable to the power law fluid. He extended his analysis to FRW turbulence in rough pipes which for the power law fluid gives

$$\frac{1}{\sqrt{f}} = \frac{4.07}{n} \log \frac{R}{k} + 6.0 - \frac{2.65}{n} \tag{5.101}$$

where R is the pipe radius.

4.25 The Lord, Hulsey, and Melton Scale-Up Procedure

Lord, Hulsey, and Melton (1967) have recently developed a generalized, although completely empirical, approach to the analysis of the turbulent flow in smooth tubes of any time-independent viscous or viscoelastic fluid which exhibits linear relationship in a logarithmic plot of $D\Delta P/4L$ versus $8V/D$ under turbulent flow conditions. The method is applicable to such diverse systems as aqueous solutions or suspensions of bentonite clay, calcium carbonate, sodium carboxymethyl cellulose (CMC), and guar gum, the latter two exhibiting viscoelastic effects. Lord *et al.* follow the suggestion of Melton and Malone (1964) by assuming a generalized form of the Blasius equation

$$f = B\left(\frac{DV\rho}{\mu}\right)^{a} \tag{5.102}$$

and combining it with the Fanning equation

$$\frac{\Delta P}{L} = \frac{2f\rho V^2}{g_c D} \tag{4.24}$$

to obtain

$$\frac{D\Delta P}{4L} = AD^{e}\left(\frac{8V}{D}\right)^{s} \tag{5.103}$$

where A = constant incorporating the density and viscosity terms, $\text{lb}_f\ \text{sec}^s/\text{ft}^{e+2}$,

e and s are dimensionless constants equal to 1.60 and 1.80, respectively, for a Newtonian fluid.

Lord *et al.* analyzed their data in pipes ranging from 1/8 to 1 inch in diameter to obtain values of the constants A, e, and s. The net diameter exponent $(1 - e + s)$ was expected and found to be more or less independent of fluid properties permitting Equation 5.103 to be rearranged to

$$\frac{D^{1.20}\Delta P}{4L} = A(8V)^{s} \tag{5.104}$$

The constants A and s were determined from the experimental data. Figures 5.27 and 5.28 show typical data of Lord *et al.* for 0.25 percent CMC, and Table 5.3 gives the values of the constants A and s for the systems they studied. This generalized approach is a powerful empirical one for scale-up of data obtained in a small diameter pipe, although it is clearly dependent upon the constancy of the diameter exponent. Hall (1969) has drawn attention to the actual variation in this exponent for the systems reported by Lord *et al.* and cautions against scale-up over too extensive a range.

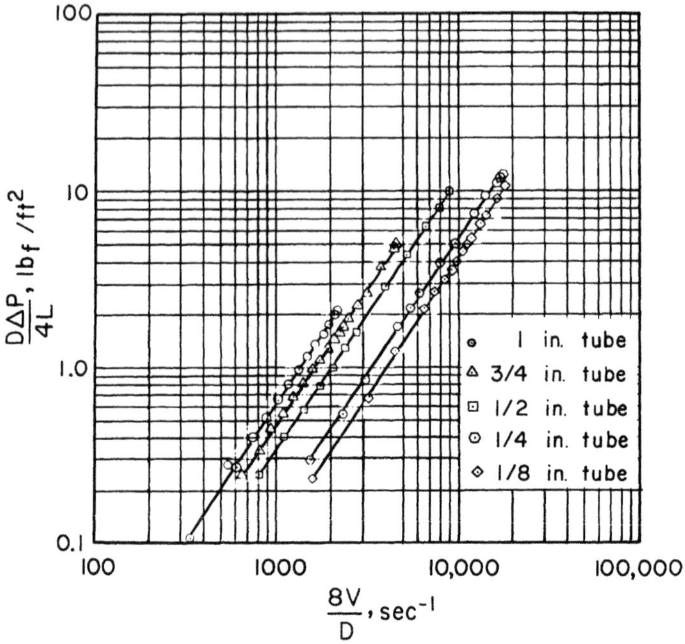

Figure 5.27. Stress-flow diagram for 0.25 percent CMC. (From Lord, D. L., B. W. Hulsey, and L. L. Melton, *Soc. Petrol. Engrs. J.*, **7**, 252, 1967.)

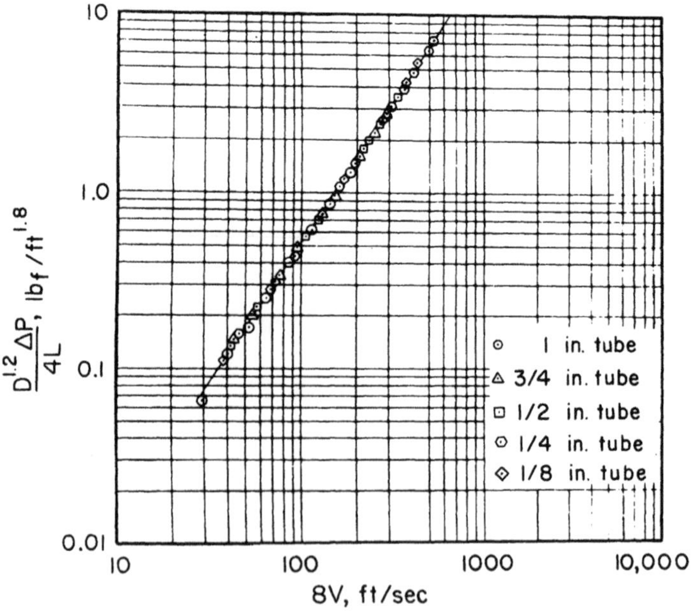

Figure 5.28. Multiple pipe correlation for 0.25 percent CMC. (From Lord, D. L., B. W. Hulsey, and L. L. Melton, *Soc. Petrol. Engrs. J.*, **7**, 252, 1967.)

TABLE 5.3
Constants[a] for Equation 5.104

Water Mixture with	Concentration, Weight %	$A \times 10^4$	s	Standard Deviation, Percent
Water		1.176	1.780	4.19
Bentonite	8.0	5.250	1.615	2.67
Bentonite	6.0	2.525	1.707	4.17
Bentonite	4.0	2.167	1.707	3.77
$CaCO_3$	30.0	3.491	1.658	4.35
$CaCO_3$	27.5	1.736	1.771	3.68
$CaCO_3$	25.0	1.710	1.768	2.96
$CaCO_3$	22.5	1.677	1.769	2.86
$CaCO_3$	20.0	1.639	1.771	3.02
$CaCO_3$	17.5	1.476	1.785	3.29
$CaCO_3$	15.0	1.498	1.779	3.72
CMC	1.0	30.01	1.208	2.84
CMC	0.50	8.092	1.431	3.14
CMC	0.25	3.428	1.588	3.88
Guar gum	0.48	7.993	1.221	3.30
Guar gum	0.36	7.913	1.220	3.90
Guar gum	0.06	3.745	1.385	3.71

[a] D. L. Lord, B. W. Hulsey, and L. L. Melton, *Soc. Petrol. Engrs. J.*, **7**, 252, 1967. The units of A are $lb_f\ sec^s/ft^{s+1.8}$

4.3 The Relations for a Bingham Fluid

Although Dodge and Metzner (1959) suggest that their approach to the turbulent flow in smooth pipes of non-Newtonian fluids, whose behavior may be expressed by

$$\frac{D\Delta P}{4L} = K'\left(\frac{8V}{D}\right)^{n'} \tag{5.61}$$

is also applicable to Bingham plastics, it is clearly not attractive to apply what is essentially a power law equation to a system that behaves as a Bingham fluid. On the basis of data of Govier and Winning (1948), Alves, Boucher, and Pigford (1952), Hedstrom (1952), Worster (1955), and others, it seems well established from an empirical point of view that, at Re_B values above $(Re_B)_c$, and following some transition range, the friction factor for Bingham plastics is dependent only on Re_B and the pipe roughness. For smooth pipes the $f - Re_B$ relationship is close to that for Newtonian fluids.

Figure 5.29, modified after Govier and Winning, shows turbulent flow data obtained in a smooth tube for Bingham-like clay in water suspensions. These data confirm the lack of dependence of f on Y at the higher Reynolds numbers.

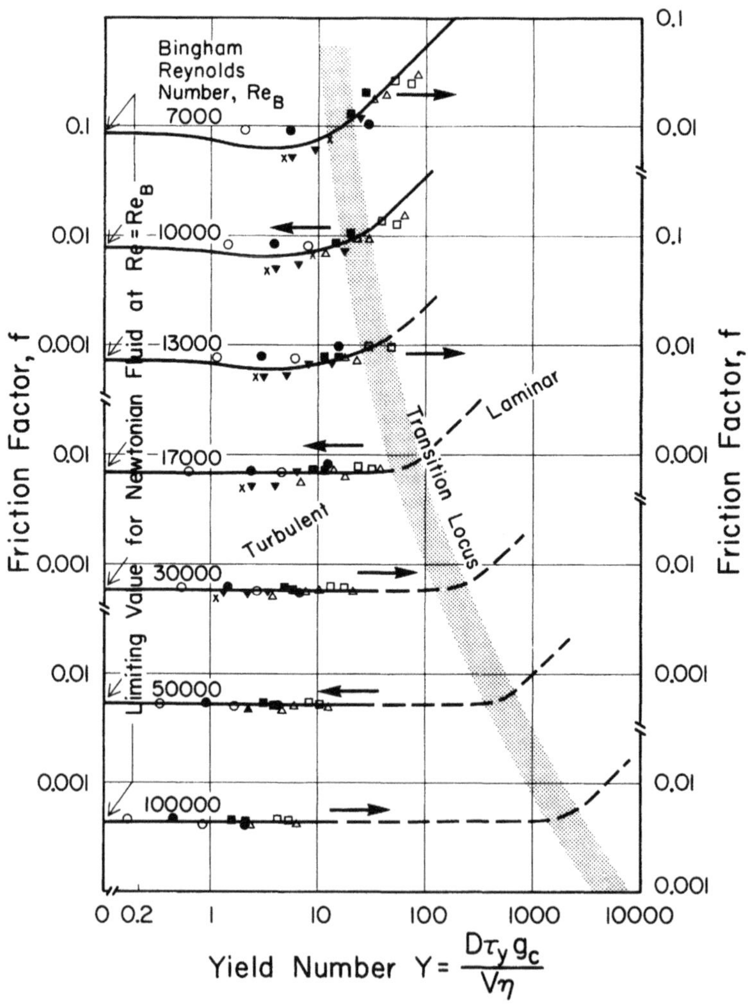

Figure 5.29. Friction factor—yield number correlation for Bingham plastics. (Modified after Govier, G. W., and M. D. Winning, The Flow of Non-Newtonian Fluids. Paper presented at the Meeting of the American Institute of Chemical Engineers, Montreal, Quebec, Sept. 1948.)

4.31 The Tomita Relations

Tomita (1959) has also applied the Prandtl mixing length concept to the turbulent flow in smooth tubes of Bingham plastics to obtain

$$\sqrt{\frac{1}{f_{BT}}} = 4 \log \mathrm{Re}_{BT}\sqrt{f_{BT}} - 0.40 \tag{5.105}$$

where

$$\mathrm{Re}_{BT} = \frac{DV\rho}{\eta}\left[\frac{(1-x)(x^4 - 4x + 3)}{3}\right] \tag{5.106}$$

and

$$f_{BT} = \frac{D\Delta P g_c}{2\rho L V^2(1-x)} \tag{5.107}$$

In these relations $x = \tau_y/\tau_w$ is calculated as though the flow were laminar, i.e., from Equation 5.51 or the equivalent. We may note the relations between Tomita's modified friction factor and the conventional friction factor,

$$f_{BT} = \frac{f}{1-x} \tag{5.108}$$

and between his modified Reynolds number and the Bingham Reynolds number

$$\mathrm{Re}_{BT} = \frac{(1-x)(x^4 - 4x + 3)}{3}\mathrm{Re}_B \tag{5.109}$$

Tomita's equation is confirmed over the range $2000 < \mathrm{Re}_{BT} < 100{,}000$ by twenty data points covering a good range of diameter, a sevenfold range of yield stress, but only a slight variation in coefficient of rigidity.

4.32 The Torrance Relations

Torrance's analysis (1963) for the turbulant flow in smooth pipes of yield-pseudoplastic fluids discussed in the next section reduces to the following when $n = 1$, corresponding to the Bingham plastic case:

$$\frac{1}{\sqrt{f}} = 4.53 \log (1 - x) + 4.53 \log (\mathrm{Re}_B\sqrt{f}) - 2.3 \tag{5.110}$$

We see that if the quantity x is small, which is ordinarily the case for the turbulent flow of Bingham fluids, Torrance's equation for the Bingham fluid reduces to

$$\frac{1}{\sqrt{f}} = 4.53 \log (\mathrm{Re}_B\sqrt{f}) - 2.3 \tag{5.111}$$

which, with Re_B replacing Re, is the same as his equation for the Newtonian fluid. This offers some confirmation of the experimental observations that, for smooth tubes, f is a function of Re_B alone. For values of x exceeding 0.1 or so, Equation 5.110 predicts higher friction factors for Bingham than for Newtonian fluids, as illustrated by Figure 5.33 discussed later.

Torrance's analysis for FRW turbulence in rough pipes is also applicable to the Bingham case and reduces to the same relationship for Newtonian fluids,

$$\frac{1}{\sqrt{f}} = 4.07 \log \frac{R}{k} + 3.36 \tag{5.112}$$

Although unconfirmed by experimental data, there is every reason to expect this equation to be applicable to the Bingham fluid.

4.33 The Thomas Correlation

Thomas (1963) presents data on the SW turbulent pipe flow of suspensions of thorium oxide, kaolin, and titanium oxide in water in tubes varying in diameter from 1/8 to 1 inch. These materials exhibited Bingham behavior under laminar flow conditions. Thomas confirms that the appropriate form of the Reynolds number is $DV\rho/\eta$ and that the data conform to a Blasius type relationship. His data, however, fall on a line some 20–30 percent below the normal Blasius (or Nikuradse) line. Thomas indicates that the coefficients, B and b, in the Blasius type equation

$$f = B\left(\frac{DV\rho}{\eta}\right)^{-b} \tag{5.113}$$

are functions of the rheological properties of the fluid and suggests

$$B = 0.079\left[\left(\frac{\mu}{\eta}\right)^{0.48} + \left(\frac{g_c \rho \tau_y \kappa^2}{\mu^2}\right)^2\right] \tag{5.114}$$

and

$$b = 0.25\left[\left(\frac{\mu}{\eta}\right)^{0.15} + \left(\frac{g_c \rho \tau_y \kappa^2}{\mu^2}\right)^2\right] \tag{5.115}$$

where $\kappa = 6 \times 10^{-6}$ ft.

Thomas' data showing the variation of B and b with yields stress are given in Figure 5.30. Thomas' turbulent flow data were mostly taken within less than one-half a cycle of Reynolds number above the end of the laminar region and we would expect some gradual transition to the Nikuradse line. Particularly at low yield numbers, this would cause the data to fall below the Nikuradse line with a lower negative slope and, consequently, the Blasius coefficients to be low.

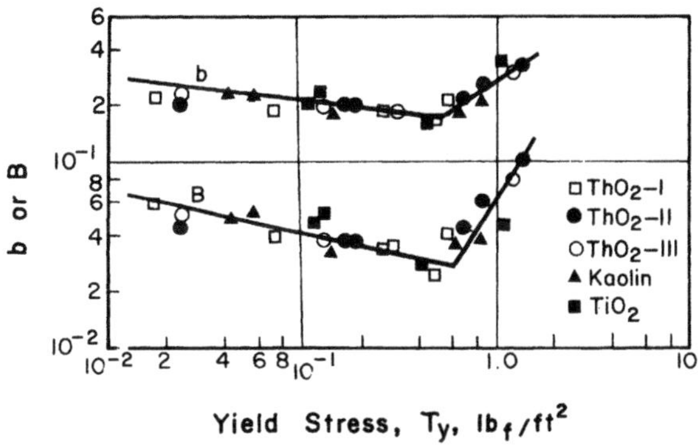

Figure 5.30. Effect of yield stress on the coefficients of the Blasius type equation (Eq. 5.113). (From Thomas, D. G., *Ind. Eng. Chem.*, **55**, 27, 1963.)

4.34 The Hanks and Dadia Analysis

Hanks and Dadia (1969) have extended the analysis of Hanks (1968) for the turbulent flow of Newtonian fluids in smooth pipes (as discussed in Chapter 4) to the case of Bingham plastics. The analysis incorporates the relationship between $(\mathrm{Re}_B)_c$ and $(\mathrm{He})_c$ developed by Hanks and Pratt (1967). Starting from the equivalent of Equation 3.46,

$$\tau = \tau_l + \tau_t$$

where $\tau_l = \tau_0 - \eta \dfrac{du}{dr}$ for $\tau > \tau_0$,

$$\frac{du}{dr} = 0 \qquad \text{for } \tau \leq \tau_0,$$

$$\tau_t = \rho(l')^2(du/dr)^2,$$

$$l' = l(R - r)\left[1 - \exp\left(-\phi_B \frac{(R-r)}{R}\right)\right]$$

They express the parameter ϕ_B as

$$\phi_B = \frac{\mathrm{Re}\sqrt{f} - \mathrm{Re}_c\sqrt{f_c}}{2\sqrt{2}B} \tag{5.116}$$

where $B = B$ (He) to be determined by analysis of experimental data. A consideration of thirteen sets of turbulent flow data from four different investigators led them to the following empirical relationship

$$B = 157.5\ \mathrm{He}^{-0.151} \tag{5.117}$$

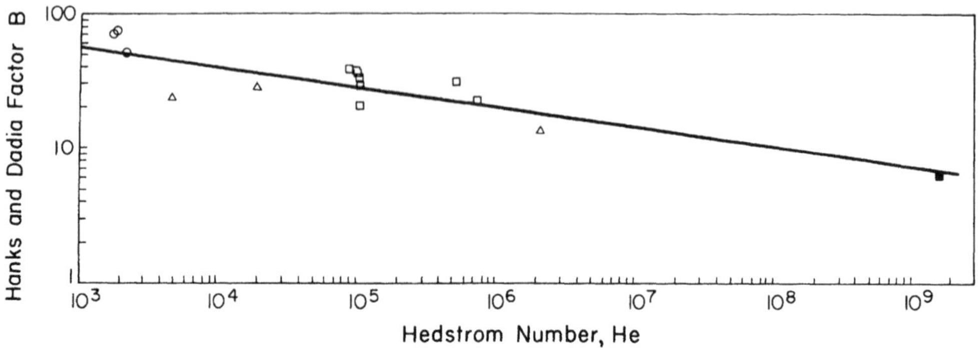

Figure 5.31. Hanks and Dadia factor B. Data from selected references; line represents Eq. 5.117. (From Hanks, R. W., and B. H. Dadia, Private Communication, 1969, *A.I.Ch.E. Journal*, **17**, 554, 1971.)

applicable in the range $10^3 \le \text{He} \le 2 \times 10^9$. The amount of scatter of the derived values of B about this line is fairly substantial, as shown in Figure 5.31 from Hanks and Dadia. In fact, one wonders about the reliability of the data and if these values of B would not be almost as well represented by B taken as a constant equal to its Newtonian value of 22.

In any case from these relationships, including that for B, and by the procedure used by Hanks (1968) for Newtonian fluids, Hanks and Dadia developed the relationships for f versus Re_B at each of several values of He above $(\text{He})_c$. These are shown in Figure 5.32 taken from their work. The figure also shows the f–Re_B lines in the laminar region and the Hanks and Pratt (1967) relationship between f_c and He_c.

The work of Hanks and Dadia indicates values of f in the turbulent range lower than for Newtonian fluids at $\text{Re}_B = \text{Re}$ for all values of $\text{He} < 5 \times 10^5$, equivalent values at $\text{He} = 5 \times 10^5$, and higher values at $\text{He} > 5 \times 10^5$. Also, at $\text{He} > 10^6$ and in the range of $\text{Re}_B \simeq 10^6$, their prediction shows maxima in the f–Re_B lines. Other observations that we have discussed confirm the lower f values (values of $B > 22$) immediately beyond $(\text{Re}_B)_c$ but, apart from Equation 5.117, on which Figure 5.32 depends, we have no direct confirmation of the continued lower values of f for $\text{He} < 10^6$, of the maxima, and of the higher values of f for $\text{He} > 10^6$. Figure 5.32 probably reflects the maximum likely deviation of the f–Re_B relationship from the f–Re relationship for Newtonian fluids. The use of the lower f values, except immediately beyond $(\text{Re}_B)_c$ is not recommended for design purposes.

We may conclude that the turbulent flow of Bingham plastics in a smooth pipe may conservatively be represented by the Blasius or Nikuradse line with Re replaced by Re_B. It is important to note that laminar flow extends to higher Reynolds numbers than in the case of Newtonians (see Figure 5.21)

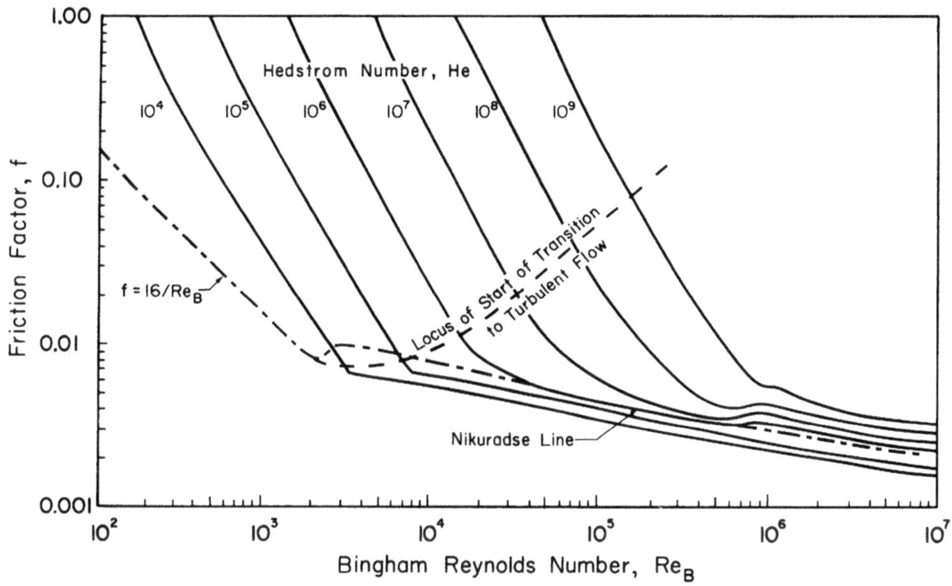

Figure 5.32. Hanks and Dadia friction factor correlation for Bingham fluids. (From Hanks, R. W., and B. H. Dadia, Private Communication, 1969, *A.I.Ch.E. Journal,* **17**, 554, 1971.)

and that a gradual transition to turbulent flow may result in friction factors somewhat below the Nikuradse line in the Reynolds number range immediately above the critical Re_B. In this region Thomas' modified Blasius equation may be helpful and the Hanks and Dadia analysis provides a theoretically based method for estimation of the friction factor. For values of $\mathrm{He} > 10^6$, the Hanks and Dadia analysis indicates values of f greater than for Newtonian fluids.

4.4 The Relations for a Yield-Pseudoplastic Fluid; the Torrance Relations

Little quantitative or analytical work has yet been directed to the case of the yield-pseudoplastic fluid, even the simple power law variety of this complex fluid. The notable exception to this is the work of Torrance (1963) who extended Clapp's (1961) treatment to the power law yield-pseudoplastic.

Torrance (1963) has treated the case of the turbulent flow in smooth pipes of a yield-pseudoplastic fluid described by Equation 1.48:

$$\tau = \tau_y + \frac{K}{g_c}\dot{S}^n \tag{1.48}$$

by following the method of analysis of Clapp (1961) based on Prandtl's mixing length concept. Torrance obtains the relationship

$$\sqrt{\frac{2}{f}} = (A - 1.5nB) + B\,2.303\log(1-x)$$

$$+ B\,2.303\log(\mathrm{Re}_{PLC}\sqrt{f^{2-n}}) + 0.347(5n-8)B \qquad (5.118)$$

where Re_{PLC} is the Reynolds number previously defined by Clapp (see Equation 5.97) and A and B are constants dependent upon n. From Clapp's work Torrance took

$$A = \frac{3.8}{n} \qquad \text{and} \qquad B = \frac{2.78}{n}$$

enabling Equation 5.118 to be written

$$\frac{1}{\sqrt{f}} = \left[\left(\frac{2.69}{n}\right) - 2.95\right] + \frac{4.53}{n}\log(1-x)$$

$$+ \left(\frac{4.53}{n}\log \mathrm{Re}_{PLC}\sqrt{f^{2-n}}\right) + \frac{0.68}{n}(5n-8) \qquad (5.119)$$

This general equation is represented graphically in Figures 5.33, 5.34, 5.35, and 5.36. For a power law pseudoplastic, i.e., $\tau_y = 0$, Torrance's equation simplifies identically to Clapp's equation, i.e., Equation 5.99. For a Bingham plastic, i.e., $n = 1$ and $\mathrm{Re}_{PLC} = \mathrm{Re}_B$, Torrance's equation becomes

$$\frac{1}{\sqrt{f}} = 4.53\log(1-x) + 4.53\log(\mathrm{Re}_B\sqrt{f}) - 2.3 \qquad (5.110)$$

as discussed earlier. For a Newtonian fluid, i.e., $n = 1$, $\tau_y = 0$, and $\mathrm{Re}_{PLC} = \mathrm{Re}$, it reduces to

$$\frac{1}{\sqrt{f}} = 4.53\log(\mathrm{Re}\sqrt{f}) - 2.3 \qquad (5.100)$$

identical with Clapp's equation.

Torrance also extended his analysis to FRW turbulence following the methods described in Chapter 4 for Newtonian fluids. Incorporating a velocity profile described by

$$u^+ = A\,2.303\log\left(\frac{y}{k}\right) + E \qquad (5.120)$$

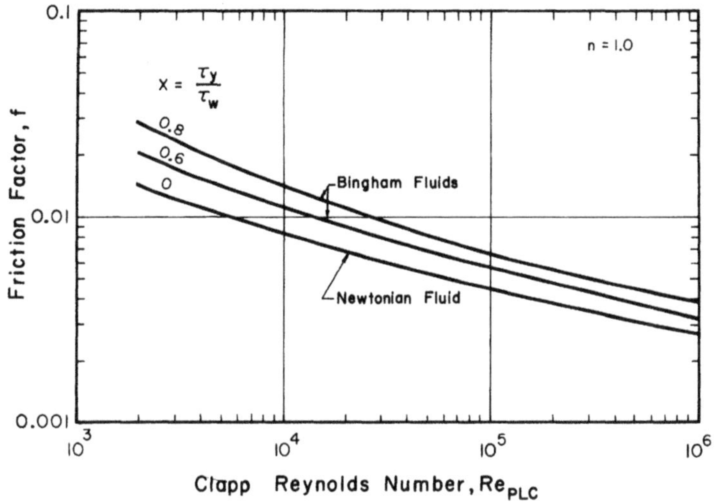

Figure 5.33. Friction factors for turbulent flow of Newtonian and Bingham fluids—Torrance equation (Eq. 5.119).

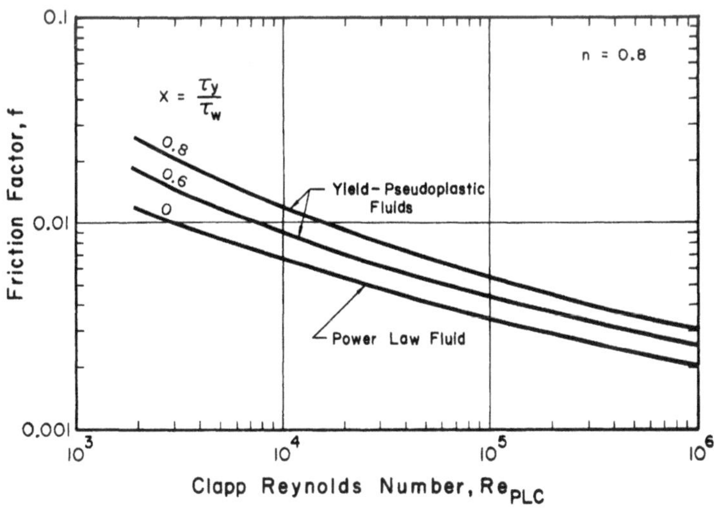

Figure 5.34. Friction factors for turbulent flow of yield-pseudoplastic fluids—Torrance equation (Eq. 5.119).

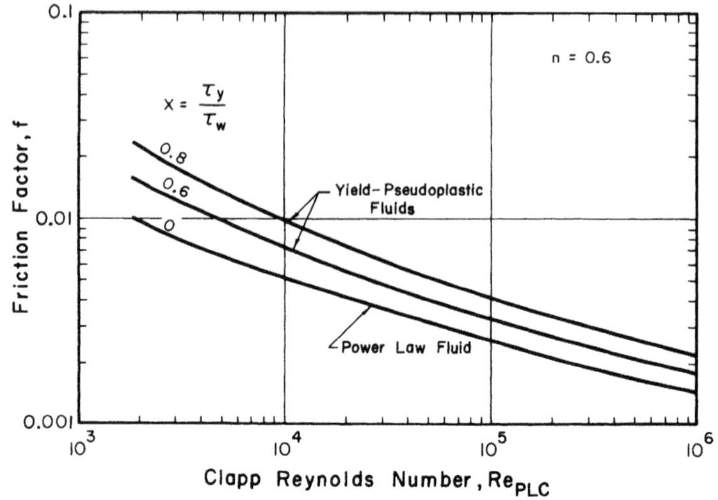

Figure 5.35. Friction factors for turbulent flow of yield-pseudoplastic fluids—Torrance equation (Eq. 5.119).

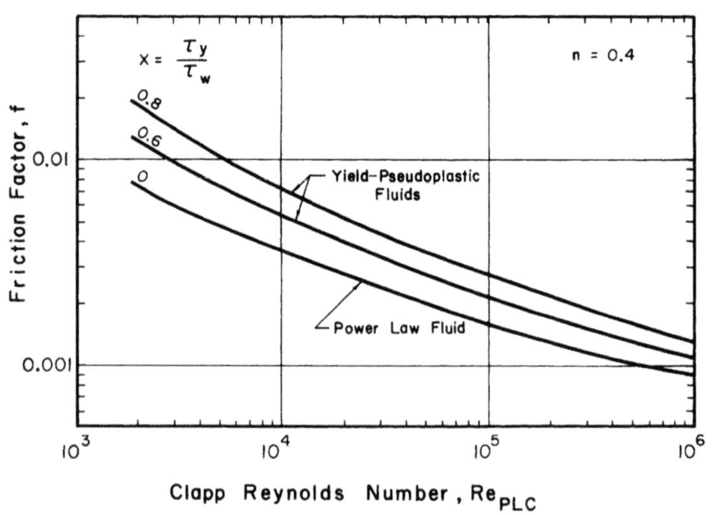

Figure 5.36. Friction factors for turbulent flow of yield-pseudoplastic fluids—Torrance equation (Eq. 5.119).

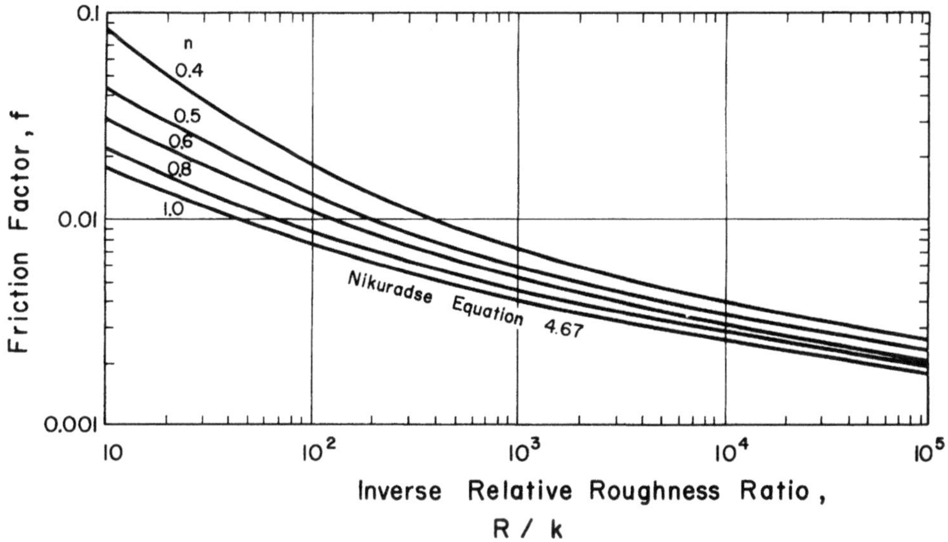

Figure 5.37. Friction factors for turbulent flow of yield-pseudoplastics in the fully rough wall region—Torrance equation (Eq. 5.121).

where A and E are constants, he used a value of $A = 2.5/n$ based on the work of Nikuradse and Nikuradse's actual value of $E = 8.5$ to obtain by integration

$$\frac{1}{\sqrt{f}} = 4.07 \log\left(\frac{R}{k}\right) + 6.0 - \frac{2.65}{n} \tag{5.121}$$

Equation 5.121 is not confirmed by experimental data but would appear suitable for the prediction of friction factors for the fully rough turbulent flow of power law yield-pseudoplastics, Bingham fluids, or simple pseudoplastics in rough pipes. Equation 5.121 is represented graphically in Figure 5.37.

4.5 The Relations for Viscoelastic Fluids

4.51 Introduction

The quantitative effects of viscoelasticity in turbulent flow are to reduce the pressure gradient or friction factor below what it would be in the absence of elastic effects. This phenomenon is referred to as "drag reduction." Although a number of solute-solvent combinations are known to exhibit drag reduction, the phenomenon is most commonly encountered in water solutions of long-chain polymers. In the design of piping systems of viscoelastic materials it is essential for the engineer to have at least a qualitative understanding of the major and diverse effects that are known to occur. This understanding is necessary in order realistically to evaluate empirical analyses and to avoid errors in extrapolating available data or employing empirical correlations beyond the conditions from which they were developed.

The drag reduction phenomenon appears first to have been mentioned in the engineering literature by Toms (1949), who noted it with solutions of polymethyl-methacrylate in monochlorobenzene. Subsequently drag reduction was reported in independent studies by Dodge and Metzner (1959) and by Shaver and Merrill (1959), both with dilute solutions of carboxymethyl cellulose (CMC) in water. Shaver and Merrill presented a correlation of their results based entirely on the viscous properties of the solutions. Theirs and other correlations based entirely on viscous properties are of limited utility owing to the fact that it has been demonstrated in a number of instances (e.g., Savins, 1964; and Wells, 1965) that drag reduction may occur in systems whose viscous properties are not measurably different from those of a Newtonian fluid. Following the work of Shaver and Merrill, a large volume of data has become available, defining drag reduction for a variety of materials, molecular weights, and concentrations. A review paper by Patterson, Zakin, and Rodriguez (1969) discusses the most significant of the recent studies. Figure 5.38 from Seyer and Metzner (1969a) illustrates the reduction in pressure gradient that occurs in small tubes with a typical viscoelastic additive. It is of interest to note that, with the 0.1 percent solution, the illustrated reduction in drag is achieved in a solution whose apparent viscosity is nearly ten times that of water and therefore has a pressure drop nearly ten times greater than the water under laminar flow conditions.

The 0.01 percent data of Figure 5.38 and other data of Seyer and Metzner (1967) are plotted in friction factor form in Figure 5.39. At low concentration (0.01 percent in this case) the behavior of the viscoelastic fluid is identical with that of a Newtonian fluid until a certain Reynolds number is reached. At higher flow rates the material shows distinct, although modest, drag reduction. Capillary measurements of the viscosity of the 0.01 percent solution showed Newtonian behavior with a viscosity of about 1.16 centipoises. For this material Seyer and Metzner indicate that the properties of the turbulence, including the intensities and time-averaged velocity profile in the core of the tube, do not differ markedly from those of a Newtonian fluid at similar conditions. At the much higher concentration of 0.40 percent, the data follow the laminar relation to quite high Reynolds numbers and, in contrast to the known behavior of purely viscous materials, depart only gradually from the laminar relation. This region is in fact a broad transition region, beginning at a Reynolds number where departure from the laminar relation first occurs and extending to a Reynolds number in excess of 10,000. The boundaries of the

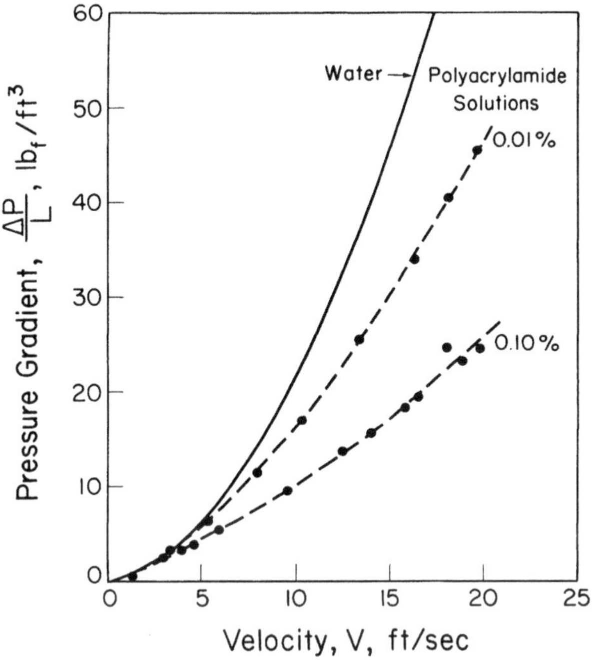

Figure 5.38. Pressure gradient reduction in a 1-inch pipe caused by a typical viscoelastic fluid—aqueous solution of polyacrylamide. (From Seyer, F. A., and A. B. Metzner, *Can. J. Chem. Eng.*, **47**, 525 (1969.)

transition region are not yet defined but depend strongly on the diameter, and the type and concentration of the polymer under consideration.

The diameter effect in drag reduction is also illustrated in Figure 5.39 by the data for different pipe sizes. The amount of drag reduction decreases significantly as the tube size increases. Although no data have been published for large diameter tubes (1 foot or larger in diameter) the trend of the data is to suggest that little or no drag reduction would occur in large diameter pipes. This matter is considered in more detail later.

It is often convenient to think of a drag reduction efficiency representing the fractional reduction in friction factor from the normally expected turbulent flow quantity toward the hypothetical laminar flow quantity, i.e.,

$$F = \frac{f_t - f}{f_t - f_l} \tag{5.122}$$

where f, f_t, and f_l are, respectively, the actual, the normally expected turbulent, and the hypothetical laminar friction factors all at the actual Reynolds number. This was proposed by Seyer and Metzner (1967), who reasoned that F would be dependent only on Re_{MR}, n' and the Deborah number, De (see Chapter 1, Section 4.42). From data on water solutions of polyacrylamide

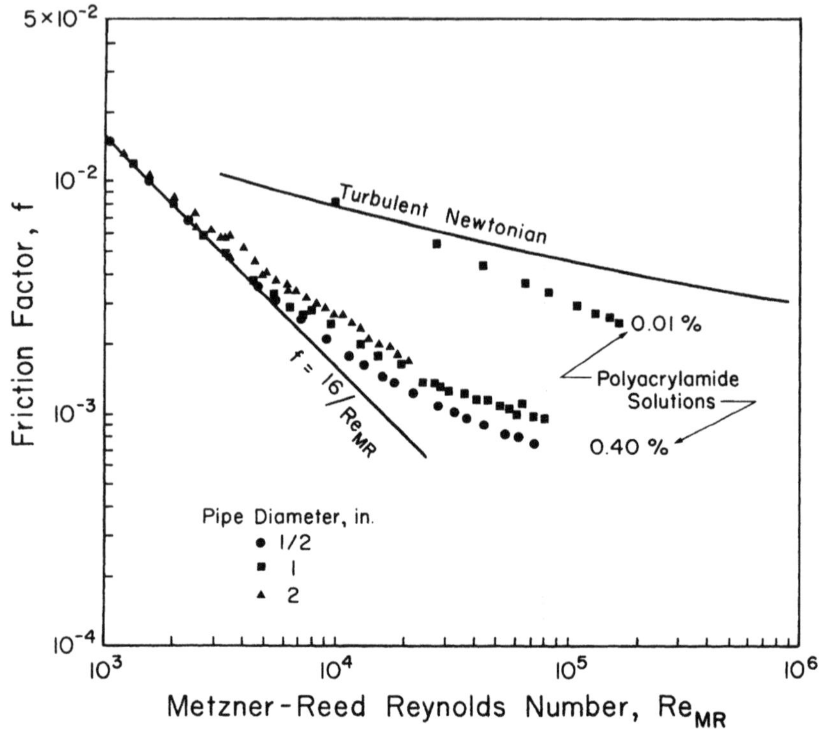

Figure 5.39. Friction factors for typical viscoelastic fluids—aqueous solutions of polyacrylamide. (From the data of Seyer, F. A., and A. B. Metzner, *Can. J. Chem. Eng.*, **45**, 121, 1967; **47**, 525, 1969.)

(essentially constant n') they developed an empirical correlation which is a graphical representation of $F = \phi(\text{Re}_{MR}, \text{De})$.

White (1970) quotes data of several workers which offer support for a maximum drag reduction effect described by the equation

$$f_{\min} = 0.126\ \text{Re}^{-0.43} \tag{5.123}$$

This corresponds to a drag reduction efficiency, F, of 0.82 to 0.83.

A number of strictly empirical relationships for correlating and predicting drag reduction effects have been developed. These include those of Meter and Bird (1964), Kilbane and Greenkorn (1966), Lord, Hulsey, and Melton (1967), Seyer and Metzner (1967), Virk, Merrill, Mickley and Smith (1967), and Whitsitt *et al.* (1969). Virk *et al.* (1967) have had some success in relating the onset of drag reduction to molecular parameters of the polymer. Their work also includes a correlation of the amount of drag reduction with concentration for a homologous series of polyethylene oxides in water. These relationships are restricted in their application but are of interest because of their simplicity and qualitative significance. We will discuss the Kilbane and

Greenkorn correlation as representative of these and then deal with the developments of Meyer (1966), Seyer and Metzner (1969b), and Astarita, Greco, and Nicodemo (1969), which have a stronger theoretical basis and are of more general applicability.

4.52 The Kilbane and Greenkorn Correlation

Kilbane and Greenkorn (1966) propose a correlation of the Blasius form

$$f = \bar{K} f'' \,\mathrm{Re}_{KG}{}^{-0.284} \tag{5.124}$$

where f'' is the friction factor predicted from the Dodge and Metzner viscous fluid turbulent flow correlation for the appropriate value of n' and for a modified Reynolds number

$$\mathrm{Re}_{KG} = (1 + \mathrm{Ws})\,\mathrm{Re}_{MR} \tag{5.125}$$

where Ws = the Weissenberg number $= (P_{11} - P_{22})/\tau_w$. The quantity $\bar{K}$ is a constant for a given fluid found by fitting a straight line of slope $= -0.284$ to a log-log plot of f/f'' versus Re_{MR} for turbulent flow data. The evaluation of $\bar{K}$ for the data of Metzner and Park (1964) and Meter (1964) is illustrated in Figure 5.40; for the solutions tested, $\bar{K}$ ranges from 10.16 to 13.65. The Kilbane and Greenkorn correlation fits the data of Seyer (1965) with a single value of $\bar{K} = 13.1$.

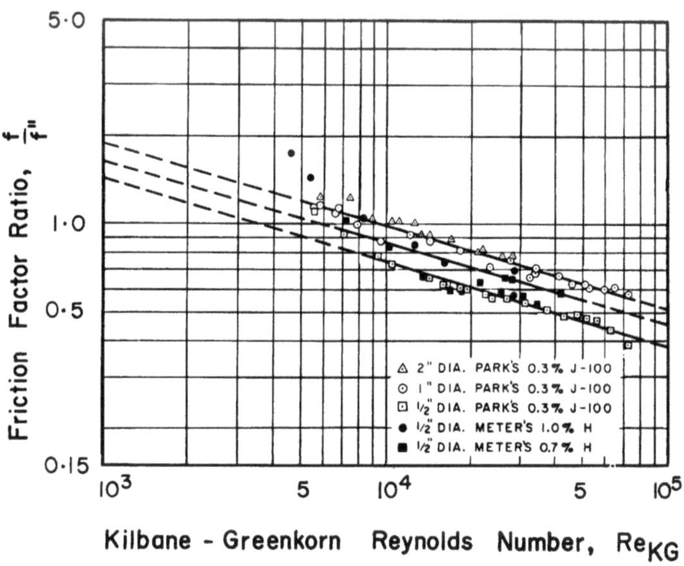

Figure 5.40. Evaluation of $\bar{K}$ (Eq. 5.124) for experimental data. (From Kilbane, J. K., and R. A. Greenkorn, Soc. Petrol. Engrs., A.I.M.E., Paper SPE 1679, 1966.)

4.53 Logarithmic Relationships (Meyer, and Seyer and Metzner)

A number of analyses are available which correlate friction factor data for drag reducing systems using modifications of the well known logarithmic relationships for Newtonian fluids. The advantage of these relationships lies in the fact that their form may be deduced in part by theoretical reasoning.

As we have seen in Chapter 4, the velocity profile in the turbulent core for Newtonian fluids is of the form

$$u^+ = A \ln y^+ + B$$

in which A and B are constants. Pitot-tube measurements reported by Elata, Lehrer, and Kahanovitz (1966), Ernst (1966), and Meyer (1966) indicated that this equation, with a modified constant B, is applicable to viscoelastic fluids. Meyer's analysis of the data obtained by Ernst indicated

$$B = B_N + X' \log \frac{U^*}{U_c^*}; \qquad U^* > U_c^* \tag{5.126}$$

where $B_N = 5.5$, the value of B for Newtonian fluids,
X' = parameter dependent upon the viscoelastic properties,
U^* = friction velocity $= \sqrt{\tau_w g_c/\rho} = V\sqrt{f/2}$
U_c^* = the value of U^* below which the viscoelastic effects are not observed.

The data of Ernst (1965) and those of Elata and Tirosh (1965) both showed U_c^* to be 0.23 ft/sec. From these relations Meyer developed the following equation for the friction factor

$$\frac{1}{\sqrt{f}} = 4.0 \log (\text{Re}\sqrt{f}) - 0.394 + \frac{\alpha'}{\sqrt{2}} \log \frac{U^*}{U_c^*} \tag{5.127}$$

in which the shear dependent viscosity is evaluated at the rate of shear at the wall, and α' is a constant related to X'. The applicability of Equation 5.127 to the data of Ernst on a 0.05 percent solution of carboxymethyl cellulose (CMC) is shown in Figure 5.41. Although this equation adequately represents the CMC data, two shortcomings should be noted. The very dilute CMC solutions exhibit only a slight amount of drag reduction relative to that presented in Figure 5.39, and therefore the correlated effects are small ones. Additionally the local velocity data were obtained with Pitot-tube measurements which have been subjected to serious questions by a number of investigators. Savins (1965), Astarita and Nicodemo (1966), and others have shown how the Pitot-tube measurement depends directly on the magnitude of the first normal stress difference, a fact that has not been accounted for in the work of Ernst,

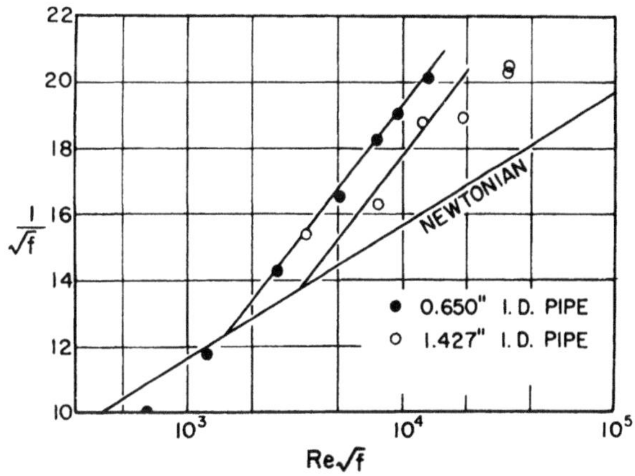

Figure 5.41. Evaluation of Meyer's friction factor relationship for viscoelastic fluids—Equation 5.126. (From Meyer, W. A., *A.I.Ch.E. Journal*, **12**, 522, 1966.)

and Elata and Tirosh. In this regard, however, the modest level of the normal stresses for CMC solutions would tend to reduce the possible errors.

Seyer and Metzner (1969b) have formulated the logarithmic similarity laws for drag reducing systems using theoretical arguments similar to those of Millikan (1939) for Newtonian fluids, and of Dodge and Metzner (1959) for purely viscous non-Newtonian fluids. Their analysis is based primarily on the observation that viscoelastic fluids may offer an extremely large resistance to elongational or stretching deformations, depending on the magnitude of a Deborah number. In turbulent flow one of the primary processes affecting the drag or pressure drop is the formation and action of large eddies near the wall which eject low momentum fluid toward the center of the tube.

Based upon data for Newtonian fluids Seyer and Metzner have shown that the deformations within the large eddies are essentially stretching deformations as discussed in Chapter 1. Furthermore their calculations show that for known drag reducers the Deborah number, as defined by Equation 1.63 and determined by the relaxation time of the fluid, is sufficiently large to inhibit the motion of the large eddies. Thus it is assumed that the velocity in the wall region but beyond the laminar sublayer is determined by:

$$u = \phi_1(\tau_w, y, \rho, \mu_a, \theta)$$

and in the turbulent core by

$$u_m - u = \phi_2(\tau_w, y, \rho, R)$$

These equations are identical to those applicable to Newtonian fluids except for the addition in the first equation of the fluid relaxation time to account for elastic effects. In the core the flow is assumed to be essentially inertial and independent of both elastic and viscous effects.

From dimensional analysis, and requiring that the above equations overlap in some region near the wall but outside the viscous sublayer, the velocity profile for the turbulent core may be shown to be

$$u^+ = A \ln y^+ + B(J) \tag{5.128}$$

where $A = 2.5$ is the same constant as for Newtonian fluids and $B(J)$ is an arbitrary function of the dimensionless group

$$J = \frac{\theta(U^*)^2\rho}{\mu_a} \tag{5.129}$$

which must be determined experimentally. Determining the average velocity by integration of Equation 5.128 across the core and assuming a linear velocity profile within the viscous sublayer we obtain

$$\sqrt{\frac{2}{f}} = 2.5(1-\zeta_l)^2 \ln \mathrm{Re}\sqrt{f} + (1-\zeta_l)^2[B(J) - A\ln 2\sqrt{2}] - 3 \tag{5.130}$$

in which $\zeta_l = y_l/R$ is the dimensionless thickness of the laminar sublayer. The quantity ζ_l is defined arbitrarily as the point at which the linear velocity profile and that given by Equation 5.128 intersect.

For most systems, as in the case for Newtonian fluids, the ζ_l term may be neglected and Equation 5.130 reduces to Meyer's Equation (Equation 5.127) except for the form of the function $B(J)$.

Elata *et al.* (1966) have also derived an equation similar to Equation 5.130 in which they have assumed

$$B(J) = 5.5 + \alpha'' \log J \tag{5.131}$$

where α'' is an empirical parameter which was found to be a complicated function of concentration. It should be noted that Elata's determination of B is based on Pitot-tube measurements and is therefore subject to some doubt.

Seyer and Metzner found that Equation 5.130 would reproduce the available friction factor data to within a mean deviation of 4.2 percent. Their evaluation of the function $B(J)$ from data on several concentrations of a polyacrylamide is reproduced in Figure 5.42. These data have been obtained using an optical technique for measuring the velocities and are therefore free of the limitations imposed on data obtained from Pitot-tube measurements. In this regard the close agreement between the independent experimental determinations of B through Equation 5.128 with velocities, and through Equation 5.130 by measurement of the pressure drop should be noted. Although the Seyer and Metzner analysis appears fundamentally sound, additional data with other viscoelastic fluids are necessary to confirm the relationship between $B(J)$ and the group J.

The effects of diameter and solute concentration on drag reduction are of

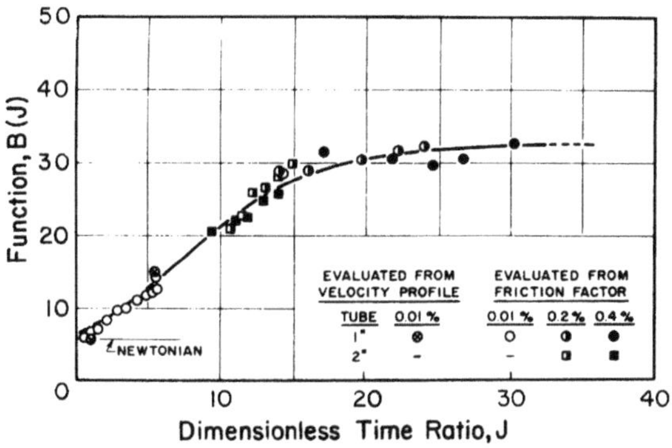

Figure 5.42. Evaluation of the function $B(J)$ from behavior of viscoelastic solutions of polyacrylamide in water. (From Seyer, F. A., and A. B. Metzner, *A.I.Ch.E. Journal*, **15**, 426, 1969.)

great importance. We may discuss these within the framework of Equations 5.129 and 5.130. The effect of diameter on the friction factor for turbulent flow of a viscoelastic fluid will depend upon the way in which the group J in Equation 5.129 changes with diameter. At a given Reynolds number the wall shear stress (and hence U^*) varies inversely with the diameter, and therefore the group J tends to decrease with increasing pipe diameter. From Figure 5.42 and Equation 5.130 we see that the drag reduction effect will decrease accordingly. The dependence of the group J on shear stress is somewhat reduced, however, since the relaxation time, θ, generally increases with decreasing shear stress. The data of Seyer and Metzner (1967) indicates that for 0.2 percent and higher concentrations of polyacrylamide, J is directly proportional to τ_w, whereas for the very dilute, 0.01 percent, polyacrylamides J is directly proportional to $\sqrt{\tau_w}$. Thus the quantitative effect of diameter depends directly on the details of the rheological properties.

Table 5.4 illustrates reductions in drag, as a percentage of the pressure drop for water, for a 0.01 percent solution of polyacrylamide. Entries for the 1-inch pipe are experimental values, and the values for the 12-inch pipe have been obtained from Equation 5.130. These calculations indicate that, for the polymer considered, only modest drag reduction can be expected in large pipes under conditions of commercial interest.

Qualitatively, increases in concentration of a polymeric additive result in increased elastic effects in a solution and thereby increased drag reduction. The changes in drag reducing effectiveness with concentration are illustrated in Table 5.5, where J at a given set of flow conditions may be taken as a measure of the drag reducing effectiveness. It is apparent that the initial amounts of polymeric material result in large effects and, although further increases in the concentration (to 0.1 percent and in excess in this case) increase the elasticity of the system, these increases are offset by rapid increases in the viscosity.

TABLE 5.4

Drag Reduction in Small and Large Diameter Tubes[a]

Reynolds Number	Drag Reduction Percent	
	1-Inch Tube	12-Inch Tube
10,000	0	0
50,000	20	0
100,000	30	0
500,000	61	14
1,000,000	78	25

[a] F. A. Seyer and A. B. Metzner, *Can. J. Chem. Eng.*, **47**, 525 (1969).

TABLE 5.5

Effect of Concentration of a Polyacrylamide Solution in Water on Drag Reduction at Constant Wall Shear Stress, $\tau_w = 1.0\ \mathrm{lb_f/ft^2}$ [a]

Concentration, Weight percent	Apparent Kinematic Viscosity, μ_a/ρ, $10^5 \mathrm{ft^2/sec}$	θ, 10^4 sec	J, dimensionless
0	1.00	10^{-13}	10^{-12}
0.01	1.60	1.3	5.8
0.10	2.35	8.5	18.7
0.20	3.62	12.0	17.1
0.40	8.57	21.0	12.8
0.60	15.20	23.0	7.8

[a] F. A. Seyer and A. B. Metzner, *Can. J. Chem. Eng.*, **47**, 525 (1969).

4.54 The Astarita *et al.* Relationships

Astarita, Greco, and Nicodemo (1969) have also developed a method for estimating drag reduction effects based in part on theoretical considerations. They reason that the friction factor should be a function of both the Reynolds number and a Deborah number

$$f = \phi(\mathrm{Re}, \nu T) \tag{5.132}$$

where ν = the smallest frequency of energy dissipating eddies,
T = a characteristic time for the process.

Following Astarita (1965) and Seyer and Metzner (1967) they express v as

$$v = \frac{V}{D}\,\mathrm{Re}^{0.75} \qquad (5.133)$$

For convenience and anticipating the possibility of separately expressing the role of the Deborah number, they define a fractional friction factor

$$\beta = \frac{f}{f_t} \qquad (5.134)$$

in terms of f and f_t defined following Equation 5.122.

The quantity β lies between the limits of unity (no drag reduction) and a value of f/f_l, where $f_l = 16/\mathrm{Re}$. Astarita *et al.* assume that

$$\beta = \beta(vT) \qquad (5.135)$$

where T is a characteristic constant for any drag reducing fluid but one that is difficult to predict or measure. To avoid the difficulty they re-express Equation 5.135 as

$$\beta = \beta'\left(\frac{v}{v_{0.5}}\right) \qquad (5.136)$$

by taking a ratio of the expression for β and the expression for $\beta = 0.5$, at which v has the value $v_{0.5}$.

They anticipated that drag reduction data for different solutions and different diameters should be correlated by a single curve if plotted in the form of Equation 5.136. The method does, however, require sufficient experimental data for each solution to determine $v_{0.5}$, i.e., the value of $(V/D)\mathrm{Re}^{0.75}$ at which $\beta = 0.5$.

Astarita *et al.* obtained drag reduction data with a series of water solutions of a commercial polymer (ET597) covering concentrations ranging from 125 to 2000 ppm, in tube diameters of 0.69, 1.10, and 2.05 cm and over a range of Re of 2×10^3 to 2×10^5. Since the higher concentration polymer solutions were non-Newtonian, they obtained their values of f_t (for calculating β) from Dodge and Metzner's (1959) friction factor correlation (see Figure 5.23).

The final correlation of Astarita *et al.* is given in Figure 5.43, where we see that a single curve correlates data for all their polymer concentrations and tube diameters. Although we might expect other polymers to yield the same relationship, the correlation may only be considered proved for the polymer tested. Nonetheless, given sufficient measurements to determine $v_{0.5}$ for any particular solution, the correlation should give good order of magnitude predictions of drag reduction at different values of Re and tube diameter.

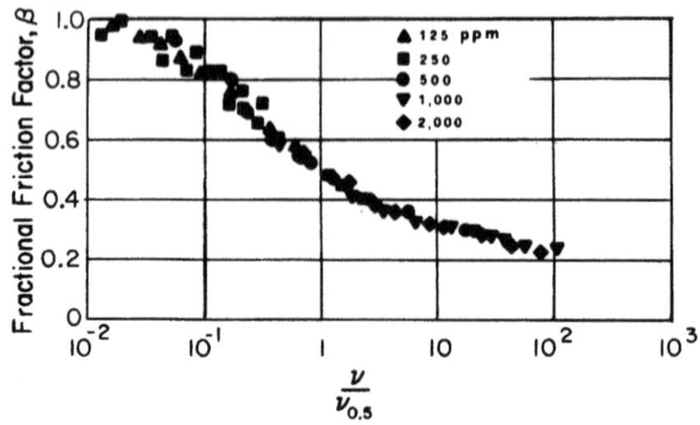

Figure 5.43. Fractional friction factors for viscoelastic fluids—data on aqueous solutions of polyacrylamide. (From Astarita, G., G. Greco, Jr., and L. Nicodemo, *A.I.Ch.E. Journal*, **15**, 564, 1969.)

4.6 Bizarre Fluids

In a series of papers, Savins (1967, 1968) and White (1967) have explored the flow properties of a variety of soap solutions. Reductions in drag as large as those observed for the polymer solutions have been noted. Typical data, taken from White, are presented in Figure 5.44 for a dilute water solution

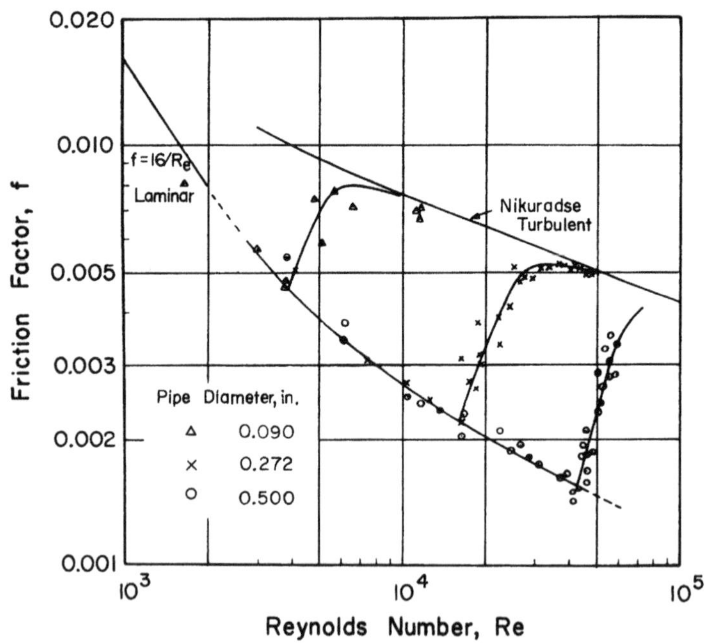

Figure 5.44. Friction factors for a typical soap solution. (From White, A., *Nature*, **214**, 585, 1967.)

of equimolar amounts of cetyltrimethylammonium bromide and 1-naphthol. These solutions are known to form micelles, which are disrupted at sufficiently high stresses and which subsequently reform on removal of the stress. This breakdown feature, coincident with the start of the high positive slope lines in Figure 5.44, has been noted in the work of Savins. Although little is known of the physical properties of these materials, it is conceivable that the micelles may be effective in resisting sudden deformations and thereby in reducing the drag in much the same way as viscoelastic materials. Savins (1969) gives a detailed discussion of the various types of drag reduction and the various materials that exhibit this phenomenon.

4.7 Entrance Effects

Few data are available concerning the travel distance from the entrance of a tube required for the establishment of fully developed turbulent flow or concerning the pressure gradient in the entrance region in the turbulent flow of non-Newtonian fluids. Dodge and Metzner (1959) present limited data for purely viscous, non-Newtonian materials suggesting that

$$\frac{X_E}{D} \leq 15$$

which is substantially less than for Newtonian fluids. General indications of Bogue and Metzner (1963) and Weltmann and Keller (1957) are to the effect that the excess pressure gradient in the entrance region is much the same as for Newtonian fluids. This is not unexpected since, in turbulent flow, most deviations from Newtonian behavior caused by viscous, non-Newtonian fluids are modest.

Limited data for viscoelastic systems indicate that, at least under some conditions, major deviations from Newtonian behavior may occur. Dodge (1957) and Seyer (1965) have noted entry lengths much greater than for Newtonians with drag reducing fluids. Recent data obtained by Seyer (1969) for a 0.2 percent solution of a polyacrylamide (Separan AP30) in water are plotted in Figure 5.45, which shows typical pressure profiles for water and the polymer solution flowing from a large reservoir into a sharp-edged $\frac{1}{2}$-inch tube. Similar data have also been obtained in a 2-inch pipe. As a drag reducer the polymer behaves similarly to the 0.4 percent polyacrylamide solution of Figure 5.39. The entry length for the polymer solution is in excess of 150 diameters, whereas that for the water, at the same velocity, is roughly 50 to 60 diameters. These data are in the transition region from laminar to turbulent flow where, for Newtonian fluids, very large entry lengths are observed. Thus the qualitative features of Figure 4.1 are retained, but the sharp peak

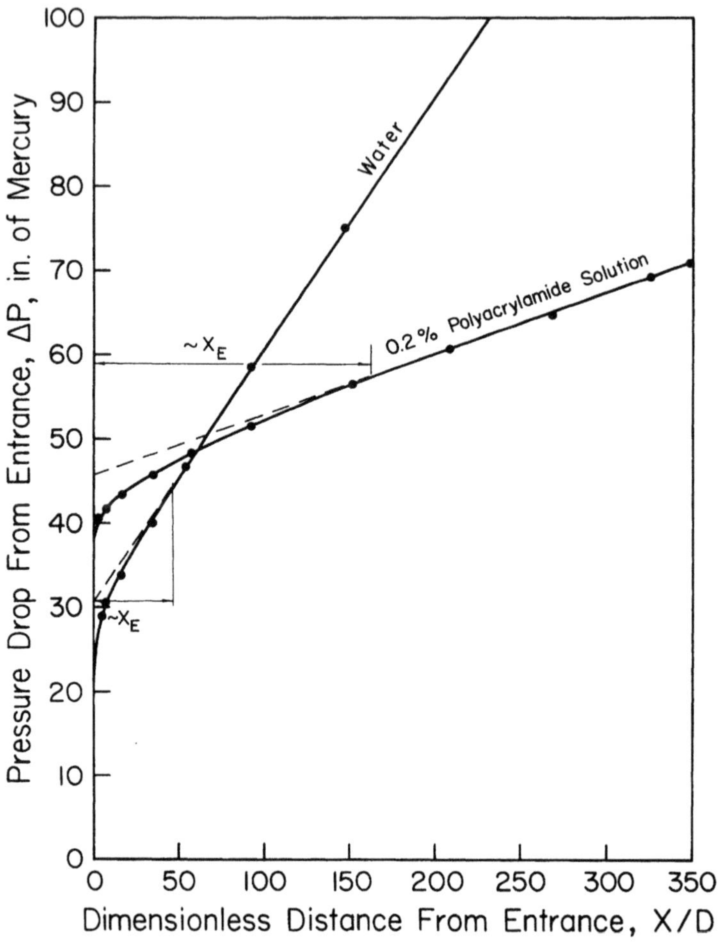

Figure 5.45. Pressure profiles for water and a typical viscoelastic fluid—0.20 percent polyacrylamide in water.

Constant flow rate of 34 U.S. gallons/min
Tube diameter, 1/2 inch
Reynolds number for water, 1.9×10^5
Metzner-Reid Reynolds number for solution, 3.5×10^4
(From Seyer, F. A., Private Communication, 1969.)

is broadened to include the very large range of transitional Reynolds numbers that occur for the more concentrated solutions.

Few data are available for sudden contraction losses for viscoelastic fluids and no data are available for pressure losses for flow through various fittings. Figure 5.45 shows, however, that the contraction loss for a polyacrylamide solution is significantly larger than for the water and suggests therefore that major effects may occur in the flow of viscoelastic fluids through fittings and changes in cross section.

4.8 General Form of the Pressure Drop Equation

In steady flow the mechanical energy balance for the turbulent flow of non-Newtonian materials takes the same form as for Newtonian fluids. The velocity profile measurements of Bogue and Metzner (1963) for purely viscous materials and those of Seyer (1969) for viscoelastic materials indicate that, since the velocity profiles are very nearly the same as for Newtonians, the numerical value of α must be much the same as for Newtonian fluids.

For purely viscous materials the effect of entrance region and of the presence of pipe fittings appears to be little different from that of Newtonian materials and the reader is, therefore, referred to the relations for incompressible Newtonian fluids in Chapter 4. With viscoelastic materials the few data available suggest that major effects may occur under some conditions and that the usual relations for purely viscous materials may yield erroneous estimates of the pressure losses. Additionally, if major elastic effects occur, for example, in the entrance of a tube, the energy balance should include an additional term to account for the storage of elastic energy. Present knowledge, however, is insufficient to provide a quantitative evaluation of these effects.

5 THE FLOW OF SUSPENSIONS OF FIBROUS MATERIALS

5.1 Introduction and Typical Data

Liquid suspensions of wood pulp and other fibrous materials exhibit flow behavior that in some ranges of flow is similar to that of some of the pseudo-homogeneous, non-Newtonian materials we have discussed, in some ranges is more like that of a Newtonian fluid, and in still other ranges is similar to that of solid-liquid suspensions to be discussed in Chapter 11. The behavior is complex, not yet well understood, and there are no completely satisfactory and general methods available for the prediction of velocity profiles or pressure gradients. For these reasons we will not endeavor to treat the subject in any depth but we will look at some typical data and relate the observed performance to the earlier discussion based upon the various rheological types.

The pipe flow behavior of water suspensions of wood-pulp and synthetic fibers has been the subject of many investigations. A great number of these have originated from the Pulp and Paper Research Institute of Canada, The Technical Association of the Pulp and Paper Industry (T.A.P.P.I.) who sponsored research at the Massachusetts Institute of Technology, and the Eastman Kodak Company.

Wood-pulp fibers are flexible, their density is only slightly greater than that of water, they range from less than 1 to 5 mm in length, and they are of the order of 25 to 50 microns in width. In suspension they do not act independently but may flocculate in large masses. At extremely low flow rates, below those of commercial interest, the fibers may form a bed on the bottom of the pipe and flow in the moving bed pattern which we will discuss in connection with granular solids in Chapter 11. They may also form into loose aggregates, which slide and roll on the pipe bottom. At modest flow rates the fibers tend to congregate in a central core, which moves much as the plug of a Bingham plastic in laminar flow. The plug is surrounded by an annulus of essentially clear liquid in which the velocity gradient is high. As the flow rate is increased, the higher shear and turbulent stresses gradually break up the plug and the fibers are more or less symmetrically distributed. This is analogous to, but also exhibits an important difference from, the symmetric suspension flow pattern which we discuss in Chapter 11.

The flow of flexible fibers in liquid suspension, as might be expected from the difference in the flow patterns observed, results in pressure gradient curves

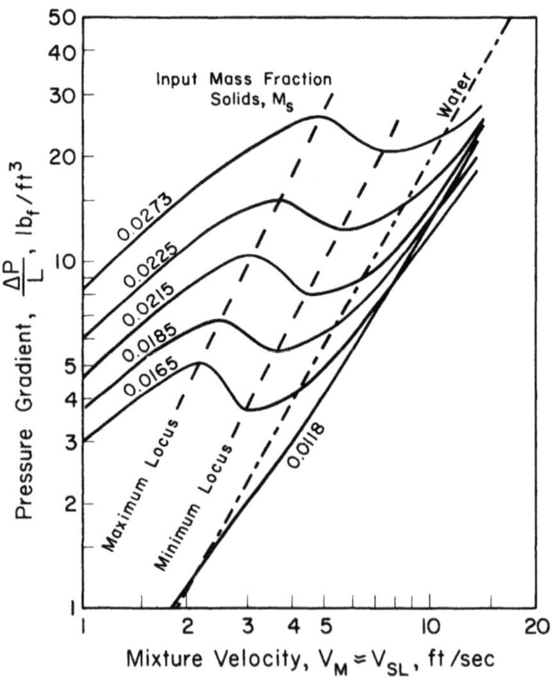

Figure 5.46. Pressure gradient relationship for the flow of water suspensions of unbleached sulfite pulp fibers in a 1-inch smooth copper pipe. (From data of Guenter, T. E., and N. H. Ceaglske, *T.A.P.P.I.*, **34**, 140, 1951.)

quite different from those of Newtonian fluids. Data of Guenter and Ceaglske (1951) on the flow of unbleached sulfite pulp fibers in water are typical and are given in Figure 5.46. For each line of constant composition, above the line for the mass fraction of 0.0118, the data show a maximum and a minimum in the pressure gradient curve. Further, and most significant, with increasing flow velocity the curves cross the line for water tending to parallel it at higher velocities and remaining below it.

Many workers (Forrest, 1949; Daily and Bugliarello, 1958; Condolios and Constans, 1964; Serwinski, Kemblowski, Kielbasa, and Iciek, 1968; and others) have related visually observed flow patterns to the pressure gradient curves.

At flow rates below the maximum point on the curve the fibers move as a sliding bed or as clusters rolling on the pipe wall. From the maximum to the minimum point the plug flow pattern with the clear water annulus is observed, the plug decreasing in diameter as the flow rate is increased. As the velocity is increased beyond the minimum point, the turbulence is increased and the

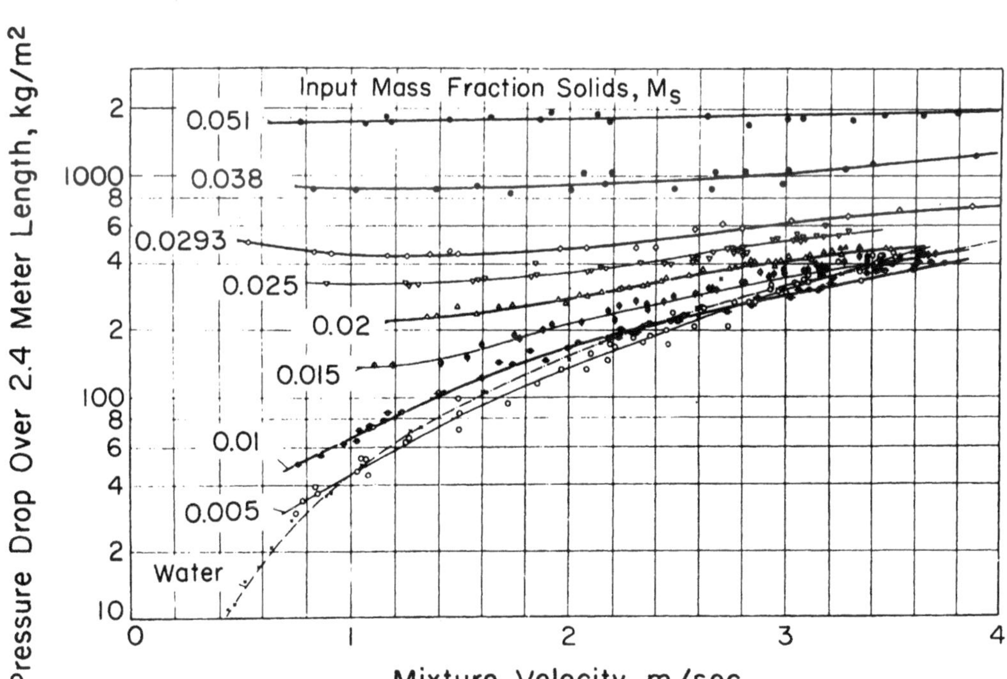

Figure 5.47. Pressure drop relationship for the flow of water suspensions of bleached, nonbeated sulfite pulp in a 60-mm pipe. (From Serwinski, M., Z. Kemblowski, J. Kielbasa, and J. Iciek, *Int. Chem. Eng.*, **8**, 453, 1968.)

fibers become distributed more and more uniformly until a pseudohomogeneous mixture results. From this point the pressure gradient tends to become proportional approximately to the square of the velocity. The fact that the pressure gradient is lower than that observed for water is attributed to a suppression of the turbulence caused by the presence of the fibers.

Typical data of Serwinski *et al.* covering a wider range of fiber concentration, but not such low velocities, are given in Figure 5.47. (Note the arithmetic scale on the abscissa as compared with Figure 5.46.) Here we see only a slight indication of the minimum point and a wide range of velocity over which the pressure gradient is essentially constant. This reflects the greater shear resistance and stability of the fiber plug with the more concentrated suspensions.

5.2 Prediction of Pressure Gradient

From our knowledge of the behavior of homogeneous non-Newtonian fluids, we might expect that the plug flow region of fiber suspensions could be described as that of a Bingham or near-Bingham plastic with a coefficient of rigidity equal to the viscosity of water and with a yield stress, or apparent yield stress, dependent upon the nature and concentration of the fibers. This concept has been proposed and tested by Baines (1958), Daily and Bugliarello (1958), and others. It is found to have some qualitative merit but not to be entirely suitable as a basis for correlation. One problem is the difficulty of obtaining meaningful measurements by laboratory instrument of the yield value as it would be effective in pipe flow. Nonetheless, as indicated in Figure 5.48 taken from Daily and Bugliarello (1958), the friction factor curves are similar to those for Bingham fluids as discussed in Section 2.3. Baines (1958) has also demonstrated that shear stress data interpreted from pipe flow and concentric cylinder tests for 0.95 weight percent suspensions of bleached sulfite pulp fibers align with one another on the equivalent of an f versus Re plot at constant He.

Baines (1958), Forgacs, Robertson, and Mason (1958), and others have shown that the plug remains stable only so long as laminar flow persists in the annulus. Forgacs *et al.* indicate that the end of stable plug flow occurs at

$$\mathrm{Re}_c = \left(\frac{DV_M\rho}{\mu}\right)_c = 310 \tag{5.137}$$

It is possible that further analysis of data of this sort would confirm the applicability of the Buckingham equation over the plug flow region as long as laminar flow persists in the annulus surrounding the plug.

Although not shown in Figure 5.48, the transition to turbulent flow appears in some cases to take place through an intermediate or mixed flow stage, as reflected in Figure 5.49 taken from Daily and Tsuchiya (1959). Similar data have been shown by Forgacs *et al.* Beyond this and as shown in Figure 5.48

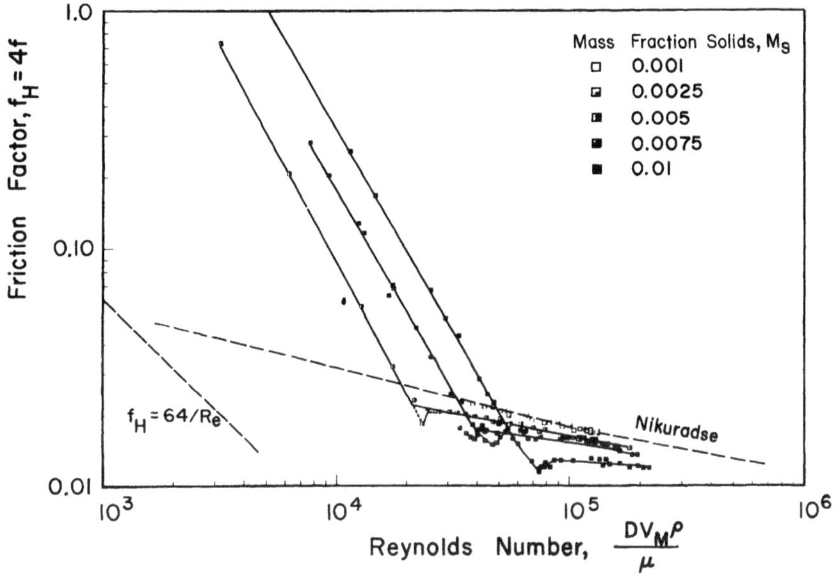

Figure 5.48. Friction factor data for the flow of water suspensions of ground wood poplar pulp in a 2-inch pipe. (From Daily, J. W., and G. Bugliarello, M. I. T. Hydrodynamics Laboratory, Technical Report No. 30, Oct. 1958.)

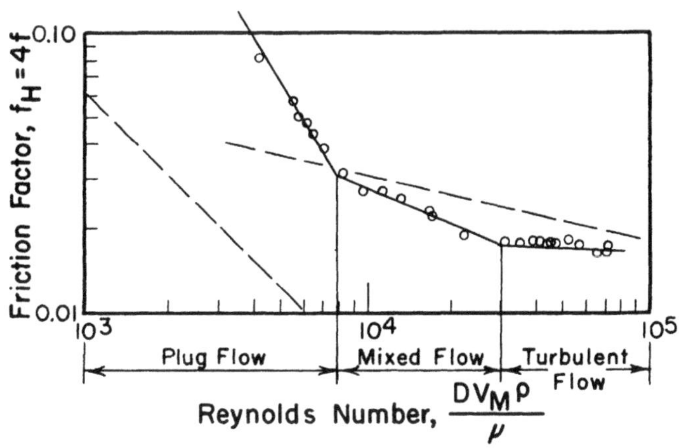

Figure 5.49. Friction factor data for the flow of water suspensions of Kraft softwood pulp (65 percent spruce, 35 percent jack pine) in a 3/4-inch pipe. Fiber concentration (mass fraction solids) = 0.005. (From Daily, J. W., and A. Tsuchiya, M. I. T. Hydrodynamics Laboratory, Technical Note No. 4, Feb. 1959.)

the f–Re lines, which fall below the turbulent flow line for water, tend slowly toward it, with f declining somewhat less with Re than in the Newtonian case. This again is qualitatively similar to the turbulent flow of Bingham fluids as discussed in Section 4.3. It is interesting also to compare this behavior with the turbulent flow of viscoelastic fluids discussed in Section 4.5. Considering the flexibility characteristics of typical fibers we might expect energy storage possibilities and some elastic behavior. This is discussed by Forgacs *et al.*

Guthrie (1959) has applied the generalized method of Metzner and Reed discussed in Section 2.6 to data on a variety of pulps and a range of consistencies in $\frac{1}{2}$- and 1-inch pipes. The rheological properties of the pulps expressed in the Metzner and Reed terminology ranged from $0.36 < n' < 1.0$ and $0.10 \times 10^3 < K' < 17.5 \times 10^3$. The units of K' are $\mathrm{lb_f\ sec}^{n'}/\mathrm{ft}^2$. Guthrie's results are shown in Figure 5.50. For $\mathrm{Re}_{MR} < 2100$ they show good average agreement, although considerable scatter, with the laminar line, $f = 16/\mathrm{Re}_{MR}$. At higher values of Re_{MR} the data fall well below the Newtonian line which was Guthrie's basis of comparison. Actually, if the Dodge and Metzner, or Clapp or Torrance lines (see Figure 5.26) for $n' < 1$ are superimposed on Guthrie's Figure 5.50, the data are seen to be reasonably well bracketed between the lines $n' = 0.4$ and $n' = 1.0$. Guthrie's work is not sufficiently

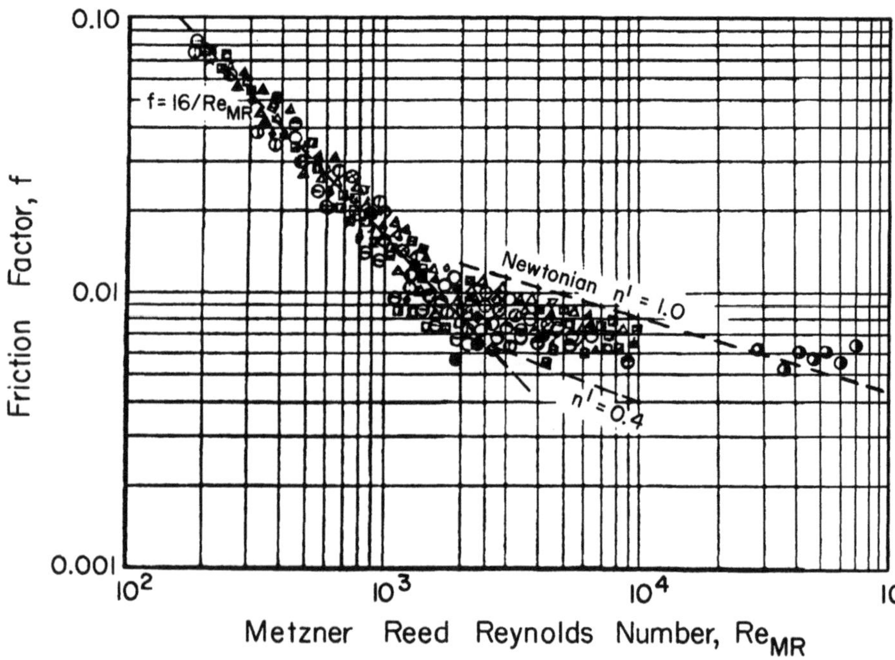

Figure 5.50. Correlation of friction factors for the flow of fiber suspensions. (Modified after Guthrie, W. E., *T.A.P.P.I.*, **42**, 232, 1959.)

extensive to confirm fully the applicability of the Metzner–Reed and Dodge and Metzner correlations for laminar and turbulent flow, but it is strongly indicative of their suitability at least for an approximate interpretation of pipe flow data of fiber suspensions.

In addition to the prediction methods which relate the flow behavior to the Bingham plastic or to the generalized Metzner and Reed expression, there have been many attempts to interpret pressure gradient data by strictly empirical means. We will not discuss these but refer the reader to the work of Forgacs *et al.* (1958), Daily and Bugliarello (1958, 1961), Daily, Bugliarello, and Troutman (1959) and Manteufel (1965). The last-mentioned reference reviews an empirical prediction method proposed earlier by the Hydraulics Committee of the Technical Association of the Pulp and Paper Industry (T.A.P.P.I.).

Our discussion indicates that the phenomena underlying the observed behavior of the pipe flow of fiber suspensions are still far from being fully understood. Much work has recently been directed to the fundamentals of the flow. Typical is the work of Mason (1954) and Robertson and Mason (1954, 1956, 1957); that on the flow of model fibers by Bobkowicz and Gauvin (1965); and that on rigid particles by Daily and Hardison (1964), and Daily and Chu (1961).

6 RECOMMENDED DESIGN METHODS

6.1 Introduction

Design methods for the pipe flow of time-independent non-Newtonian fluids are more complex than for Newtonians. It is first necessary to identify that the fluid or mixture is pseudo-homogeneous and behaves as a time independent non-Newtonian fluid. The next step is to classify it as to type and to determine its rheological properties.

The steps preliminary to the pipeline calculations are:

1. Following the method suggested in Section 3.5 of Chapter 1, determine that the mixture will be essentially nonsettling in its pipe flow. If otherwise, the problem should be handled by the methods of Chapter 9 or 11.
2. From actual rheological data, or generalizations discussed in Chapters 1 and 2, determine that the fluid behavior is time-independent or that its time dependency may be neglected for the problem at hand. If neither is

the case, i.e. if time dependency is significant, the problem should be handled by the methods of Chapter 6.

3. Preferably from actual rheological data interpreted as discussed in Chapter 1, or from generalizations as discussed in Chapter 2, determine the rheograms for the mixture of interest at the temperature of interest and over at least a decade of shear rate up to a shear stress not less than the maximum anticipated wall shear stress in the pipe.
4. Fit the simplest appropriate rheological equation to the data and determine the relevant constants. The discussion of Sections 4.2 and 5.2 of Chapter 1 will serve as a guide.

In what follows we assume that at least the following rheological constants are known, having been evaluated over the shear stress range of interest:

For a pseudoplastic:
a. the power law constants, n, K,
b. the Metzner-Reed constants, n', K',

For a Bingham or near Bingham fluid:
a. the Bingham or apparent Bingham constants, τ_y, η or τ_{ya}, η_a,
b. the Metzner-Reed constants, n', K',

For a yield-pseudoplastic:
a. the yield–power law constants, τ_y, η, K,
b. the Metzner-Reed constants, n', K',

For a viscoelastic fluid:
a. the viscous behavior property parameters; μ if appropriate, otherwise n and K, and n' and K',
b. the principal normal stress difference, $P_{11} - P_{22}$, as a function of the rate of shear, or some equivalent measure of the viscoelastic or drag reducing behavior of the fluid (see Section 6.6, step 5).

6.2 Entrance Effects

The pipe travel distance required for the attainment of stabilized flow, X_E, depends upon whether the flow is laminar or turbulent, as well as upon the rheological properties of the mixture. The nature of the flow may be determined as follows:

1. Calculate the appropriate modified Reynolds number
   For a power law pseudoplastic:
   Re_{PL1} from Equation 5.29

Re_{PL2} from Equation 5.30
Re_{MR} from Equation 5.63
For a Bingham or near Bingham fluid:
Re_B, He, and Y from the definitions following Equation 5.53
Re_{MR} from Equation 5.63
For a more complex but still viscous fluid:
Re_{MR} from Equation 5.63
For a viscoelastic fluid:
Re, Re_{PL1}, or Re_{MR} depending upon the viscous behavior of the fluid.

2. Determine whether the flow will be laminar or turbulent by comparison of the calculated Reynolds number with the transition Reynolds number
   $(Re_{PL1})_c$ as given by Figure 5.4
   $(Re_{PL2})_c = (Re_{MR})_c = 2100$
   $(Re_B)_c$ as given by Figure 5.9 or 5.21.
3. If the flow is turbulent, we have no reliable way of estimating X_E or the pressure drop over the entrance length. See the discussion in Section 5 of Chapter 4.
4. If the flow is laminar, X_D or X_E may be estimated as follows:
   a. For power law pseudoplastic fluids, calculate X_D from Figure 5.12; for Bingham fluids, estimate X_E from Figure 5.15; for more complex but not viscoelastic fluids, estimate X_D from Figure 5.12 taking $n' = n$ and $K' = (K/g_c)(1 + 3n/4n)^n$.
   b. No methods are available for viscoelastic fluids but it must be expected that the entrance length (or the total entrance effect) will be greater than for a corresponding viscous fluid.
5. Where the entrance length is found to be less than 10 percent of the pipe length, its effect may be neglected; otherwise follow step 6.
6. Determine the pressure drop over the entrance length for a power law pseudoplastic from Equation 5.64 and Figure 5.14; for a Bingham fluid, from Equation 5.65 and Figure 5.16; for a yield-pseudoplastic, from Equation 5.64 and Figure 5.14, assuming that $n' = n$ and $K' = (K/g_c) \times (1 + 3n/4n)^n$.

6.3 Effect of Bends and Fittings

As discussed in Section 2.8 we have no specific methods for non-Newtonian fluids to account for the effect in a pipe system of bends and fittings. The methods for Newtonian fluids discussed in Section 5 of Chapter 4 may be applied to estimate L_E which may then be added to the actual pipe length for purposes of the calculation of the pressure drop due to friction.

6.4 Potential Energy or Hydrostatic Head Effects

No complications arise in the calculation of ΔP_{PE} for non-Newtonian fluids and the quantity is determined directly from the net elevation difference and the mixture density as

$$\Delta P_{PE} = \frac{g}{g_c} \rho \Delta z$$

as shown in Equation 5.66.

6.5 Acceleration or Kinetic Energy Effects

Effects of acceleration or kinetic energy are usually small with the essentially incompressible non-Newtonian fluids we are discussing. The pressure change ΔP_{KE} is given by $(\rho/2g_c)[(V_2^2/\alpha_2)-(V_1^2/\alpha_1)]$ as shown in Equation 5.66. For laminar flow the α terms should be evaluated having regard to Equation 5.68 or Figure 5.17 for pseudoplastics and Equation 5.69 for Bingham or yield-pseudoplastic fluids. For turbulent flow the α terms may be taken as 0.8.

6.6 Pressure Gradient for Steady Flow

The component of the pressure drop attributable to irreversibilities or friction may be estimated as follows. We consider first the case when the fluid does not exhibit viscoelastic effects, or, if it does, when the flow is laminar.

1. For a pseudoplastic fluid:
   a. First assume power law behavior and the power law constants, K and n, for τ less than the estimated value of $\tau_w = D\Delta P/4L$.
   b. If the flow is laminar as determined in Section 6.2, determine the pressure gradient due to friction from Equation 5.25 or from Equation 4.24 and a friction factor determined from Figure 5.4. This will be sufficiently accurate for most purposes. However, if greater accuracy, or a check, is desired and if the rheological behavior of the fluid or mixture may be better represented by the Ellis, Powell-Eyring, Meter, or Sisko equation, the pressure gradient due to friction may also be calculated by the applicable methods discussed in Sections 2.22 to 2.25.
   c. If the pipe is smooth and the flow is turbulent, calculate Re_{PL2} from Equation 5.30 and determine the pressure gradient due to friction from Equation 4.24 and a friction factor determined from Figure 5.26.

d. If the flow is turbulent and the pipe is rough, multiply the pressure gradient obtained in step c by the ratio of the friction factors, f_{rough}/f_{smooth}, as obtained from Figure 4.16 at $Re = Re_{PL2}$.
e. If the pipe roughness and flow are such that FRW turbulence will occur, estimate the friction factor from Equation 5.121 or Figure 5.37.

2. For a Bingham or near Bingham plastic:
   a. If the flow is laminar as determined in Section 6.2, determine the pressure gradient due to friction from Equation 4.24 with f determined from either of Figures 5.9 or 5.10.
   b. If the flow is turbulent and the pipe is smooth, calculate the pressure gradient due to friction from Equation 4.24 with f determined from the appropriate Figure 5.32 or 5.33.
   c. If the flow is turbulent and the pipe is rough, multiply the pressure gradient calculated in step b by the ratio, f_{rough}/f_{smooth}, as determined from Figure 4.16 at $Re = Re_B$.
   d. If the pipe roughness is such that FRW turbulence may be expected, estimate the friction factor from Figure 5.37, assuming $n = 1$.
3. For a power law yield-pseudoplastic, or a fluid best described by the power law yield-pseudoplastic equation (Equation 1.48):
   a. If the flow is laminar as determined in Section 6.2, determine the pressure gradient due to friction from Equation 5.59 by trial and error as indicated in Section 2.4.
   b. If the flow is turbulent and the pipe is smooth, determine the pressure gradient due to friction from Equation 4.24 and a friction factor based on Torrance's work as in Figures 5.33 to 5.36.
   c. If the flow is turbulent and the pipe is rough, multiply the pressure gradient obtained in step b by the ratio of the friction factors, f_{rough}/f_{smooth}, as obtained from Figure 4.16 at $Re = Re_{PLC}$.
   d. If the pipe roughness and flow are such that FRW turbulence will occur, estimate the friction factor from Equation 5.121 or Figure 5.37 using the appropriate value of n.
4. If the fluid is not well described by any of the power law, the Bingham, or the power law yield-pseudoplastic equation, the generalized approach of Metzner and Reed, and Dodge and Metzner, is useful.
   a. If the flow is laminar as determined in Section 6.2, calculate the pressure gradient due to friction from Equation 4.24 and a friction factor determined by Equation 5.62. The values of K' and n' should be evaluated for $\tau \le \tau_w$. This involves some trial and error.
   b. If the flow is turbulent and the pipe is smooth, calculate the pressure gradient due to friction from Equation 4.24 and a friction factor determined from Figure 5.23 or 5.26, again using values of K' and n' appropriate for $\tau \le \tau_w$.

c. If the flow is turbulent and the pipe is rough, multiply the pressure gradient obtained in step b by the ratio of the friction factors, $f_{\text{rough}}/f_{\text{smooth}}$, as obtained from Figure 4.16 at $\text{Re} = \text{Re}_{MR}$.

d. If the pipe roughness and flow are such that FRW turbulence will occur, estimate the friction factor from Equation 5.121 or Figure 5.37, taking $n = n'$.

5. If the fluid is viscoelastic (drag reducing) and the flow is turbulent, experimental data on the elastic properties of the fluid or the magnitude of the drag reduction effect are necessary to permit a good estimate of the pressure gradient. The data may be such as to permit the estimation of $B(J)$ of Equation 5.131 or of $\nu_{0.5}$ of Equation 5.136.

a. If data are available to estimate $B(J)$ (see Section 4.53) the pressure gradient due to friction for turbulent flow in a smooth pipe may be estimated from Equation 4.24 and a friction factor determined from Equation 5.130.

b. If data are available to estimate $\nu_{0.5}$ (see Section 4.54), the pressure gradient due to friction for turbulent flow in a smooth pipe may be estimated from Equation 4.24, Figure 5.26 (for the determination of f), and Figure 5.43.

c. If data are not available to estimate $B(J)$ or $\nu_{0.5}$, but $P_{11} - P_{22}$ is known or may be estimated, the pressure gradient due to friction in a smooth pipe may be estimated from Equation 4.24, and a friction factor determined from Equation 5.124 (using a value of $\overline{K}$ of 13.1) as discussed in Section 4.52.

d. If the flow is turbulent and the pipe is rough, the actual pressure gradient will exceed that calculated in steps a, b, or c but, owing to the thickened boundary layer with the viscoelastic fluid, will be less than that which would be obtained by multiplying the pressure gradient obtained in steps a, b, or c by the ratio of the friction factors, $f_{\text{rough}}/f_{\text{smooth}}$, as obtained from Figure 4.16 at the appropriate Reynolds number.

6. Regardless of whether steps *1*, *2*, *3*, *4* or *5* are followed, the total pressure gradient is estimated having regard for Equation 5.66 and the discussion in Sections 6.2, 6.3, 6.4, and 6.5 relating to entrance effects, effects of bends and fittings, potential energy effects, and acceleration effects.

Bibliography

Allis-Chalmers Manufacturing Co., Bulletin No. 1649.

Alves, G. E., D. F. Boucher, and R. L. Pigford, *Chem. Eng. Progress*, **48**, 385 (1952).

Astarita, G., *Ind. Eng. Chem. Fundam.*, **4**, 354 (1965).

Astarita, G., G. Greco, Jr., and L. Nicodemo, *A.I.Ch.E. Journal*, **15**, 564 (1969).

Astarita, G., G. Greco, and L. Peluso, *Ind. Eng. Chem. Fundam.*, **7**, 595 (1968).

Astarita, G., and L. Nicodemo, *A.I.Ch.E. Journal*, **12**, 478 (1966).

Baines, W. D., Laminar Flow of Dilute Fibre Suspensions, Report MH-73, National Research Council of Canada, Ottawa, August, 1958.

Bobkowicz, A. J., and W. H. Gauvin, *Can. J. Chem. Eng.*, **43**, 87 (1965).

Boger, D. V., and A. V. Ramamurthi, *A.I.Ch.E. Journal*, **16**, 1088 (1970).

Bogue, D. C., *Ind. Eng. Chem.*, **51**, 874 (1959).

Bogue, D. C., and A. B. Metzner, *Ind. Eng. Chem. Fundam.*, **2**, 143 (1963).

Buckingham, E., *A.S.T.M. Proc.*, **21**, 1154 (1921).

Caldwell, D. H., and H. F. Babbitt, *Ind. Eng. Chem.*, **33**, 249 (1941).

Chen, S. S., L. T. Fan, and C. L. Hwang, *A.I.Ch.E. Journal*, **16**, 293 (1970).

Christiansen, E. B., N. W. Ryan, and W. E. Stevens, *A.I.Ch.E. Journal*, **1**, 544 (1955).

Clapp, R. M., Conf. on Int. Developments in Heat Transfer, Univ. Colorado, Boulder, Col., 1961, *Proc.* p. 652.

Collins, M., and W. R. Schowalter, *A.I.Ch.E. Journal*, **9**, 804 (1963).

Condolios, E., and J. Constans, *Allgemeine Papier-Rundschau*, **3**, 136 (1964).

Daily, J. W., and G. Bugliarello, The Effects of Fibers on Velocity Distribution, Turbulence and Flow Resistance of Dilute Suspensions, Technical Report No. 30 Hydrodynamics Laboratory, Massachusetts Institute of Technology, Cambridge, Mass., October, 1958.

Daily, J. W., and G. Bugliarello, *T.A.P.P.I.*, **44**, 497 (1961).

Daily, J. W., G. Bugliarello, and W. W. Troutman, Measurement and Analysis of Turbulent Flow of Wood Pulp Fiber Suspensions, Technical Report No. 35, Hydrodynamics Laboratory, Massachusetts Institute of Technology, Cambridge, Mass., September, 1959.

Daily, J. W., and T. K. Chu, Rigid Particle Suspensions in Turbulent Shear Flow: Some Concentration Effects, Technical Report No. 48, Hydrodynamics Laboratory, Massachusetts Institute of Technology, Cambridge, Mass., October, 1961.

Daily, J. W., and R. L. Hardison, Rigid Particle Suspensions in Turbulent Shear Flow: Measurement of Total Head, Velocity and Turbulence with Impact Tubes, Report No. 67, Hydrodynamics Laboratory, Massachusetts Institute of Technology, Cambridge, Mass., April, 1964.

Daily, J. W., and A. Tsuchiya, Flow of Nylon Fiber Suspensions, Technical Note No. 4, Hydrodynamics Laboratory, Massachusetts Institute of Technology, Cambridge, Mass., February, 1959.

Deissler, R. G., *Trans. A.S.M.E.*, **73**, 101 (1951).

Dodge, D. W., Doctoral Dissertation, University of Delaware, Newark, Del., 1957.

Dodge, D. W., and A. B. Metzner, *A.I.Ch.E. Journal*, **5**, 189 (1959).

Elata, C., J. Lehrer, and A. Kahanovitz, *Israel J. Technol.*, **4**, 87 (1966).

Elata, C., and J. Tirosh, *Israel J. Technol.*, **3**, 1 (1965).

Ernst, W. D., *A.I.Ch.E. Journal*, **11**, 940 (1965); **12**, 581 (1966).

Forgacs, O. L., A. A. Robertson, and S. G. Mason, Hydrodynamic Behaviour of Paper Making Fibres, 44th Annual Meeting of the Technical Section, Canadian Pulp and Paper Assoc., Montreal, Que., 1958, *Proc.* p. 79.

Forrest, F. R., *Paper Mill News*, **72** (21), 53 (1949).

Govier, G. W., *Chem. Eng.*, **66** (24 Aug.), 138 (1959).

Govier, G. W., and M. D. Winning, The Flow of Non-Newtonian Fluids, Paper presented at the Meeting of the American Institute of Chemical Engineers, Montreal, Quebec, September, 1948.

Griskey, R. G. and R. G., Green, *A.I.Ch.E. Journal*, **17**, 725 (1971).

Guenter, T. E., and N. H. Ceaglske, *T.A.P.P.I.*, **34**, 140 (1951).

Guthrie, W. E., *T.A.P.P.I.*, **42**, 232, (1959).

Hall, C. D., *Soc, Petrol. Engrs J.*, **9**, 131 (1969).

Hanks, R. W., *A.I.Ch.E. Journal*, **9**, 306 (1963); **14**, 691 (1968).

Hanks, R. W., and E. B. Christiansen, *A.I.Ch.E. Journal*, **8**, 464 (1962).

Hanks, R. W., and B. H. Dadia, Theoretical Analysis of the Turbulent Flow of Non-Newtonian Slurries in Pipes. Private Communication, 1969. Published: *A.I.Ch.E. Journal*, **17**, 554 (1971).

Hanks, R. W., and D. R. Pratt, *Soc. Petrol. Engrs. J.*, **7**, 342 (1967).

Hedstrom, B. O. A., *Ind. Eng. Chem.*, **44**, 651 (1952).

Herzog, R. L., and K. Weissenberg, *Kolloid-Z.*, **46**, 277 (1928).

Hornbeck, R. W., Non-Newtonion Laminar Flow in the Inlet of a Pipe, A.S.M.E. Paper 65-WA/FE-4, November, 1965.

Kilbane, J. K., and R. A. Greenkorn, Correlation of Friction Factors for Viscoelastic Fluids in Tubes, Soc. Petrol. Engrs., A.I.M.E., Paper No. SPE 1679, 1966.

Kreith, F., and R. Eisenstadt, *Trans. A.S.M.E.*, **79**, 1070 (1957).

Lord, D. L., B. W. Hulsey, and L. L. Melton, *Soc. Petrol. Engrs. J.*, **7**, 252 (1967).

Manteufel, R. J., *T.A.P.P.I.*, **48**, 631 (1965).

Mason, S. G., *T.A.P.P I*, **37**, 494 (1954).

McMillen, E. L., *Chem. Eng. Progr.*, **44**, 537 (1948).

Melton, L. L., and W. T. Malone, *Soc. Petrol. Engrs. J.*, **4**, 56 (1964).

Meter, D. M., Doctoral Dissertation, University of Wisconsin, Madison, Wisconsin, 1963.

Meter, D. M., *A.I.Ch.E. Journal*, **10**, 881 (1964).

Meter, D. M., and R. B. Bird, *A.I.Ch.E. Journal*, **10**, 878 (1964).

Metzner, A. B., in *Advances in Chemical Engineering*, edited by T. B. Drew and J. W. Hoopes, Vol. 1, p. 77, Academic Press, New York, 1956.

Metzner, A. B., *Ind. Eng. Chem.*, **49**, 1429 (1957).

Metzner, A. B., Tube and Annular Flow Problems, Course Notes, Univ. Delaware, Newark, Del., 1968.

Metzner, A. B., and M. G. Park, *J. Fluid Mech.*, **20**, 291 (1964).

Metzner, A. B., and J. C. Reed, *A.I.Ch.E. Journal*, **1**, 434 (1955).

Meyer, W. A., *A.I.Ch.E. Journal*, **12**, 522 (1966).

Michiyoshi, L., K. Mizuno, and Y. Hoshiai, *Int. Chem. Eng.*, **6**, 373 (1966).

Millikan, C. B., Fifth Int. Congress for Applied Mechanics, *Proc.* p. 386, John Wiley & Sons Inc., New York; 1939.

Mooney, M. *J. Rheol.*, **2**, 210 (1931).

Patterson, G. K., J. L. Zakin, and J. M. Rodriguez, *Ind. Eng. Chem.*, **61**, 22 (1969).

Perkins, A., and J. J. Glick, B. Chem. Eng. Thesis, University of Delaware, Newark, Del., 1954.

Powell, R. E., and H. Eyring, *Nature*, **154**(3909), 427 (1944).

Rabinowitsch, B., *Z. phisik. Chem., Ser. A*, **145**, 1 (1929).

Robertson, A. A., and S. G. Mason, *Pulp Paper Mag. Can.*, **55**(3), 263 (1954); **57**(6), 121 (1956).

Robertson, A. A., and S. G. Mason, *T.A.P.P.I.*, **40**, 326 (1957).

Ryan, N. W., and M. M. Johnson, *A.I.Ch.E. Journal*, **5**, 433 (1959).

Savins, J. G., *Soc. Petrol. Engrs. J.*, **4**, 203 (1964).

Savins, J. G., *A.I.Ch.E. Journal*, **11**, 673 (1965).

Savins, J. G,. *Rheol. Acta*, **6**, 323 (1967); **7**, 87 (1968).

Savins, J. G., in *Viscous Drag Reduction*, edited by C. S. Wells, Plenum Press, Inc., New York, 1969, p. 183.

Savins, J. G., R. F. Burdyn, and G. C. Wallick, *Soc. Petrol. Engrs. J.*, **3**, 197 (1963).

Savins, J. G., and D. B. Cox, *Fuel*, **62**, 363 (1963).

Savins, J. G., G. C. Wallick, and W. R. Foster, *Soc. Petrol. Engrs. J.*, **2**, 211 (1962).

Serwinski, M., Z. Kemblowski, J. Kielbasa, and J. Iciek, *Int. Chem. Eng.*, **8**, 453 (1968).

Seyer, F. A., M.Sc. Thesis, Univ., Delaware, Newark, Del., 1965.

Seyer, F. A., Private Communication, 1969.

Seyer, F. A., and A. B. Metzner, *Can. J. Chem. Eng.*, **45**, 121 (1967).

Seyer, F. A., and A. B. Metzner:
a. *Can. J. Chem. Eng.*, **47**, 525 (1969).
b. *A.I.Ch.E. Journal*, **15**, 426 (1969).

Shaver, R. G., and E. W. Merrill, *A.I.Ch.E. Journal*, **5**, 181 (1959).

Sisko, A. W., *Ind. Eng. Chem.*, **50**, 1789 (1958).

Skelland, A. H. P., *Non-Newtonian Flow and Heat Transfer*, John Willey & Sons Inc., New York, 1967.

Stevens, W. E., Doctoral Dissertation, University of Utah, Salt Lake City, Utah, 1953.

Sylvester, N. D., and S. L. Rosen, *A.I.Ch.E. Journal*, **16**, 964 (1970).

Thomas, D. G., *Ind. Eng. Chem.*, **55**, 27 (1963).

Tomita, Y., *Bulletin J.S.M.E.*, **2**, 469 (1959); **4**, 77 (1961).

Toms, B. A., First Int. Rheological Congress (1948), *Proc.* Part 2, p. 135, North-Holland Publishing Co., Amsterdam, 1949.

Torrance, B. McK., *South African Mechanical Engineer*, **13**, 89 (1963).

Virk, P. S., E. W. Merrill, H. S. Mickley, and K. A. Smith, *J. Fluid Mech.*, **30**, 305 (1967).

Wells, C. S., Jr., *A.I.A.A. Journal*, **3**, 1800 (1965).

Weltmann, R. N., and T. A. Keller, N.A.C.A., T.M. 3889, 1957.

White, A., *Nature*, **214**, 585 (1967).

White, D. A., *Chem. Eng. Sci.*, **25**, 1127 (1970).

Whitsitt, N. F., L. J. Harrington, and H. R. Crawford, in *Viscous Drag Reduction*, edited by C. S. Wells, Plenum Press, Inc., New York, 1969, p. 265.

Wilhelm, R. H., D. M. Wroughton, and W. F. Loeffel, *Ind. Eng. Chem.*, **31**, 622 (1939).

Worster, R. C., Flow in Pipes of Fly Ash and Other Suspensions of Finely Divided Solids, British Hydromechanics Research Association, Publication RR 515, Cranfield, Bedford, November, 1955.

The Flow of Time-Dependent Non-Newtonian Fluids in Pipes

6

1 INTRODUCTION

The problems encountered in the design of pipe flow systems for time-dependent fluids have received relatively little attention, notwithstanding the gradual improvements in our understanding of the rheology of such fluids. This is not as serious as it may at first appear, however, because time-dependent fluids approach time-independent behavior after having been subjected to shear for a reasonable period of time. Despite this, it is of real interest to know how long a time-dependent fluid must flow in a pipe to reach this condition of approximate time-independent behavior and how the pressure gradient changes with time and travel distance. These are the problems we will examine in this chapter. We will confine our attention to the case of thixotropic fluids, although much the same approach may be employed for rheopectic fluids.

The discussion will be restricted to purely viscous time-dependent fluids. Viscoelastic fluids display a type of time dependency of a different nature and usually of much shorter duration than that associated with thixotropy or rheopexy.

In a thixotropic fluid, as flow progresses from start-up, the structural breakdown of the fluid results in a decrease in the pressure gradient at a constant flow rate or an increase in the flow rate at constant pressure gradient. Although the mechanisms of thixotropy have been investigated for many years, none of the proposed constitutive equations are readily applicable to the prediction of flow behavior in pipes. This difficulty results from the inadequacy and/or the complexity of the rheological equation. In our discussion we will assume the availability of a series of shear stress–shear rate–shear duration, (τ–$\dot{S}$–θ), curves as described in Chapter 1, covering the range of shear duration of interest. We will base our analysis on such data rather than on any assumed rheological equation.

We must remember that the rheological behavior and the shape of the τ–$\dot{S}$–θ curves may, and usually do, depend upon the previous shear and thermal history of the fluid. Accordingly, it is important that the curves represent the behavior of a sample of fluid that has experienced a shear and thermal history approximating as closely as possible that anticipated for the fluid whose pipe flow behavior is to be analyzed. When the fluid moves into the pipe slowly and with a minimum of mixing, such as from a large storage tank directly into the pipe, the τ–$\dot{S}$–θ curves should be based upon tests on a sample that has been stored at zero shear for an appropriate time and has not been significantly sheared in its transfer to the test equipment. At the other extreme, when a fluid undergoes intense agitation, as in the passage through a turbine or centrifugal pump, the curves should be obtained using a highly premixed or sheared fluid. In each case the thermal history of the sample should be the same as that of the fluid involved in the application. For any given thermal history and regardless of the shear history, the rheological behavior of a fluid element at a particular rate of shear will gradually approach a limiting condition corresponding to the infinite shear duration rheogram.

In our analysis of the effect of the duration of shear on the pipe flow of thixotropic fluids, we will assume that we have the τ–$\dot{S}$–θ curves based on suitably preconditioned samples and that the measurement of the duration of shear, θ, starts from the time of commencement of flow in the pipe.

2 LAMINAR FLOW

2.1 Qualitative Description

It is helpful first to consider qualitatively the changes that occur in individual fluid elements as they enter a pipe and flow from the pipe inlet in laminar motion. For simplicity, in this discussion we assume that the fluid entering the pipe, and the fluid filling the pipe at the start of flow, have not been subjected to previous significant shear and may be described by the zero shear duration curve as illustrated by the upper curve of Figure 1.21. Also the velocity profile, appropriate for the properties of the entering fluid, is assumed to be developed instantaneously, i.e., entrance effects due to boundary layer development are neglected. The volumetric flow rate is taken to be constant for the entire period of flow from start-up.

Generally, thixotropic fluids, and more particularly thixotropic oils, display yield values. That is, after the material has been at rest for a sufficient length of time, a "gelled" structure exists throughout the fluid and a finite stress is required to break the structure and to initiate flow. If an applied stress or pressure drop is not sufficient to break the gel or exceed the yield stress, no flow will occur. If the applied stress is greater than the gel or yield strength, the gel will be broken from the wall to a point where the yield strength is not exceeded, and the annulus of fluid will flow. The flow will involve a solid cylindrical plug carried in the moving annular fluid in a manner similar to that of a Bingham plastic. However, in many thixotropic fluids, once the gel is broken by simple shear or by premixing, the sheared fluid, unlike a Bingham plastic, may no longer display a finite yield strength because of the slow rate of redevelopment of the structure.

Let us consider the start-up of flow of a fluid that has been at rest in a pipe for sufficient time for a gel to develop. We assume also that the fresh fluid entering the pipe has not been premixed and that it has the same gel structure as that in the pipe. At the commencement of flow, the velocity profile is that characteristic of a yield-pseudoplastic; the shear duration is zero throughout the length and cross section of the pipe; the rate of shear is a maximum at the wall decreasing to zero in the pluglike core.

Behavior Immediately Downstream from the Entrance, $L < L_c$ After a short time of flow, say 1 minute, the region from the entrance of the pipe to a distance L_c, corresponding with that traveled by the maximum velocity elements of fluid in 1 minute, will contain fluid which, at the wall, has been subjected to shear for 1 minute and, elsewhere in the sheared annulus, has been subject to shear for a time θ that depends upon the radial position and the distance from the inlet, but is within the range $0 < \theta < 1$. In an idealized situation the central plug remains unsheared, its properties remain unchanged, and its diameter remains nearly constant. Because of the variations in both shear duration and rate of shear, the fluid properties in the annulus vary with the radial position and distance. In our illustration, the consistency of the fluid will decrease, being lowest at the wall, and elsewhere lowest at the points farthest from the pipe inlet. As a result, the initial velocity profile will change, with the fluid elements nearest the wall and farthest from the entrance accelerating, and the elements nearest the plug and the plug itself decelerating. The pressure gradient, dP/dL, will decrease with distance from the pipe inlet to the distance L_c.

As the time of flow is increased, say to 10 minutes, the penetration of the entering fluid increases, i.e., L_c increases, and a greater length of pipe contains fluid which has been subject to varying shear duration. The range of shear duration is $0 < \theta < 10$; the consistency of the sheared fluid decreases still further; the velocity profile becomes steeper at the wall; the plug, still retaining its initial diameter, decelerates further. The pressure gradient is now everywhere lower than after 1 minute of flow and decreases with distance from the inlet to the new value of L_c.

Eventually, when the time of flow is sufficiently increased, and if the pipe is long enough, L_c reaches its maximum value, L_{cm}, and at L_{cm} and beyond an equilibrium velocity profile is reached and dP/dL becomes constant with time and position. The maximum value of L_c corresponds with the travel distance of the fastest moving elements in a time equal to that for the entering unsheared material to reach its minimum consistency. In the section from the entrance to L_{cm} the velocity profile and dP/dL vary with the travel distance but no longer with time.

Behavior Remote from the Entrance, $L > L_c$ In the region beyond L_c, where the freshly entering fluid has not penetrated, all elements of fluid within the flowing annulus are subjected to a constant duration of shear equal to the time from the start of flow, and they exhibit properties corresponding with a duration of shear equal to the time from the start of flow. At any time of flow, therefore, the velocity profile and dP/dL are both constant with travel distance beyond L_c. As the time of flow is increased from 1 to 10 minutes and longer, the fluid elements in the annular region undergo their thixotropic reduction in consistency and dP/dL decreases further. The fluid in the central plug remains unsheared and the plug diameter remains constant.

For a thixotropic fluid that does not exhibit a yield value, or for one in which any gel in the entering fluid has been broken and the gel in the fluid standing in the pipe has not yet reformed, the overall situation is similar to that just described except that the sheared region extends to the center line of the pipe and there is no unsheared plug.

A more complex situation results when a thixotropic fluid that exhibits a gel after standing in a pipe is displaced by an entering fluid that is premixed, as by the action of a pump. Here the gelled fluid originally in the pipe results in a central plug throughout the length of the pipe containing original material. This is as in the first case discussed. Over the length of the pipe containing the entering premixed material, no plug exists and the situation is as discussed in the preceding paragraph.

Ritter and Batycky (1967) discuss in detail the case of a premixed material displacing a gelled material. Figure 6.1, taken from their work, shows calculated velocity profiles for a crude oil (the Pembina crude oil at 44.5°F described

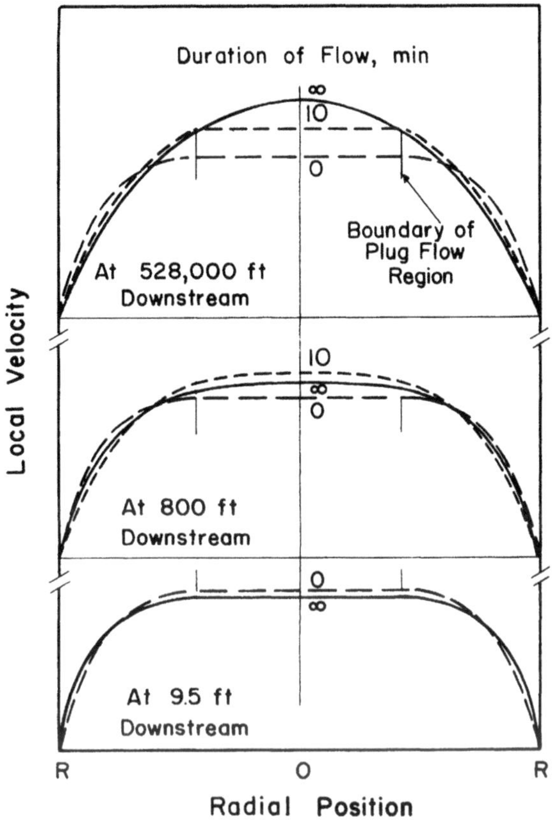

Figure 6.1. Illustrative velocity profiles at downstream positions for a typical thixotropic fluid. Pembina crude oil, temperature = 44.5°F, constant flow rate = 15,000 barrels/day (average velocity, $V = 1.24$ ft/sec), $D = 12$ in., duration of flow: — — — start-up, – – – – – – 10 minutes, ———— steady state. (From Ritter, R. A., and J. P. Batycky, *Soc. Petrol. Engrs. J.*, **7**, 369, 1967.)

in Chapter 1). The change in the profile with duration of flow and distance from the pipe inlet is illustrated.

Cheng, Ray, and Valentin (1965) report studies on the flow of thixotropic bentonite suspensions in $\frac{1}{2}$-, 1-, and $1\frac{1}{2}$-inch pipes. Under conditions of laminar flow they found broad agreement with their qualitative predictions (similar to those we have discussed) of the effect of pipe travel–distance on pressure gradient, but they did not present quantitative relations. Actually the observed effects due to thixotropy were modest, and Cheng *et al.* found their pressure drop data to be consistent with their equilibrium consistency curves, i.e., $\tau = f(\dot{S})$ with $\theta \simeq \infty$.

2.2 General Equations

In the simplest case of a time-dependent fluid, the constitutive equation, or the τ–$\dot{S}$–θ curves, may be expressed, as

$$f_H(\tau, \theta) = \dot{S} \tag{6.1}$$

where f_H designates a functional relationship dependent upon the previous shear and thermal history. For a fluid that has been at constant temperature and not subject to shear at any significant rate for a long period of time, the relationship becomes

$$f_{T,0}(\tau, \theta) = \dot{S} \tag{6.2}$$

where $f_{T,0}$ represents a reproducible function dependent only upon the nature of the fluid and the constant historical temperature, T.

Similarly, for a fluid subject to a constant temperature and a constant high rate of shear (say 1000 sec^{-1} or greater) for a long period of time, we have

$$f_{T,\infty}(\tau, \theta) = \dot{S} \tag{6.3}$$

To develop the general equations applicable to the laminar pipe flow of time-dependent fluids, we assume a case where the τ–$\dot{S}$–θ relations are known for the fluid as it enters the pipe and that the duration of shear is measured from this time. Let the relations be those indicated by Equation 6.1. For convenience, we assume that the fluid contained in the pipe at the time of start-up has the same rheological properties as the entering fluid. If we now express $\dot{S}$ as $(-du/dr)$, we have, from Equation 6.1, 6.2, or 6.3,

$$\tau = f\left(-\frac{du}{dr}, \theta\right) \tag{6.4}$$

We may now recall Equation 5.6,

$$\tau = -\frac{r}{2}\frac{dP}{dL} \tag{5.6}$$

developed without assumption for the case of time-independent fluids. This equation may also be applied to time-dependent fluids if we assume that the radial flow causing the changing velocity profile and producing the corresponding changes in stresses is negligible compared with the axial flow. This is a reasonable assumption.

In principle, the velocity profile, the average velocity, and the relationship between Q and dP/dL may now be obtained by combination and integration of Equations 6.4 and 5.6. The problem is complicated by the presence of the term θ, since the shear duration is dependent both upon travel distance and radial position. Referring to Figure 6.2 we may consider a shell of fluid bounded by stream lines within the radii r and $(r + dr)$ moving at a velocity $u_{r,0}$ at the pipe inlet changing its radial position to r' and $(r' + dr)$ and its velocity to $n_{r,L}$ at a downstream distance L. Then

$$L = \int_0^{\theta_r} u_r \, d\theta \tag{6.5}$$

or

$$\theta_r = I(L, r)$$

where θ_r is the shear duration for the fluid in the cylindrical shell. For a given discharge, Q, we also have the relation

$$Q = -\pi \int_0^{R^2} r^2 du \tag{5.4}$$

With Equation 6.4 or the equivalent τ–$\dot{S}$–θ curves, and Equations 5.6, 6.5, and 5.4, we may now, in principle at least, perform the integration to obtain the velocity profile at any travel distance L, i.e.,

$$u_{L,r} = \phi(L, r, \dot{S}, \theta, Q) \tag{6.6}$$

and the pressure gradient at any travel distance, i.e.,

$$\left(\frac{dP}{dL}\right)_L = \phi_1(L, \dot{S}, \theta, Q) \tag{6.7}$$

In practice, the complexity of the function represented by Equation 6.4 and the need for simultaneous solution of the equations renders this task difficult. An approximate solution to the problem by Govier and Ritter (1963) and a more comprehensive analysis by Ritter and Batycky (1967) are presented in the next section.

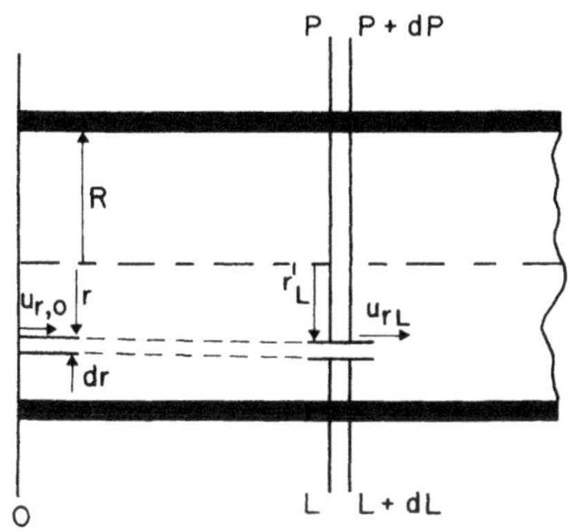

Figure 6.2. Schematic diagram of basis for Equations 6.5 and 6.6.

2.3 Integration of the General Equation

2.31 The Method of Govier and Ritter

Govier and Ritter (1963) presented an approximate method upon which the design of pipe flow systems for laminar flow of thixotropic fluids may be based. The method is applicable to a fluid that does not exhibit a significant yield value or, if it does, to conditions after any gel is broken.

They approached the problem by assuming an average time of shearing, θ_{av}, to be applicable and then demonstrated how the relationship between the pressure gradient, dP/dL, and the volumetric flow rate, Q, varied with this average shearing time. They did not relate the average time of shearing to the travel distance in the pipe. Actually, from Equation 6.5 we can see that, at any finite distance from the pipe entrance, the fluid elements adjacent to the wall have been subject to a duration of shear equal to the total time of flow whereas, within the travel distance L_c, those at the pipe center will have been exposed to a time of shear corresponding to the travel time of the center-line element moving at the velocity, u_0. The extent to which any meaningful average duration of shear may be defined within the travel distance L_c will depend on a complex interrelationship of Equations 6.4, 6.5, 5.4, and 5.6. On the other hand, the average time of shear, at any finite distance up to L_c from the pipe entrance, must lie between the aforementioned limits. Beyond the travel distance L_c from the pipe entry the fluid is essentially unaffected by the entry of fresh fluid, and the duration of shear for all fluid elements will be constant and equal to the duration of flow.

We may summarize these ideas in terms of three time values defined as:

θ_0 = time of shearing of central fluid element within the travel distance $L < L_c$,

θ_{av} = average time of shear of all fluid elements across a section at a travel distance L,

θ_F = duration of flow.

Then for $L < L_c$,

$$\theta_{av} = \phi\left(\theta_F, \frac{L}{\bar{u}_0}\right) \tag{6.8}$$

and for $L \geq L_c$,

$$\theta_{av} = \theta_F = \frac{L_c}{\bar{u}_{0c}} \tag{6.9}$$

where

$$\bar{u}_0 = \frac{1}{L}\int_0^L u_0 \, dL \tag{6.10}$$

and

$$\bar{u}_{0c} = \frac{1}{L_c}\int_0^{L_c} u_0\,dL \tag{6.11}$$

Recognizing the complexities inherent in the solution of Equations 6.8 and 6.10, we may use the form of Equation 6.8 to define an approximate average time of shear of the fluid elements at travel distances $L < L_c$ as

$$\theta'_{\mathrm{av}} = \theta_E + \frac{L}{V} \tag{6.12}$$

where θ_E = the rheological age of the entering fluid,
V = the cross-sectional average velocity of the fluid.

Govier and Ritter's method involves choosing values of θ_{av} between zero and infinity and, from the applicable τ–$\dot{S}$ curve, chosen from the τ–θ–$\dot{S}$ set, evaluating the relationship between τ and $\dot{S}$ with a power law fit to the curve. Following Dodge and Metzner (1959), they choose values of the power law coefficients, K and n, to fit the τ–$\dot{S}$ curve at $\dot{S}$ values corresponding with the rate of shear at the wall, $(du/dr)_w$, and lower values. This is equivalent to evaluating the coefficients at and below the wall shear stress, τ_w. The relations for $(du/dr)_w$ and τ_w have been given previously as

$$-\left(\frac{du}{dr}\right)_w = \frac{8V}{D}\left(\frac{1+3n}{4n}\right) \tag{5.24}$$

and from Equation 5.7, with $\Delta P/L$ replacing $-dP/dL$,

$$\tau_w = \frac{D\Delta P}{4L}$$

Trial-and-error solution of the equations is required, since $(du/dr)_w$ is dependent upon n and this, in turn, is not determinable until $(du/dr)_w$ or τ_w is known. The procedure is to assume a value of $(-du/dr)_w = 8V/D$ or somewhat greater; to determine the power law exponent, n, to fit the τ–$\dot{S}$ curve at and below the assumed $(du/dr)_w$; to recompute $(du/dr)_w$ from Equation 5.24; and to repeat the process until the assumed and recomputed values of $(du/dr)_w$ check. Once the proper value of n, and then of K, are determined the pressure gradient is calculated as for a time-independent power law fluid, i.e., the power law Reynolds number, Re_{PL1}, is calculated and from it the friction factor and the pressure gradient, all at the chosen θ_{av}. The whole procedure is then repeated at other values of θ_{av}. Typical results of such calculations for the laminar flow of a thixotropic fluid are shown in the left-hand portion of Figure 6.3. These results are for a thixotropic crude oil at 44.5°F flowing in a horizontal 6-inch pipe.

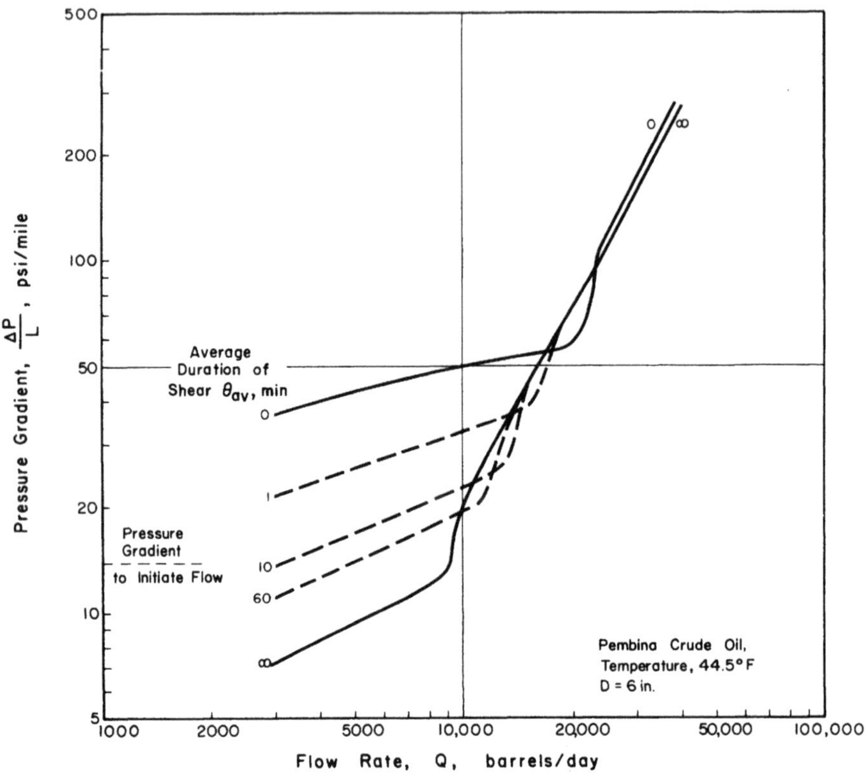

Figure 6.3. Pressure gradient relationships for a typical thixotropic fluid. (Modified after Govier, G. W., and R. A. Ritter, Sixth World Petroleum Congress, Section VII, Paper No. 1, Frankfurt, 1963.)

The Govier and Ritter method is exact beyond the travel distance L_c, where $\theta_{av} = \theta_F$ and dP/dL is independent of L. For travel distances less than L_c, it does not provide a relationship between dP/dL and the travel distance, but an approximation for this may be obtained from Equation 6.12. For a given flow rate, we estimate θ_{av} as θ'_{av} as a function of θ_E and L. The adequacy of this approximation is considered later.

2.32 The Method of Ritter and Batycky

A procedure developed by Ritter and Batycky (1967) does not retain the conceptual simplicity and ease of computation embodied in that of Govier and Ritter, but it does provide a reasonably accurate description of the variations of pressure gradient and velocities along the pipe and with duration of flow in the region of $L < L_c$.

The starting point of the analysis is to express the τ–$\dot{S}$–θ data in the form of an interpolation polynomial. This is necessary both to permit the τ–$\dot{S}$ relationship to be evaluated at any value of θ and to facilitate the numerical analysis required.

The use of Equation 5.6 provides, as before, a linear distribution of shear stress with radius expressible in terms of either the wall shear stress or the pressure gradient. (We assume the velocity profile to be fully developed at the pipe entrance. This is discussed in Section 2.4.) The distribution of the duration of shear is not so easily expressed because of the effect of introduction of new material into the pipe and the continuous aging of the fluid in the pipe. It is therefore necessary to maintain a complete description of the shear duration of all fluid elements in the flow system in order to apply Equation 6.4 properly.

The procedure by which the ages of fluid elements are obtained and mapped and that which enables the incorporation of the appropriate rheological properties across the entire cross section involve the introduction of radial increments. By restricting consideration to a constant volumetric flow rate, it is possible also to define increment boundaries as lying along flow streamlines. Then, to maintain the flow rates within the increment or streamline boundaries constant along the pipe and with duration of flow, the boundaries are allowed to shift radially. Equation 5.2 must be written as two equations: one in terms of the incremental flow rate in increment i between r_i and r_{i+1}, as follows

$$q_i = 2\pi \int_{r_i}^{r_{i+1}} urdr \tag{6.13}$$

where u is a continuous function of radius, obtained by discrete integration of the shear rate followed by interpolation to make it continuous; and the second equation providing the total flow rate which must be maintained constant

$$Q = \sum_{i=1}^{N} q_i \tag{6.14}$$

where N is the number of radial increments. In order that the distribution of shear duration may be expressed in terms of the changing velocity distribution, and also that the pressure gradient may be computed at intervals along the pipe, longitudinal increments are also introduced. Then, θ_j, the duration of shear at the leading edge of the jth longitudinal increment at a particular streamline within the ith radial increment is defined as

$$\theta_j = \frac{2(l_j - l_{j-1})}{u_j + u_{j-1}} + \theta_{j-1} \tag{6.15}$$

where $l_j - l_{j-1}$ is the longitudinal increment length, $\frac{1}{2}(u_j + u_{j-1})$ is the average velocity on the particular streamline within the ith radial increment,

θ_{j-1} is the duration of shear at the leading edge of the previous increment at the same streamline, and θ_1 is the age associated with the fluid entering the pipe. For the solution at any duration of flow after start-up, Equation 6.15 is applied along each streamline until the duration of flow has been reached, at which point no further aging occurs along the streamline.

There are two sequences of calculation. The first scheme, used when there is a constant duration of shear across the pipe, involves altering the shear stress at the wall, integrating the shear rate and velocity gradients until the flow rate corresponds to the specified value. The incremental flow rates derived for the entrance are the ones used in the second calculation sequences. The second procedure is similar to the first but the sequence of calculations involves adjustments to satisfy Equations 6.13, 6.14, and 6.15. This procedure is used at any duration of flow down the entire aging section.

Figure 6.4, taken from Ritter and Batycky, shows the change in pressure gradient along the pipe and with duration of flow for the constant volumetric flow of a thixotropic oil in a pipe in which the fluid present initially was a gel displaying a yield stress and the material which entered had no previous shear

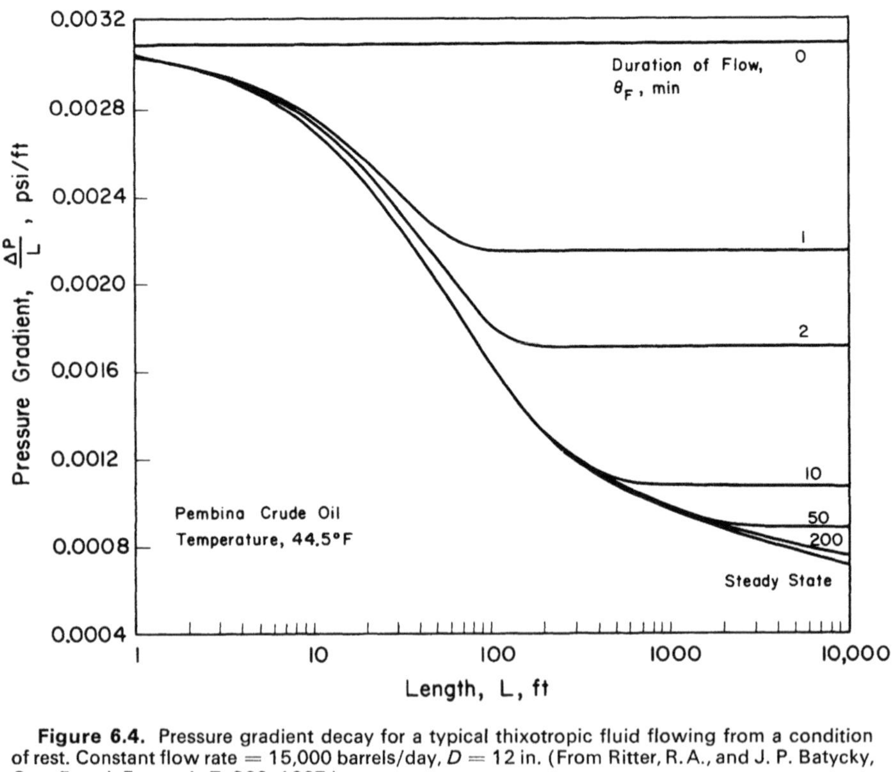

Figure 6.4. Pressure gradient decay for a typical thixotropic fluid flowing from a condition of rest. Constant flow rate = 15,000 barrels/day, D = 12 in. (From Ritter, R. A., and J. P. Batycky, *Soc. Petrol. Engrs. J.*, **7**, 369, 1967.)

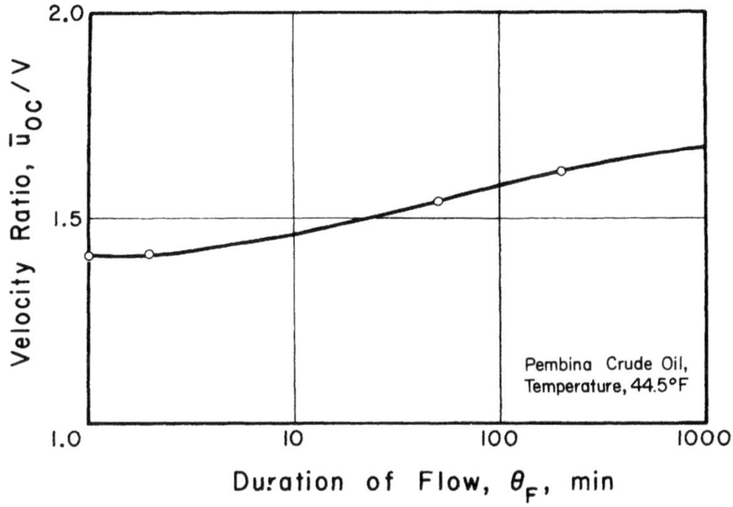

Figure 6.5. Variation of average effective velocity of the central fluid element with shear duration. Constant flow rate = 15,000 barrels/day, D = 12 in. (From data of Ritter, R. A., and J. P. Batycky, *Soc. Petrol. Engrs. J.*, **7**, 369, 1967.)

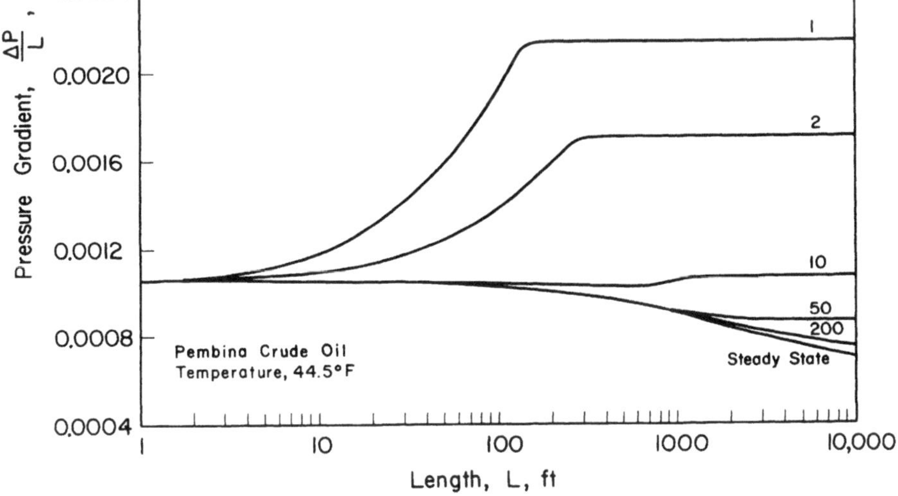

Figure 6.6. Pressure gradient build-up and decay for a typical thixotropic fluid flowing from a previously sheared condition. Constant flow rate = 15,000 barrels/day, D = 12 in. (From the data of Batycky, J. P., M.Sc. Thesis, University of Alberta, Edmonton, Alberta, 1966.)

history. This example is for a flow rate of 15,000 barrels per day in a 12-inch I.D. pipe; the average velocity is 1.24 ft/sec. We may note the clear demonstration of the existence of the travel distance L_c beyond which the gradient is constant with travel distance. From the indicated values of L_c at each of the durations of shear, we may calculate $\bar{u}_{0c}$ from Equation 6.9. The values so determined, expressed as a fraction of V, are shown in Figure 6.5. The nearly constant value of $\bar{u}_{0c}/V$ at approximately 1.5 is consistent with the value of u_0/V predicted from Equation 5.21 for a pseudoplastic having an n value in the range of those exhibited by the Pembina crude oil.

The uniform gradients at $L > L_c$ are identical with those calculated by the Govier and Ritter method with the duration of shear taken as the duration of flow.

In Figure 6.6, taken from Batycky (1966), we show the effect of assigning a shear duration of 10 minutes to the entering fluid. Comparison of the gradients in the two figures serves to emphasize the necessity of assessing the age of the entering fluid fairly accurately. In Figure 6.7 we compare, for travel distances less than L_c, an approximate calculation method based upon Govier

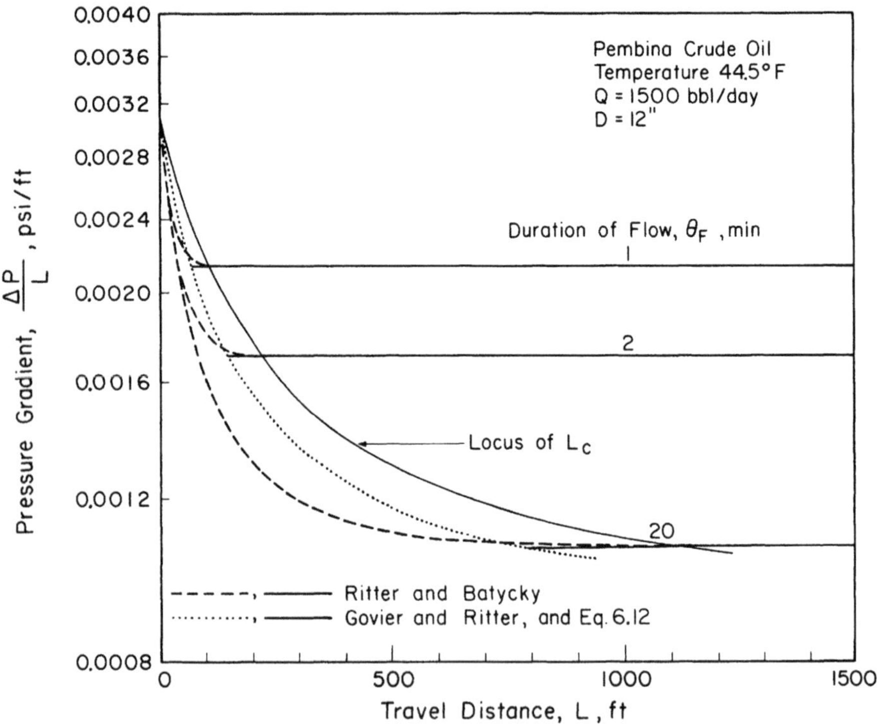

Figure 6.7. Comparison of pressure gradient decay calculations for a typical thixotropic fluid flowing from a condition of rest.

and Ritter's procedure and θ_{av} estimated by Equation 6.12, with $\theta_E = 0$, with the more nearly exact method of Ritter and Batycky. The approximate method produces a single curve falling consistently but not markedly above the gradients predicted by Ritter and Batycky. As each θ_F is reached, this curve is intersected by the level gradient corresponding with θ_F. This approximation therefore provides, by simple calculation, a reasonable estimate of the changing pressure gradient for $L < L_c$.

2.4 Entrance Effects

Having in mind the analytical difficulties in describing the entrance effect for time-independent fluids (see Chapter 5, Section 2.7) it is not surprising that the situation with time-dependent fluids is even more complex. If, at the inlet, the duration of shear is zero and the velocity profile is flat, no shearing occurs except adjacent to the wall. In the regions along the pipe where the velocity profile is flat, the rheological age will remain zero. As the fluid flows down the pipe and the boundary layer increases in thickness, there is a distribution of duration of shear across the boundary layer ranging from that of the entering age to that of the entering age plus the duration of flow. Further, a fluid element experiences a range of shear rates as it becomes immersed relatively further into the boundary layer on its way down the pipe. Similarly, if the fluid entering the pipe has been subject to previous shear, then, as the flow proceeds down the pipe, a structural growth could occur in the non-boundary layer region similar to that which would occur in a fluid at rest. Given the appropriate τ–$\dot{S}$–θ curves, however, there is no difficulty in obtaining a simple bracket solution of the entrance length and the pressure drop in it.

First, we may obtain a wide-range bracket solution by determining the entrance length and the pressure gradient for each of the zero and infinite time rheograms of the fluid, using the methods discussed in Chapter 5 for simple pseudoplastics. The results of such calculations are presented, for illustrative purposes only, in Figures 6.8 and 6.9. These results correspond to the growth of the boundary layer to the center of the pipe and are based on the methods of Collins and Schowalter (1963), discussed in Chapter 5, for the previously discussed Pembina crude oil flowing in a 6-inch horizontal pipe.

Secondly, if desired we could "narrow in" with an approximate trial-and-error bracket solution by evaluations for the zero time rheogram and for rheograms corresponding with θ'_{av} determined from the equivalent of Equation 6.12, i.e.,

$$\theta'_{av} = \theta_E + \frac{(X'_E)_{av}}{V} \tag{6.16}$$

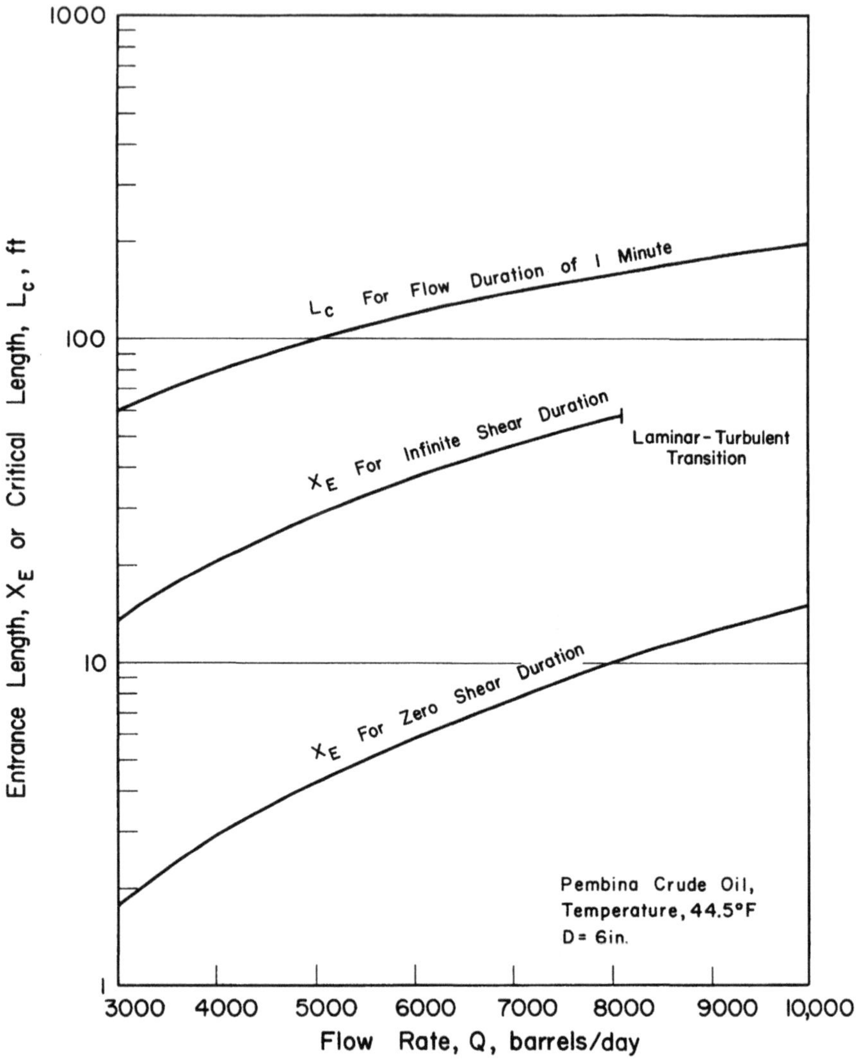

Figure 6.8. Comparison of entrance lengths and travel distance required for flow stabilization for a typical thixotropic fluid.

where $(X_E')_{av}$ is the entrance length calculated using the rheological properties assuming a duration of shear $\theta = \theta_{av}'$. Actually, if this is done, we find that the contribution to θ_{av}' of $(X_E')_{av}/V$ is so small that $\theta_{av}' \simeq \theta_E$. Accordingly we may interpret the curves for zero and infinite shear durations in Figure 6.8 as curves for the range of entrance lengths which would be encountered with entrance fluid having been subject to shear for times ranging from zero to infinity.

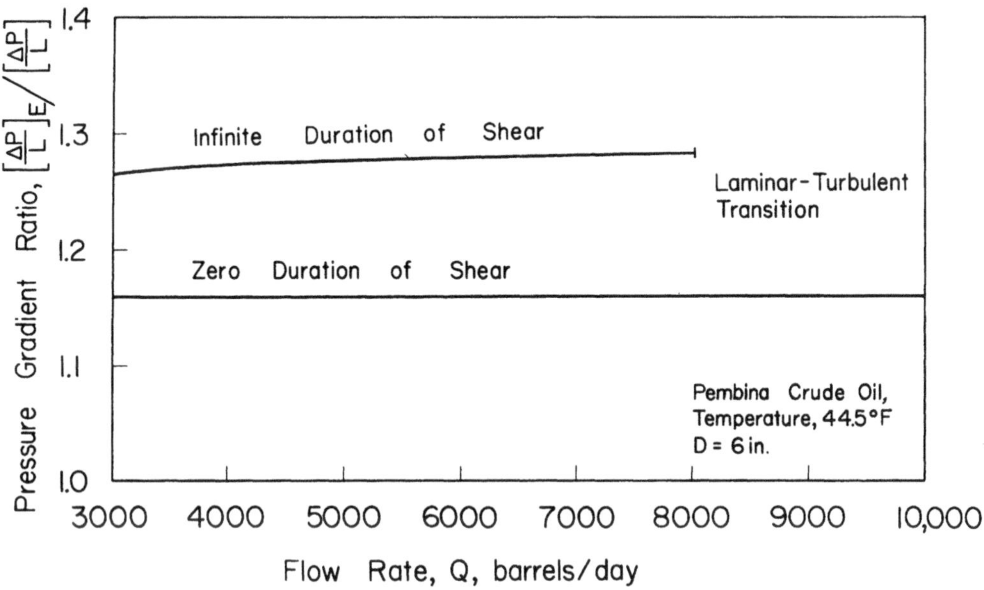

Figure 6.9. Entrance length pressure gradient for a typical thixotropic fluid.

To put matters in perspective we may compare the entrance length calculations based on the two values of θ with the values of L_c discussed earlier and as estimated from Equation 6.9 and Figure 6.5. These values are also shown in Figure 6.8. It is apparent, in this case at least, that the entrance length, whether taken to correspond with $\theta = 0$, or $\theta = \infty$, is very short compared with the travel distance to obtain steady-state rheological properties. Indeed, at normal pipe velocities ($1 < V < 7$ ft/sec) and diameters, the average travel times associated with the entrance lengths X_{E0}, $(X_E')_{av}$, $X_{E\infty}$ are of the order of seconds, whereas the time required to attain equilibrium rheological properties is of the order of tens or hundreds of minutes. It is therefore reasonable, in most applications, to assume that the boundary layer develops instantaneously to the center line, that is to say, that the entering velocity is fully developed.

2.5 General Form of the Pressure Drop Equation

The general form of the pressure drop equation is the same as that discussed for the laminar flow of time-independent fluids (see Chapter 5, Section 2.8) except that one such equation may be written for each average duration of flow that may be assigned. Further, over a distance of the order of L_c, as given by Equation 6.9 and Figure 6.5, from the pipe inlet, the pressure gradient will vary with position and should be estimated having regard to the variation of shear duration with radial and axial position. This may be done with good accuracy by the method of Ritter and Batycky or approximately through the use of Equation 6.12.

3 TRANSITION FROM LAMINAR TO TURBULENT FLOW

The transition from laminar to turbulent flow for time-independent non-Newtonians was shown in Chapter 5 to depend upon the rheological properties, as well as on the diameter of the pipe and on the density and velocity of the fluid. For simple pseudoplastics that could be described by the power law, we found the transition to occur approximately at a value of $Re_{PL2} = 2100$ or more exactly at a value of $f = f_c$, given by Figure 5.4 as a function of Re_{PL1}, and n. Further, a thixotropic liquid has been shown to behave, over time, as a series of rheologically different fluids, so that an element of fluid may display a different pseudoplastic or yield-pseudoplastic behavior at each duration of shear. As a consequence, even though laminar flow may prevail immediately following the entrance to a pipe, transition from laminar to turbulent flow

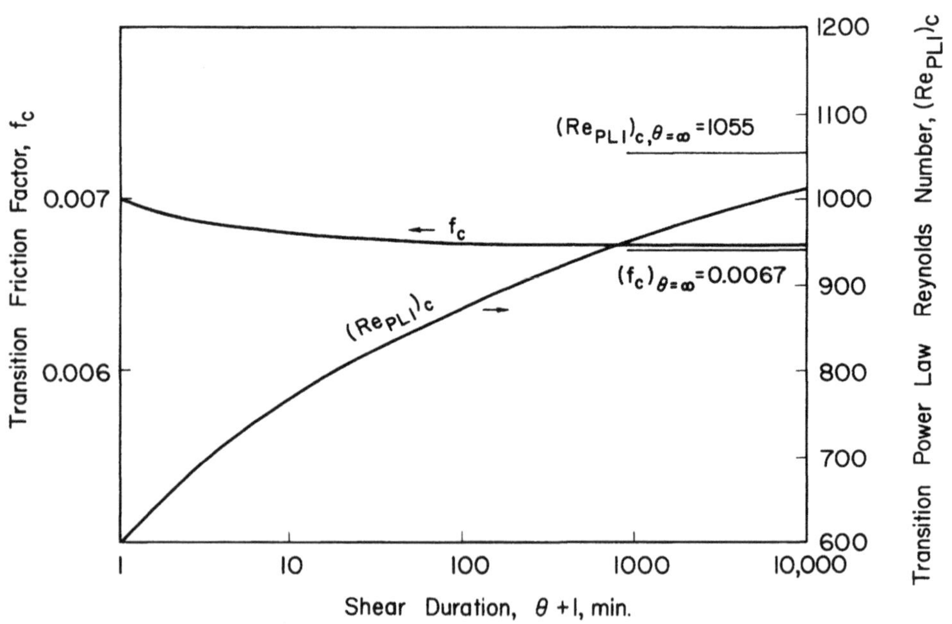

Figure 6.10. Effect of shear duration on transition friction factor and Reynolds number for a typical thixotropic fluid.

may occur at some distance down the pipe, since the consistency of the fluid decreases, causing an increase in the Reynolds number with the possibility that it may reach the transition value. An exact solution is most difficult because of the changing consistency of the fluid with flow duration. On the other hand, the solution may be bracketed by considering the extremes of the rheological behavior at zero and infinite time of shear and applying the methods discussed in Chapter 5 for power law pseudoplastics.

We may illustrate these methods by applying them to the case of the Pembina crude oil. Here we obtain the critical friction factors, and values of Re_{PL1} shown in Figure 6.10. The changing transition values of Re_{PL1} may be observed. It may be noted in this illustration that the transition Reynolds number at zero shear duration is only about one-half what it is at infinite shear duration. The slight change in f with shear duration reflects the relative insensitivity of f_c to changes in Re_{PL1} as shown in Figure 5.4.

4 TURBULENT FLOW

4.1 Introduction

Immediately beyond the laminar-turbulent transition point the distribution of rheological ages across the pipe becomes most complex with velocity fluctuations mixing the various fluid elements. This mixing grows in intensity as the transition point is passed and the turbulence becomes fully developed. A knowledge of the rheological behavior that results with the mixing of fluid elements, which have been exposed to different levels and duration of shear, would therefore be necessary for a complete understanding of the problem. Once turbulence is fully developed, it is reasonable to assume that the consistency at any cross section will be uniform. However, questions remain concerning the level of the rate of shear and the duration of shear at which it is appropriate to evaluate the consistency.

4.2 Approximate Treatment

Data of Cheng, Ray, and Valentin (1965) on the turbulent flow of thixotropic bentonite suspensions offer support to the idea that in turbulent flow it is the τ–$\dot{S}$ curve at infinite shear duration that is most significant. This may well be so, and certainly will be so after a fluid has traversed a long length of pipe, but it seems appropriate at least to consider the limiting cases of the turbulent flow of the zero and the infinite shear duration fluid.

This is the approach of Govier and Ritter who show how to provide a bracket solution through the use of the power law equation with previously unsheared and infinitely sheared fluids by trial using an appropriate friction factor–Reynolds number relationship. The method involves an assumption of the value of τ_w; the use of a power law fit to the appropriate shear duration curve in the range of τ immediately below τ_w; the use of (say) the Metzner and Reed (1955) f versus Re_{MR} correlation with the selected values of K and n to obtain f; the calculation of τ_w from f; and finally a repeat of the calculations until the assumed and the calculated values of τ_w check.

Using the data previously presented for the Pembina crude oil, we may illustrate the effect upon the pressure gradient in turbulent flow of assuming zero and infinite average duration of shear. The results of the calculations, using a trial-and-error procedure and Metzner and Reed's friction factor–Reynolds number relationship, are shown on the right hand side of Figure 6.3. We may note, as is to be expected, that the effect of duration of shear on the pressure gradient is much less in the turbulent flow region than in the previously discussed laminar flow region. In fact, the modest effect between even zero and infinite shearing duration confirms that the bracket solution is adequate for any practical purpose.

4.3 Entrance Effects

Turbulence may originate in the developing flow region as the fluid proceeds from a flat profile at the entry point through the growth of a laminar boundary layer leading to development finally of a turbulent boundary layer. As with time-independent fluids, no methods are available for the accurate prediction of either the entrance length or the pressure drop in the entrance region. In general, however, we would expect the entrance effect itself, as distinct from the effect due to thixotropy, to be small and of negligible importance in practice.

4.4 General Form of the Pressure Drop Equation

The general form of the pressure drop equation is the same as that discussed for the turbulent flow of time-independent non-Newtonians in Section 4.8 of Chapter 5, except that, at least in principle, up to travel distances of the order of $L_c \simeq V\theta_\infty$ the equation should be written for a series of values of $\theta_{av} \simeq L/V$

to give the pressure gradient as a function of pipe travel distance. In practice, the use of a single solution based upon the rheological properties at an infinite duration of shear gives a good approximation for all but those thixotropic fluids that reach their equilibrium consistencies most slowly.

5 RECOMMENDED DESIGN METHODS

5.1 Introduction

The peculiarities of time-dependent non-Newtonian fluids introduce two matters of design importance. The first is the variation of the pressure gradient with travel distance in the pipe over a length L_c wherein the properties of the fluid vary both radially and axially. The second is the variation of the pressure gradient with flow duration beyond the distance L_c. In the following discussion we assume the availability of suitable τ–$\dot{S}$–θ data for a sample of the fluid properly representative of the condition of the entering fluid.

5.2 Entrance Effects

As discussed in Section 2.4, the normal entrance effects associated with the growth of the boundary layer are usually of no importance compared with those caused by the radial and axial variation of rheological properties. The latter effect may be estimated as follows.

1. Determine whether the flow in the entrance length will be laminar or turbulent by application of the methods discussed in Section 6.2 of Chapter 5, using the τ–$\dot{S}$ relations for both $\theta = 0$ and $\theta = \infty$ or some other appropriate value of θ.
2. If the flow is turbulent, the entrance length will probably not be significant.
3. If the flow is laminar, the entrance length, L_c, associated with different durations of flow, and the radial and axial variation of rheological properties, may be determined
   a. by the approximate method of Govier and Ritter using Equation 6.12 and a range of values of θ'_{av} covering the range of thixotropic decay of the fluid, or
   b. by the more accurate method of Ritter and Batycky involving the detailed procedure discussed in Section 2.32.
4. If L_c so determined is 10 percent or less of the total pipe length, L, its effect may usually be neglected and the problem may be dealt with entirely as discussed in Section 5.6.

5. If L_c exceeds $0.10L$ the pressure gradient profile from $L = 0$ to $L = L_c$ should be evaluated
   a. by the approximate method of Govier and Ritter based upon the discussion of Section 2.31, or
   b. by the more nearly exact method of Ritter and Batycky involving the detailed procedure of Section 2.32.

5.3 Effect of Bends and Fittings

See Chapter 5, Section 6.3.

5.4 Potential Energy or Hydrostatic Head Effects

See Chapter 5, Section 6.4.

5.5 Acceleration Effects

See Chapter 5, Section 6.5.

There are no special problems associated with time-dependent fluids provided, with respect to Sections 5.3 and 5.5, that the average rheological age of the fluid is determinable and the fluid of this age may be characterized by a simple constitutive equation.

5.6 Pressure Gradient for Steady Flow Beyond the Stabilization Distance, L_c

1. If the flow is laminar the pressure gradient may be calculated for each of several durations of shear (equal to the duration of flow) by the method of Govier and Ritter discussed in Section 2.31. The method of Ritter and Batycky gives the same results for $L > L_c$.
2. If the flow is turbulent the pressure gradient will not vary greatly with duration of flow and may be narrowly bracketed by applying the procedure of Govier and Ritter discussed in Section 4.2.

For any duration of flow, the total pressure drop is obtained by summing the pressure drop over the length L_c, the effects of elevation changes, acceleration effects, and the pressure drop over the length $(L - L_c + \sum L_E)$. The final design prediction would be a set of pressure gradient profiles such as those of Figure 6.4 for a range of flow rates and each of a selection of pipe sizes.

Bibliography

Batycky, J. P., M. Sc. Thesis, University of Alberta, Edmonton, Alberta, 1966.

Cheng, D. C.-H., D. J. Ray, and F. H. H. Valentin, *Trans. Inst. Chem. Engrs.*, **43**, T176 (1965).

Collins, M., and W. R. Schowalter, *A.I.Ch.E. Journal*, **9**, 804 (1963).

Dodge, D. W., and A. B. Metzner, *A.I.Ch.E. Journal*, **5**, 189 (1959).

Govier, G. W., and R. A. Ritter, Sixth World Petroleum Congress, Section VII, Paper No. 1, Frankfurt, 1963.

Metzner, A. B., and J. C. Reed, *A.I.Ch.E. Journal*, **1**, 434 (1955).

Ritter, R. A., and J. P. Batycky, *Soc. Petrol. Engrs. J.*, **7**, 369 (1967).

Fundamental Concepts of the Flow of Multiphase Mixtures

7

1 INTRODUCTION

The fundamental concepts common to the pipe flow of gas-liquid, liquid-liquid, gas-solid, and liquid-solid mixtures are reviewed in this chapter. Basically these are the relationships, for each phase and for the interface, derivable by an approach similar to the one used for the fundamental equations discussed in Chapter 3. These concepts constitute a common framework within which the flow of all two-phase systems may be viewed and interrelated.

In this chapter we will develop equations that apply to multiphase flow systems with adsorption and surface tension effects neglected.* The microscopic balance equations can be applied to some simplified models of multiphase flow, whereas the macroscopic equations are completely general and apply to all multiphase flow systems. The treatment of the laws of conservation in this chapter is based largely on the works of Standart (1964) and Delhaye (1967, 1968). Other investigators who have contributed toward an understanding of the basic concepts of multiphase systems are: Teletov (1958), van Deemter and van der Laan (1961), Birkhoff (1964), Kraiko and Sternin (1965), Sparenberg (1965), Van Wijngaarden (1968), and Panton (1968).

The application of the fundamental equations to the flow of multiphase mixtures is complicated by a number of factors not present in the case of single-phase systems.

* Delhaye (1967) takes account of the interfacial tension in his development.

a. The flow of multiphase mixtures may not be characterized merely as laminar, or as a combination of laminar and turbulent flow, but the relative quantities and the distribution of the phases (i.e., the flow pattern) must be considered (Govier, Radford, and Dunn, 1957). Moreover, even where the flow may be known to be laminar or turbulent, the free boundaries of the phases add to the difficulties of applying the fundamental relations.
b. Because of the differences in phase densities normally encountered, the flow pattern in horizontal or inclined flow is not symmetrical about the flow axis.
c. The presence of the interface between the phases itself adds to the complexity of the problem, and the fundamental equations must be written for it as well as for the individual phases.
d. In cases involving a large amount of interfacial surface, the energy associated with the creation of the surface must be considered in the total energy equation.
e. In the general case, the phases may not be in equilibrium with respect to temperature or component concentrations. Here the heat and mass transfer relationships must be considered. Fortunately, in multiphase mixtures, the heat and mass transfer rates are usually high because of the large interfacial area, the good mixing, and the frequent presence of turbulence. As a result, in most cases of the pipe flow of such mixtures, the phases may be assumed to reach an equilibrium condition rather rapidly and to remain in equilibrium, during their passage through the pipe.
f. In most instances the separate phases move at different average velocities and the *in situ* concentrations are not the same as the concentrations in which the phases are introduced or removed from the system. The variation of the *in situ* concentrations from the supply concentrations is referred to as the holdup phenomenon.

In the following discussion, a general two-phase system in flow is described. The continuity equation is developed for the multiphase system as a whole and for the individual components. The simplified form of the equation for the common case of steady flow is then presented. The general form of the momentum equation is developed, and reference is made to the instances where it may be solved. We treat the energy equation in a similar manner. The mechanical energy balance equation is then introduced, in two different ways, leading to a final overall form involving a friction or irreversibility term. This equation is especially useful in practical applications.

For steady-state flow, and equilibrium between phases, the holdup phenomenon is then treated in some detail.

2 DESCRIPTION OF A GENERAL TWO-PHASE SYSTEM

Before we proceed with the development of the conservation laws it is desirable to describe the system and introduce some useful notation and generally applicable relationships. Although the treatment to follow is applicable to any number of phases and to their flow through a duct of any shape, for simplicity in presentation we will discuss the flow of a two-phase multicomponent system in a circular pipe. The situation is illustrated schematically in Figure 7.1. Let α and β refer, respectively, to the less and the more dense phase, with A_b as the interphase boundary. The external boundary of the system is A_e comprising the inlet boundary A_1, the outlet boundary A_2, and the pipe wall A_w, all of which are fixed. The volume occupied by the two phases is denoted by $\mathscr{V}_\alpha$ and $\mathscr{V}_\beta$, and the total system volume by $\mathscr{V}$. Then

$$\mathscr{V} = \mathscr{V}_\alpha + \mathscr{V}_\beta \tag{7.1}$$

Any extensive quantity of the system such as mass, momentum, energy, or enthalpy, say F, is given by

$$F = \int_{\mathscr{V}_\alpha} \rho_\alpha \mathfrak{f}_\alpha \, d\mathscr{V} + \int_{\mathscr{V}_\beta} \rho_\beta \mathfrak{f}_\beta \, d\mathscr{V} \tag{7.2}$$

where $\mathfrak{f}$ is the local specific value of F, and ρ is the local density of the fluid. The time rate of change of F is given by

$$\frac{dF}{dt} = \frac{d}{dt} \int_{\mathscr{V}_\alpha} \rho_\alpha \mathfrak{f}_\alpha \, d\mathscr{V} + \frac{d}{dt} \int_{\mathscr{V}_\beta} \rho_\beta \mathfrak{f}_\beta \, d\mathscr{V}$$

This relationship may be expanded to (Truesdell and Toupin, 1960)

$$\frac{dF}{dt} = \int_{\mathscr{V}_\alpha} \frac{\partial(\rho_\alpha \mathfrak{f}_\alpha)}{\partial t} d\mathscr{V} + \int_{\mathscr{V}_\beta} \frac{\partial(\rho_\beta \mathfrak{f}_\beta)}{\partial t} d\mathscr{V}$$
$$+ \int_{A_b} (\rho_\alpha \mathfrak{f}_\alpha \mathbf{v}_b \cdot \mathbf{n}_\alpha + \rho_\beta \mathfrak{f}_\beta \mathbf{v}_b \cdot \mathbf{n}_\beta) dA \tag{7.3}$$

where $\mathbf{n}_\alpha$ and $\mathbf{n}_\beta$ are the external unit normals from the α and β phases, respectively, and $\mathbf{v}_b$ is the local velocity of the interface.

For an arbitrary vector or tensor quantity, say $\mathbf{X}$, and from the Gauss–Ostrogradsky theorem (Standart, 1964), we may write

$$\int_{A_{e\alpha}} \mathbf{X}_\alpha \cdot d\mathbf{A} + \int_{A_{e\beta}} \mathbf{X}_\beta \cdot d\mathbf{A} = \int_{\mathscr{V}} \nabla \cdot \mathbf{X} d\mathscr{V} - \int_{A_b} (\mathbf{X}_\alpha \cdot \mathbf{n}_\alpha + \mathbf{X}_\beta \cdot \mathbf{n}_\beta) dA \tag{7.4}$$

where $d\mathbf{A}$ is a vector of magnitude dA, with a direction outward and normal to the plane of the area, i.e., $d\mathbf{A} = \mathbf{n} dA$.

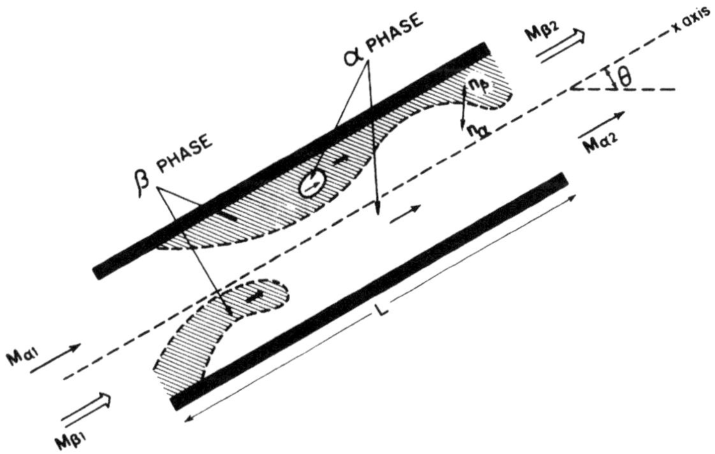

Figure 7.1. Schematic diagram of basis for equations for the flow of a two-phase system.

3 THE CONTINUITY EQUATION

The continuity equation representing the conservation of mass may be developed both for the entire flowing system and for the individual components.

3.1 Total System Continuity Equation

Macroscopic In a multicomponent multiphase system, the rate of accumulation of mass must equal the influx of mass through the external boundaries:

$$\frac{d}{dt}\left(\int_{\mathscr{V}_\alpha} \rho_\alpha \, d\mathscr{V} + \int_{\mathscr{V}_\beta} \rho_\beta \, d\mathscr{V}\right) = -\int_{A_{e\alpha}} \rho_\alpha \mathbf{v}_\alpha \cdot d\mathbf{A} - \int_{A_{e\beta}} \rho_\beta \mathbf{v}_\beta \cdot d\mathbf{A} \tag{7.5}$$

This is the macroscopic total system equation of continuity. It may also be expressed as

$$\frac{\Delta \mathscr{M}}{\Delta t} = M_{\alpha 1} + M_{\beta 1} - M_{\alpha 2} - M_{\beta 2} \tag{7.6}$$

where $\mathscr{M}$ is the mass of the material in the system,

$$\mathscr{M} = \int_{\mathscr{V}_\alpha} \rho_\alpha \, d\mathscr{V} + \int_{\mathscr{V}_\beta} \rho_\beta \, d\mathscr{V} \tag{7.7}$$

The mass flow rates of the individual phases at the inlet or the outlet are given by such equations as

$$M_{\alpha 1} = \int_{A_{\alpha 1}} \rho_\alpha \mathbf{v}_\alpha \cdot d\mathbf{A} \tag{7.8}$$

for the α phase at the inlet.

For one-dimensional (in a macroscopic sense) flow parallel to the axis, as illustrated in Figure 7.1, the following equations result:

$$M_{\alpha 1} = \rho_{\alpha 1} V_{\alpha 1} E_{\alpha 1} A = \text{rate of mass influx of the } \alpha \text{ phase} \tag{7.8a}$$

$$M_{\beta 1} = \rho_{\beta 1} V_{\beta 1} E_{\beta 1} A = \text{rate of mass influx of the } \beta \text{ phase} \tag{7.8b}$$

$$M_{\alpha 2} = \rho_{\alpha 2} V_{\alpha 2} E_{\alpha 2} A = \text{rate of mass efflux of the } \alpha \text{ phase} \tag{7.8c}$$

$$M_{\beta 2} = \rho_{\beta 2} V_{\beta 2} E_{\beta 2} A = \text{rate of mass efflux of the } \beta \text{ phase} \tag{7.8d}$$

where V_α and V_β are the cross-sectional area average velocities,
E_α is the average volume fraction of the α phase over a differential length of the pipe and therefore is the fractional cross-sectional area occupied by the α phase, i.e.,

$$E_\alpha = \frac{A_\alpha}{A} \qquad \text{and} \qquad E_\alpha + E_\beta = 1$$

In the special case where the inflows $M_{\alpha 1}$ and $M_{\beta 1}$ are constant with time and where the flow in the pipe is stabilized, resulting in a constant mass in the pipe, $\Delta\mathcal{M}/\Delta t = 0$, the continuity equation reduces to the following:

$$M_{\alpha 1} + M_{\beta 1} = M_{\alpha 2} + M_{\beta 2} \tag{7.9}$$

or

$$M_{\alpha 1} = M_{\alpha 2} - \Delta M_\alpha \tag{7.9a}$$

and

$$M_{\beta 1} = M_{\beta 2} - \Delta M_\beta \tag{7.9b}$$

where $\Delta M_\alpha = -\Delta M_\beta$. Defining the volume average *in situ* volume fractions of phases in the pipe as $\bar{E}_\alpha$ and $\bar{E}_\beta$, where $\bar{E} = (1/L)\int_0^L E dL$, and the volume average *in situ* velocities as $\bar{V}_\alpha$ and $\bar{V}_\beta$, Equations 7.9a and 7.9b, can be rewritten as

$$M_{\alpha 1} = A\bar{E}_\alpha \bar{V}_\alpha \rho_{\alpha 2} - \Delta M_\alpha \tag{7.10}$$

$$M_{\beta 1} = A\bar{E}_\beta \bar{V}_\beta \rho_{\beta 2} - \Delta M_\beta \tag{7.11}$$

For a two-phase system,

$$\bar{E}_\alpha + \bar{E}_\beta = 1 \tag{7.12}$$

Where ΔM_α is zero, as in the case of complete insolubility, or where it is small, as with limited solubility or with presaturated phases, and where

$\rho_{\alpha 1} = \rho_{\alpha 2} = \rho_\alpha$ and $\rho_{\beta 1} = \rho_{\beta 2} = \rho_\beta$, these equations reduce to the volume balances

$$\overline{V}_{S\alpha} = \overline{E}_\alpha \, \overline{V}_\alpha \tag{7.13}$$

$$\overline{V}_{S\beta} = \overline{E}_\beta \, \overline{V}_\beta \tag{7.14}$$

where $\overline{V}_{S\alpha}$ and $\overline{V}_{S\beta}$ are the superficial volume average velocities of the phases α and β.

Microscopic Starting with Equation 7.5 and using the Gauss–Ostrogradsky theorem, we can also derive an equation of continuity for each phase (Standart, 1964)

$$\frac{\partial \rho_\alpha}{\partial t} + \nabla \cdot (\rho_\alpha \mathbf{v}_\alpha) = 0 \tag{7.15}$$

$$\frac{\partial \rho_\beta}{\partial t} + \nabla \cdot (\rho_\beta \mathbf{v}_\beta) = 0 \tag{7.16}$$

The relations may also be derived directly from mass balances on differential volume elements in each phase.

In addition, Equation 7.5 leads to the following mass balance for the interface itself:

$$\rho_\alpha (\mathbf{v}_\alpha - \mathbf{v}_b) \cdot \mathbf{n}_\alpha + \rho_\beta (\mathbf{v}_\beta - \mathbf{v}_b) \cdot \mathbf{n}_\beta = 0 \tag{7.17}$$

where $\mathbf{v}_b$ is the phase boundary velocity. Equation 7.17 simply states that no mass can accumulate in the surface. The term $\rho_\alpha (\mathbf{v}_\alpha - \mathbf{v}_b) \cdot \mathbf{n}_\alpha$ represents the mass flow rate per unit area from the α phase, $\rho_\beta (\mathbf{v}_\beta - \mathbf{v}_b) \cdot \mathbf{n}_\beta$ represents the mass flow rate per unit area from the β phase, and $\mathbf{v}_b \cdot \mathbf{n}$ is the displacement velocity of the interface. If there is no transfer of mass at the interface,

$$(\mathbf{v}_\beta - \mathbf{v}_b) \cdot \mathbf{n}_\beta = 0$$

and

$$(\mathbf{v}_\alpha - \mathbf{v}_b) \cdot \mathbf{n}_\alpha = 0$$

In addition, if the interface position does not change ($\mathbf{v}_b = 0$), we obtain the "no slip" condition at the interface, i.e.,

$$\mathbf{v}_\alpha = \mathbf{v}_\beta = \mathbf{v}_b$$

When there is mass transfer across the interface and the assumption of "no slip" is valid, the tangential component of $(\mathbf{v}_\beta - \mathbf{v}_b)$ and $(\mathbf{v}_\alpha - \mathbf{v}_b)$ must be equal. The vector $(\mathbf{v}_\alpha - \mathbf{v}_\beta)$ should then be perpendicular to the interface.

Under steady-state conditions without mass transfer and with flow parallel to the axis of the pipe, the microscopic mass balances reduce to

$$\frac{d(\rho_\alpha v_\alpha)}{dx} = 0 \tag{7.18}$$

and

$$\frac{d(\rho_\beta v_\beta)}{dx} = 0 \tag{7.19}$$

which are the local equivalents of Equations 7.10 and 7.11 with ΔM_α and ΔM_β equalling zero.

Pseudohomogeneous Fluid In the special case of a two-phase system, in which one phase is finely and uniformly dispersed in the other, the continuity equations may be derived in a different way (see van Deemter and van der Laan, 1961), and the resulting equations are

$$\frac{\partial(\varepsilon_\alpha \rho_\alpha)}{\partial t} + \nabla \cdot (\varepsilon_\alpha \rho_\alpha \mathbf{v}_\alpha) = 0 \tag{7.20}$$

$$\frac{\partial(\varepsilon_\beta \rho_\beta)}{\partial t} + \nabla \cdot (\varepsilon_\beta \rho_\beta \mathbf{v}_\beta) = 0 \tag{7.21}$$

These equations are simpler in form than Equations 7.15, 7.16, and 7.17. The sum of Equations 7.20 and 7.21 gives the total system continuity equation,

$$\frac{\partial(\varepsilon_\alpha \rho_\alpha + \varepsilon_\beta \rho_\beta)}{\partial t} + \nabla \cdot (\varepsilon_\alpha \rho_\alpha \mathbf{v}_\alpha + \varepsilon_\beta \rho_\beta \mathbf{v}_\beta) = 0 \tag{7.22}$$

Note that no equation is required for the interface because the form of the interface, which is "finely and uniformly dispersed," is assumed in the derivation of this form of the continuity equation. [See Panton (1968) for an excellent discussion of these equations for one-dimensional flow.] The steady-state equivalents of these equations for flow parallel to the axis of the pipe are

$$\frac{d(\varepsilon_\alpha \rho_\alpha v_\alpha)}{dx} = 0 \tag{7.23}$$

and

$$\frac{d(\varepsilon_\beta \rho_\beta v_\beta)}{dx} = 0 \tag{7.24}$$

3.2 Individual Component Continuity Equations

Macroscopic If the system consists of n components then, in addition to the total continuity equation, $(n - 1)$ further independent mass conservation equations may be derived. Assuming no chemical reactions within the phases or at the interface, we have, for each component,

$$\frac{d}{dt}\left(\int_{\mathscr{V}_\alpha} \rho_{\alpha i}\, d\mathscr{V} + \int_{\mathscr{V}_\beta} \rho_{\beta i}\, d\mathscr{V}\right) = -\int_{A_{e\alpha}} \rho_{\alpha i}\mathbf{v}_{\alpha i} \cdot d\mathbf{A} - \int_{A_{e\beta}} \rho_{\beta i}\mathbf{v}_{\beta i} \cdot d\mathbf{A} \quad (7.25)$$

where ρ_i is the mass of component i per unit volume of the system, and $\mathbf{v}_i$ is the velocity of component i. The overall density and velocity of a phase in the mixture are defined by

$$\rho = \sum_{i=1}^{n} \rho_i \quad (7.26)$$

$$\mathbf{v} = \frac{1}{\rho} \sum_{i=1}^{n} \rho_i \mathbf{v}_i \quad (7.27)$$

Summing Equation 7.25 for n components yields the continuity equation (Chapter 3, Equation 3.6). This equation may also be written as

$$\frac{\Delta \mathscr{M}_i}{\Delta t} = M_{\alpha 1 i} + M_{\beta 1 i} - M_{\alpha 2 i} - M_{\beta 2 i} \quad (7.28)$$

for each component.

Microscopic The local laws of conservation for the individual components may also be derived as they were for the overall system. The resulting equations are

$$\frac{\partial \rho_{\alpha i}}{\partial t} + \nabla \cdot (\rho_{\alpha i}\mathbf{v}_{\alpha i}) = 0 \quad (7.29)$$

$$\frac{\partial \rho_{\beta i}}{\partial t} + \nabla \cdot (\rho_{\beta i}\mathbf{v}_{\beta i}) = 0 \quad (7.30)$$

for each component. At the phase boundary we have n relations of the form

$$\rho_{\alpha i}(\mathbf{v}_{\alpha i} - \mathbf{v}_b) \cdot \mathbf{n}_\alpha + \rho_{\beta i}(\mathbf{v}_{\beta i} - \mathbf{v}_b) \cdot \mathbf{n}_\beta = 0 \quad (7.31)$$

For a single-component system, Equations 7.29, 7.30, and 7.31 reduce to corresponding equations for the overall system.

4 THE MOMENTUM EQUATION

The momentum equation developed in Chapter 3 may be applied in a macroscopic sense, and locally or in a microscopic sense, to a flowing, two-phase, multicomponent system.

Macroscopic The momentum equation for a two-phase system contained within the pipe may be written (Standart, 1964) as

$$\frac{d}{dt}\left(\int_{\mathscr{V}_\alpha} \rho_\alpha \mathbf{v}_\alpha \, d\mathscr{V} + \int_{\mathscr{V}_\beta} \rho_\beta \mathbf{v}_\beta \, d\mathscr{V}\right) = -\int_{A_\alpha} \{\rho_\alpha \mathbf{v}_\alpha \mathbf{v}_\alpha + g_c P\mathscr{U} + g_c \boldsymbol{\tau}\} \cdot d\mathbf{A}$$
$$-\int_{A_\beta} \{\rho_\beta \mathbf{v}_\beta \mathbf{v}_\beta + g_c P\mathscr{U} + g_c \boldsymbol{\tau}\} \cdot d\mathbf{A} + \sum_{i=1}^{n} g_c \left(\int_{\mathscr{V}_\alpha} \rho_{\alpha i} \mathbf{F}_{\alpha i} \, d\mathscr{V} + \int_{\mathscr{V}_\beta} \rho_{\beta i} \mathbf{F}_{\beta i} \, d\mathscr{V}\right) \tag{7.32}$$

where $\mathscr{U}$ is the unit tensor, $\boldsymbol{\tau}$ is the stress tensor, and $\mathbf{F}_{\alpha i}$ is the external force per unit mass of i, ordinarily the force of gravity, in the α-phase. Equation 7.32 is the macroscopic momentum equation, and it is written in the vector-tensor notation. The quantities enclosed in the braces, { }, are all tensors (3 × 3 matrices) and may be written as

$$\rho\mathbf{v}\mathbf{v} = \begin{bmatrix} \rho v_1^2 & \rho v_1 v_2 & \rho v_1 v_3 \\ \rho v_2 v_1 & \rho v_2^2 & \rho v_2 v_3 \\ \rho v_3 v_1 & \rho v_3 v_2 & \rho v_3^2 \end{bmatrix} \tag{7.33}$$

$$P\mathscr{U} = \begin{bmatrix} P & 0 & 0 \\ 0 & P & 0 \\ 0 & 0 & P \end{bmatrix} \tag{7.34}$$

$$\boldsymbol{\tau} = \begin{bmatrix} \tau_{11} & \tau_{12} & \tau_{13} \\ \tau_{21} & \tau_{22} & \tau_{23} \\ \tau_{31} & \tau_{32} & \tau_{33} \end{bmatrix}, \quad \tau_{ij} = \tau_{ji} \tag{7.35}$$

The left-hand side of Equation 7.32 represents the rate of change of the system momentum, which must equal the resultant of all the forces acting on the system. The first two terms on the right-hand side represent the rate of influx of momentum through the external surface, and the last two terms represent the external forces. Additional terms must be included in this equation to account for interfacial forces (see Delhaye, 1968).

When the inlet and outlet flows may be taken as parallel to the axis of the pipe and the only external force acting is that of gravity, Equation 7.32 may be simplified to

$$\frac{dJ}{dt} = (MV)_{\alpha 1} + (MV)_{\beta 1} - (MV)_{\alpha 2} - (MV)_{\beta 2}$$
$$+ g_c(P_1 - P_2)\,A - g_c F_w - g \sin\theta \, \mathscr{M} \tag{7.36}$$

where

$$J = \mathbf{I} \cdot \left[\int_{\mathscr{V}_\alpha} \rho_\alpha \mathbf{v}_\alpha \, d\mathscr{V} + \int_{\mathscr{V}_\beta} \rho_\beta \mathbf{v}_\beta \, d\mathscr{V} \right]$$

$\mathbf{I}$ = a unit vector in the flow direction,
θ = the angle of inclination (Figure 7.1).

The contribution of the frictional stress over the inlet and outlet sections have been neglected, and the pressure is assumed to be constant over the cross-section. We can simplify the equation further by writing

$$\mathbf{F}_w = \mathbf{F}_{wP} + \mathbf{F}_{w\tau}$$

and the component of this force in the flow direction is given by

$$\mathbf{I} \cdot \mathbf{F}_w = \mathbf{I} \cdot \mathbf{F}_{wP} + \mathbf{I} \cdot \mathbf{F}_{w\tau} \tag{7.37}$$

where $\mathbf{I} \cdot \mathbf{F}_{wP}$ is the pressure force on the pipe surface in the direction of flow, equals zero for a pipe of constant cross section.

Substituting Equation 7.37 into Equation 7.36, we obtain

$$\frac{dJ}{dt} = (MV)_{\alpha 1} + (MV)_{\beta 1} - (MV)_{\alpha 2} - (MV)_{\beta 2} + g_c (P_1 - P_2)A - g_c F_{w\tau} - g \sin\theta \, \mathscr{M} \tag{7.38}$$

The wall shear force, $F_{w\tau}$, cannot be evaluated in the general case, but when the flow geometry is simple (annular, stratified, or pseudohomogeneous), and when the flow is laminar, it may be determined with the aid of the constitutive equation of the fluid. We will deal with such cases later in other chapters.

Microscopic The microscopic momentum equation may be derived from Equation 7.32 through the Gauss-Ostrogradsky theorem (Standart, 1964). The resulting equations are:
for the α-phase,

$$\frac{\partial}{\partial t}(\rho_\alpha \mathbf{v}_\alpha) + \nabla \cdot (\rho_\alpha \mathbf{v}_\alpha \mathbf{v}_\alpha + g_c P \mathscr{U} + g_c \boldsymbol{\tau}) - g_c \sum_i \rho_{\alpha i} \mathbf{F}_i = 0 \tag{7.39}$$

for the β-phase,

$$\frac{\partial}{\partial t}(\rho_\beta \mathbf{v}_\beta) + \nabla \cdot (\rho_\beta \mathbf{v}_\beta \mathbf{v}_\beta + g_c P \mathscr{U} + g_c \boldsymbol{\tau}) - g_c \sum_i \rho_{\beta i} \mathbf{F}_i = 0 \tag{7.39a}$$

and at the phase boundary,

$$\{\rho_\alpha \mathbf{v}_\alpha \mathbf{v}_\alpha + g_c P \mathscr{U} + g_c \boldsymbol{\tau}\} \cdot \mathbf{n}_\alpha + \{\rho_\beta \mathbf{v}_\beta \mathbf{v}_\beta + g_c P \mathscr{U} + g_c \boldsymbol{\tau}\} \cdot n_\beta = 0 \tag{7.40}$$

5 THE TOTAL ENERGY EQUATION

The total energy equation in its most general form for a two-phase flowing system may be derived in a manner analogous to that used for a single-phase system. Standart (1964) presents such a derivation. The resulting equation is complex and of little practical utility. As in the case of single-phase flow, however, if axial conduction, diffusion, frictional stresses at the inlet and outlet are neglected, and if the only field force is that of gravity, a simpler but still quite general form of the equation results. This macroscopic energy balance is

$$\frac{d\mathscr{E}}{dt} = \left[M\left(H + \frac{V^2}{2g_c} + \frac{g}{g_c}z\right)\right]_{\alpha 1} + \left[M\left(H + \frac{V^2}{2g_c} + \frac{g}{g_c}z\right)\right]_{\beta 1} - \left[M\left(H + \frac{V^2}{2g_c} + \frac{g}{g_c}z\right)\right]_{\alpha 2} - \left[M\left(H + \frac{V^2}{2g_c} + \frac{g}{g_c}z\right)\right]_{\beta 2} - W + \mathscr{Q} \tag{7.41}$$

where

$$W = \int_{A_{w\alpha}} (\mathbf{v}_\alpha \cdot \boldsymbol{\tau}_\alpha) \cdot d\mathbf{A} + \int_{A_{w\beta}} (\mathbf{v}_\beta \cdot \boldsymbol{\tau}_\beta) \cdot d\mathbf{A} \tag{7.42}$$

$$\mathscr{E} = \int_{\mathscr{V}_\alpha} \rho_\alpha \left(U + \frac{V^2}{2g_c} + \frac{g}{g_c}z\right)_\alpha d\mathscr{V} + \int_{\mathscr{V}_\beta} \rho_\beta \left(U + \frac{V^2}{2g_c} + \frac{g}{g_c}z\right)_\beta d\mathscr{V} \tag{7.43}$$

$$\mathscr{Q} = -\int_{A_{w\alpha}} \mathbf{q}_\alpha \cdot d\mathbf{A} - \int_{A_{w\beta}} \mathbf{q}_\beta \cdot d\mathbf{A} \tag{7.44}$$

$$H = U + vP \tag{7.45}$$

In Equation 7.41 $d\mathscr{E}/dt$ is the rate of accumulation of total energy in the system, and the terms within the square brackets represent the sums of the internal energy, the kinetic energy, and the potential energy of the inlet and outlet streams. The terms W and $\mathscr{Q}$ are, respectively, the work done by the system on the surroundings, and the heat absorbed by the system from the surroundings. The work done by the system on the surroundings, in the case being considered, is restricted to that resulting from the wall shear stresses.

In order that we may apply this equation, the internal energy of the streams must be evaluated at the inlet and outlet conditions, and the wall shear stresses and the applicable portion of the wall area must be determined for each phase. Further, the internal energy of the outlet streams is influenced by the degradation of mechanical energy due to shear within the phases. As a result, the total energy equation in this form is applicable to laminar flow only, and where the flow geometry is known and simple. In the limiting case where the wall shear stresses are low and the work done by the system, W, may be neglected, the equation expresses the relationship between heat absorbed by the system and changes in the mechanical and internal forms of energy. If the changes in the mechanical forms of energy are otherwise determined, e.g., from the momentum equation, the equation relates the heat absorbed to changes in internal energy and, consequently, to changes in temperature. For computation involving a changing heat flux over the length of a system, it is convenient to express Equation 7.41 in differential form.

Under steady-state conditions, where W equals zero, and where there is no solubility between phases, or in the more common case when the phases are mutually saturated and the change in solubility is negligible, the equation becomes

$$d\mathcal{Q} = (M_\alpha + M_\beta)\frac{g}{g_c}\,dz + M_\alpha\,dH_\alpha + M_\beta\,dH_\beta + \frac{M_\alpha V_\alpha\,dV_\alpha}{\alpha_\alpha g_c} + \frac{M_\beta V_\beta\,dV_\beta}{\alpha_\beta g_c} \tag{7.46}$$

where the terms α_α and α_β are introduced to account for the velocity profile in the evaluation of the kinetic energy terms from the average velocities. This is the same equation as that developed for gas-liquid flow by Vohr (1962), and it is the total energy equivalent of the equation employed, for example, by Govier *et al.* (1957). For cases not involving a gas phase, or involving a gas phase at low velocities or under small fractional pressure drop, the last two terms of the equation are negligible. The changes in enthalpy may be expressed in terms of the specific heats of the phases and the temperature and pressure changes:

$$dH = C_P\,dT + \left[v - T\left(\frac{\partial v}{\partial T}\right)_P\right]dP$$

where $v = 1/\rho$ = specific volume. In the case of gases, the term in the brackets is small, and it is zero for ideal gases.

6 THE MECHANICAL ENERGY EQUATION

The mechanical energy equation is not independent of the conservation laws already dealt with. It may be considered either a consequence of the momentum equation or a reduced form of the total energy equation. It is instructive to present both these views.

6.1 Derivation from the Momentum Equation

Following the method illustrated in Chapter 3 for single-phase systems, we may derive from Equation 7.32 the macroscopic mechanical energy balance for a two-phase system

$$\frac{d}{dt}\int_{\mathscr{V}} \rho(\tfrac{1}{2}\mathbf{v}\cdot\mathbf{v})d\mathscr{V} = -\oint_{A_e} [\rho(\tfrac{1}{2}\mathbf{v}\cdot\mathbf{v})(\mathbf{v}) + g_c P\mathbf{v} + g_c \mathbf{v}\cdot\boldsymbol{\tau}]\cdot d\mathbf{A}$$
$$-\int_{A_b}\left\{\sum_{\alpha}^{\beta}[\rho(\tfrac{1}{2}\mathbf{v}\cdot\mathbf{v})(\mathbf{v}-\mathbf{v}_b) + g_c P\mathbf{v} + g_c\mathbf{v}\cdot\boldsymbol{\tau}]\cdot\mathbf{n}\right\}dA$$
$$+\int_{\mathscr{V}} g_c(P\nabla\cdot\mathbf{v} + \boldsymbol{\tau}:\nabla\mathbf{v})d\mathscr{V} + \sum_i g_c\int_{\mathscr{V}}\rho_i(\mathbf{F}_i\cdot\mathbf{v})d\mathscr{V} \quad (7.47)$$

Equation 7.47 may be put in a more convenient form by defining the interface mass flux as

$$\mathbf{w} = \rho(\mathbf{v}-\mathbf{v}_b) = \sum_i \mathbf{w}_i$$

and

$$\mathbf{w}_i = \rho_i(\mathbf{v}_i - \mathbf{v}_b) = \frac{\rho_i}{\rho}\mathbf{w}$$

This leads to the following form of the mechanical energy balance (Standart, 1964)

$$\frac{d}{dt}\int_{\mathscr{V}} \rho(\tfrac{1}{2}\mathbf{v}\cdot\mathbf{v}+\Omega)d\mathscr{V} = -\oint_{A_e}\{\rho(\tfrac{1}{2}\mathbf{v}\cdot\mathbf{v}+\Omega)\mathbf{v} + g_c P\mathbf{v} + g_c\mathbf{v}\cdot\boldsymbol{\tau}\}\cdot d\mathbf{A}$$
$$-\int_{A_b}\left\{\sum_{\alpha}^{\beta}\left[\mathbf{w}g_c P\upsilon + \upsilon\mathbf{w}\cdot\left(g_c\boldsymbol{\tau} + \frac{\upsilon}{2}\mathbf{w}\mathbf{w}\right)\right]\cdot\mathbf{n}\right\}dA$$
$$+\int_{\mathscr{V}} g_c(P\nabla\cdot\mathbf{v} + \boldsymbol{\tau}:\nabla\mathbf{v})d\mathscr{V} \quad (7.48)$$

where $\Omega \equiv gz$ is the gravitational potential. Now neglecting the frictional stress contribution over the inlet and outlet, we have

$$\frac{d}{dt}\int_{\mathscr{V}} \rho(\tfrac{1}{2}\mathbf{v}\cdot\mathbf{v}+\Omega)d\mathscr{V} = [M(g_c P\upsilon + \tfrac{1}{2}\mathbf{v}\cdot\mathbf{v}+\Omega)]_{\alpha 1} + [M(g_c P\upsilon + \tfrac{1}{2}\mathbf{v}.\mathbf{v}+\Omega)]_{\beta 1}$$
$$-[M(g_c P\upsilon + \tfrac{1}{2}\mathbf{v}\cdot\mathbf{v}+\Omega)]_{\alpha 2} - [M(g_c P\upsilon + \tfrac{1}{2}\mathbf{v}\cdot\mathbf{v}+\Omega)]_{\beta 2}$$
$$-\int_{A_b}\left[\sum_{\alpha}^{\beta} g_c\mathbf{w}P\upsilon\cdot\mathbf{n}\right]dA + \int_{\mathscr{V}} g_c P\nabla\cdot\mathbf{v}d\mathscr{V}$$
$$-\left[\int_{A_b}\left\{\sum_{\alpha}^{\beta}[\upsilon\mathbf{w}\cdot(g_c\boldsymbol{\tau} + \tfrac{1}{2}\upsilon\mathbf{w}\mathbf{w})]\cdot\mathbf{n}\right\}dA + \int_{A_w} g_c\mathbf{v}\cdot\boldsymbol{\tau}\cdot d\mathbf{A} + \int_{\mathscr{V}} g_c\boldsymbol{\tau}:\nabla\mathbf{v}d\mathscr{V}\right]$$
$$(7.49)$$

The final term in square brackets includes all irreversible effects. It is defined as $g_c \mathscr{F}$, the total friction loss or irreversible energy degradation.

The integral over the entire system volume may be simplified if we assume one-dimensional flow. The equation of continuity then yields

$$\nabla \cdot \mathbf{v} = \rho \frac{dv}{dt} = \rho\left(\frac{\partial v}{\partial t} + V\frac{\partial v}{\partial L}\right) \tag{7.50}$$

In addition, we have

$$d\mathscr{V}_\alpha = E_\alpha A dL \tag{7.51}$$

$$d\mathscr{V}_\beta = E_\beta A dL \tag{7.52}$$

thus

$$\int_{\mathscr{V}} P\nabla \cdot \mathbf{v} d\mathscr{V} = \int_L \rho EAP \frac{\partial v}{\partial t} dL + \int_L MP \frac{\partial v}{\partial L} dL \tag{7.53}$$

where the line integrals are to be summed over both phases. For the steady-state case, the mechanical energy balance may now be written as

$$\left[M\left(\frac{V^2}{2\alpha_\alpha} + \Omega\right)\right]_{\alpha 2} + \left[M\left(\frac{V^2}{2\alpha_\beta} + \Omega\right)\right]_{\beta 2} - \left[M\left(\frac{V^2}{2\alpha_\alpha} + \Omega\right)\right]_{\alpha 1}$$
$$- \left[M\left(\frac{V^2}{2\alpha_\beta} + \Omega\right)\right]_{\beta 1} = g_c\left[W - \int_L Mv\frac{dP}{dL}dL - \mathscr{F}\right] \tag{7.54}$$

M_α and M_β are each constants; Equation 7.54 may be written in differential form as

$$M_\alpha\left(\frac{VdV}{\alpha_\alpha g_c} + \frac{g}{g_c}dz + vdP\right)_\alpha + M_\beta\left(\frac{VdV}{\alpha_\beta g_c} + \frac{g}{g_c}dz + vdP\right)_\beta = -d\mathscr{F} \tag{7.55}$$

and after some rearranging as

$$-dP = \frac{1}{Q_\alpha + Q_\beta}\left[(M_\alpha + M_\beta)\frac{g}{g_c}dz + \frac{M_\alpha V_\alpha dV_\alpha}{\alpha_\alpha g_c} + \frac{M_\beta V_\beta dV_\beta}{\alpha_\beta g_c} + d\mathscr{F}\right] \tag{7.56}$$

where $Q_\alpha = M_\alpha v_\alpha$ and $Q_\beta = M_\beta v_\beta$. Defining $dP_f = d\mathscr{F}/(Q_\alpha + Q_\beta)$ results in the final form of the equation

$$-dP = \frac{1}{Q_\alpha + Q_\beta}\left[(M_\alpha + M_\beta)\frac{g}{g_c}dz + \frac{M_\alpha V_\alpha dV_\alpha}{\alpha_\alpha g_c} + \frac{M_\beta V_\beta dV_\beta}{\alpha_\beta g_c}\right] - dP_f \tag{7.57}$$

6.2 Derivation from the Total Energy Equation

The derivation of the mechanical energy equation follows that of the total energy equation, except that only the "mechanical" forms of energy are included, and this requires the addition of terms that refer to the mechanical energy degraded to heat (see Govier *et al.*, 1957). Thus, in place of the terms $M_\alpha\, dU_\alpha + M_\beta\, dU_\beta - d\mathcal{Q}$ in Equation 7.46, the terms $M_\alpha\, d\mathcal{F}_\alpha + M_\beta\, d\mathcal{F}_\beta$ appear. The irreversibility terms $d\mathcal{F}_\alpha$ and $d\mathcal{F}_\beta$ represent the net effect, arbitrarily attributed to unit masses of phases α and β, respectively, of all irreversibilities or of all degradations of mechanical energy to internal energy. Clearly this net effect must be the net change in the internal energy of the system, less any energy absorbed from the surroundings, i.e., $d\mathcal{Q}$. Introducing the substitution, the mechanical energy equivalent of Equation 7.46 is

$$(M_\alpha + M_\beta)\frac{g}{g_c}\,dz + (M_\alpha v_\alpha + M_\beta v_\beta)dP + \frac{M_\alpha V_\alpha\, dV_\alpha}{\alpha_\alpha g_c} + \frac{M_\beta V_\beta\, dV_\beta}{\alpha_\beta g_c} + M_\alpha\, d\mathcal{F}_\alpha + M_\beta\, d\mathcal{F}_\beta = 0 \quad (7.58)$$

assuming the system does no useful work on the surroundings, i.e., $W = 0$. Rearranging, we have

$$-dP = \frac{M_\alpha + M_\beta}{Q_\alpha + Q_\beta}\frac{g}{g_c}\,dz + \left(\frac{M_\alpha V_\alpha\, dV_\alpha}{\alpha_\alpha g_c} + \frac{M_\beta V_\beta\, dV_\beta}{\alpha_\beta g_c}\right)\frac{1}{Q_\alpha + Q_\beta} + \frac{M_\alpha\, d\mathcal{F}_\alpha + M_\beta\, d\mathcal{F}_\beta}{Q_\alpha + Q_\beta} \quad (7.59)$$

The distinction between $d\mathcal{F}_\alpha$ and $d\mathcal{F}_\beta$ is arbitrary, and it is preferable to substitute a single irreversibility term. This is most conveniently done by defining

$$-dP_f = \frac{M_\alpha\, d\mathcal{F}_\alpha + M_\beta\, d\mathcal{F}_\beta}{Q_\alpha + Q_\beta}$$

where dP_f is the change due to all irreversibilities, and to them alone. This results in the general form of the mechanical energy balance previously given as Equation 7.57.

Equation 7.57 may also be applied to a finite length of pipe, L, inclined at an angle θ from the horizontal provided the properties of the mixture are relatively constant over this distance. The resulting equation is

$$\Delta P = \left(\frac{M_\alpha + M_\beta}{Q_\alpha + Q_\beta}\right)\frac{g}{g_c}L\sin\theta + \left[\frac{M_\alpha(V_{\alpha 2}^2 - V_{\alpha 1}^2)}{2\alpha_\alpha(Q_\alpha + Q_\beta)g_c} + \frac{M_\beta(V_{\beta 2}^2 - V_{\beta 1}^2)}{2\alpha_\beta(Q_\alpha + Q_\beta)g_c}\right] + \Delta P_f \quad (7.60)$$

where, for convenience in sign, we define $\Delta P = (P_1 - P_2)$. The three terms on the right-hand side are the contributions due potential energy change (ΔP_{PE}), kinetic energy change (ΔP_{KE}), and friction (ΔP_f).

6.3 *The Irreversibility Term, dP_f*

Except in laminar flow, the term dP_f is not predictable from theory and must be determined by measurements or from generalizations based upon measurements. In the development of generalizations based upon measurements, three distinct approaches have been taken. These are the methods based upon the pressure drop ratio, the dimensionless excess pressure drop, and the two-phase friction factor.

Martinelli and coworkers (1944), in the field of the horizontal flow of gas-liquid mixtures, have defined quantities expressing the ratio of the two-phase pressure drop due to friction, to the pressure drop that would be expected if either of the phases were flowing alone in a pipe of the same diameter and length. These are

$$\phi_\alpha = \sqrt{\frac{dP_f}{dP_{fS\alpha}}} \tag{7.61}$$

$$\phi_\beta = \sqrt{\frac{dP_f}{dP_{fS\beta}}} \tag{7.62}$$

Martinelli's approach was to seek generalizations by which these ratios could be predicted. Multiplication of either of the ratios by the appropriate single-phase pressure drop then gives the two-phase pressure drop.

Durand (1953) and others working on the pipe flow of solid-liquid mixtures and Hughmark and Pressburg (1961) working on the vertical flow of gas liquid mixtures have used an "excess dimensionless pressure drop," and an "excess pressure gradient" defined as

$$\frac{dP_f - dP_{fS\alpha}}{dP_{fS\alpha}} \qquad \text{(Durand)}$$

or

$$-\frac{dP_f - dP_{fS\alpha}}{dL} \qquad \text{(Hughmark and Pressburg)}$$

and have developed correlations in terms of it.

Govier and coworkers (1961) and others have defined two-phase friction factors in terms of one, or the other, or both of the flowing phases and have correlated these with the variables affecting them. Among the various Fanning type friction factors that may be defined, the following are the most common:

$$f_{tp\alpha} = -\frac{g_c D}{2\rho_\alpha V_{S\alpha}^2}\frac{dP_f}{dL} \tag{7.63}$$

$$f_{tp\beta} = -\frac{g_c D}{2\rho_\beta V_{S\beta}^2}\frac{dP_f}{dL} \tag{7.64}$$

$$f_{tpM} = -\frac{g_c D}{2\rho_M V_M^2}\frac{dP_f}{dL} \tag{6.65}$$

In the above equations $V_{S\alpha}$, $V_{S\beta}$ are, respectively, the superficial velocities of the α and β phases defined as the volume per unit time of the respective phase entering the pipe divided by the cross-sectional area of the pipe; V_M is the mixture velocity $= V_{S\alpha} + V_{S\beta}$; and ρ_M is the density of a mixture of the composition entering the pipe, i.e.,

$$\rho_M = \rho_\alpha \frac{Q_\alpha}{Q_\alpha + Q_\beta} + \rho_\beta \frac{Q_\beta}{Q_\alpha + Q_\beta}$$

$$= \rho_\alpha C_\alpha + \rho_\beta C_\beta \tag{7.66}$$

Although the choice of friction factor definition is arbitrary, that based on the superficial velocity and the density of one of the phases is simpler, and therefore probably preferable.

The relationships among the ϕ terms, the dimensionless excess pressure drop, and the various two-phase friction factors are readily derivable from Equations 7.61 through 7.66 and the Fanning equation for single-phase pressure drop. In terms of $f_{tp\alpha}$ these are

$$\phi_\alpha^2 = \frac{f_{tp\alpha}}{f_\alpha} \tag{7.67}$$

$$\phi_\beta^2 = \frac{f_{tp\alpha}\rho_\alpha V_{S\alpha}^2}{f_\beta \rho_\beta V_{S\beta}^2} \tag{7.68}$$

$$\frac{dP_f - dP_{fS\alpha}}{dP_{fS\alpha}} = \frac{f_{tp\alpha}}{f_\alpha} - 1 \tag{7.69}$$

$$f_{tp\beta} = f_{tp\alpha}\frac{\rho_\alpha V_{S\alpha}^2}{\rho_\beta V_{S\beta}^2} \tag{7.70}$$

$$f_{tpM} = f_{tp\alpha}\frac{\rho_\alpha V_{S\alpha}^2}{\rho_M V_M^2} \tag{7.71}$$

The meaning of Equation 7.57 or 7.60 should be recognized. Although the derivation of the equation from the momentum equation indicates the nature of the term, dP_f or ΔP_f, it does not provide means of predicting it. The derivation of the equation from the total energy equation gives only the broadest insight into the dP_f or ΔP_f term and again offers no means for its prediction. Notwithstanding the assumptions involved in its derivation, for practical purposes the validity of Equation 7.57, or its more convenient form 7.60, is not subject to question because the equation, in effect, defines dP_f or ΔP_f as the quantity required to balance it.

The pressure drop due to the potential energy effect, ΔP_{PE}, which in the flow of a single-phase fluid is the "hydrostatic head" effect, does not, in the two-phase case, truly represent the pressure due to a static column of the fluid mixture of the concentration and configuration that may exist under all the possible flow patterns. The term is based only upon the inlet (or outlet) quantities and densities and does not account for the fact that, because of holdup, the *in situ* mixture may differ greatly in composition. The irreversible effect, ΔP_f, as here defined, includes the pressure drop due to the effects of holdup.

It is possible, however, to write an equation similar to Equation 7.60 substituting for the potential energy term a hydrostatic head term, i.e.,

$$\Delta P_{HH} = (\rho_\alpha E_\alpha + \rho_\beta E_\beta) \frac{g}{g_c} \Delta z \tag{7.72}$$

The equation is

$$\Delta P = (\rho_\alpha E_\alpha + \rho_\beta E_\beta) \frac{g}{g_c} \Delta z + \frac{1}{Q_\alpha + Q_\beta} \left(\frac{M_\alpha \Delta V_\alpha^2}{2\alpha_\alpha g_c} + \frac{M_\beta \Delta V_\beta^2}{2\alpha_\beta g_c} \right) + \Delta P_f^\bullet \tag{7.73}$$

This, in effect, defines a different pressure drop due to irreversibilities which excludes the effect of holdup on mixture density. This pressure drop is

$$\Delta P_f^\bullet = \Delta P_f + \left[\frac{M_\alpha + M_\beta}{Q_\alpha + Q_\beta} - (\rho_\alpha E_\alpha + \rho_\beta E_\beta) \right] \frac{g}{g_c} \Delta z \tag{7.74}$$

Several investigators have endeavored to evaluate ΔP_{HH}, the actual weight per unit area of the column of the *in situ* fluid mixture. These include Ros (1961) who investigated the vertical flow of gas-liquid mixtures, Jones and coworkers (1967) who worked with gas-solid mixtures, and Newitt, Richardson, and Gliddon (1961) who studied the flow of liquid-solid mixtures. This involves a determination of the *in situ* composition or the holdup. There is nothing wrong with this procedure; from a mechanistic point of view it may have advantages. On the other hand, where a strictly empirical approach is necessary, the use of ΔP_f and ΔP_{PE} seems preferable because it does not require an empirical correlation for ΔP_{HH} before the $\Delta P_f^{\bullet}$ term may be evaluated. In some cases, notably with the vertical flow of fluid-solid mixtures, the ΔP_{HH} approach has been more successful.

7 THE SLIP OR HOLDUP EFFECT

For reasons that will be more fully discussed later, when the phases differ in density and/or viscosity, one of them—usually the less dense phase—tends to flow at a higher *in situ* average velocity than does the other. This gives rise to an all-important characteristic of two-phase flow, the existence of "slip" of one phase past the other, or "holdup" of one phase relative to the other. The phenomenon may be measured in many ways but, however measured, it involves the *in situ* volume fractions of the phases that may be, and usually is, different from the input volume fraction.

It is convenient to recall the three measures of the *in situ* volume fraction of a phase as previously defined:

$\varepsilon_\alpha =$ the local *in situ* volume fraction of the α-phase (see discussion following Equation 7.22),

$E_\alpha = A_\alpha/A = 1/A \int_0^A \varepsilon_\alpha \, dA =$ the cross-sectional average *in situ* fraction of the α-phase (see discussion following Equation 7.8d),

$\bar{E}_\alpha = 1/\mathscr{V} \int_0^{\mathscr{V}} \varepsilon_\alpha \, d\mathscr{V} = 1/L \int_0^L E_\alpha dL =$ the volume average *in situ* volume fraction of the α phase (see discussion following Equation 7.10).

In most instances the holdup phenomenon is measured or described over a cross-sectional area or a finite system volume and E_α or $\bar{E}_\alpha$ is the appropriate volume fraction.

The volume fraction of the α phase supplied to or delivered from a system is

$$C_\alpha = \frac{Q_\alpha}{Q_\alpha + Q_\beta}$$

with an appropriate subscript for C_α if the supplied and delivered fractions differ due to mass transfer.

7.1 Average Slip Velocity

One measure of the holdup phenomenon is the average slip velocity at a given cross section defined as

$$S = V_\alpha - V_\beta \tag{7.75}$$

This is not to be confused with the local relative velocity of the phases, $v_\alpha - v_\beta$; it is merely the difference between the two average *in situ* velocities as defined by Equations 7.13 and 7.14, i.e.,

$$S = \frac{V_{S\alpha}}{E_\alpha} - \frac{V_{S\beta}}{E_\beta} \tag{7.76}$$

Assuming $V_{S\alpha}$ and $V_{S\beta}$ to be defined through the input quantities, S and E_α are directly interrelated, and a knowledge of one determines the other.

Equation 7.76 may be solved for E_α or E_β and expressed in terms of the mixture velocity,

$$V_M = V_{S\alpha} + V_{S\beta} = \frac{Q_\alpha + Q_\beta}{A} \tag{7.77}$$

and C_α or C_β as defined earlier. The results are

$$E_\alpha = \left(\frac{V_M + S}{2S}\right) - \left[\left(\frac{V_M + S}{2S}\right)^2 - \frac{V_M C_\alpha}{S}\right]^{1/2} \tag{7.78a}$$

and

$$E_\beta = -\left(\frac{V_M - S}{2S}\right) + \left[\left(\frac{V_M - S}{2S}\right)^2 + \frac{V_M C_\beta}{S}\right]^{1/2} \tag{7.78b}$$

Slip velocities vary widely but are highest with gas-liquid and gas-solid flow where they may reach levels of 50 ft/sec or more.

Note that if the average slip velocity is zero, there is no holdup of one phase relative to the other,

$$\frac{V_{S\alpha}}{E_\alpha} = \frac{V_{S\beta}}{E_\beta}$$

and

$$E_\alpha = \frac{V_{S\alpha}}{V_{S\alpha} + V_{S\beta}} = \frac{Q_\alpha}{Q_\alpha + Q_\beta} = C_\alpha$$

which means that E_α and E_β are determined solely by the input quantities. This is true for the flow of pseudohomogeneous mixtures.

7.2 Ratio of the Average in situ Velocities (Holdup Ratio or Slip Ratio)

A second measure of the holdup phenomenon is the ratio of the average *in situ* velocities at any cross section

$$H = \frac{V_\alpha}{V_\beta}$$

which, from Equations 7.13 and 7.14, is equal to

$$H = \frac{V_{S\alpha}}{V_{S\beta}} \frac{E_\beta}{E_\alpha} = \frac{C_\alpha E_\beta}{C_\beta E_\alpha} \tag{7.79}$$

and we see that H is also the ratio, at any cross section, of

a. the *in situ* volume fraction ratio of the heavier to the lighter phase, E_β/E_α, to
b. the input volume fraction ratio of the heavier to the lighter phase $V_{S\beta}/V_{S\alpha} = C_\beta/C_\alpha$.

The ratio H has been called the "holdup ratio" by Govier and coworkers (1957), and the "slip ratio" by Marchaterre and Hoglund (1962) and by Petrick and Kudirka (1966). As in the case of the average slip velocity, for fixed input quantities the holdup ratio and the average *in situ* volume fraction are interrelated and one fixes the other. Rearrangement of Equation 7.79 yields

$$E_\alpha = \frac{V_{S\alpha}}{V_{S\alpha} + HV_{S\beta}} = \frac{C_\alpha}{C_\alpha + HC_\beta} \tag{7.80}$$

and

$$E_\beta = \frac{HV_{S\beta}}{V_{S\alpha} + HV_{S\beta}} = \frac{HC_\beta}{C_\alpha + HC_\beta} \tag{7.81}$$

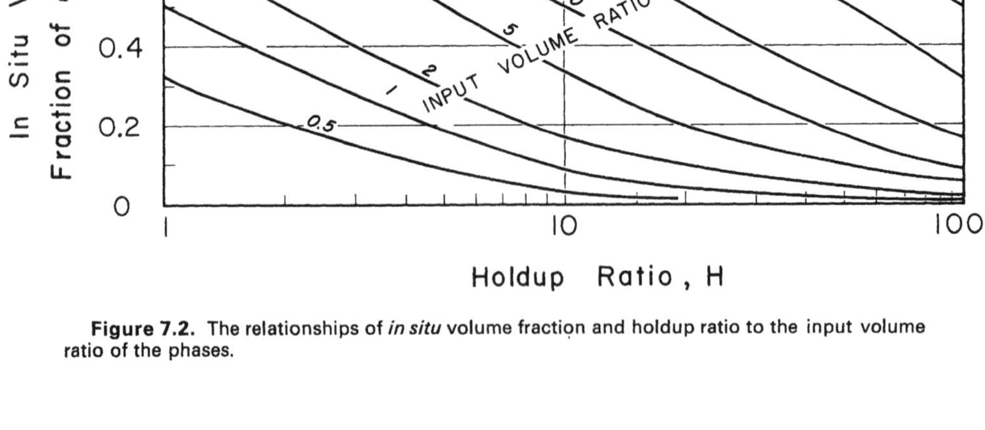

Figure 7.2. The relationships of *in situ* volume fraction and holdup ratio to the input volume ratio of the phases.

We may note that for a two-phase system, whereas S is defined only by a knowledge of E_α, $V_{S\alpha}$, and $V_{S\beta}$, H is defined through E_α and the ratio of $V_{S\alpha}$ to $V_{S\beta}$. This latter permits a simple graphical representation of the relationship between E_α, H, and the input volume or the concentration ratio of the phases, $V_{S\alpha}/V_{S\beta} = C_\alpha/C_\beta$, as given in Figure 7.2. From Figure 7.2 we may see that the holdup ratio is a sensitive measure of the holdup phenomenon.

The relationship between the average slip velocity and the holdup ratio may be determined from Equations 7.76 and 7.80 as

$$S = \frac{H-1}{H}(V_{S\alpha} + HV_{S\beta}) \tag{7.82}$$

The holdup ratio may vary from unity (flow of a pseudohomogeneous mixture with no average slip) to 100 or more.

7.3 Bankoff K Factor, Armand C Factor

Another measure of holdup is the K factor used by Bankoff (1960) in connection with the bubble flow pattern in gas-liquid flow. (This is the same as a flow parameter, C, proposed by Armand, 1946.) The K factor is defined by

$$K = E_\alpha \frac{V_{S\alpha} + V_{S\beta}}{V_{S\alpha}} = \frac{E_\alpha}{C_\alpha} \tag{7.83}$$

i.e., as the ratio of the *in situ* to the input volume fraction of phase α. From Equations 7.79 and 7.83, we obtain

$$H = \frac{E_\beta}{K - E_\alpha}$$

or

$$K = \frac{E_\beta + E_\alpha H}{H} \tag{7.84}$$

which is an equation developed by Bankoff (1960).

As for other measures of holdup, the K or C ratio is not independent, but is related to the *in situ* and the input volume fractions. The range of K has been shown by Bankoff to be from about 0.6 to 1.0. The relationship between H and K in terms of the superficial velocities and the input and the *in situ* volume fractions should be noted. From Equations 7.80 and 7.84,

$$K = \frac{V_{S\alpha} + V_{S\beta}}{V_{S\alpha} + HV_{S\beta}} = \frac{1}{C_\alpha + HC_\beta} = \frac{E_\beta + E_\alpha H}{H} \tag{7.85}$$

The relationship between K, E_α, and H is shown in Figure 7.3. From Equation 7.85 it is seen that K approaches E_α as H becomes large, and as H approaches 1 (no slip) K also becomes 1.

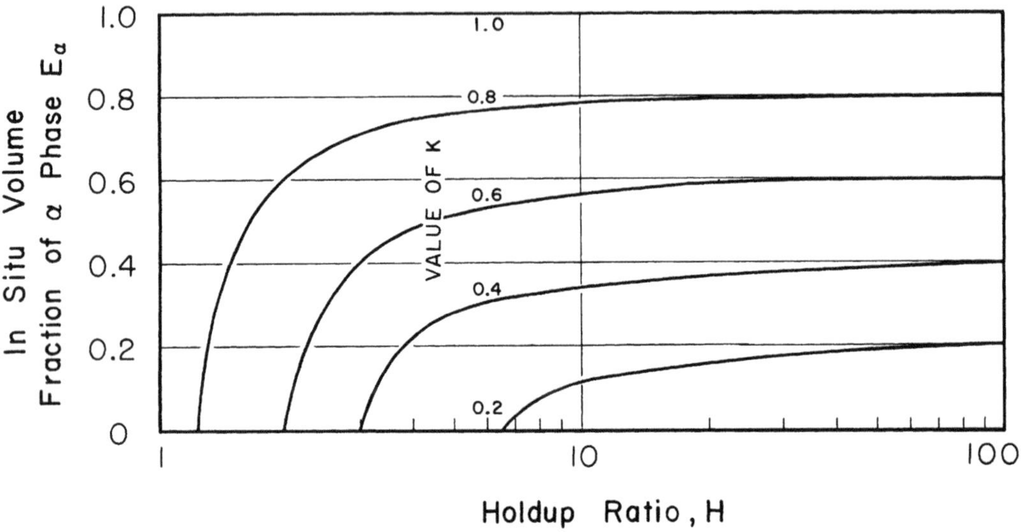

Figure 7.3. The relationship of the *in situ* volume fraction and the holdup ratio to K.

Equations 7.76 to 7.85 show the interrelation of the commonly used measures of the phenomenon of holdup or slip for the important case of a two-phase system with no mass transfer between phases. This case is closely approximated for isothermal gas-solid, liquid-solid, liquid-liquid, gas-capsule, and liquid-capsule flow. For gas-liquid flow the variations with pressure and temperature of gas solubility in the liquid phase may be significant, thus rendering Equations 7.13 and 7.14 approximate. The equivalent accurate relations for the case of mass transfer may be derived from Equations 7.10 and 7.11.

7.4 Factors Influencing Holdup

It is now clearly recognized that three factors may contribute to holdup or slip. These are

a. the existence of a velocity profile across the cross section;
b. the existence of a concentration profile across the cross section, and;
c. the local relative velocity between phases caused by gravitational effects.

Bankoff (1960), Neal (1963), Zuber and Findlay (1965), Petrick and Kudirka

(1966), and Nassos and Bankoff (1967) have all investigated these effects as they apply to the vertical flow of gas-liquid mixtures. Certain of the conclusions of these studies have application to any two-phase system. These will be introduced here, and the conclusions applicable only to the vertical flow of gas-liquid mixtures will be discussed later.

7.41 Effect of Velocity and Concentration Profile

Let us first consider the limiting case of the flow in a pipe of a mixture of one phase dispersed in another, under conditions where there is no local relative velocity between the phases, i.e., where at any radial position the two phases flow at the same local velocity, v. This case is encountered where the settling or rising velocity of the phases is small. The situation is represented by Figure 7.4, which illustrates the type of velocity profile that may be encountered and shows that there can also be a "concentration profile" or a variation of the *in situ* volume fraction of the phases with position. If the phases were uniformly mixed, as in a fine emulsion, the concentration profile would be flat, but in general the phase distribution is not uniform. In vertical flow the concentration profile is reasonably symmetrical. In inclined or horizontal flow even modest gravitational effects cause a skewing of the profile.

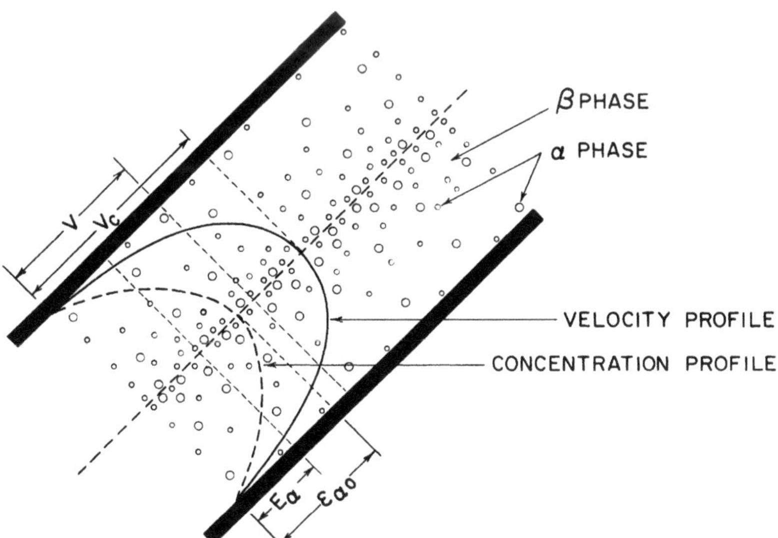

Figure 7.4. Illustrative velocity and concentration profiles.

The effect of the velocity and concentration profiles is made clear by expressing the average *in situ* velocities of the phases in terms of the local velocities and concentrations

$$V_\alpha = \frac{Q_\alpha}{AE_\alpha} = \frac{V_{S\alpha}}{E_\alpha} = \frac{\overline{\varepsilon_\alpha v}}{E_\alpha} \tag{7.86}$$

$$V_\beta = \frac{Q_\beta}{AE_\beta} = \frac{V_{S\beta}}{E_\beta} = \frac{\overline{\varepsilon_\beta v}}{E_\beta} \tag{7.87}$$

$$V_{S\alpha} + V_{S\beta} = \frac{Q_\alpha + Q_\beta}{A} = V_M \tag{7.88}$$

where the overlining indicates cross-sectional averages, $V_M = 1/A \int_0^A vdA$ and $\overline{\varepsilon v} = 1/A \int_0^A \varepsilon v dA$. It is proper to define the cross-sectional average *in situ* velocities as in Equations 7.86 and 7.87 so that the product of the velocity, the appropriate average volume fraction, and the area will yield the true volumetric flow rate. We see that V_α and V_β are volume fraction weighted cross-sectional average velocities (Zuber and Findlay, 1965).

From these relations for V_α, V_β, and V_M, we see that only if one or both the concentration and the velocity profile are flat would $V_\alpha = V_\beta = V_M$, or would $S = 0$ and $H = 1$. In other words, holdup or slip will occur as a result of the interplay of one of the profiles on the other. The exact effect of the profiles on holdup may be predicted only if the profiles are known.

7.42 Effect of Local Relative Velocity Between Phases

Many workers, especially in the flow of gas-liquid mixtures, have recognized the effect of the local relative velocity between phases most commonly caused by gravitational forces. Zuber and Findlay (1965) have done much to clarify the situation. We will follow their analysis.

Recalling that v_α and v_β are the local velocities of the phases we may define local volumetric flux densities or local superficial velocities for each of the phases and the mixture as

$$j_\alpha = \varepsilon_\alpha v_\alpha \tag{7.89}$$

$$j_\beta = \varepsilon_\beta v_\beta \tag{7.90}$$

$$j_M = j_\alpha + j_\beta = \varepsilon_\alpha v_\alpha + \varepsilon_\beta v_\beta \tag{7.91}$$

From these definitions we may express local drift velocities, or local phase velocities relative to j_M, as

$$v_{\alpha j} = v_\alpha - j_M \tag{7.92}$$

$$v_{\beta j} = v_\beta - j_M \tag{7.93}$$

and the local relative velocity between the phases as

$$v_r = v_\alpha - v_\beta \tag{7.94}$$

Combining Equations 7.91, 7.92, 7.93, and 7.94 we may express the local drift velocities as

$$v_{\alpha j} = (1 - \varepsilon_\alpha)\, v_r \tag{7.95}$$

$$v_{\beta j} = -\, \varepsilon_\alpha\, v_r \tag{7.96}$$

Now if the local relative velocity v_r is zero, then $v_{\alpha j} = v_{\beta j} = 0$ and $v_\alpha = v_\beta = j_M$.

Similar and more useful relationships may be expressed in terms of the velocities averaged over the cross-sectional area. Thus we have from Equation 7.89

$$\overline{j_\alpha} = \frac{1}{A}\int_0^A j_\alpha\, dA = \overline{\varepsilon_\alpha v_\alpha} \tag{7.97}$$

and substituting for v_α from Equation 7.92

$$\overline{j_\alpha} = \overline{\varepsilon_\alpha j_M} + \overline{\varepsilon_\alpha v_{\alpha j}} = V_{S\alpha} \tag{7.98}$$

Similarly,

$$\overline{j_\beta} = \overline{\varepsilon_\beta j_M} + \overline{\varepsilon_\beta v_{\beta j}} = V_{S\beta} \tag{7.99}$$

and

$$\overline{j_M} = V_{S\alpha} + V_{S\beta} = V_M \tag{7.100}$$

Dividing Equation 7.98 by E_α, we obtain

$$\frac{V_{S\alpha}}{E_\alpha} = V_\alpha = \frac{\overline{\varepsilon_\alpha j_M}}{E_\alpha} + \frac{\overline{\varepsilon_\alpha v_{\alpha j}}}{E_\alpha} \tag{7.101}$$

Noting that the first term on the right-hand side of Equation 7.101 would equal V_M if the concentration profile were uniform, it is convenient to define a concentration effect factor, or a distribution coefficient,

$$C_0 = \frac{\overline{\varepsilon_\alpha j_M}}{E_\alpha V_M} \tag{7.102}$$

Also we may define a volume fraction weighted average drift velocity

$$V_{\alpha j} = \frac{\overline{\varepsilon_\alpha v_{\alpha j}}}{E_\alpha} \tag{7.103}$$

Equation 7.101 may now be written as

$$\frac{V_{S\alpha}}{E_\alpha} = V_\alpha = C_0 V_M + V_{\alpha j} \tag{7.104}$$

and we see that the average absolute velocity of the α phase may be expressed as the sum of a term proportional to the mixture velocity and the weighted average drift velocity. Equation 7.104 may also be written in the form

$$\frac{V_{S\alpha}}{E_\alpha V_M} = \frac{C_\alpha}{E_\alpha} = C_0 + \frac{V_{\alpha j}}{V_M} \tag{7.105}$$

It is of interest to note that under the special conditions where $V_{\alpha j} = 0$, Equation 7.105 shows that C_0 is the reciprocal of Bankoff's K, i.e.,

$$C_0 = \frac{C_\alpha}{E_\alpha} = \frac{1}{K}$$

Equation 7.104 indicates a linear relationship between $V_{S\alpha}/E_\alpha$ and V_M with a slope of C_0 and an intercept of $V_{\alpha j}$. Zuber and Findlay (1965) find that data for the flow of gas-liquid mixtures confirm this linearity within individual flow patterns, and they obtain values of C_0 and $V_{\alpha j}$ for several different flow patterns, including slug and annular flow. Actually the derivation of Equation 7.104 is based upon the continuity equation for dispersed systems (Equations 7.20 and 7.21) and there is no theoretical basis for the application of the equation to other than dispersed systems. Nonetheless the linear form of the equation is confirmed empirically and, with experimentally determined values of C_0 and $V_{\alpha j}$, it has broad application in the interpretation of holdup data.

Equation 7.105 is in general form; from it and the appropriate definitions (Equations 7.75 and 7.79), we may write general equations for the average *in situ* volume fraction of the α phase, the slip velocity, and the holdup ratio. These equations are

$$E_\alpha = \frac{C_\alpha}{C_0 + \dfrac{V_{\alpha j}}{V_M}} \tag{7.106}$$

$$S = C_0 V_{S\alpha} + \left(C_0 - \frac{1}{E_\beta}\right) V_{S\beta} + V_{\alpha j} \tag{7.107}$$

$$H = \frac{1}{C_\beta}\left(C_0 + \frac{V_{\alpha j}}{V_M}\right)\left(1 - \frac{C_\alpha}{C_0 + \dfrac{V_{\alpha j}}{V_M}}\right) \tag{7.108}$$

Note that Equation 7.108 reduces to Equation 7.85 when $V_{\alpha j} = 0$ and C_0 is equal to $1/K$.

These equations assist in our understanding of the holdup phenomenon and in clarifying the roles of the velocity and the concentration profiles, and of the relative velocity between the phases. To apply them in a particular case, however, we must know both C_0 and $V_{\alpha j}$. The average drift velocity, $V_{\alpha j}$, depends upon the specific circumstances, including the size of the particles, bubbles or droplets of the dispersed phase, the densities of the phases, etc., and its discussion is best left to the consideration of the various two-phase systems. The precise value of C_0 also depends upon the system under consideration, but here it is possible to draw some general conclusions for certain simplified but not unreasonable cases.

Zuber and Findlay (1965) illustrate the effect of nonuniform flow and concentration distribution for a simplified case in a circular pipe when both profiles are axially symmetric, and they assume power law velocity and concentration profiles of the form

$$\frac{j_M}{j_{M0}} = 1 - \left(\frac{r}{R}\right)^m \tag{7.109}$$

where j_{M0} is the value of j_M at the center line of the pipe, and

$$\frac{\varepsilon_\alpha - \varepsilon_{\alpha w}}{\varepsilon_{\alpha 0} - \varepsilon_{\alpha w}} = 1 - \left(\frac{r}{R}\right)^n \tag{7.110}$$

where $\varepsilon_{\alpha w}$ and $\varepsilon_{\alpha 0}$ are, respectively, the values of ε_α at the wall and at the center line of the pipe. Combining these relations with the definition of C_0 (Equation 7.102) and performing the necessary integration, we obtain

$$C_0 = 1 + \frac{2}{m + n + 2}\left(1 - \frac{\varepsilon_{\alpha w}}{E_\alpha}\right) \tag{7.111}$$

or

$$C_0 = \frac{m + 2}{m + n + 2}\left[1 + \frac{\varepsilon_{\alpha 0}}{E_\alpha}\left(\frac{n}{m + 2}\right)\right] \tag{7.112}$$

Thus, for the conditions assumed (power law velocity and concentration profiles) we have a way of predicting C_0 from the profile exponents m and n. This relationship may be expected to yield good results for the flow of dispersed systems provided gravitational effects do not cause a skewness in the concentration profile. From these relations it is seen that, regardless of the velocity profile, i.e., of the value of m,

a. $C_0 = 1$ if the concentration is uniform, i.e., $\varepsilon_{\alpha w} = \varepsilon_\alpha = \varepsilon_{\alpha 0}$;
b. $C_0 > 1$ if the concentration at the center exceeds that at the wall, i.e., $\varepsilon_{\alpha 0} > \varepsilon_{\alpha w}$; and
c. $C_0 < 1$ if the concentration at the wall exceeds that at the center, i.e., $\varepsilon_{\alpha w} > \varepsilon_{\alpha 0}$.

Values of C_0 for various values of m and n and for ratios of $\varepsilon_{\alpha w}/\varepsilon_{\alpha 0}$ varying from 0 to unity are given in Figure 7.5 modified after Zuber and Findlay (1965).

We may now summarize by noting that for any given flow situation, as characterized by the fluid properties, the pipe size, and the quantities of phases flowing, the holdup phenomenon is determined by the local velocity and volume fraction profiles and the local relative velocity between the phases. With the effect of the concentration and velocity profiles reflected by the coefficient C_0 and the effect of local relative velocity incorporated in $V_{\alpha j}$, any of the measures of holdup may be expressed in terms of these quantities and the input velocities or concentrations. Table 7.1 summarizes the principal relationships for the limiting and general values of C_0 and $V_{\alpha j}$.

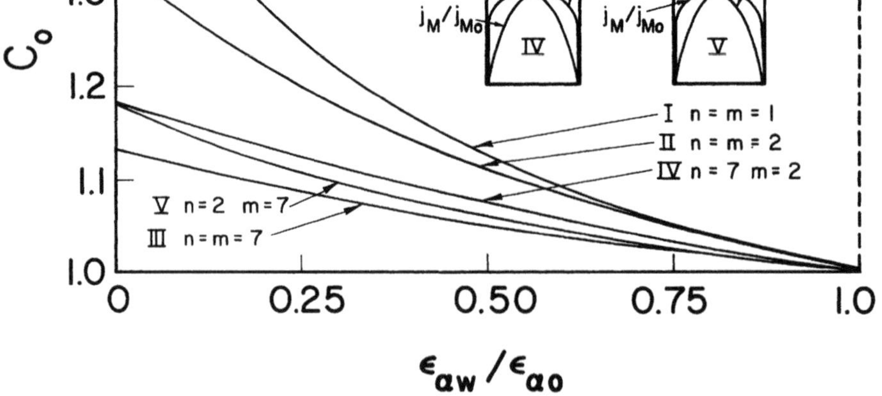

Figure 7.5. Range of values of the distribution coefficient C_0. (Modified after Zuber, N. and J. A. Findlay, *J. Heat Transfer, Trans. A.S.M.E., Ser. C.*, **87**, 453, 1965.)

TABLE 7.1
Summary of Holdup Relations[a]

Characteristics of Flow Pattern			Measure of Holdup Phenomenon			
Distribution coefficient, C_0	Local relative velocity, v_r	Weighted average drift velocity, $V_{\alpha j}$	E_α	H	S	K
$C_0 = 1$ Uniform velocity or concentration profile or both	$v_r = 0$	$V_{\alpha j} = 0$	$E_\alpha = C_\alpha$	$H = 1$	$S = 0$	$K = 1$
	$v_r \neq 0$	$V_{\alpha j} \neq 0$	$E_\alpha = \dfrac{C_\alpha V_M}{V_M + V_{\alpha j}}$	$H = \dfrac{C_\alpha}{C_\beta}\left(\dfrac{1 + \frac{V_{\alpha j}}{V_M}}{C_\alpha} - 1\right)$	$S = \left(1 + \dfrac{V_{\alpha j}}{V_M}\right)\left(V_M - \dfrac{C_\beta V_M}{C_\beta + \frac{V_{\alpha j}}{V_M}}\right)$	$K = \dfrac{V_M}{V_M + V_{\alpha j}}$
$C_0 \neq 1$ variable velocity or concentration profile or both	$v_r = 0$	$V_{\alpha j} = 0$	$E_\alpha = \dfrac{C_\alpha}{C_0}$	$H = \dfrac{C_\alpha}{C_\beta}\left(\dfrac{C_0}{C_\alpha} - 1\right)$	$S = C_0\left(V_M - \dfrac{C_\beta V_M}{C_0 - C_\alpha}\right)$	$K = \dfrac{1}{C_0}$
	$v_r \neq 0$	$V_{\alpha j} \neq 0$	$E_\alpha = \dfrac{C_\alpha V_M}{C_0 V_M + V_{\alpha j}}$	$H = \dfrac{C_\alpha}{C_\beta}\left(\dfrac{C_0 + \frac{V_{\alpha j}}{V_M}}{C_\alpha} - 1\right)$	$S = \left(C_0 + \dfrac{V_{\alpha j}}{V_M}\right)\left(V_M - \dfrac{C_\beta V_M}{C_0 + \frac{V_{\alpha j}}{V_M} - C_\alpha}\right)$	$K = \dfrac{V_M}{C_0 V_M + V_{\alpha j}}$

[a] V_M, C_α, and C_β in which the holdup measures are expressed are functions of the known quantities D, Q_α, and Q_β.

7.43 Variation of Holdup with Length

Throughout the foregoing discussion of the measures of holdup and the influence of velocity and concentration profiles and relative velocity we have, in effect, been considering the situation at a single cross section in the flow stream. The values of E_α, S, H, K, and C_0 are, therefore, values applicable at a given distance from the inlet of the pipe. When the fluid properties or the flow pattern varies with travel distance, the holdup, and all these measures of it, will vary with distance. In principle, this variation may be taken into account by integration over the length. For small finite lengths or when the changes are not too large, volume average values, $\bar{E}_\alpha$, $\bar{S}$, $\bar{H}$, $\bar{K}$, and $\bar{C}_0$ may be used in place of E_α, S, H, K, and C_0 in the equations developed in this section.

Bibliography

Armand, A. A., *Izv. Vsesou Teplotek Inst.*, No. 1, p. 16 (1946), *AERE Libr. Transl.* Vol. 828, 1959.

Bankoff, S. G., *J. Heat Transfer, Trans, A.S.M.E., Ser. C*, **82**, 265 (1960).

Bird, R. B., W. E. Stewart, and E. N. Lightfoot, *Transport Phenomena*, John Wiley & Sons, Inc., New York, 1960.

Birkhoff, G., *J. Math. Anal. Applic.*, **8**, 66 (1964).

Delhaye, J.-M., *Houille Blanche*, **22**, 559 (1967).

Delhaye, J.-M., *Compt. Rend. Acad. Sci., Paris, Ser. A*, **267**, 660 (1968).

Durand, R., Minnesota Int. Hydraulics Convention, *Proc.* p. 89, Int. Assoc. for Hydraulic Research, 1953.

Govier, G. W., B. A. Radford, and J. S. C. Dunn, *Can. J. Chem. Eng.*, **35**, 58 (1957).

Govier, G. W., G. A. Sullivan, and R. K. Wood, *Can. J. Chem. Eng.*, **39**, 67 (1961).

Hughmark, G. A., and B. S. Pressburg, *A.I.Ch.E. Journal*, **7**, 677 (1961).

Jones, J. H., W. G. Braun, T. E. Daubert, and H. D. Allendorf, *A.I.Ch.E. Journal* **13**, 608 (1967).

Kraiko, A. N., and L. E. Sternin, *Prikladnaya Matematika i Mekhanika*, **29**, 418 (1965), English translation *J. Appl. Math. Mech.*, **29** :482 (1965).

Marchaterre, J. F., and B. M. Hoglund, *Nucleonics*, **20**, 142 (1962).

Martinelli, R. C., L. M. K. Boelter, T. H. M. Taylor, E. G. Thomsen, and E. H. Morrin, *Trans. A.I.M.E.*, **66**, 139 (1944).

Nassos, G. P. and S. G. Bankoff, *Chem. Eng. Sci.*, **22**, 661 (1967).

Neal, L. G., An Analysis of Slip in Gas-Liquid Flow Applicable to the Bubble and Slug Flow Regimes. Report K.R.-62, Kjeller Research Establishment, Kjeller, Norway, Dec. 1963.

Newitt, D. M., J. F. Richardson, and B. J. Gliddon, *Trans. Inst. Chem. Engrs.*, **39**, 93 (1961).

Panton, R., *J. Fluid Mech.*, **31**, 273 (1968).

Petrick, M., and A. A. Kudirka, Third Int. Heat Transfer Conf., *Proc.*, **4**, 184, Chicago, Ill., 1966.

Ros, N. C. J., *J. Petrol Technol.*, **13**, 1037 (1961).

Sparenberg, J. A., *Appl. Sci. Res., Sec. A*, **15**, 221 (1965).

Standart, G., *Chem. Eng. Sci.*, **19**, 227 (1964).

Teletov, C. G., *Vestn. Mosk. Univ. Ser.* **2**, (1958).

Truesdell, C., and Toupin, R. A., Classical Field Theories, in *Encyclopedia of Physics*, Vol. 3, Springer, Berlin, 1960.

van Deemter, J. J., and E. T. van der Laan, *Appl. Sci. Res., Sec. A*, **10**, 102 (1961).

van Wijngaarden, L., *J. Fluid Mech.*, **33**, 465 (1968).

Vohr, J., *A.I.Ch.E. Journal*, **8**, 280 (1962).

Zuber, N., and J. A. Findlay, *J. Heat Transfer, Trans. A.S.M.E., Ser. C*, **87**, 453 (1965).

The Vertical Flow of Gas-Liquid and Liquid-Liquid Mixtures in Pipes

8

1 INTRODUCTION

The vertical pipe flow of gas-liquid and liquid-liquid mixtures is widely encountered in the petroleum industry, the chemical process industry, steam generation equipment, and nuclear reactor design.

In the petroleum industry the production of gas or oil through wells almost invariably involves the flow of mixed fluid phases. In the case of a well producing gas, frequently at least small amounts of water in the liquid phase and/or light liquid hydrocarbons are produced simultaneously and the flow mixture is one of two or even three phases. The gas-liquid ratio is high but the presence of even a small amount of liquid has a significant effect on the flow. With many oil wells gas is produced simultaneously, and often water as well. Again the flow mixture is one of two or even three phases. In both these applications the flow column is ordinarily the "production tubing," a jointed pipe of uniform diameter, but occasionally flow through the annulus between the "casing" and the production tubing is also encountered. The length of the vertical flow column is the depth of the well—which may range from less than 1000 to 15,000 feet or more. Typical inside diameters are from $1\frac{1}{4}$ to about 4 inches. The solubility of the gas in the liquid and its change with pressure are important factors.

In the chemical process industry, in steam generation equipment, and in nuclear reactor design, gas-liquid flow is frequently encountered with simultaneous heat transfer and arises from the partial vaporization of an initially single phase liquid. Here the problem is complicated by the continuously changing proportion of the mixture that is present in the gas phase. In such applications the flow column is relatively short—often no more than 20 feet and seldom in excess of 100 feet.

2 TYPICAL FLOW PATTERNS, HOLDUP, AND PRESSURE GRADIENT

Experimental work has been carried out extensively since 1950 with many different gas-liquid, but only a few liquid-liquid, systems, with the result that the various flow patterns and the general holdup and pressure drop behavior which may be encountered are well known. The typical situation may most easily be described by referring to a simple case and considering the behavior over a short length of flow column such that effects due to solubility changes need not concern us. For the gas-liquid case, the air-water system flowing in a 1-inch inside diameter tube is suitable, and data of Govier, Radford, and Dunn (1957) taken at a constant temperature and a constant average pressure (such that the average density of the air is constant) are convenient to depict typical behavior. For the vertical flow of two immiscible liquids, the data of Govier, Sullivan, and Wood (1961) for a water and a 20-centipoise oil system in a 1-inch-diameter tube are as comprehensive as any and are suitable to illustrate the general behavior.

2.1 Flow Patterns

Figure 8.1 and Table 8.1 summarize in scale drawings and tabulated form the results of photographic observations of the flow patterns observed at low, medium, and high water rates, and over a range of air rates. Some interpolation and a modest amount of extrapolation of original data have been made for convenience in this representation, but all diagrams reflect *in situ* volume fractions of the phases consistent with measured holdup data. The drawings depict the flow patterns as they would appear in a short time-exposure photograph: thus all bubbles or droplets show when they are widely spaced, but less than one half are visible when the clearance between them becomes small.

At the lowest water rates Govier *et al.* (1957) distinguished six different flow patterns: bubble, slug, froth, ripple (or semiannular), film (or annular), and mist, each occurring at successively higher air rates. We will treat the last

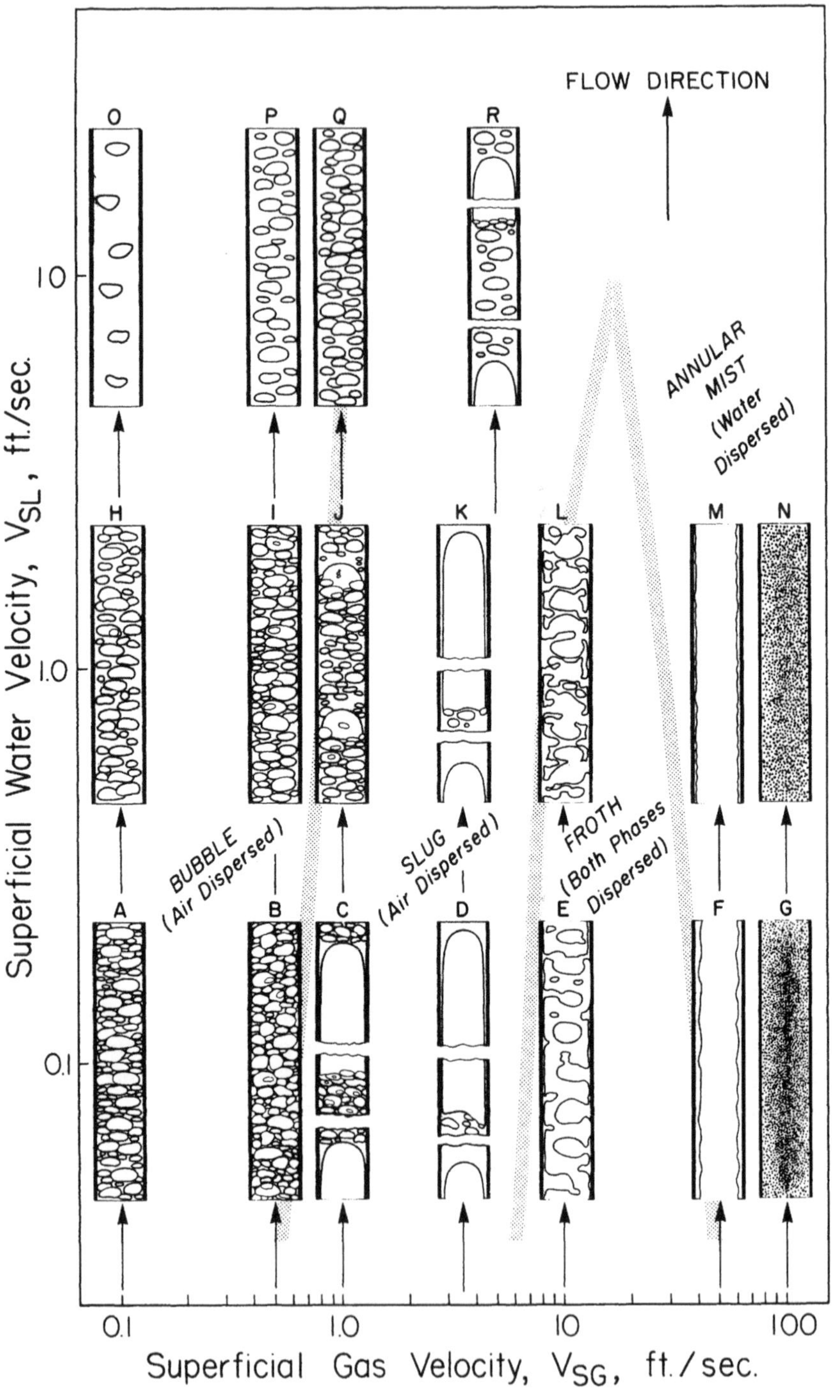

Figure 8.1. Flow patterns for air-water mixtures flowing in a 1.025-inch pipe based on observations of Govier, Radford, and Dunn, 1957, and calculations.

TABLE 8.1

Flow Pattern Details for Air-Water Mixture Flowing in 1.025 Inch I.D. Vertical Pipe (See Figure 8.1)

Sketch Identification	Flow Pattern	Continuous Phase	Superficial Velocity Water, V_{SL} ft/sec	Superficial Velocity Air, V_{SG} ft/sec	Volume Fraction Air, E_G	Holdup Ratio, H	Slip Velocity, S ft/sec
A	bubble	water	0.1	0.1	0.47	1.15	0.03
B	bubble	water	0.1	0.5	0.67	2.50	0.45
C	slug	water	0.1	1.0	0.70	4.30	1.10
D	slug	water	0.1	3.5	0.75	11.6	4.26
E	froth	neither	0.1	10.0	0.77	29.0	12.5
F	annular mist	air	0.1	50.0	0.96	20.5	49.5
G	annular mist	air	0.1	100.0	0.988	12.5	93.1
H	bubble	water	1.0	0.1	0.09	1.05	0.05
I	bubble	water	1.0	0.5	0.294	1.20	0.28
J	slug	water	1.0	1.0	0.43	1.35	0.61
K	slug	water	1.0	3.5	0.62	2.15	3.02
L	froth	neither	1.0	10.0	0.69	4.40	11.1
M	annular mist	air	1.0	50.0	0.877	7.0	48.9
N	annular mist	air	1.0	100.0	0.93	7.0	91.7
O	bubble	water	10.0	0.1	0.01	1.01	0.10
P	bubble	water	10.0	0.5	0.04	1.1	1.04
Q	bubble	water	10.0	1.0	0.08	1.13	1.41
R	slug	water	10.0	5.0	0.27	1.35	4.80

Sketch Identification	Flow Pattern	Approximate Only: Diameter of Bubbles or Droplets, d_s, in.	Number of Bubbles or Droplets, n_b[a]	Average Film Thickness, δ_{av}, in.	Length of Liquid Slug, L_s, in.	Length of Taylor Bubble, L_b, in.
A	bubble	0.25	540	—	—	—
B	bubble	0.25	780	—	—	—
C	slug	0.25	350	—	8	18
D	slug	0.2–0.25	150	—	8	32
E	froth	0.2–0.5	30	0.025	—	—
F	annular mist	—	—	0.08	—	—
G	annular mist	0.027	2100	0.0025	—	—
H	bubble	0.25	100	—	—	—
I	bubble	0.25	350	—	—	—
J	slug	0.25	400	—	3	0.6
K	slug	0.2–0.25	70	—	8	16
L	froth	0.2–0.5	80	0.035	—	—
M	annular mist	—	—	0.015	—	—
N	annular mist	0.05	1300	0.015	—	—
O	bubble	0.25	12	—	—	—
P	bubble	0.25	50	—	—	—
Q	bubble	0.25	95	—	—	—
R	slug	0.25	45	—	8	3.7

[a] Per foot of 1-inch pipe; in the slug flow pattern it refers to the small bubbles in the liquid slug only.

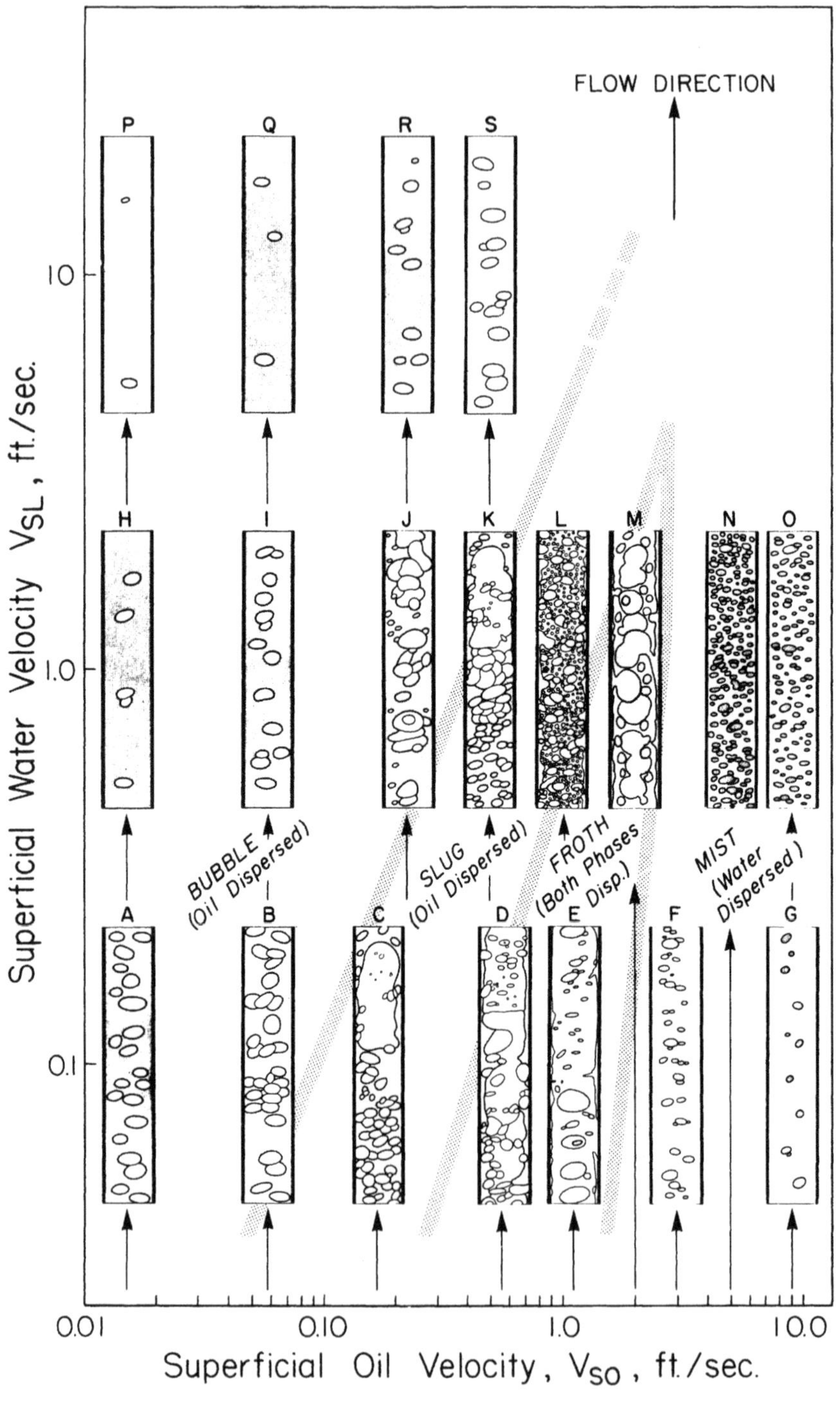

Figure 8.2. Flow patterns for 20.09-centipoise viscosity, 0.851-specific gravity oil, and water mixtures in a 1.04-inch pipe based on observations of Govier, Sullivan, and Wood, 1961.

TABLE 8.2

Flow Pattern Details for 20.09-Centipoise Viscosity, 0.851-Specific Gravity Oil Flowing with Water in 1.04-Inch I.D. Vertical Pipe (See Figure 8.2)

Sketch Identification	Flow Pattern	Continuous Phase	Superficial Velocity Water, V_{SL} ft/sec	Superficial Velocity Oil, V_{SO} ft/sec	Volume Fraction Oil, E_O	Holdup Ratio, H	Slip Velocity, S ft/sec
A	bubble	water	0.1	0.015	0.08	1.70	0.08
B	bubble	water	0.1	0.058	0.17	2.85	0.22
C	slug	water	0.1	0.167	0.32	3.60	0.38
D	froth	neither	0.1	0.562	0.67	2.80	0.54
E	froth	neither	0.1	1.11	0.87	1.62	0.49
F	mist	oil	0.1	2.94	0.97	0.84	−0.58
G	mist	oil	0.1	9.0	0.99	0.72	−3.53
H	bubble	water	1.0	0.015	0.01	1.26	0.26
I	bubble	water	1.0	0.058	0.035	1.60	0.62
J	bubble	water	1.0	0.22	0.114	1.71	0.80
K	slug	water	1.0	0.49	0.24	1.52	0.69
L	froth	neither	1.0	1.61	0.58	1.17	0.40
M	froth	neither	1.0	2.0	0.65	1.10	0.28
N	mist	oil	1.0	5.0	0.85	0.92	−0.52
O	mist	oil	1.0	9.0	0.909	0.90	−1.10
P	bubble	water	10.0	0.015	0.0014	1.05	0.50
Q	bubble	water	10.0	0.058	0.005	1.10	1.00
R	bubble	water	10.0	0.22	0.02	1.10	1.02
S	bubble	water	10.0	0.49	0.04	1.10	1.04

Sketch Identification	Flow Pattern	Approximate Only: Diameter of Bubbles or Droplets, d_s, in.	Number of Bubbles or Droplets, n_b[a]	Average Film Thickness, δ_{av}, in.	Length of Liquid Slug, L_s, in.	Length of Taylor Bubble, L_b, in.
A	bubble	0.35	35	—	—	—
B	bubble	0.37	60	—	—	—
C	slug	0.22	110	—	8.0	2.3
D	froth	—	—	0.082	—	—
E	froth	0.2–0.46	40	0.015	—	—
F	mist	0.20	65	—	—	—
G	mist	0.20	20	—	—	—
H	bubble	0.30	8	—	—	—
I	bubble	0.30	25	—	—	—
J	bubble	0.40	30	—	—	—
K	slug	0.20	140	—	8.0	1.8
L	froth	0.1–0.25	1500	0.06	—	—
M	froth	0.2–0.5	75	0.045	—	—
N	mist	0.20	365	—	—	—
O	mist	0.20	215	—	—	—
P	bubble	0.20	3	—	—	—
Q	bubble	0.25	6	—	—	—
R	bubble	0.30	15	—	—	—
S	bubble	0.35	20	—	—	—

[a] Per foot of 1 inch pipe; in the slug flow pattern it refers to the small bubbles in the liquid slug only.

three flow patterns as one under the heading of annular mist. At the higher water rates the froth flow pattern is missing and slug flow gives way to annular-mist flow at higher air velocities.

The following detailed description of the various flow patterns is taken with some modification from Govier *et al.* (1957).

At the low liquid velocities, as the gas rate is increased from zero to superficial gas velocities up to about 1 or 2 ft/sec, the gas is dispersed as discrete bubbles which increase steadily in number and, with coalescence, may increase in size. This is the bubble flow pattern. With a further increase in gas flow some of the bubbles coalesce to form larger, cap-shaped bubbles nearly spanning the tube. This marks the beginning of the slug flow pattern. As the gas rate is further increased, these bubbles become longer and assume a bullet shape. The large bubbles are called Taylor bubbles (see Section 5.1). The slug flow pattern is characterized by alternating bullet-shaped bubbles of gas which are surrounded by a thin liquid annulus and interspersed with slugs of liquid containing small bubbles of gas. As the gas velocity is increased still further, the large gas bubbles grow in length until they become unstable and collapse under the pressure gradient created by the flowing liquid film. At this point the bubbles and the slugs become highly distorted and seem to merge and degenerate into a froth. The froth pattern is highly turbulent, with coarse agitation of the liquid phase being its most predominant characteristic.

As the gas velocity is further increased, the liquid distributes itself between an upward moving wavy or ripply layer of liquid on the tube wall and coarse droplets carried in the continuous gas phase. This is the beginning of the annular-mist flow pattern. At slightly higher gas velocities, the liquid film becomes relatively smooth as a result of attrition of liquid into the faster moving gas and a greater fraction of the liquid is carried in droplet form by the gas. Eventually, at some sufficiently high gas rate, a sufficient number of water droplets is captured from the liquid interface to remove the liquid layer and to result in a region of pure mist flow.

The above description of flow patterns does not apply in its entirety to superficial water rates in excess of 7 or 8 ft/sec. At these higher water rates the flow pattern of the liquid phase seems much the same as at the lower rates, but the gas phase pattern changes markedly. In the slug flow region the gas phase, as before, is characterized by an agitated pattern that extends the entire length of the tube, with fast moving slugs traveling intermittently through the system. As the gas rate increases these slugs move faster but decrease in relative length. No froth flow pattern is observed and annular-mist flow develops directly at superficial gas velocities of about 20 ft/sec.

Figure 8.2 and Table 8.2 show in scale drawings and tabulated form the results of photographic observations of the flow patterns observed by Govier *et al.* (1961) for a liquid-liquid system. The general comments made with respect to Figure 8.1 also apply to Figure 8.2.

We notice a striking similarity between the two sets of drawings and see that the same sequence of flow patterns is observed with the liquid-liquid system as with the gas-liquid system. Such differences as are apparent in the two flow patterns are because of the differences in the viscosities, densities, and interfacial tensions. The difference is most apparent in the case of annular-mist flow where the oil-water system exhibits much fewer but larger water droplets than are shown in the air-water system. It is interesting to see that at any water rate the air-water system shows the same flow pattern as the oil-water system when the air rate is about ten times the oil rate.

With respect to both Figures 8.1 and 8.2 and others to be shown later, we should note that the bubble and droplet sizes and numbers are not uniquely determined by the flow rates and properties but are influenced by the entrance device and the travel distance from it. For example, bubble flow could result in patterns with, say, 50 percent fewer or more bubbles 50 percent larger or smaller than those shown. Additionally, as is discussed later, the average size of the bubbles may be increased and their number decreased as a result of bubble collision, adherence, and coalescence.

2.2 Holdup

Turning to the holdup phenomenon, which is closely related to the flow pattern, data of Govier *et al.* (1957) for the air-water system are shown in Figure 8.3 along with the approximate flow boundaries. The marked increase in holdup ratio with decrease in superficial water velocity is to be noted, together with the maximum in the curves at constant water rates which occurs in the middle of the froth flow region. We also observe the tendency of the holdup ratio to decline at the highest air rates where annular-mist flow occurs.

The data of Govier *et al.* (1961) illustrate the phenomena for a liquid-liquid system. These are shown in Figure 8.4. Although the magnitude of the holdup ratio is much less than that observed with the air-water system, the general trends shown in Figure 8.4 are similar to those of Figure 8.3. One notable difference is the decline of the holdup ratio to below unity at the highest oil rates. This corresponds with the change in flow pattern to one in which the oil phase becomes continuous.

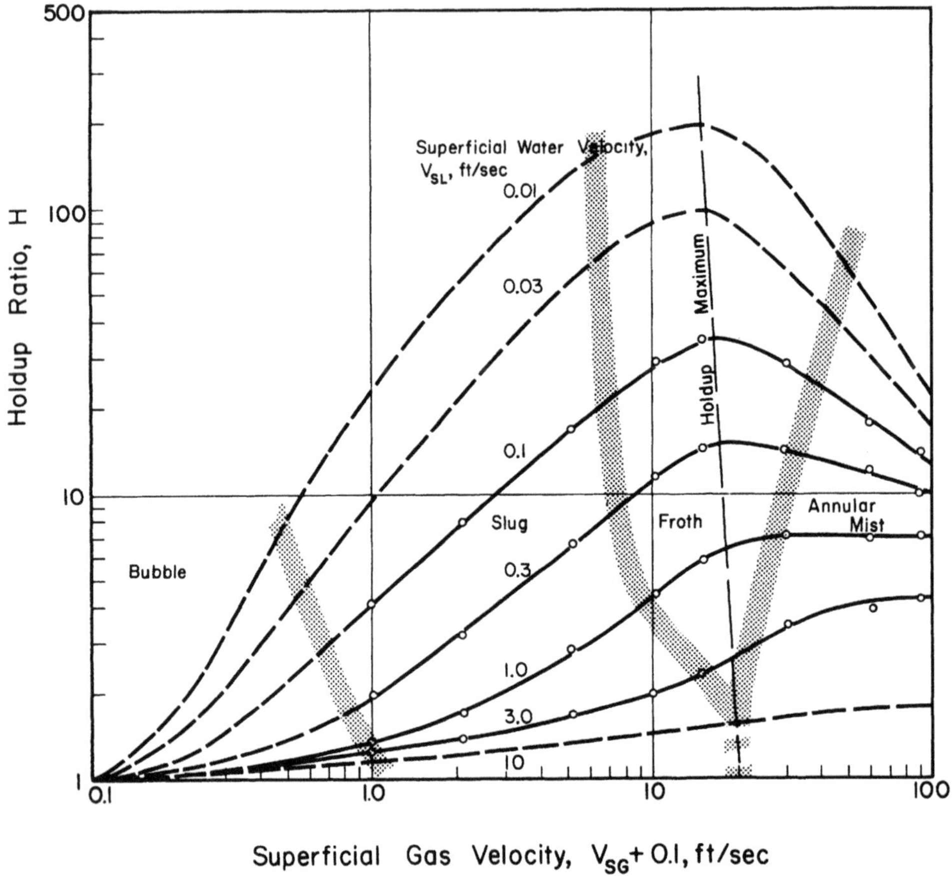

Figure 8.3. Holdup data for air-water mixtures in a 1.025-inch pipe. (Replotted from Govier, Radford, and Dunn, 1957.)

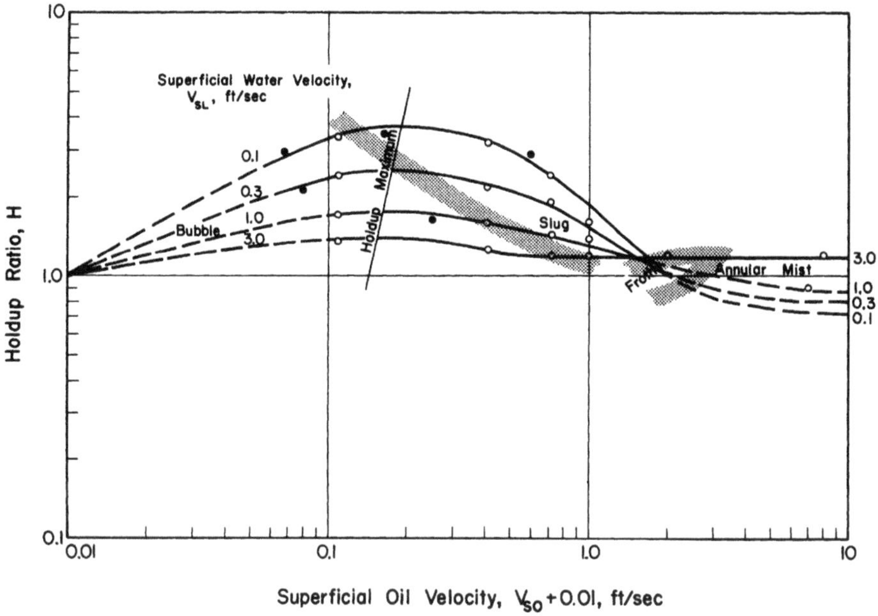

Figure 8.4. Holdup data for 20.09-centipoise viscosity, 0.851-specific gravity oil, and water mixtures in a 1.04-inch pipe. (Replotted from Govier, Sullivan, and Wood, 1961.)

2.3 Pressure Gradient

To obtain an appreciation of the trends of the pressure gradient with the velocities of the two phases, it is useful to examine both the total pressure gradient, including the potential energy component, and that due to irreversibilities alone. For this purpose the kinetic energy effects, which are usually small in any event, may be neglected and Equation 7.60 may be written

$$\Delta P = \frac{M_L + M_G}{Q_L + Q_G}\left(\frac{g}{g_c}\right)\Delta z + \Delta P_f \tag{8.1}$$

where ΔP = total pressure drop over the length Δz,
ΔP_f = pressure drop due to irreversibilities over the length Δz,
M_L = mass flow rate of the liquid phase,
M_G = mass flow rate of the gas phase,
Q_L = volumetric flow rate of the liquid phase,
Q_G = volumetric flow rate of the gas phase.

Multiplying by $1/\rho_L \Delta z$ gives an expression for the gradients in terms of feet of liquid (or of the more dense liquid of a liquid-liquid mixture) per foot of height

$$\frac{\Delta P}{\rho_L \Delta z} = \frac{M_L + M_G}{Q_L + Q_G} \frac{g}{g_c \rho_L} + \frac{\Delta P_f}{\rho_L \Delta z} \tag{8.2}$$

The data of Govier *et al.* (1957) on the total pressure gradient, $\Delta P/\rho_W \Delta z$, for the air-water system in a 1-inch pipe are shown by the solid lines in Figure 8.5. Extrapolated and interpolated data are shown by heavy dashed lines. The light dashed line shows the calculated potential energy term, i.e.,

$$\frac{M_L + M_G}{Q_L + Q_G} \left(\frac{g}{\rho_W g_c} \right)$$

The flow pattern boundaries, as observed visually and shown in Figure 8.1, are depicted by bands reflecting an uncertainty of plus or minus about 10 percent. Several features of the pressure gradient phenomenon are apparent. First, at low gas and liquid rates, the potential energy term accounts for most of the total pressure gradient. Secondly, at low and intermediate liquid

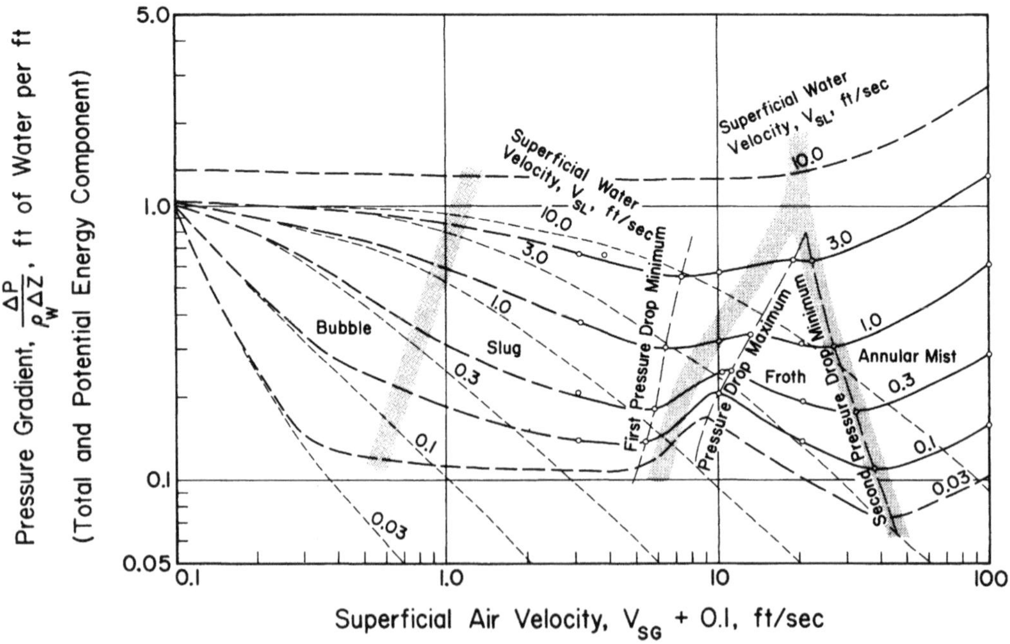

Figure 8.5. Pressure gradient data for air-water mixtures in a 1.025-inch pipe. (Replotted from Govier, Radford, and Dunn, 1957.)

rates, the total pressure gradient goes through a minimum, a maximum, and a second minimum with increasing gas rates; at higher liquid rates the maximum disappears and a single minimum occurs while at the highest liquid rates the gradient increases steadily, at increasing rate, with gas velocity. The loci of these maximum and minimum points are shown as light dashed lines. Govier *et al.* (1957) defined pressure drop regimes I, II, III, and IV by the loci lines such that regime I extends to gas flow rates defined by the first pressure drop minimum, regime II from the first pressure drop minimum to the pressure drop maximum, regime III from the pressure drop maximum to the second minimum, and regime IV beyond the second minimum. Thirdly, the loci lines in the total pressure gradient curve are related to changes in the flow patterns as illustrated by Figure 8.1. These relationships enabled Govier *et al.* (1957) to support their visual observations of flow pattern changes with the more quantitative detection of the pressure gradient minimum and maximum points.

The pressure gradient due to the irreversibilities, $\Delta P_f/\rho_W \Delta z$, i.e., the difference between the total gradient and that due to the potential energy term is shown in Figure 8.6. We should note, as discussed in Chapter 7, that the

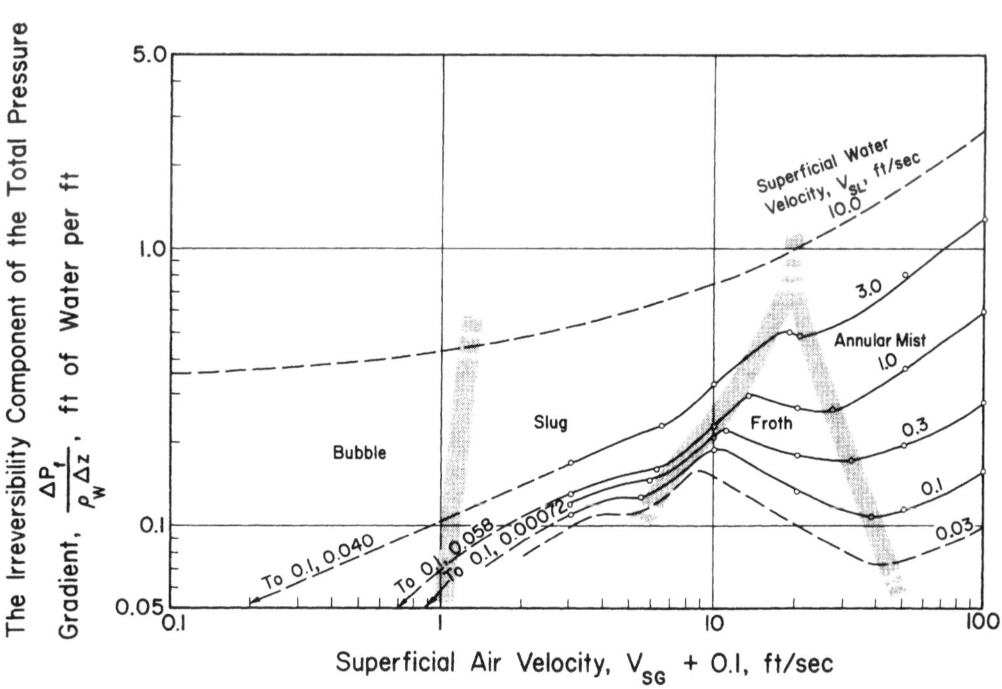

Figure 8.6. Irreversibility component of the pressure gradient for air-water mixtures in a 1.025-inch pipe. (Replotted from Govier, Radford, and Dunn, 1957.)

term $\frac{M_L + M_G}{Q_L + Q_G}\left(\frac{g}{\rho_L g_c}\right)$ is not the same as the hydrostatic head unless the holdup ratio is unity and therefore $\Delta P_f/\rho_L \Delta z$ or $\Delta P_f/\rho_W \Delta z$ includes the excess pressure gradient due to the holdup effect.

The pressure gradient characteristics of a liquid-liquid system are illustrated by the data of Govier *et al.* (1961) for water and a 20-centipoise oil flowing in a vertical, 1-inch tube. Figure 8.7 is the counterpart of Figure 8.5. Qualitatively, the picture obtained is much the same as for the air-water system. Quantitatively, because of the much lower density contrast, the minima reached in the low water rate curves are less distinct.

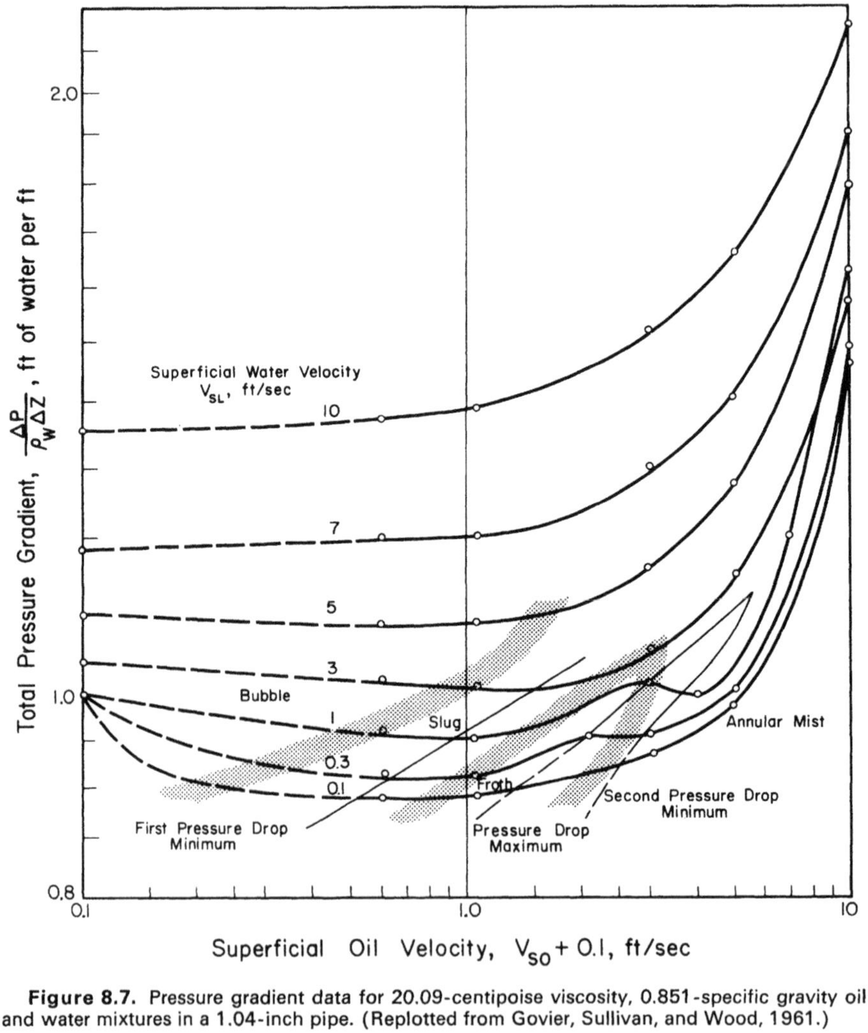

Figure 8.7. Pressure gradient data for 20.09-centipoise viscosity, 0.851-specific gravity oil and water mixtures in a 1.04-inch pipe. (Replotted from Govier, Sullivan, and Wood, 1961.)

3 EMPIRICAL OVERALL CORRELATIONS

A number of strictly empirical correlations of data of the type described in the preceding section have been developed. These cover, to varying degrees, fluid systems of different properties, different pipe diameters, and different ranges of flow rates. No completely satisfactory such correlations have yet been developed and, in view of the extreme complexity of the flow mechanism and its changes with flow rate, it is doubtful if there ever will be. Nonetheless, the empirical overall correlations are useful for orientation purposes, and where an analysis based upon more fundamental considerations cannot be made, they are all that are available to us.

3.1 Flow Patterns

Galegar, Stovall, and Huntington (1954), Kozlov (1954), Govier *et al.* (1957, 1961), Griffith and Wallis (1961), Duns and Ros (1963) and others have all made empirical attempts to define the conditions under which the various flow patterns may be expected. This has been done by proposing "flow pattern maps" of various kinds.

The flow pattern diagram of Griffith and Wallis (1961) is reproduced in Figure 8.8. Their regions, I, II, and III correspond to the bubble, slug and froth (or semiannular), and annular-mist regions; the froth flow pattern is not separately identified but it would be the transition from region II to III. The Duns and Ros diagram is presented later as Figure 8.19 in connection with the discussion of their pressure drop correlation.

It seems clear that for any given fluid system the major factors in determining the flow pattern are the flow velocities. The fluid densities, viscosities, and

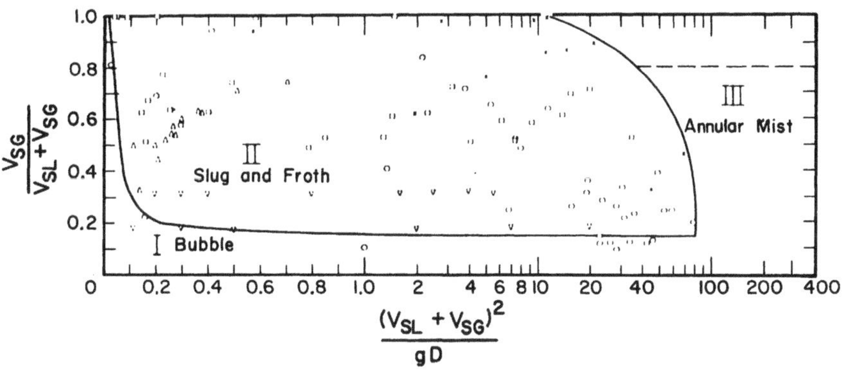

Figure 8.8. Griffith and Wallis flow pattern map for the flow of gas-liquid (air-water) mixtures. (From Griffith, P., and G. B. Wallis, *J. Heat Transfer, Trans. A.S.M.E., Ser. C.*, **83**, 307, 1961.)

interfacial tension and the pipe diameter are the other factors involved and their effects are not well known, although Duns and Ros (1963) claim that the viscosities of the phases are not important in determining the flow pattern, and that the effect of liquid density and of interfacial tension is not great. Their correlation contains the term $(\rho_L/\sigma)^{1/4}$ in both ordinate and abscissa. Govier *et al.* (1961) confirm that the viscosity of the phase not in contact with the wall is not important. Brown, Sullivan, and Govier (1960) show the effect of gas-phase density to be proportional to $(\rho_G)^{1/3}$ over the small density range of 0.092 to 0.552 lb_m/ft^3. The effect of diameter on flow pattern is claimed to be negligible by Duns and Ros. Brown *et al.* (1960) confirmed that it was small in the bubble flow region but observed some effect of diameter in the transitions from slug to froth and from froth to annular-mist flow.

All matters considered, it would appear that the best, simple, empirical correlation of the flow pattern transitions at least for gas-liquid systems might be expected on coordinates, YV_{SL} versus XV_{SG}, where

$$Y = (\rho_L \sigma_{WA}/\rho_W \sigma)^{1/4},$$
$$X = (\rho_G/\rho_A)^{1/3}\, Y,$$

ρ_L = density of the liquid under flowing conditions,
ρ_G = density of the gas under flowing conditions,
σ = interfacial tension of the liquid-gas system at flowing conditions,
σ_{WA} = interfacial tension of the water-air system at standard atmospheric conditions (taken as 60°F and 14.65 psia)
ρ_A = density of air at standard atmospheric conditions,

and where we would anticipate the boundaries to be bands whose widths are somewhat influenced by the tube diameter. Note that such a correlation gives a direct reading in V_{SL} and V_{SG} for the air-water system at atmospheric conditions.

A flow pattern map based on these considerations and the data of Govier *et al.* (1957), and generally compatible with the diagrams of Duns and Ros (1963), and Griffith and Wallis (1961) is presented in Figure 8.9. The relationships between the loci of the pressure gradient minimum and maximum points and the flow pattern boundaries have been used as a guide in the preparation of the correlation. The shaded bands reflect estimated uncertainties of about plus or minus 10 percent. The boundaries on this diagram are well substantiated by visual observations and the pressure gradient maximum and minimum points for the air-water system. They also show good order of magnitude agreement with the flow pattern boundaries determined by Govier *et al.* (1961) for their oil-water systems with oil viscosities of 0.936, 20.1, and 150 centipoises.

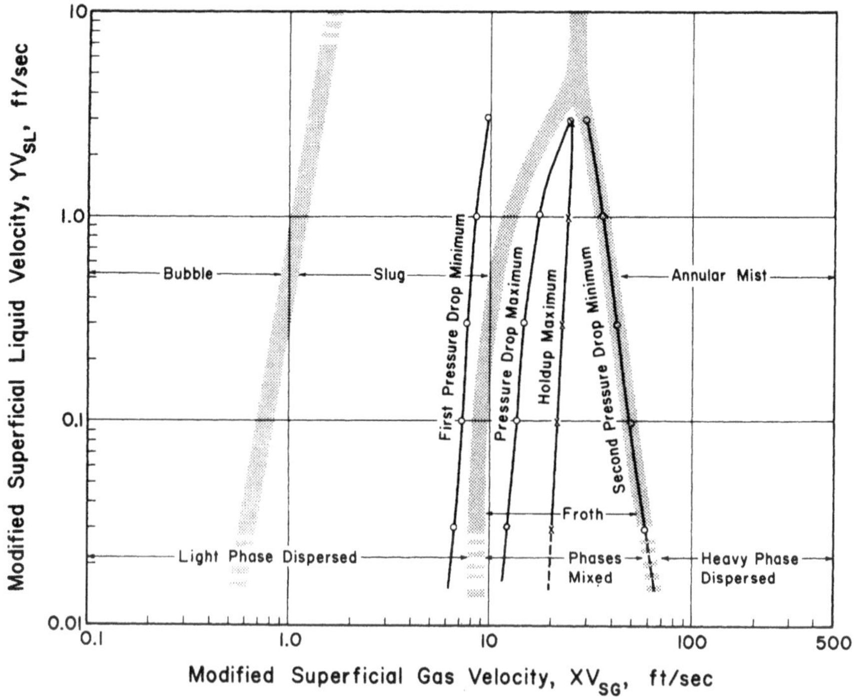

Figure 8.9. Generalized flow pattern map for flow of gas-liquid mixtures.

The boundaries of Figure 8.9 are not directly comparable with those proposed by Griffith and Wallis (1961) or Duns and Ros (1963). The bubble-slug transition is considerably different from that suggested by Griffith and Wallis, but they themselves expressed doubt about the location of their boundary, whereas the transition in Figure 8.9 is well confirmed. The froth to annular-mist transition agrees closely with that of Griffith and Wallis at superficial liquid velocities below 1.5 ft/sec but, at higher liquid velocities, it indicates higher superficial gas velocities than does their correlation. Duns and Ros define their region I to include the bubble and part of the slug region and, considering this, their correlation of region I is compatible with Figure 8.9. If the froth to annular-mist transition is taken as the equivalent of Duns and Ros' region II to region III transition, agreement exists at superficial liquid velocities below about 0.5 ft/sec, but Figure 8.9 predicts lower superficial gas velocities at higher rates.

Figure 8.9 is recommended for a simple overall assessment of the flow pattern to be expected under particular conditions. The correlation may be used with reasonable confidence for all gas-liquid systems and as a rough approximation for liquid-liquid systems.

3.2 Holdup

Empirical correlations for holdup (or the other directly related quantities, i.e., *in situ* volume fraction of the gas phase or light phase and the slip velocity) have also been developed by a number of workers. Galegar *et al.* (1954) presented one of the earliest, although restricted, correlations covering their air-water and air-kerosene data in a 2-inch pipe. Only the most recent correlations and those proposed as applicable over the full range of flow patterns will be discussed here.

Hughmark and Pressburg (1961) have developed probably the best general holdup correlation for gas-liquid mixtures covering the widest range of physical properties and diameters. Their correlation was based on their own data on the vertical flow in a 1-inch pipe of air and water, saturated aqueous solution of sodium carbonate, kerosene, trichlorethylene, and two oils of viscosities 5.8 and 28.6 centipoises, and carefully selected data of others. They developed a relationship between $\bar{E}_G$, the volume average *in situ* volume fraction of the gas phase, and a quantity

$$X' = 6 \times 10^4 \left(\frac{V_{SL}\rho_L}{V_{SG}\rho_G}\right)^{0.9} \left(\frac{\mu_L^{0.19}\sigma^{0.205}\rho_G^{0.70}\mu_G^{2.75}}{G_M^{0.435}\rho_L^{0.72}}\right) \tag{8.3}$$

where $G_M = (V_{SL}\rho_L + V_{SG}\rho_G)$, mass velocity is in $lb_m/ft^2 sec$,
μ and σ are in gram-centimeter-second units, the velocities are in ft/sec, and the densities are in lb_m/ft^3.

The correlation given in Figure 8.10 reproduces the data upon which it was based with an average deviation of 0.022 in $\bar{E}_G$. The correlation does not include the influence of diameter, although it reproduces the data of Govier *et al.* (1957) for an air-water system in tubes of diameters ranging from 0.4 to 2.34 inches to an average deviation of 0.038 in $\bar{E}_G$.

Hughmark (1962) developed another empirical holdup correlation in which the Bankoff K factor, as discussed in Chapter 7 and related to $\bar{E}_G$ through

$$K = \bar{E}_G \frac{V_M}{V_{SG}} = \frac{\bar{E}_G}{C_G} \tag{7.83}$$

is correlated with

$$\mathscr{X} = \mathrm{Re}_H^{1/6}\,\hat{\mathrm{Fr}}_M^{1/8}\,C_L^{-1/4} \tag{8.4}$$

where

$$\mathrm{Re}_H = \frac{DG_M}{\bar{E}_G\mu_G + (1-\bar{E}_G)\mu_L} \tag{8.5}$$

$$\hat{\mathrm{Fr}}_M = \frac{V_M^2}{gD}$$

$$V_M = V_{SL} + V_{SG}$$

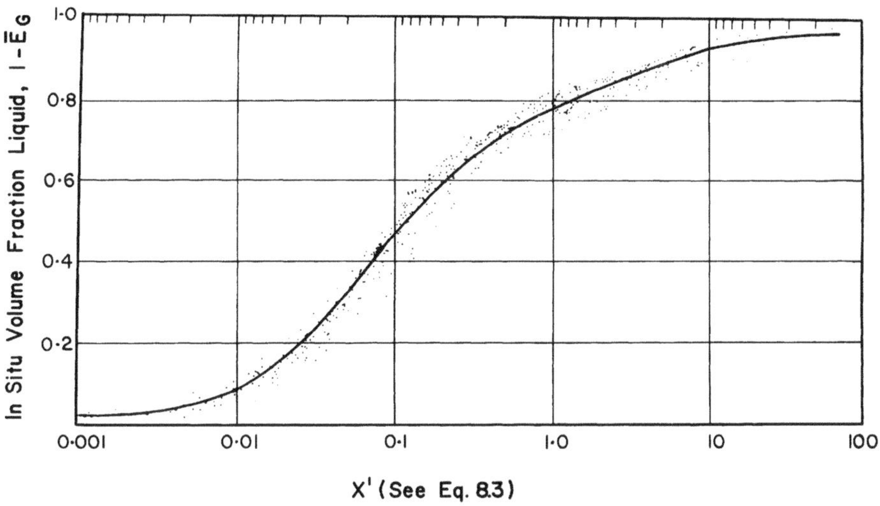

Figure 8.10. Hughmark and Pressburg holdup correlation for gas-liquid mixtures. (From Hughmark, G. A., and B. S. Pressburg, *A.I.Ch.E. Journal*, **7**, 677, 1961.)

Hughmark expressed Equation 7.83 in the equivalent form

$$\frac{M_L + M_G}{M_G} = 1 - \frac{\rho_L}{\rho_G}\left(1 - \frac{K}{\bar{E}_G}\right) \tag{8.6}$$

The correlation presented in Figure 8.11 was based on data for the vertical flow of air-liquid systems in tubes of diameters 0.63 to 2.5 inches. The average absolute deviation of results of the correlation when compared with other measured data was 0.04 in the value of $\bar{E}_G$ derived from K, although individual deviations were as high as 0.15. Consideration of the net exponents of the components of $\mathscr{L}$ shows the correlation to be insensitive to each of the components and that it contains a term involving $\bar{E}_G$ to about the same power as the other variables. With $\bar{E}_G$ being present, in effect, in both the ordinate and abscissa, and with the low sensitivity to what are known to be important variables, the correlation should not be counted upon for more than order of magnitude accuracy.

Ros (1961) and Duns and Ros (1963) have also developed an empirical correlation of holdup. Their correlation, which is in terms of slip velocity, is discussed in Section 3.32 in conjunction with their pressure drop correlation.

Brown *et al.* (1960) present a graphical correlation based upon their quite extensive air-water data for pressures to 110 psia and for diameters from $\frac{3}{4}$ to $2\frac{1}{2}$ inches. Although limited to air-water or closely similar systems at low pressures, it closely reproduces actual data and may be considered reliable for such systems.

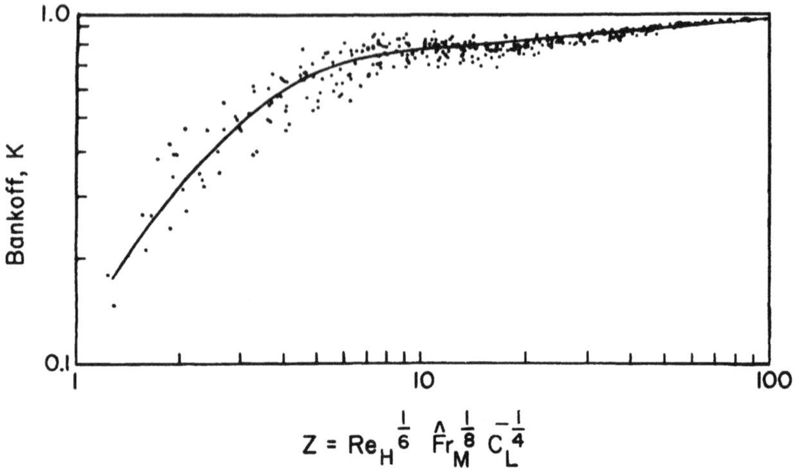

Figure 8.11. Hughmark holdup correlation for gas-liquid mixtures. (From Hughmark, G. A., *Chem. Eng. Progress,* **58**(4), 62, 1962.)

The holdup correlation of Chawla (1969) discussed in Section 3.2 of Chapter 10 may also be applied to vertical flow.

3.3 Pressure Gradient

A number of overall correlations have been developed for the prediction of the pressure gradient in the vertical flow of gas-liquid mixtures. These may be divided into two groups depending upon whether the analysis or correlation is based upon or related to Equation 3.22 or 7.60 involving only the input quantities and ΔP_{PE}, or upon Equation 7.73 involving ΔP_{HH} and an estimate of the density of the *in situ* mixture. In both categories we will consider first the pressure drop exclusive of any contribution due to kinetic energy or acceleration effects and account for these later.

3.31 Analyses or Correlations Not Involving Estimates of the *in situ* Concentration or Density

Correlations of this type include the pioneer work of Poettmann and Carpenter (1952) for the flow of natural gas, and crude oil with or without water; the extensions and modifications to the Poettmann-Carpenter correlation by Baxendell and Thomas (1961), Fancher and Brown (1963), and Hagedorn and Brown (1964); the correlation of Galegar *et al.* (1954); the work of Govier *et al.* (1957), Govier and Short (1958), Brown *et al.* (1960) on the air-water system; the correlation of Hughmark and Pressburg (1961); the correlation of Tek (1961) for the flow of natural gas, crude oil and water; the work of Gaither, Winkler, and Kirkpatrick (1963) on the flow of natural gas-water mixtures; and the work of Yocum (1959) on natural gas and crude oils.

The more important of these general empirical correlations are discussed in the following.

The Poettmann-Carpenter and Related Correlations The Poettmann-Carpenter correlation is based upon actual field data from 49 wells producing mixtures of oil, gas, and water at rates varying from 5 barrels of oil per day with a gas-oil ratio of 35,000 standard cubic feet per barrel to more than 1400 barrels of oil per day at a gas-oil ratio of 342. Most of the oil rates were less than 300 barrels per day. The gravity of the oil varied from about 30° to 54° API. The data were mostly for 2.0- and 2.5-inch flow tubing with two points for 3-inch tubing.

The starting point for the correlation is the mechanical energy balance for a single-phase fluid. The correlation assumes, in effect, that the flowing mixture is homogeneous, with a composition equal to that of the produced mixture, but whose density varies because of the variation with pressure and temperature of the solubility of the gas in the oil. Kinetic energy effects are assumed to be small and are neglected. No account is taken of the holdup phenomenon, and the hydrostatic head term is evaluated as for the assumed homogeneous flowing mixture.

The mechanical energy balance for a single-phase fluid with kinetic energy effects neglected (see development in Chapter 3) may be applied to the steady flow of the assumed homogeneous mixture

$$\int_{P_1}^{P_2} \frac{1}{\rho_M} dP + \frac{g}{g_c} \Delta z + \frac{\Delta P_f}{\rho_M} = 0 \tag{8.7}$$

where ρ_M is the density of the mixture (assuming no holdup). Expressing $\Delta P_f / \rho_M$ in terms of a friction factor, f_{PC}, through a Fanning type equation

$$\frac{\Delta P_f}{\rho_M} = \frac{2 f_{PC} \bar{V}^2 \Delta z}{g_c D} \tag{8.8}$$

where $\bar{V}$ = an assumed constant average value of the velocity V, taken as the integrated average value of V between the pressure limits of P_1 and P_2, where V is based upon the ratio of fluids flowing into and out of the flow tubing. Substitution of Equation 8.8 into Equation 8.7 yields

$$\Delta z = \frac{\int_{P_2}^{P_1} \frac{1}{\rho_M} dP}{1 + \frac{2 \bar{V}^2 f_{PC}}{gD}} \left(\frac{g_c}{g} \right) \tag{8.9}$$

which Poettmann and Carpenter also expressed in terms of "field units" as

$$\Delta z = \frac{\dfrac{1}{M'}\displaystyle\int_{P_2}^{P_1} v_M \, dP}{1 + \dfrac{f_{PC} Q'^2 \bar{v}_M^2}{7.413 \times 10^{10} D^5}} \tag{8.10}$$

where v_M = cubic feet of mixed gas, oil, and water at pressure P per barrel of stock-tank oil based on the ratio of the fluids flowing into and out of the flow tubing,

M' = total mass of gas, oil, and water, in pounds, associated with one barrel of stock-tank oil flowing into and out of the flow tubing,

Q' = barrels of stock-tank oil produced per day,

$\bar{v}_M$ = arbitrarily defined as constant and equal to the integrated average value of v_M between pressure limits of P_1 and P_2,

D = inside diameter of tubing, in feet.

The volume v_M associated with one barrel of produced stock-tank oil may be expressed as follows:

$$v_M = 5.61F + \frac{P_0 T_{av} Z}{P T_0}(S_p - S_s) + v_W \tag{8.11}$$

where F = formation-volume factor of the oil at pressure P, barrels of reservoir oil per barrel of stock-tank oil,

S_p = producing gas-oil ratio, cubic feet per barrel of stock-tank oil,

S_s = solubility of gas in oil at pressure P, cubic feet per barrel of oil,

v_W = cubic feet of water produced per barrel of stock-tank oil,

T_{av} = arithmetic average temperature in the flow tubing, degrees Rankine,

P = pressure, psia,

P_0 = reference or standard pressure, psia,

T_0 = reference or standard temperature, degrees Rankine,

Z = compressibility factor of the gas in the tubing at temperature, T_{av}, and pressure, P.

The total mass of flowing fluid associated with one barrel of stock-tank oil may be calculated from the following expression:

$$M' = 350G_{sto} + 0.0764G_g S_p + 62.4G_b v_W \tag{8.12}$$

where G_{sto} = specific gravity of stock-tank oil,

G_g = separator-gas gravity (air = 1),

G_b = specific gravity of produced water.

The term $\int_{P_2}^{P_1} v_M\,dP$ of Equation 8.10 is evaluated by graphical integration of a plot of v_M, calculated from Equation 8.11 versus P. The value of $\bar{v}_M$ is also determined graphically by evaluation of

$$\bar{v}_M = \frac{\int_{P_2}^{P_1} v_M dP}{P_1 - P_2} \tag{8.13}$$

Poettmann and Carpenter go further in the combination and rearrangement of Equations 8.10, 8.11, 8.12, and 8.13 and present tables applicable to hydrocarbon systems for the evaluation of the integrals involving the compressibility factor, Z. The original work should be consulted for these.

The final step in the development of the correlation was a calculation, from the field data mentioned, of the friction factor, f_{PC}, using the equivalent of Equations 8.10, 8.11, 8.12, and 8.13 and the appropriate relationships for F, S_s, and Z. This friction factor was then correlated with a reduced form of the Reynolds number, $DV\rho_M$, which in field units is

$$\mathrm{Re}_{PC} = 1.4737 \times 10^{-5}\left(\frac{M'Q'}{D}\right) \tag{8.14}$$

Poettmann and Carpenter reasoned that, in the highly turbulent flow normally encountered in oil and gas wells, the role of viscosity would be negligible. The final correlation is presented, as the line labeled "Poettmann-Carpenter" in Figure 8.12, a plot of f_{PC} versus Re_{PC}.

Despite its theoretical shortcomings and the rather limited data on which it is based, the Poettmann-Carpenter correlation has been successfully used in the petroleum industry and is generally considered capable of predicting pressure gradients within about 20 percent for tubing sizes, flow rates, and oil properties comparable to those on which it was based.

Baxendell and Thomas (1961) proposed the first modification to the Poettmann-Carpenter correlation in the form of an extension of the correlation to higher total mass flow rates. They conducted experiments in $2\frac{7}{8}$- and $3\frac{1}{2}$-inch tubings in a 6250-foot well at flow rates up to 5000 barrels per day, and at a nearly constant gas-oil ratio of some 750 cubic feet per barrel. Their data enabled them to adjust the lower end of the Poettmann-Carpenter f_{PC} versus Re_{PC} curve, and to extend it to higher values of Re_{PC}. The Baxendell-Thomas extension of the correlation is indicated by the line labeled "Baxendell and Thomas" in Figure 8.12.

Fancher and Brown (1963) conducted extensive field tests in an 8000 foot well equipped with $2\frac{3}{8}$-inch tubing. Their flow rates varied up to 1000 barrels per day with gas-liquid ratios up to 9400 cubic feet per barrel. In general, they

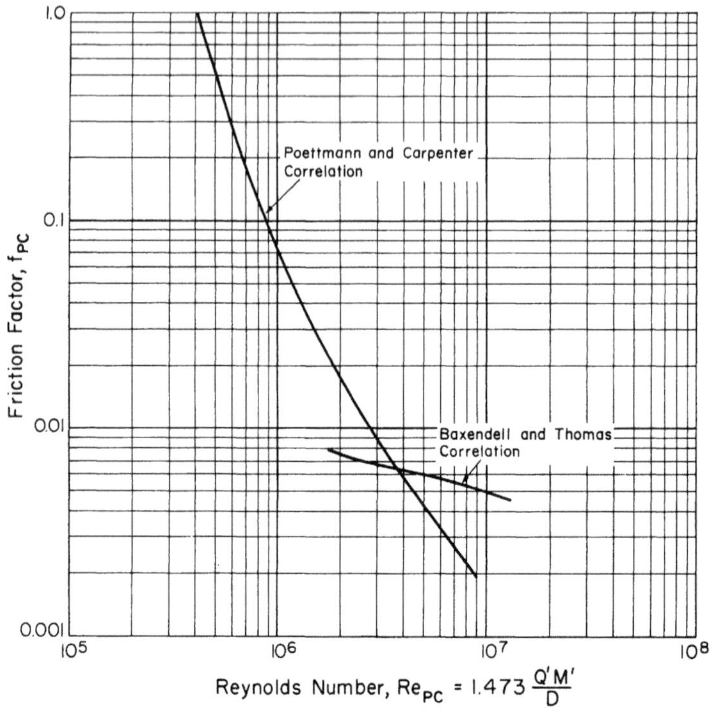

Figure 8.12. Poettmann-Carpenter and Baxendell-Thomas friction factor correlations. (Redrawn from Baxendell, P. B., and R. Thomas, *J. Petrol. Technol.*, **13**, 1023, 1961.)

confirmed the Poettmann-Carpenter correlation at low gas-liquid ratios but found consistent deviations exceeding 20 percent in the calculated pressure gradient at flow rates below 200 barrels per day at gas-oil ratios exceeding 1000 cubic feet per barrel. The correlation proposed by Fancher and Brown, with gas-liquid ratio as a parameter, is given in Figure 8.13.

Hagedorn and Brown (1964) propose a further modification of the Poettmann-Carpenter correlation, specifically for flow in small diameter tubing. They conducted tests with water, oils of various viscosities up to 110 centipoises, and gas in a 1500-foot experimental well equipped with $1\frac{1}{4}$-inch tubing. They demonstrated that the gradients were independent of liquid viscosity up to about 10 centipoises, but showed that higher liquid viscosities resulted in significantly higher pressure gradients. Following Fancher and Brown (1963) they also introduced the gas-liquid ratio as a correlating parameter to

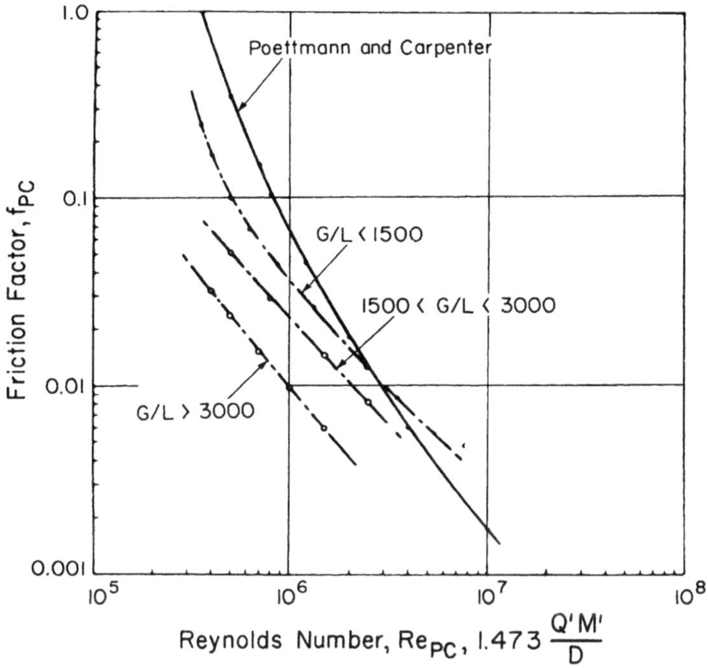

Figure 8.13. Fancher and Brown modification of the Poettmann-Carpenter friction factor correlation. (*G/L* in cubic feet per barrel.) (Redrawn from Fancher, G. H., and K. E. Brown, *Soc. Petrol. Engrs. J.*, **3**, 59, 1963.)

develop their final correlations for a $1\frac{1}{4}$-inch tubing, as shown in Figures 8.14 and 8.15.

Tek and Chan (1959) and Tek (1961) have taken an approach different from the others who have attempted to improve the Poettmann-Carpenter correlation. They applied the concept of a "two-phase Reynolds number," conceived and applied to horizontal flow by Bertuzzi, Tek, and Poettmann (1956), and developed a correlation of f_{PC} based upon it. Tek and Chan base their correlation on the test data for 31 of the 49 wells studied by Poettmann and Carpenter. These cover liquid flow rates up to 500 barrels per day, with gas-liquid ratios to 40,000 standard cubic feet per barrel in 2.0- and 2.5-inch tubings. Reasoning that the shortcomings of the basic Poettmann-Carpenter correlation were related to the omission of the viscosity effect and the lack of generality because of the dimensional character of Re_{PC}, they concluded that the two-phase Reynolds number offered possibilities for an improved correlation of f_{PC}. The two-phase Reynolds number first proposed by Bertuzzi, Tek, and Poettmann is defined as

$$Re_{BTP} = (Re_{SG})^{k/(1+k)}(Re_{SL})^{1/e^{0.1k}} \tag{8.15}$$

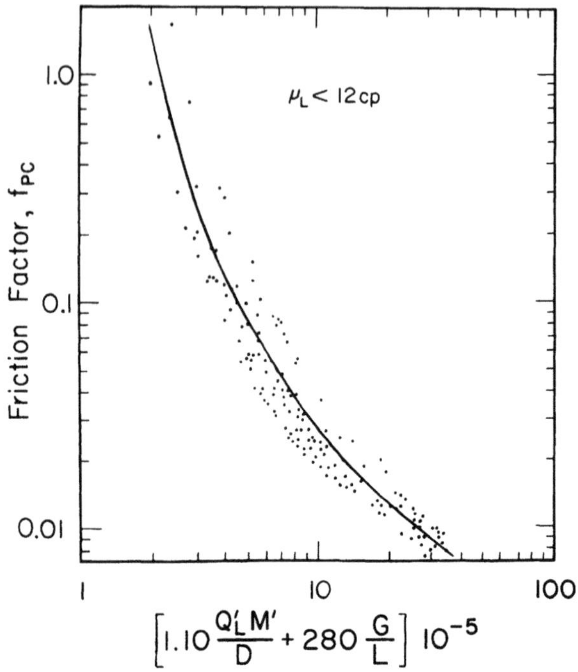

Figure 8.14. Hagedorn and Brown friction factor correlation for low viscosity liquids. (From Hagedorn, A. R., and K. E. Brown, *J. Petrol. Technol.*, **16**, 203, 1964.)

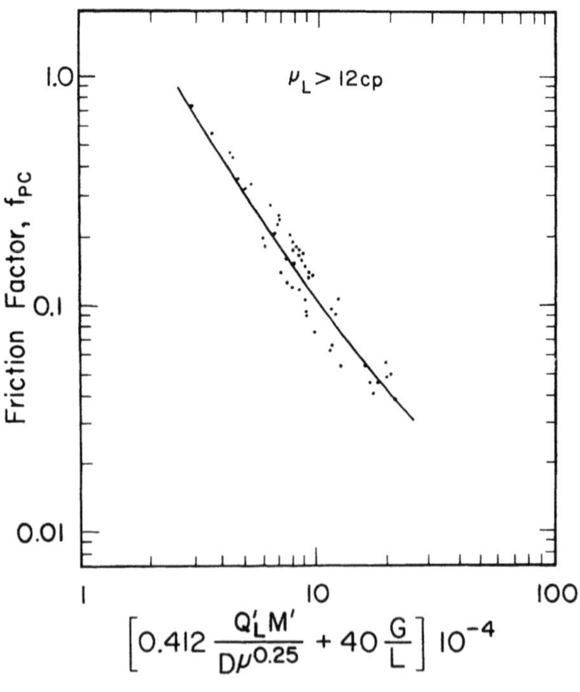

Figure 8.15. Hagedorn and Brown friction factor correlation for high viscosity liquids. (From Hagedorn, A. R., and K. E. Brown, *J. Petrol. Technol.*, **16**, 203, 1964.)

where Re_{BTP} = two-phase Reynolds number function defined by Equation 8.15, dimensionless,

$$Re_{SG} = \frac{DV_{SG}\rho_G}{\mu_G} = \text{superficial Reynolds number for the gas phase,}$$

$$Re_{SL} = \frac{DV_{SL}\rho_L}{\mu_L} = \text{superficial Reynolds number for the liquid phase,}$$

$$k = \frac{V_{SG}\rho_G}{V_{SL}\rho_L} = \text{mass ratio of gas to liquid, based on separator conditions and stock-tank quantities.}$$

The exponents on the individual Reynolds numbers were chosen by Bertuzzi, Tek, and Poettmann so that the two-phase Reynolds number function reduces to the well known single-phase Reynolds number whenever k is equal to zero or infinity.

Tek (1961) presents the results of the correlation in two equivalent ways; one is a plot of f_{PC} versus Re_{BTP} with the gas-liquid mass ratio as a parameter, and the other is a plot of f_{PC} versus a modified version of Re_{BTP} designed to bring the lines of various k values together. The first of the two forms is presented in Figure 8.16. Tek indicates that the data of the 31 wells are correlated by Figure 8.16 to within a bias of 2.2 percent and a standard deviation

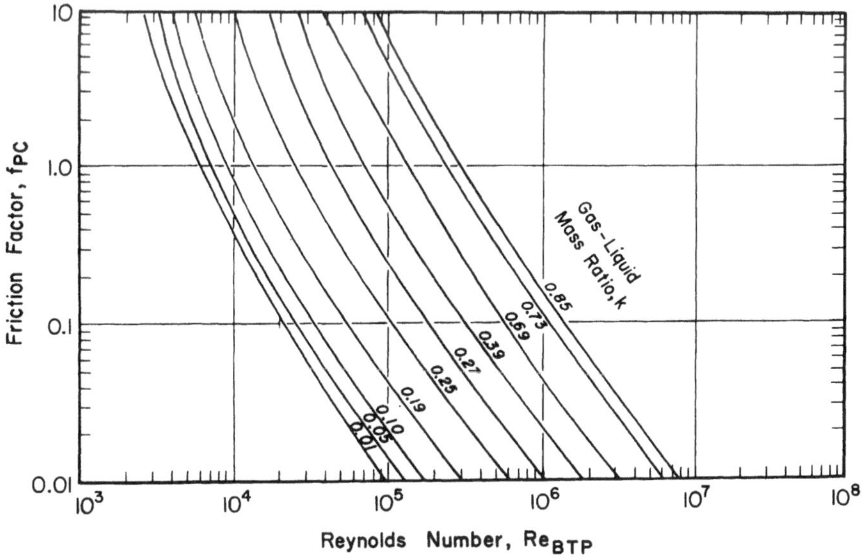

Figure 8.16. Tek and Chan friction factor correlation. (From Tek, M. R., *J. Petrol. Technol.*, **13**, 1029, 1961.)

of 7.3 percent. Individual deviations exceeded 20 percent. Despite the use of the additional parameter it seems questionable whether the Tek-Chan correlation is any better than the original Poettmann-Carpenter correlation as extended by Baxendell and Thomas, and others, which is based upon somewhat more data.

The Poettmann-Carpenter based correlations are useful for routine calculations in wells producing mixtures of water, oil, and gas. The work of Fancher and Brown, Hagedorn and Brown, and Tek and Chan confirms that the underlying assumptions of the method are severely limiting, and indicates that the interrelated effects of liquid viscosity, total mass flow rates, gas-liquid ratio, and tubing diameter cannot be properly accounted for. The several modifications are helpful in the range of conditions where they are applicable, but they fall far short of complete explanation. All the correlations in this group are based upon measured or estimated data at the bottom and the top of the well and therefore are of the integral form employing average velocities and properties (defined in special ways) and basically suited for predicting the overall pressure gradient in a single-step calculation. The correlations may be expected to yield predictions within about 20 percent and better in ranges when the case under consideration closely matches the test data.

The Govier* et al. *Correlation Govier *et al.* (1957), Govier and Short (1958), and Brown *et al.* (1960) developed a pressure gradient correlation for their data on an air-water system based upon the two-phase mechanical energy balance (Equation 7.60) neglecting kinetic energy effects, in the form

$$\frac{\Delta P}{\Delta z} = \frac{V_{SG}\rho_G + V_{SL}\rho_L}{V_M} + \frac{\Delta P_f}{\Delta z} \tag{8.16}$$

and Equation 7.64

$$\frac{\Delta P_f}{\Delta z} = \frac{2\rho_L V_{SL}^2}{g_c D} f_{tpL} \tag{7.64}$$

where f_{tpL} = two-phase friction factor. Their data cover the tube diameters 0.63, 1.025, 1.50, and 2.50 inches, air densities from 0.092 to 0.552 lb_m/ft^3, and a range of air and water rates covering all flow patterns except the annular mist. The correlation is based upon laboratory data taken at constant average pressure and temperature and it may therefore be considered a point condition correlation suitable for use in an integration procedure for significantly long pipe lengths. The correlation is in the form of the two-phase friction factor defined by Equation 7.64 versus a reduced form of the liquid Reynolds number (being the product of diameter and superficial water velocity, DV_{SL}) with a dimensional parameter, $D^{3.5}\,V_{SG}$. The correlation is reproduced in Figure 8.17. Although based solely on air-water data it should

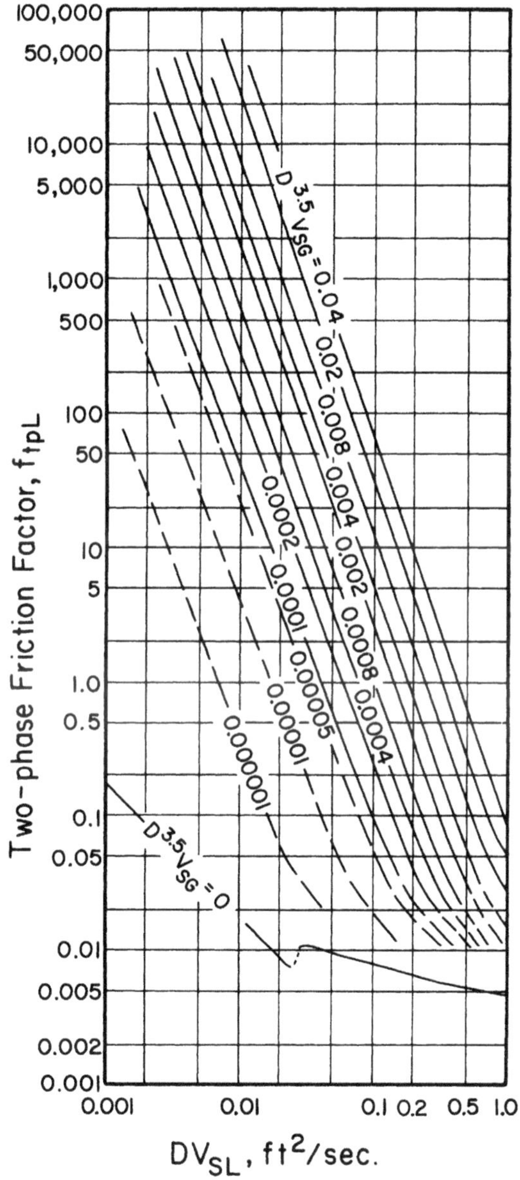

Figure 8.17. Govier *et al.* friction factor correlation for air-water mixtures. Solid lines based on data points, dashed lines interpolated. (From Brown, R. A. S., G. A. Sullivan, and G. W. Govier, *Can. J. Chem. Eng.*, **38**, 62, 1960.)

be applicable to reasonably similar gas-liquid systems by interpreting the abscissa as

$$DV_{SL}\left(\frac{\rho_L}{\rho_W}\right)\left(\frac{\mu_W}{\mu_L}\right)$$

where ρ_L and μ_L are the density and viscosity, respectively, of the liquid,
ρ_W and μ_W are the density and viscosity of water at approximately 70°F.

This correlation closely reproduces actual data for the air-water system and may be used with confidence for water with any gas at low or medium pressure, for all flow patterns except annular mist. No effect on friction factor was found for gas densities ranging from 0.092 to 0.552 lb_m/ft^3.

The Hughmark and Pressburg Correlation Hughmark and Pressburg (1961) have also developed a pressure gradient correlation based upon the mechanical energy balance applied to a two-phase system. They used Equation 7.60 in exactly the way it was used by Govier *et al.* to calculate the pressure gradient due to irreversibilities, but instead of defining a friction factor, they dealt with the excess pressure gradient. They correlated their own pressure gradient data on air with several liquids, together with data of others, in terms of the slip velocity (derived from their holdup correlation referred to in Section 3.2), the flow rates, the liquid and gas properties, and the diameter of the tube. The slip velocity, S, is related to $\bar{E}_G$ given by their holdup correlation through Equation 7.76:

$$S = \frac{V_{SG}}{\bar{E}_G} - \frac{V_{SL}}{1 - \bar{E}_G} \tag{7.76}$$

Their final correlation is $(\Delta P_{tp} - \Delta P_L)/\Delta z$ versus S with the following parameter:

$$\psi = \frac{1}{G_G^{0.7}\mu_L^{0.147}\sigma^{0.194}}\left(\frac{D}{0.0873}\right)^{0.5} \tag{8.17}$$

where ΔP_L = the pressure drop that would exist with liquid flowing alone at the total mass velocity of the mixture,
$G_G = V_{SG}\,\rho_G$, mass velocity of the gas phase, lb_m/ft^2 sec,
μ and σ are in gram-centimeter-second units and D is in feet.

The correlation is reproduced in Figure 8.18. To accommodate different pipe roughnesses, they proposed that ΔP_{tp} determined from the correlation be multiplied by the ratio, $f_L/(f_L)_{0.002}$, i.e., the friction factor that would be expected for the liquid flowing at the total mass velocity of the system in the actual pipe to that which would be expected if it were flowing in a pipe of

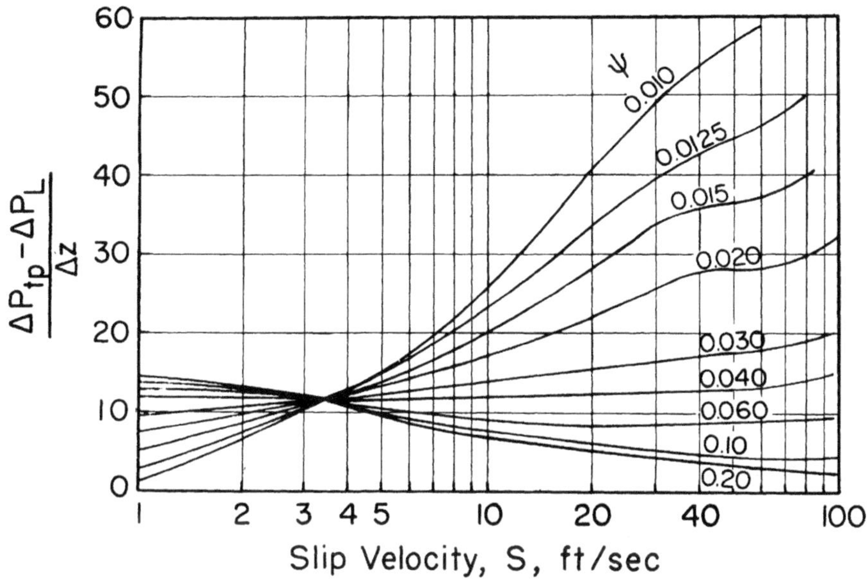

Figure 8.18. Hughmark and Pressburg pressure gradient correlation. (From Hughmark, G. A., and B. S. Pressburg, *A.I.Ch.E. Journal*, **7**, 677, 1961.)

relative roughness equal to 0.002. Like that of Govier *et al.*, the correlation is of the point condition type. The authors of the correlation indicated it to be accurate to within 15 percent for the range of data investigated. This covers a substantial range of liquid properties and pipe diameters from 0.5 to 2.0 inches. Larger deviations from measured values are indicated by Orkiszewski (1967). An unattractive feature of the method is its dependence upon a separate empirical correlation for slip velocity.

3.32 Analyses or Correlations Involving Estimates of the *in situ* Concentration or Density

Here the pioneer work is that of Ros (1961) extended by Duns and Ros (1963). Another well known correlation based on this approach is that of Hagedorn and Brown (1965). Several workers, including Griffith and Wallis (1961), have developed methods of this type for a single flow pattern. These are discussed later.

The Ros Correlation Ros (1961) and Duns and Ros (1963) have developed what is probably the best, general empirical correlation for the pressure drop in vertical gas-liquid flow. They start with the equivalent of Equation 7.73 written as

$$\Delta P = \Delta P_{HH} + \Delta P_{KE} + \Delta P_f^{\bullet} \qquad (8.18)$$

where

$$\Delta P_{HH} = [\bar{E}_G \rho_G + (1 - \bar{E}_G)\rho_L] \frac{g}{g_c} \Delta z,$$

$$\Delta P_f^\bullet = \frac{2 f_R V_{SL}^2 \rho_L}{g_c D} \left(1 + \frac{V_{SG}}{V_{SL}}\right) \Delta z \text{ when the liquid phase is continuous,}$$

$$\Delta P_f^\bullet = \frac{2 f_R V_{SG}^2 \rho_G}{g_c D} \left(1 + \frac{V_{SL}}{V_{SG}}\right) \Delta z \text{ when the gas phase is continuous,}$$

ΔP_{KE} = the pressure drop due to kinetic energy or acceleration effects,
f_R = Ros friction factor.

From a dimensional analysis, Ros determined that ten dimensionless groups are required to describe fully the two-phase flow phenomenon, and therefore the pressure gradient in dimensionless form may be expressed in terms of nine such groups. Of the nine, he reasoned that only four would be really significant in a vertical flow problem. These are N_{VL}, the liquid velocity number; N_{VG}, the gas velocity number; N_D, the diameter number; and N_L the liquid viscosity number, defined as follows:

$$N_{VL} = V_{SL} \sqrt[4]{\rho_L/g\sigma} \quad \text{liquid velocity number}$$
$$N_{VG} = V_{SG} \sqrt[4]{\rho_L/g\sigma} \quad \text{gas velocity number}$$
$$N_D = D \sqrt{\rho_L g/\sigma} \quad \text{pipe diameter number}$$
$$N_L = \mu_L \sqrt[4]{g/\rho_L \sigma^3} \quad \text{liquid viscosity number}$$

Ros carried out some 4000 test runs involving 20,000 data points in a laboratory column with a long stabilizing section and a test section some 33 feet long. The tube diameter was varied from 3.2 to 14.23 cm, liquid specific gravity from 0.828 to 1.000, liquid viscosity from 1 to 337 centistokes, and interfacial tension from 24.5 to 72 dynes/cm. Superficial gas and liquid velocities were varied from 0 to 328 and 0 to 10.5 ft/sec, respectively. The correlations of Ros, and the slightly modified form by Duns and Ros (1963) are based upon the recognition that the relationships controlling $\bar{E}_G$ and f_R will be different for each flow pattern. Thus, the first step is the determination of the flow "region" from the correlation with N_{VL} and N_{VG} previously mentioned. This correlation is reproduced in Figure 8.19. Ros' flow regions may be approximately related to the flow patterns we have discussed as follows:

Region I Bubble and slug flow at low liquid velocities; froth flow at moderate liquid velocities; liquid phase continuous.

Region II Froth flow; "heading" or pulsating flow; neither phase continuous.

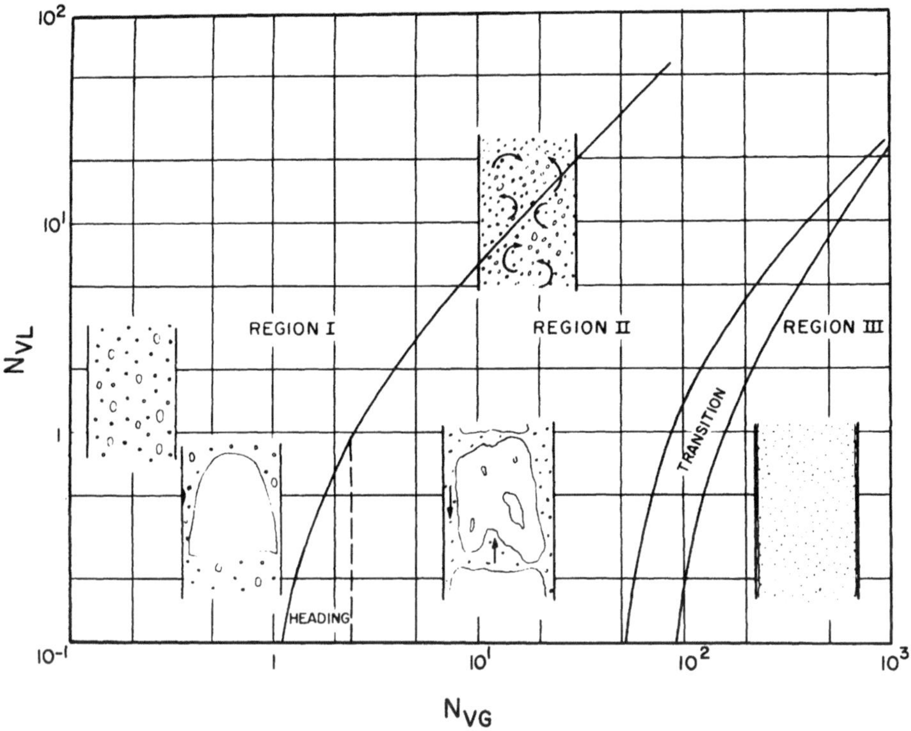

Figure 8.19. Regions of validity of Ros' correlation. (From Duns, H. Jr., and N. C. J. Ros, Sixth World Petroleum Congress, 1963.)

Region III Annular-mist flow; gas phase continuous.

The *in situ* volume fraction gas is correlated indirectly through the related slip velocity, S, defined by Equation 7.76.

Ros defined a slip velocity number

$$N_S = S\sqrt[4]{\rho_L/g\sigma} \tag{8.19}$$

and correlated his holdup data in terms of it, the other dimensionless numbers and a series of "F" factors. His relationships for N_S are:

Region I: $N_{VG} \leq L_1 + L_2 N_{VL}$

$$N_S = F_1 + F_2 N_{VL} + F_3'\left(\frac{N_{VG}}{1 + N_{VL}}\right)^2 \tag{8.20}$$

where

$$F_3' = F_3 - \frac{F_4}{N_D}$$

The factors F_1, F_2, F_3, F_4, are given as functions of N_L in Figure 8.20, L_1, and L_2 are given as functions of N_D in Figure 8.21.

Region II: $L_1 + L_2 N_{VL} \leq N_{VG} \leq 50 + 36N_{VL}$

$$N_S = (1 + F_5)\frac{N_{VG}^{0.982} + F_6'}{(1 + F_7 N_{VL})^2} \tag{8.21}$$

where

$$F_6' = 0.029N_D + F_6$$

The factors F_5, F_6, and F_7 are given as functions of N_L in Figure 8.22.

Region III: $N_{VG} \geq 75 + 84N_{VL}^{0.75}$. Here Ros assumed homogeneous flow with $S = 0$.

For the friction factor, f_R, Ros postulated that Regions I and II, i.e. where

$$N_{VG} \leq 50 + N_{VL}$$

might be treated together since both involved a continuous liquid phase. He assumed a relationship of the form

$$f_R = f_1\left(\frac{f_2}{f_3}\right) \tag{8.22}$$

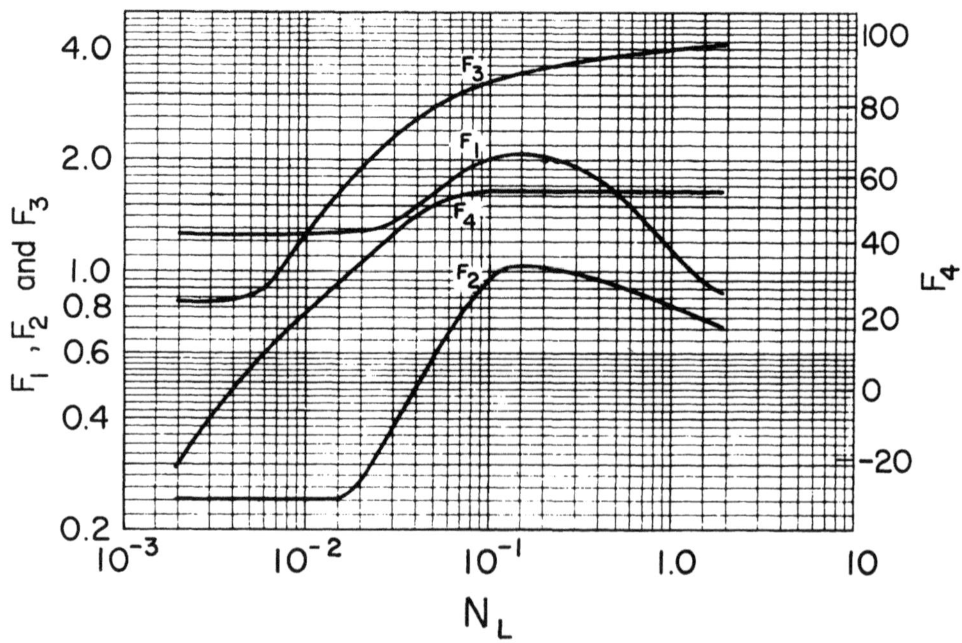

Figure 8.20. Ros factors F_1, F_2, F_3, and F_4. (From Ros, N. C. J., *J. Petrol. Technol.*, **13**, 1037, 1961.)

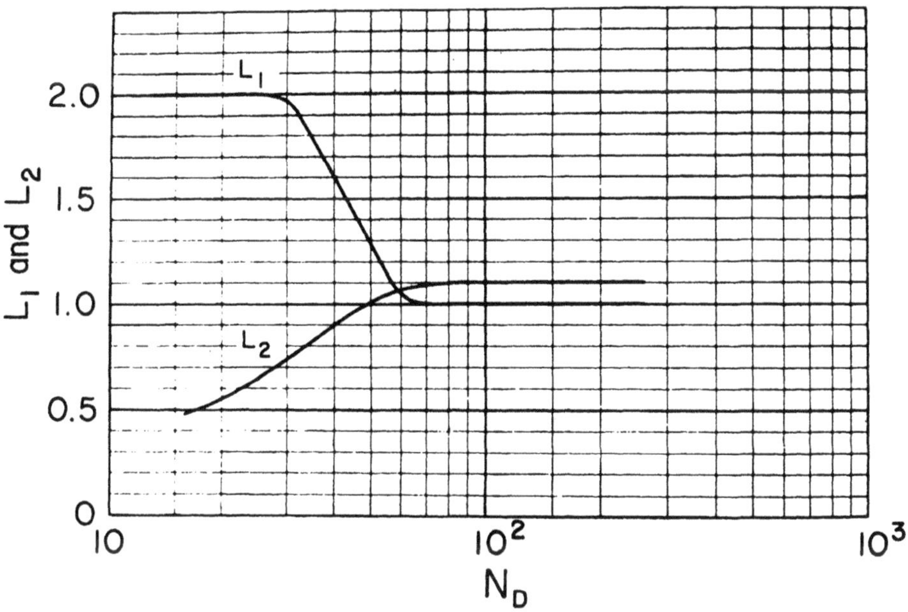

Figure 8.21. Ros factors L_1, and L_2. (From Ros, N. C. J., *J. Petrol. Technol.*, **13**, 1037, 1961.)

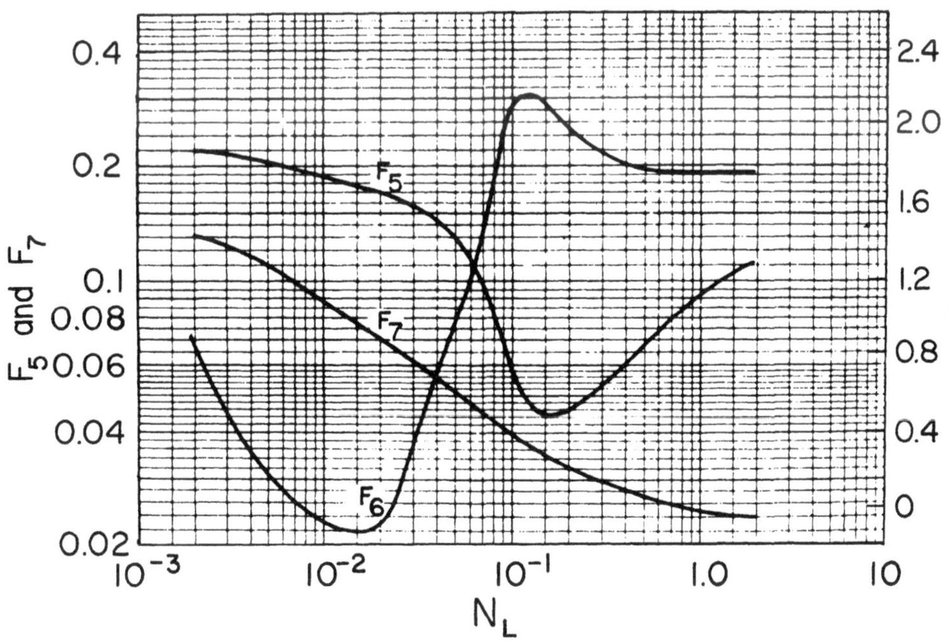

Figure 8.22. Ros factors F_5, F_6, and F_7. (From Ros, N. C. J., *J. Petrol. Technol.*, **13**, 1037 1961.)

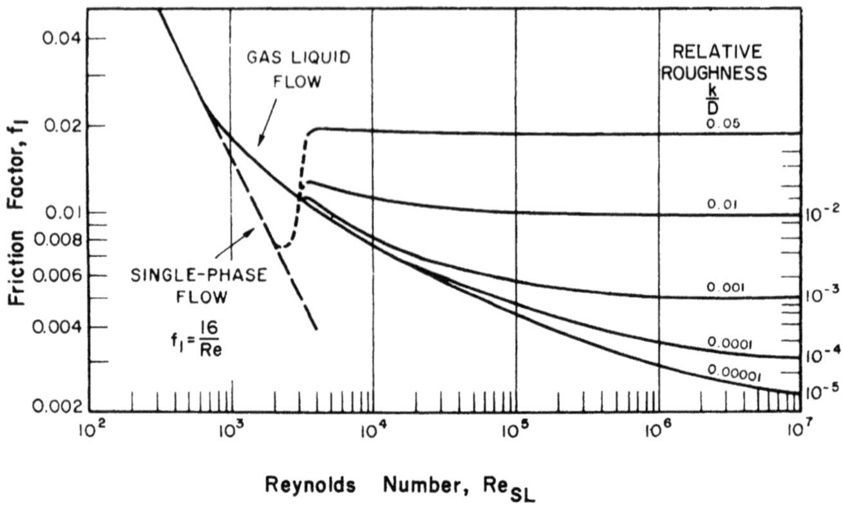

Figure 8.23. Ros friction factor f_1. (From Duns, H., Jr., and N. C. J. Ros, Sixth World Petroleum Congress, 1963.)

where f_1 is the conventional single-phase friction factor dependent upon the single-phase Reynolds number and the relative roughness of the pipe wall but modified in the range of $700 < \text{Re} < 3000$. Ros evaluated the Reynolds number as $\text{Re}_{SL} = DV_{SL}\rho_L/\mu_L$. His modified $f_1 - \text{Re}_{SL}$ – relative roughness relationship is given in Figure 8.23. The factors f_2 and f_3 are intended to correct for the effects of holdup.

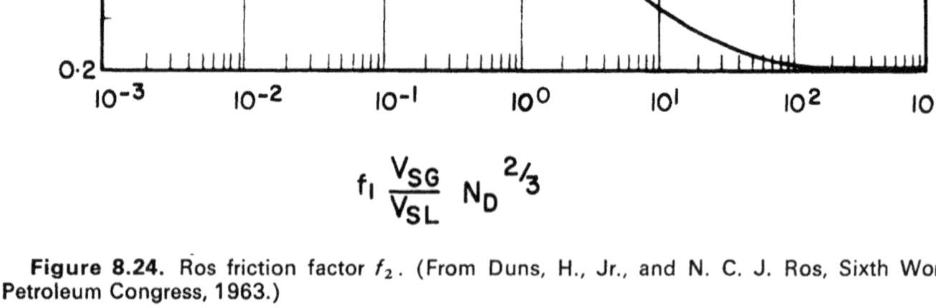

Figure 8.24. Ros friction factor f_2. (From Duns, H., Jr., and N. C. J. Ros, Sixth World Petroleum Congress, 1963.)

The factor f_2, as determined from the data, is correlated with f_1, V_{SG}/V_{SL} and $N_D^{2/3}$, as shown in Figure 8.24. The factor f_3 is a second-order correction factor important only where the liquid viscosity exceeds 50 centistokes. It is given by the equation

$$f_3 = 1 + f_1\sqrt{V_{SG}/50V_{SL}} \tag{8.23}$$

In region III, where $N_{VG} \geq 75 + 84N_{VL}^{0.75}$ and annular-mist flow prevails, Duns and Ros found that $f_R = f_1$ provided that f_1 was evaluated at the superficial gas Reynolds number, $\mathrm{Re}_{SG} = DV_{SG}\rho_G/\mu_G$ and the "wall roughness," k, was properly representative of the roughness of the liquid film. They correlated the film roughness through the relation given in Figure 8.25. They also suggested that, at gas rates near the transition from froth or slug flow to annular-mist flow, an improvement in accuracy might be obtained by substituting $(D - k)$ for D throughout the calculation of the friction gradient.

Although the pressure gradient due to acceleration is often negligible, this is not always the case in the annular-mist region. Duns and Ros evaluate dP_{KE}/dz as

$$\frac{dP_{KE}}{dz} = -(\rho_L V_{SL} + \rho_G V_{SG})\frac{dV_{SG}}{dz} \tag{8.24}$$

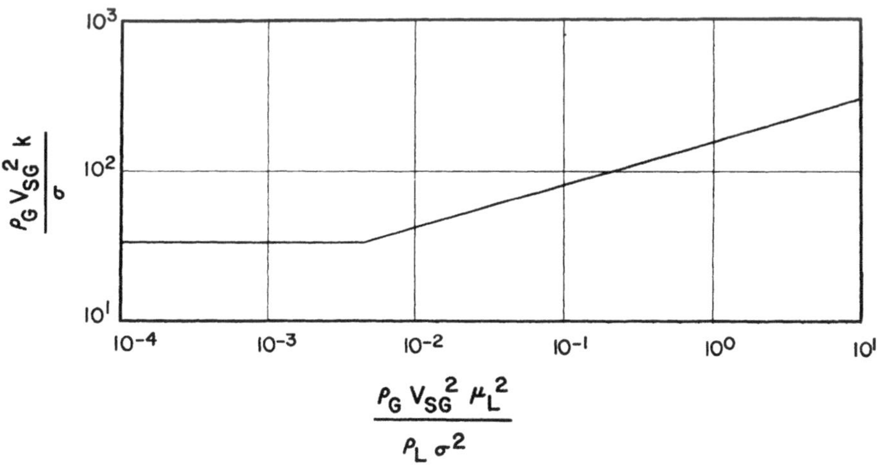

Figure 8.25. Duns and Ros film roughness correlation for annular-mist flow. (From Duns, H., Jr., and N. C. J. Ros, Sixth World Petroleum Congress, 1963.)

and taking the gas expansion to be isothermal, such that PV_{SG} is constant,

$$V_{SG}\frac{dP}{dz} + P\frac{dV_{SG}}{dz} = 0 \tag{8.25}$$

from which

$$\frac{dP_{KE}}{dz} = (\rho_L V_{SL} + \rho_G V_{SG})\frac{V_{SG}}{P}\frac{dP}{dz} \tag{8.26}$$

Writing Equation 8.26 over a finite length of pipe, Δz, and combining it with Equation 8.18 we obtain

$$\frac{\Delta P}{\Delta z} \quad \frac{\dfrac{\Delta P_{HH}}{\Delta z} + \dfrac{\Delta P_f^{\bullet}}{\Delta z}}{1 - (\rho_L V_{SL} + \rho_G V_{SG})\dfrac{V_{SG}}{P}} \tag{8.27}$$

The Ros correlation applies over a short length of pipe Δz, similar in that respect to that of Govier *et al.* and of Hughmark and Pressburg. For deep wells or situations involving a pressure drop significant with respect to the bottom hole pressure, it should be used with a series of step calculations. The correlation is probably based upon more extensive laboratory data than any other and has met the test of field comparison extremely well. Ros shows that his correlation gives essentially identical results to that of Baxendell and Thomas (1961) for high flow rates in wells. Orkiszewski (1967) finds an average error of 2.4 percent and a standard deviation of 27 percent in comparing the Ros predicted pressure gradient with measured gradients in 148 wells. Kern and Nicholson (1965) also demonstrate the good engineering accuracy attainable by the Ros correlation.

We may conclude that the Ros correlation represents about the best that can be expected from empirical methods. One serious disadvantage is its complexity in application.

The Hagedorn and Brown Correlation Hagedorn and Brown (1965) have developed an independent correlation based upon an extended form of the single-phase mechanical energy balance (to take into account the kinetic energy effects) and incorporating an estimate of the density of the *in situ* mixture. Their version of Equation 4.31, written in field units, is

$$144\frac{\Delta P}{\Delta z} = \bar{\rho}_M + \frac{4fQ_L'^2 M^2}{2.9652 \times 10^{11} D^5 \bar{\rho}_M} + \frac{\bar{\rho}_M \Delta\left(\dfrac{V_M^2}{2g_c}\right)}{\Delta z} \tag{8.28}$$

where Q'_L = total liquid production rate, barrels per day. They calculate $\bar{\rho}_M$, the effective mean density in the *in situ* mixture, as

$$\bar{\rho}_M = (1 - \bar{E}'_G)\rho_L + \bar{E}'_G \rho_G \tag{8.29}$$

where $\bar{E}'_G$ is an effective average volume fraction gas of the *in situ* mixture. A further major assumption is that the friction factor, f, is related to a Reynolds number and the relative roughness in the same way as it is for single-phase flow. This introduces the problem of a definition of the Reynolds number for two-phase flow. Hagedorn and Brown arbitrarily define it in field units as

$$\mathrm{Re}_{HB} = 2.2 \times 10^{-2} \frac{Q'_L M}{D\mu_L^{(1-\bar{E}'_G)}\mu_G^{\bar{E}'_G}} \tag{8.30}$$

Using this definition for the Reynolds number, and the conventional relationship between f and Re for single-phase fluids, Hagedorn and Brown calculated $\bar{E}'_G$ from Equations 8.28 and 8.29 for their own data on water, and 10-, 30-, 35-, and 110-centipoise oils flowing with gas in 1-, $1\frac{1}{4}$-, and $1\frac{1}{2}$-inch tubings in a 1500 foot experimental well. They emphasize that $\bar{E}'_G$ may not be the same as the true average volume fraction gas of the *in situ* mixture. They then sought a correlation of the values of $\bar{E}'_G$ so obtained using as a guide the dimensionless ratios proposed by Ros (1961), i.e., N_{VL}, N_{VG}, N_D, and N_L. They also introduced a pressure ratio term, P/P_a, to reduce scatter due to pressure variation, and two correlating coefficients C and ψ.

Figure 8.26 gives the relationship between the product CN_L and N_L, whereas Figure 8.27 shows the relationship between the correlating factor, ψ, and a parameter involving the gas velocity number, the liquid viscosity number, and the diameter number. Figure 8.28 is the final correlation for the

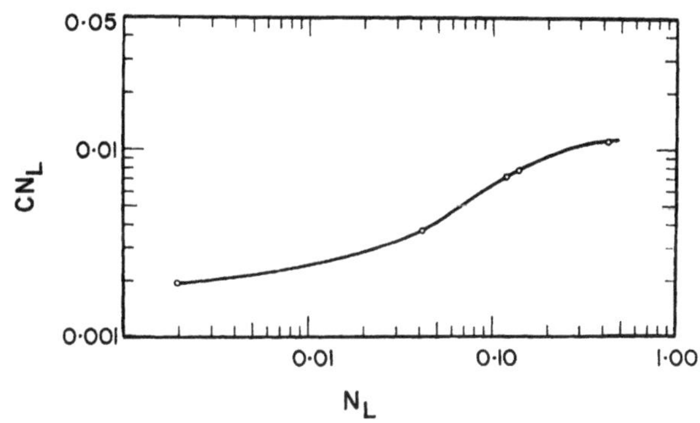

Figure 8.26. Hagedorn and Brown factor CN_L. (From Hagedorn, A. R., and K. E. Brown, *J. Petrol. Technol.*, **17**, 475, 1965.)

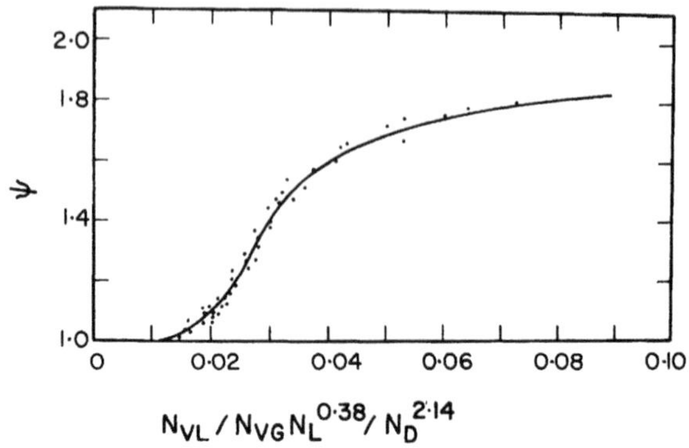

Figure 8.27. Hagedorn and Brown correlating factor ψ. (From Hagedorn, A. R., and K. E. Brown, *J. Petrol. Technol.*, **17**, 475, 1965.)

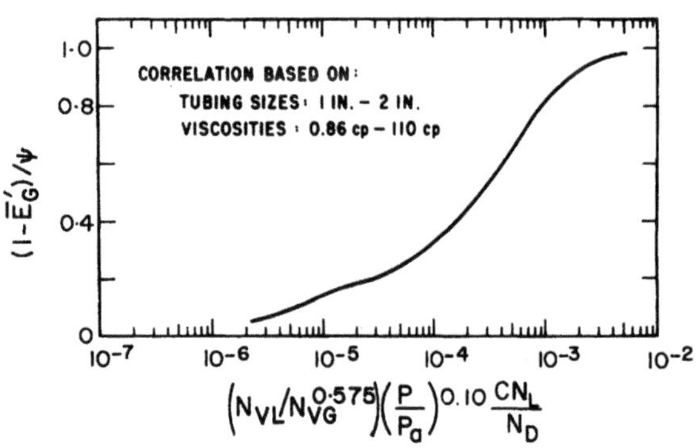

Figure 8.28. Hagedorn and Brown correlation for effective holdup. (From Hagedorn, A. R., and K. E. Brown, *J. Petrol. Technol.*, **17**, 475, 1965.)

effective holdup yielding $\bar{E}'_G$ in terms of ψ, the other dimensionless ratios, and C. The effective volume fraction gas, $\bar{E}'_G$, may therefore be estimated by proceeding from Figure 8.26 through Figures 8.27 and 8.28.

De Gance and Atherton (1970) express the relationships of Figures 8.26, 8.27, and 8.28 by equations explicit in CN_L, ψ, and $1 - \bar{E}'_G$.

Hagedorn and Brown have tested their correlation with data reported by Fancher and Brown (1963), Baxendell (1961), and Gaither *et al.* (1963). They report average errors below 6 percent, and standard deviations below 9 percent. For the combined data they indicate an average error of -1.1 percent with a standard deviation of 6.5 percent. This is a good accuracy considering the range of diameters (1 to $3\frac{1}{2}$ inches) and the range of liquid viscosities (0.8 to 110 centipoises) covered.

The correlation is similar to those discussed in Section 3.31 in that it is based upon measured data at the bottom and top of the well and is, therefore, of the integral form. The correlation is entirely empirical and in effect attaches to $\bar{\rho}_M$ and consequently to $\bar{E}'_G$ any discrepancies in the evaluation of the friction factor by the single-phase correlation and the arbitrarily determined two-phase Reynolds number. The effective *in situ* volume fraction gas is therefore one defined to balance Equation 8.28 with the irreversibility term evaluated in the Hagedorn-Brown manner. Despite this, under some circumstances the method gives surprisingly good answers as independently attested to by Orkiszewski (1967) who found an average error of only 0.7 percent and a standard deviation of 24.2 percent in comparing predicted and measured gradients in 148 different well flow conditions.

3.33 The Prediction Scheme of Orkiszewski

Orkiszewski (1967) proposed, and tested against data for 148 well conditions, a pressure gradient prediction scheme involving:

1. The flow pattern discrimination
   a. between bubble and slug flow in accordance with the method of Griffith and Wallis (1961), see Section 3.1, and
   b. among the other flow patterns as suggested by Ros (1961) and Duns and Ros (1963), see Section 3.32;
2. the estimation of the pressure gradient
   a. for bubble and the low velocity slug flow patterns, by the method of Griffith and Wallis based upon their analysis of the flow mechanisms, see Sections 4.13, 5.6, and 5.9,
   b. for high velocity slug flow, by an empirical modification of the Griffith and Wallis mechanistic analysis which adjusts their predicted value of E_G, and
   c. for flow in the transition from the slug to the annular-mist pattern and in the annular-mist pattern itself, by the method of Ros, see Section 3.32.

Orkiszewski's comparisons show that his combination of methods gives improved average error and standard deviation over the generally applied methods of Ros and of Hagedorn and Brown (1965). Espanol, Holmes, and Brown (1969) confirm this with independent data. We should note, however, that out of Orkiszewski's 148 test results none were for flow in the annular-mist region, 7 were partly in the transition from slug to annular-mist flow, 26 were partly in slug and partly in bubble flow, 4 were completely in bubble flow, and the remaining 111 were entirely in slug flow. Most of the 44 well test data points used by Espanol *et al.* also represented bubble, slug, or bubble and slug flow.

Our conclusions relative to the work of Orkiszewski, and Espanol *et al.* are that, for the prediction of pressure gradient,

1. the empirical method of Ros is superior to the empirical method of Hagedorn and Brown in the transition to the annular-mist flow pattern;
2. for the bubble or the low velocity slug flow regions, a method based upon the mechanism of flow (such as that of Griffith and Wallis, although see Sections 4.13, 5.6, 5.9, 5.10 and 5.11 for improvements to this) is superior to the Ros or the Hagedorn and Brown empirical methods; and
3. the empirical adjustments proposed by Orkiszewski to the Griffith and Wallis analysis of slug flow for situations where the friction component is significant results in improvements in the prediction but really represent compensations for what are now known to be flaws in the Griffith and Wallis model of the mechanism. The analyses presented in Sections 5.6, 5.9, 5.10 and 5.11 do not require such adjustments as has been confirmed by calculations by Aziz, Fortems, and Settari (1971) for 26 of the well data points reported by Espanol. They found the results based upon the methods presented in the sections referred to be somewhat better than the predictions by Orkiszewski's procedure.

Gould and Tek (1970) have developed a flow pattern map [coordinates $V_{SL}(\rho_L/g\sigma)^{1/4}$ and $V_{SG}(\rho_L/g\sigma)^{1/4}$] and a computer program based upon the Orkiszewski design method. They also discuss several applications of the method to practical problems relating to the performance of oil and gas wells.

4 THE BUBBLE FLOW PATTERN

4.1 Introduction

The bubble flow pattern occurs at all rates of flow of the continuous phase, and from infinitesimal to modest rates of flow of the dispersed phase. It may result from the separate admission to a flowing liquid of a stream of bubbles or from the generation of bubbles within the system itself through addition of heat, the reduction of pressure, electrolysis, or chemical reaction. We will not discuss the internal generation of bubbles which is extensively covered in the literature of evaporation and boiling (Tong, 1965; Westwater, 1956, 1958; Rohsenow, 1963; Leppert and Pitts, 1964; Davies, 1963). Our attention will be restricted to the generation of bubbles by some type of orifice device. So far as the actual pipe flow behavior is concerned, the basic phenomena are similar regardless of the method of origin of the bubbles.

If the gas or an immiscible liquid is introduced through an orifice device, the number of bubbles increases with increasing gas rate to the point where large cap-shaped bubbles are formed by collision, adherence, and coalescence. The diameters of these bubbles nearly equal the tube diameter. This marks the end of the bubble flow pattern and the start of slug flow.

Distinction needs to be drawn between the number, size, and shape of bubbles introduced into a flowing, continuous liquid phase and the number, size, and shape of bubbles resulting after any interaction among the bubbles, or between the bubbles and the continuous phase. Of greatest interest is the pseudo-equilibrium condition where the number, size, and shape of the bubbles have become relatively stabilized and are no longer affected by entrance conditions. If the bubbles may collide, a bubble mixture is essentially unstable against coalescence, since there is always a tendency for a reduction of the total surface area. (Anderson and Quinn, 1970a, have shown that bubbles of gas in a highly purified liquid tend to coalesce more readily than in a liquid containing trace contaminants.) The bubble flow pattern is, therefore, wholly stable only when bubbles are generated continuously under conditions where, after entrance effects are no longer significant, no collisions occur among the bubbles. Anderson and Quinn (1970b) suggest, on the basis of an idealized analysis for a bubble column with no net liquid flow, that small bubbles may be expected to coalesce to form cap-shaped bubbles after a travel distance of the order of $D/E_G^{4/3}$, where D is the tube diameter.

Some investigators (Zuber *et al.*, 1967) subdivide the overall bubble flow region into a bubbly, and a bubbly-slug region, the former consisting of small dispersed bubbles only, whereas the latter includes part of what could be termed the transition to slug flow and consists of both small dispersed bubbles and larger aggregates of bubbles.

In the next subsection we consider the question of bubble generation at orifices and present some empirical correlations for the prediction of bubble frequency or bubble volume from a given orifice. This is followed by a discussion of limiting rise velocity of bubbles under various conditions. This development allows us in Section 4.11 to predict, at least approximately, the stabilized bubble flow pattern from given flow variables. A mechanistic approach to holdup and pressure drop calculations is presented in Sections 4.12 and 4.13. Finally we treat the transition to slug flow in Section 4.14.

4.2 Bubble Generation at Submerged Orifices

The starting point for a discussion of the mechanics of the bubble flow pattern is a consideration of the factors influencing the frequency and size of bubbles generated at orifices submerged in a liquid.

The volume of a gas bubble generated at an orifice submerged in a liquid is known (Coppock and Meiklejohn, 1951; Datta *et al.*, 1950; Siemes, 1954a;

Hughes and Evans, 1955; Mahoney and Wenzel, 1963; van Krevelen and Hoftijzer, 1950) to be dependent upon the orifice diameter, the properties of the gas and the liquid including the gas-liquid interfacial tension, the condition of the solid surface of the orifice, the liquid-solid and the gas-solid interfacial tension, and the average volumetric rate of flow of the gas through the orifice. When gas is introduced at low rates through an orifice of given size, the volume of the bubbles generated remains relatively constant and the frequency of generation increases with the volumetric rate of flow of the gas. With continuing increase in the flow rate, the frequency reaches a limiting value, and therefore the volume of the individual bubbles increases to accommodate the increased gas flow rate. This continues until discrete bubbles are no longer formed, and the gas is emitted from the orifice in a more or less continuous jet.

In the constant volume region

$$Q_G = n\omega v_{bo} \tag{8.31}$$

where Q_G is the volumetric flow rate of the gas, n is the number of orifices, ω is the frequency (number of bubbles generated per second), and v_{bo} is the average individual bubble volume at the generation site. Under ideal conditions of pure fluids, uniform solid surface, and a bubble growing at equilibrium (at an infinitesimally slow rate), a simple force balance indicates that

$$v_{boi} = \frac{\pi D_o \sigma}{g \Delta\rho} \tag{8.32}$$

where D_o = orifice diameter,
σ = gas-liquid interfacial tension,
$\Delta\rho = (\rho_L - \rho_G)$, density difference.

This is the "force balance" bubble volume which, ideally, would be expected as the bubble generation frequency approaches zero and inertial effects are negligible.

The actual bubble volume, even at a generation frequency approaching zero, will differ from v_{boi} because of the nonideality of the surface conditions and of bubble deformation. In practice, bubbles are released at lower volumes because of inertial effects within the gas or in the surrounding liquid, and contamination of the solid surfaces.

As the gas flow rate increases such that the volume of the bubbles must increase, Equation 8.32 predicts bubble volumes that are too small and reliance must be placed on empirical relationships. Of the many such relationships which have been proposed we will refer only to two.

Siemes (1954a) and Siemes and Kauffmann (1956) have made a careful theoretical and experimental study of the problem. The latter propose two equations for the prediction of bubble volume in gas-liquid systems. For liquid viscosities less than 20 centipoises, their equation is

$$v_{bo} = 2.55 \times 10^{-3} \frac{\sigma}{\rho_L} D_o + 3.68 \times 10^{-2} Q_G^3 D_o^{1/2} \tag{8.33}$$

For liquid viscosities in excess of 20 centipoises they give

$$v_{bo} = 2.55 \times 10^{-3} \frac{\sigma}{\rho_L} D_o + 2.23 \times 10^{-4} \mu_L^2 D_o^{1/2} + (0.015 + 10^{-5} \mu_L^{1.41} + 0.045 D_o) Q_G \tag{8.34}$$

Both equations apply only under conditions where

$$Q_G < 7.5 \sigma D_o^2$$
$$Q_G < 0.2\sigma$$

and, in Equation 8.34, Q_G has to be greater than 0.2–0.3 cm^3/sec.

For the air-water system Yip *et al* (1970) proposed the following empirical equation where inertial effects are negligible. The equation reflects their data on the air water system and is consistent with data of van Krevelen and Hoftijzer (1950)

$$v_{bo}^* = 0.0942 \sinh (1.64\ D_o) \tag{8.35}$$

where v_{bo}^* = actual bubble volume at generating orifice at zero frequency. Yip (1967) has also explored the effects of gas phase inertia, and for the air-water system has correlated the actual bubble diameter with the zero frequency diameter by

$$v_{bo} = v_{bo}^*(1 + 0.075\omega) \tag{8.36}$$

Equations 8.33 through 8.36 are expressed in centimeter-gram-second units.

4.3 The Rise Velocity of Single Bubbles in a Stagnant Liquid

The velocity with which bubbles of gas rise in a stagnant liquid depends upon the fluid properties and rather strikingly upon the bubble volume or equivalent diameter. Figure 8.29 shows some data of Peebles and Garber (1953) indicating the behavior of air bubbles rising in a 1-inch tube in water and in an 11-centipoise mixture of ethyl acetate and cottonseed oil.

There are two approaches to the question of the rise velocity of bubbles in a stagnant liquid. First, under essentially ideal conditions of spherical bubbles rising in laminar motion the problem is amenable to theoretical analysis

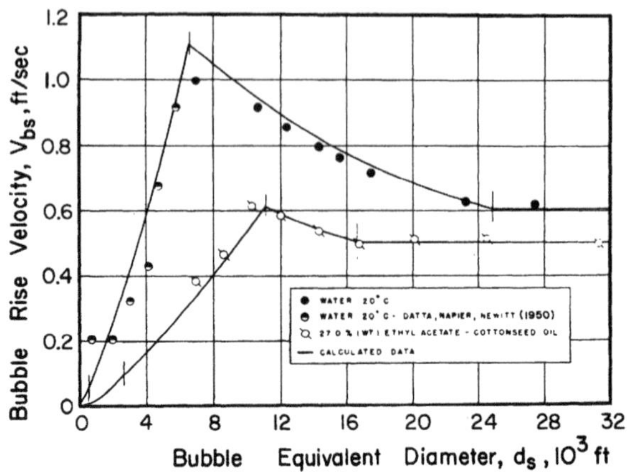

Figure 8.29. Rise velocity of bubbles. The lines represent values predicted by the appropriate relations of Peebles and Garber. (From Peebles, F. N., and H. J. Garber, *Chem. Eng. Progress,* **49**, 88, 1953.)

based upon the continuity and momentum equations. Secondly, under any reasonably steady state condition the problem may be approached, as in the case of the settling of solid particles, from the viewpoint of a balance between the buoyant and the drag force acting on the bubble; in most cases, the latter are determined empirically through a drag coefficient. We shall follow the second approach making reference to the theoretical analysis where appropriate.

From Chapter 1, Equations 1.1 and 1.2, under conditions where the buoyant and drag forces are equal, we have for spherical bubbles

$$\frac{\pi d_s^3}{6} g\Delta\rho = C_D \frac{1}{2} V_{bs}^2 \rho_L \left(\frac{\pi d_s^2}{4}\right) \tag{8.37}$$

and

$$V_{bs} = 1.155\sqrt{\frac{g\Delta\rho d_s}{\rho_L C_D}} \tag{8.38}$$

where d_s = equivalent spherical bubble diameter = $(6v_b/\pi)^{1/3}$,
v_b = volume of bubble,
C_D = drag coefficient,
V_{bs} = steady-state terminal rise velocity of bubble in a stagnant liquid.

Peebles and Garber (1953) and others have shown that for bubble diameters of the order of 1 mm and less or for bubble Reynolds numbers below about 2, bubbles are, in fact, spherical and they rise in a rectilinear path at velocities corresponding with

$$C_D = \frac{24}{\text{Re}_b} \tag{8.39}$$

where $\mathrm{Re}_b = d_s V_{bs} \rho_L/\mu_L$. This gives

$$V_{bs} = \frac{g\Delta\rho d_s^2}{18\mu_L} \tag{8.40}$$

which is identical with Stokes' Law as discussed in Chapter 1.

At Re_b values above 2 but still below a critical value determined by the properties of the fluids, bubbles remain spherical and rise in more or less straight paths, but circulation within the bubble becomes important. This causes a decrease in the drag coefficient and an increase in the rise velocity. The region is defined by

$$2 \leq \mathrm{Re}_b \leq 4.02 G_1^{-0.214}$$

where

$$G_1 = \frac{g\mu_L^4}{\rho_L \sigma^3} \tag{8.41}$$

Here the drag coefficient is empirically found to be

$$C_D = \frac{18.7}{\mathrm{Re}_b^{0.68}} \tag{8.42}$$

and the equation for the rise velocity is

$$V_{bs} = 0.33 g^{0.76} \left(\frac{\rho_L}{\mu_L}\right)^{0.52} \left(\frac{d_s}{2}\right)^{1.28} \tag{8.43}$$

This may be compared with

$$C_D = \frac{16}{\mathrm{Re}_b} \tag{8.44}$$

and

$$V_{bs} = \frac{g\Delta\rho d_s^2}{12\mu_L} \tag{8.45}$$

which is the modified Stokes' Law discussed in Chapter 1.

For the higher Re_b range given by

$$4.02 G_1^{-0.214} \leq \mathrm{Re}_b \leq 3.10 G_1^{-0.250}$$

or

$$16.32\, G_1^{0.144} \leq G_2 \leq 5.75$$

the bubbles are deformed, flattened horizontally and ellipsoidal in cross section, move in spiraling, zigzag paths, and the drag coefficient which increases sharply with Re_b is

$$C_D = 0.0275 G_1\, \mathrm{Re}_b^4 \quad \text{or} \quad C_D = 0.44 G_2 \tag{8.46}$$

where

$$G_2 = \frac{g d_s^4 V_{bs}^4 \rho_L^3}{16\sigma^3} \tag{8.47}$$

and the rise velocity is

$$V_{bs} = 1.35\left(\frac{2\sigma}{\rho_L d_s}\right)^{0.50} \tag{8.48}$$

At still higher Reynolds numbers the bubbles, greatly deformed, mushroom- or cap-shaped, tend to fill the flow tube and to rise in a rectilinear path at a constant velocity, independent of bubble size. Peebles and Garber define this region by

$$3.10G_1^{-0.25} \leq \text{Re}_b$$

or

$$5.75 \leq G_2$$

and find the drag coefficients to be given by

$$C_D = 0.82G_1^{0.25}\ \text{Re}_b$$

or (8.49)

$$C_D = 1.64G_2^{0.25}$$

and the bubble rise velocity by

$$V_{bs} = 1.18\left(\frac{\sigma g}{\rho_L}\right)^{0.25} \tag{8.50}$$

We will see later that the conditions where these last relations apply are normally not considered to be within the bubble flow pattern. Equations 8.41 through 8.50 are expressed in centimeter-gram-second units, converted from the British units used by Peebles and Garber, to be consistent with the remainder of this section.

Peebles and Garber conducted all their experiments in a 1.03 inch I.D. pipe. No attempt was made by these investigators to evaluate the effect of tube wall on the rise velocity.

Harmathy (1960) has also dealt with the rise velocities of oblate spheroidal and cap-shaped bubbles rising in turbulent motion. These bubbles correspond with those covered by Peebles and Garber's Equations 8.48 and 8.50. His relationship for such flow in an infinite medium is based upon Newton's settling equation for spheres as discussed in Chapter 1. He demonstrates that the bubble shape is largely determined by the Eotvos number, Eo, defined as

$g\Delta\rho d_s^2/\sigma$. For Eo $\leq$ 15 the fluid particles are spheroidal in shape, for 15 $\leq$ Eo $\leq$ 40 they are irregular spheroid like tending toward mushroom- or cap-shaped, and for Eo $\geq$ 40 they are cap-shaped (as in the slug flow pattern discussed in Section 5). For Eo < 13 Harmathy gives

$$V_{bs} = 0.88\ \mathrm{Eo}^{-1/4} V_{bsN} \tag{8.51}$$

where

$$V_{bsN} = 1.74\sqrt{\frac{g\Delta\rho d_s}{\rho}} \tag{1.11}$$

Combining Equations 1.11 and 8.51 we have

$$V_{bs} = 1.53\left(\frac{g\Delta\rho\sigma}{\rho^2}\right)^{1/4} \tag{8.52}$$

an equation in which the bubble size itself does not appear. For the mushroom- or cap-shaped bubble region (Eo > 40) Harmathy gives

$$V_{bs} = 0.715\sqrt{\frac{g\Delta\rho d_s}{\rho}} \tag{8.53}$$

Siemes (1954b) published the results of his extensive theoretical and experimental studies almost simultaneously with Peebles and Garber. He classifies bubble flow into essentially the same four regions as do Peebles and Garber and gives similar equations.

Mendelson (1967) developed an analogy between the rise velocity of large bubbles in liquids of infinite extent and the velocity of surface waves over infinitely deep liquids. He found that, in the regions where Equations 8.48 and 8.50 are applicable, the wave equation correlates data for bubbles rising in pure liquids quite well.

4.4 The Effect of the Tube Wall

Uno and Kintner (1956) measured rise velocities of a variety of shapes of air bubbles in four different liquids and six different tube sizes ranging from 2.09 to 15.25 cm in diameter. They proposed an empirical relation involving the quantity $(1 - d_s/D)$ and a constant dependent on D and the surface tension of the liquid for the retarding effect of the tube wall on the bubble rise velocity.

Haberman and Sayre (1958) have conducted an analytical study of the motion of rigid and fluid spheres in stationary and moving liquids inside cylindrical tubes. Their final equation applicable in cases of laminar particle flow, i.e., where the particle or bubble Reynolds number <2, in a stagnant liquid, may be expressed as

$$V_{bs} = \frac{d_s^2 g \Delta\rho}{18\mu_L K_1}\left[\frac{1 + (\mu_L/\mu_G)}{1 + \frac{2}{3}(\mu_L/\mu_G)}\right] \tag{8.54}$$

where K_1 is a correction factor for the wall effect. This equation reduces to Stokes' Law for $\mu_L/\mu_G = 0$, and $K_1 = 1.0$. These are precisely the conditions for a rigid sphere in an infinite medium. The equation reduces to the modified Stokes Law for gas bubbles ($\mu_G \ll \mu_L$) of diameter much smaller than the tube diameter. Haberman and Sayre show that their theoretical analysis gives results in good agreement with experiments for both solid and fluid particles. Their results are shown in graphical form in Figure 8.30, which gives their wall correction factor, K_1, in terms of the ratios d_s/D and μ_L/μ_G. Figure 8.30 is applicable only to the laminar flow of spherical and near spherical solid and fluid particles in a stationary fluid.

Harmathy (1960) has extended his analysis of the rise of irregularly shaped bubbles at bubble Reynolds numbers exceeding 500 to cover the effect of

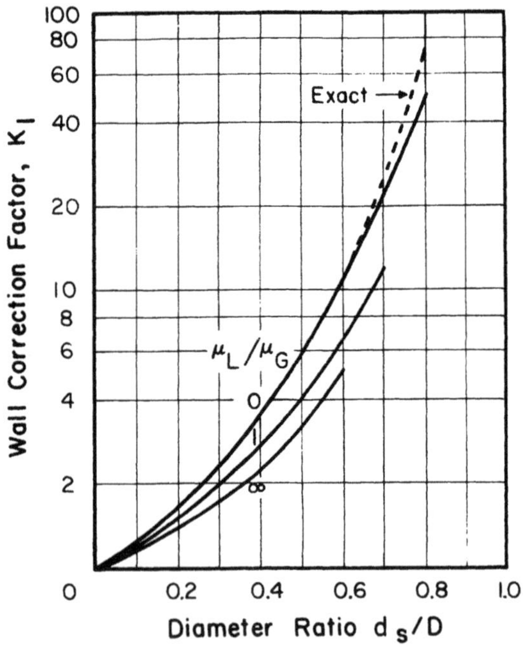

Figure 8.30. Wall correction factor, K_1, for spheres concentrically within a cylinder. (From Haberman, W. L., and R. M. Sayre. Dept. of the Navy, Hydromechanics Lab. Res. and Dev. Report No. 1143, Oct. 1958.)

proximity to the wall. As a general equation including the effects of shape, properties, and wall proximity, he proposes

$$V_{bs} = 1.53\left(\frac{g\Delta\rho\sigma}{\rho^2}\right)^{1/4} K_2 \tag{8.55}$$

where

$$K_2 = \frac{0.88\left[1 - \left(\frac{d_s}{D}\right)^2\right]}{\mathrm{Eo}^{1/4}\sqrt{1 + \left(\frac{d_s}{D}\right)^4}}\left[1 - \frac{2}{3}\left(\frac{\mathrm{Eo}}{1+\mathrm{Eo}}\right)\left(\frac{d_s}{D}\right)^{3/2}\right]$$

Figure 8.31 is a plot of the factor K_2 as a function of d_s/D and Eo. Harmathy found good agreement between his correlation and his own and other data, including that of Uno and Kintner. The correlation is valid under the following conditions:

$$\frac{d_s}{D} < 1.0 \qquad \text{for Eo} \leq 4.5$$

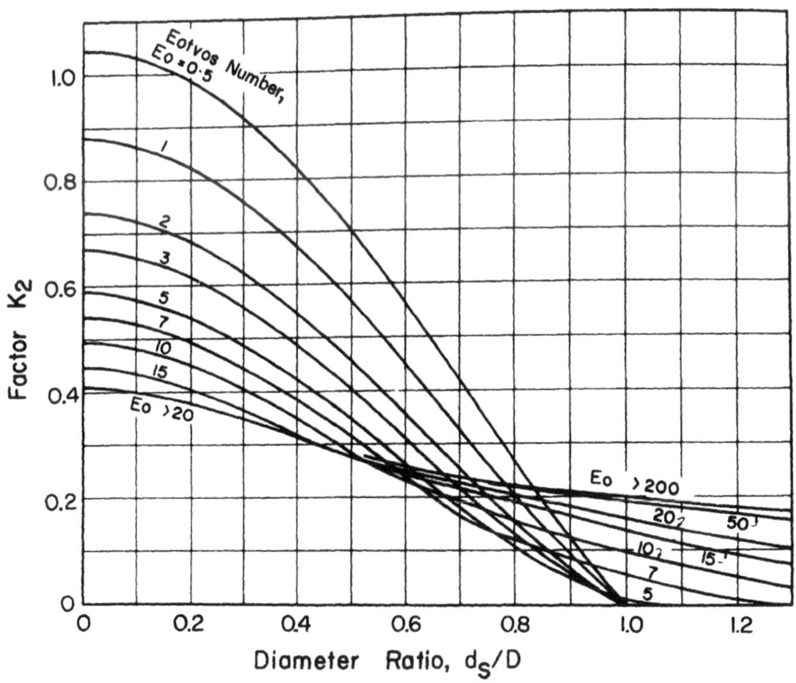

Figure 8.31. Harmathy factor K_2, Equation 8.55. (From Harmathy, T. Z., *A.I.Ch.E. Journal*, **6**, 281, 1960.)

$$\frac{d_s}{D} < 1 - 0.175\sqrt{\text{Eo} - 4.5} \qquad \text{for } 4.5 < \text{Eo} < 20$$

$$\frac{d_s}{D} < 0.31 \qquad \text{for Eo} \geq 20$$

Manery and Mendelson (1968) extended the wave analogy (Mendelson, 1967) to take into account the effect of wall proximity by relating the bubble rise velocity in bounded liquids to wave velocities over liquids of finite depth. Their equation for the bubble rise velocity is in terms of the diameter ratio d_s/D, the Eotvos number, and the bubble rise velocity in infinite media. Comparison with Uno and Kintner's experimental data shows good agreement.

4.5 Effect of Radial and Axial Interaction Between Bubbles

4.51 Radial Interaction

Relatively little of a quantitative nature is known about the influence on a rising bubble of adjacent bubbles in approximately the same horizontal plane. Limited theoretical and experimental studies of the settling of solid, spherical particles in a fluid have some relevance to the problem.

Theoretical analysis of the effect of particles on the motion of a single, solid, spherical particle in the same horizontal plane has been restricted to pairs of such particles. Kynch (1959), by a technique involving the superposition of the motion of two isolated spheres, expressed the velocity of the doublet in terms of the ratio of center-to-center separation distance to the particle diameter. He found that two equal spheres in contact, with their line of centers horizontal, would have a falling velocity 40 percent higher than that of a single sphere.

Jayaweera *et al.* (1964) have studied the behavior of small clusters of solid spheres experimentally over a particle Reynolds number range from 10^{-4} to 10. They confirmed the higher rate of fall of the doublets, found that below $\text{Re} = 0.03$ the spheres remained adjacent to one another with no tendency to separate or rotate, and above $\text{Re} = 0.05$ observed both separation and inward rotation of the spheres until a maximum separation distance was reached (eight diameters decreasing to about three diameters with increasing Reynolds numbers). The behavior of clusters of three to six spheres was similar in that their rate of fall also exceeded that of single spheres, and rotation occurred up to a certain separation distance (six diameters at $\text{Re} \simeq 1$, and three diameters at $\text{Re} \simeq 7$).

The work of Kynch and of Jayaweera *et al.* deals with solid spheres. Similar interactions are to be expected among bubbles but there are several complicating factors: internal circulation may occur; the shape may be spherical, oblate spheroidal, or other; oscillation often occurs; spiral and zigzag travel paths are observed; and, most important, when bubbles come into contact they may adhere to one another and, if they adhere, they may coalesce. These matters do not lend themselves to theoretical analysis and their complexity suggests that the overall velocity effect due to radial interaction, i.e., ΔV_r, must be determined by direct observation of differences in rise velocities of single and multiple bubbles.

Yip *et al.* (1970) investigated the radial interaction in an air-water system and found, using a 5.04 cm tube, that it was significant only for bubbles of equivalent diameter greater than 8 percent of the tube diameter. Their empirical correlation for the effect of horizontally adjacent bubbles on bubble rise velocity is

$$\Delta V_r = 4.4(n - 1)(d_s - 0.39) \qquad \text{for } d_s > 0.45 \tag{8.56a}$$

and

$$\Delta V_r = 0 \qquad \text{for } d_s \leq 0.45 \tag{8.56b}$$

where ΔV_r = the increment in bubble velocity over the velocity of a single bubble, cm/sec,
n = the number of bubbles at a horizontal plane,
d_s = the equivalent bubble diameter, cm.

Little is known about the radial separation distance below which the radial interaction becomes important, but from the work on solid spheres it would appear that bubbles would be sensitive to the presence of others within as much as eight diameters at low Reynolds numbers, and that their rise velocity would be increased from that of single bubbles by small amounts increasing to perhaps 40 percent. A related matter is the minimum, stable separation distance. Clearly below a certain minimum separation distance proportional to the bubble diameter, the effect of oscillation, spiraling, and random motion could cause bubble collision with possible adherence and coalescence. If the minimum, stable, radial separation (face to face) is taken as $N \times d$, then simple geometrical considerations indicate that for $d \ll D$,

$$n_{\max} \simeq \frac{\pi}{4}\left(\frac{D}{d + Nd}\right)^2 \simeq \frac{\pi}{4}\frac{D^2}{d^2}\left(\frac{1}{1 + N}\right)^2 \tag{8.57}$$

where $n_{\max}$ is the maximum number of stable bubbles in an approximately horizontal plane. The factor N probably lies in the range of 0.5 to 1.5.

4.52 Axial Interaction—Approach to Stabilization

As in the case of the radial interaction, not too much is yet known of the effect of the wake of a leading bubble on a trailing one. Again some of our qualitative understanding comes from studies of the motion of solid spheres.

Jayaweera *et al.* (1964) observed the axial interaction between pairs of solid spheres settling one behind the other. They showed that the wake effect becomes significant even at center-to-center distances greater than ten diameters. Up to a separation of eight to ten diameters the interaction, as measured by the relative velocity between the spheres, varies inversely with center-to-center separation distance, Δz. At lesser separations the relative velocity is nearly constant, increasing only with the bubble Reynolds number. Figure 8.32, taken from Jayaweera *et al.*, shows these effects.

Many investigators (Datta *et al.*, 1950; Coppock and Meiklejohn, 1951; Guyer and Pfister, 1946; Garner and Hammerton, 1954) have observed that when a bubble is followed by another within less than some critical distance there is an axial interaction between the bubbles which results in an increase in the rise velocity. The interaction is due to a combination of several factors, including the vortex disturbance created by a preceding bubble, the additional liquid velocity component induced by the replacement of liquid following passage of the bubble, and the oscillation in bubble shape and path. For large cap-shaped bubbles Moissis and Griffith (1962) have reported that axial interaction becomes significant when bubbles follow one another more closely than five to six tube diameters and is perceptible to about eight tube diameters. As in the case of the radial interaction, it seems that the best

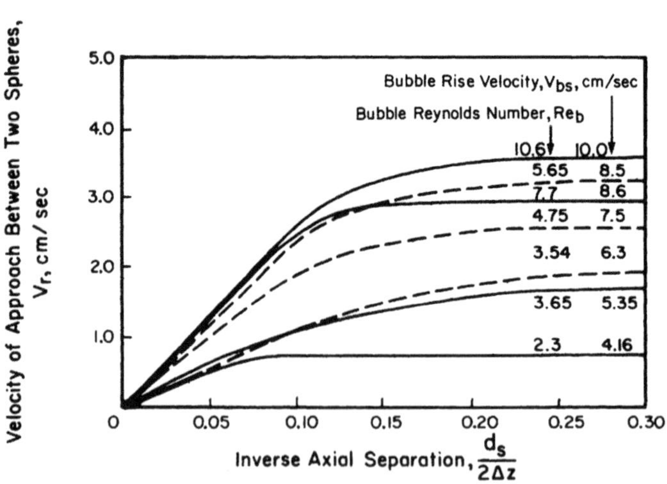

Figure 8.32. Effect of axial interaction on bubble rise velocity. (From Jayaweera, K.O.L.F., B. J. Mason, and G. W. Slack, *J. Fluid Mech.*, **20**, 121, 1964.)

approach to the problem of the axial interaction is by direct observation of the effect of bubble separation distance on bubble rise velocity.

Yip *et al.* (1970) found no axial interaction and a stable flow pattern if the axial bubble separation distance exceeded a critical value. For air bubbles rising in water in a 5.04 cm tube they determined the critical center-to-center separation distance Δz_c, cm, to be

$$\Delta z_c = \frac{n v_b D^2}{0.00067 + 0.625\, n v_b}$$

where n = number of bubbles in approximately the same horizontal plane,
v_b = bubble volume, cm^3,
D = tube diameter, cm.

The limiting value, for large cap-shaped bubbles, was found from this equation to be 41.8 cm compared with the 40.0 cm obtained by Moissis and Griffith (1962).

4.6 Rise Velocity of a Continuous Swarm of Bubbles in a Stagnant Liquid

If closely spaced, uniformly sized bubbles are continuously generated in a swarm, as might issue from a porous plate bubble generator, certain further effects are observed.

First, under laminar flow conditions a situation may exist where a myriad of uniformly distributed small bubbles, each in contact or almost in contact with its neighbors, rise at more or less constant velocity. Second, these bubbles may remain distinct for a considerable travel distance despite their proximity and the normal tendency for agglomeration or coalescence. Zuber *et al.* (1967) refer to this situation as the laminar bubbly regime. Finally, the velocity of such a swarm of bubbles is significantly less than that of a single bubble or a small group of bubbles of the same size. The situation is analogous to hindered settling, which we discussed in Chapter 1. The rising continuous swarm of bubbles may be considered stationary and the liquid may be thought of as draining through a bed of bubbles whose permeability decreases as the concentration of bubbles increases. Qualitatively the effect is described by the Carman–Kozeny relationship (Figure 1.8).

The phenomenon is illustrated by data of Nicklin (1962) reproduced in Figure 8.33 showing the effect of superficial gas velocity (and therefore the volume fraction gas) on the rise velocity in a stagnant liquid of the bubble swarm. We may derive Figure 8.34 from Figure 8.33, and show directly, through a factor $F_H = (V_{bs} \text{ at } E_G)/(V_{bs} \text{ at } E_G = 0)$, the effect of the *in situ* volume fraction gas on the ratio of the stagnant liquid rise velocity of a bubble swarm to the stagnant liquid rise velocity of a single bubble.

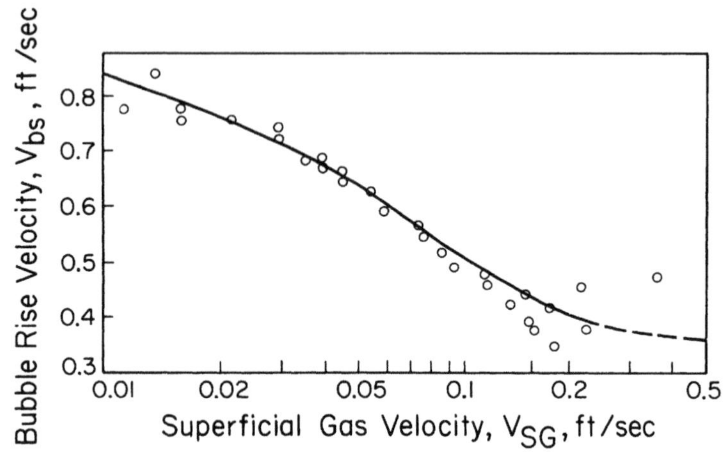

Figure 8.33. Effect of superficial gas velocity on bubble rise velocity. (Redrawn from Nicklin, D. J., *Chem. Eng. Sci.*, **17**, 693, 1962.)

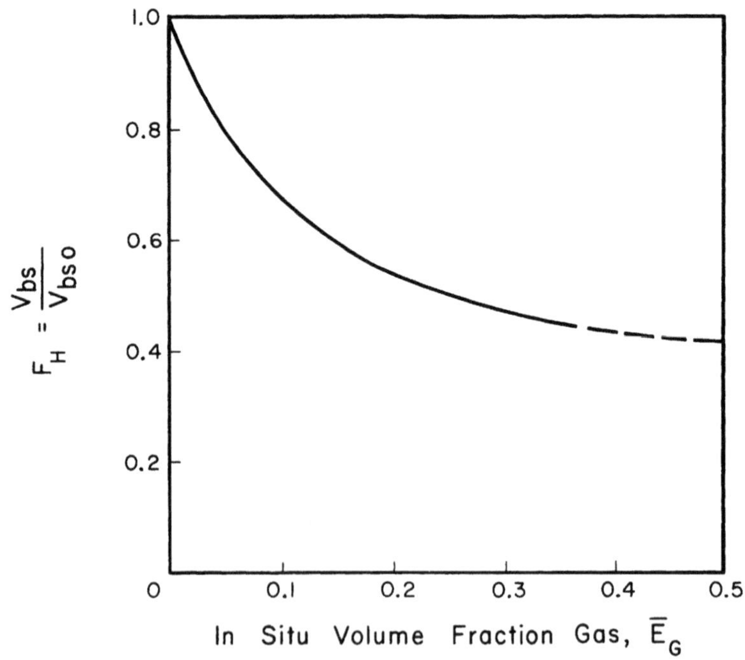

Figure 8.34. Effect of gas phase concentration on bubble rise velocity. (From the data of Nicklin, D. J., *Chem. Eng. Sci.*, **17**, 693, 1962.)

The work of Zuber and Hench (1962) and Zuber (1964) provides an alternate and probably more accurate way of predicting the rise velocity of bubble swarms. Under conditions where the bubble swarm rises in a stagnant liquid in laminar motion these authors propose

$$V_{bs} = \frac{g\Delta\rho d^2}{18\mu_L}(1 - E_G)^3 \qquad \text{for } d < 0.05 \text{ cm} \tag{8.59}$$

$$V_{bs} = 1.41\left(\frac{\sigma g\Delta\rho}{\rho_L^2}\right)^{1/4}(1 - E_G)^{3/2} \quad \text{for } 0.05 < d < 2.0 \text{ cm} \tag{8.60}$$

It is to be noted that these two equations may be expressed as

$$V_{bs} = V_{bs0}(1 - E_G)^K \tag{8.61}$$

where V_{bs0} is the value of V_{bs} at $E_G = 0$, with K reducing from 3.0 for very small bubbles to 1.5 for larger bubbles.

At higher bubble Reynolds numbers the bubbles lose their separate identity and a region which Zuber *et al.* (1967) describe as the churn turbulent bubbly regime is encountered. In this region Equation 8.50 derived by Peebles and Garber is applicable, although Zuber *et al.* recommend the following equation:

$$V_{bs} = 1.41\left(\frac{\sigma g\Delta\rho}{\rho_L^2}\right)^{1/4} \tag{8.62}$$

The constant 1.41 is that proposed by Levich (1962), and it is intermediate between the value 1.18 proposed by Peebles and Garber (1953) and the constant 1.53 suggested by Harmathy (1960).

4.7 Cross-Sectional Distribution of Bubbles in a Flowing Stream

A train of single bubbles in a flowing stream tends to occupy the central position with minor deviations in all directions. Two bubbles initially separated tend to remain so, although, depending upon their size relative to the tube diameter, they may collide, adhere, and coalesce. As the number of bubbles increases, by analogy with the case of solid spheres, we might expect an arrangement approaching a regular polygon for three to six, or so, bubbles, but thereafter about the only order to be expected in the arrangement is that the time average distribution might be described statistically.

A number of experimental studies relating to the distribution of bubbles across a cross section have been carried out. Commonly the distribution is expressed in terms of ε_G, the time average local volume fraction of the gas phase (often referred to as the local "void ratio"). Typical profiles are given in Figure 8.35 taken from Malnes (1966). In this figure the separate lines

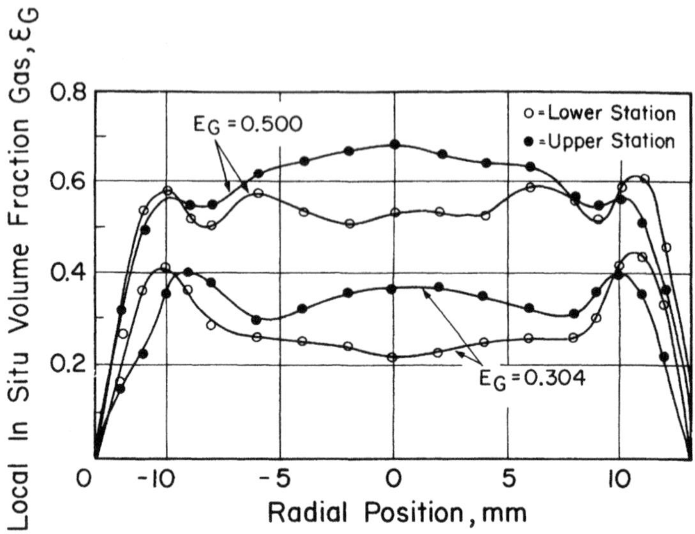

Figure 8.35. Typical concentration profiles in the bubble flow pattern. (From Malnes, D., Institutt for Atomenergi, Kjeller Research Establishment, Kjeller, Norway, May 1966.)

represent profiles taken simultaneously at the lower and upper end of the test section, and the spread between them is indicative of the time variation (and uncertainties) in the concentration profile. On the basis of data such as those of Malnes, several workers, (Bankoff, 1960; Zuber and Findlay, 1965; and others) have found it expedient to approximate the time average distribution of bubbles by the power law expression discussed in Chapter 7, i.e.,

$$\frac{\varepsilon_G - \varepsilon_{Gw}}{\varepsilon_{G0} - \varepsilon_{Gw}} = 1 - \left(\frac{r}{R}\right)^n \tag{7.110}$$

It is clear that such an expression, even with the proper values of ε_{Gw}, ε_{G0}, and n can only simulate the more regular profiles and is incapable of reflecting the peaking phenomenon observed in Figure 8.35.

4.8 Cross-Sectional Distribution of Bubble Velocity in a Flowing Stream

When more than a single bubble is rising in a flowing stream, the location of bubbles within the cross section influences their velocities and we may think not only of the velocity profile of the flowing liquid, but also of the (leading) velocity profile of the bubbles. Although theoretically one might hope to be able to compute the bubble velocity profile from that of the liquid, the bubble concentration profile, the rise velocity of individual bubbles, the wall effects, and the radial interactions, in practice this is not possible because the

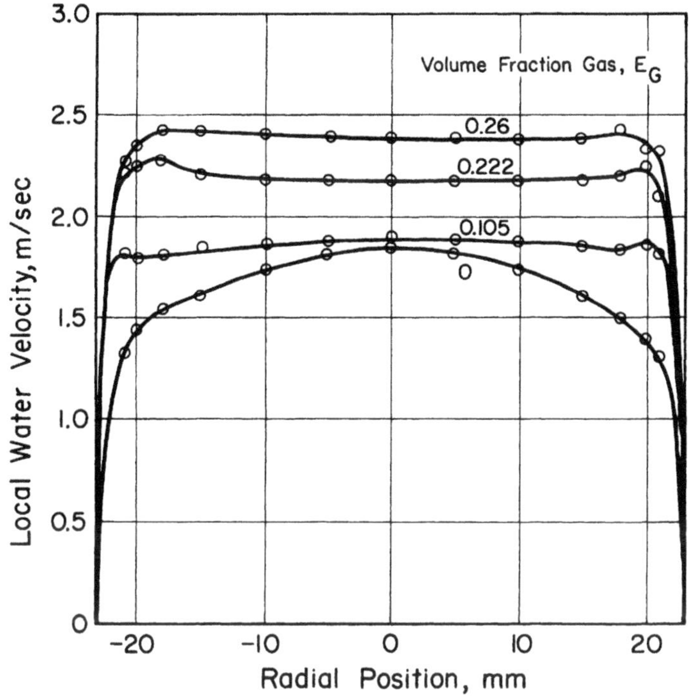

Figure 8.36. Typical velocity profiles in the bubble flow pattern. (From Malnes, D., Institutt for Atomenergi, Kjeller Research Establishment, Kjeller, Norway, May 1966.)

presence of the bubbles influences the velocity profile of the liquid. Our real knowledge of both the liquid and the bubble velocity profiles comes from experimental work.

The work of Malnes (1966) on an air water system reveals the effect of the presence of bubbles, distributed as discussed earlier, on the profile of the flowing liquid. This is shown in Figure 8.36, which demonstrates that, with average volume fractions of the gas phase in excess of approximately 0.10, the normal turbulent velocity profile changes to a plug-shaped profile. We would expect the bubble velocity profiles to be similar to these but leading the liquid profile.

4.9 The Average Absolute Rise Velocity of Bubbles in a Flowing Stream

The absolute rise velocity of bubbles in a flowing stream would be expected to be the simple vector sum of the bubble rise velocity in a stagnant liquid and the local absolute velocity of the flowing liquid. That such is the case has been extensively confirmed by experiments. Where we are dealing with a single, very small bubble or a train of such bubbles traveling in a known rectilinear path, and where the velocity profile of the liquid is known, there is no problem, and the local value of V_{bf} is

$$u_{bf} = u_L + V_{bs} \tag{8.63}$$

where u_L is the appropriate local liquid velocity. Haberman and Sayre (1958) approached the problem through an addition to their equation (Eq. 8.54) of a term related to the velocity of the liquid at the center line of the tube, i.e.,

$$V_{bf} = \frac{d^2 g \Delta\rho}{18\mu_L K_1}\left(\frac{1 + \mu_L/\mu_G}{1 + \frac{2}{3}\mu_L/\mu_G}\right) + K_2' u_{LO} \tag{8.64}$$

Their values for K_2' are given in Figure 8.37.

Similarly, in the case of multiple bubbles uniformly distributed across the cross section with a uniform liquid velocity profile the absolute bubble rise velocity is given by

$$V_{bf} = V_M + V_{bs} \tag{8.65}$$

where V_M is the constant liquid velocity above and below the bubbles. From continuity considerations, as illustrated in Figure 8.38,

$$V_M = \frac{Q_G + Q_L}{A} = V_{SG} + V_{SL} \tag{8.66}$$

Also it may be shown that $V_M > V_L$.

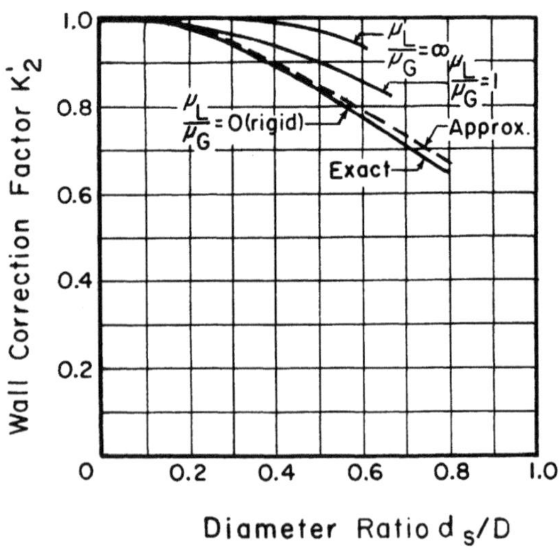

Figure 8.37. The Haberman and Sayre wall correction factor, K_2', Equation 8.64. (From Haberman, W. L., and R. M. Sayre, Dept. of the Navy, Hydromechanics Lab. Res. and Dev. Report No. 1143, Oct. 1958.)

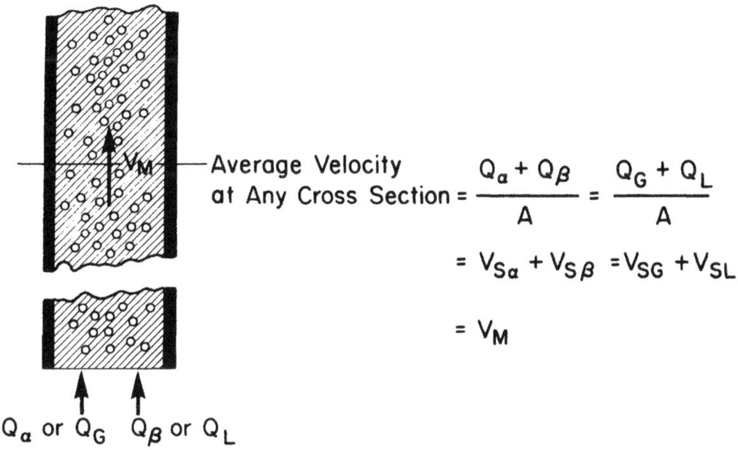

Figure 8.38. Overall volumetric relationships in vertical bubble flow.

The difficulty in the application of the vector summing concept to determine the average, absolute rise velocity of bubbles in a flowing stream arises from the nonuniformity of the bubble concentration, and the liquid velocity across the cross section. It is just in this situation that the analysis of Zuber and Findlay (1965) discussed in Chapter 7 is so helpful. The weighted average drift velocity $V_{\alpha j}$ of Equation 7.104 may be interpreted as V_{bs}, the rise velocity of the bubbles in a stagnant liquid, and we obtain

$$V_{bf} = C_0 V_M + V_{bs} \tag{8.67}$$

where V_{bf} is the same as V_α of Equation 7.104 and C_0 is the distribution coefficient related to the velocity and the concentration profiles.

For the cases of concern to us we may assume $\varepsilon_{Gw} = 0$, i.e., a solid wall with no heat addition such that the bubble concentration drops to zero at the wall. Then, if we assume power law velocity and concentration profiles we have Equation 7.112,

$$C_0 = \frac{m+2}{m+n+2}\left[1 + \frac{\varepsilon_{G0}}{E_G}\left(\frac{n}{m+2}\right)\right] \tag{7.112}$$

Zuber *et al.* (1967) have analyzed extensive data of their own and others with the aid of Equation 7.112. Typical results are shown in Figure 8.39 dealing with the air water data of Petrick (1962) in the bubble flow pattern. It is observed that the data conform to Equation 8.67 with $V_{bs} = 1.41(\sigma g \Delta\rho/\rho_L^2)^{1/4}$ and $C_0 = 1.6$. Zuber *et al.* have found that

1. C_0 is not very sensitive to n and m;

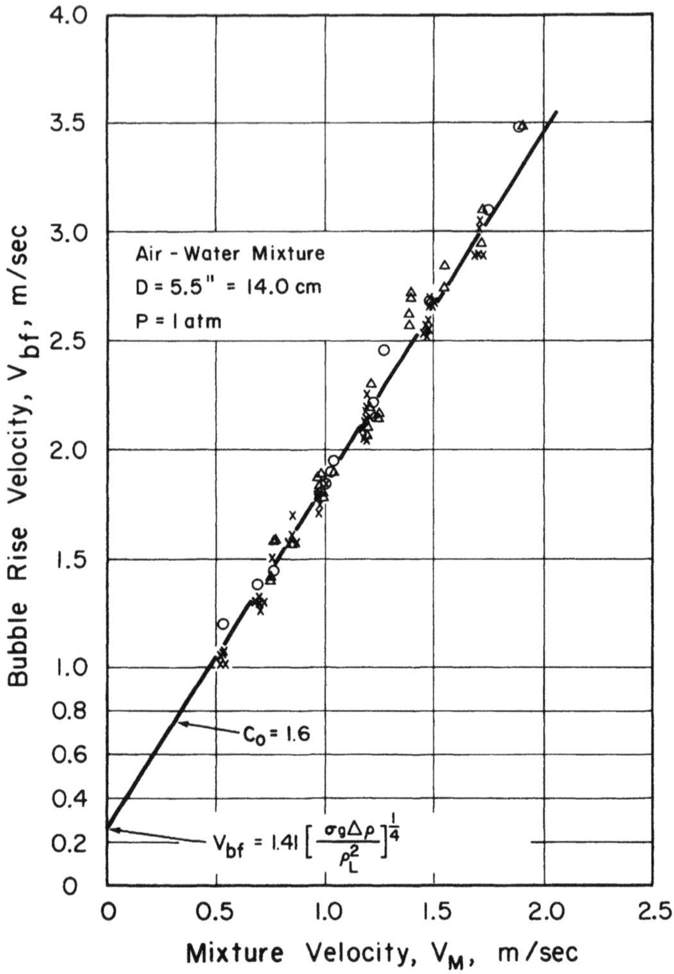

Figure 8.39. Confirmation of Equation 8.67. (From N. Zuber *et al.*, E.U.R.A.E.C. Report No. 1949, General Electric Co., San José, California, 1967.)

2. where values of both n and m are not known they may be taken as equal, in which case

$$C_0 = \frac{n+2}{n+1}$$

3. values of C_0 vary from about 1.60 to 1.10 corresponding (if n and m are assumed to be equal) to n ranging from below unity (high bubble concentration and liquid velocity at center line) to $n = 9$ (very flat concentration and velocity profile).

From these considerations it is clear that the absolute average bubble rise velocity may be determined from Equation 8.67 provided one can calculate or estimate the distribution coefficient, C_0. This requires a knowledge of, or assumptions concerning, the concentration and velocity profiles (i.e., n and m). Fortunately, approximately correct values for n and m yield good values for C_0. Figure 7.5 is helpful in this respect.

From the foregoing and the most probable profiles we may summarize the relations for the absolute average rise velocity of bubbles in the bubble flow regime as follows:

Discrete Bubbles in an Axial Train

$\mathrm{Re}_b < 2$

$$V_{bf} = (1.45 \pm 0.05) V_M + \frac{g \Delta\rho d_s^2}{18 \mu_L} \tag{8.68}$$

$2 < \mathrm{Re}_b < 4.02 G_1^{-0.214}$

$$V_{bf} = (1.45 \pm 0.05) V_M + 0.33 g^{0.76} \left(\frac{\rho_L}{\mu_L}\right)^{0.52} \left(\frac{d_s}{2}\right)^{1.28} \tag{8.69}$$

$4.02 G_1^{-0.214} < \mathrm{Re}_b < 3.10 G_1^{-0.250}$

$$V_{bf} = (1.45 \pm 0.05) V_M + 1.35 \left(\frac{2 \sigma g_c}{\rho_L d_s}\right)^{0.50} \tag{8.70}$$

Continuous Swarms of Bubbles

$\mathrm{Re}_b < 2$ (laminar bubbly)
$d_s < 0.05$ cm

$$V_{bf} = (1.15 \pm 0.05) V_M + \frac{g \Delta\rho d_s^2}{18 \mu_L} (1 - E_G)^3 \tag{8.71}$$

$2.0 > d_s > 0.05$ cm

$$V_{bf} = (1.15 \pm 0.05) V_M + 1.41 \left(\frac{\sigma g \Delta\rho}{\rho_L^2}\right)^{1/4} (1 - E_G)^{3/2} \tag{8.72}$$

$\mathrm{Re}_b > 2$ (churn-turbulent, bubbly)

$$V_{bf} = (1.25 \pm 0.05) V_M + 1.41 \left(\frac{\sigma g \Delta\rho}{\rho_L^2}\right)^{1/4} \tag{8.73}$$

4.10 The Overall Mechanics of the Bubble Flow Pattern

The various factors previously discussed lead to an overall understanding of the bubble flow pattern and its stability. The situation may best be described by starting with a train of single or a few (n) bubbles introduced into a flowing liquid and rising axially. The average axial spacing of the bubbles is determined by the gas flow rate, the average individual bubble volume, and the bubble velocity, i.e.,

$$\Delta z_b = \frac{V_{bf} v_b n}{Q_G} \tag{8.74}$$

Notwithstanding this, there are random variations in bubble volume, velocity, and spacing. As the gas rate is increased, with constant average individual bubble volume, the bubble spacing decreases and eventually axial interaction occurs between the closest bubbles with consequent bubble acceleration. Following this, bubble collision may occur. Under laminar flow conditions the situation is reasonably stable, bubbles resist agglomeration or coalescence, and ideal bubbly flow takes place. Under turbulent conditions bubble adherence may result and coalescence may also take place depending upon the interfacial tension. Bubble adherence, with or without coalescence, results in bubbles of larger effective size and more widely spaced. Flow stability is again obtained at a bubble relative velocity corresponding with the new bubble volume and shape. This velocity may be higher or lower than the original velocity, depending on the part of the drag coefficient curve that is applicable. During this time the increase in the *in situ* volume fraction gas of the flowing mixture has the effect of reducing the relative bubble velocity in a manner analogous to the reduction of the settling velocity of particles with increase in particle concentration. This effect, modest at first, becomes quite important as the volume fraction of gas increases beyond about 0.05.

With further increase in the gas rate initially small bubbles transform even more quickly to fewer larger ones, and eventually the point may be reached where some enlarged bubbles stretch nearly across the tube. The bubble then is cap-shaped with a near-spherical nose and with a breadth-to-depth ratio of about four (Rosenberg, 1950, Harmathy, 1960). The development of this type of bubble marks the beginning of the slug flow pattern.

The foregoing sequence considers the flow pattern changes resulting from increasing the gas flow rate in a train of single or of a relatively few bubbles of initially constant volumes. Suppose that instead of such a train of bubbles there were a myriad of small bubbles of varying size as might be issued from a porous plate bubble generator. Initially these bubbles might be discharged at a low frequency, but as the gas rate is increased the frequency would increase. The situation is not too different from the one described for the single train, except that in the initial bubble clusters axial and radial interactions and random oscillations will give early rise to bubble collision, adherence, and possible coalescence. In fact, the probability is that, given sufficient opportunity (i.e., travel time), a swarm of small bubbles will change over to a single train of larger bubbles or bubble aggregates. Nonetheless, under laminar conditions, the swarm may be highly persistent. Thus, regardless of the manner of initial introduction of the gas and liquid, if the gas volumes are such as to result in bubbles within axial, radial, or oscillatory influence of one another, the ultimate stable flow pattern will tend to be the same. The travel distance required for the stable pattern to develop depends upon the flow rates, as well as the interfacial and other properties, of the fluids.

4.11 Flow Pattern Details

The details of the bubble flow pattern may be predicted at least approximately from a knowledge of the characteristics of the bubble generating device, the properties of the gas and liquid, the gas and liquid flow rates, the distance from the bubble generator and the static pressure at the bubble generator or any other known point.

1. From the nature of the bubble generator (number and size of orifices), the fluid properties and the gas flow rate, we may estimate the average volume and frequency of bubbles as released from the bubble generator from Equation 8.31 and the most appropriate of Equations 8.33, 8.34, and 8.36.
2. A check on the probable stability of the n bubbles in a cross section may be made by calculating the bubble Reynolds number and comparing n with n_{max} as given by Equation 8.57. If Re_b exceeds 2 and n exceeds n_{max}, then bubble collision, adherence, and possible coalescence must be anticipated, resulting in an approximate doubling of the average bubble volume and halving of n. This process might be expected to continue until the resulting n is less than n_{max}. The following calculations should be made with the adjusted n and v_b values.
3. The absolute rise velocity of the bubbles may be estimated from the appropriate of Equations 8.68 to 8.70. Corrections for wall effect and radial and axial interaction may be made where appropriate.
4. The initial average axial spacing, Δz_b, of the bubbles may be estimated from Equation 8.74.
5. If the axial spacing exceeds the critical spacing Δz_c, as determined from Equation 8.58, the flow pattern may be considered stable and is defined by Δz_b, n, v_b, ω, and V_{bf}.

6. An always present complication results from a variation in bubble size and consequently in bubble velocity. Smaller, slower moving bubbles tend to be absorbed by larger, more rapidly moving ones; thus the number of bubbles tends to decrease and their average size to increase. Qualitative account may be taken of this in the estimation of n and v_b, i.e., it may be assumed that the smallest bubbles in the distribution will have a short life.
7. If the axial spacing of the bubbles is less than the critical spacing given by Equation 8.58, the flow pattern is probably not stable and a transition to double average bubble volumes with double spacing and half frequency is to be expected. If the estimated spacing for the doubled-volume bubble is still below the critical spacing for that volume, further bubble enlargement through collision, adherence, and possible coalescence will likely occur. The stable pattern will be that when the final spacing $\Delta z_b > \Delta z_c$ for the final bubble volume and frequency. Although bubble adherence and coalescence must eventually be expected if the nearness and velocities of bubbles may result in collision, in many practical situations the bubble size and the interfacial properties are such that coalescence does not occur rapidly, and adhering bubbles are eventually separated by other collisions. Thus, it is possible, given sufficiently high interfacial tension, to have a relatively stable bubble pattern with a swarm of rather closely spaced bubbles. Under these conditions the bubbles tend to rise as a group, i.e., hindered rising takes place. The relative rise velocity of the group cannot be predicted precisely, but it may be estimated from the free rise velocity of the individual bubbles and the factor of Figure 8.34 based upon the data of Nicklin (1962) or alternately from the appropriate of Equations 8.71 to 8.73. The flow pattern is defined through n, v_b, Δz_b, ω, and V_{bf} as before.

When the details of the stable, or relatively stable flow pattern are known, the holdup and the pressure gradient may be calculated.

4.12 Holdup

The *in situ* volume fraction gas is calculated as follows. Consider a tube of length L:

$$v_t = \frac{\pi D^2}{4} L \tag{8.75}$$

where v_t = the volume of the tube

$$\Delta z_b = \frac{V_{bf}}{\omega} \tag{8.76}$$

$$\bar{E}_G v_t = \frac{L}{\Delta z_b} n v_b = \frac{L n v_b \omega}{V_{bf}} \tag{8.77}$$

$$\bar{E}_G = \frac{4 n \omega v_b}{\pi D^2 V_{bf}} = \frac{4 Q_G}{\pi D^2 V_{bf}} = \frac{V_{SG}}{V_{bf}} \tag{8.78}$$

With V_{bf} determined from the appropriate of Equations 8.68 to 8.73, the *in situ* average volume fraction gas may be determined from Equation 8.78, and the holdup ratio, the slip velocity, and other measures of the holdup may be calculated from the appropriate equations discussed in Chapter 7. Orkiszewski (1957) reports that Griffith (1962) expresses E_G for bubble flow in terms of the slip velocity through Equation 7.78a and recommends that a good approximation to E_G may be obtained by using $S = 0.8$ ft/sec.

4.13 Pressure Gradient

The pressure gradient may now be estimated by essentially single-phase methods, i.e., by recognizing that the average mixture density is

$$\rho_M = (1 - E_G)\rho_L + E_G \rho_G \tag{8.79}$$

or, where the second phase is a gas phase at low pressure,

$$\rho_M \simeq (1 - E_G)\rho_L \tag{8.80}$$

and that the shear stress at the wall is essentially that due to the liquid phase flowing at the average mixture velocity, $V_M = V_{SL} + V_{SG}$. Applying the mechanical energy balance, we obtain

$$\frac{\Delta P}{\rho_L \Delta z} = (1 - \bar{E}_G)\frac{g}{g_c} + \frac{2 f_L V_M^2}{g_c D} \tag{8.81}$$

where f_L is determined from the conventional Reynolds number–friction factor plot (Figure 4.16) using a Reynolds number

$$\mathrm{Re}_M = \frac{D V_M \rho_L}{\mu_L} \tag{8.82}$$

Equation 8.81 is very similar to that proposed by Griffith and Wallis (1961) and discussed by Orkiszewski (1967). The difference lies in the use of V_M in Equations 8.81 and 8.82 rather than $V_L = V_{SL}/(1 - E_G)$ as proposed by Griffith and Wallis. From a consideration of the flow mechanism, V_M should be the better quantity to use.

4.14 The Transition from Bubble to Slug Flow

The transition from bubble to slug flow occurs when n becomes unity and the bubble, either as initially generated or after bubble agglomeration and coalescence, increases in size to the point where it behaves like a Taylor bubble. The minimum size bubble that behaves this way is a cap-shaped bubble with a spherical nose of diameter about $0.75D$. Although the transition to slug flow might be said to start at the point of the formation of such bubbles, it seems more practical to define the start of slug flow as when Taylor bubbles, having a length at least equal to the diameter of the tube, are formed. Such bubbles would be readily identified visually.

As we will see in the next section, slug flow is characterized not only by the Taylor bubbles, but also by a myriad of small bubbles in the intervening liquid slug. In fully developed slug flow the average volume fraction of gas in the intervening liquid slug is of the order of 0.10 (Akagawa and Sakaguchi, 1966), but at the transition from bubble to slug flow the fraction gas in the liquid slug is much higher. We may take it to be of the order of 0.25. The volume of a Taylor bubble may be approximated by Equation 8.122 (Griffith and Wallis, 1961) discussed later. For Taylor bubbles having a length equal to their diameter, this becomes

$$v_b = \frac{\pi D^2}{4}(0.913D - 0.526D)$$

and consequently the volume of gas in a minimum Taylor bubble and the liquid slug associated with it is

$$v_b' = \frac{\pi D^2}{4}(0.913D - 0.526D) + 0.25\Delta z_c \tag{8.83}$$

The minimum spacing for stable slugs is in the range of six to eight diameters (Moissis and Griffith, 1962). Substituting $7D$ for Δz_c, we obtain, approximately,

$$v_b' \simeq 0.52\pi D^3 \tag{8.84}$$

Now if we define V_{SG}' as the transition superficial gas velocity, we may write from Equation 8.74

$$V_{SG}' = \frac{4V_{bf}v_b'}{\Delta z_c \pi D^2} \tag{8.85}$$

The rise velocity of Taylor bubbles in a flowing liquid is given by

$$V_{bf} = 1.2(V_{SG}' + V_{SL}) + 0.35\sqrt{gD\frac{\Delta\rho}{\rho}} \tag{8.86}$$

where ρ is the density of the continuous phase. This equation is derived as Equation 8.116 in the next section. Substituting for V_{bf}, v_b', and Δz_c in Equation 8.85, we obtain, on rearrangement,

$$V_{SG}' = 0.56V_{SL} + 0.16\sqrt{gD\frac{\Delta\rho}{\rho}} \tag{8.87}$$

Equation 8.87 should permit good order of magnitude estimates of the bubble-slug transition velocity.

Orkiszewski (1967) gives an equation for Griffith and Wallis' transition from bubble to slug flow. The relationship is entirely empirical however, representing the flow boundary line of Figure 8.8 which, according to Griffith and Wallis, was not too well defined.

5 THE SLUG FLOW PATTERN

5.1 Introduction

The slug flow pattern occurs at all rates of flow of the continuous phase and over a modest intermediate range of flow of the discontinuous phase. Dealing with the gas-liquid case, and with increasing gas rate at constant liquid rate, the pattern commences when the rate of gas flow is such that cap-shaped bubbles, which almost bridge the tube, are formed. (See discussion of the Transition from Bubble to Slug Flow, Section 4.14.) The wake of these bubbles contains anywhere from a few to many much smaller, more or less spherical bubbles, but the region between bubbles is mostly liquid filled. The liquid filled region is referred to as the slug.

With increasing gas flow rate the cap-shaped bubbles become enlarged and elongated, taking on a distinctive shape with a near-spherical nose and a tapered transition to a near-cylindrical body. Surrounding the bubble is a relatively thin film of liquid flowing downward. The interior of the bubble is a region of essentially constant pressure, and consequently the pressure in the adjacent film is also essentially constant. This means that the gravitational force acting on the film must be entirely balanced by the shear stress at the wall and the (small) upward shear stress at the bubble film interface. Thus, at the velocities encountered in the slug flow regime, the liquid film flows downward generating an upward wall stress. The situation is illustrated in Figure 8.40. As the bubbles become elongated the slug length decreases, and a point may be reached where the wake of a bubble influences the velocity of the trailing bubble. This causes acceleration of the trailing bubble and ultimately its collision and coalescence with the leading bubble, resulting in an approximate doubling of the length of both the bubble and the liquid slug.

At low and intermediate liquid velocities as the gas rate and the bubble length are further increased, waves and instabilities in the liquid film surrounding the bubble cause a penetration of liquid into the bubble, a destruction of the characteristic shape, and an alternate merging and separation of fragments of the bubble with the liquid slug. The resulting pattern, exhibiting much agitation, is the froth pattern, really a broad transition to the more stable annular-mist flow pattern. On the other hand, at superficial liquid velocities of the order of 7 or 8 feet per second and greater, further increase in the gas flow rate results in very long stable bubbles and a direct transformation to annular-mist flow.

The above discussion and that following throughout Section 5 relate primarily to gas-liquid systems. We have little specific knowledge of slug flow in liquid-liquid systems. Nonetheless, the overall mechanics for the two types of systems should be the same and the relations derived for gas-liquid mixtures should be generally applicable to liquid-liquid mixtures except in those cases in which the density or the viscosity of the gas phase is neglected.

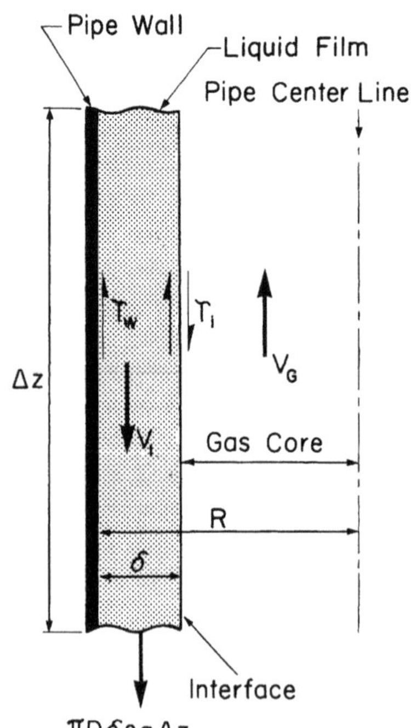

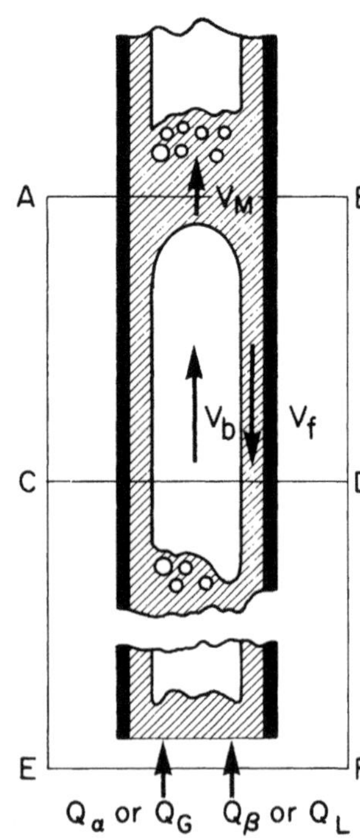

Figure 8.40 (left). Schematic illustration of stresses in the liquid film in slug flow.

Figure 8.41 (right). Overall volumetric relationships in vertical slug flow.

5.2 Overall Continuity Considerations

A consideration of the inflow and outflow into a typical element ABCD, as illustrated in Figure 8.41, is revealing as to the relationships between the bubble, film, and slug velocities. We may note immediately that overall continuity requirements over section ABEF, assuming incompressible fluids, indicate that the average absolute velocity of the liquid ahead of the bubble, i.e., the average slug velocity is

$$V_M = \frac{Q_G + Q_L}{A} = V_{SG} + V_{SL} \tag{8.88}$$

The compressibility of the gas phase may be accounted for by expressing Q_G, and hence V_{SG}, as the volume or velocity corrected from the inlet pressure and temperature conditions to the pressure and temperature at the section under consideration.

Defining V_b and V_f as the average absolute velocities of the bubble and the film, each being positive in the sense shown in the figure, overall continuity considerations require that

$$V_f A_f = V_b A_b - V_M A \tag{8.89}$$

where $A = \frac{\pi}{4} D^2$,

$A_b = \frac{\pi}{4} (D - 2\delta)^2$,

$A_f = \pi\delta D - \pi\delta^2 \simeq \pi\delta D$,

δ = film thickness.

Note that V_f is taken as positive in the downward direction. From these relations

$$V_f = \frac{V_b(D - 2\delta)^2 - V_M D^2}{4\delta(D - \delta)} \tag{8.90}$$

or

$$V_b = \frac{V_M D^2 + 4V_f \delta(D - \delta)}{(D - 2\delta)^2} \tag{8.91}$$

It is convenient to define $\xi = 2\delta/D$, which permits a simplification of Equations 8.90 and 8.91 to

$$V_f = \frac{V_b(1-\xi)^2 - V_M}{2\xi - \xi^2} \tag{8.92}$$

$$V_b = \frac{V_M + V_f(2\xi - \xi^2)}{(1-\xi)^2} \tag{8.93}$$

If in Equations 8.92 and 8.93 we neglect the second-order term, ξ^2, we obtain

$$V_f = \frac{V_b(1-2\xi) - V_M}{2\xi} \tag{8.92a}$$

$$V_b = \frac{V_M + V_f\,2\xi}{1-2\xi} \tag{8.93a}$$

Equation 8.92 is essentially the same as the equation derived by Nicklin and Davidson (1962). Equations 8.92 and 8.93 show that the average film velocity may be expressed in terms of the bubble velocity, the average slug velocity, and the film thickness; or the bubble velocity may be determined from the film velocity, the average slug velocity, and the film thickness.

From Equation 8.92 or 8.92a, we observe that V_f may be positive or negative, depending upon the relative values of the terms in the numerator. For V_f to be negative and the flow in the liquid film to be upward,

$$V_b(1-2\xi) < V_M$$

and since (see Equation 8.116 given subsequently)

$$V_b = 1.2V_M + V_{bs}$$

we have

$$2\xi > 1 - \frac{V_M}{1.2V_M + V_{bs}}$$

or

$$\frac{\delta}{D} > \frac{1}{4}\left(\frac{0.2 + V_{bs}/V_M}{1.2 + V_{bs}/V_M}\right) \tag{8.94}$$

This relationship is plotted in Figure 8.42, which shows the range of conditions where upward and downward flow occurs in the film over the range of

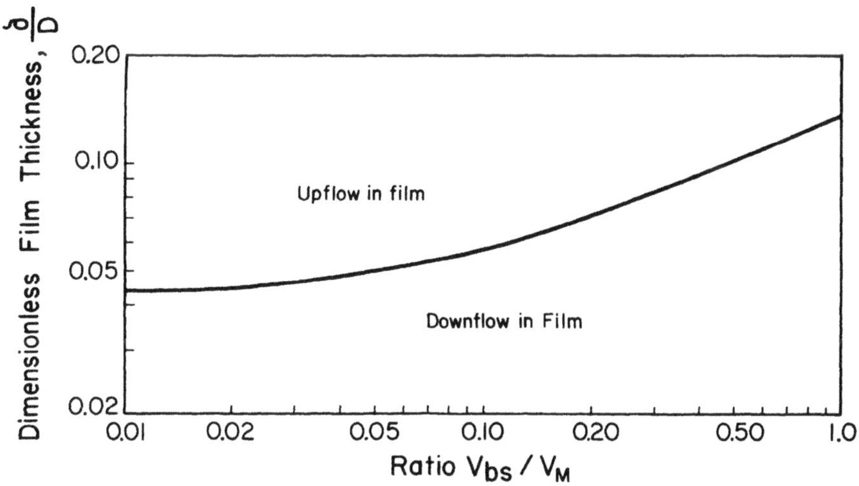

Figure 8.42. Conditions for upflow and downflow in the liquid film in slug flow.

V_{bs}/V_M normally encountered in the slug flow region. We observe that for the film to be moving downward, as is the case in slug flow, continuity considerations indicate that the film thickness has a maximum value in the range of $0.04D$ to $0.14D$.

In arriving at the inequality of Equation 8.94, we have not considered the conditions under which the film may become unstable. The relationship is based on simple volumetric balance considerations for the film.

For the limiting case where the volume of the small bubbles within the liquid slug is negligible with respect to the volume of the large bubble, a macroscopic mass balance on the gas phase is expressed by equating the average gas flow rate to the frequency and volume of the gas bubbles

$$Q_G = \omega v_b \tag{8.31}$$

where

$$\omega = \frac{V_b}{L_b + L_s} \tag{8.95}$$

Equations 8.31 and 8.95 lead to

$$Q_G = \frac{v_b V_b}{L_b + L_s} \tag{8.96}$$

which is the gas continuity equation. This relationship has been given by Griffith and Wallis (1961) and Moissis (1963).

Substituting Equation 8.96 into Equation 8.88, we obtain the liquid volume or mass balance

$$Q_L = AV_M - \frac{v_b V_b}{L_b + L_s} \tag{8.97}$$

With $v_b = L_b A_b$, $v_s = L_s A$, and from Equation 8.89, Equation 8.97 may be rearranged to the form

$$Q_L = \frac{v_s V_M}{L_s + L_b} - \frac{A_f L_b V_f}{L_s + L_b} \tag{8.98}$$

Here we see that the liquid flow rate is the sum of two terms, the first being the volume of the liquid slug, v_s, multiplied by an effective frequency, $V_M/(L_s + L_b)$; and the second term, which is inherently negative in slug flow, represents the downflow of the liquid in the film. The second term is equivalent to the volume of the liquid in the film, $A_f L_b$ or v_f multiplied by a second effective frequency $V_f/(L_s + L_b)$. Equation 8.97 may also be written as

$$Q_L = \frac{v_s V_M - v_f V_f}{L_b + L_s} \tag{8.99}$$

indicating that the liquid flow rate may be expressed as the product of a volume-weighted effective frequency, $(v_s V_M - v_f V_f)/[(v_s + v_f)(L_b + L_s)]$, and the total liquid volume, $(v_s + v_f)$. We note that the true frequency of the system, ω, previously defined as $V_b/(L_s + L_b)$ differs from all of these effective frequencies.

Akagawa and Sakaguchi (1966) have pointed out that in many cases 10 to 20 percent of the total gas volume is contained in the small bubbles, and that this volume should not be neglected. If the total average volume fraction of the small bubbles in the liquid slug is defined as E_{GSL}, and the average volume of the liquid slug is taken as v_s, then Equation 8.96 may be modified to

$$Q_G = \frac{v_b V_b}{L_s + L_b} + \frac{E_{GSL} v_s V_M}{L_s + L_b} \tag{8.100}$$

where the second term is based upon the volume and effective frequency of the liquid slug as discussed in connection with Equation 8.98.

This leads to a modification of Equations 8.97 and 8.98 as follows:

$$Q_L = AV_M - \frac{v_b V_b + E_{GSL} v_s V_M}{L_b + L_s} \tag{8.101}$$

$$Q_L = \frac{(1 - E_{GSL})v_s V_M - A_f L_b V_f}{L_b + L_s} \tag{8.102}$$

5.3 The Flow of a Taylor Bubble in Stagnant Liquids

The starting point for an analysis of slug flow (Griffith and Wallis, 1961; Nicklin, Wilkes, and Davidson, 1962; and Brown, 1965) is the rise of large bubbles in stagnant liquids. The initial studies of the rate of rise of such bubbles were by Dumitrescu (1943) and Davies and Taylor (1950), and bubbles of the characteristic shape are, therefore, often referred to as Dumitrescu or Taylor bubbles. The latter term will be used here consistent with the suggestion of Griffith and Wallis (1961). Dumitrescu, and Davies and Taylor proposed that the behavior of gas bubbles rising through actual liquids in tubes could be approximated by the hypothetical case of a weightless bubble rising through an inviscid liquid. In addition, they neglected surface tension effects.

Let us consider first a gas-liquid system, where the density of the bubble fluid can be neglected relative to the density of the liquid, and the surface of the bubble is, therefore, a surface of constant pressure. It is convenient to treat the bubble as stationary, and to consider the liquid to be flowing down and around it. The liquid velocity at the center line and on the nose of the bubble is zero. We neglect viscosity effects, assume steady-state conditions, and apply Equation 3.11 to find that the liquid velocity elsewhere on the bubble-liquid interface is given by

$$V_i^2 = 2gh \tag{8.103}$$

where V_i is the tangential velocity at the interface at a vertical distance, h, below the nose of the bubble.

Dumitrescu assumed that the bubble would have a spherical nose, and solved simultaneously the flow around the bubble and the asymptotic film to obtain the bubble velocity and the frontal radius of curvature as

$$V_{bs} = 0.35\sqrt{gD} \tag{8.104}$$

and

$$D_{nb} = 0.75D \tag{8.105}$$

where D_{nb} is the diameter of the assumed spherical nose of the bubble.

Davies and Taylor (1950) solved the problem with somewhat different assumptions and obtained

$$V_{bs} = 0.328\sqrt{gD} \tag{8.106}$$

Nicklin *et al.* (1962) showed that the Davies and Taylor solution was not unique but tended to the limiting value

$$V_{bs} = 0.346\sqrt{gD} \tag{8.107}$$

Despite the approximations made in obtaining these solutions, they come surprisingly close to predicting the velocities of air bubbles rising in liquids of low viscosities, the case which would be expected to resemble most closely the behavior in inviscid fluids. Equation 8.104 is significantly better than Equation 8.106 because of the less severe truncation of the series equations arising in the solutions.

Equations 8.104, 8.105, and 8.106 assume the Taylor bubble to be formed of a gas of zero density. Where the bubble density is significant, e.g., immiscible liquids, it may be shown (Neal, 1963) that

$$V_{bs} = c\sqrt{gD\left(\frac{\Delta\rho}{\rho_L}\right)} \tag{8.108}$$

where c is a constant having a value near 0.35, and $\Delta\rho = \rho_L - \rho_G$ or, for liquid-liquid systems, $\Delta\rho = \rho_\beta - \rho_\alpha$.

Brown (1965) found experimentally that, although the solutions of Dumitrescu, and Davies and Taylor described the behavior of gas bubbles in low viscosity liquids quite well, they were not suitable for high viscosity liquids. (Dumitrescu's equation predicated a velocity 19 percent too high for a 142-centipoise liquid.) He devised a modified model by superimposing a developing boundary layer on the potential flow concept. Brown's model is idealized, but it does account for the effect of liquid viscosity on the flow in the film, and it is an improvement on the inviscid model.

Brown analyzed the portion of the liquid film in laminar flow and developed the following equation for the average velocity in the film

$$V_f = \frac{\rho_L g D^2 \xi_0^2 (1 - \xi_0)}{6\mu_L (2 - \xi_0)} \tag{8.109}$$

where $\xi_0 = \dfrac{2\delta_0}{D}$

δ_0 = the limiting thickness of the liquid film surrounding the cylindrical portion of the bubble.

Combining Equation 8.109 with the continuity expresion, Equation 8.93, with $\xi = \xi_0$ and with $V_M = 0$ (i.e., $V_b = V_{bs}$), he obtained

$$V_{bs} = \frac{\rho_L g D^2 \xi_0^3}{6\mu_L(1 - \xi_0)} \tag{8.110}$$

Equation 8.110 is a relationship between the bubble velocity and the thickness and properties of the film surrounding the cylindrical part of the bubble.

Brown has pointed out that experimental data confirm the applicability of the assumption of laminar flow in the film over a range of film Reynolds numbers, $\mathrm{Re}_f = 4\delta V_f \rho_L/\mu_L$, from 0.01 to 4000. (Note that Re_f is based upon an equivalent diameter of 4δ for the film.) Noting that the bubble nose shapes in different liquids were nearly identical when expressed as a portion of a sphere having a diameter, $D - 2\delta_0$, Brown calculated δ_0 from Equation 8.110 from experimental data on several systems and found a good correlation for V_{bs} with the equation

$$V_{bs} = 0.35\sqrt{g(D - 2\delta_0)} \tag{8.111}$$

where

$$\delta_0 = \frac{D\sqrt{1 + N_{LB}} - D}{N_{LB}} \tag{8.112}$$

and

$$N_{LB} = \left(\frac{14.5\rho_L^2 D^3 g}{\mu_L^2}\right)^{1/3} \quad \text{a liquid viscosity number} \tag{8.113}$$

Equation 8.111 indicates good agreement with data in the literature covering a range of liquid properties. The limits of applicability, as proposed by Brown, are

a. For surface tension: $\mathrm{Eo}(1 - \xi_0)^2 > 20$
where Eo is the Eotvos number referred to the pipe diameter, $\rho_L g D^2/\sigma$.
b. For viscosity: $N_{LB} > 60$.

At lower values of these parameters the assumption of potential flow at the nose of the bubble is considered invalid.

White and Beardmore (1962) have carried out an extensive experimental investigation of the rise velocity of Taylor-type air bubbles in a variety of liquids covering a wide range of properties. From dimensional analysis they reasoned that the behavior of cylindrical bubbles might be correlated in terms of three dimensionless ratios and they selected as the most convenient the Froude number, $\mathrm{Fr}_b = V_b/\sqrt{gD}$, a viscosity–surface tension number, $Y = g\mu_L^4/\rho_L\sigma^3$, and the Eotvos number referred to the pipe diameter. They expected the following forms of correlation:

a. Where the relative magnitude of the viscous, inertial and interfacial forces are all significant,

$$\mathrm{Fr}_b = \phi(\mathrm{Eo}, Y)$$

b. where viscous forces are unimportant,

$$\mathrm{Fr}_b = \phi(\mathrm{Eo})$$

c. where interfacial forces are unimportant,

$$\mathrm{Fr}_b = \phi(\mathrm{Eo}^3/Y)$$

d. where both viscous and interfacial forces are unimportant,

$$\mathrm{Fr}_b = \text{constant}.$$

The last situation is covered by the theoretical work of Dumitrescu, Davies and Taylor, and others. The work of Brown is a semitheoretical analysis of the second last situation.

White and Beardmore's final correlation is given in Figure 8.43. Where both viscous and interfacial forces may be neglected the correlation reduces to Equation 8.107, in excellent agreement with the work of others. The conditions required for this simplification are $\mathrm{Eo} > 70$ and $\sqrt{\mathrm{Eo}^3/Y} = \rho_L^2 g D^3/\mu_L^2 > 3 \times 10^5$. (We may note that $\sqrt{\mathrm{Eo}^3/Y}$ is proportional to N_{LB}^3 and is the square of Wallis' N_{LW} defined later by Equation 8.115). Where viscous forces may be neglected but interfacial forces are important, it reduces to a single relationship between the Froude and the Eotvos numbers, indicated by the line labeled $Y < 10^{-8}$ in Figure 8.43; this applies where $\rho_L^2 g D^3/\mu_L^2 > 3 \times 10^5$.

The White and Beardmore correlation may be taken to sum up present empirical knowledge and theoretical understanding of the flow of Taylor-type bubbles. Their empirical correlation confirms Brown's more theoretical analysis for values of $N_{LB} > 60$, and supplements it for lower values of N_{LB}. A comparison of Brown's equation, the White and Beardmore empirical curves, and experimental data is given in Figure 8.44.

Wallis (1969) presents an empirical relationship for the factor c of Equation 8.108 which, with this equation, permits the prediction of V_{bs} over the full

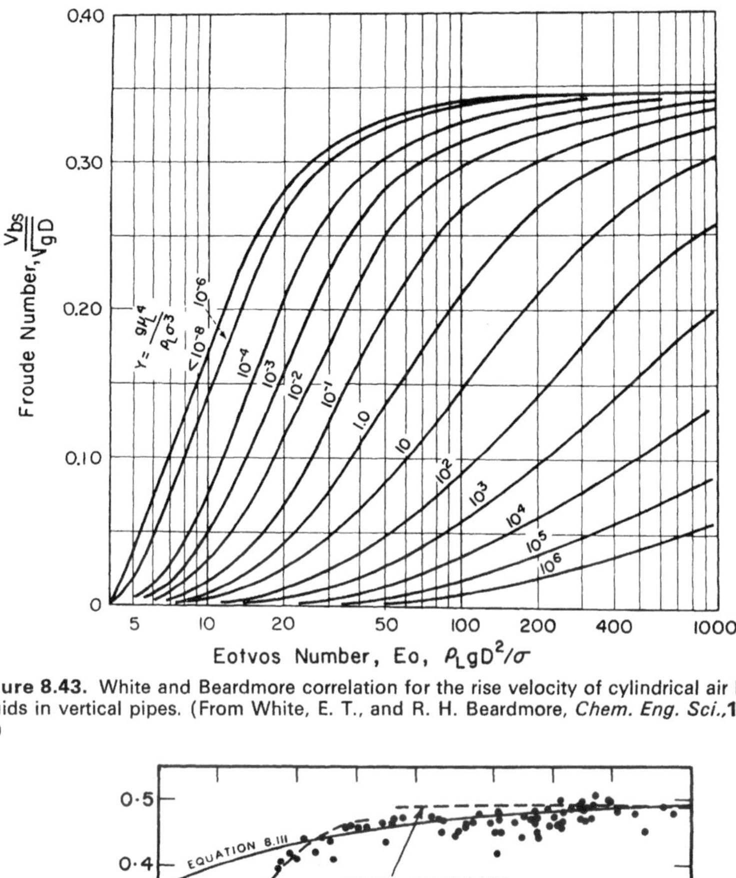

Figure 8.43. White and Beardmore correlation for the rise velocity of cylindrical air bubbles in liquids in vertical pipes. (From White, E. T., and R. H. Beardmore, *Chem. Eng. Sci.*,**17**, 351, 1962.)

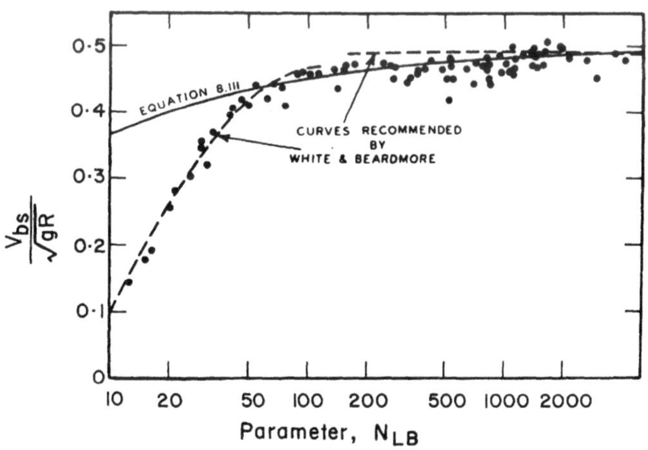

Figure 8.44. Comparison of the Brown and Govier, and the White and Beardmore correlations for bubble rise velocity.
Note:

$$V_{bs} = 0.496\sqrt{gR}\sqrt{1 - \frac{-2 + 2\sqrt{1 + N_{LB}}}{N_{LB}}}$$

(From Brown, R. A. S., *Can. J. Chem. Eng.*, **43**, 217, 1965.)

range of viscous, inertial, and interfacial forces shown in Figure 8.43. His equation is

$$c = 0.345\left[1 - \exp\left(-\frac{0.01N_{LW}}{0.345}\right)\right]\left[1 - \exp\left(\frac{3.37\ \mathrm{Eo}}{m}\right)\right] \tag{8.114}$$

where N_{LW} is a liquid viscosity number defined by

$$N_{LW} = \frac{[(\rho_L - \rho_G)\rho_L g D^3]^{1/2}}{\mu_L} = \left(\frac{\mathrm{Eo}^3}{Y}\right)^{1/2} \simeq \frac{\rho_L g^{1/2} D^{3/2}}{\mu_L} \tag{8.115}$$

and $m = 10$ for $N_{LW} > 250$
$m = 69\ N_{LW}^{-0.35}$ for $18 < N_{LW} < 250$
$m = 25$ for $N_{LW} < 18$

Wallis indicates that this relationship provides a good approximation to the correlation of Figure 8.43.

5.4 The Flow of a Taylor Bubble in Flowing Liquids

From general considerations one would expect Taylor-type bubbles in a liquid, which itself is in motion, to rise at a velocity equal to the sum of their absolute rise velocity in a stagnant liquid and the effective absolute velocity of the liquid. This situation is similar to the rise of smaller bubbles, except that the effective velocity of the liquid extends over a greater portion of the pipe cross section.

Nicklin *et al.* (1962) and Neal (1963) have studied the rise velocity of Taylor bubbles in flowing water, at liquid Reynolds numbers from zero to more than 60,000. Their experimental data confirm that, as in the case of bubble flow, the absolute rise velocity of a Taylor bubble is the sum of the rise velocity in a stagnant liquid and a term proportional to the velocity of the liquid, i.e.,

$$V_b = C_0 V_M + V_{bs} \tag{8.67}$$

where V_M = the average velocity of the liquid above and below the bubble, i.e., the average velocity of the slug,
C_0 = the correction to the average velocity for the velocity profile as discussed in Chapter 7.

Nicklin *et al.* found that for liquid Reynolds numbers higher than 8000, $C_0 = 1.20$, whereas at lower Reynolds numbers C_0 approached the limiting value of 2.0. Apparently the stable Taylor bubble, unlike oscillating small bubbles, finds itself in a flow field essentially equivalent to that existing at the center line of the pipe where the center line velocity is 2.0 and 1.2 times the average velocity for laminar and turbulent flow, respectively. Under normal circumstances of turbulent flow in the liquid slug

$$V_b = 1.2V_M + V_{bs} \tag{8.116}$$

Akagawa and Sakaguchi (1966) confirm the applicability of Equation 8.116 to the air-water system in a 1-inch pipe and point out that the effect of the term V_{bs} is negligible except at very low gas and liquid velocities.

It is interesting to note that Equation 8.116 has been derived by Neal (1963) assuming the definition of C_0 discussed in Chapter 7 (see Equation 7.102) and a liquid velocity profile ahead of the bubble given by

$$\frac{u_L}{u_{LO}} = \left(1 + \frac{r}{R}\right)^{1/n} \tag{8.117}$$

where u_{LO} is the local liquid velocity at the center line of the tube.

5.5 Small Bubbles in the Liquid Slug

The only definitive analysis of the small bubbles carried within the liquid slug is that of Akagawa (1964) and Akagawa and Sakaguchi (1966). Their studies with the air-water system in a tube 1.09 inches in diameter provide some empirical insight into the size, distribution, and velocity of these bubbles.

Akagawa and Sakaguchi's data covering superficial air and water velocities to 4.9 and 2.6 ft/sec, respectively, show the average proportion of the total gas within the system present in the liquid slugs to be essentially independent of water rate, and to vary linearly with air rate, from 30 percent at very low air rates to about 15 percent at 4.9 ft/sec. This corresponds with a nearly constant gas volume fraction (based on the total system volume) contained in the liquid slugs of approximately 10 percent. They correlate their data by

$$\bar{E}_{GB} = \bar{E}_G^{1.3} \tag{8.118}$$

$$\bar{E}_{GS} = \bar{E}_G - \bar{E}_G^{1.3} \tag{8.119}$$

$$\bar{E}_{GSL} = \bar{E}_G^{1.8} \tag{8.120}$$

where $\bar{E}_G$ = average volume fraction gas in the system as a whole,
$\bar{E}_{GB}$ = average volume fraction gas (based on the total system volume) present in the Taylor bubble region,
$\bar{E}_{GS}$ = average volume fraction gas (based on the total system volume) present in the liquid slug region,

$\bar{E}_{GSL}$ = average volume fraction gas (based on the volume of the liquid slug) present in the liquid slug region.

The data on the effect of V_{SL} and V_{SG} on $\bar{E}_{GSL}$ are shown in Figure 8.45, which is useful in making a preliminary estimate of $\bar{E}_{GSL}$.

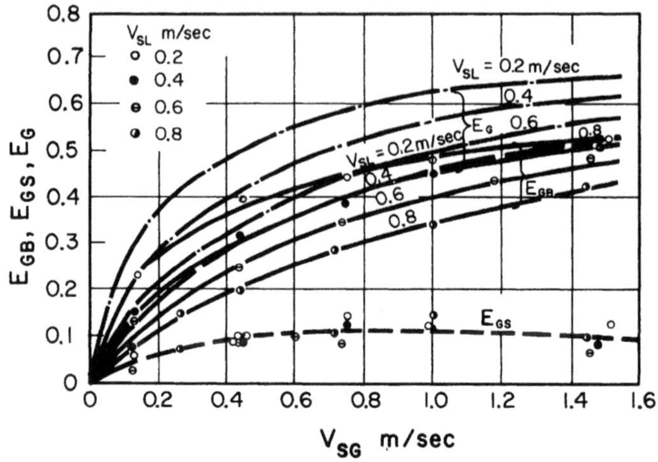

Figure 8.45. The volume fraction of gas in the bubble, liquid slug, and total system regions. (From Akagawa, K., and T, Sakaguchi, *Bulletin J.S.M.E.*, **9**, 104, 1966.)

The small bubbles were found to range in diameter from about 1.5 to 3.5 mm and were distributed more or less uniformly in the liquid slug. The work of Akagawa and Sakaguchi suggests that the presence of small bubbles in the liquid slug has a slight effect on the rise velocity of the Taylor bubbles and their data indicate that C_0 (Equation 8.67) is in the range of 1.25–1.35. They also give the following empirical equation for estimating the average velocity of the small bubbles in the liquid slug, V_{bLS},

$$V_{bLS} = 1.10 V_M \tag{8.121}$$

and note that in the presence of small bubbles the average velocity of the liquid in the slug is $0.96 V_M$ rather than V_M.

5.6 Relationship Between Bubble Length and Slug Length (Bubble Separation Distance)

Griffith and Wallis (1961) have clarified the relations, required by continuity considerations, between the lengths of the Taylor bubbles and the intervening liquid slugs. However, they neglected the presence of the small bubbles and also they assumed the absolute velocity of the Taylor bubble, V_b, to be $V_{bs} + V_M$ instead of $V_{bs} + 1.2V_M$, which is now established for turbulent flow. Adjusting for the latter, but otherwise following their approach, we may substitute in Equation 8.96 the value for V_b from Equation 8.67 and an empirical relationship for v_b,

$$v_b = A(mL_b - nD) \tag{8.122}$$

where m and n are dimensionless constants, to obtain

$$L_b = \frac{V_{SG}L_s + nD(V_{bs} + C_0 V_M)}{m(V_{bs} + C_0 V_M) - V_{SG}} \tag{8.123}$$

Griffith and Wallis have also shown that for Taylor bubbles, in potential flow, the constants in Equation 8.122 are

$$m = 0.913, \qquad n = 0.526 \quad \text{for} \quad L_b/D < 20$$

It should be noted from Equation 8.123 that there is no unique set of bubble and slug lengths for a given combination of V_{SG}, V_{SL}, D, and V_{bs}, but continuity considerations may be satisfied, for example, by short closely spaced bubbles or long widely spaced bubbles. The actual situation is controlled by the inlet conditions and momentum considerations.

5.7 *Axial Interaction Between Bubbles*

Evidence of interaction between two adjacent Taylor bubbles and the acceleration of the trailing bubble due to the wake effect of the leading bubble has been observed by many investigators. Moissis and Griffith (1962) and Moissis (1963) have made a detailed experimental analysis of the problem and found that when Taylor bubbles are separated by a distance less than $\Delta z_c = 8D$ the trailing bubble moves at a velocity given by the Dumitrescu equation (Eq. 8.104) with a modified coefficient as follows:

$$V_{bs} = [0.35 + 2.8\exp(-1.06\,\Delta z/D)]\sqrt{gD} \tag{8.124}$$

Moissis and Griffith's work is partially confirmed by that of Akagawa and Sakaguchi (1966) who report that the liquid slug length is nearly independent of the gas and liquid rates, averaging about $7.5D$, and varying about 30 percent either way. Yip (1967) also offers confirmation when he extrapolates his data on small bubbles to indicate a minimum separation distance for cap-shaped bubbles of about $8D$.

5.8 *The Flow of the Film Surrounding the Taylor Bubble*

In Sections 7.21 and 7.31 of this chapter we deal in detail with the continuity and momentum relationships for the flow of an annular film of liquid. The relationships there derived, with the simplification resulting from the approximation that $\tau_i = 0$, are directly applicable to the film surrounding the Taylor bubble. Also where the Taylor bubble is of a gas we may neglect the density of the gas or bubble phase.

For laminar flow in the film Equation 8.174 and 8.176 simplify to

$$u = \frac{g\rho_L}{2\mu_L}(2\delta y - y^2) \tag{8.125}$$

$$Q_f = \frac{\pi D g \rho_L \delta^3}{3\mu_L} \quad \text{or} \quad V_f = \frac{g\rho_L \delta^2}{3\mu_L} \tag{8.126}$$

Now the film Reynolds number, $\text{Re}_f = 4\delta V_f \rho_L/\mu_L$, may also be expressed as $DV_{Sf}\rho_L/\mu_L$, where

$$V_{SF} = \frac{4Q_f}{\pi D^2} = \frac{4\delta V_f}{D}$$

Introducing the definition of Re_f, Equation 8.126 may be rearranged to the dimensionless form

$$\delta^* = 0.909\ \text{Re}_f^{1/3} \tag{8.127}$$

where

$$\delta^* = \delta\left(\frac{g\rho_L^2}{\mu_L^2}\right)^{1/3}$$

This equation was derived by Nusselt (1916). The film thickness, δ, may be calculated from either Equation 8.126 or 8.127 provided the total flow in the film is known.

When the flow in the film is turbulent the analysis involves the assumption of a suitable turbulent velocity profile through the film—or the equivalent. The force balance on the film, given subsequently in general form by Equation 8.167, simplifies to

$$\tau = \frac{\rho_L g(\delta - y)}{g_c} \tag{8.128}$$

At $y = 0$,

$$\tau = \tau_w = -\frac{DdP}{4dz} = \frac{\rho_L f V_f^2}{2g_c}$$

where f is the liquid film friction factor. Substitution of these relations in Equation 8.128 leads to

$$\delta = \frac{fV_f^2}{2g} \tag{8.129}$$

which in dimensionless form may be expressed as

$$\delta^* = 0.315(\mathrm{Re}_f \sqrt{f})^{2/3} \tag{8.130}$$

As a check on Equation 8.130 we note that for laminar flow between parallel plates,

$$f = \frac{24}{\mathrm{Re}} \tag{8.131}$$

and the substitution of this value for f into Equation 8.130 yields Equation 8.127 as expected.

Experimental determinations of the friction factor for turbulent flow of liquids between parallel plates separated by a distance 2δ are correlated by Walker, Whan, and Rothfus (1957) in the form $\mathrm{Re}_f \sqrt{f} = \phi(\mathrm{Re}_f)$ as shown in Figure 8.46. Belkin, Macleod, and Rothfus (1959) have made use of this

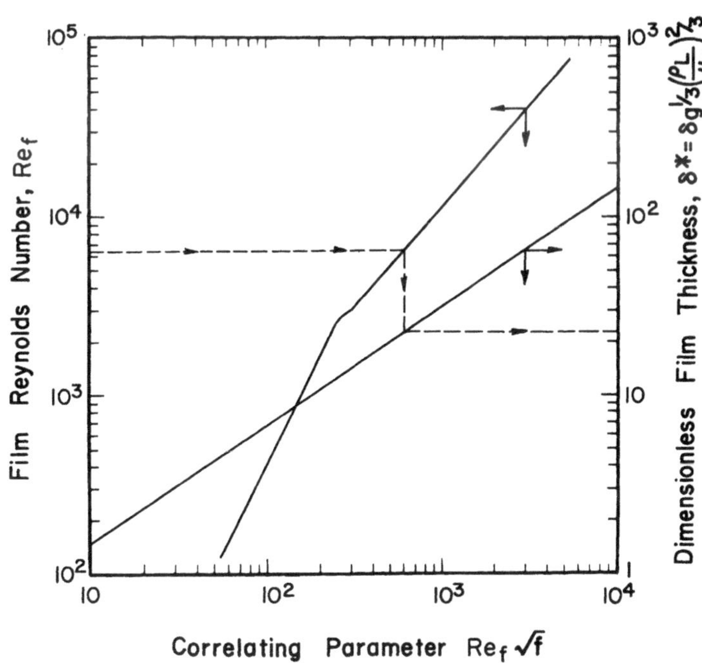

Figure 8.46. The Walker-Whan-Rothfus liquid film friction factor correlation and the Belkin *et al.* film thickness correlation. (From Belkin, H. H., A. A. Macleod, C. C. Monrad, and R. R. Rothfus *A.I.Ch.E. Journal*, **5**, 245, 1959.)

friction factor correlation to predict film thicknesses from Equation 8.130. Figure 8.46 gives a convenient graphical method of calculating film thickness from a knowledge of the film Reynolds number, Re_f. Figure 8.47 shows a comparison of the film thickness calculated from Figure 8.46 with the experimental data of several investigators. The agreement is quite good, thus supporting the simple analysis presented here. The experimental data show the critical film Reynolds number, $(Re_f)_c$, to be about 2700.

In their more approximate analysis of turbulent film flow, Collier and Wallis (1966) use a constant value of $f = 0.008$ in Equation 8.130 to arrive at

$$\delta^* = 0.063\ Re_f^{2/3} \tag{8.132}$$

In addition to the momentum considerations discussed in the foregoing, the continuity equations developed earlier must also be satisfied. The calculation of film thickness and film velocity involves a simultaneous solution of the appropriate momentum equation (Equation 8.127 or 8.130) along with the continuity equation (Equation 8.92 or 8.92a). When the bubble velocity may be expressed by

$$V_{bs} = 0.345\sqrt{gD\frac{\Delta\rho}{\rho_L}} \tag{8.133}$$

(as compared with Equation 8.108), Collier and Wallis have presented a graphical method of solving for film thickness. Their method is based on the following equations

$$Re_f = 0.345\frac{V_{Sf}}{V_{bs}}N_{LW} \tag{8.134}$$

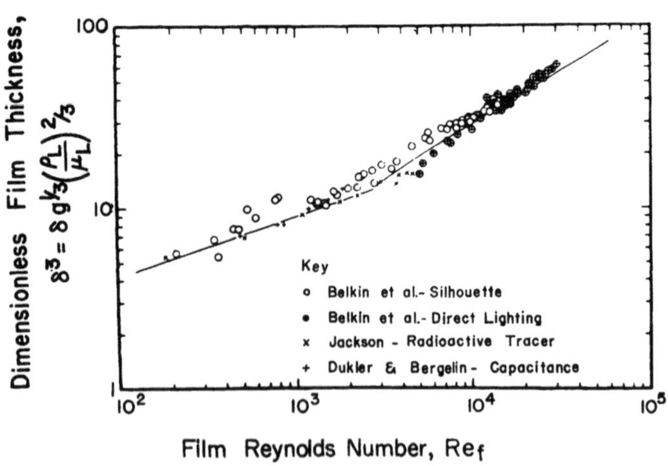

Figure 8.47. Comparison of predicted and measured film thickness. (From Belkin, H. H., A. A. Macleod, C. C. Monrad, and R. R. Rothfus, *A.I.Ch.E., Journal,* **5**, 245, 1959.)

$$\frac{V_{Sf}}{V_{bs}} = 3.85 N_{LW}\left(\frac{\delta}{D}\right)^3 \qquad \text{for } Re_f < 3500 \tag{8.135}$$

$$\frac{V_{Sf}}{V_{bs}} = 190\left(N_{LW}\frac{\delta}{D}\right)^{3/2} \qquad \text{for } Re_f > 3500 \tag{8.136}$$

and

$$\frac{V_{Sf}}{V_{bs}} = \left(1.2\frac{V_M}{V_{bs}} + 1\right)(1 - 2\xi)^2 - \frac{V_M}{V_{bs}} \tag{8.137}$$

where N_{LW} is the liquid viscosity number defined by Equation 8.115.

$$N_{LW} = \frac{D\rho_L^{1/2}[Dg(\rho_L - \rho_G)]^{1/2}}{\mu_L} \simeq \frac{\rho_L g^{1/2} D^{3/2}}{\mu_L}$$

Equations 8.134 through 8.137 may be easily derived by combining various equations already discussed in this section. We should note that Equation 8.136 for the turbulent flow case assumes the approximate value of $f = 0.008$ mentioned earlier. A graphical procedure for solving these equations is shown in Figure 8.48. At the intersection of the curves for appropriate values of

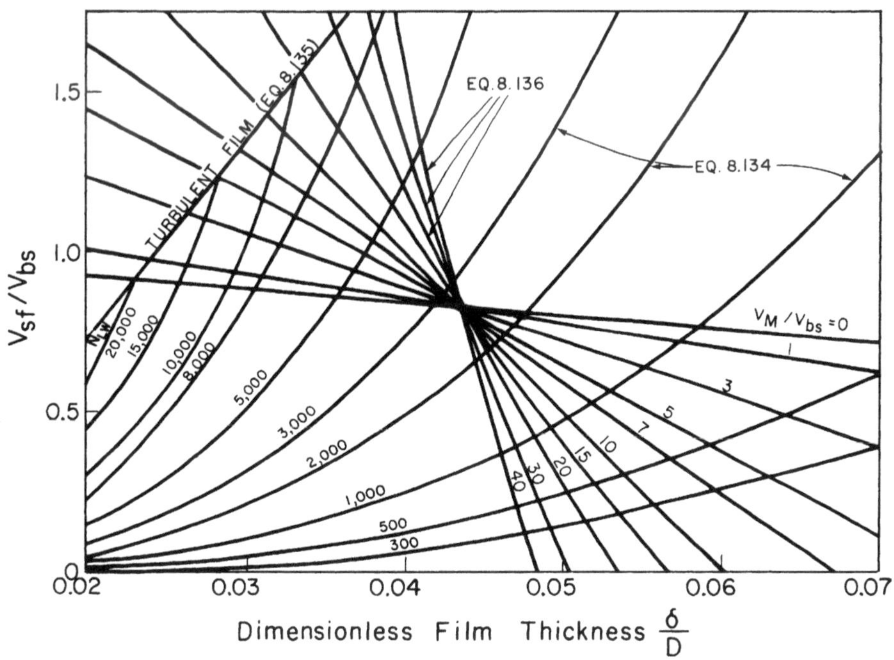

Figure 8.48. Graphical solution for film thickness. (From Collier, J. G., and G. B. Wallis, Two-Phase Flow and Heat Transfer, 1966, Vol. 2, p. 482.)

N_{LW} and V_M/V_{bs}, one may read both δ/D and V_{Sf}/V_b from the figure. The use of Figure 8.48 should be adequate for normal calculations but the correlation of Figure 8.46 used with Equation 8.92 or 8.92a will provide somewhat more accurate results.

5.9 Flow Pattern Details

The details of the slug flow pattern, i.e., the bubble length and spacing, the bubble and liquid flow rates, and the film thickness may be predicted, at least approximately, from the manner in which the gas is introduced into the flow stream, the gas and liquid flow rates and properties, and the diameter of the flow tube.

1. A preliminary estimate of the volume fraction of gas contained within the liquid slug may be made from V_{SG}, V_{SL}, and Figure 8.45.
2. The absolute velocity of the Taylor bubble may be calculated by the methods discussed in Sections 5.3 and 5.4.
3. If the gas is initially dispersed, but cap-shaped or fully developed Taylor bubbles are formed by collision and coalescence and, where sufficient detail is known, the approximate volume and spacing of the bubbles and the slugs may be estimated from the characteristics of the bubble generating device and the methods discussed in Section 4 on the bubble flow pattern, and Equation 8.122.
4. If the gas is introduced into the flow stream intermittently, the gas "bursts," or the approximate bubble volume, and the length of the bubble may be estimated by the rate of gas flow during the burst, the duration of the burst, and Equation 8.122.
5. The consistency of the bubble and slug lengths estimated in step *3* or *4* may be confirmed, and necessary adjustments made from the gas continuity relationship, Equation 8.96 or 8.100 and the volume of the bubble, or directly from Equation 8.123.
6. The stability of the flow pattern may be checked, first by the general criterion of Moissis and Griffith, $L_s > 8D$, and, further, by estimates based on Equation 8.124 of the rate, time, and travel distance for a following bubble to close on its leader.
7. If the closing velocity of adjacent bubbles is significant in the particular case, the stable pattern may be estimated by assuming the bubble volume to be doubled and recalculating as appropriate.
8. Usually, our knowledge of the entrance flow conditions is insufficient to permit a good estimate of the bubble volume; therefore, recourse must be

made to a bracketing procedure. Combining Equations 8.96 and 8.122, and using the Griffith and Wallis constants, we have

$$\frac{0.913L_b - 0.526D}{L_b + L_s} = \frac{V_{SG}}{V_b} \tag{8.138}$$

Now the minimum value of L_s for stable slug flow is about $8D$. This means that, if the final stable slugs have resulted from agglomeration of originally more closely spaced bubbles, L_s will be between $8D$ and $16D$ (the lower value prevailing if the predecessor bubbles were $4D$ apart leading to stable slugs just over the minimum spacing; and the higher value prevailing if predecessor bubbles were nearly $8D$ apart, just below the minimum spacing). The slug length may therefore be taken as $12D$ with an uncertainty of some 30 percent. For a given V_{SG}/V_b, L_b may then be calculated from Equation 8.138. Since the ratio $L_b/(L_b + L_s)$ is fixed by the ratio V_{SG}/V_b, the uncertainty in the calculated value of L_b will be essentially the same as that in L_s. Notwithstanding the possible uncertainties in each, L_s and L_b, the ratio $L_b/(L_b + L_s)$ is known with fair accuracy. This is important from the viewpoint of determining the holdup and pressure drop.

9. The thickness of the liquid film surrounding the bubble may be estimated from continuity relations and Equation 8.126 or 8.127 if the film flow is laminar, or Equation 8.129 or 8.130 if the film flow is turbulent. The trial-and-error procedure involved in the solution of these equations may be eliminated by use of Figure 8.48.

5.10 Holdup

Notwithstanding that the bubble and slug lengths are not both uniquely defined by the flow rates, fluid properties, and diameter, the holdup may be closely estimated.

As measured by the average *in situ* volume fraction of the gas phase, the holdup may be determined in two different ways. First, we can express, in terms of the volume of the Taylor bubble, the volume of the small bubbles in the liquid slug and the total volume over the length of one average bubble and one average slug, i.e.,

$$E_G = \frac{v_b + E_{GSL} A L_s}{A(L_b + L_s)} \tag{8.139}$$

When the bubble volume, the bubble and slug lengths, and the gas volume fraction in the liquid slug are known or can be reasonably estimated, this is a suitable method.

An alternate method is based upon the volume of gas admitted to a given length of flow tube compared with the total volume of the length of the flow tube. Consider a length, L, being traversed by bubbles moving at a velocity, V_b. Then in the time, L/V_b, a given bubble will traverse the length L and a fresh quantity of gas in the volume $Q_G L/V_b$ will enter the tube and be contained in the length L. The *in situ* volume fraction gas will therefore be

$$\bar{E}_G = \frac{Q_G L/V_b}{LA} = \frac{V_{SG}}{V_b} \tag{8.140}$$

Clearly, Equations 8.139 and 8.140 must yield the same result, when applied to the same finite system, and this is confirmed by the fact that

$$\bar{E}_G = \frac{v_b + \bar{E}_{GSL} v_s}{A(L_b + L_s)} = \frac{V_{SG}}{V_b} \tag{8.141}$$

is identical with Equation 8.96, the gas continuity equation.

For a gas flowing with a low viscosity liquid, and where interfacial forces are small, the simple Dumitrescu equation applies, the volume fraction gas may be computed from Equation 8.140, and the applicable relation for V_b,

$$\bar{E}_G = \frac{V_{SG}}{0.35\sqrt{gD} + 1.2V_M} \tag{8.142}$$

At low liquid rates the volume fraction gas tends to a value of 0.833 with high gas rates. The validity of this relationship for the air-water system has been confirmed by Nicklin *et al.* (1962).

5.11 Pressure Gradient

As in the case of bubble flow, since the liquid phase is continuous, the pressure gradient may be estimated by essentially single-phase methods. The average *in situ* density of the flowing mixture is

$$\rho_M = (1 - E_G)\rho_L + E_G \rho_G \tag{8.79}$$

When the second phase is a gas phase at low pressure, we may write

$$\rho_M \simeq (1 - E_G)\rho_L \tag{8.80}$$

The wall shear adjacent the liquid slugs is that due to a liquid flowing at an average velocity of V_M. The frictional pressure drop is essentially that over the liquid slugs since the bubble region is nearly one of constant pressure. There is also a pressure drop at the ends of the bubble, but this term is small compared to the pressure drop in the liquid slug.

Applying the mechanical energy balance neglecting kinetic energy effects (Equation 7.73)

$$\frac{\Delta P}{\rho_L \Delta z} = \frac{(1 - \bar{E}_G)g}{g_c} + \frac{2f_{LM} V_M^2 L_s}{(L_b + L_s)g_c D} \tag{8.143}$$

Here f_{LM} is determined at a Reynolds number $D V_M \rho_L/\mu_L$ as for single-phase flow in a pipe (Figure 4.16).

Nicklin (1961) proposed the use of a simplified version of this equation in which $L_s/(L_b + L_s)$ is approximated by $1 - \bar{E}_G$

$$\frac{\Delta P}{\rho_L \Delta z} = (1 - \bar{E}_G)\left(\frac{g}{g_c} + \frac{2f_{LM} V_M^2}{g_c D}\right) \tag{8.143a}$$

Actually the presence of liquid in the film reduces what otherwise would be the slug length and it may be shown that

$$\frac{L_s}{L_s + L_b} = \frac{1 - \bar{E}_G}{(1 - \xi)^2}$$

With this substitution we have (Brown and Govier, 1965)

$$\frac{\Delta P}{\rho_L \Delta z} = \frac{1 - \bar{E}_G}{(1 - \xi)^2}\left(\frac{g}{g_c} + \frac{2f_{LM} V_M^2}{g_c D}\right) \tag{8.144}$$

The term $(1 - \xi)^2$ has a minimum value of 0.52 (corresponding with the maximum value of δ of $0.14D$).

Experimental data (Govier, Radford, and Dunn, 1957) show both these approaches to be reasonable; the former predicts gradients somewhat high and the latter somewhat low. If L_s, L_b, and δ are all calculated, there seems to be no reason for not using Equation 8.163 or 8.144.

5.12 *Transition from Slug to Froth Flow*

The transition from slug to froth flow at any constant liquid rate with increasing gas rate occurs as a result of a gradual decrease in the downward velocity and in the stability of the liquid film surrounding the increasingly rapid movement of the bubbles. Waves develop at the film-bubble interface which eventually cause a breakdown of the film and a penetration of the liquid into the bubble, and generally a coarse, highly agitated mixture of gas and liquid develops. The breakdown of the film in the transition from slug to froth flow is analogous to the flooding phenomenon observed in wetted wall columns.

The flooding phenomenon in vertical tubes has been reviewed by Hewitt and Wallis (1963), Collier and Wallis (1966), and Wallis (1969). These investigators correlate flooding data (in $\frac{3}{4}$- to $1\frac{1}{4}$-inch pipes) of several workers in terms of two dimensionless superficial velocities defined as

$$V_{Sf}^* = V_{Sf}\left(\frac{\rho_L}{gD(\rho_L-\rho_G)}\right)^{1/2} \simeq \frac{V_{Sf}}{\sqrt{gD}} \tag{8.145}$$

$$V_{SG}^* = V_{SG}\left(\frac{\rho_G}{gD(\rho_L-\rho_G)}\right)^{1/2} \simeq V_{SG}\left(\frac{\rho_G}{\rho_L gD}\right)^{1/2} \tag{8.146}$$

For the slug flow pattern we can easily relate V_G to the bubble velocity V_b, i.e.,

$$V_G = (1-\xi)^2 V_b$$

Collier and Wallis find an approximate correlation in the form

$$\sqrt{V_{SG}^*} + \sqrt{V_{Sf}^*} = C \tag{8.147}$$

where C ranges from 0.88 to 1.0. Hewitt and Wallis have shown that the data of Nicklin and Davidson (1962) may be correlated by Equation 8.147 with $C = 1.0$.

Wallis (1969) has developed a more general flooding correlation that has been tested for fluids of viscosities from 1 to 3000 centipoises. The correlation is

$$\sqrt{V_{SG}^*} + m\sqrt{V_{Sf}^*} = C \tag{8.148}$$

where both m and C are functions of a liquid viscosity number, N_{LW} defined by Equation 8.115. Wallis' correlation for m and C are reproduced as Figures 8.49 and 8.50.

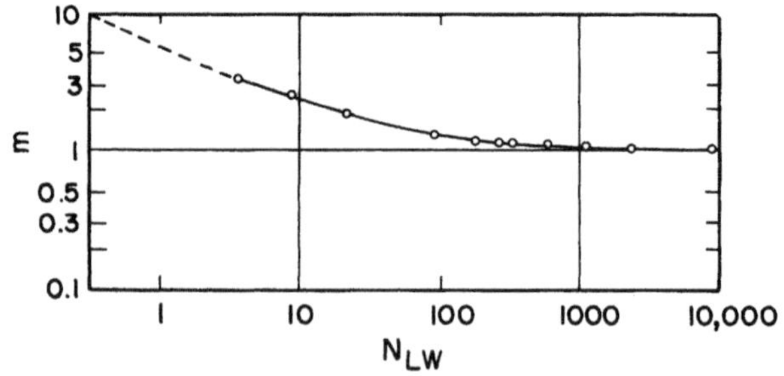

Figure 8.49. Wallis factor *m*. (From Wallis, G. B., *One-Dimensional Two-Phase Flow,* McGraw-Hill Book Co., New York, 1969.) (Used with permission of McGraw-Hill Book Co.)

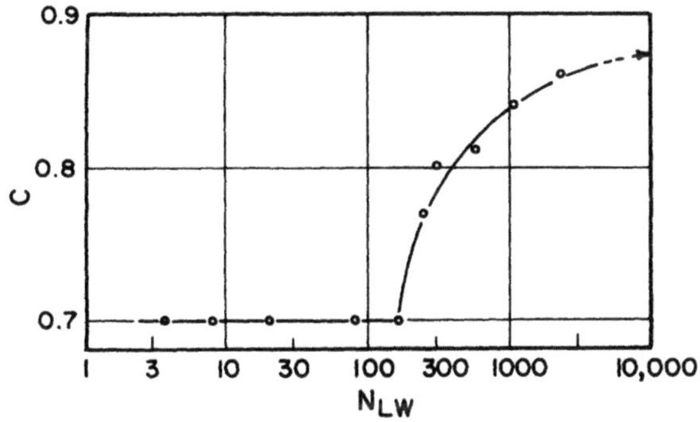

Figure 8.50. Wallis factor *C*. (From Wallis, G. B., *One-Dimensional Two-Phase Flow* McGraw-Hill Book Co., New York, 1969.) (Used with permission of McGraw-Hill Book Co.)

The procedure for predicting the transition to froth flow is to calculate the film thickness from one of the methods discussed in Section 5.8 and then to calculate the left-hand side of Equation 8.147 or 8.148. If the calculated value of C is above about 0.88, the film is not stable and transition to froth flow will take place. The results of these calculations, using a critical value of $C = 1$ and $m = 1$, are shown in Figure 8.51 taken from Collier and Wallis.

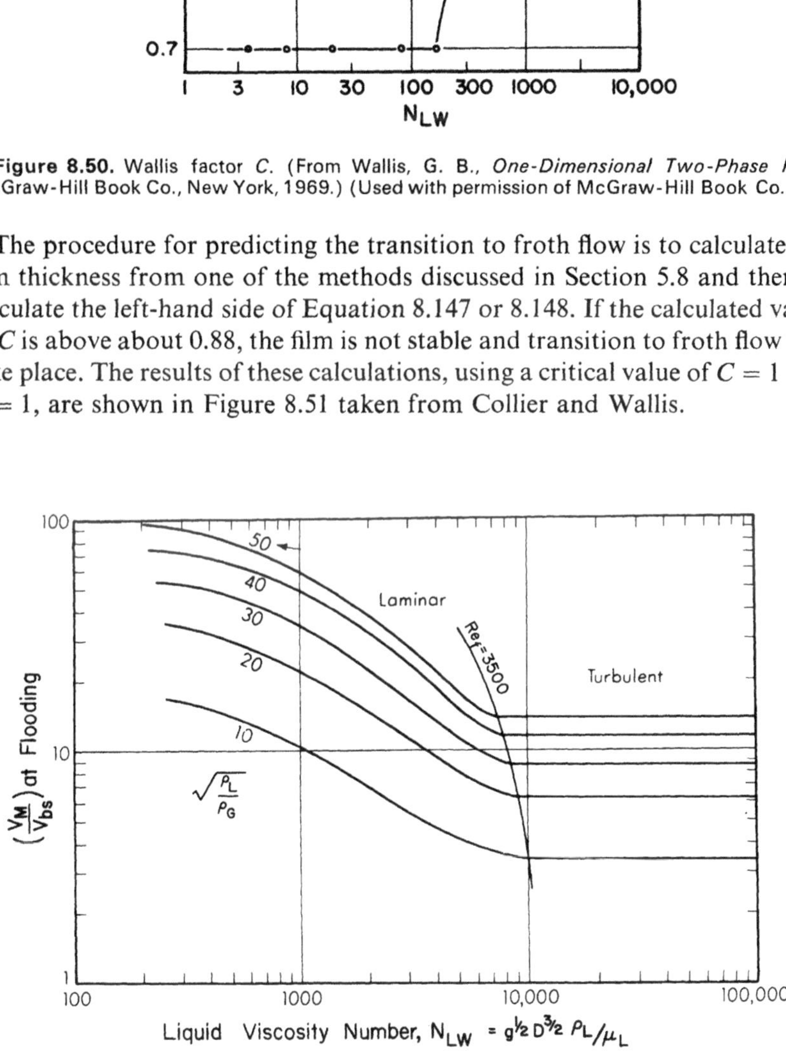

Figure 8.51. Collier and Wallis correlation for transition from slug to froth flow. (From Collier, J. G., and G. B. Wallis, Two-Phase Flow and Heat Transfer, 1966, Vol. 2, p. 494.)

Moissis (1963) has studied the transition from slug to froth flow through an analysis of the development of instabilities in the film. These instabilities depend upon the relative velocity between the bubble and the film, the thickness of the film, and the criterion of instability in the propagation of waves in the film. Considering the conditions under which waves may be propagated in the film and the most unstable wave length in the geometry involved, Moissis developed the criterion for the limiting stability condition. Noting that the bubble length was not defined by any of the relationships and that it was determined by other factors including the entrance conditions, Moissis solved the stability and the continuity and momentum relationships for a 1-inch tube and for each of several values of L_b/D. He presented his results as a plot of V_{SG}/V_M versus V_M^2/gD, as shown in Figure 8.52.

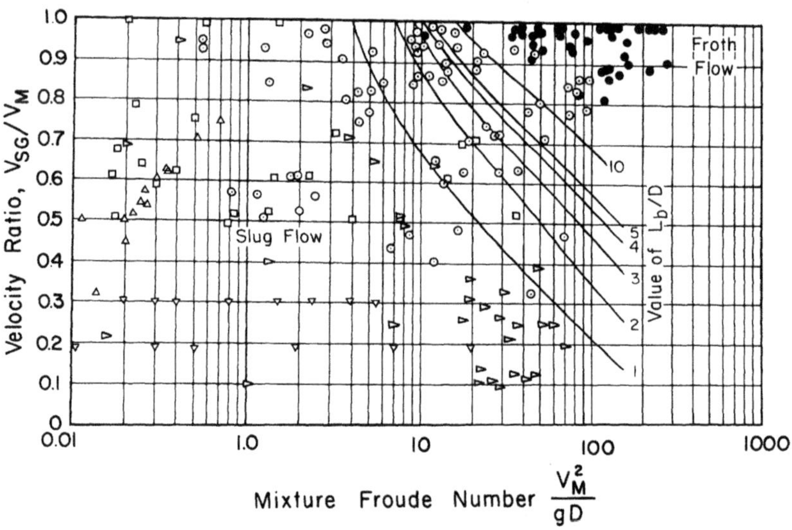

Figure 8.52. Moissis correlation for transition from slug to froth flow. All white points denote slug flow observations, all black points denote froth flow observations.
○ ● Govier *et al.* (1957) △ Brancart (1958) ▽ Griffith and Wallis (1961)
□ Kozlov (1952) ▷ Schwartz (1954)
(From Moissis, R., *J. Heat Transfer, Trans. A.S.M.E., Ser. C.*, **85**, 366, 1963.)

6 THE FROTH FLOW PATTERN

The froth flow pattern has been variously called churn, froth, wave entrainment, dispersed plug, and semiannular. It is similar to the slug pattern in that the flow is pulsating and there are alternate slugs of gas and liquid. At the higher flow rates characteristic of froth flow, the bubble wakes become more agitated, a large number of small bubbles are torn off at the tail, and the whole wake becomes richer in bubbles. The froth pattern differs from the regular orderly slug flow pattern in that neither the bubbles of gas nor the slugs of liquid maintain their identity as they move up the tube. Regions within the tube are continually being filled and emptied of liquid and there is much to-and-fro motion. The flow pattern is characterized by an abrupt increase in the pressure gradient and a maximum in the holdup ratio (see Figures 8.3 and 8.5.). Nicklin and Davidson (1962) relate froth flow to the flooding phenomenon in a wetted wall column and attribute the development of the pattern to an instability in the liquid film. This causes a reduced downward flow of liquid and a liquid "starvation" of the zone below, resulting in continual breakdown and buildup of the liquid zones.

The pattern occurs over a modest range of superficial gas velocities and apparently only up to a certain critical superficial liquid velocity (see Figure 8.9). It is a transition pattern from the regular slug flow to annular-mist flow. At any given liquid rate the pattern extends from the point of breakdown of the regular liquid slugs and gas bubbles to a gas velocity sufficient to carry the bulk of the liquid up the wall of the tube by the surface drag exerted.

The equivalent of the froth flow pattern occurs with liquid-liquid systems, but our knowledge of it is very scant and caution must be used even in making qualitative comparisons with the froth pattern of gas-liquid mixtures.

The froth flow pattern is not amenable to theoretical analysis and has not been the subject of any extensive experimental study. At the present state of knowledge it seems only possible to treat it as a modified form of the slug flow pattern.

Although discrete bubbles are not identifiable in the froth flow pattern, and the flowing mixture is made up of an extremely turbulent nonhomogeneous mixture of gas and liquid, still some success has been had (Harmathy, 1960; Peebles and Garber, 1953; Zuber and Findlay, 1965) in predicting effective average local drift velocities (equivalent to the velocity in an infinite stagnant liquid).

Equation 8.73 may be used to get a rough idea of the effective bubble rise velocity, V_{bf}. With this value of V_{bf}, an indication of E_G may be obtained through Equation 8.140. The pressure drop might be expected to be at least that of the hydrostatic head of the mixture plus a friction term based upon the liquid properties and the mixture velocity, i.e.,

$$\frac{\Delta P}{\rho_L \Delta z} = (1 - \bar{E}_G)\frac{g}{g_c} + \frac{2fV_M^2}{g_c D} \tag{8.149}$$

where f is determined as for single-phase liquid flow at a Reynolds number, $DV_M \rho_L/\mu_L$.

Duns and Ros (1963) suggest that froth flow may be treated by a weighting of the E_G's and the pressure gradients calculated for the slug and annular-mist regions. Orkiszewski (1967) also follows this procedure. These two references should be consulted for details.

7 THE ANNULAR-MIST FLOW PATTERN

7.1 Introduction

The annular-mist flow pattern is widely encountered in the flow of gas-liquid mixtures at high gas rates and gas-liquid ratios. The equivalent pattern is also encountered, although not so commonly, with liquid-liquid mixtures. In this case, in the limit where there is no film, the pattern is really the same as the dispersed bubble flow pattern with an inversion of the phases. In the discussion to follow we restrict our consideration to gas-liquid systems. Certain qualitative analogies may be applied to liquid-liquid systems, but we have little to aid our quantitative understanding. The annular-mist flow pattern in gas-liquid systems is characterized by an upward moving, continuous, smooth to wavy film of liquid on the tube wall and a much more rapidly moving central core of gas, containing entrained droplets of liquid in a concentration which may vary from low to high. The liquid film may be wholly in laminar motion or it may be laminar only nearest the wall, and turbulent nearest the gas-liquid interface. The gas phase is in turbulent motion and, at steady state, any entrained droplets remain constant in amount and move at a velocity somewhat less than that of the gas.

Some authors (e.g., Shearer and Nedderman, 1965) divide the annular-mist flow pattern into the "small ripple" regime and the "disturbance wave" regime. In the small ripple regime, small waves develop on the liquid surface and move at velocities of the order of the interface velocities and then lose their identity. At higher liquid flow rates the waves are larger, and they travel at velocities two to five times the interfacial velocity. These waves are known as disturbance waves.

Telles and Dukler (1970) present the results of a statistical analysis of varied data on the vertical flow of thin wavy liquid films. They show that, when waves develop, up to 50 percent of the total flow may be in large "lumps" of liquid which are spaced as widely as 1000 times the mean film thickness.

The annular and the mist flow patterns are often treated as two separate patterns. This does not seem appropriate since, except in the two extreme or limiting cases, the liquid is distributed between the film and the entrained droplets, and there is a gradual change in its distribution without any abrupt change in mechanism. Nonetheless, the theoretical analysis is considerably simplified when mist-free annular flow or annulus-free mist flow is assumed. As will be seen later, under certain limiting conditions this approach yields reasonable results.

At any given liquid rate, decreasing the gas rate causes more of the liquid to be present in the film, the liquid film velocity to decrease, and its thickness to

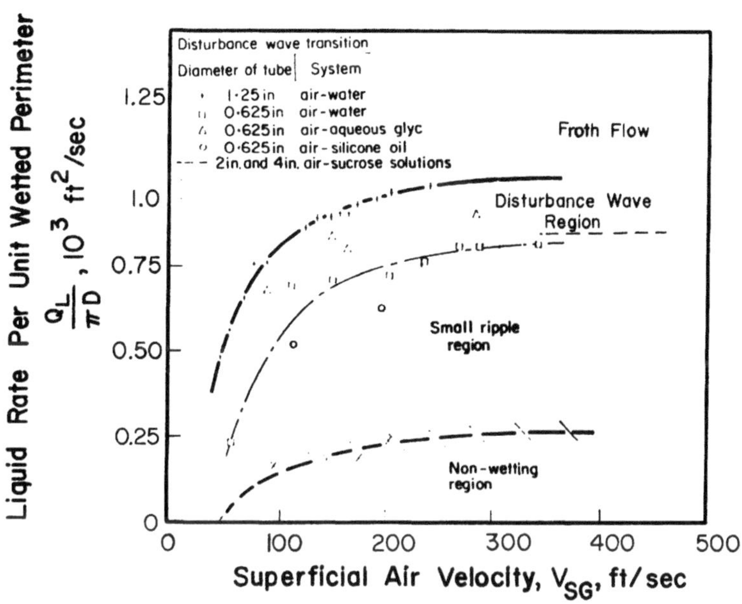

Figure 8.53. Regions of annular-mist flow. (From Shearer, C. J., and R. M. Nedderman, *Chem. Eng. Sci.*, **20**, 671, 1965.)

increase. At a certain critical gas rate the liquid film velocity becomes zero, and below this rate the liquid film thickens rapidly, has a negative velocity near the wall, penetrates the gas phase, and froth flow occurs. As the gas rate is increased, turbulence takes place in the liquid film, the thickness of the film decreases, waves develop at the interface, an increasing number of droplets are torn from the film and entrained in the gas. Eventually the continuous film is destroyed and almost all the liquid is transported as entrained droplets in the gas phase. Figure 8.53, taken from Shearer and Nedderman (1965), shows the nonwetting region when all the liquid is in the gas core, the small ripple region, and the disturbance wave region.

7.2 Overall Continuity Considerations

7.21 Pure Annular Flow—No Entrainment

Let us first consider the limiting case where no liquid is entrained in the gas core. The overall material balance under steady-state conditions indicates that the average velocity in the gas core, V_G, is related to the gas flow rate by

$$Q_G = V_G A_c = V_G \frac{\pi}{4} (D - 2\delta)^2 \tag{8.150}$$

or

$$Q_G = V_G \frac{\pi D^2}{4} (1 - \xi)^2 \tag{8.151}$$

where δ = the mean film thickness,
$\xi = 2\delta/D$.

The liquid flow rate may be expressed in terms of the average liquid velocity by

$$Q_L = Q_f = V_f A_f = V_f \frac{\pi D^2}{4} (2\xi - \xi^2) \tag{8.152}$$

The total flow rate is given by

$$Q_M = Q_G + Q_L = \frac{\pi D^2}{4} [V_G(1 - \xi)^2 + V_f(2\xi - \xi^2)] \tag{8.153}$$

Assuming the film to be thin allows us to neglect terms of the order ξ^2 in the above discussion. The resulting simplified equations are

$$Q_G = V_G \frac{\pi D^2}{4} (1 - 2\xi) \tag{8.154}$$

$$Q_L = V_f \frac{\pi D^2}{4} 2\xi \tag{8.155}$$

and

$$Q_M = \frac{\pi D^2}{4} [V_G(1 - 2\xi) + 2V_f \xi] \tag{8.156}$$

Alternate forms of some of these equations and other related equations are given in Section 8.21 of Chapter 10.

7.22 Annular-Mist Flow

When part of the liquid is carried in the form of entrained droplets, e denoting the volume fraction of these droplets in the gas phase, and V_e denoting their average absolute velocity, the equations for liquid, gas, and total mass balance are modified to

$$Q_L = Q_f + Q_e$$

$$= V_f A_f + V_e A_c e = \frac{\pi D^2}{4} [(2\xi - \xi^2)V_f + e(1 - \xi)^2 V_e] \tag{8.157}$$

$$Q_G = V_c A_c(1 - e) = \frac{\pi D^2}{4} [V_c(1 - \xi)^2(1 - e)] \tag{8.158}$$

$$Q_M = Q_G + Q_L = \frac{\pi D^2}{4}[(1-\xi^2)(V_c + eV_e - eV_c) + (2\xi - \xi^2)V_f] \tag{8.159}$$

where V_c = the average velocity of the mixture in the core. (Again these equations appear in alternate forms in Section 8.22 of Chapter 10.) In many cases the slip between the core and the entrained drops may be neglected, i.e., $V_c = V_e$, and the second-order terms containing ξ^2 may also be neglected, which further simplifies Equations 8.157, 8.158, and 8.159.

At very high gas flow rates all the liquid is carried in the form of a mist; ξ becomes zero, and the equations are reduced to

$$Q_L = Q_e = eV_e \frac{\pi D^2}{4} \tag{8.160}$$

$$Q_G = \frac{\pi D^2}{4} V_c(1-e) \tag{8.161}$$

$$Q_G + Q_L = \frac{\pi D^2}{4}(V_c + eV_e - eV_c) \tag{8.162}$$

It is convenient to define F_e representing the fraction of the flowing liquid which is entrained in the gas phase

$$F_e = \frac{Q_e}{Q_L} \tag{8.163}$$

Then F_e, V_e, and e are interrelated by

$$Q_e = F_e Q_L = eV_e(1-\xi)^2 \frac{\pi D^2}{4} \tag{8.164}$$

If the average diameter of the entrained droplets is denoted by d_e, and their number per unit volume by n_e, then

$$e = \frac{n_e \pi d_e^3}{6} \tag{8.165}$$

7.3 Momentum Considerations

7.31 Pure Annular Flow—No Entrainment

The momentum equation for the steady flow of the liquid film in the annular flow pattern has been written in several approximate forms by Calvert and Williams (1955), Dukler (1959), Willis (1965), Shearer and Nedderman (1965), and others. Our development here closely parallels the treatment by Shearer and Nedderman (1965).

The Force Balance on the Film Consider Figure 8.54, which shows the portion, dz, of a liquid film of thickness δ flowing upward at an average velocity, V_f. Under steady flow conditions, the momentum equation reduces to a force balance. Consider the forces acting on the liquid film bounded by the cylindrical surfaces at radii $(R - y)$ and $(R - \delta)$. The forces are:

at the surface $(R - y)$ distant from the center line acting downward

$$\tau 2\pi(R - y)dz$$

at the surface $(R - \delta)$ distant from the center line (the interface), acting upward

$$\tau_i 2\pi(R - \delta)dz$$

as a result of gravitational forces, acting downward

$$F_g = \frac{\pi}{4}[(D - 2y)^2 - (D - 2\delta)^2]dz\rho_L \frac{g}{g_c}$$

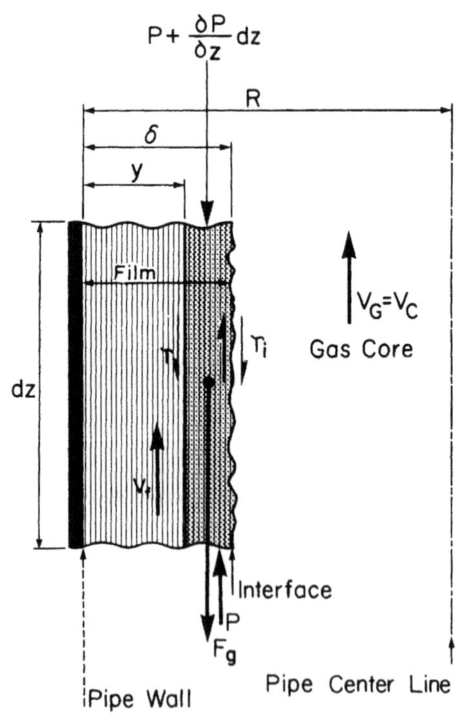

Figure 8.54. Forces on an element of the liquid film in annular flow.

as a result of the pressure gradient, acting upward

$$-\frac{\pi}{4}[(D-2y)^2-(D-2\delta)^2]dP$$

The force balance for the portion of the film being considered yields

$$\tau=\tau_i\left(\frac{D-2\delta}{D-2y}\right)-\left(\frac{dP}{dz}+\rho_L\frac{g}{g_c}\right)\left[\frac{(2D-2y-2\delta)(2\delta-2y)}{4(D-2y)}\right] \quad (8.166)$$

This expression may be simplified by expanding the term contained within the square brackets and neglecting terms of the second order (y^2, δy, and δ^2). The resulting equation is

$$\tau=\tau_i\left(\frac{D-2\delta}{D-2y}\right)-\left(\frac{dP}{dz}+\rho_L\frac{g}{g_c}\right)(\delta-y) \quad (8.167)$$

The Force Balance on the Gas Core The shear stress at the interface may be related to the pressure gradient by a second force balance for the gas phase in the core, where the forces are:

at the surface $(R-\delta)$ distant from the center line (the interface), acting downward

$$\tau_i 2\pi(R-\delta)dz$$

as a result of gravitational forces, acting downward

$$\pi(R-\delta)^2\rho_G\frac{g}{g_c}dz$$

as a result of the pressure gradient, acting upward

$$-\pi(R-\delta)^2dP$$

The force balance is

$$\tau_i=\frac{D-2\delta}{4}\left(-\frac{dP}{dz}-\rho_G\frac{g}{g_c}\right) \quad (8.168)$$

The Combined Force Balance Equations 8.167 and 8.168 may be combined, with second-order terms again neglected, to yield

$$\tau=\left(-\frac{dP}{dz}-\rho_G\frac{g}{g_c}\right)\left(\frac{D-2\delta}{4}\right)+\left(-\frac{dP}{2dz}-\rho_L\frac{g}{g_c}+\frac{\rho_G}{2}\frac{g}{g_c}\right)(\delta-y) \quad (8.169)$$

This final equation for the shear stress distribution in the liquid film differs from that derived by Calvert and Williams (1955). As pointed out by Shearer and Nedderman, the difference between Equation 8.169 and the equation of Calvert and Williams is due to an inconsistency in the analysis of the latter, who considered the liquid film to be planar for one part of the analysis and annular for another.

Dukler (1959) and Hewitt (1961) simplify the combined force balance to

$$\tau = \frac{R}{2}\left(-\frac{dP}{dz} - \frac{g\rho_G}{g_c}\right) - (\delta - y)\frac{g\rho_L}{g_c} \tag{8.170}$$

by recognizing that δ is small* with respect to R, and that dP/dz is ordinarily small with respect to $g\rho_L/g_c$.

Willis (1965) simplifies Equation 8.170 further by neglecting the weight of the gas, i.e., the term $g\rho_G/g_c$, and bases his analysis on the equivalent of the following expression

$$\tau = -\frac{R}{2}\frac{dP}{dz} - (\delta - y)\frac{g\rho_L}{g_c} \tag{8.171}$$

Equations 8.169, 8.170, and 8.171 are all normally acceptable forms of the momentum balance for a steady-state film with a core of single-phase gas, but Equation 8.169 certainly is to be preferred under conditions of high pressure gradients or long flow columns.

From Equation 8.170 the wall shear stress may be written as

$$\tau_w = \frac{R}{2}\left(-\frac{dP}{dz} - \frac{g\rho_G}{g_c}\right) - \delta\frac{g\rho_L}{g_c} \tag{8.172}$$

and

$$\frac{\tau}{\tau_w} = 1 + \frac{y\rho_L g}{g_c \tau_w} \tag{8.173}$$

Equation 8.173 is probably the most concise, although an approximate, expression of the momentum balance.

Before we can further develop useful relationships from the momentum equation, we must relate τ to the velocity gradient within the film. This may be done precisely for the case of laminar flow, but only approximately when the liquid film is turbulent.

The Velocity Profile, Total Flow, and Thickness of the Liquid Film Willis (1965) and Shearer and Nedderman (1965) have made an analysis of the situation assuming laminar flow throughout the film. Starting with Equation 8.169, and incorporating the constitutive equation for the Newtonian liquid and the boundary conditions $u = 0$ at $y = 0$, we obtain by integration

*The film thickness, is normally less than 5 percent of the radius, R.

$$u = \frac{g_c}{\mu_L}\left[\frac{y(D-2\delta)}{4}\left(-\frac{dP}{dz} - \rho_G\frac{g}{g_c}\right) + \left(\frac{dP}{2dz} + \rho_L\frac{g}{g_c} - \rho_G\frac{g}{2g_c}\right)\frac{(\delta-y)^2-\delta^2}{2}\right] \tag{8.174}$$

The liquid flow rate is given by

$$Q_L = \int_0^\delta \pi(D-2y)u\,dy \simeq \pi D\int_0^\delta u\,dy \tag{8.175}$$

Substituting Equation 8.174 into 8.175 and integrating, we obtain

$$Q_L = \frac{\pi D g_c}{\mu_L}\left[-\frac{dP}{dz}\frac{D\delta^2}{8} + \frac{dP}{dz}\left(\frac{\delta^3}{12}\right) - \rho_L\frac{g}{g_c}\frac{\delta^3}{3} - \rho_G\frac{g}{g_c}\frac{D\delta^2}{8} + \frac{5}{12}\rho_G\frac{g}{g_c}\delta^3\right] \tag{8.176}$$

Willis (1965), using the equivalent of Equation 8.171, obtained the following expressions for the liquid velocity and flow rate:

$$u = \frac{yg}{g_c}\left[-\frac{D-2\delta}{4}\frac{dP}{dz} + \left(\frac{dP}{2dz} + \rho_L\frac{g}{g_c}\right)\left(\frac{2\delta-y}{2}\right)\right] \tag{8.177}$$

$$Q_L = \frac{\pi D g_c}{\mu_L}\left[\left(-\frac{dP}{dz}\right)\frac{D\delta^2}{8} - \frac{1}{4}\delta^3\left(-\frac{dP}{dz}\right) - \frac{1}{3}\delta^3\rho_L\frac{g}{g_c} + \frac{5}{12}\frac{\delta^4}{D}\rho_L\frac{g}{g_c}\right] \tag{8.178}$$

We note that the first term on the right-hand side of Equation 8.178 is identical with the first term in Equation 8.176, but the remaining three terms differ because of the approximations of Willis' development.

Shearer and Nedderman (1965) have shown that, for the film thickness values encountered in the small ripple regime of annular flow, the last four terms on the right-hand side of Equation 8.176 may be neglected without causing errors in excess of 1 percent. The resulting equation for the liquid flow rate is

$$Q_L = -\frac{\pi D^2\delta^2 g_c}{8\mu_L}\frac{dP}{dz} \tag{8.179}$$

Also for very thin films and low gas density Equation 8.168 may be simplified to

$$\tau_i = -\frac{dP}{dz}\frac{D}{4} = \tau_w \tag{8.180}$$

Shearer and Nedderman claim that in the small ripple regime the assumption that $\tau_i = \tau_w$ (constant shear stress in the film) does not lead to any more than 1 percent error. If we define a dimensionless film thickness as

$$\delta^+ = \frac{\delta\rho_L}{\mu_L}\sqrt{\tau_i g_c/\rho_L} = \frac{\delta}{\mu_L}\sqrt{\tau_i g_c \rho_L} \tag{8.181}$$

then from Equations 8.179 and 8.180 we have

$$\delta^+ = \sqrt{\frac{2\rho_L Q_L}{\pi D \mu_L}} = \sqrt{\frac{\text{Re}_f}{2}} \tag{8.182}$$

We see from the above discussion that the thickness of the liquid film can be predicted from the overall liquid flow rate, the liquid properties, the tube radius, and the pressure gradient. As will be seen later, the pressure gradient may be evaluated through the gas core.

The situation is significantly more complex in the general case when part of the film is in turbulent motion. The essence of the problem is the development of a relationship between τ and u over the range of y of interest. To do this, Calvert and Williams (1955) assumed that the mixing length distribution in the film was the same as that for full pipe flow and that the portion of the film in laminar motion could be determined as for single-phase flow. Anderson and Mantzouranis (1960) assumed the applicability of the universal velocity profile without considering the actual shear stress distribution in the film.

Turner (1966) has shown that it is reasonable to assume that the wall shear stress is the same as if the liquid film were part of a liquid stream flowing at the average film velocity and filling the whole pipe. Wallis (1970) has developed a simplified analysis incorporating the same approximation. We discuss Turner's analysis in Chapter 10 and defer discussion of Wallis' treatment to later in this section, since he applied his force balances to the entire contents of the pipe and to the gas core, not singling out a force balance on the liquid film. Levy's (1966) analysis of annular flow in the presence of entrainment incorporates the universal velocity profile for the turbulent liquid film. We deal with this in Section 7.32.

Probably the most detailed analysis of pure annular flow with a turbulent liquid film is that by Hewitt (1961) based upon the analysis previously made by Dukler (1959) for downward flow. Hewitt's analysis involves the Deissler (1950) equation for relating shear stress to the velocity distribution. The method is cumbersome, and it requires numerical integration (or the use of tables). For these reasons, and since there is no evidence that it gives significantly better results than the simpler analyses of Turner, Wallis, and Levy we will not consider it here.

Estimation of the Interfacial Shear Stress For the application of the momentum equation in the absence of entrainment, whether by the Willis (1965) or the Shearer and Nedderman (1965) method for laminar flow; or the Calvert and Williams (1955), Anderson and Mantzouranis (1960), Hewitt (1961), or other methods for a laminar-turbulent film, it is necessary to evaluate the interfacial shear stress, τ_i. This may be done by considering and applying single-phase methods to the gas core. A difficulty is that the liquid film may be smooth (low liquid and gas rates), ripply (intermediate rates), or relatively large waves may occur in it. Although other approaches have been used (e.g., Calvert and Williams) the best seems to be to liken the effect of the liquid film on the gas flow to roughness in a pipe wall and to seek an empirical relationship between a friction factor for the gas phase, the gas phase Reynolds number, and measures of the "roughness" of the liquid film.

Anderson and Mantzouranis correlated their data through a friction factor, f_{AM}, defined through the equation

$$\tau_i = \frac{f_{AM} \rho_G V_G^2}{g_c} \tag{8.183}$$

the gas phase Reynolds number, $D_G V_G \rho_G/\mu_G$, and the parameter

$$Q^* = \frac{W^+}{\delta^+}$$

where $W^+ = Q_L \rho_L/\pi D \mu_L$, a dimensionless liquid film flow rate,
δ^+ = the previously defined dimensionless film thickness (Equation 8.181).

The friction factor, f_{AM}, is equivalent to one-half the conventional Fanning friction factor, f. Collier and Hewitt (1961) and Hewitt, King, and Lovegrove (1961) have followed this approach. Figure 8.55 is a composite, giving the data of Anderson and Mantzouranis (1960), Hewitt *et al.* (1961) and of Bennett and Thornton (1961). The lack of full agreement of all data is in part at least due to the lack of fully established equilibrium in some of the test data. This is reflected by the fact that different results have been obtained with different methods of liquid introduction (Collier and Hewitt, 1961). Nonetheless, Figure 8.55 probably reflects the best general present relationship between the shear stress at the interface and the variables affecting it.

For thin film (small ripple region) Shearer and Nedderman have been able to get a rather good correlation between a friction factor, f_{SN}, and an effective relative roughness, k/D, related to the dimensionless film thickness, δ^+, and the relative velocity, $V_{SG} - u_i$. The friction factor is defined by

$$\tau_i = \frac{f_{SN} \rho_G (V_{SG} - u_i)^2}{2 g_c} \tag{8.184}$$

Shearer and Nedderman defined δ^+ in terms of τ_i, but since they assume $\tau_i \simeq \tau_w$ their definition is the same as that of Hewitt *et al.*

Shearer and Nedderman express the relationship for the single-phase friction factor in FRW (fully rough wall) turbulence in the form

$$\frac{1}{\sqrt{f_{SN}}} = 4 \log \frac{D}{2(k - 5y^*)} + 3.48 \tag{8.185}$$

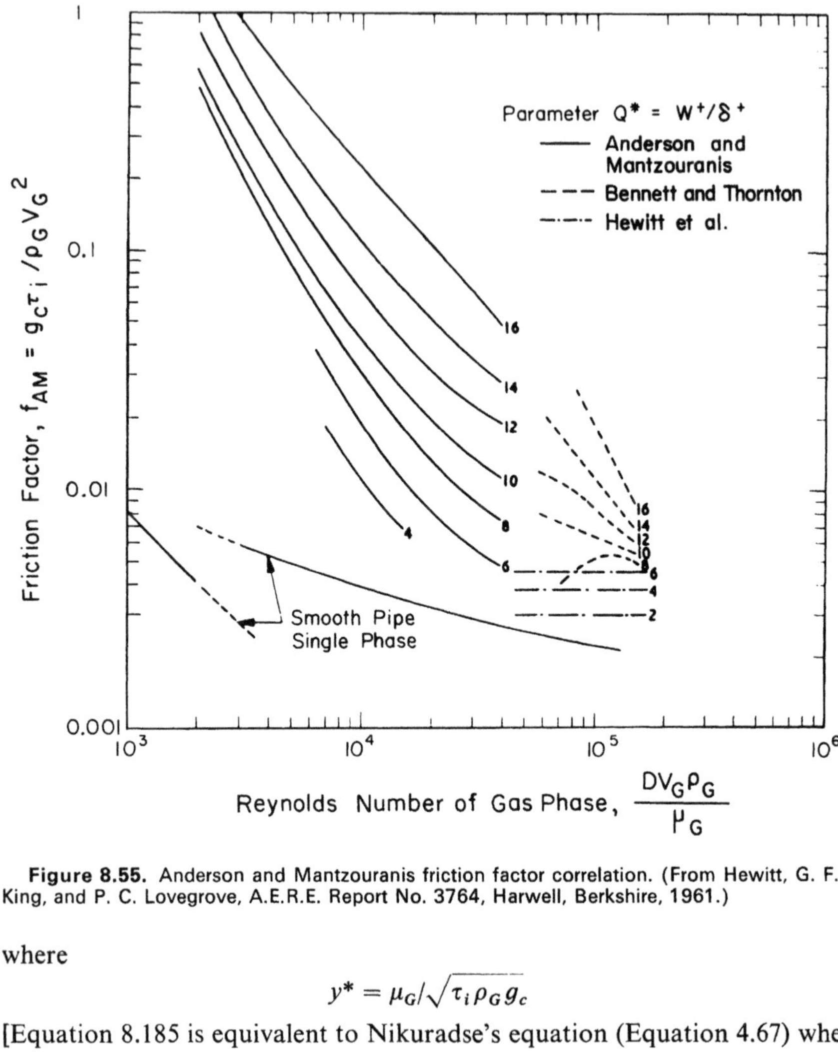

Figure 8.55. Anderson and Mantzouranis friction factor correlation. (From Hewitt, G. F., I. King, and P. C. Lovegrove, A.E.R.E. Report No. 3764, Harwell, Berkshire, 1961.)

where

$$y^* = \mu_G/\sqrt{\tau_i \rho_G g_c}$$

[Equation 8.185 is equivalent to Nikuradse's equation (Equation 4.67) where $k \gg 5y^*$.] They correlate values of k, calculated from actual data through Equation 8.185, by the empirical relation

$$\delta^+ = K_1 D^2 \sigma + K_2 \left(\frac{V_{SG} - u_i}{\mu_L}\right) \sqrt{\rho_G}\,\frac{k}{D} \tag{8.186}$$

where σ is the surface tension, and the empirical values of the constants K_1 and K_2 are

$$K_1 = 2.52 \times 10^4 \text{ sec}^2/\text{lb}_m \text{ ft}^2$$

$$K_2 = 0.098\ (\text{lb}_m/\text{ft})^{1/2}$$

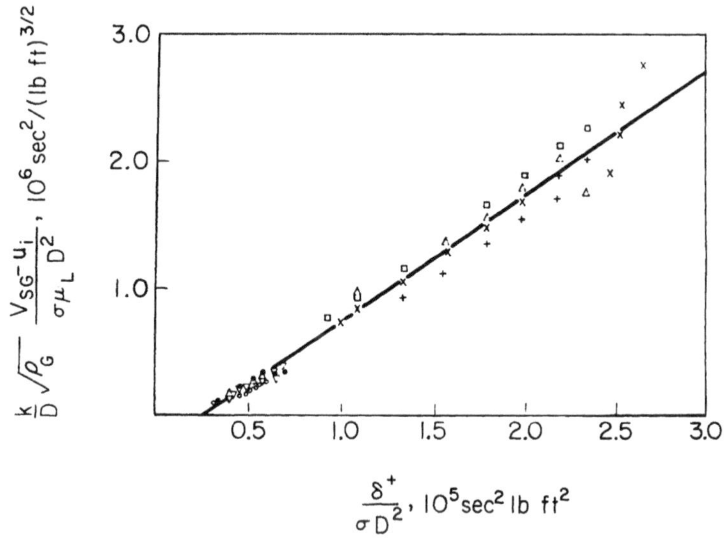

Figure 8.56. Shearer and Nedderman pressure drop correlation for "small ripple" flow regime. (From Shearer, C. J., and R. M. Nedderman, *Chem. Eng. Sci.*, **20**, 671, 1965.)

Figure 8.56 shows a comparison of experimental data of Shearer and Nedderman with Equation 8.136. The agreement is good for $\delta^+/D^2\sigma < 2 \times 10^{-5}$ and becomes increasingly poor as the film thickness increases.

Pogson, Roberts, and Waibler (1970), in connection with their experimental study of the annular-mist flow of steam-water mixtures in a vertical pipe 1.335 inches in diameter, combined force balance relationships for the liquid film and the gas core and developed the following empirical equation for the prediction of film thickness

$$\frac{\delta}{D} = 126\left(\frac{\text{Re}_{SL}^{0.42}}{\text{Re}_G^{1.186}}\right)\left(\frac{\mu_L \rho_G}{\mu_G \rho_L}\right)\left(\frac{D\sigma g_c \rho_G}{\mu_G^2}\right)^{1/4} \tag{8.187}$$

The equation reproduces their data, in the absence of heat transfer and covering film thicknesses of 3 to 13 × 10^{-3} inch, extremely well. Accordingly, it accounts implicitly for the variation with flow rates of the nature of the interface and the effect of ripples and waves on the interfacial shear stress. It may be useful for preliminary estimates with other systems.

Wallis (1970) presents a simplified analysis for the interfacial shear stress and the shear stress in the film. He defines an interface friction factor, f, by analogy with single-phase flow as

$$\tau_i = \frac{f\rho_G V_c^2}{2g_c}$$

where V_c is the mean velocity of the gas core. The mean diameter of the gas core is given by

$$D\sqrt{E_G} = D - 2\delta \tag{8.188}$$

where E_G is the fraction of the cross section occupied by the gas phase. Combining these two equations with a force balance over the gas core rearranging, and substituting for V_c in terms of the superficial gas velocity, V_{SG}, we obtain

$$f = \frac{-(dP/dz)DE_G^{5/2}}{2\rho_G V_{SG}^2} g_c \tag{8.189}$$

Here we have neglected $\rho_G g/g_c$ in comparison to dP/dz. Wallis calculated values of δ and of f from experimental data in the annular flow region and found a simple approximate correlation in the form

$$f = 0.005\left(1 + 300\frac{\delta}{D}\right) \tag{8.190}$$

or, since $\delta/D \simeq (1 - E_G)/4$,

$$f = 0.005[1 + 75(1 - E_G)] \tag{8.191}$$

Comparing Equation 8.190 or 8.191 with approximate equations for single-phase flow in rough pipes, one can show that a wavy annular film is about equivalent to a "sand roughness" of four times the film thickness. Combination of Equations 8.189 and 8.191 leads to the following form of the force balance over the gas core:

$$-\left(\frac{dP}{dz} + \rho_G\frac{g}{g_c}\right) = 10^{-2}\frac{\rho_G V_{SG}^2}{g_c D}\frac{[1 + 75(1 - E_G)]}{E_G^{5/2}} \tag{8.192}$$

Wallis writes a force balance for the total contents of the pipe incorporating an estimate of the wall shear stress as if the liquid film were a part of a liquid stream filling the whole pipe and flowing at an average velocity V_L. For turbulent flow in the film he assumes a constant value of the friction factor, f, of 0.005 and obtains

$$-\left(\frac{dP}{dz} + \rho_G\frac{g}{g_c}\right) = (1 - E_G)(\rho_L - \rho_G)\frac{g}{g_c} + \frac{10^{-2}}{g_c}\frac{\rho_L V_{SL}^2}{D(1 - E_G)^2} \tag{8.193}$$

For complete laminar flow of the film the force balance becomes

$$-\left(\frac{dP}{dz} + \rho_G\frac{g}{g_c}\right) = \frac{32}{g_c}\frac{V_{SL}\mu_L}{D^2(1 - E_G)^2} + \left[1 - \frac{2E_G^2 \ln E_G}{(1 - E_G)^2} - \frac{2E_G}{1 - E_G}\right]\frac{g}{g_c}(\rho_L - \rho_G) \tag{8.194}$$

Equation 8.194 may be simplified by inserting the following approximation:

$$1 - \frac{2E_G^2 \ln E_G}{(1 - E_G)^2} - \frac{2E_G}{1 - E_G} \simeq 0.684(1 - E_G)$$

which is applicable over the range $0 < (1 - E_G) < 0.2$.

Wallis (1969) expresses Equations 8.192, 8.193, and 8.194 in dimensionless form by defining the quantities

$$\Delta P^* = \frac{\dfrac{g_c}{g}\dfrac{\Delta P}{\Delta z} + \rho_G}{(\rho_L - \rho_G)} \tag{8.195}$$

$$\Delta P_G^* = \frac{g_c \Delta P_{SG}}{\Delta z g(\rho_L - \rho_G)} \tag{8.196}$$

$$\Delta P_L^* = \frac{g_c \Delta P_{SL}}{\Delta z g(\rho_L - \rho_G)} \tag{8.197}$$

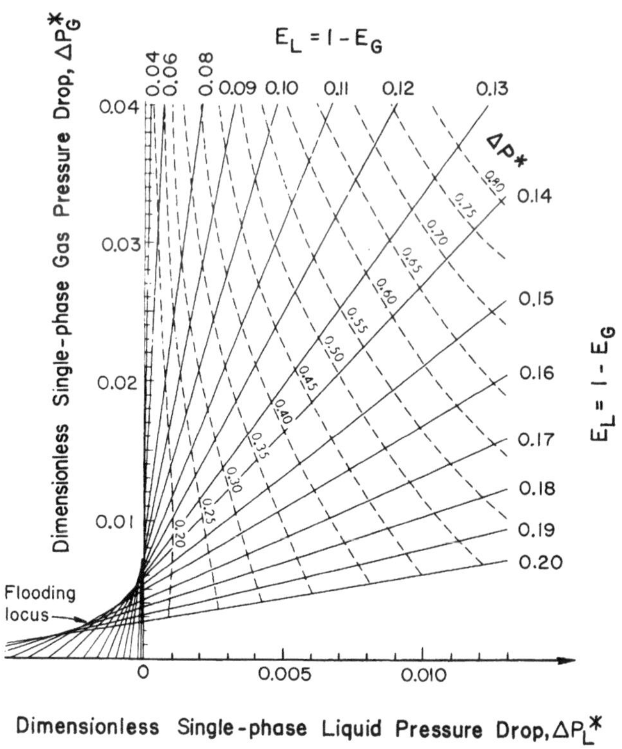

Figure 8.57. Wallis' pressure drop correlation for annular-mist flow. ΔP^* = dimensionless two-phase pressure drop. (From Wallis, G. B., *One-Dimensional Two-Phase Flow,* McGraw-Hill Book Co., New York, 1969.) (Used with permission of McGraw-Hill Book Co.)

where ΔP^* is a dimensionless two-phase pressure drop, ΔP_G^* is the dimensionless pressure drop that would be encountered if the gas were to flow alone at the velocity V_{SG}, and ΔP_L^* is the dimensionless pressure drop that would be encountered if the liquid were to flow alone at the velocity V_{SL}.

Figure 8.57 gives Wallis' graphical solution of the equations. It is useful for a quick, approximate estimate of the pressure gradient and volume fraction gas in mist-free annular flow.

Turner (1966) has presented a detailed analysis for pure annular flow of a gas-liquid mixture in a pipe inclined at any angle to the horizontal. This treatment, which is discussed in some detail in Section 8.31 of Chapter 10, may also be applied to the vertical flow case.

7.32 Annular-Mist Flow—the Effect of Entrainment

To this point we have assumed that all of the liquid is carried in the film and there is no entrainment. The data of many workers confirm that except at low liquid rates a significant portion of the liquid may be carried as droplets in the gas core. The data of Collier and Hewitt (1961) are typical. The results of their measurements on the vertical flow of air-water mixtures in a $1\frac{1}{4}$-inch plastic tube are shown in Figure 8.58. The data show that the flow in the liquid film initially increases with liquid flow rate, but eventually levels out at a value largely dependent on the gas flow rate. At the highest gas and liquid flow rates approximately 90 percent of the liquid is carried in entrained droplets.

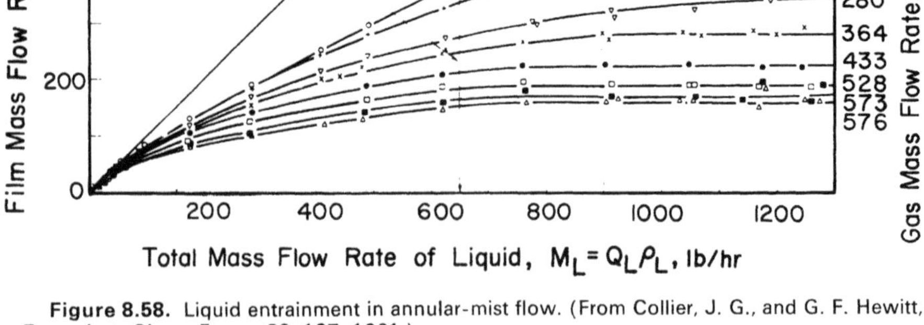

Figure 8.58. Liquid entrainment in annular-mist flow. (From Collier, J. G., and G. F. Hewitt, *Trans. Inst. Chem. Engrs.*, **39**, 127, 1961.)

Gill, Hewitt, and Lacey (1964) have measured liquid droplet concentration profiles for air-water flow in a $1\frac{1}{4}$-inch tube and observed essentially flat profiles over a substantial range of liquid velocities. Typical data from their work are given in Figure 8.59 in terms of the local volume fraction of the liquid, ε_L.

There are no methods known for the prediction of the size and size distribution of the entrained droplets in annular-mist flow. We would expect, however, that the droplet size would be somewhat less than the film thickness. Pogson *et al.* (1970) are among the very few who have measured droplet size

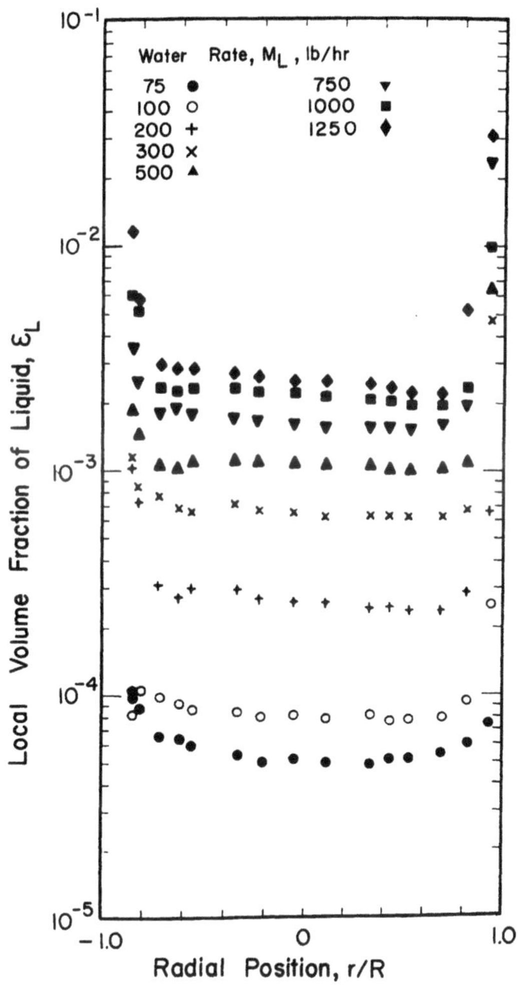

Figure 8.59. Typical liquid droplet concentration profiles. Air rate = 300 lb/hr. (From Gill, L. E., G. F. Hewitt, and P. M. C. Lacey, *Chem. Eng. Sci.*, **19**, 665, 1964.)

and size distribution. In their work with the steam-water system they found the arithmetic mean droplet diameter to be insensitive to the total liquid rate and to vary only from 39 to 44 microns under a threefold variation in liquid rate and under circumstances where the liquid film thickness ranged from about 5 to 8.5×10^{-3} inch (127 to 216 microns). The droplet size distribution covered the range from about 10 to 110 microns and was skewed toward the lower size.

The entrainment data of Gill *et al.* (1964) are shown in Figure 8.60 where the volume ratio of the liquid in the gas core is given as a function of the total liquid flow rate. This diagram should not be used as a general correlation but is useful as a guide.

Gill, Hewitt, Hitchon, and Lacey (1963), Gill, Hewitt, and Lacey (1964), and Gill and Hewitt (1968) have studied the gas core in the vertical annular flow of air-water mixtures using a sampling probe. These authors have observed dramatic and persistent differences between axial and peripheral methods of liquid injection. The experimental pressure gradients obtained by the two methods are compared with results calculated by several methods in Figure 8.61. This figure also shows that in the $1\frac{1}{4}$-inch-diameter tube used by these investigators there is a tendency for the pressure gradient curves to approach each other, but the entrance effect persists for a large distance. The experimental value of the film thickness toward the end of the $17\frac{1}{2}$-foot test section differs by a factor of 2 for the two different methods of injecting liquid. As expected, the film is thinner in the case of axial gas injection.

Entrainment affects the previously described analysis of the momentum equation in two ways. First, the quantity of liquid flowing in the film is not the total quantity flowing but must be calculated from Equation 8.157 from a knowledge of the quantity of liquid entrained. Second, the interfacial shear calculation must reflect the greater effective density of the mixture in the "gas" core. Equation 8.168 is modified to

$$\tau_i = \frac{D - 2\delta}{4}\left(-\frac{dP}{dz} - \rho_c \frac{g}{g_c}\right) \tag{8.198}$$

where

$$\rho_c = (1 - e)\rho_G + e\rho_L \tag{8.199}$$

These two corrections are not difficult in principle but they do require a knowledge of the amount of entrainment, i.e., of the volume fraction of the entrained droplets in the gas phase, e, or the equivalent. At the present state of knowledge we must resort to empirical correlations based on data such as shown in Figures 8.58, 8.59, 8.60, and 8.61.

Collier and Hewitt (1961) and Gill, Hewitt, and Lacey (1965) have followed an approach suggested by Wicks and Dukler (1960) and have correlated the

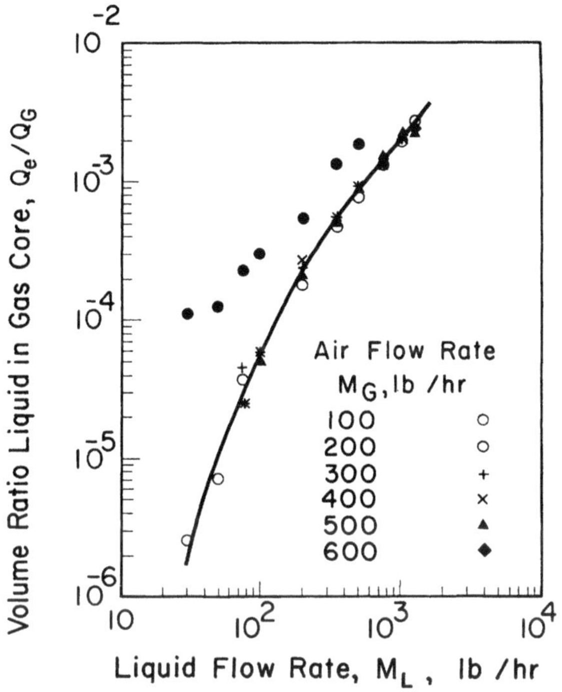

Figure 8.60. Entrainment data of Gill *et al.* (From Gill, L. E., G. F. Hewitt, and P. M. C. Lacey, *Chem. Eng. Sci.*, **19**, 665, 1964.)

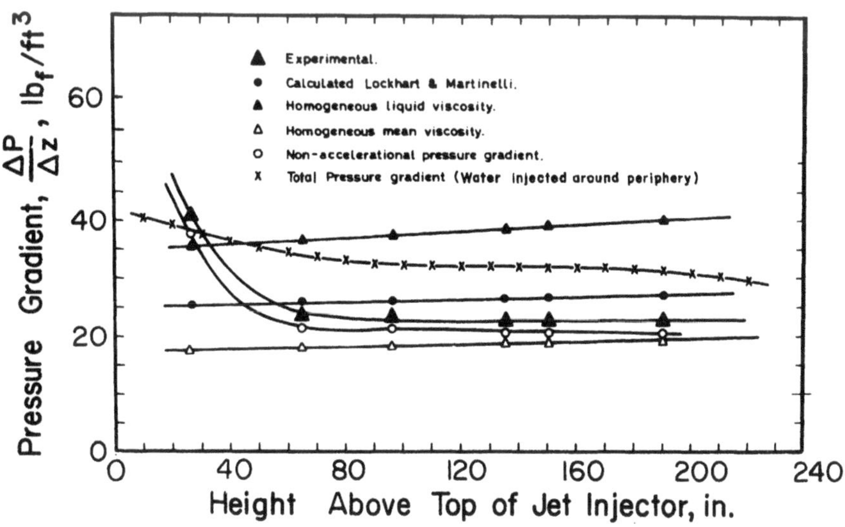

Figure 8.61. Comparison of observed and calculated pressure gradients. (From Gill, L. E., and G. F. Hewitt, *Chem. Eng. Sci.*, **23**, 677, 1968.)

mass rate of flow of the entrained liquid through the Lockhart and Martinelli (1949) parameter (see Section 3.2 of Chapter 10)

$$X = \frac{(\Delta P/L)_{SL}}{(\Delta P/L)_{SG}}$$

where $\left(\frac{\Delta P}{L}\right)_{SL}$ = pressure gradient for the liquid flowing alone in a horizontal pipe at the velocity V_{SL}

$\left(\frac{\Delta P}{L}\right)_{SG}$ = pressure gradient for the gas flowing alone in a horizontal pipe at the velocity V_{SG}

and an entrainment number

$$\mathscr{R}_E = \frac{(Q_L/Q_G)\mathrm{We}_c M_e}{\left(\frac{\Delta P}{L}\right)_{SG}}$$

where $M_e = Q_e \rho_L$ = mass flow rate of entrained liquid,

$\mathrm{We}_c = \left[\frac{\rho_G (V_G - V_L)^2 D}{\sigma g_c}\right]_c$

= critical value of the Weber number dependent on the method of introduction of the liquid,

V_G, V_L = the average entering velocities of the gas and the liquid,

σ = interfacial tension.

Collier and Hewitt and Gill *et al.* have taken Wicks and Dukler's values for We_c as follows:

$\mathrm{We}_c = 13$ when the liquid is introduced abruptly as with a tee entry,

$\mathrm{We}_c = 22$ when the liquid is introduced smoothly as with an annular entrance device.

Figure 8.62 taken from Collier and Hewitt shows data of Bennett and the mean lines of Wicks and Dukler for horizontal flow. Gill *et al.* (1965) confirm Collier and Hewitt's empirical relation for $X > 0.08$, namely

$$X = 0.069 \mathscr{R}_E^{0.39} \tag{8.200}$$

and show that $X > 0.08$ approximately corresponds with the condition where $\delta^+ > 18$. For $X < 0.08$ or $\delta^+ < 18$, they recommend

$$X = k \mathscr{R}_E^{0.16} \tag{8.201}$$

where k, depending on the gas velocity, ranges from 0.03 at gas flow rates (in a $1\frac{1}{4}$-inch tube) of the order of 600-700 lb/hr to 0.08 at rates of the order of 165 lb/hr.

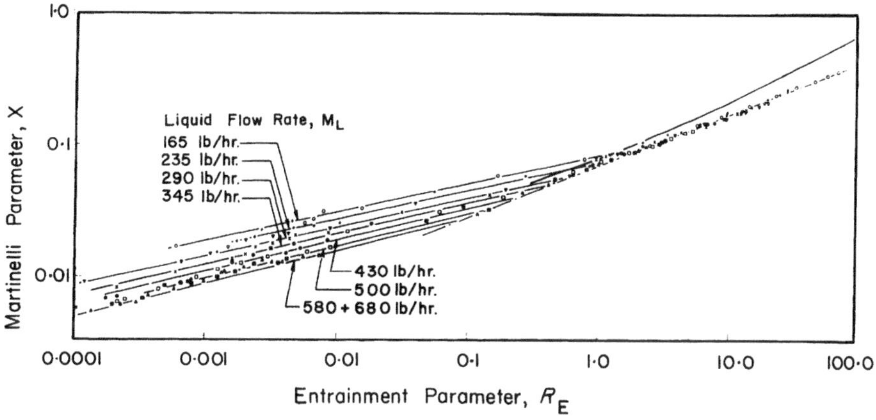

Figure 8.62. Wicks and Dukler entrainment correlation. (From Collier, J. G., and G. F. Hewitt, *Trans. Inst. Chem. Engrs.*, **39**, 127, 1961.)

If M_e is estimated from this correlation, the related quantity, e, may be determined from a modification of Equation 8.164, i.e.,

$$M_e = \rho_L \frac{e\pi D^2}{4}(1 - \xi)^2 V_e \tag{8.202}$$

where V_e is taken as the average gas velocity, V_G, less the effective average "settling" velocity of the droplets. At the gas velocities encountered in annular flow, the average settling velocity of the droplets is small relative to the gas velocity, permitting the approximation

$$M_e = \rho_L \frac{e\pi D^2}{4}(1 - \xi)^2 V_G \tag{8.203}$$

and immediate estimation of e from M_e, ξ, and V_G.

Levy (1966) has developed a method for estimating the distribution of the liquid between the film and the entrained droplets in the usual situation where the entrainment of liquid in the gas core is accompanied by turbulence in part of the film. The method requires an independent estimate of the pressure drop. Although developed for vertical upward flow, the analysis may also be applied with minor modifications to horizontal and to vertical downward flow. We will summarize Levy's treatment here.

The local shear stress in turbulent flow with variable density is given by

$$g_c \tau = -\overline{(\rho u)' v'} \tag{8.204}$$

where the prime (′) denotes the fluctuating component (see Chapters 3 and 4). The fluctuating components may be expressed in terms of the appropriate mixing lengths, i.e.,

$$\sqrt{(v'^2)} \simeq \sqrt{(u'^2)} = l_u \left(\frac{du}{dy}\right)$$

and

$$\sqrt{\overline{(\rho u)'^2}} = l_{\rho u} \left| \left(\frac{d\rho u}{dy}\right) \right|$$

where l_u = mixing length for velocity,
$l_{\rho u}$ = mixing length for momentum.

The local shear stress, τ, becomes

$$g_c \tau = \rho l_u l_{\rho u} \left(\frac{du}{dy}\right)^2 + u l_u l_{\rho u} \frac{du}{dy} \left| \left(\frac{d\rho}{dy}\right) \right| \tag{8.205}$$

We note that the shear stress is made up of a momentum term and a mass transfer term. The mass transfer term involving $d\rho/dy$ is usually negligible in single-phase flow but cannot be neglected in annular flow with entrainment. In order to simplify the treatment, Levy approximates the derivative in Equation 8.205 by first-order differences.

For the core side we have

$$\frac{du}{dy} \simeq \frac{V_c - V_f}{D/4}$$

and

$$-\frac{d\rho}{dy} \simeq \frac{\rho_f - \rho_c}{D/4}$$

where V and ρ are the mean values of velocity and density and the subscripts c and f refer to the core and the film, respectively. The shear stress on the core side of the interface, τ_c, may be approximated as

$$g_c \tau_c \simeq l_u l_{\rho u} V_c \left(\frac{V_c - V_f}{D/4}\right)\left(\frac{\rho_f - \rho_c}{D/4}\right)\left[1 + \frac{\rho_c}{\rho_f - \rho_c}\left(\frac{V_c - V_f}{V_c}\right)\right] \tag{8.206}$$

For most practical systems, both $\rho_c/(\rho_f - \rho_c)$ and $(V_c - V_f)/V_c$ are much less than unity and the momentum term becomes negligible compared with the mass transfer term. Equation 8.206 may therefore be simplified to

$$g_c \tau_c \simeq l_u l_{\rho u} V_c \left(\frac{V_c - V_f}{D/4}\right)\left(\frac{\rho_f - \rho_c}{D/4}\right) \tag{8.207}$$

For the liquid or film side of the interface we may write

$$\frac{du}{dy} \simeq \frac{V_f}{\delta/2}$$

and

$$\frac{d\rho}{dy} \simeq \frac{\rho_L - \rho_f}{\delta/2}$$

which yields the following expression for the shear stress on the film side of the interface:

$$g_c \tau_f = \rho_f l_u l_{\rho u} \left(\frac{2V_f}{\delta}\right)^2 \left(1 + \frac{\rho_L - \rho_f}{\rho_f}\right) \qquad (8.208)$$

where ρ_L is the density of the pure liquid. Here $\rho_L \simeq \rho_f$ and the mass transfer component of the shear stress may be neglected, yielding

$$g_c \tau_f = \rho_f l_u l_{\rho u} \left(\frac{2V_f}{\delta}\right)^2 \qquad (8.209)$$

The film flow rate may be expressed in dimensionless form as

$$Q_L^+ = \frac{D V_{Sf} \rho_L}{4\mu_L} \qquad (8.210)$$

where

$$V_{Sf} = \frac{4Q_{Lf}}{\pi D^2} = \text{superficial film velocity}$$

Note that Q_L^+ is 1/4 times the Reynolds number for the flow in the film. Levy assumes that the universal velocity relationships, (Equations 4.41 4.46, and 4.52) apply in the liquid film and integrates these to obtain the total flow in the film

$$Q_L^+ = -64 + 3\delta^+ + 2.5\delta^+ \ln \delta^+ \qquad \text{for } \delta^+ \geq 30 \qquad (8.211)$$

$$Q_L^+ = 12.5 - 8.05\delta^+ + 5\delta^+ \ln \delta^+ \qquad \text{for } 5 < \delta^+ < 30 \qquad (8.212)$$

$$Q_L^+ = 0.5\delta^{+2} \qquad \text{for } \delta^+ \leq 5 \qquad (8.213)$$

Equations 8.211 and 8.212 are the same as those derived by McManus (1961) for concentric horizontal annular flow and discussed in Section 8.33 of Chapter 10.

If G_{Sf}, G_{SL}, and G_{SG} represent the superficial mass velocities of the film liquid, the total liquid, and the gas, then the mass fraction of the gas in the core, X_c, is

$$X_c = \frac{G_{SG}}{G_{SG} + G_{SL} - G_{Sf}} \qquad (8.214)$$

Treating the core as pseudohomogeneous and assuming additive volumes, we have

$$\frac{1}{\rho_c} = \frac{1}{\rho_L}(1 - X_c) + \frac{1}{\rho_G} X_c \qquad (8.215)$$

The average film and core velocities are

$$V_f = \frac{G_{Sf}}{\rho_L(2\xi - \xi^2)} \qquad (8.216)$$

$$V_c = \frac{G_{SG}}{X_c \rho_c(1 - 2\xi - \xi^2)} \qquad (8.217)$$

Following single-phase methods and based on Equation 8.207, Levy postulates that the shear stress on the core side of the interface may be related to the film thickness by

$$\sqrt{\frac{g_c \tau_c}{V_c(\rho_L - \rho_c)(V_c - V_f)}}\, R_L = F' \qquad \text{for } -\frac{dP}{dz} \geq \rho_L g/g_c \qquad (8.218)$$

$$\sqrt{\frac{g_c \tau_c}{V_c(\rho_L - \rho_c)(V_c - V_f)}} \left(\frac{\rho_L g/g_c}{-dP/dz}\right)^{-n} R_L = F' \qquad \text{for } \frac{-dP}{dz} < \rho_L g/g_c \qquad (8.219)$$

where R_L = function of the density ratio ρ_L/ρ_c,
F' = function of ξ,
n = exponent on pressure drop grouping, dimensionless.

Levy has calculated F' using experimental pressure drop and film thickness data for several systems. Good agreement was obtained with $n = \frac{1}{3}$. Figure 8.63 shows F' as a function of ξ. The relationship between R_L and ρ_L/ρ_G is shown in Figure 8.64.

Levy has also developed a simplified treatment which he finds to represent experimental data equally well. Assuming that $\rho_G/\rho_L \ll 1$ and that $\xi \ll 1$, he simplifies Equations 8.215 and 8.216, takes $\tau_c \simeq \tau_w$, and reduces Equations 8.218 and 8.219 to

$$\sqrt{\frac{\left(\frac{-dP}{dz}\right)\frac{D}{4}}{\rho_L/g_c}\left(\frac{\rho_G}{G_{SG}}\right)}\left[\frac{\rho_L g/g_c}{\left(\frac{-dP}{dz}\right)}\right]^{-n} R_L = F' \qquad (8.220)$$

where $n = 0$ for $(-dP/dz) \geq \rho_L g/g_c$ and
$n = \frac{1}{3}$ for $(-dP/dz) \leq \rho_L g/g_c$.

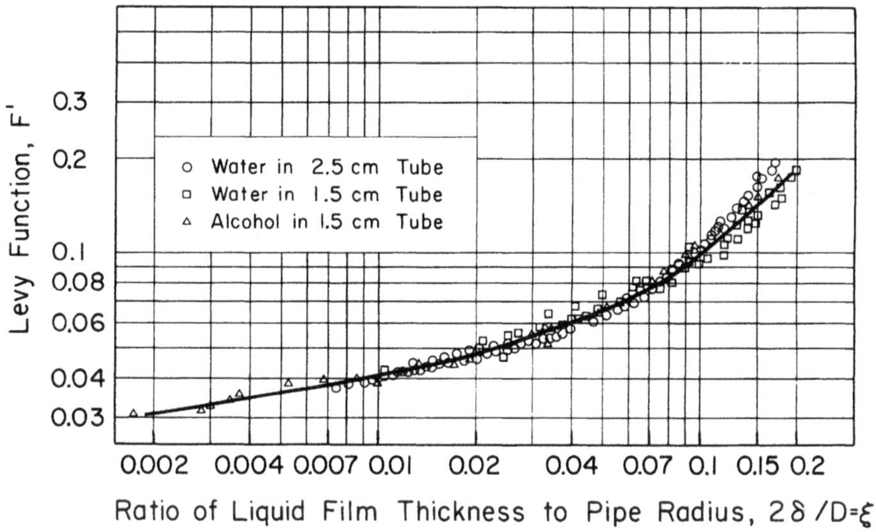

Figure 8.63. Levy function F'. (From Levy, S., *Int. J. Heat and Mass Transfer,* **9**, 171, 1966.)

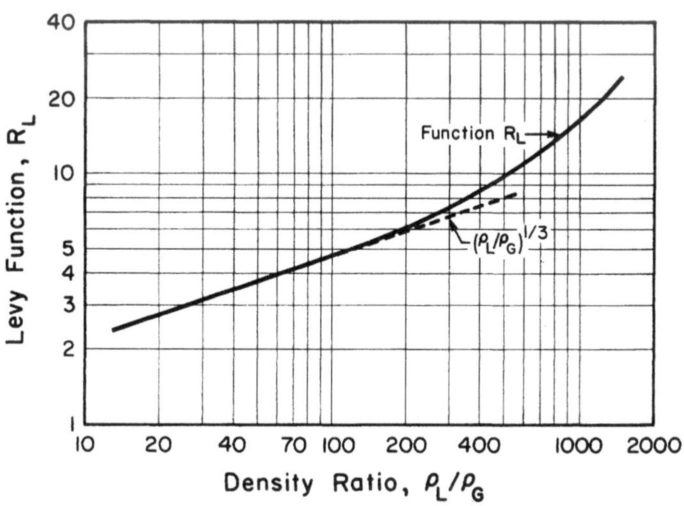

Figure 8.64. Levy function R_L. (From Levy, S., *Int. J. Heat and Mass Transfer,* **9**, 171, 1966.)

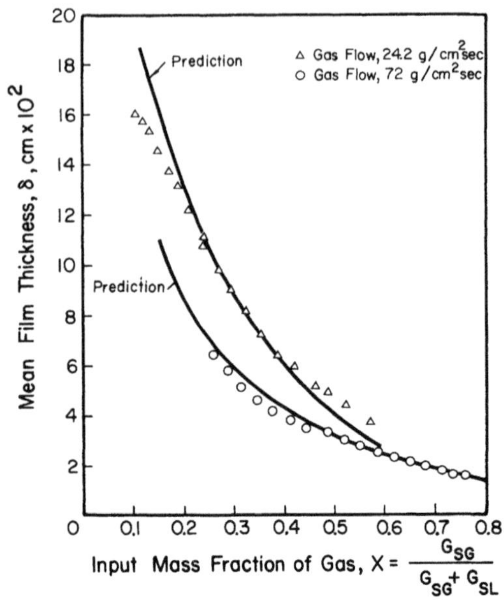

Figure 8.65. Comparison of Levy's prediction method for film thickness with actual data on the flow of gas-water mixtures in a 2.5-cm tube, $\rho_G = 0.0361$ g/cm³. (From Levy, S., *Int. J. Heat and Mass Transfer*, **9**, 171, 1966.)

Equation 8.220 with Figures 8.63 and 8.64 and an independent estimate of dP/dz permit a direct solution for the film thickness without the iterations required in the use of Equations 8.218 and 8.219. Levy uses an empirical correlation (Equation 8.225 given subsequently) applicable to annular-mist flow developed by Adorni *et al.* (1961, 1963) for estimating the pressure gradient.

Figure 8.65 from Levy compares his predictions of film thickness with actual data in a 1-inch tube. Although further confirmation of Levy's method is required, and it is dependent upon a reliable independent estimate of the pressure gradient, it appears soundly based on a rational analysis of the flow mechanism.

7.4 *Flow Pattern Details*

In the annular-mist region a complete description of the flow pattern involves a knowledge of the distribution of the liquid between the entrained droplets and the film; the thickness, velocity, and surface structure of the film; the volume fraction, size, and velocity of the entrained droplets; and the velocity of the gas. The prediction of certain of the characteristics of the pattern depends upon empirical correlations, and most of these have been based on air-water systems flowing in relatively small diameter tubes. For these reasons the prediction methods may be considered reasonably reliable for air-water and closely similar systems, but are relatively untested for others.

7.41 Pure Annular Flow—No Entrainment

The flow pattern details, including the pressure gradient, may be estimated approximately following Wallis, as discussed in Section 7.31. The method is as follows:

1. The film structure may only be approximated but Figure 8.53 will indicate whether small ripples or disturbance waves are to be expected.
2. Calculate ΔP_G^* and ΔP_L^* using Equations 8.196 and 8.197.
3. Using Figure 8.57, estimate ΔP^* and $(1 - E_G)$. ΔP^* may be converted to $\Delta P/\Delta z$ from Equation 8.195, and the film thickness δ may be calculated from Equation 8.188.
4. The actual gas velocity may be calculated from Equation 8.151.
5. The average film velocity is determined from Equation 8.152.
6. The interface velocity, u_i, may be estimated from Equation 8.177 and the film thickness and the pressure gradient determined in step *3*.

The above calculations are approximate only and may be refined if greater accuracy is desired by applying the method of Levy and continuing the procedure as follows:

7. From the estimate of the pressure gradient and Figures 8.63 and 8.64 and Equation 8.180, an estimate of τ_i may be obtained.
8. The film thickness may now be recalculated from Equation 8.179.
9. Steps *4*, *5* and *6* may now be repeated using the revised values of the film thickness.

7.42 Annular-Mist Flow

1. The volume of liquid entrained in the gas core may be estimated from Figure 8.60 (Gill *et al.*, 1964) or Figure 8.58 (Collier and Hewitt, 1961). The liquid flow rate in the film is obtained by difference.
2a. The pressure gradient may be estimated as in Section 7.41.
2b. In place of steps *1* and *2a*, or as a check, we may use the method proposed by Levy, which has been discussed in detail in Section 7.32.
3. The film thickness may be estimated, using the film flow rate, from Equation 8.179.
4. Steps *4*, *5*, and *6* of Section 7.41 may be followed to obtain the actual gas velocity, the average film velocity, and the interface velocity.

7.5 Holdup

Two different holdup quantities are of interest in annular-mist flow. These are the holdup in the gas core due to the presence of entrained liquid and the overall system holdup due to both the liquid film and any entrained liquid.

The volume fraction of liquid entrained in the gas core, e, is related to F_e, the fraction of the flowing liquid entrained in the gas phase, and the other variables through Equation 8.164. This equation may be solved for e with the aid of the values of F_e, V_e, and δ as determined in Section 7.42.

The overall system holdup may be expressed in terms of E_G (or $\bar{E}_G$ for a finite system length), the overall volume fraction gas. From volumetric considerations this may be calculated as

$$E_G = (1 - \xi)^2(1 - e) = (1 - \xi)^2 - F_e\left(\frac{V_L}{V_e}\right) \tag{8.221}$$

Note that e, rather than the overall E_G, is the significant factor in affecting the pressure gradient.

Collier and Wallis (1966) have obtained an empirical correlation for $\bar{E}_G$ using the same dimensionless velocities used in the flooding correlation discussed in Section 5.12. Their correlation is given by

$$\frac{V_{SG}^*}{1 - 3.1(1 - \bar{E}_G)} - \frac{V_{SL}^*}{3.1(1 - \bar{E}_G)} = 1 \tag{8.222}$$

They have evaluated their correlation with the data of Govier *et al.* (1957), Govier and Short (1958), Brown *et al.* (1960), Hewitt and Lovegrove (1963), and Hewitt *et al.* (1961). The agreement is quite good and this correlation is recommended for the direct empirical estimation of $\bar{E}_G$.

7.6 Pressure Gradient

As has been indicated in the discussion of the momentum equation and in the prediction of the flow pattern details, the pressure gradient may be calculated by the application of single-phase methods to the gas core. Where entrainment occurs we must use the average *in situ* density of the core mixture as given by Equation 8.199. The mixture is flowing at an average velocity relative to the moving interface,

$$V_r = V_c - u_i = \frac{V_{SG}}{(1 - \xi)^2} - u_i \tag{8.223}$$

For most practical purposes, u_i may be neglected and we may apply the mechanical energy balance, Equation 7.57, to obtain

$$\frac{\Delta P}{\Delta z} = \rho_c\left(\frac{g}{g_c}\right) + \frac{2f_f\rho_c V_G^2}{g_c D} + \frac{\Delta V_G^2}{2\alpha g_c}(1-e)\rho_G + \frac{\Delta(V_G - S)^2}{2\alpha g_c} e\rho_L \quad (8.224)$$

The major problem is the determination of f_f, the friction factor appropriate for the flow of the gas core, with any entrained liquid, in the presence of the liquid film which may be smooth, ripply, or wavy. In the absence of serious entrainment, the correlation of Shearer and Nedderman (Equation 8.185) enables us to predict f_f in the small ripply region, whereas that of Wallis (Equation 8.190 or 8.191) is another approximation.

There is no reliable method, based upon an analysis of the flow mechanism, known for the prediction of the pressure gradient in the presence of serious entrainment. Adorni *et al.* (1961, 1963) have developed an empirical correlation for the pressure gradient in annular-mist flow based upon their data on the vertical flow of argon-water mixtures in a 1-inch pipe. Their equation is

$$-\frac{dP}{dz} = \frac{g}{\dfrac{G_{SL}}{G_{SL}+G_{SG}}\left(\dfrac{1}{\rho_L}\right) + \dfrac{G_{SG}}{G_{SL}+G_{SG}}\left(\dfrac{1}{\rho_G}\right)} + \frac{0.43}{D^{1.2}}\left(\frac{\sigma}{73}\right)^{0.4}\left[(G_{SL}+G_{SG})^2\left(\frac{G_{SL}}{G_{SL}+G_{SG}}\frac{1}{\rho_L} + \frac{G_{SG}}{G_{SL}+G_{SG}}\frac{1}{\rho_G}\right)\right]^{0.75} \quad (8.225)$$

Equation 8.225 is in centimeter-gram-second units, and σ is the surface tension. Levy has apparently had success with the use of this method. Also the method of Ros (1961) and Duns and Ros (1963) for their flow region III is available and is broadly applicable whether or not there is serious entrainment. This was discussed in detail in Section 3.32.

The last two terms in Equation 8.224 represent the change in kinetic energy of the gas and the entrained liquid. Ordinarily the slip velocity, S, is small and the two kinetic energy terms may be approximated as

$$\Delta P_{KE} = \frac{\Delta V_G^2}{2\alpha g_c}\left[(1-e)\rho_G + e\rho_L\right] \quad (8.226)$$

with α taken as 0.80 for turbulent flow.

8 RECOMMENDED DESIGN METHODS

8.1 Introduction

If there is any doubt whether the gas-liquid or liquid-liquid system may behave as a pseudohomogeneous system (in which case the method of Chapter 4 or 5 will be applicable), the procedure outlined in Chapter 1 should be followed. The procedure discussed here is for fluid mixtures which may not be classed as pseudohomogeneous. The method of application of the relations discussed in this chapter to actual design problems depends upon whether or not the flowing mixture may be assumed incompressible and, if not, upon whether the relationship to be applied is expressed in differential or integral form. Liquid-liquid mixtures generally, and gas-liquid mixtures where the anticipated pressure drop is less than about 10 percent of the upstream pressure, may be treated as incompressible. In these cases all the relations for flow pattern, holdup, and pressure gradient may be applied directly to the finite actual system.

For gas-liquid mixtures where pressure drops exceeding some 10 percent of the upstream pressure occur or are expected, the differential relationships applicable to point conditions must be integrated or applied successively in step calculations with average properties and flow rates over pipe sections such that the pressure drop in each step is of the order of 10 percent of the upstream pressure. It is to be noted that the flow pattern itself, as well as the holdup and the pressure gradient, may change over the length of the pipe. In such cases the stepwise procedure is the only way in which full account of the change may be taken. Certain of the relationships of this chapter, however, notably the Poettmann and Carpenter (1952) and related correlations, are expressed in integral form and may be applied directly to the total system.

In the following outline of the recommended design methods we will describe the procedure as for the incompressible case or for a single step in the calculations for a compressible mixture. Where alternative methods are suggested, and where no clear recommendation is given as to the best method, the reader must use his judgment based upon the discussion of the characteristics of the methods within the chapter.

For simplicity the procedure is set forth only for the case where the flow rates, V_{SG} and V_{SL}; the flow properties, ρ_G, μ_G, ρ_L, μ_L, and σ; the pipe diameter and length, D and L; and the upstream pressure, P_1 are known or are estimated in accordance with the methods of Chapter 2, and the flow pattern, holdup, and pressure drop are to be estimated. For alternate design problems, such as estimating the flow rate, the change to the procedure should be apparent, although trial-and-error calculations may be involved.

8.2 Entrance Effects

There are no confirmed methods to predict reliably the length of pipe in which the flow pattern develops or the changes in the flow pattern, holdup, and pressure gradient with the entrance region.

1. For a rough estimate of the entrance length and its relative importance, follow the methods described in Section 5.2 of Chapter 4 assuming that the continuous phase is flowing at the mixture velocity. (The continuous phase is defined by the flow pattern, determined as in Section 8.6.)
2. If the length estimated in step *1* is less than about 10 percent of the total pipe length, entrance effects may be neglected.
3. If the estimated entrance length exceeds about 10 percent of the total pipe length, a rough idea of the flow pattern, holdup, and pressure gradient within it may be obtained by assuming that conditions change linearly from those at the entrance to those under stable flow condition.

8.3 Effect of Bends and Fittings

Precise methods to account for bends and fittings are not available but an approximate account may be taken by increasing the pipe length by ΣL_E, the sum of the equivalent lengths of the fittings in the pipe system. This may be estimated by the methods discussed in Chapter 4, using as the criterion of flow a Reynolds number $DV_M \rho_c/\mu_c$, where ρ_c and μ_c refer to the continuous phase. Care must be taken not to increase the elevation change on which ΔP_{HH} or ΔP_{PE} is calculated.

8.4 Potential Energy or Hydrostatic Head Effects

For the vertical flow considered in this chapter the pressure change due to potential energy or hydrostatic head effects is often of great importance. The equations presented and referred to in Section 8.6 all include terms for ΔP_{PE} or ΔP_{HH}.

8.5 Acceleration or Kinetic Energy Effects

Acceleration effects are usually negligible with liquid-liquid systems but may be significant with gas-liquid mixtures, particularly in the annular-mist flow pattern. The pressure drop due to acceleration, ΔP_{KE}, may be calculated from Equation 8.24 or 8.26 or the appropriate term of Equation 8.224 or 8.226.

8.6 Flow Pattern, Holdup, and Pressure Gradient for the Steady Flow of Incompressible Mixtures (Liquid-Gas Mixtures when $\Delta P < 0.10P_1$)

8.61 Prediction by General Methods

1. Determine the probable flow pattern from Figure 8.9. If the conditions are close to a flow pattern boundary the actual pattern may be on either side of the boundary.

2. Estimate the holdup or $\bar{E}_G$ from Figure 8.10 or 8.11 as appears most applicable.

3. Estimate the pressure drop due to elevation change and irreversibilities as follows.

   a. If the system is air-water or gas of density comparable to air at 30 to 100 psia and water, determine f_{tpL} from Figure 8.17 and calculate the pressure gradient from Equation 8.16.

   b. If the system is natural gas-oil,

      i. determine f_{PC} from Figure 8.12, 8.13, 8.14, 8.15, or 8.16 as appears most appropriate and calculate the pressure gradient from Equation 8.10, or

      ii. estimate the "effective" *in situ* volume fraction gas from Figures 8.26, 8.27, and 8.28; calculate $\bar{\rho}_M$ from Equation 8.29, Re_{HB} from Equation 8.30, obtain f from Figure 4.16, and estimate $\Delta P/\Delta z$ from Equation 8.28, or

      iii. use the correlations of Duns and Ros given in Section 3.32 and Figures 8.19 through 8.25, or

      iv. calculate the pressure drop as in all of items (i), (ii), and (iii), compare, and make a suitable choice.

8.62 Confirmation by Specific Flow Pattern Methods

(A) Bubble Flow Pattern

1. A check that the flow is in the bubble flow region may be obtained by determining whether $V_{SG} < V'_{SG}$ with V'_{SG} estimated from Equation 8.85.

2. If the details of the bubble flow pattern are of interest, they may be predicted, approximately only, as described in items *1* through *7* of Section 4.11.

3. The holdup may be predicted as described in Section 4.12 from Equation 8.78, with V_{bf} determined from the appropriate of Equations 8.68 to 8.73.

4. The pressure gradient is calculated as described in Section 4.13 from Equation 8.81.

(B) Slug Flow Pattern

1. A check that the flow is in the slug flow region may be made by estimating the transition gas velocity to bubble flow as described in Section 4.14 and to froth flow as described in Section 5.12.

2. If the details of the slug flow pattern are of interest, they may be predicted, approximately only, as described in items *1* through *9* of Section 5.9.

3. The holdup may be predicted as described in Section 5.10 from Equation 8.139 with the terms in the equation determined as indicated in Section 5.9, or from Equation 8.142.

4. The pressure gradient may be calculated from Equation 8.143 or 8.144 using E_G as determined in step *3*, and L_s and L_b as determined in step *2*.

(C) Froth Flow Pattern

A check that the flow is in the froth flow region may be obtained by determining the transition velocities to slug flow (Equation 8.85) and to annular-mist flow as approximately indicated by Figure 8.53.

Only very approximate estimates of holdup and pressure gradient may be made from mechanistic considerations in the froth flow pattern. Such estimates as are possible are described in Section 6.

(D) Annular-Mist Flow Pattern

1. The details of the annular-mist flow pattern may be estimated as described in Section 7.41 (items *1* through *9*) for pure annular flow, and Section 7.42 (items *1* through *4*) for annular-mist flow.

2. The holdup may be predicted as described in Section 7.5 from Equations 8.221 or 8.222.

3. The pressure gradient due to hydrostatic head and irreversible effects may be determined as described in Section 7.6 using Equation 8.224.

Bibliography

Adorni, N., I. Casagrande, L. Cravarolo, A. Hassid, and M. Silvestri, Experimental Data on Two-Phase Adiabatic Flow: Liquid Film Thickness, Phase and Velocity Distribution, Pressure Drops in Vertical Gas-Liquid Flow, Report EURAEC-150, U.S. Atomic Energy Commission, 1961.

Adorni, N., I. Casagrande, L. Cravarolo, A. Hassid, E. Pedrocchi, and M. Silvestri, Further Investigations in Adiabatic Dispersed Two-Phase Flow: Pressure Drop and Film Thickness Measurements with Different Channel Geometrics—Analysis of the Influence of Geometrical and Physical Parameters, CISE Report R-53, 1963.

Akagawa, K., *Bulletin J.S.M.E.*, **7**, 122 (1964).

Akagawa, K., and T. Sakaguchi, *Bulletin J.S.M.E.*, **9**, 104 (1966).

Anderson, G. H., and B. G. Mantzouranis, *Chem. Eng. Sci.*, **12**, 233 (1960).

Anderson, J. L., and J. A. Quinn, *Chem. Eng. Sci.:*
a. **25**, 373 (1970).
b. **25**, 338 (1970).

Aziz, K., C. C. Fortems, and A. Settari, Interaction of Wellbore Conditions with flow in the Reservoir in the Mathematical Simulation of Petroleum Reservoirs. Paper presented at the Eighth World Petroleum Congress, Moscow, June 1971.

Bankoff, S. G., *J. Heat Transfer, Trans. A.S.M.E., Ser. C.*, **82**, 265 (1960).

Baxendell, P. B., and R. Thomas, *J. Petrol. Technol.*, **13**, 1023 (1961).

Belkin, H. H., A. A. MacLeod, C. C. Monrad, and R. R. Rothfus, *A.I.Ch.E. Journal*, **5**, 245 (1959).

Bennett, J. A. R., and J. D. Thornton, *Trans Inst. Chem. Engrs.*, **39**, 101 (1961).

Bertuzzi, A. F., M. R. Tek, and F. H. Poettmann, *Petrol. Trans.*, **207**, 17 (1956).

Brancart, C., B. Sc. Thesis, Dept. Mechanical Engineering, Massachusetts Institute of Technology, Cambridge, Mass., 1958.

Brown, R. A. S., *Can. J. Chem. Eng.*, **43**, 217 (1965).

Brown, R. A. S., and G. W. Govier, *Can. J. Chem. Eng.*, **43**, 224 (1965).

Brown, R. A. S., G. A. Sullivan, and G. W. Govier, *Can. J. Chem. Eng.*, **38**, 62 (1960).

Calvert, S., and Williams, B., *A.I.Ch.E. Journal*, **1**, 78 (1955).

Chawla, J. M., *Chem. Ing. Technik*, **41**, 328, (1969).

Collier, J. G., and G. F. Hewitt, *Trans. Inst. Chem. Engrs.*, **39**, 127 (1961).

Collier, J. G., and G. B. Wallis, Two-Phase Flow and Heat Transfer, Notes for a summer course, Vol. 2, Thayer School, Dartmouth College, Hannover, N.H., July 1966.

Coppock, P. D., and G. T. Meiklejohn, *Trans. Inst. Chem. Engrs.*, **29**, 75 (1951).

Davies, J. T., in *Advances in Chemical Engineering*, Vol. 4, p. 1 edited by T. B. Drew, J. W. Hoopes, Jr., and T. Vermeulen, Academic Press, New York, 1963.

Davies, R. M., and Sir Geoffrey Taylor, *Proc. Royal Soc. London, Ser. A*, **200**, 375, (1950).

Datta, R. L., D. H. Napier, and D. M. Newitt, *Trans. Inst. Chem. Engrs.*, **28**, 14, (1950).

DeGance, A. E., and R. W. Atherton, *Chem. Eng.*, **77**, (21), 87 (1970).

Deissler, R. G., Analytical and Experimental Investigation of Adiabatic Turbulent Flow in Smooth Tubes, Nat. Advisory Committee Aeronautics, Tech. Note No. 2138, July 1950, 41 pp.

Dukler, A. E., *Chem. Eng. Progress*, **55**(10), 62 (1959).

Dukler, A. E., and O. P. Bergelin, *Chem. Eng. Progress*, **48**, 557 (1952).

Dumitrescu, D. T., *Z. angew. Math. u. Mech.*, **23**, 139 (1943).

Duns, H., Jr., and N. C. J. Ros, Sixth World Petroleum Congress, Section 2, Paper No. 22, Frankfurt, 1963.

Espanol, J. H., C. S. Holmes, and K. E. Brown, Comparison of Existing Multiphase Flow Methods for the Calculation of Pressure Drop in Vertical Wells, Soc. Petrol. Engrs., A.I.M.E., Paper No. SPE 2553, 1969.

Fancher, G. H., and K. E. Brown, *Soc. Petrol. Engrs J.*, **3**, 59 (1963).

Gaither, O. D., H. W. Winkler, and C. V. Kirkpatrick *J. Petrol. Technol.*, **15**, 309 (1963).

Galegar, W. C., W. B. Stovall, and R. L. Huntington, *Petrol. Refiner*, **33** (11), 208 (1954).

Garner, F., and D. Hammerton, *Chem. Eng. Sci.*, **3**, 1 (1954).

Gill, L. E., and G. F. Hewitt, *Chem. Eng. Sci.*, **23**, 677 (1968).

Gill, L. E., G. F. Hewitt, J. W. Hitchon, and P. M. C. Lacey, *Chem. Eng. Sci.*, **18**, 525 (1963).

Gill, L. E., G. F. Hewitt, and P. M. C. Lacey, *Chem. Eng. Sci.*, **19**, 665 (1964).

Gill, L. E., G. F. Hewitt and P. M. C. Lacey, *Chem. Eng. Sci.*, **20**, 71 (1965).

Gould, T. L., and M. R. Tek, Steady and Unsteady State Two-Phase Flow Through Vertical Flow Strings, Soc. Petrol. Engrs., A.I.M.E., Paper No. SPE 2804, 1970.

Govier, G. W., B. A. Radford, and J. S. C. Dunn, *Can. J. Chem. Eng.*, **35**, 58 (1957).

Govier, G. W., and W. L. Short, *Can. J. Chem. Eng.*, **36**, 195 (1958).

Govier, G. W., G. A. Sullivan, and R. K. Wood, *Can. J. Chem. Eng.*, **39**, 67 (1961).

Griffith, P., Two-Phase Flow in Pipes, Special summer program, Massachusetts Institute of Technology, Cambridge, Mass., 1962.

Griffith, P., and G. B. Wallis, *J. Heat Transfer, Trans. A.S.M.E., Ser. C.*, **83**, 307 (1961).

Guyer, A., and X. Pfister, *Helv. Chim. Acta*, **29**, 1173 (1946).

Haberman, W. L., and R. M. Sayre, Motion of Rigid and Fluid Spheres in Stationary and Moving Liquids inside Cylindrical Tubes, Report No. 1143, Dept. of the Navy, David Taylor Model Basin, October 1958.

Hagedorn, A. R., and K. E. Brown, *J. Petrol. Technol.*, **16**, 203 (1964); **17**, 475 (1965).

Harmathy, T. Z., *A.I.Ch.E. Journal*, **6**, 281 (1960).

Hewitt, G. F., Analysis of Annular Two-Phase Flow: Application of the Dukler Analysis to Vertical Upward Flow in a Tube, Report AERE-R 3680, Atomic Energy Research Establishment, Harwell, Berkshire, England, 1961.

Hewitt, G. F., I. King, and P. C. Lovegrove, Holdup and Pressure Drop Measurements in the Two-Phase Annular Flow of Air-Water Mixtures, Report AERE-R 3764, Atomic Energy Research Establishment, Harwell, Berkshire, England, 1961.

Hewitt, G. F., and P. C. Lovegrove, Comparative Film Thickness and Holdup Measurements in Vertical Annular Flow, Report AERE-M 1203, Atomic Energy Research Establishment, Harwell, Berkshire, England, 1963.

Hewitt, G. F., and G. B. Wallis, Multiphase Flow Symposium, *Proc.* p. 62, American Society of Mechanical Engineers, New York, 1963.

Hughmark, G. A., *Chem. Eng. Progress*, **58**(4), 62 (1962).

Hughmark, G. A., and B. S. Pressburg, *A.I.Ch.E. Journal*, **7**, 677 (1961).

Hughes, R. R., and H. D. Evans, *Chem. Eng. Progress*, **51**, 557 (1955).

Jackson, M. L.,, *A.I.Ch.E. Journal*, **1**, 231 (1955).

Jayaweera, K. O. L. F., B. J. Mason, and G. W. Slack, *J. Fluid Mech.*, **20**, 121 (1964).

Kern, C. P., and F. R. Nicholson, Jr., *J. Petrol. Technol.*, **17**, 1373 (1965).

Kozlov, B. K., *Zhur. Tekh. Fiz.*, **24**, 656 (1952); *J. Tech. Phys.*, **24**, 2285 (1954).

Kynch, G. J., *J. Fluid Mech.*, **5**, 193 (1959).

Leppert, G., and C. C. Pitts in *Advances in Heat Transfer*, Vol. 1, edited by T. F. Irvine, Jr. and J. H. Hartnett, Academic Press Inc., New York, 1964.

Levich, V. G., *Physicochemical Hydrodynamics*, Prentice-Hall, Inc., Englewood Cliffs, N.J., 1962.

Levy, S., *Int. J. Heat and Mass Transfer*, **9**, 171 (1966).

Lockhart, R. W., and R. C. Martinelli, *Chem. Eng. Progress*, **45**, 39 (1949).

Mahoney, J. F., Jr., and L. A. Wenzel, *A.I.Ch.E. Journal*, **9**, 641 (1963).

Malnes, D., Slip Ratios and Friction Factors in the Bubble Flow Regime in Vertical Tubes, Report KR 110, Kjeller Research Establishment, Kjeller, Norway, May 1966.

Manery, C. C., and H. D. Mendelson, *A.I.Ch.E. Journal*, **14**, 295 (1968).

Mendelson, H. D., *A.I.Ch.E. Journal*, **13**, 250 (1967).

Moissis, R., *J. Heat Transfer, Trans, A.S.M.E., Ser. C.*, **85**, 366 (1963).

Moissis, R., and P. Griffith, *J. Heat Transfer, Trans. A.S.M.E., Ser. C.*, **84**, 29 (1962).

Neal, L. G., Analysis of Slip in Gas-Liquid Flow Applicable to the Bubble and Slug Flow Regimes, Report KR-62, Kjeller Research Establishment, Kjeller, Norway, December 1963.

Nicklin, D. J., Doctoral Dissertation, University of Cambridge, England, 1961.

Nicklin, D. J., *Chem. Eng. Sci.*, **17**, 693 (1962).

Nicklin, D. J., and J. F. Davidson, Symposium on Two-Phase Flow, *Proc.* p. 29, Institution of Mechanical Engineers, London, February 1962.

Nicklin, D. J., J. O. Wilkes, and J. F. Davidson, *Trans. Inst. Chem. Engrs.*, **40**, 61 (1962).

Nusselt, W., *Z. Ver. deutsch. Ing.*, **60**, 541 (1916).

Orkiszewski, J., *J. Petrol. Technol.*, **19**, 829 (1967).

Peebles, F. N., and H. J. Garber, *Chem. Eng. Progress*, **49**, 88 (1953).

Petrick, M., A Study of Vapor Carry-under and Associated Problems, Report ANL 6581, Argonne National Laboratories, July 1962, also *A.I.Ch.E. Journal*, **9**, 253 (1963).

Poettmann, F. H., and P. G. Carpenter, *Amer. Petrol. Inst.*, *Drilling and Production Practice*, p. 257 (1952).

Pogson, J. T., J. H. Roberts, and P. J. Waibler, *J. Heat Transfer, Trans. A.S.M.E., Ser. C*, **92**, 651 (1970).

Rohsenow, W. M., in *Modern Development in Heat Transfer*, edited by W. Ibele, Academic Press, New York, 1963.

Ros, N. C. J., *J. Petrol. Technol.*, **13**, 1037 (1961).

Rosenberg, B., Drag and Shape of Air Bubbles Moving in Liquids. The David W. Taylor Model Basin, Report 727, Navy Dept., Washington D. C., September 1950.

Schwartz, K., *VDI-Forschungsheft*, **20** (445), 1 (1954).

Shearer, C. J., and R. M. Nedderman, *Chem. Eng. Sci.*, **20**, 671 (1965).

Siemes, W., *Chemie-Ing.-Techn.*:
a. **26**, 479 (1954);
b. **26**, 614 (1954).

Siemes, W., and J. F. Kauffmann, *Chem. Eng. Sci.*, **5**, 127 (1956).

Tek, M. R., *J. Petrol. Technol.*, **13**, 1029 (1961).

Tek, M. R., and W. J. Chan, Simultaneous Flow of Liquid and Gas Through Vertical Pipe, Paper Presented at the Annual Meeting of the American Institute of Chemical Engineers, San Francisco, December 1959.

Telles, A. S., and A. E. Dukler, *Ind. Eng. Chem. Fundam.*, **9**, 412 (1970).

Tong, L. S., *Boiling Heat Transfer and Two-Phase Flow*, John Wiley & Sons Inc., New York, 1965.

Turner, J. M., Doctoral Dissertation, Dartmouth College, Hanover, New Hampshire, 1966.

Uno, S. and R. C. Kintner, *A.I.Ch.E. Journal*, **2**, 420 (1956).

Van Krevelen, D. W., and P. J. Hoftijzer, *Chem. Eng. Progress*, **46**, 29 (1950)

Walker, J. E., G. L. Whan, and R. R. Rothfus, *A.I.Ch.E. Journal*, **3**, 484 (1957).

Wallis, G. B., *J. Basic Eng., Trans. A.S.M.E., Ser. D.*, **92**, 59 (1970).

Wallis, G. B., *One-Dimensional Two-Phase Flow*, McGraw-Hill Book Co., New York, 1969.

Westwater, J. W., in *Advances in Chemical Engineering*, Vol. 1, p. 1, (1956), Vol. 2 p. 1, (1958), edited by T. B. Drew and J. W. Hoopes Jr., Academic Press Inc., New York.

White, E. T., and R. H. Beardmore, *Chem. Eng. Sci.*, **17**, 351 (1962).

Wicks, M., and A. E. Dukler, *A.I.Ch.E. Journal*, **6**, 463 (1960).

Willis, I. J., *Chem. Eng. Sci.*, **20**, 895 (1965).

Yip, K. C., M.Sc. Thesis, Dept. Mechanical Engineering, The University of Calgary Calgary, Alberta, 1967.

Yip, K. C., J. E. S., Venart, and G. W. Govier, *Can. J. Chem. Eng.*, **48**, 229 (1970).

Yocum, B. T., *The Petroleum Engineer*, **31**, November (1959), p. B40.

Zuber, N., *Chem. Eng. Sci.*, **19**, 897 (1964).

Zuber, N., and J. A. Findlay, *J. Heat Transfer*, *Trans. A.S.M.E.*, *Ser. C.*, **87**, 453 (1965).

Zuber, N., and J. Hench, Steady State and Transient Void Fraction of Bubbling Systems and Their Operating Limits, Part 1, Report No. 62 GL 100, General Electric Co., Schenectady, N.Y., 1962.

Zuber, N., F. W. Staub, G. Bijwaard, and P. G. Kroeger, Steady State and Transient Void Fraction in Two-Phase Flow Systems, Vol. 1, Report EURAEC-GEAP-5417, General Electric Co., San Jose, California, January 1967.

The Vertical Flow of Gas-Solid and Liquid-Solid Mixtures in Pipes

9

1 INTRODUCTION

The vertical pipeline transport of fluid-solid mixtures is an old technique that is being encountered with greater and greater frequency in the modern industrial plants, grain elevators, rocket propulsion systems, nuclear reactor cooling, and the mining industry. Liquid-solid flow systems are also finding new applications in the technology of large-scale desalination plants. The solids are added to the flowing fluid streams to control scale deposition and to enhance convective transport by increasing the intensity of turbulence.

Much has been published on the flow of fluid-solid mixtures since the 1950's. Some of this information has been consolidated in the works of Leva (1959), Zenz and Othmer (1960), Orr (1966), Soo (1967), Brodkey (1967), and Wallis (1969). Zenz (1957) and Frantz (1962) have published review articles on the subject. In spite of the extensive literature, however, our knowledge of flow pattern details and especially our ability to predict the holdup and pressure drop attending the vertical flow of fluid-solid mixtures is limited. Most of the useful correlations are entirely empirical and result in poor accuracy when applied to systems other than the ones for which correlations are developed. Some of the more recent studies have involved attempts at a mechanistic treatment of the problem (Julian and Dukler, 1965, Shook and Daniel, 1969). These studies improve our understanding of the phenomena, but only in a few cases do they result in any useful design procedures.

It is useful, as introductory to the vertical pipe flow of fluid-solid mixtures, to consider first what takes place when a fluid passes upward through a bed of initially stationary and finely divided particles.

At low fluid velocities the solid particles are undisturbed and the fluid flows in a tortuous path through the bed of unconsolidated particles. An understanding of such "flow through porous media" is important in filtration operations, fixed bed catalytic operations, the movement of groundwater, and the flow of oil, gas, and water in underground petroleum reservoirs. The topic is dealt with in detail in Muskat (1949), Scheidegger (1960), Collins (1961), and the recent technical literature. We will not treat it here.

As the fluid flow rate is increased, a velocity is reached where the pressure drop equals the weight per unit area of the bed. At this point the particles become suspended in the fluid but do not travel with it; their slip velocity equals the fluid velocity. This is known as the "fluidized state," which is so important in fluid-solid contacting in the process industries. Fluidization has been dealt with extensively in the work of Zenz and Othmer (1960) and Leva (1959) and is also a subject often considered in the recent technical literature. Again this topic is not included in our treatment.

At fluid velocities above that at which the particles are fully suspended, the particles attain a finite upward velocity and true fluid-solid transport begins. The holdup of solids is high at this minimum transport velocity. With mixed size particles, the smaller particles are transported at lower velocities and the fluid velocity must be increased to beyond the slip velocity of the largest particles before true transport takes place. Under true transport conditions there is no accumulation of material in the system and the composition of the mixture flowing into the system is the same as that of the mixture flowing out.

If a fluid-solid mixture is introduced in a system at a velocity below the velocity at which a steady-state condition is possible, the system will eventually choke with the accumulation of solids. The prediction of the choking or minimum transport velocity is important in the design of transport systems.

If, as is usual, the solid particles are introduced at a lower axial velocity than the flowing fluid, there is a considerable travel distance over which the particles are accelerating before a steady velocity is reached. When the fluid is a gas and the pressure gradient is significant, this condition is never reached and the velocity of both the gas and the solid particles steadily increases. The acceleration of the solid particles, particularly in the upstream reaches of the pipe, causes significant changes in the concentration profile, holdup, and pressure gradient. Accordingly great care must be taken in the interpretation of laboratory data obtained in relatively short pipes.

In our treatment of fluid-fluid systems we referred to the superficial velocities of each of the phases and characterized the mixture introduced into a pipe through these two quantities. The same may be done with fluid-solid systems, where the superficial velocity of the solid phase is defined as

$$V_{SS} = \frac{4Q_S}{\pi D^2} = \frac{4M_S}{\pi D^2 \rho_S} \tag{9.1}$$

where Q_S = volume rate of flow of solids,
M_S = mass rate of flow of solids,
ρ_S = density of solids.

Another often used method of specifying the quantity of solids present is by the so-called "loading" or input mass ratio of solids to fluid, $\mathscr{R}$, i.e.,

$$\mathscr{R} = \frac{M_S}{M_F} = \frac{Q_S \rho_S}{Q_F \rho_F} \tag{9.2}$$

where M_F is the mass rate of flow of the fluid. Alternately, and even more commonly, a fluid-solid system is identified through the input volume fraction or concentration of solids, C_S, where

$$C_S = \frac{Q_S}{Q_S + Q_F} = \frac{V_{SS}}{V_{SS} + V_{SF}} \tag{9.3}$$

and either the mixture velocity, V_M, or the superficial fluid velocity, V_{SF}, V_{SL}, or V_{SG}. We shall use all these methods even though it makes direct comparison between fluid-solid and fluid-fluid systems more difficult.

Our objective in this chapter is to present the available information from the point of view of the engineer concerned with the design of vertical gas-solid and liquid-solid pipe transport systems. The treatment here will parallel as far as possible the treatment of vertical gas-liquid and liquid-liquid systems introduced in Chapter 8. There is really only one flow pattern in this case. Furthermore, we rely mainly on generalized empirical or semi-empirical correlations. In Section 2 we will discuss the flow pattern encountered in fluid-solid systems. This will be followed by a presentation of typical holdup and pressure drop data. In Section 3 we will discuss the available empirical correlations and the limited theoretical relationships for predicting some aspects of the flow pattern and the holdup and pressure drop.

2 TYPICAL DATA

The extensive data of Newitt, Richardson, and Gliddon (1961) on the vertical flow in a 1-inch pipe of a variety of fine and medium sized solids in water and the data of Worster and Denny (1955) for the flow of water suspensions of coal and gravel in 3-, 4,- and 6-inch pipes are representative for the liquid-solid system. For gas-solid systems we will refer to the data of Belden and Kassel (1949) on air suspensions of fine catalyst particles flowing in a 1-inch vertical pipe.

Unfortunately only limited data are available on the details of the flow pattern or on holdup, and we will have to satisfy ourselves with rather general observations concerning the former and with derived results based upon simplifying assumptions for the latter.

2.1 Flow Pattern

The flow pattern observed in the vertical upward flow of fluid-solid mixtures involves the solid particles in a dispersed state in the fluid. The concentration profile is symmetrical about the pipe axis, but only under limiting conditions is it uniform across the pipe. Zenz (1957) has reported that under some conditions with gas-solid suspensions a slug-like pattern is observed, with much of the gas phase traveling in elongated bubbles analogous to the Taylor bubbles of vertical gas-liquid flow. Generally, however, with both liquid-solid and gas-solid systems, the dispersion of the solid is relatively uniform with travel distance, and the flow pattern may be described as one of axi-symmetric suspension. Few photographic observations are available and, although flow pattern diagrams could be developed from concentration profile measurements, the simplicity of the situation is such that they would have little value.

Turning to the actual details of the symmetric concentration flow pattern we find that there are at least limited data relating to the fluid velocity profile, the solids velocity profile, and the solids concentration profile.

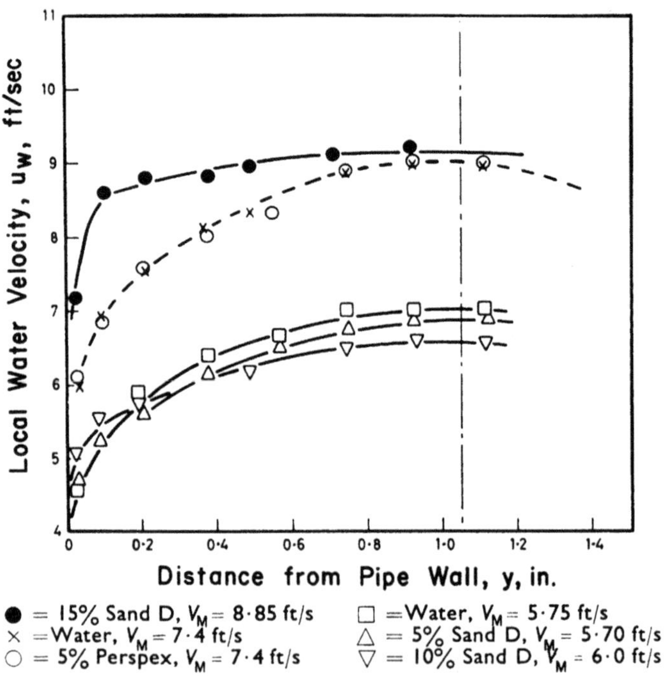

Figure 9.1. Water velocity profiles. Particle diameter, $d = 0.05$ in. (sand) and 0.048 in. (perspex); $D = 2$ in. (From Newitt, D. M., J. F. Richardson, and B. J. Gliddon, *Trans. Inst. Chem. Engrs.*, **39**, 93, 1961.)

For water suspensions, Durand (1953) found no effect on the water velocity profile of solids up to concentrations of 8 percent. Newitt *et al.* (1961) confirm this but show a flattening effect on the profile with 15 volume percent sand in water. Their data are given in Figure 9.1. Data of Toda, Konno, Saito, and Maeda (1969) (see Figure 9.2) confirm the observations of Durand and Newitt *et al.* at relatively high flow rates but suggest a significant effect even at low concentrations when the flow rate is low (see Figure 9.3).

Measurements of the velocity profile of solids suspended in liquids are scarce, but what data are available confirm our expectation that the average velocity of the solids would lag that of the liquid by an amount of the order of V_0, the terminal settling velocity. Actually, because of particle diffusion, and as will be illustrated with data for the gas-solid system, we would expect the particle velocity profile to be blunter than that of the fluid, lagging the fluid velocity by more than V_0 at the axis but leading it near the wall.

Data on the solids concentration profile are also limited. Newitt *et al.* (1961)

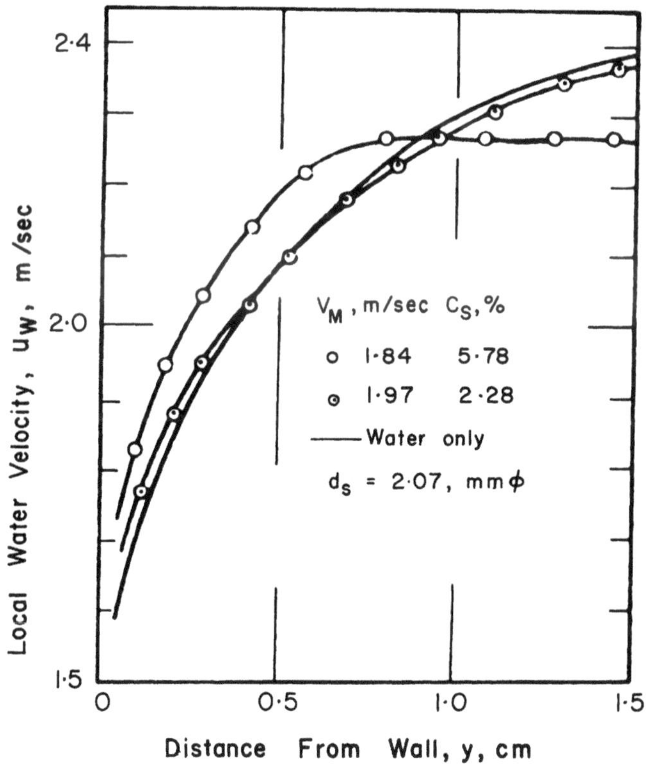

Figure 9.2. Water velocity profiles. Glass particles, $d = 2.07$ mm; $D = 30.2$ mm. (From Toda, M., H. Konno, S. Saito, and S. Maeda, *Int. Chem. Eng.*, **9**, 553, 1969.)

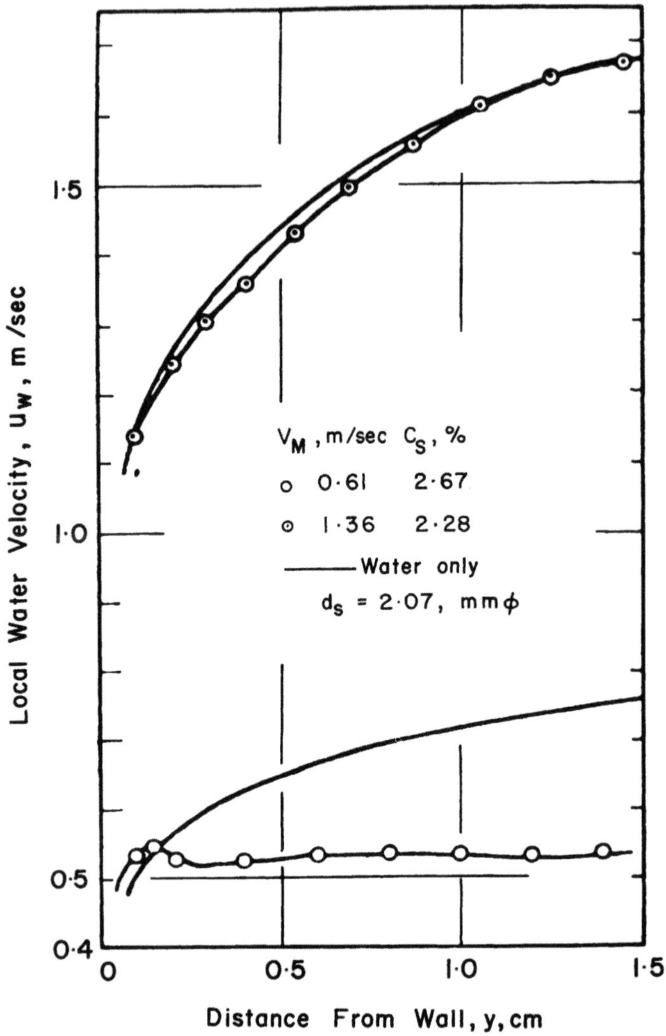

Figure 9.3. Water velocity profiles. Glass particles, $d = 2.07$ mm; $D = 30.2$ mm. (From Toda, M., H. Konno, S. Saito, and S. Maeda, *Int. Chem. Eng.*, **9**, 553, 1969.)

found that at low velocities (about 4 ft/sec) sand and pebbles flowing in water were fairly uniformly distributed. At higher velocities (about 11 ft/sec), however, they observed a marked concentration of the particles into a central core with an almost clear surrounding annulus of water. These observations are confirmed by Toda *et al.* (1969).

With gas-solids suspensions the situation is similar, although in the case of fine particles it may be complicated by electrostatic effects (Soo *et al.*, 1964).

Most available data relate to horizontal flow; however, since they have been taken under conditions where the relative effect of the gravitational force is small, they are also applicable to the vertical case.

Data concerning the effect of solid particles on the velocity profile of the carrier gas are somewhat conflicting; Soo, Trezek, Dimick, and Hohnstreiter (1964) indicated no effect up to gas-solid input mass ratios of about 8, whereas Peskin and Dwyer (1965) observed a considerable flattening of the profile with solids loadings in excess of 3. Data of Doig and Roper (1967) for 756-micron glass particles in air confirm the observations of Peskin and Dwyer.

As with liquid-solid systems, the particle velocity profile is somewhat blunter than the fluid velocity profile, lagging it at the center line and leading it near the wall. Figure 9.4 taken from Doig and Roper (1967) illustrates the phenomenon.

Data relating to the distribution of particles in vertically flowing gas-solid systems indicate the importance of particle size, loading, gas velocity, and electrostatic effects. In the absence of the latter, or where such effects are small, and at low loadings (say less than 0.20) and high gas velocities (say greater than 30 ft/sec)), data of Doig (1965) and Reddy and Pei (1969 indicate essentially uniform concentration profiles. On the other hand, Soo and Regalbuto (1960) measured concentration profiles with 230-micron glass particles in air which showed a significant peaking at the pipe axis. Their data are

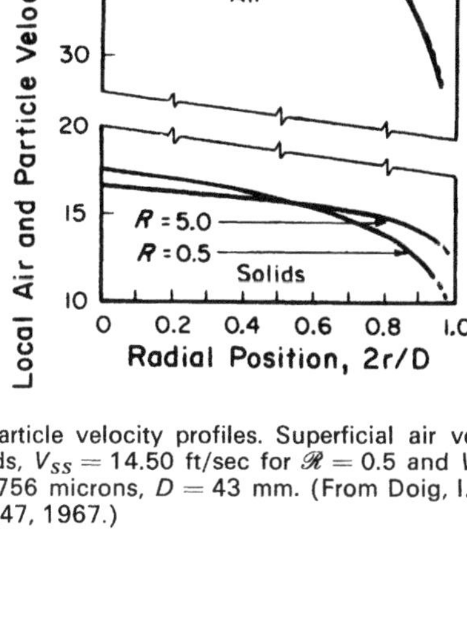

Figure 9.4. Air and particle velocity profiles. Superficial air velocity, $V_{SA} = 36.0$ ft/sec; superficial velocity of solids, $V_{SS} = 14.50$ ft/sec for $\mathscr{R} = 0.5$ and $V_{SS} = 14.56$ ft/sec for $\mathscr{R} = 5.0$; glass particles, $d = 756$ microns, $D = 43$ mm. (From Doig, I. D., and G. H. Roper, *Ind. Eng. Chem. Fundam.*, **6**, 247, 1967.)

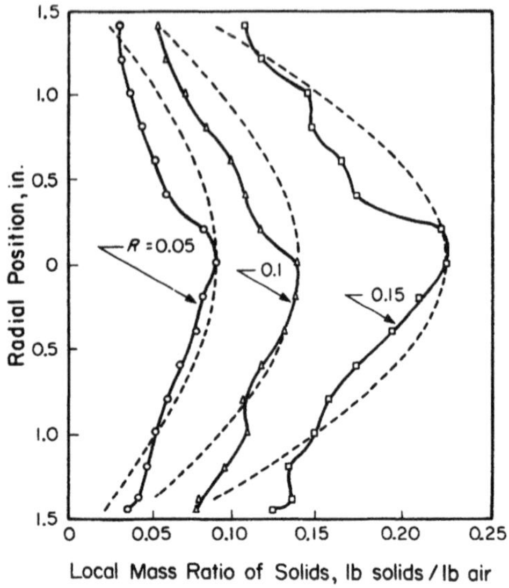

Figure 9.5. Solids concentration profiles. Superficial air velocity, $V_{SA} = 75$ ft/sec; glass particles, $d = 230$ microns, $D = 3$ in. (From Soo, S. L., and J. A. Regalbuto, *Can. J. Chem. Eng.*, **38**, 160, 1960.)

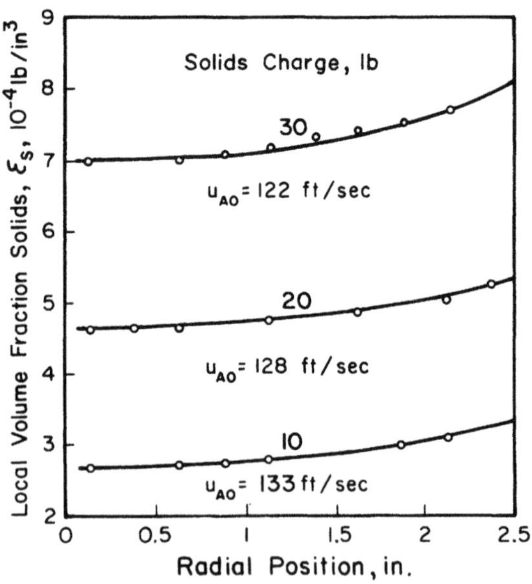

Figure 9.6. Solids concentration profiles. Glass particles, $d = 50$ microns, $D = 5$ in.; u_{A0} refers to air velocity at the center of the pipe. (From Soo, S. L., G. J. Trezek, R. C. Dimick, and G. F. Hohnstreiter, *Ind. Eng. Chem. Fundam.*, **3**, 98, 1964.)

illustrated in Figure 9.5 by the points and the solid lines. (The dashed lines represent their approximate theoretically based relationship for the profiles.)

Soo *et al.* (1964) show that electrostatic charges may develop on the particles resulting in self-repulsion and an increase in concentration near the wall. Typical data are shown in Figure 9.6 taken form their work.

2.2 Holdup

Reliable actual measurements of the holdup or the equivalent for the vertical flow of liquid-solid and gas-solid systems are scarce, and none covering a significant range of the variables is known to the authors. Newitt *et al.* (1961) did measure the terminal settling velocities of the particles they studied. If we assume that the average slip velocity, S is equal to the measured terminal settling velocity, V_0, we may calculate the holdup ratio H from the relationship

$$S = \frac{H-1}{H}(V_{SL} + HV_{SS}) \tag{7.82}$$

Actually, as we saw in Chapter 7, the assumption that $S = V_0$ is valid only where the concentration or the velocity profile, or both, are uniform. With solids dispersed in a fluid, neither profile is really uniform and $S \neq V_0$, although the assumption should not be seriously in error. We will discuss this further in Section 3. Calculating H in this way, we find that it is very close to unity for the Newitt *et al.* data on 0.0040-inch sand, for which the terminal settling velocity is 0.033 ft/sec. For the pebbles whose terminal settling velocity is 1.06 ft/sec there is significant holdup as shown in Figure 9.7. It is interesting to note that the holdup ratio, at least as calculated from the data of Newitt *et al.* assuming $S = V_0$, is essentially independent of solids concentration. This is a consequence of S being taken independent of C_S and of V_{SS} being much smaller than V_{SL}.

For gas-solid systems we may calculate the holdup ratio for the data of Belden and Kassel (1949) again by assuming, as they did, that $S = V_0$ and estimating V_0 from Newton's settling equation (Equation 1.11). Holdup ratios so calculated are shown in Figure 9.8. We again observe no significant effect of solids concentration on the holdup ratio and the expected approach toward unity at the highest gas velocities. It should be noted that the assumption of $S = V_0$ and a more or less uniform concentration profile would not be valid if electrostatic effects were significant. Actual holdup data of Mehta, Smith, and Comings (1957) for the flow of air with 36- and 97-micron glass particles in 1/2-inch commercial steel pipe show trends quite unlike the holdups of Figure 9.8 and are probably seriously affected by electrostatic charges causing high particle concentrations near the wall.

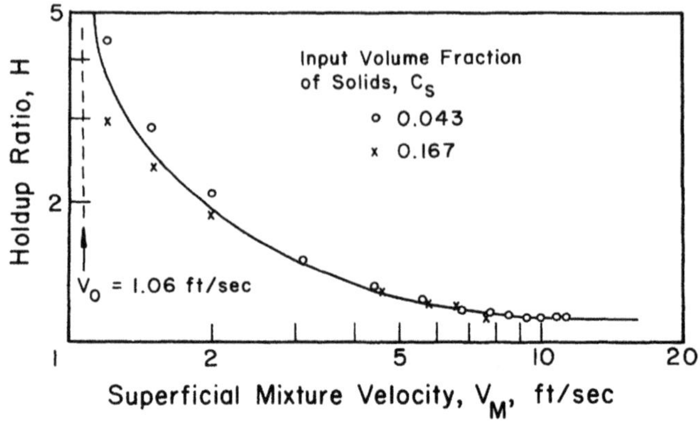

Figure 9.7. Holdup of solids; water-pebble mixture. Particle diameter, $d_{av} = 0.15$ in., $D = 1$ in. (From the data of Newitt, D. M., J. F. Richardson, and B. J. Gliddon, *Trans. Inst. Chem. Engrs.*, **39**, 93, 1961.)

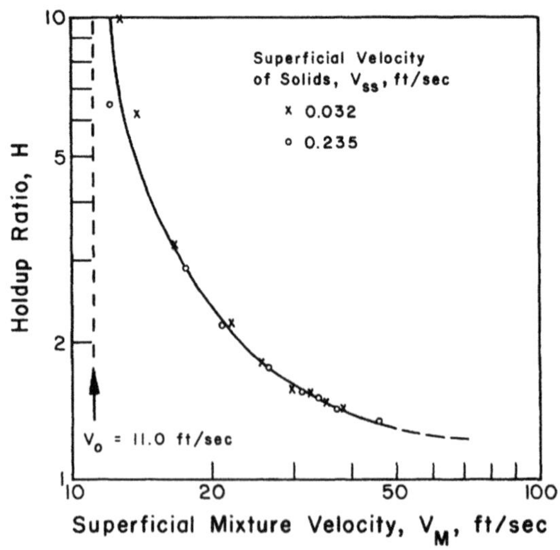

Figure 9.8. Holdup of solids; catalyst particles in air. Particle diameter, $d_{av} = 0.0379$ in., $D = 1.023$ in. (From the data of Belden, D. H., and L. S. Kassel, *Ind. Eng. Chem.*, **41**, 1174, 1949.)

2.3 Pressure Gradient

The mechanical energy balance developed in Chapter 7 (Equation 7.57) may be written for the vertical flow of a fluid-solid mixture as

$$\Delta P_T = \frac{1}{Q_F + Q_S}\left[(M_F + M_S)\frac{g}{g_c}\Delta z + \int_1^2 \frac{M_F V_F\, dV_F}{\alpha_F g_c} + \int_1^2 \frac{M_S V_S\, dV_S}{\alpha_S g_c}\right] + \Delta P_f \tag{7.60}$$

or

$$\Delta P_T = \Delta P_{PE} + \Delta P_{KEF} + \Delta P_{KES} + \Delta P_f \tag{9.4}$$

Alternately, and from Equation 7.73, we may write

$$\Delta P_T = (\rho_F E_F + \rho_S E_S)\frac{g}{g_c}\Delta z + \frac{1}{Q_F + Q_S}\left[\int_1^2 \frac{M_F V_F\, dV_F}{\alpha_F g_c} + \int_1^2 \frac{M_S V_S\, dV_S}{\alpha_S g_c}\right] + \Delta P_f^{\bullet} \tag{7.73}$$

or

$$\Delta P_T = \Delta P_{HH} + \Delta P_{KEF} + \Delta P_{KES} + \Delta P_f^{\bullet} \tag{9.5}$$

The kinetic energy term related to the solids, ΔP_{KES}, may be quite significant, especially if the pressure drop is measured over a section of pipe closely downstream of the point of introduction of the solids. Kolpakov and Donat (1970) have made a detailed experimental study of the pressure drop associated with the acceleration of solid particles in pneumatic conveying in verticle pipes. They conclude that the simple expression for ΔP_{KES} of the form included in Equation 7.60 or 7.73 is not adequate and correlate their data through a complex empirical relationship involving the terminal settling velocity of the particles as well as the mean gas velocity and the solids concentration. The original paper should be consulted for details.

In our discussion of typical pressure drop data we consider the gradient at a point in the line where the solids are fully accelerated and the contribution of kinctic energy effects to the total pressure gradient is negligible. Also, for convenience we will express all pressure gradient data in units of feet of water per foot of height.

Whether we think in terms of Equation 9.4 or 9.5 we may consider the total pressure drop, less any kinetic energy effects, to be

$$\Delta P_T - \Delta P_{KEF} - \Delta P_{KES} = \Delta P_{PE} + \Delta P_f \tag{9.6}$$

or

$$\Delta P_T - \Delta P_{KEF} - \Delta P_{KES} = \Delta P_{HH} + \Delta P_f^{\bullet} \tag{9.7}$$

and the pressure gradient to be

$$\frac{\Delta P_T - \Delta P_{KEF} - \Delta P_{KES}}{\rho_W \Delta z} = \frac{\Delta P_{PE}}{\rho_W \Delta z} + \frac{\Delta P_f}{\rho_W \Delta z} \tag{9.8}$$

or

$$\frac{\Delta P_T - \Delta P_{KEF} - \Delta P_{KES}}{\rho_W \Delta z} = \frac{\Delta P_{HH}}{\rho_W \Delta z} + \frac{\Delta P_f^{\bullet}}{\rho_W \Delta z} \tag{9.9}$$

where ρ_W is the density of water.

In presenting typical pressure gradient data we will express the term $(\Delta P_T - \Delta P_{KEF} - \Delta P_{KES})/\rho_W \Delta z$ simply as $\Delta P/\rho_W \Delta z$.

Newitt *et al.* (1961) have obtained extensive pressure drop data for the vertical flow of liquid-solid mixtures in a 1-inch smooth pipe. Their data on 0.004-inch (102 micron) sand particles in water and 0.15-inch pebbles in water are typical for fine to medium and for coarse particles. These are presented in Figures 9.9, 9.10, and 9.11. Figures 9.9 and 9.10 show the identical data for

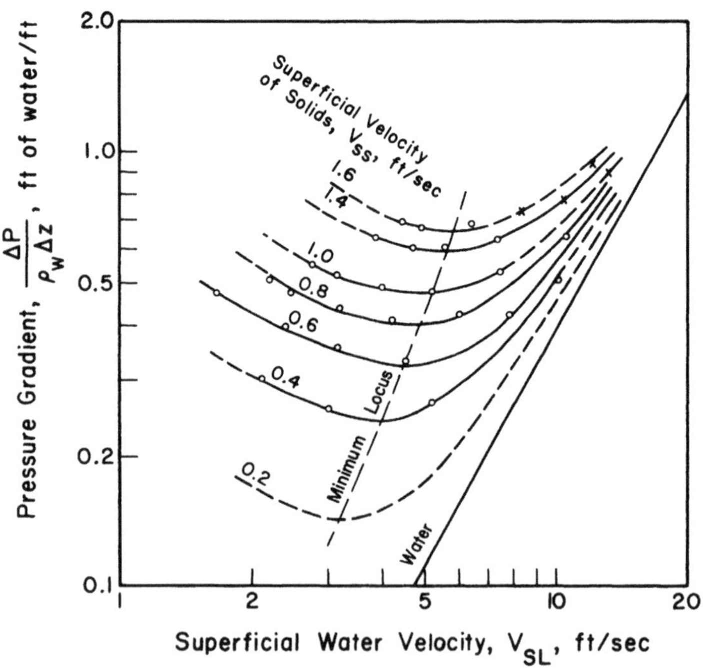

Figure 9.9. Pressure gradients, water-sand system, in terms of V_{SS} and V_{SL}. Particle diameter, $d_{av} = 0.004$, $D = 1$ in. (From data of Newitt, D. M., J. F. Richardson, and B. J. Gliddon, *Trans. Inst. Chem. Engrs.*, **39**, 93, 1961.)

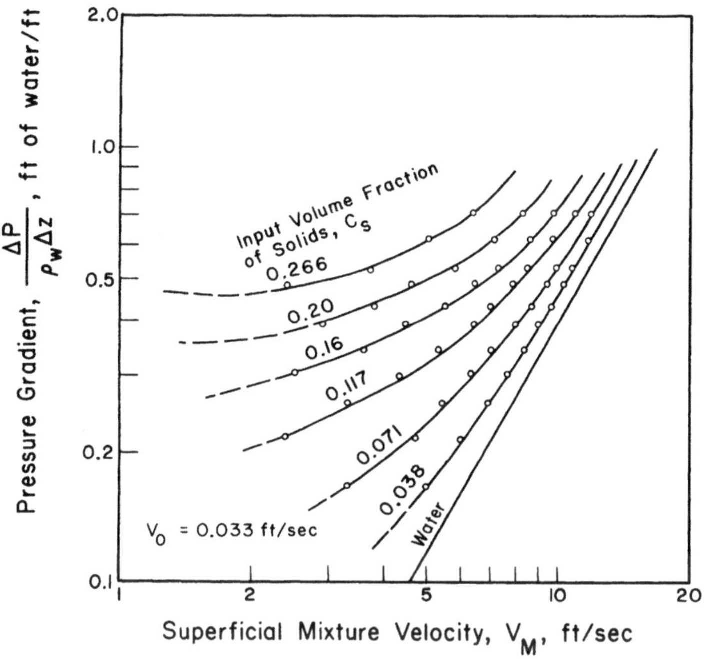

Figure 9.10. Pressure gradients, water-sand system, in terms of C_S and V_M. Particle diameter $d_{av} = 0.004$, $D = 1$ in. (From data of Newitt, D. M., J. F. Richardson, and B. J. Gliddon, *Trans, Inst. Chem Engrs.* **39**, 93, 1961.)

the water-sand system but plotted in terms of V_{SS} and V_{SL} in Figure 9.9 and in terms of C_S and V_M in Figure 9.10. The curves of Figure 9.9 show clearly defined minima. These curves are directly comparable with those of Figure 8.5 for the vertical flow of gas-liquid systems. The reduction in the pressure gradient at a given value of V_{SS} with decrease in V_{SL} is due to the decrease in the friction component of the total pressure gradient. As V_{SL} is decreased beyond the value corresponding with the pressure gradient minimum, the gradient increases as a result of increasing holdup and hydrostatic head. Figure 9.10 shows the data in the more customary manner. No minima appear within the range of data shown in these curves but minima would occur as the mixture velocity is decreased further. We note that the curves tend to parallel each other and the line for water at the lower concentrations and higher velocities. The pressure gradient data for the water-pebble system are shown in Figure 9.11. The curves have the same shape as those in Figure 9.10 and indicate that at a given concentration in the low concentration range the pebble system has a pressure gradient about the same as the sand system at a concentration 40 percent higher. At the highest concentrations the pressure gradients for the two systems are nearly the same. In Figure 9.11 we illustrate the relationship between the shape of the pressure gradient curve

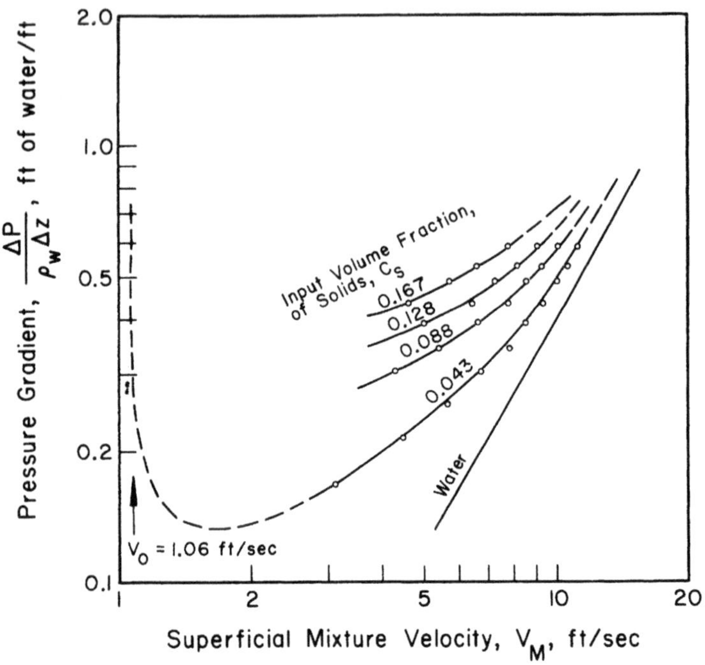

Figure 9.11. Pressure gradient, water-pebble system, in terms of C_S and V_M. Particle diameter, $d_{av} = 0.15$ in., $D = 1$ in. (From data of Newitt, D. M., J. F. Richardson, and B. J. Gliddon, *Trans. Inst. Chem. Engrs.*, **39**, 93, 1961.)

and the terminal settling velocity by extrapolating the former to its asymptotic relationship with the latter. The terminal settling velocity corresponds with the minimum transport or choking velocity and the point in Figure 9.11 where the pressure gradient rises rapidly.

Worster and Denny (1955) studied the vertical and horizontal flow of water suspensions of very coarse solids (1/10- to 3/16-inch gravel and 1/2- to 1-inch coal) in a 4-inch pipe. They noted that in vertical flow the *in situ* concentration of solids exceeded the input concentration. They do not present actual pressure gradient data for vertical flow but state that within experimental error their observations indicate that

$$\frac{\Delta P_f^{\bullet}}{\rho_W \Delta z} = \frac{\Delta P_{fW}}{\rho_W \Delta z} \tag{9.10}$$

and

$$\frac{\Delta P_{HH}}{\rho_W \Delta z} \simeq [C_S(s-1)+1]\frac{g}{g_c} \tag{9.11}$$

where

$$s = \frac{\rho_S}{\rho_W}$$

We note that Equation 9.11 is really only valid if there is no holdup, in which case $\Delta P_{HH} = \Delta P_{PE}$.

The data of Belden and Kassel obtained with spherical catalyst particles (0.027 to 0.052 inch) suspended in air and flowing in a 1.023-inch commercial steel pipe are typical for a gas-solid system. The data are shown in Figure 9.12 in terms of V_{SS}. The shapes of the curves are generally similar to those of Figure 9.9, but the minima are more pronounced in Figure 9.12 than in Figure 9.9.

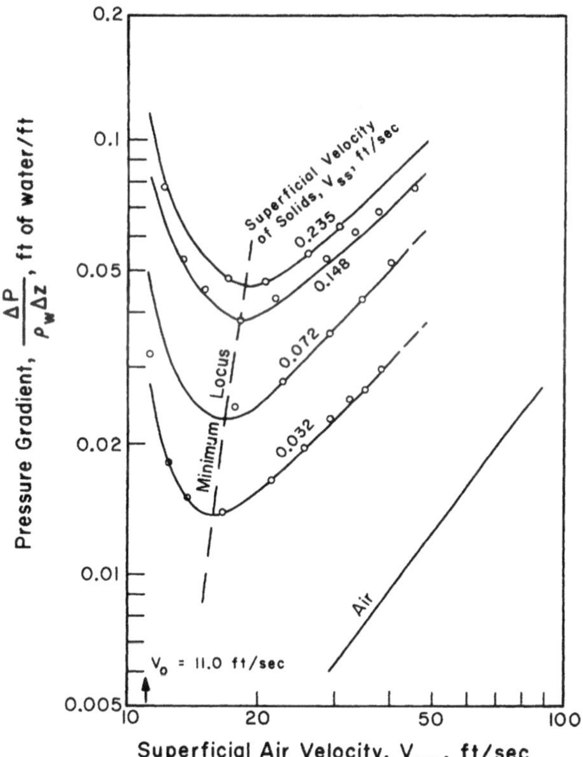

Figure 9.12. Pressure gradients, catalyst particles in air, in terms of V_{SS} and V_{SG}. Particle diameter, $d_{av} = 0.0379$ in., $D = 1.023$ in. (From data of Belden, D. H., and L. S. Kassel, *Ind. Eng. Chem.*, **41**, 1174, 1949.)

3 GENERAL CORRELATIONS

In other chapters we have dealt with the generally applicable empirical correlations for the prediction of flow pattern, holdup, and pressure gradient, and then in separate sections have dealt with mechanistic and theoretical matters for each of the flow patterns. We will follow the same approach here, although we have really only one flow pattern.

3.1 Flow Pattern

Since in vertical pipes, all fluid-solid mixtures flow in a symmetrically dispersed manner at velocities beyond the choking or minimum transport velocity, our concern with flow pattern prediction is mainly that of predicting the minimum transport velocity. In principle, this velocity is V_0, the terminal settling velocity of the solid particles (or the largest of them if there is a range of size) in the carrier fluid. In practice, a somewhat greater velocity is required; otherwise the particles would be suspended but motionless. Any carrier fluid velocity in excess of V_0 will theoretically result in solids transport, although velocities too close to V_0 would be risky and, from the typical data we have seen, would result in a higher than necessary pressure gradient. A suitable velocity, for purposes of defining the minimum transport velocity in a practical sense, would be twice V_0. For both liquid- and gas-solid systems this would also correspond with a holdup ratio of about 2 (see Figures 9.7 and 9.8) and be not far away from the pressure gradient minimum. Thus, if we define $(V_M)_{\min} = 2V_0$ as the practical minimum mixture transport velocity, we have from Equations 1.1 and 1.2 with $F_g = F_D$

$$(V_M)_{\min} = 2\sqrt{\frac{4gd(s-1)}{3C_D}} \tag{9.12}$$

The approach just discussed should be entirely suitable for determining the minimum transport velocity in vertical flow. Zenz (1957), however, has developed an empirical relationship between $(V_{SF})_{\min}$, the solid and fluid properties, and the mass flow rate of the solids. Zenz shows separate correlations for liquid-solid and gas-solid systems, but the data actually suggest a single correlation. We present this in Figure 9.13 redrawn from his work. The scatter of the data points is considerable.

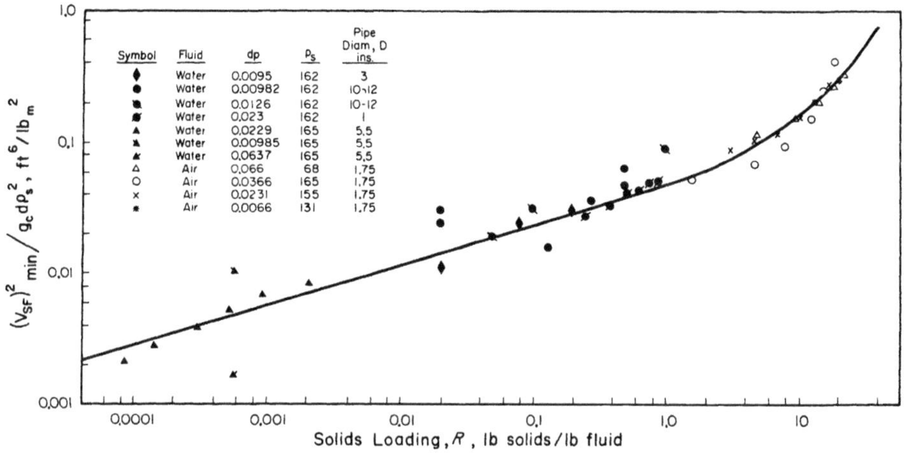

Figure 9.13. Zenz correlation for the minimum transport velocity. (Modified after Zenz, F. A., *Hydrocarbon Processing,* **36**(6), 133, 1957. (Copyright © 1957 by Gulf Publishing Co. Houston, Texas.)

3.2 Holdup

The holdup phenomenon in the vertical flow of fluid-solid mixtures may be predicted having regard to the analysis of Zuber and Findlay (1965) discussed in Chapter 7. From Equation 7.104 we may write

$$V_S = C_0 V_M - V_{\alpha j} \tag{9.13}$$

where V_S = the average velocity of the solids,
C_0 = a distribution coefficient,
$V_{\alpha j}$ = the volume fraction weighted average drift velocity of the solids relative to the mixture (positive downward).

Now, since $H = V_F/V_S$ (see Section 7.2 of Chapter 7), the holdup ratio may be estimated provided C_0 and $V_{\alpha j}$ can be determined. We recall that C_0 is a measure of the uniformity or otherwise of the distribution of the dispersed phase. For most fluid-solid mixtures the distribution tends to be approximately uniform with a modest peaking in the centre of the pipe. Also, as we have seen, the velocity profile, which is normally that for turbulent flow, tends to be flattened somewhat by the presence of the solids. Data for fluid-solid systems are very limited; but from data of Zuber and Findlay (1965) for dispersed fluid-fluid systems we may estimate that $1 \leq C_0 \leq 1.2$, and we may approximate C_0 at a value of 1.1. The weighted average drift velocity, $V_{\alpha j}$, is probably best estimated as the terminal settling velocity, V_0, of the particles under the prevailing conditions of turbulence. Under normal turbulent flow conditions, Newton's equation is applicable and

$$V_{\alpha j} = V_0 = 1.74[gd(s-1)]^{1/2} \tag{1.11}$$

From these relations and the following,

$$V_M = V_{SS} + V_{SF}$$

$$V_M = V_S E_S + V_F(1 - E_S)$$

$$H = \frac{V_{SF}}{V_{SS}}\left(\frac{E_S}{1 - E_S}\right) \tag{7.79}$$

we may calculate E_S or H.

Most often in the literature relating to the vertical flow of fluid-solid mixtures, C_0 of Equation 9.13 is implicitly assumed to be unity leading to

$$V_M - V_S = V_0 \tag{9.14}$$

If we express V_M in terms of E_S, V_S, and V_F, Equation 9.14 may be expressed as

$$S = \frac{V_0}{1 - E_S} \tag{9.15}$$

where S is the average slip velocity, i.e., $(V_F - V_S)$. If E_S is small, $S \simeq V_0$. The equality of S and V_0 is frequently assumed even under conditions where neither C_0 nor $(1 - E_S)$ is close to one.

The quantity E_S is frequently calculated from Equation 7.78b applied to a fluid-solid system (Newitt *et al.*, 1961)

$$E_S = -\left(\frac{V_M - S}{2S}\right) + \left[\left(\frac{V_M - S}{2S}\right)^2 + \frac{V_M C_S}{S}\right]^{1/2} \tag{9.16}$$

Equation 9.16 is perfectly valid in itself but the circumstances when V_0 may properly be substituted for S are decidedly limited.

On the assumptions the $C_0 = 1$ and $S = V_0$, Worster and Denny (1955) calculate the relationship between

a. $\dfrac{E_S}{C_S}$ and $\dfrac{V_M}{V_S}$, and

b. $\dfrac{V_L}{V_M}$ and $\dfrac{V_M}{V_S}$

for a range of values of C_S and for each of upward and downward vertical flow of water suspensions of coal in a 6-inch pipe. Their relationships, based

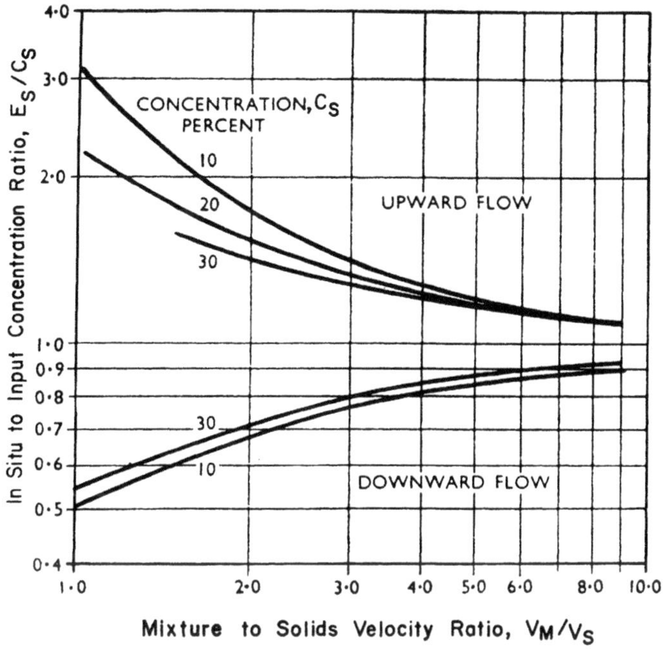

Figure 9.14. Effect of holdup on concentration ratio. (From Worster, R. C., and D. F. Denny, *Proc. Inst. Mech. Engrs.*, **169**, 563, 1955.)

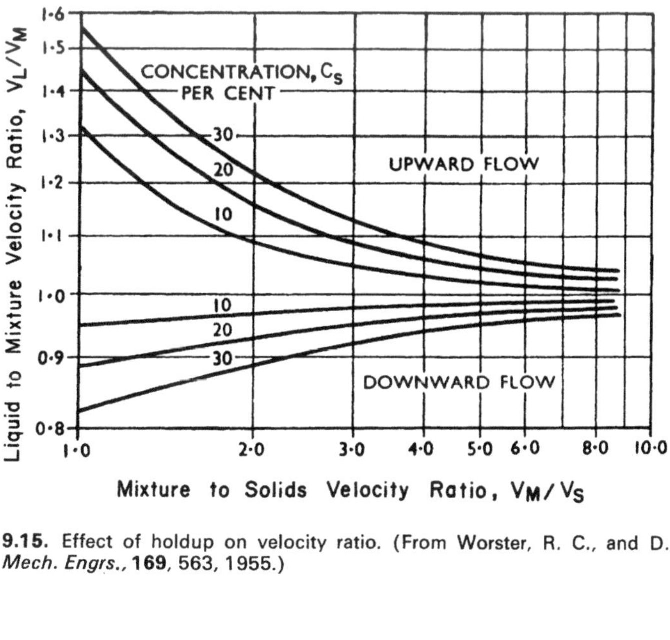

Figure 9.15. Effect of holdup on velocity ratio. (From Worster, R. C., and D. F. Denny, *Proc. Inst. Mech. Engrs.*, **169**, 563, 1955.)

upon the continuity equations discussed in Section 3 of Chapter 7, are presented in Figures 9.14 and 9.15. The curves illustrate the approach of E_S to C_S, and of V_L to V_M as V_M/V_S increases. They also show that E_S is within 10 percent of C_S, and V_L is within 5 percent of V_M when $V_M/V_S = 10$.

Oedjoe and Buchanan (1966) studied the flow of water suspensions of coal, gravel, and sand with wide particle-size distribution in 2-, 3-, and 4-inch pipes. For the coal and gravel suspensions they found

$$V_0 = 1.44\sqrt{\bar{d}(s-1)} \tag{9.17}$$

and for the sand mixtures

$$V_0 = \frac{1.15 \times 10^{-4}}{\bar{d}}\sqrt{1 + 1.56 \times 10^6 \bar{d}(s-1)} - 1 \tag{9.18}$$

where $\bar{d}$ = the weighted average diameter calculated from

$$\bar{d} = \sum_i^n x_i d_{ai},$$

x_i = weight on screen i,

d_{ai} = average opening of screens $(i - 1)$ and i.

Equations 9.17 and 9.18 provide an alternate to Newton's law for estimating the terminal settling velocity of coal, gravel, and sand suspensions of the size range tested by Oedjoe and Buchanan.

Cloete, Miller, and Streat (1967) reported actual holdup data for the flow in 1/2- and 3/4-inch pipes of water suspensions of sand and glass particles in the size range 0.007 to 0.0275 inch. They confined their attention to suspensions of concentration in excess of 40 volume percent and mostly in the range of 45 to 53 percent. With these concentrated suspensions they observed essentially constant holdup ratios (range 1.31 to 1.44) reflecting *in situ* solids concentrations of 53 to 61 percent. They noted that the *in situ* solids concentrations fell between the free-settled solids fraction (51 to 56 percent) and the "tap-settled" (settled under vibrating conditions) solids fraction (61 to 62 percent) for the solids in water. We may note that these suspensions have input concentrations close to those falling in the nonsettling or pseudohomogeneous category. Holdup occurs to the limiting condition where the *in situ* mixture is essentially nonsettling.

For gas-solid flow Soo *et al.* (1964), Doig (1965), Reddy and Pei (1969), and others have proposed various empirical relations for the prediction of the solids velocity profile and the relative velocity between the gas and the solids. These relations provide an alternative approach to calculating the holdup for gas-solid systems. Following are the equations for predicting the solids velocity profile.

Soo et al.

$$u_S = u_{Sw} + (u_{S0} - u_{Sw})\left(\frac{2y}{D}\right)^{1/m} \tag{9.19}$$

where

$$0.8 < m < 1.5$$

Doig

$$u_S = u_{S0}\left(\frac{2y}{D}\right)^{1/n} \tag{9.20}$$

where

$$2.13 < n < 14.3$$

Reddy and Pei

$$\frac{u_S - u_{Sw}}{u_{S0} - u_{Sw}} = \left(\frac{2y - d}{D - d}\right)^{1/m} \tag{9.21}$$

$$u_{Sw} = u_{S0}\left(\frac{d}{D}\right)^{1/n} \tag{9.22}$$

where

$$m = 4.15 - \frac{1}{350 \times 10^{-5}}\left(\frac{d}{D}\right) \tag{9.23}$$

and

$$n \simeq 7$$

In Equations 9.19 to 9.23 the symbols used are defined as follows:

d = particle diameter,
D = diameter of the pipe,
u_S = local velocity of solids,
u_{S0} = velocity of solids at the center of the pipe,
u_{Sw} = velocity of solids at the pipe wall,
y = distance from pipe wall.

Reddy an Pei also correlate the relative velocity of the solids at the axis of the pipe, u_{r0}, with their terminal settling velocity V_0 as

$$\frac{u_{r0}}{V_0} = 1 + 1.59\mathscr{R}^{2/3} \tag{9.24}$$

where $u_{r0} = u_{G0} - u_{S0}$.

If we assume an undisturbed turbulent gas velocity profile and Newton's terminal settling equation, the above relationships permit us, for any given average superficial gas velocity, particle size, and loading, to estimate u_{r0}, u_{S0}, u_{Sw}, and u_S as functions of y and, by integration, to obtain $(V_G - V_{\alpha j})$ and $V_{\alpha j}$. From Equation 9.13, assuming a value of C_0, we could then predict V_S and H. Actually this refinement is hardly warranted in the light of the uncertainties in Equations 9.21 to 9.24 and our lack of detailed knowledge of the concentration profile (and consequently of C_0).

Barth (1958), in a correlation he developed for the prediction of pressure gradient in vertical pneumatic transport of solids, arrives at a relationship for the prediction of the actual average solids velocity, V_S (Equation 9.55). This equation provides another method for the estimation of H or E. We defer discussion of Equation 9.55 to Section 3.33.

Rose and Duckworth (1969), in connection with their overall analysis of the flow of fluid-solid suspensions which we will discuss in Section 3.3, developed the following equation for V_S in vertical flow systems

$$\frac{V_S}{V_{SF}} = 0.63\phi_4 \tag{9.25}$$

where ϕ_4 is an empirical function of V_{SF}^2/gds^2 as shown in Figure 9.16. From Equation 9.25 we may write

$$H = \frac{V_F}{V_S} = \frac{V_F}{0.63 V_{SF}\,\phi_4} \tag{9.26}$$

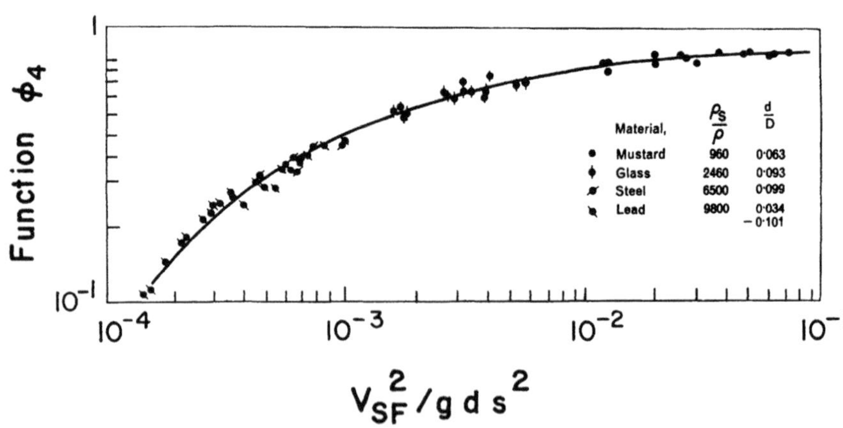

Figure 9.16. Rose and Duckworth function ϕ_4. (From Rose, H. E., and R. A. Duckworth, *The Engineer*. **227**, (5904), 430, 1969.)

and from Equation 7.79

$$E_S = \frac{1.6 V_{SS}}{V_{SF}\phi_4} \tag{9.27}$$

Equation 9.27 is probably suitable for estimating E_S within the range of variables upon which Equation 9.25 is based, i.e.,

$$960 < s < 9800$$
$$0.03 < d/D < 0.101$$
$$10^{-4} < (V_{SF}^2/gd) < 10^{-1}$$

As V_{SF}^2/gds^2 approaches the maximum value covered by Figure 9.16, ϕ_4 approaches 1.0, E_S approaches $1.6V_{SS}/V_{SF}$, and H approaches $1.6V_{SF}/(V_{SF} - 1.6V_{SS})$. These latter two limits do not appear correct, indicating that the correlation should be used cautiously beyond the actual range of the data.

3.3 Pressure Gradient

3.31 Introduction

Efforts to correlate the pressure gradient for fluid-solid mixtures flowing in vertical pipes originated with the work of Gasterstadt (1924), who proposed that data be interpreted in terms of the ratio of the pressure gradient for the two-phase flow to that for the flow of the fluid phase alone. Since that time a number of other methods have been proposed, and today we have a confusing variety of correlations, some for only gas-solid or liquid-solid mixtures, others for both. We will follow the procedure of Chapter 8 and deal with these in two groups, depending upon whether the analysis or correlation is based upon, or related to, Equation 7.60, involving only the input quantities, or Equation 7.73, involving an estimate of the density of the *in situ* mixture. Unless specifically noted otherwise, we will deal only with ΔP, and its components, as defined in Section 2.3, i.e., the pressure drop after deduction of any applicable kinetic energy effects. In addition, we will observe that, as discussed in Chapter 7, certain of the correlations are based upon a pressure drop ratio, $\Delta P/\Delta P_F$, $\Delta P_f/\Delta P_{fF}$, or $\Delta P_f^\bullet/\Delta P_{fF}$, where ΔP_F and ΔP_{fF} refer to ΔP and ΔP_f for the pure carrier fluid. Others are based on an excess or a dimensionless excess pressure drop, $\Delta P - \Delta P_F$ or $(\Delta P - \Delta P_F)/\Delta P_F$, and still others express ΔP_f or $\Delta P_f^\bullet$ in terms of a friction factor or the equivalent. Many workers in the field of hydraulic or pneumatic transport of solids use the Fanning equation in the form

$$\frac{\Delta P_f}{\Delta z} = \frac{f_H \rho V^2}{2g_c D} \tag{4.25}$$

defining f_H, which is four times the f we have been using (see Section 3.2 of Chapter 4). For the sake of consistency we have converted all such relations to the equivalent form in terms of f defined by

$$\frac{\Delta P_f}{\Delta z} = \frac{2 f \rho V^2}{g_c D} \tag{4.24}$$

The analyses that involve estimates of the density of the *in situ* mixture use various forms of, or approximations to, the basic relationship for mixture density, Equation 2.44 of Chapter 2,

$$\rho_M = E_S \rho_S + (1 - E_S)\rho_F \tag{9.28}$$

Alternate and identically equivalent forms of this expression are

$$\rho_M = [E_S(s - 1) + 1]\rho_F \tag{9.28a}$$

and

$$\rho_M = \rho_F + \frac{V_{SS}}{V_S}(\rho_S - \rho_F) \tag{9.28b}$$

It is useful also to list the various approximations to ρ_M.

If ρ_F is neglected,

$$\rho_M \simeq E_S \rho_S = \left(\frac{V_{SS}}{V_S}\right)\rho_S = \frac{G_{SS}}{V_S} = \frac{\mathscr{R} V_{SF} \rho_F}{V_S} \tag{9.29}$$

If holdup itself is neglected, or if reference is made to the input mixture, we may write Equation 2.44 as

$$\rho_M = C_S \rho_S + (1 - C_S)\rho_F \tag{9.30}$$

which is identical with

$$\rho_M = \frac{M_S + M_F}{Q_S + Q_F} \tag{9.30a}$$

the term that enters Equation 7.57.

If, along with the holdup, ρ_F is also neglected

$$\rho_M \simeq C_S \rho_S \simeq \frac{V_{SS}}{V_M}\rho_S \simeq \frac{G_S}{V_M} \simeq \frac{\mathscr{R} V_{SF} \rho_F}{V_M} \tag{9.31}$$

Further, if V_{SF} and V_M are taken as equal,

$$\rho_M \simeq \mathscr{R}\rho_F \tag{9.32}$$

An alternative approximation often used is to neglect the volume occupied by the solids. This leads from Equation 9.30a to

$$\rho_M \simeq \frac{M_S + M_F}{M_F}\rho_F = (1 + \mathscr{R})\rho_F \tag{9.33}$$

3.32 Analyses or Correlations Not Involving Estimates of the *in situ* Concentration or Density

These correlations are based upon Equation 7.60 (see Equation 9.4) or a closely related relationship which may involve some approximation in the treatment of the kinetic energy terms but which, in any case, deal with ΔP_{PE} and ΔP_f, as we have previously discussed.

Vogt and White (1948) proposed the first such correlation. They wrote the equivalent of Equation 7.60 as

$$M_S \int \frac{dP}{\rho_S} + M_F \int \frac{dP}{\rho_F} + \frac{M_S \Delta V_S^2}{2g_c} + \frac{M_F \Delta V_F^2}{2g_c} + (M_S + M_F)\frac{g}{g_c}\Delta z + W_f = 0 \tag{9.34}$$

where W_f is the energy dissipated per unit of time due to friction. They defined α, a dimensionless, fractional increase in the frictional pressure drop, such that

$$W_f = \frac{\alpha M_F 2 f G_{SF}^2 \Delta z}{g_c \rho_F^2 D} \tag{9.35}$$

where f = the conventional friction factor for the single-phase carrier fluid,
G_{SF} = superficial mass velocity of the fluid = $V_{SF}\rho_F$.

For their data for the vertical flow, in 1/2-inch commercial steel pipe, of air with 24-, 35-, 48-, and 80-mesh sand, 35 mesh steel shot, and with wheat and clover seed, they find a correlation of the form

$$\alpha - 1 = A\left(\frac{D}{d}\right)^2\left(\frac{\mathscr{R}}{s\,\mathrm{Re}_{SF}}\right)^K \tag{9.36}$$

where

$$\mathrm{Re}_{SF} = \frac{D V_{SF} \rho_F}{\mu_F}$$

A and K are functions of the quantity $[\frac{1}{3}(\rho_S - \rho_F)\rho_F g d^3]^{1/2}/\mu_F$.

Vogt and White proposed their correlation as preliminary and promising, but indicated that it was based upon limited ranges of the variables.

Gopichand, Sarma, and Rao (1959) confirm that the Vogt and White correlation does not have general applicability and that V_{SF} plays a role beyond that accounted for in α. Our interest in the Vogt and White correlation is primarily for its historical significance.

Bhattacharya and Roy (1955) followed the lead of Vogt and White in analyzing their data on the vertical flow of low concentration water and oil suspensions of kieselguhr, alkalized iron-copper kieselguhr, and alkalized iron-copper catalyst particles of diameter range 40 to 80 microns in 10- and 19-mm tubes. They were unable to obtain a satisfactory correlation of α with Vogt and White's method but did find that α could be expressed by the relation

$$\alpha - 1 = A\left(\frac{D}{d}\right)^2\left(\frac{1}{s}\right)\frac{\mathscr{R}}{\mathrm{Re}_M} \tag{9.37}$$

where $\mathrm{Re}_M = DV_M\rho_M/\mu_M$.

A is a function of the quantity $[\frac{1}{3}(\rho_S - \rho_L)\rho_L g d^3]^{1/2}/\mu_L$ as given in Figure 9.17 redrawn from Bhattacharya and Roy. Their correlation is based on limited data for fine particles in low concentration and flowing at low velocities. Like the Vogt and White correlation it should only be used under conditions close to those for which it was developed.

For convenience we also include in this category of analyses that of Farbar (1949). He analyzed his data for catalyst particles (10 to 220 microns) in a 17-mm I.D. glass tube by the method originally proposed by Gasterstadt (1924). The correlation is of the form

$$\frac{\Delta P}{\Delta P_{SF}} = \phi(\mathscr{R}) \tag{9.38}$$

where ΔP and ΔP_{SF} are the pressure drops, respectively, for the two-phase system and for air alone with $V_G = V_{SG}$. Farbar found that the data could not be correlated by a straight line as originally proposed by Gasterstadt and, in fact, suggested that a family of curves might be required for a precise representation of the behavior of even a single system. We should note that this approach, which does not separate the effects of ΔP_f and ΔP_{PE} or ΔP_{HH}, cannot be expected to be successful where both terms are significant, since each could be affected differently by $\mathscr{R}$. Even for the system investigated by Farbar, this correlation results in up to 50 percent error in the pressure drop.

Korn (1950) extends the type of treatment given by Farbar by considering

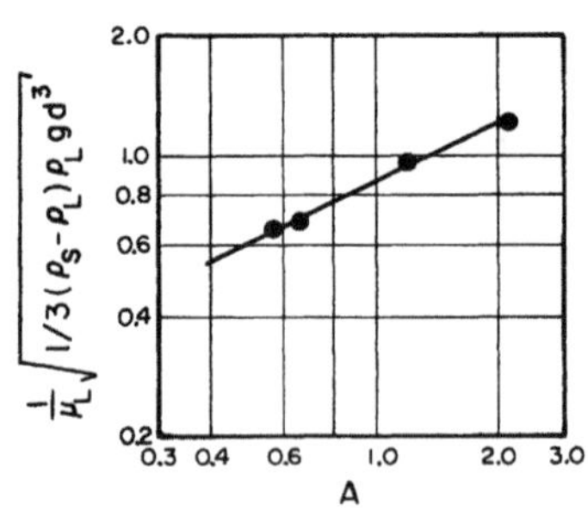

Figure 9.17. Effect of solid and liquid properties on the constant A in Equation 9.37. (Redrawn from Bhattacharya, A., and A. N. Roy, *Ind. Eng. Chem.*, **47**, 268, 1955.)

the two principal components of the pressure drop. In our notation he expresses the pressure drop as

$$\Delta P = \Delta P_f + \Delta P_{PE} \tag{9.39}$$

where

$$\Delta P_f = \frac{2 f_{FM} \rho_M V_M^2}{g_c D} \Delta z \tag{9.40}$$

$$\Delta P_{PE} = \rho_M \Delta z \frac{g}{g_c} \tag{9.41}$$

and f_{FM} = the conventional single-phase friction factor (Figure 4.16) based upon $\mathrm{Re} = D V_M \rho_F / \mu_F$ and the appropriate relative roughness. Korn approximates ρ_M as

$$\rho_M = (1 + \mathscr{R}) \rho_F \tag{9.42}$$

from which we see that he neglects holdup (and the volume occupied by the solids). Combining Equations 9.39 through 9.42, we obtain

$$\Delta P = \left(\frac{2 f_{FM} V_M^2}{g_c D} + \frac{g}{g_c} \right) \rho_F (1 + \mathscr{R}) \Delta z \tag{9.43}$$

The use of ρ_M in both Equations 9.40 and 9.41 presupposes that the presence of the solids affects ΔP_f and ΔP_{PE} in the same way and Korn's approach reduces essentially to that of Farbar. Korn did not actually demonstrate the validity or utility of Equation 9.43 and it is doubtful if it could lead to a useful general correlation.

Of direct interest, however, is Korn's observation that the data of several previous workers could be reasonably well correlated by a strictly empirical equation of the form

$$\frac{\Delta P}{\left(\frac{\rho_F V_F^2 \Delta z}{2 g_c} \right)} = 4 f_{FM} + b \mathscr{R}^n \tag{9.44}$$

(Note that the numeral 4 appears in the right-hand side of the equation because Korn worked in terms of $f_H = 4f$; see discussion in Section 3.31.) The best fit values of the constants b and n for several sets of data are given in Table 9.1.

Equation 9.44 with the values of b and n proposed by Korn offers a reliable method of calculating the pressure drop for systems for which b and n have been determined.

Boothroyd (1966) has considered the flow of zinc particles under 40

TABLE 9.1

Factors *b* and *n* for Equation 9.44[a]

Investigator	Material	b	n
Vogt and White (1948)	wheat	0.008	1.0
	clover	0.020	0.855
	clover	$0.04\,(V_0/V_F)^{0.6}$	0.762
	sand	0.05	$(V_0/V_F)^{0.624}$
Hariu and Molstad (1949)	sand and cracking catalyst	0.022 (erratic)	1.0
Belden and Kassel (1949)	catalyst	$0.62\left(\frac{2Dg}{V_F^2}\right)^{0.3}\left(\frac{V_0}{V_F}\right)^{0.27}$	1.0
Korn (1950)	any	$4f+\left(\frac{2Dg}{V_F^2}\right)\left(1+\frac{1}{\mathscr{R}}\right)$	1.0

[a] A. H. Korn, *Chem. Eng.*, **57**, 108 (1950). (Reprinted by special permission of McGraw-Hill, Inc., New York.)

microns in air in 1-, 2-, and 3-inch pipes. He has separated the total pressure drop according to Equation 9.4 as

$$\Delta P = \Delta P_{PE} + \Delta P_f$$

where ΔP_{PE} is defined by Equations 9.41 and 9.42,

$$\Delta P_f = \frac{2f_{tpG}\rho_G V_{SG}^2}{g_c D}\,\Delta z \tag{9.45}$$

f_{tpG} is a two-phase friction factor defined by Equation 9.45 (see also Equation 7.63).

Boothroyd's approach, like that of Korn, neglects the holdup effect and the volume occupied by the solids. Boothroyd expresses his data in terms of the ratio of f_{tpG}, as determined from Equation 9.45 to f_G corresponding with the single-phase flow of the carrier gas of density ρ_G and at a velocity V_{SG}. He finds that f_{tpG}/f_G is always less than one for the flow in the one-inch pipe but may be less or greater than one for the flow in the 2- or 3-inch pipes. Additionally Boothroyd reports that

$$\Delta P_f < \frac{2f_G\rho_M V_M^2}{gD}\,\Delta z$$

where ρ_M is defined by Equation 9.42. The fine particles tend, on the one hand, to suppress turbulence and to cause a reduction in friction factor and, on the other hand, to result in a mixture whose density and velocity exceeds ρ_G and V_{SG}, thereby increasing the friction factor calculated from Equation 9.45. Whether f_{tpG}/f_G is less or greater than one depends on the balance of these two effects. The investigations of Boothroyd show that conservatively safe estimates of ΔP_f may be made using single-phase methods with the mixture velocity and density.

3.33 Analyses or Correlations Involving Estimates of the *in situ* Concentration or Density

These correlations are based either upon Equation 3.29 adapted to a mixed-phase system, or upon Equation 7.73 or a closely related form. They involve various approximations to the kinetic energy terms, but all take some account of holdup and deal with ΔP_{HH} and $\Delta P_f^{\bullet}$ as we have previously discussed.

The correlations for liquid-solid systems of Durand (1953), Newitt *et al.* (1961), of Oedjoe and Buchanan (1966), and of Toda *et al.* (1969) are all of this type, as are the works of Worster and Denny (1955) and Cloete *et al.* (1967). Newitt *et al.* (1961) correlate the quantity, $(\Delta P - \Delta P_{HH} - \Delta P_F)/\Delta P_F$; Toda *et al.*, Oedjoe and Buchanan, and the others express their correlations in terms of $\Delta P - \Delta P_{HH}$ and Fanning type friction factors.

Belden and Kassel (1949), Uspenskii (1951), Mehta, Smith, and Comings (1957), Barth (1958), Gopichand, Sarma, and Rao (1959), Razumov (1962), and Jones, Braun, Daubert, and Allendorf (1967), base their correlations for gas-solid mixtures on the same separation of the hydrostatic component, i.e., they relate $\Delta P - \Delta P_{HH}$ to some type or combination of friction factor. Reddy and Pei (1969), although recognizing the role of ΔP_{HH}, correlate $(\Delta P/\Delta z) - (\Delta P_F/\Delta z)$ in terms of the mass flow rate of solids. Hinkle (1953) developed an equation for the calculation of the pressure drop from the mechanical energy balance requiring the use of the Fanning friction factor and a drag coefficient correlation.

Rose and Duckworth (1969) present a detailed analysis and correlation applicable both to gas-solid and liquid-solid mixtures.

In this section we will present the details of several of the correlations for liquid-solid systems, of two correlations applicable to gas-solid mixtures, and of the Rose and Duckworth analysis.

Liquid-Solid Systems Most of the correlations for the hydraulic transport of solids in vertical pipes are based on single-phase methods with the hydrostatic effects evaluated using a mixture density based on estimated *in situ* conditions. The friction term is calculated using the single-phase friction factor with either the fluid or mixture velocity.

Durand (1953) seems to be the first one to carry out a careful experimental study of the matter. He studied both the ascending and descending flow of water suspensions of sand and fine gravel in a vertical pipe 150 mm in diameter. The sand particle sizes were 0.18 and 1.0 mm; the fine gravel was 4.36 mm. He assumed the pressure drop due to friction to be independent of flow direction and interpreted his data through the equivalent of

$$\Delta P_{\text{up}} = \Delta P_{HH} + \Delta P_f^{\bullet} \qquad \text{(for upflow)} \tag{9.46}$$

$$\Delta P_{\text{down}} = \Delta P_{HH} - \Delta P_f^{\bullet} \qquad \text{(for downflow)} \tag{9.47}$$

resulting in

$$\Delta P_f^{\bullet} = \frac{\Delta P_{\text{up}} - \Delta P_{\text{down}}}{2} \tag{9.48}$$

Durand found that $\Delta P_f^{\bullet}$ so determined was essentially the same as for pure water flowing at a velocity equal to that of the two-phase mixtures. He did not discuss methods of estimating ΔP_{HH}.

Newitt, Richardson, and Gliddon (1961) have conducted tests using 1- and 2-inch diameter pipes and four different sands (with mean diameter ranging from 0.0040 to 0.05 inch), pebbles (mean diameter 0.15 inch), zircon (mean diameter 0.054 inch) and perspex (mean diameter 0.048 inch) in water suspension. They calculated the hydrostatic head by

$$\Delta P_{HH} = \rho_M \frac{g}{g_c} \Delta z \tag{9.49}$$

where

$$\rho_M = [E_S(s - 1) + 1]\rho_L \tag{9.28a}$$

Estimating E_S from the equivalent of Equation 7.76 by assuming that $S = V_0$. Their findings with respect to $\Delta P_f^{\bullet}$ may be divided into two categories:

1. For slurries of particles of diameter over 0.01 inch they confirm the findings of Durand that

$$\Delta P_f^{\bullet} - \Delta P_{fM} = 0 \tag{9.50}$$

where

$$\Delta P_{fM} = \frac{2 f \rho_L V_M^2}{g_c D} \Delta z$$

$$V_M = V_{SL} + V_{SS}$$

2. For slurries of particles of diameter less than 0.01 inch, Newitt *et al.* find that $\Delta P_f^{\bullet}$ considerably exceeds ΔP_{fM}. They therefore correlate their own data and those of Bhattacharya and Roy (1955) in terms of the dimensionless excess pressure drop. The correlation involves the Froude number, the pipe to particle diameter ratio, the solid to fluid density ratio, and the delivered solids concentration, C_S, as follows:

$$\frac{\Delta P_f^{\bullet} - \Delta P_{fM}}{\Delta P_{fM}} = 0.0037\, C_S \left(\frac{gD}{V_M^2}\right)^{1/2} \left(\frac{D}{d}\right) s^2 \tag{9.51}$$

The correlation appears to be excellent for the range of variables tested.

Kostuik (1966) has found that the Newitt *et al.* correlation applies for the vertical hoisting of crushed magnetite provided $V_M/S \geq 10$, where S is taken as V_0. This means the equation applies under circumstances where the *in situ* concentration, E_S, is approximately the same as the input concentration, C_S, as shown by Figures 9.14 and 9.15 discussed earlier.

Cloete, Miller, and Streat (1967), using their own data for the flow of water suspensions of sand and glass particles in 1/2- and 3/4-inch pipes, point out that the assumption that $\Delta P_f^{\bullet}$ may be calculated as for the single-phase carrier liquid at the mixture velocity is not valid for concentrated suspensions. Figure 9.18 modified from their work compares some of their data for $\Delta P_f^{\bullet}/\rho_W$ with such a prediction. The figure also gives the predicted pressure gradient based on the assumption of a single-phase fluid of density equal to the *in situ* density of the mixture, of viscosity equal to that of water, and flowing at the mixture velocity. The latter prediction is much closer than the former at $V_M > 3$ ft/sec, but neither fits the data well.

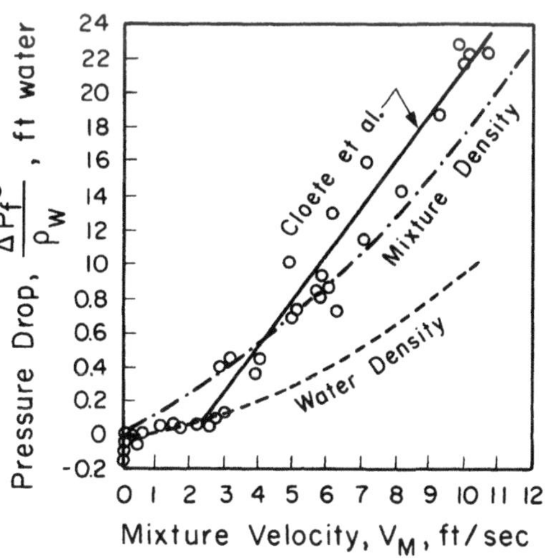

Figure 9.18. Pressure drop data of Cloete *et al.* (Redrawn from Cloete, F. L. D., A. I. Miller, and M. Streat, *Trans. Inst. Chem. Engrs.*, **45**, 392, 1967.)

The pressure drop correlations discussed in the foregoing have been developed for uniformly sized particles in low concentrations. Oedjoe and Buchanan (1966) investigated the pressure drop relationships for high concentration slurries (C_S up to 0.40) of large solid particles with a wide range of size distribution. The solids used in their study included coal and gravel particles of various sizes in 2-, 3-, and 4-inch diameter pipes. Velocities did not exceed 5.0 ft/sec. They calculated ΔP_{HH} in the same manner as Newitt *et al.* (using Equation 9.49) except that they calculated V_0 from their own correlations, i.e., Equations 9.17 and 9.18. They expressed $\Delta P_f^\bullet$ through the Fanning type equation in terms of V_L. Oedjoe and Buchanan compared their method of pressure drop prediction with their extensive experimental data and observed errors up to 10 percent with improved accuracy of prediction for uniformly sized particles.

Toda *et al.* (1969) report data on the vertical flow, to velocities of about 6 ft/sec, of water suspension of glass particles of diameters ranging from 0.44 to 2.78 mm in a 30.2-mm diameter pipe. They expressed the pressure drop as

$$\Delta P = \Delta P_{HH} + \Delta P_f^\bullet \tag{9.7}$$

The hydrostatic head term was evaluated through the equivalent of Equation 9.49 with ρ_M defined by Equation 9.28a. Toda *et al.* made actual measurements on the *in situ* volume fraction solids and used these in the calculation of ΔP_{HH}. They offered no method of predicting E_S or of correlating their data on E_S. In the range of concentrations which they studied (apparently about 6 volume percent and less) they found $\Delta P_f^\bullet$ to be correlated by Equation 9.50.

For conditions where holdup is clearly negligible the work of Graf and Acaroglu (1967), mentioned subsequently in Chapter 11, confirms the applicability of single-phase methods to the prediction of ΔP. Here $\Delta P_{PE} = \Delta P_{HH}$ and $\Delta P_f = \Delta P_f^\bullet$ and Equations 9.30, 9.49, and 9.50 apply.

The works of Durand, Newitt *et al.*, Kostuik, Cloete *et al.*, Oedjoe and Buchanan, and Toda *et al.* confirm the suitability of the equation, $\Delta P = \Delta P_{HH} + \Delta P_f^\bullet$, and that ΔP_{HH} may be estimated from the *in situ* mixture density, ρ_M. In practice where the holdup effect is not great, ρ_M may be based upon input quantities. The frictional term, $\Delta P_f^\bullet$ may be estimated from the single-phase friction factor for the carrier fluid. At velocities below about 3 ft/sec, the use of the actual liquid velocity and density is recommended. At higher velocities the situation is less clear, and in some cases the use of a velocity and density corresponding with that of the mixture seems preferable.

Gas-Solid Systems Barth (1958) has derived equations for pneumatic transport of solids by a consideration of the balance of various forces acting on the particles, the fluid, and the pipe wall. He expresses the pressure drop as

$$\Delta P = \Delta P_G + \Delta P_S \tag{9.52}$$

where ΔP_G = the pressure drop due to the flow of gas alone in the pipe at the superficial gas velocity V_{SG},

ΔP_S = the additional pressure drop caused by the presence of solids.

In evaluating ΔP_G Barth neglects the small hydrostatic head term. Such a term, however, is included in the evaluation of ΔP_S as follows:

$$\frac{\Delta P_S}{\Delta z} = \frac{2 f_S \mathscr{R} \rho_G V_{SG}^2}{g_c D} \tag{9.53}$$

This equation defines a solids friction factor, f_S. Barth's analysis shows that f_S may be expressed as

$$f_S = \frac{V_S}{V_{SG}} f_S^* + \frac{2Dg}{V_S V_{SG}} \tag{9.54}$$

where f_S^* is a characteristic of the solid particles and the pipe wall related to to the coefficient of restitution. For the average velocity of solids Barth derives the equation

$$1 - \left(\frac{V_S}{V_{SG}}\right)^2 = \mathrm{Fr}_0^2 \frac{2f_S^*}{2} \left(\frac{V_S}{V_{SG}}\right)^2 + \frac{1}{\mathrm{Fr}_G^2} \tag{9.55}$$

where $\mathrm{Fr}_0 = \dfrac{V_0}{\sqrt{Dg}}$, the settling velocity Froude number,

$\mathrm{Fr}_G = \dfrac{V_{SG}}{\sqrt{Dg}}$, the gas velocity Froude number.

Substitution of Equation 9.54 into Equation 9.53 and some rearrangement yields

$$\frac{\Delta P_S}{\Delta z} = \frac{2\mathscr{R}\rho_G V_S V_{SG} f_S^*}{g_c D} + \frac{g}{g_c} \mathscr{R}\rho_G \frac{V_{SG}}{V_S} \tag{9.56}$$

We see from Equation 9.29 that the second term of Equation 9.56 is a hydrostatic head term based upon the *in situ* concentration of solids and neglecting the weight of the gas itself. The first term is the pressure gradient caused by the impact of the particles on the pipe wall.

From his own and other data Barth employed these relations to calculate the effective values of f_S^*. He found these to range from 0.00075 to 0.00325 with most of the materials indicating a value of approximately 0.00125. To develop a general prediction method Barth therefore assumed a constant value, $f_S^* = 1.25 \times 10^{-3}$ and calculated f_S from Equations 9.54 and 9.55. Figure 9.19 shows the derived values of f_S as a function of Fr_G for various values of Fr_0. Experimental data are also shown in this figure. Barth has tested his correlation for the transport of fine coal (*d* up to 20 mm), coke balls (*d* = 6 to 10 mm), wheat meal (*d* up to 1 mm), coke dust (*d* up to 0.05 mm), and wheat. The data show considerable scatter and the correlation is highly sensitive to the gas Froude number, Fr_G, in certain ranges. The scatter is probably due to the use of a constant value for f_S^*. Barth has also extended his work to horizontal transport and we will discuss this in Chapter 11.

The Jones *et al.* (1967) correlation is based on extensive data for air-solid systems. The solids range in size from 200 to 765 microns and the solid density from 156 to 475 lb_m/ft^3. Jones *et al.* start with (Equation 7.73)

$$\Delta P_T = \Delta P_{HH} + \Delta P_{KES} + \Delta P_{KEF} + \Delta P_f^\bullet \tag{9.57}$$

where $\Delta P_{HH} = \rho_M \dfrac{g}{g_c} \Delta z$

$$\Delta P_{KES} = \frac{\rho_M V_S^2}{2g_c}$$

$$\Delta P_{KEF} = \frac{\rho_G V_G^2}{2g_c}$$

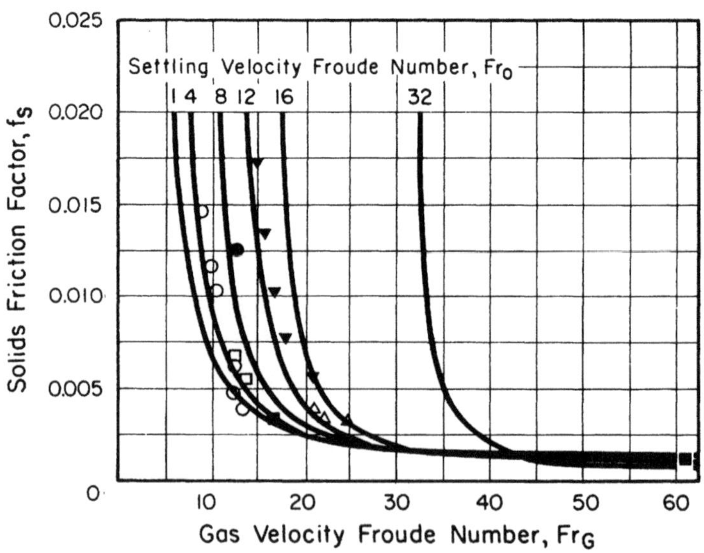

Figure 9.19. Barth's solids friction factor, f_S (based upon constant $f_S^* = 0.00125$). (Redrawn from Barth, W., *Chemie-Ing.-Techn.*, **30**, 171, 1958.)

ρ_M is based upon Equation 9.29, i.e., neglects the contribution of the mass of the gas, and the friction term, $\Delta P_f^{\bullet}$, is expressed as

$$\Delta P_f^{\bullet} = \frac{2(f_G + f_S')\rho_G V_G^2}{g_c D} \Delta z \tag{9.58}$$

where f_G = the single-phase friction factor calculated at a Reynolds number $DV_G\rho_G/\mu_G$ (Figure 4.16)
f_S' = a solids friction factor.

Jones *et al.* calculate the solids friction factor, f_S', from their data and Equation 9.58 and correlate f_S' with $\mathscr{R}$ through the relation

$$f_S' = b\mathscr{R}^n \tag{9.59}$$

The two empirical constants in Equation 9.59 are given by

$$b = 0.47 \times 10^{-6} \frac{A_S}{\sqrt{\psi}} \tag{9.60}$$

$$n = 1 \qquad A_S \leq 6300$$

$$n = \left(\frac{6300}{A_S}\right)^{1/3} \qquad A_S > 6300$$

where A_S = surface area of the spherical particles in square feet per cubic foot of material in bulk form,
ψ = particle shape factor defined by Equation 1.15 of Chapter 1.

The correlation requires a knowledge of *in situ* velocities, V_S and V_G. These may be estimated using the holdup correlations discussed in Section 3.2. Jones *et al.* do not recommend any particular method of calculating V_S and V_G. The correlation is considered to be an accurate method for the calculation of pressure drop in the pneumatic transport of solids. Average deviations from experimental values, as observed by Jones *et al.*, were, in most cases, less than 13 percent with a maximum of 25 percent in the case of certain steel shots.

Rose and Duckworth (1969) have developed the most systematically based general correlation for the prediction of the pressure gradient. Their correlation is applicable to vertical, inclined, and horizontal flow of gas-solid and liquid-solid suspensions. It does not apply to finely divided solids suspended in liquids where non-Newtonian effects occur, to gas-solids suspensions where electrostatic effects are significant, or to flow in horizontal or inclined pipes where solids are deposited on the pipe wall. The correlation is based upon the mechanical energy balance in the form

$$\Delta P_T = \Delta P_{HH} + \Delta P_{KEF} + \Delta P_{KES} + \Delta P_f^{\bullet} \tag{9.5}$$

where

$$\Delta P_{HH} = \int_{L_1}^{L_2} \rho_M \frac{g}{g_c} \sin\theta \, dL \tag{9.61}$$

$$\Delta P_{KEF} + \Delta P_{KES} = \frac{G_{SS}}{g_c}(V_{S2} - V_{S1}) + \frac{G_{SF}}{g_c}(V_{F2} - V_{F1}) \tag{9.62}$$

$$\Delta P_f^\bullet = \frac{4}{D}\int_{L_1}^{L_2} \tau_w \, dL \tag{9.63}$$

τ_w = wall shear stress,
G_{SF} = superficial mass velocity of fluid = $V_{SF}\rho_F$,
G_{SS} = superficial mass velocity of solids = $V_{SS}\rho_S$,
L = length of pipe,
ρ_M is defined without assumption, i.e., by Equation 9.28a.

Their expression of the kinetic energy terms (Equation 9.62) is not the same as in Equation 7.73, reducing to it only if C_S equals E_S. Applying dimensional analysis to the individual components of Equation 9.5, and from their own extensive data on horizontal, inclined, and vertical flow of a variety of solids in both air and water, and the data of others, they devise methods for the prediction of the travel distance and the pressure drop required to accelerate the solids to a steady velocity and each of the other factors involved in Equation 9.5.

Their results and the basic elements of their correlation are as follows:

For the acceleration distance,

$$\frac{L_A}{D} = 6\left[\left(\frac{M_S}{\rho_F g^{1/2} D^{5/2}}\right)\left(\frac{D}{d}\right)^{1/2} s^{1/2}\right]^{1/3} \tag{9.64}$$

where L_A = length of pipe required for full acceleration of the solids,
M_S = mass flow rate of solids, $\text{lb}_\text{m}/\text{sec}$;

for the acceleration pressure drop due to the solids,

$$\frac{2\Delta P_{KES}}{\rho_F V_{SF}^2} = \frac{1.12}{g_c}\mathcal{R}\phi_4\phi_5 \tag{9.65}$$

where ϕ_4 is a function of V_{SF}^2/gds^2 as given in Figure 9.16,
ϕ_5 is a function of the angle of inclination, θ, as given in Figure 9.20;

for the pressure drop due to friction and hydrostatic head,

$$\frac{\Delta P_f^\bullet + \Delta P_{HH}}{\rho_F(V_{SF}^2/2g_c)(L/D)} = 4f_{SF} + \psi_S + 2\left(\frac{gD}{V_{SF}^2}\right)\sin\theta\left[1 + \mathcal{R}\frac{V_{SF}}{V_S}\left(1 - \frac{1}{s}\right)\right] \tag{9.66}$$

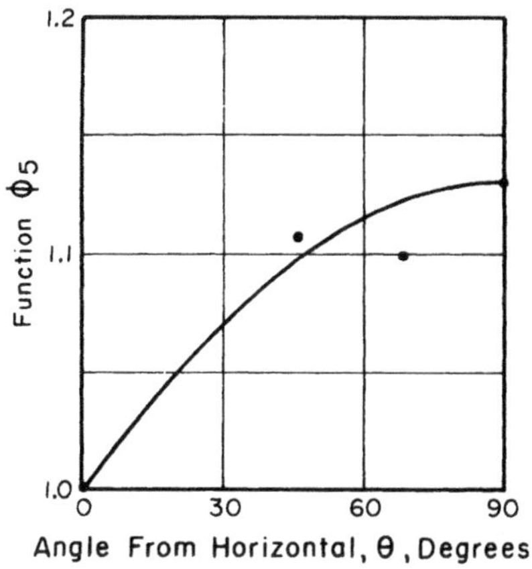

Figure 9.20. Rose and Duckworth function ϕ_5. (From Rose, H. E., and R. A. Duckworth, *The Engineer,* **227**(5904), 430, 1969.)

where f_{SF} is evaluated at $\mathrm{Re}_{SF} = \rho_F V_{SF} D/\mu$ and at the appropriate relative roughness,

$\psi_S = \psi_1\psi_2\psi_3\psi_4\psi_5\psi_6$,

ψ_1 is a function of $\mathscr{R}$ as given in Figure 9.21,

ψ_2 is a function of d/D as given in Figure 9.22,

ψ_3 is a function of ε_R, the coefficient of restitution, as given in Figure 9.23,

ψ_4 is a function of s as given in Figure 9.24,

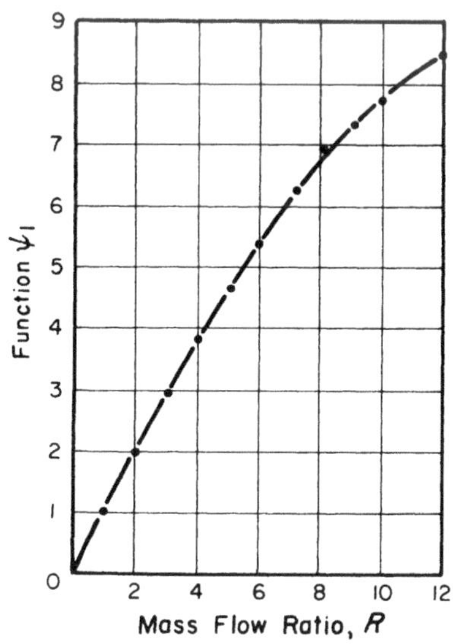

Figure 9.21. Rose and Duckworth function ψ_1. (From Rose, H. E., and R. A. Duckworth, *The Engineer,* **227**(5905), 478, 1969.)

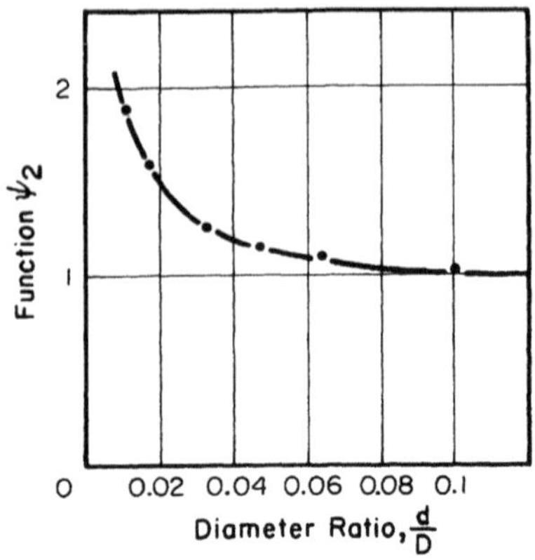

Figure 9.22. Rose and Duckworth function ψ_2. (From Rose, H. E., and R. A. Duckworth, *The Engineer,* **227**,(5905), 478, 1959.)

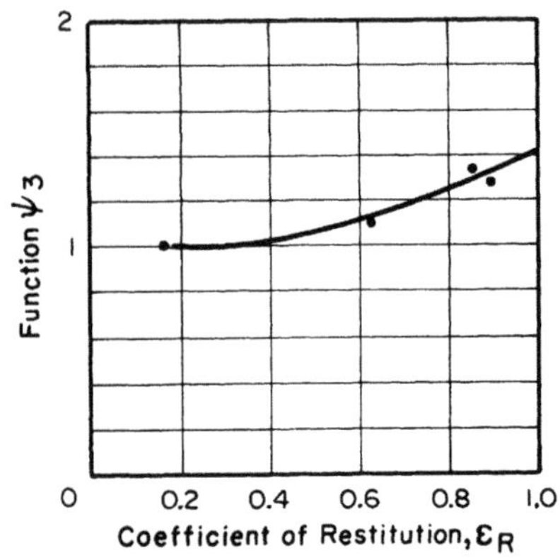

Figure 9.23. Rose and Duckworth function ψ_3. (From Rose, H. E., and R. A. Duckworth, *The Engineer,* **227**,(5905), 478, 1969.)

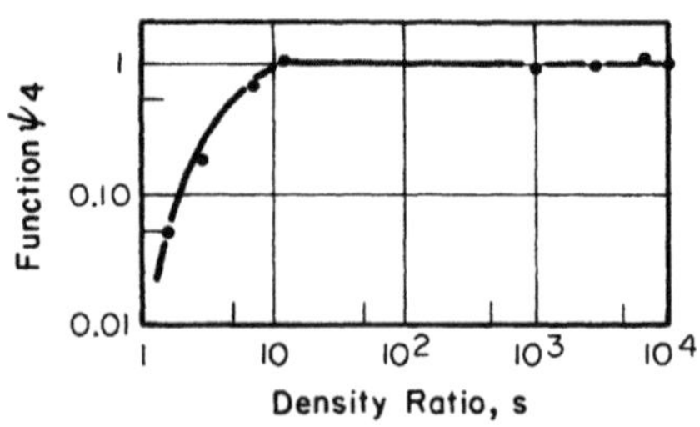

Figure 9.24. Rose and Duckworth function ψ_4. (From Rose, H. E., and R. A. Duckworth, *The Engineer,* **227**,(5905), 478, 1969.)

ψ_5 is a function of θ as given in Figure 9.25, and
ψ_6 is a function of V_{SF}^2/gD as given in Figure 9.26;

for the velocity ratio,

$$\frac{V_S}{V_{SF}} = 0.56\phi_4\phi_5 \tag{9.67}$$

for the total pressure drop,

$$\Delta P = \Delta P_{HH} + \Delta P_{KES} + \Delta P_f^\bullet + \rho_F\left(\frac{V_{SF}^2}{2g_c}\right) \tag{9.68}$$

for the minimum transport velocity,

$$\frac{(V_M)_{\min}}{V_0} = 3.2\left[\mathscr{R}^{0.2}\left(\frac{D}{d}\right)^{0.6}\left(\frac{1}{s}\right)^{0.7}\left(\frac{V_c^2}{gD}\right)^{0.25}\right] \tag{9.69}$$

where

$$V_0 = \left[\frac{4gd(s-1)}{3C_D}\right]^{1/2} \tag{9.70}$$

$(V_M)_{\min}$ = the minimum transport velocity for vertical flow and equals V_{M2} for horizontal flow (see Chapter 11).

We note that Equation 9.70 is obtained by combining Equations 1.1 and 1.2 of Chapter 1.

For comparison with some of the other correlations we may write the Rose and Duckworth correlation for vertical flow in the absence of acceleration effects as

$$\frac{\Delta P}{\rho_F \Delta z} = \frac{V_{SF}^2}{2g_c D}(4f_{SF} + \psi_S) + \left[1 + \mathscr{R}\frac{V_{SF}}{V_S}\left(1 - \frac{1}{s}\right)\right]\frac{g}{g_c} \tag{9.71}$$

where V_{SF}/V_S is given by Equation 9.67.

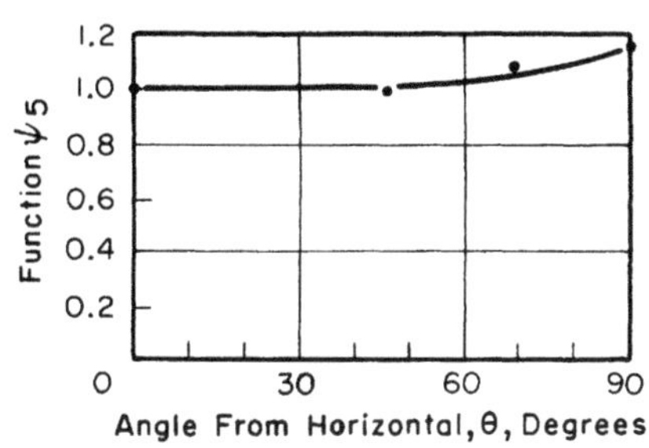

Figure 9.25. Rose and Duckworth function ψ_5. (From Rose, H. E., and R. A. Duckworth, *The Engineer*, **227**,(5905), 478, 1969.)

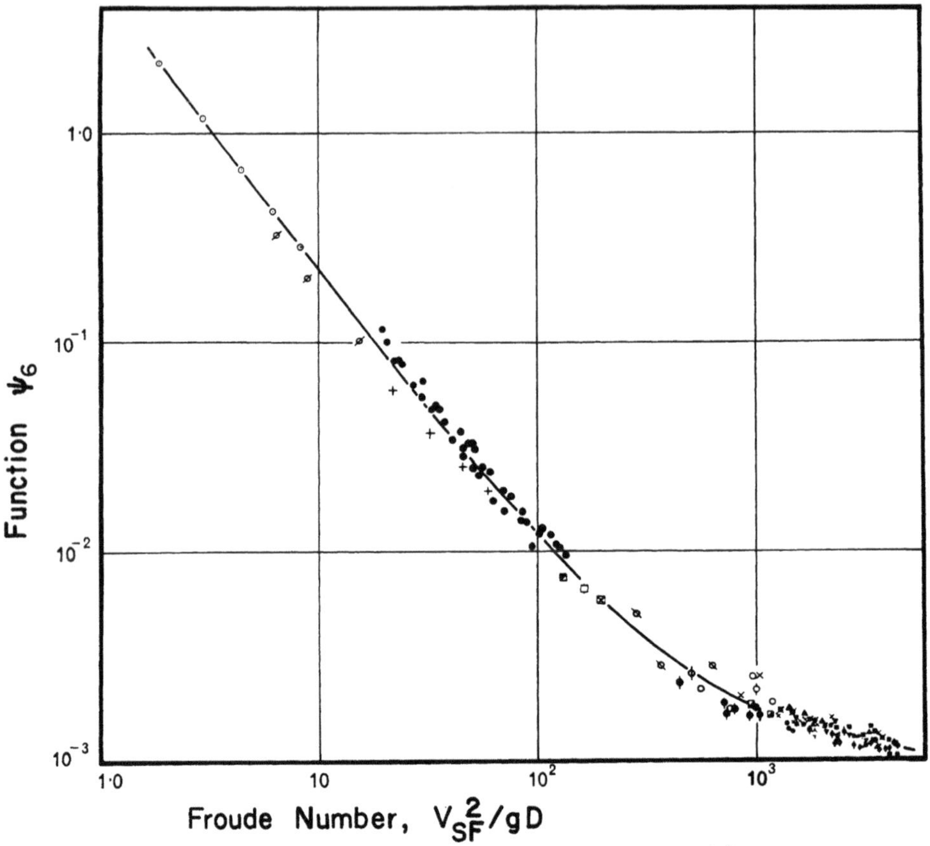

Symbol	Material	Fluid	d (in.)	D (in.)	ρ_s (lbf/ft^3)	Ref.
○	Coal	Water	0·5	3·0	94	Worster and Denny (1955)
ø	Sand	"	0·118	5·9	165	Durand and Condolios (1952)
+	Steel	"	0·125	1·265	460	Rose and Duckworth
●	Lead	"	0·095	"	695	"
ϕ	Blaud's pills	Air	0·28	2·0	114	Hitchcock and Jones (1956)
⌑	"	"	"	3·0	"	"
♦	Mustard	"	0·08	1·265	73	Rose and Duckworth
◼	Lead	"	0·128	"	695	"
◼	"	"	0·08	"	"	"
▲	"	"	0·06	"	"	"
●	"	"	0·043	"	"	"
▼	"	"	0·022	"	"	"
◥	"	"	0·015	"	"	"
■	Steel	"	0·125	"	450	"
/	Glass	"	0·118	"	187	"
♦	Limestone	"	1·25	6·0	89	Private communication
○	Coke	"	1·0	6·0	32	Private communication
◪	Iron ore	"	0·125	6·0	200	Private communication
◩	Wheat	"	0·20	16·0	87	Segler (1951)
☐	"	"		12·0	"	
⊠	"	"		10·0	"	

Figure 9.26. Rose and Duckworth function ψ_6. (From Rose, H. E., and R. A. Duckworth *The Engineer,* **227**(5905), 478, 1969.)

The suitability of the Rose and Duckworth correlation seems confirmed for a wide variety and numerous sizes of materials in a range of pipe sizes. It is recommended as being a powerful general method for estimating all components of the pressure drop in the vertical flow of fluid-solid mixtures.

3.34 Mechanistic Approach

The correlations discussed so far have been entirely empirical converting measured values of ΔP to ΔP_f or $\Delta P_f^{\bullet}$ and relating one of the latter two to the principal variables without reference to the actual mechanism of the flow. There has been one quite successful attempt to explain, from a mechanistic viewpoint, why $\Delta P_f^{\bullet}$ is generally greater than ΔP_{fF} and to develop a theoretical basis for its prediction. This is the analysis of Julian and Dukler (1965). Shook and Daniel (1969) have also made an analysis which they have tested for horizontal flow. We will describe this in Chapter 11. Both analyses involve the development of a form of the velocity distribution relationship, which is modified to take account of the solids, and the consequent integrations of the relationship to obtain an equation for the friction factor. The former analysis has been developed and tested for dilute phase ($\mathscr{R} \leq 12$) gas-solid systems, but the theory should be equally applicable to liquid-solid flow. The analysis assumes the symmetric-suspension flow pattern, but otherwise it is applicable both to vertical and horizontal flow. It applies only to circumstances where $f_M > f$, i.e., not to cases such as reported by Boothroyd for very fine particles in low concentration where the presence of the particles appears to suppress the development of turbulence and reduces the friction factor.

Julian and Dukler (1965) use data for vertical and horizontal flow in the development of their correlation. The pressure drop due to friction is calculated by applying Equations 9.5 and 9.29. They neglect the kinetic energy of the gas, i.e., ΔP_{KEF} and do not report the method of evaluating ΔP_{KES}. For the development of an expression for $\Delta P_f^{\bullet}$, they assume that in the flow of dilute phase gas-solid mixtures

> ... the solids make their influence felt primarily by modifying the local turbulence in the gas phase, increasing the turbulent fluctuations, mixing length, eddy viscosity, and thus the frictional pressure drop.

They neglect the resistance caused by form drag and the resistance caused by particle-to-wall and particle-to-particle collision, although their final correlation incorporates some empirical constants which may implicitly contain the influence of form drag and particle collisions. The analysis parallels that of Gill and Scher (1961) for single-phase flow discussed in Chapter 4.

In single-phase flow the local shear stress, τ, may be expressed as

$$\frac{\tau}{\tau_w} = \left(1 + \frac{\nu_e \rho}{\mu}\right) \frac{du^+}{dy^+} \tag{9.72}$$

where

$$u^+ = \frac{u_F}{(\tau_w g_c/\rho_F)^{1/2}}$$

$$y^+ = y\rho_F \frac{(\tau_w g_c/\rho_F)^{1/2}}{\mu}$$

$$\nu_e = \text{the eddy kinematic viscosity} = l^2 \left| \frac{du_F}{dy} \right| \tag{9.73}$$

and

$$l = \mathscr{K} y[1 - \exp(-2\Phi y/D)] \tag{4.57}$$

Substituting G_F/ρ_F for u_F in Equation 9.73 we have the starting point of Julian and Dukler's analysis, i.e.,

$$\nu_e = \frac{\mathscr{K}^2 y^2}{\rho_F} [1 - \exp(-2\Phi y/D)]^2 \frac{dG_F}{dy} \tag{9.74}$$

Equation 9.74 is valid for single-phase flow only. The basic hypothesis of Julian and Dukler is that, for gas-solid flow, G_F should be replaced by $G_F + G_S$ representing the total mass velocity, i.e.,

$$\nu_e = \frac{\mathscr{K}^2 y^2}{\rho_F} [1 - \exp(-2\Phi y/D)]^2 \frac{d}{dy}(G_F + G_S) \tag{9.75}$$

where $G_S = \rho_S u_S = r\rho_F u_F = rG_F$,
r = local mass ratio.

If we substitute $\rho_F u_F(1 + r)$ for $G_F + G_S$ in Equation 9.75 and assume that r and ρ_F are independent of y, we obtain

$$\nu_{eM} = \mathscr{K}^2 y^2 (1 + \mathscr{R})^m [1 - \exp(-2\Phi y/D)]^2 \frac{du_F}{dy} \tag{9.76}$$

where we have assumed that $\mathscr{R} = r = M_S/M_F$. In Equation 9.76, m is introduced as an empirical factor to account for the unequal contribution to the local eddy viscosity of the two phases. Equation 9.76 may be combined with

Equation 9.72 to derive the velocity distribution equation for gas-solid flow

$$u^+ = \int_0^{y^+} \frac{-1(1 + 4Cb)^{1/2}}{2C} \, dy^+ \tag{9.77}$$

where $C = (\mathscr{K} y^+)^2 \left[1 - \exp\left(-\frac{\Phi y^+}{y_m^+}\right)\right]^2$,

$y_m^+ = \mathrm{Re}\,\sqrt{f_M/8}$,
$\mathscr{K} = k(1 + \mathscr{R})^{m/2}$,
$m, k =$ constants,
$b = 1 - (y^+/y_m^+)$,
$\Phi = (y_m^+ - 60)/22$.

Combining Equation 9.77 with the momentum balance, we obtain

$$\frac{1}{\sqrt{f_M/2}} = \frac{2}{(y_m^+)^2} \int_0^{y_m^+} (y_m^+ - y^+) \int_0^{y^+} \frac{-1 + (1 + 4Cb)^{1/2}}{2C} \, dy^+ \, dy^+ \tag{9.78}$$

where

$$f_M = \frac{(\Delta P/\Delta z) D g_c}{2 \rho_G V_{SG}^2}$$

Julian and Dukler (1965) integrated this equation numerically, and calculated the best fit value of $\mathscr{K}$ for carefully screened data of several workers. They interpreted the data to show an abrupt change of slope in the plot of $\mathscr{K}$ versus $1 + \mathscr{R}$ at a value of $\mathscr{R}$ of about 12, represented by two separate equations:

$$\mathscr{K} = 0.36\,(1 + \mathscr{R})^{0.25}, \qquad \mathscr{R} \leq 12 \tag{9.79}$$

$$\mathscr{K} = 0.11\,(1 + \mathscr{R})^{0.8}, \qquad \mathscr{R} > 12 \tag{9.80}$$

Julian and Dukler suggest that dilute phase transport extends only to $\mathscr{R} \simeq 12$ and that for $\mathscr{R} > 12$ particle agglomeration may occur with a change in the flow mechanics. Actually the data are not fully definitive and an acceptable alternative interpretation applicable for the full range, to $\mathscr{R} \simeq 36$, would be a simple smooth curve through the bulk of the data. Figure 9.27 shows the data used by Julian and Dukler, their two equations, and an alternative smooth curve. For the range $\mathscr{R} \leq 12$ Julian and Dukler express their results through Figure 9.28, showing the influence of $\mathscr{R}$ on the relationship between f_M and the gas Reynolds number, $D V_{SG} \rho_G / \mu_G$, for smooth tubes. Brandon (1968) has shown that this relationship may be expressed simply as

$$\frac{f_M}{f_F} = 1 + \mathscr{R}^{0.3} \tag{9.81}$$

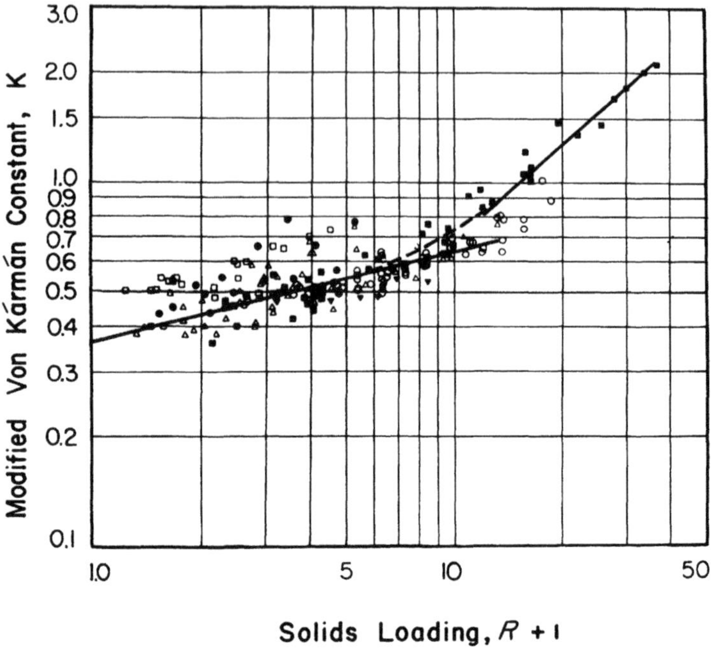

Figure 9.27. The Modified von Kármán constant. (From Julian, F. M., and A. E. Dukler, *A.I.Ch.E. Journal*, **11**, 853, 1965.)

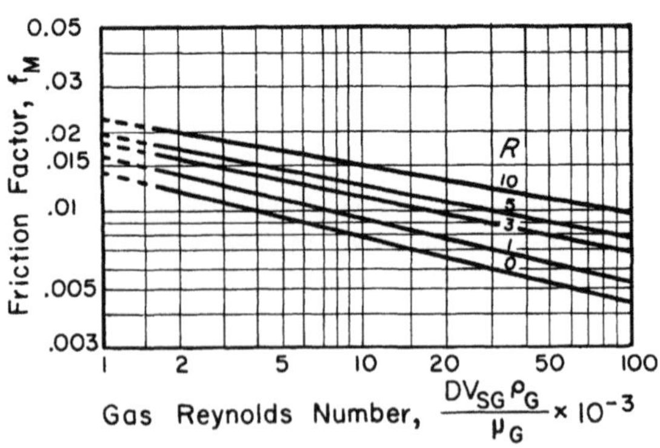

Figure 9.28. Friction factors for flow of dilute phase gas-solid mixtures. (From Julian, F. M., and A. E. Dukler, *A.I.Ch.E. Journal*, **11**, 853, 1965.)

The Julian and Dukler analysis has a reasonable theoretical basis and is supported by data of eight independent investigators covering the following ranges:

$$69.7 \text{ lb}_m/\text{ft}^3 < \rho_S < 525 \text{ lb}_m/\text{ft}^3$$
$$0.3 < \mathscr{R} < 36.0$$
$$0.001 \text{ in.} < d < 0.25 \text{ in.}$$
$$0.25 \text{ in.} < D < 2.4 \text{ in.}$$
$$3700 < \text{Re} < 178{,}000$$

Some 90 percent of the data points on the $\mathscr{K}$ versus $\mathscr{R}$ plot fall within 20 percent of the Julian and Dukler equations, a scatter not excessive, bearing in mind the difficulties involved in obtaining reliable data. The friction factor correlation of Figure 9.28 should be reliable to within about 20 percent for solids loadings up to 10 or 12.

It is of interest to note that, despite the range in solid particle size and density, the data show no systematic trend with respect to either of these properties. In the face of empirical evidence to the contrary, in liquid-solid systems we must be cautious in extrapolating the conditions of applicability of the correlation.

4 RECOMMENDED DESIGN METHODS

4.1 Introduction

We must first determine, by following the procedure outlined in Chapter 1, whether the gas-solid or liquid-solid mixture may be considered pseudohomogeneous. When the mixture may be so classified, the methods of Chapter 4, 5, or 6 apply; otherwise we follow the procedure outlined here.

As discussed in Section 5.1 of Chapter 4 and Section 8.1 of Chapter 8, either the entire system may be treated as incompressible or it may be divided into sections over each of which the mixture may be considered to be incompressible. Here we will describe the procedure for the case when the mixture may be taken as incompressible or for a single-step calculation for the compressible case. The maximum pressure drop allowed over a section of the pipe in which the system is assumed to be incompressible depends on the desired accuracy but is usually of the order of 10 percent of the upstream pressure.

We will follow the procedure of Section 8 of Chapter 8 and give methods of predicting pressure drop and holdup from a knowledge of the other system variables. In the vertical transport of solids by gas or liquids, the flow pattern is essentially the same under all practical conditions, and the only important aspect of the flow pattern detail is the prediction of the minimum transport velocity. When alternate methods are suggested, the selection of the "best" method must be based on experience and judgement. In such cases the appropriate sections of this chapter should be consulted for a discussion of the range of applicability and characteristics of the methods.

4.2 Entrance Effects

The only important entrance effect in the design of solid transport systems is the one associated with the acceleration of solids from essentially zero velocity to some stabilized pipe velocity. The following procedure is recommended.

1. The method of Rose and Duckworth and Equation 9.64 may be used to calculate the pipe length, L_A, over which the acceleration takes place.
2. The pressure drop due to the acceleration of solids over the length, L_A, i.e. $(\Delta P_{KE})_{\text{ent}}$, may be calculated from Equation 9.65. We note that the above method applies to both liquid-solid and gas-solid systems.

4.3 Effect of Bends and Fittings

No methods are available for the accurate prediction of the effect of bends and fittings. We recommend the procedure outlined in Section 8.3 of Chapter 8, which is based on the use of single-phase methods for the fluid phase flowing at the mixture velocity. (We are not dealing here with problems of local accumulation or erosion which may be encountered.)

4.4 Potential Energy or Hydrostatic Head Effects

For vertical hydraulic or pneumatic transport the hydrostatic head and potential energy effects are usually different, the appropriate one depending upon the correlation to be used for the pressure drop calculation.

1. When ΔP_{PE} is required, it may be calculated from the appropriate term of Equation 9.4.
2. When ΔP_{HH} is required, it may be calculated from the appropriate term of Equation 9.5. Various approximations for ρ_M required for the calculation of ΔP_{HH} are given in Equations 9.28 to 9.33.

4.5 Acceleration or Kinetic Energy Effects

Acceleration or kinetic energy effects may be very important in the entrance section of the pipe, and a procedure to determine them has been given in Section 4.2. Acceleration effects over the stabilized flow portion of the pipe are usually small, although they may be significant for gas-solid mixtures if the pressure drop is significant. The procedure for estimating ΔP_{KE} over the length of the pipe where the flow is stabilized is the following:

1. The average velocity of the solids, V_S, may be estimated from Equation 9.25. From Equation 9.27 the value of E_S may be calculated and from it V_F may be determined as

$$V_F = \frac{V_{SF}}{1 - E_S}$$

Other methods of calculating average phase velocities are discussed in Section 3.2.

2. The quantity, ΔP_{KE} may be calculated from the appropriate terms of Equation 9.4.

3. The total kinetic energy effect, including that for the entrance section, is obtained by adding $(\Delta P_{KE})_{ent}$ and ΔP_{KE} as determined in step *2.*

4.6 Flow Pattern, Holdup, and Pressure Gradient for the Steady Flow of Incompressible Mixtures (Liquid-Solid Mixtures and Gas-Solid Mixtures when $\Delta P < 0.10P_1$)

1. Determine the settling velocity for the solids by the methods outlined in Section 3.5 of Chapter 1 or Equations 9.17 and 9.18 as appropriate.

2. Calculate the minimum transport velocity, $(V_M)_{min}$, by
   a. the Rose and Duckworth method and Equation 9.69, or
   b. Equation 9.12, or
   c. the Zenz correlation presented as Figure 9.13.

3. If the fluid velocity is specified in the problem, it must exceed the value calculated in step *2.* If only the solid transport rate is given, step *2* gives the required minimum mass velocity and hence the maximum pipe diameter to accomplish the transport. It should be noted that the minimum transport velocity will not necessarily be the optimum fluid velocity. Usually several pressure drop calculations using different fluid rates are required to determine the overall optimum conditions.

4. The various measures of the holdup, E_S, S, and H, are related as discussed in Section 7.2 of Chapter 7 and Section 3.2 of this chapter. These quantities may be estimated by
   a. the Rose and Duckworth method, Equation 9.27, or
   b. Equations 9.15 and 9.16.

   This calculation is not always necessary for the prediction of pressure drop.
5. a. The pressure drop due to friction for either liquid-solid or gas-solid systems may be calculated by the method of Rose and Duckworth, Equation 9.71.
   b. For gas-solid transport the calculation may also be made by
      i. Equation 9.44 of Korn, or
      ii. Equation 9.58 of Jones *et al.*
   c. For liquid-solid systems an alternative procedure is that discussed in Section 3.33.
      i. Estimate the total pressure drop for coarse particles from Equation 9.7 with ΔP_{HH} determined by Equation 9.49, and $\Delta P_f^\bullet$ determined by the equivalent of Equation 9.50 based on $V = V_L$ for $V < 3$ ft/sec and $V = V_M$ for $V > 3$ ft/sec.
      ii. Use Equation 9.51 for fine particles.

Bibliography

Barth, W., *Chemie-Ing.-Techn.*, **30**, 171, (1958).

Belden, D. H., and L. S. Kassel, *Ind. Eng. Chem.*, **41**, 1174 (1949).

Bhattacharya, A., and A. N. Roy, *Ind. Eng. Chem.*, **47**, 268 (1955).

Boothroyd, R. G., *Trans. Inst. Chem. Engrs.*, **44**, 306 (1966).

Brandon, C. A., Investigation of the Interaction of Solid Particles with Fluids in Turbulent Pipe Flow. Oak Ridge National Lab. Report ORNL-TM-2124, Oak Ridge, Tenn., Sept. 1968.

Brodkey, R. S., *The Phenomena of Fluid Motions*, Addison-Wesley Publishing Co., Reading, Mass., 1967.

Cloete, F. L. D., A. I. Miller, and M. Streat, *Trans. Inst. Chem. Engrs.*, **45**, 392 (1967).

Collins, R. E., *Flow of Fluids through Porous Materials*, Reinhold Publishing Corp., New York, 1961.

Doig, I. D., Doctoral Dissertation, Univ. New South Wales, Australia, 1965.

Doig, I. D., and G. H. Roper, *Ind. Eng. Chem. Fundam.*, **6**, 247 (1967).

Durand, R., *Houille Blanche*, Special No. A, p. 124 (1953).

Durand, R., and E. Condolios, Study of the Transport of Solids in Pipes, Colloquium on Hydraulic Transport, National Coal Board, Nov. (1952).

Farbar, L., *Ind. Eng. Chem.*, **41**, 1184 (1949).

Frantz, J. F., *Chem. Eng.*, **69** (Sept. 17), 161 (1962); **69** (Oct. 1), 89 (1962); **69** (Oct. 29), 103 (1962).

Gasterstadt, J., *Zeitschrift des Vereines Deutscher Ingenieure*, **68** (24), 617 (1924).

Gill, W. N., and M. Scher, *A.I.Ch.E. Journal*, **7**, 61 (1961).

Gopichand, T., K. J. R. Sarma, and M. N. Rao, *Ind. Eng. Chem.* **51**, 1449 (1959).

Graf, W. H., and E. R. Acaroglu, *A.S.C.E. Proc., Pipeline Div.*, **93** (PL2), 63 (1967).

Hanratty, T. J., and J. M. Engen, *A.I.Ch.E. Journal*, **3**, 299 (1957).

Hariu, O. H., and M. C. Molstad, *Ind. Eng. Chem.*, **41**, 1148 (1949).

Hinkle, B. L., Doctoral Dissertation, Georgia Institute of Technology, Atlanta, Georgia, 1953.

Hitchcock, J. A., and C. Jones, Report No. 2053, Mining Research Establishment, Isleworth, 1956.

Jones, J. H., W. G. Braun, T. E. Daubert, and H. D. Allendorf, *A.I.Ch.E. Journal*, **13**, 608 (1967).

Julian, F. M., and A. E. Dukler, *A.I.Ch.E. Journal*, **11**, 853 (1965).

Kolpakov, V. M., and E. V. Donat, *Int. Chem. Eng.*, **10**, 394 (1970).

Korn, A. H., *Chem. Eng.*, **57** (March), 108 (1950).

Kostuik, S. P., *Trans. Can. Inst. Mining Met.*, **69**, 1 (1966).

Leva, M., *Fluidization*, McGraw-Hill Book Co., New York, 1959.

Mehta, N. C., J. M. Smith, and E. W. Comings, *Ind. Eng. Chem.*, **49**, 986 (1957).

Muskat, M., *Physical Principles of Oil Production*, McGraw-Hill Book Co., New York, 1949.

Newitt, D. M., J. F. Richardson, and B. J. Gliddon, *Trans. Inst. Chem. Engrs.*, **39**, 93 (1961).

Oedjoe, D., and R. H. Buchanan, *Trans. Inst. Chem. Engrs.*, **44**, 364 (1966).

Orr, C. Jr., *Particulate Technology*, The Macmillan Company, New York, 1966.

Peskin, R. L., and H. A. Dwyer, *A.S.M.E. Pub. 65-WA/FE24* (1965).

Razumov, I. M., *Int. Chem. Eng.*, **2**, 539 (1962).

Reddy, K. V. S., and D. C. T. Pei, *Ind. Eng. Chem. Fundam.*, **8**, 490 (1969).

Rose, H. E., and R. A. Duckworth, *The Engineer*, **227** (5903), 392 (1969); **227** (5904), 430 (1969); **227** (5905), 478 (1969).

Scheidegger, A. E., *The Physics of Flow Through Porous Media*, University of Toronto Press, Second Edition, 1960.

Segler, G., Pneumatic Grain Conveying, National Institute of Agricultural Eng., Silsoe, Beds., 1951.

Shook, C. A., and S. M. Daniel, *Can. J. Chem. Eng.*, **47**, 196 (1969).

Soo, S. L., *Fluid Dynamics of Multiphase Systems*, Blaisdell Publishing Co., Waltham, Mass., 1967.

Soo, S. L., and J. A. Regalbuto, *Can. J. Chem. Eng.*, **38**, 160 (1960).

Soo, S. L., G. J. Trezek, R. C. Dimick, and G. F. Hohnstreiter, *Ind. Eng. Chem. Fundam.*, **3**, 98 (1964).

Toda, M., H. Konno, S. Saito, and S. Maeda, *Int. Chem. Eng.*, **9**, 553 (1969).

Uspenskii, V. A., *Za Ekon Topliva*, **8**, 26 (1951).

Vogt, E. G., and R. R. White, *Ind. Eng. Chem.*, **40**, 1731 (1948).

Wallis, G. B., *One-Dimensional Two-Phase Flow*, McGraw-Hill Book Co., New York (1969).

Worster, R. C., and D. F. Denny, *Proc. Inst. Mech. Engrs*, **169**, 563 (1955).

Zenz, F. A., *Petroleum Refiner*, **36**(4), 173 (1957); **36**(5), 261 (1957); **36**(6), 133 (1957); **36**(7), 175 (1957); **36**(8), 147 (1957); **36**(9), 305 (1957); **36**(10), 162 (1957); **36**(11), 321 (1957).

Zenz, F. A., and D. F. Othmer, *Fluidization and Fluid Particle Systems*, Reinhold Publishing Corp., New York, 1960.

Zuber, N., and J. A. Findlay, *J. Heat Transfer*, *Trans. A.S.M.E.*, *Ser. C*, **87**, 453 (1965).

The Horizontal Flow of Gas-Liquid and Liquid-Liquid Mixtures in Pipes

10

1 INTRODUCTION

Transport of fluid mixtures in horizontal or nearly horizontal pipelines is becoming increasingly common. This is particularly evident in the field gathering of liquid and gaseous hydrocarbons. In these cases, two-phase transport, in addition to offering economies in pipeline construction itself, permits the centralization of gas processing and crude oil or condensate stabilization facilities, usually resulting in both improved processing economics and improved conservation. Two-phase flow in horizontal lines is also encountered in the operation of evaporators, chemical reactors, heat exchangers, and other types of processing equipment.

Additional applications of horizontal gas-liquid flow are found in boilers, condensers, and refrigerating and air conditioning equipment. In many of these applications pressure drop or holdup are not the sole matters of interest and the rate of heat transport may be of prime importance. The holdup and flow pattern have direct influence on the heat transfer characteristics, and a knowledge of them is required for heat transfer calculations. Our concern, however, is mainly with the actual flow mechanics of gas-liquid and liquid-liquid mixtures.

In many practical applications the pipes transporting multiphase fluids are neither vertical nor horizontal. The problem of flow in pipes inclined at various angles will also be considered.

Our treatment of horizontal flow parallels the treatment we followed for vertical flow. We deal first with typical data illustrating the general nature of the flow patterns, holdup, and pressure gradient. Following this is a discussion of empirical correlations applicable to all or several of the flow patterns. We then treat each of the flow patterns individually presenting both the applicable theory and correlations.

There has been considerable confusion in the terminology applied to the flow patterns observed in horizontal gas-liquid and liquid-liquid flow. This has been particularly true for patterns identified as plug flow and as slug flow and for their relationship to the so-called slug flow pattern in vertical flow. In our treatment we abandon the term plug and adopt terms that are less ambiguous.

2 TYPICAL FLOW PATTERNS, HOLDUP, AND PRESSURE GRADIENT

As with vertical flow, the typical situation may most easily be described by referring to the experimental data of a simple system. For the liquid-liquid case, the data of Russell, Hodgson, and Govier (1959) and of Charles, Govier, and Hodgson (1961) on oil-water systems are suitable for our purposes. For the gas-liquid case, the air-water mixture flowing in a 1-inch horizontal pipe is suitable, and the data of Govier and Omer (1962) taken at a constant temperature and a constant average pressure are convenient to depict typical behavior.

2.1 Flow Pattern

It is instructive, before considering the typical data on gas-liquid and ordinary liquid-liquid systems, to consider the flow patterns developed for two immiscible but equal density fluids. Here, because there is no density contrast, symmetry about the pipe axis exists and we see the flow patterns for horizontal flow in their simplest form. This may be done with the data of Charles *et al.* (1961) on the flow in a 1-inch pipe of water with hydrocarbon oils of the three viscosities ranging from 6.29 to 65.0 centipoises whose densities were increased to that of water through addition of carbon tetrachloride. They observed essentially the same flow patterns under similar conditions for all oil viscosities.

The flow patterns observed are illustrated in Figure 10.1 and further described in Table 10.1. The individual sketches of Figure 10.1 are based upon actual photographs supplemented by measured holdup data. At the relatively low superficial water velocity of 0.1 ft/sec, the introduction of small quantities of oil resulted in a symmetrically dispersed bubble flow pattern with initially widely separated and nearly spherical oil bubbles. The size and spacing of the bubbles are not fully fixed by the fluid properties, flow rates, and pipe geometry but, within limits, depend upon the mode of introduction of the two fluids. In the work of Charles *et al.*, large bubbles of diameter nearly that of the tube were observed.

With increase in superficial oil velocity the oil bubbles come closer together and agglomeration occurs to form a pattern of relatively symmetrical elongated or cylindrical bubbles of diameter nearly equal to that of the tube and, ultimately, a continuous central core of oil flowing inside a film of water. The stability of this particular pattern is not great, and it tends to break down into a mixed phase pattern where the two phases are coarsely intermixed. Still further increase in the superficial velocity of the oil leads to a thinning and, finally, a destruction of any remaining water film and the entrainment in the continuous oil phase of droplets of water. This flow pattern is the converse of that first described. The droplets of water tend to be small if they are formed as a result of a gradual increase in the rate of flow of the oil phase.

Much the same phenomena are observed at the intermediate and higher water rates except that, at any particular superficial oil velocity, the higher water rates result in a smaller *in situ* volume fraction of oil. Additionally the higher water rates cause a distortion of the shape of the oil bubbles and of the oil-water interface.

The flow patterns observed by Russell *et al.* (1959) in a 1-inch pipe with water and an oil of specific gravity 0.834 and viscosity of 18.0 centipoises are depicted in Figure 10.2 and Table 10.2. Again the sketches are based upon actual photographs and measurements of holdup. The flow patterns observed are generally similar to those for the equidensity system but with the distortion and eccentricity expected with the density contrast.

Hasson, Mann, and Nir (1970) have studied the flow of mixtures of water and an organic liquid of specific gravity 1.02 and viscosity 1.0 centipoise in a 12.6 mm inside diameter horizontal pipe. Although their primary attention was to the annular flow pattern, they observed the other flow patterns as well and confirm the general description given above.

Figure 10.3 and Table 10.3 summarize the results of visual and photographic observation of the flow patterns observed by Govier and Omer (1962) for an air-water system. As for vertical flow, some interpolation and a modest amount of extrapolation of original data have been made. The relationship between these flow patterns for a pair of fluids of very different densities and those for the equidensity and near equidensity systems is of interest and the differences in the observed patterns are as would be expected.

At the lowest water rate four different flow patterns may be distinguished: stratified–smooth interface, stratified–ripply interface, wave, and annular-mist flow, each occurring at successively higher air rates. At the very low gas rates, complete stratification of the fluids occurs with the gas occupying the

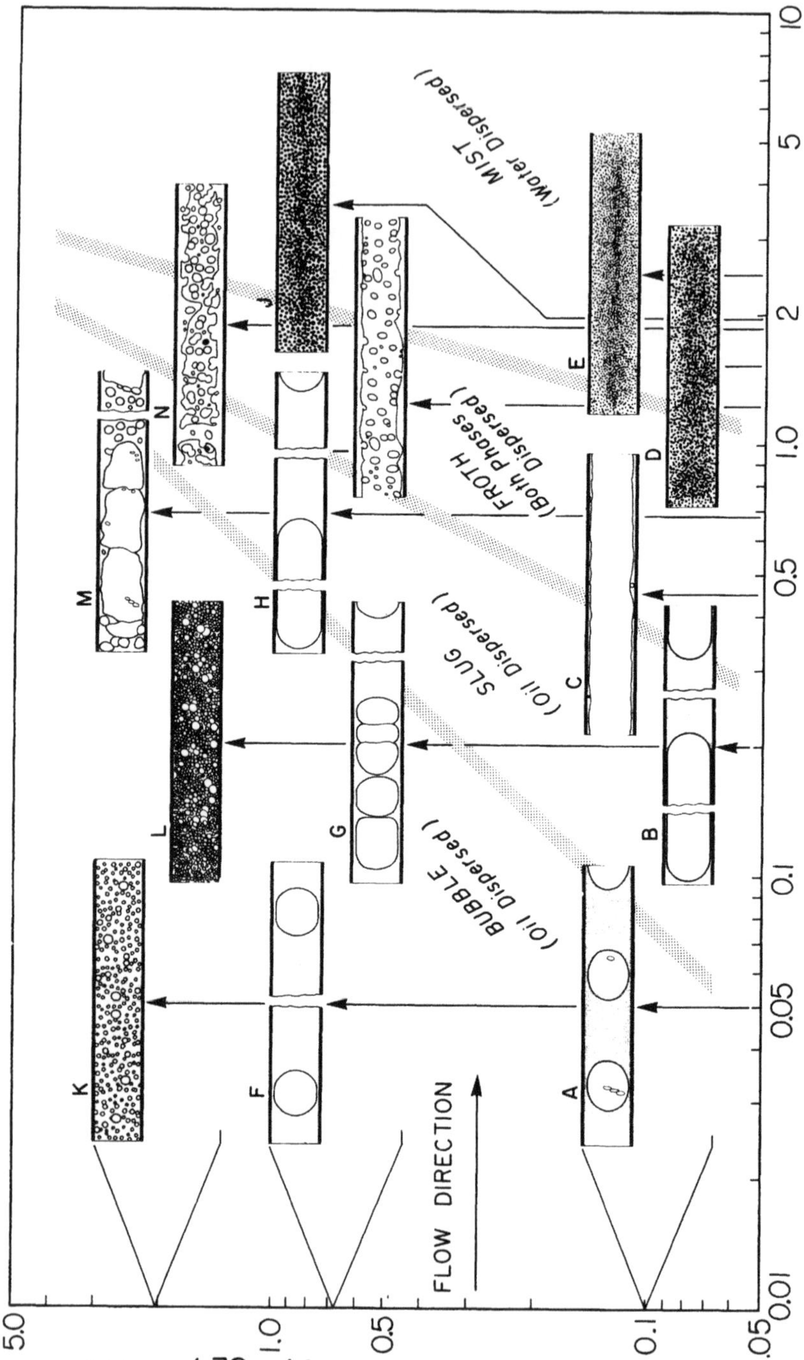

Figure 10.1. Flow patterns for 16.8-centipoise, 0.998-specific gravity oil and water mixtures in a 1.04-inch pipe based on observations of Charles, Govier, and Hodgson (1961) and calculations.

TABLE 10.1

Flow Pattern Details for 16.8-Centipoise Viscosity, 0.998-Specific Gravity Oil Flowing with Water in 1.04-Inch I.D. Horizontal Pipe (See Figure 10.1)

			Superficial Velocity					Approximate Only				
Sketch Identification	Flow Pattern	Continuous Phase	Water, V_{SL} ft/sec	Oil, V_{SO} ft/sec	Volume Fraction Oil, E_O	Holdup Ratio, H	Slip Velocity, S ft/sec	Diameter of Bubbles or Droplets, d_s, in.	Number of Bubbles or Droplets, n_b[a]	Average Film Thicknesss, δ_{av}, in.	Length of Liquid Slug, L_s, in.	Length of Taylor Bubble, L_b, in.
A	bubble	water	0.1	0.05	0.32	1.05	0.007	1.0	5.2	—	1.2	1.0
B	slug	water	0.1	0.20	0.61	1.28	0.072	1.0	2.2	—	2.1	3.3
C	oil in film[b]	both	0.1	0.45	0.73	1.64	0.24	—	—	0.07	—	—
D	mist	oil	0.1	1.36	0.91	1.39	0.42	0.03–0.06	17000	—	—	—
E	mist	oil	0.1	2.27	0.96	1.01	0.023	0.03–0.06	7800	—	—	—
F	bubble	water	0.68	0.05	0.07	1.01	0.007	1.0	1.15	—	9.3	1.1
G	bubble	water	0.68	0.20	0.22	1.04	0.035	1.0	0.85[c]	—	10.5	3.5[c]
H	slug	water	0.68	0.68	0.47	1.12	0.15	1.0	0.73	—	8.6	7.7
I	froth	neither	0.68	1.11	0.56	1.26	0.41	0.12	90	0.11	—	—
J	mist	oil	0.68	1.95	0.67	1.41	0.85	0.06–0.12	7500	—	—	—
K	bubble	water	2.04	0.05	0.02	1.01	0.02	0.07–0.19	500	—	—	—
L	bubble	water	2.04	0.20	0.09	1.03	0.067	0.07–0.19	1800	—	—	—
M	bubble[e]	water	2.04	0.68	0.23	1.11	0.29	1.0[d]	0.62	—	15.0	4.2[c,d]
N	froth	neither	2.04	1.82	0.42	1.25	0.87	0.15	300	0.14	—	—

[a] Per foot of 1-inch pipe.
[b] Corresponds to froth flow in fluids with density contrast.
[c] Bubble cluster.
[d] Large bubble only.
[e] Approaching transition to slug flow.

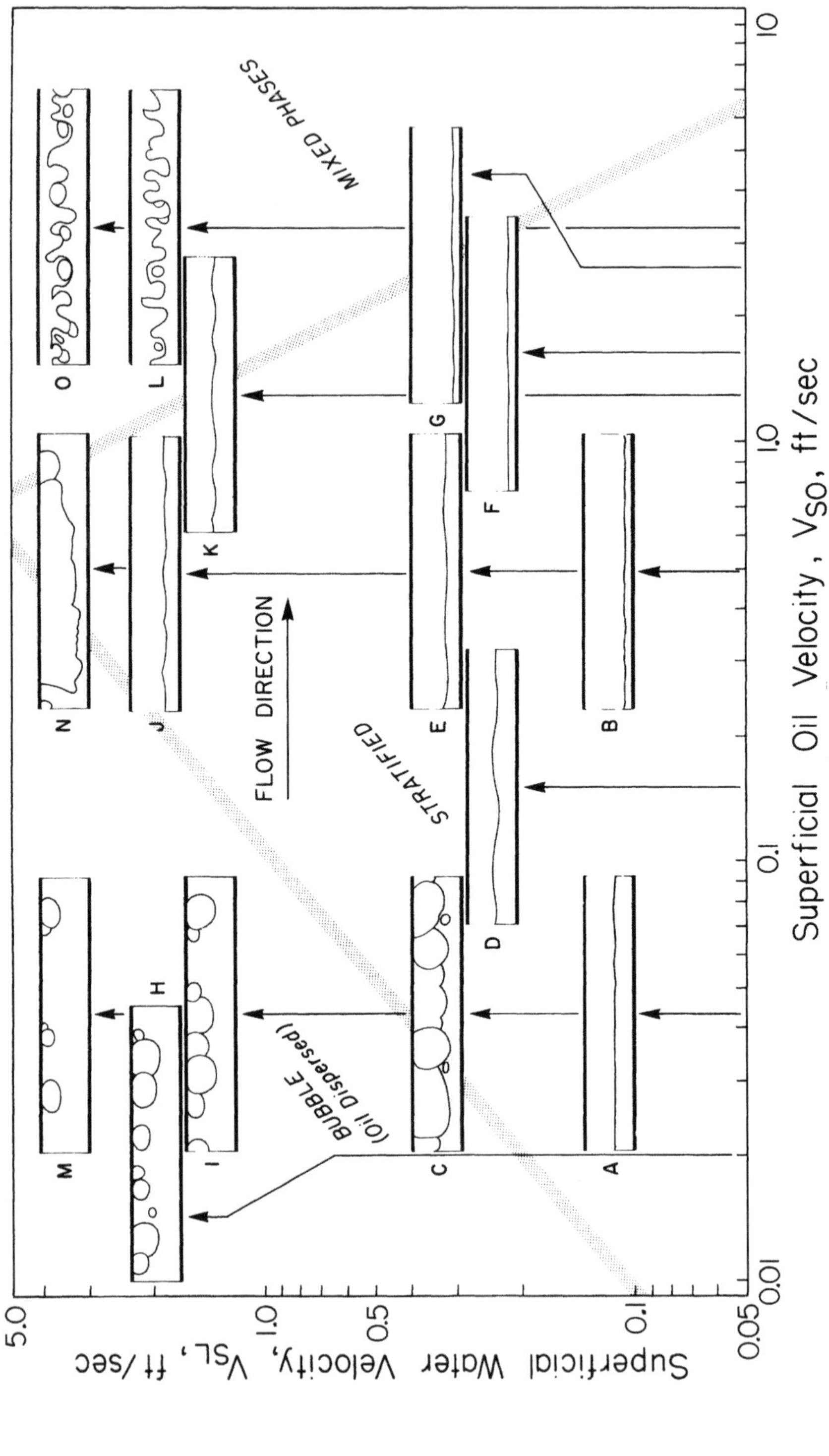

Figure 10.2. Flow patterns for 18.0-centipoise, 0.834-specific gravity oil and water mixtures in a 0.806-inch pipe based on observations of Russell, Hodgson, and Govier (1959) and calculations.

TABLE 10.2

Flow Pattern Details for 18.0-Centipoise Viscosity, 0.834-Specific Gravity Oil Flowing with Water in 0.806-Inch I.D. Horizontal Pipe (See Figure 10.2)

			Superficial Velocity					Approximate Only		
Sketch Identification	Flow Pattern	Continuous Phase	Water, V_{SL} ft/sec	Oil, V_{SO} ft/sec	Volume Fraction Oil, E_O	Holdup Ratio, H	Slip Velocity, S ft/sec	Diameter of Bubbles or Droplets, d_s, in.	Number of Bubbles or Droplets, n_b[a]	Average Height to Interface, h_{av}, in.
A	stratified	both	0.116	0.043	0.63	0.22	−0.24	—	—	0.31
B	stratified	both	0.116	0.49	0.90	0.45	−0.66	—	—	0.13
C	bubble	water	0.287	0.043	0.36	0.27	−0.32	0.5	34	0.32
D	stratified	both	0.287	0.15	0.58	0.38	−0.42	—	—	0.36
E	stratified	both	0.287	0.49	0.75	0.58	−0.48	—	—	0.24
F	stratified	both	0.287	1.63	0.884	0.75	−0.62	—	—	0.14
G	stratified	both	0.287	2.61	0.92	0.80	−0.71	—	—	0.10
H	bubble	water	1.79	0.02	0.085	0.12	−1.72	0.3–0.5	14	—
I	bubble	water	1.79	0.043	0.094	0.23	−1.52	0.3–0.5	12	—
J	stratified	both	1.79	0.48	0.232	0.89	−0.26	—	—	0.23
K	stratified	both	1.79	1.29	0.376	1.20	+0.57	—	—	0.32
L[b]	mixed phases	neither	1.79	3.26	0.55	1.50	+1.98	—	—	—
M	bubble	water	3.55	0.043	0.14	0.85	−0.54	0.3	6	—
N	slug	water	3.55	0.50	0.10	~1.30	+1.18	—	—	—
O[b]	mixed phases	neither	3.55	3.27	0.34	~1.80	+4.29	—	—	—

[a] Per foot of 0.806-inch pipe.
[b] Largely schematic.

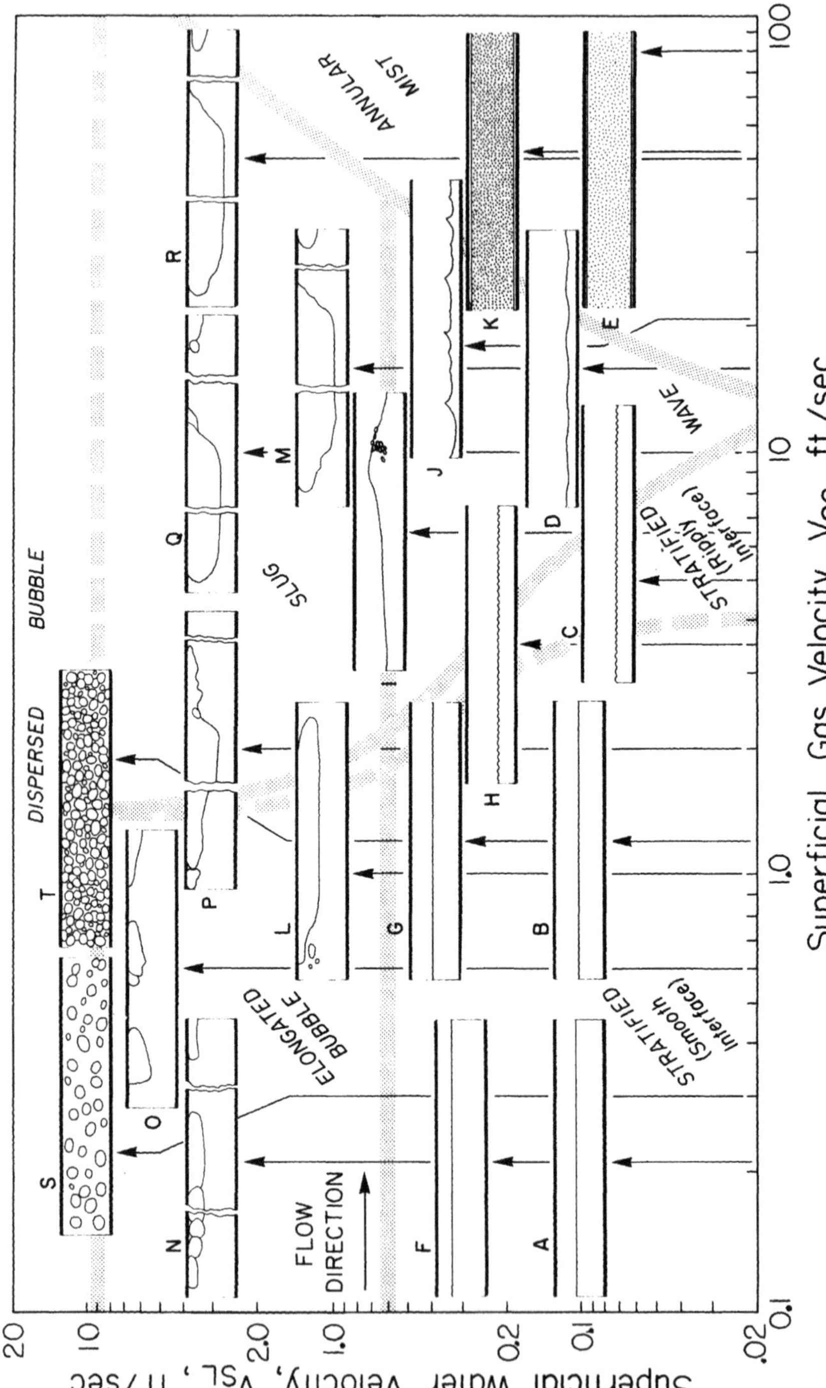

Figure 10.3. Flow patterns for air-water mixtures flowing in a 1.026-inch pipe based on observations of Govier and Omer (1962) and calculations.

TABLE 10.3

Flow Pattern Details for Air-Water Mixture Flowing in 1.026-Inch I.D. Horizontal Pipe (See Figure 10.3)

			Superficial Velocity					Approximate Only					
Sketch Identification	Flow Pattern	Continuous Phase	Water, V_{SL} ft/sec	Air, V_{SG} ft/sec	Volume Fraction Air, E_G	Holdup Ratio H	Slip Velocity S ft/sec	Diameter of Bubbles or Droplets, d_s, in.	Number of Bubbles or Droplets, n_b[a]	Average Height to Interface, h_{av}, in.	Average Film Thickness, δ_{av}, in.	Length of Liquid Slug, L_s, in.	Length of Taylor Bubble, L_b, in.
A	stratified	both	0.1	0.21	0.375	3.5	0.40	—	—	0.61	—	—	—
B	stratified	both	0.1	1.2	0.465	13.8	2.39	—	—	0.54	—	—	—
C	stratified	both	0.1	5.0	0.725	19.0	6.53	—	—	0.32	—	—	—
D	wave	both	0.1	15.9	0.90	18.0	16.7	—	—	0.16	—	—	—
E	annular mist	water	0.1	90	0.96	34.0	90.6	0.01+	60000	—	0.007	—	—
F	stratified	both	0.3	0.21	0.25	2.1	0.44	—	—	0.72	—	—	—
G	stratified	both	0.3	1.2	0.42	5.6	2.4	—	—	0.57	—	—	—
H	stratified	both	0.3	3.5	0.66	6.0	4.4	—	—	0.37	—	—	—
I	wave	both	0.4	6.5	0.75	5.5	7.1	—	—	0.28	—	—	—
J	wave	both	0.3	21	0.86	11.4	22.3	—	—	0.20	—	—	—
K	annular mist	air	0.3	52.2	0.92	16.0	53.4	0.02	40000	—	0.016	—	—
L	elongated bubble	water	1.1	1.0	0.34	1.8	1.32	—	—	0.61	—	4–5	5
M	slug	water	1.1	15.9	0.74	5.0	17.1	—	—	0.20	—	3–5	25–39
N	elongated bubble	water	3.02	0.21	0.061	1.07	0.22	—	—	0.75	—	8–9	8–9
O	elongated bubble	water	5.0	0.6	0.098	1.1	0.55	—	6	0.75	—	1	1
P	slug	water	3.02	2.0	0.32	1.4	1.78	—	—	0.64	—	3	7–8
Q	slug	water	3.02	10	0.60	2.2	9.1	—	—	0.30	—	3–5	14–23
R	slug	water	3.02	50	0.79	4.4	48.9	—	—	0.20	—	3–5	40–65
S	bubble	water	10	0.3	0.029	1.0	0	0.2	70	—	—	—	—
T	bubble	water	10	1.2	0.11	1.0	0	0.2	300	—	—	—	—

[a] Per foot of 1-inch pipe.

upper portion of the tube over a smooth gas-liquid interface. This is the stratified–smooth interface flow pattern. Increasing the gas rate produces ripples on the gas-liquid interface and the stratified–ripply interface flow pattern. Further increase in the gas rate produces waves at the interface and the wave flow pattern. As the gas flow rate is increased more, the wave action persists until waves can no longer be supported in the thinning liquid layer. Higher gas velocities result in the spreading of the liquid over the periphery of the pipe and the formation of an annular film with some liquid dispersed in the gas core. The film thickness is greater in the lower portion of the tube, as is to be expected. The amount of liquid dispersed in the gas core increases with the gas velocity. The flow is called annular when almost no liquid is dispersed in the gas core, and annular mist when liquid exists in both the film and gas core. At extremely high gas velocities, the liquid film breaks up, causing part of the tube surface to come in direct contact with the gas. This is the mist flow region.

At intermediate water rates the stratified and wave flow patterns are replaced by the elongated bubble flow pattern and the slug flow pattern. At the lowest gas velocities discrete elongated gas bubbles ride in the upper portion of the pipe. At modest gas velocities these agglomerate to form the distorted equivalent of the smooth cylindrical bubbles observed in the equidensity system. Both of these patterns are referred to as the elongated bubble flow pattern. With further increase in gas velocity the rounded nose elongated bubbles become even longer, occupy a greater portion of the pipe cross-sectional area, and suffer further distortion. They are separated by slugs of liquid which may contain entrained bubbles of gas. This is the slug flow pattern; it really is only a transition from the elongated bubble pattern to annular-mist flow. In both the elongated bubble and the slug patterns the gas travels at a noticeably higher rate than the liquid.

At the highest liquid rates and at low gas rates the gas bubbles are distributed more or less uniformly throughout the liquid. The concentration profile is somewhat asymmetric, peaking near the top of the tube. This is the dispersed-bubble or bubbly flow pattern.

2.2 Holdup

The holdup phenomenon in horizontal flow is illustrated by actual data for the three systems whose flow patterns have been discussed. Figures 10.4 and 10.5 give the data for the equal and the unequal density liquid-liquid systems. In the first case the holdup ratio is observed to pass through a maximum and

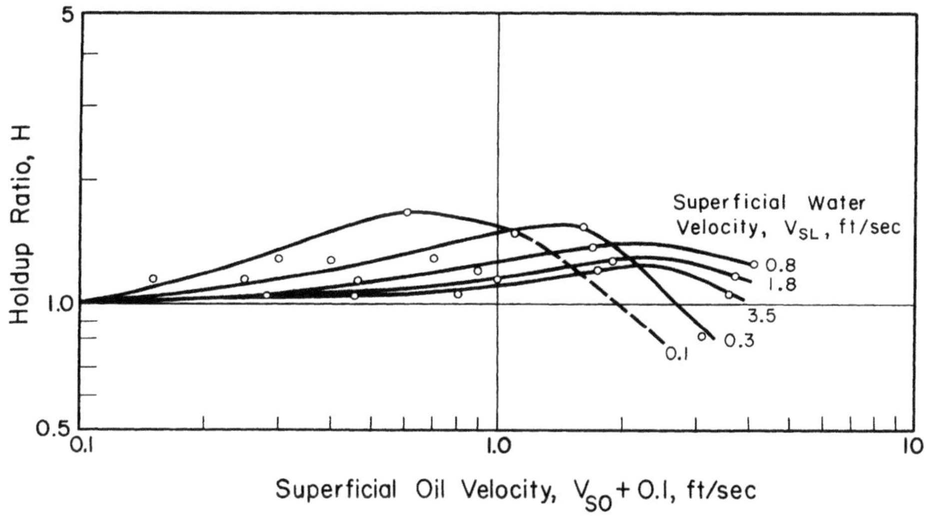

Figure 10.4. Holdup data for 16.8-centipoise 0.998-specific gravity oil and water mixtures in a 1.04-inch pipe. Replotted from Charles, Govier and Hodgson (1961).

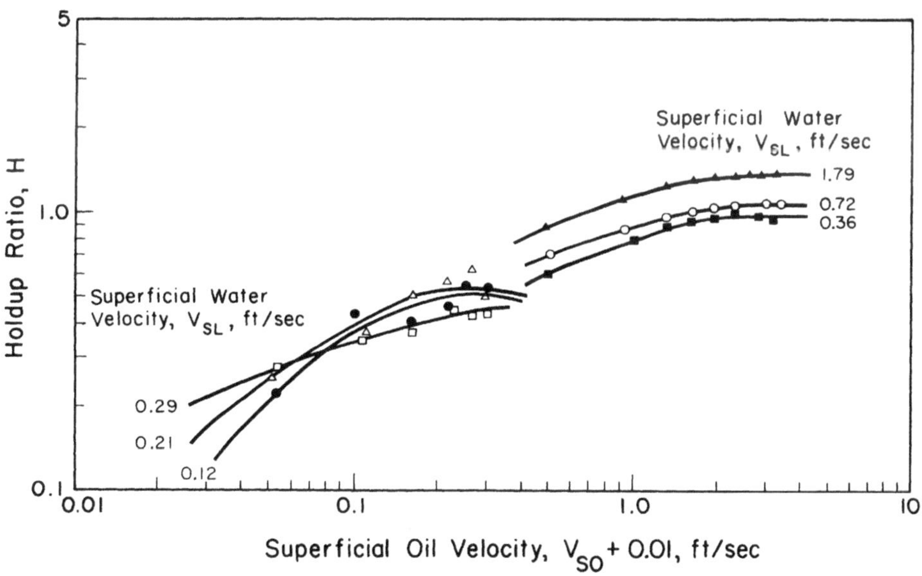

Figure 10.5. Holdup data for 18.0-centipoise, 0.834-specific gravity oil and water mixtures in a 0.806-inch pipe. Replotted from Russell, Hodgson and Govier (1959).

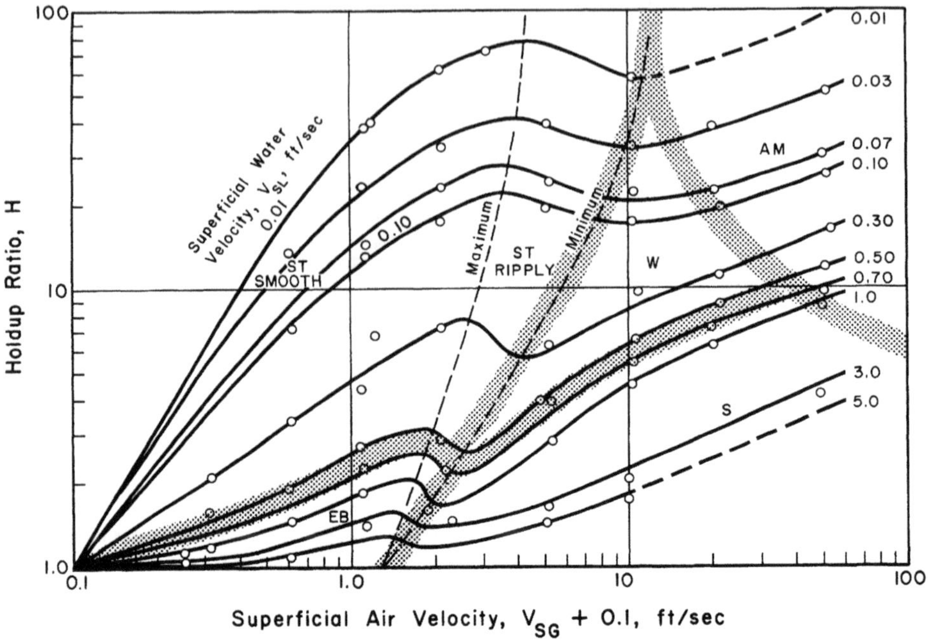

Figure 10.6. Holdup data for air-water mixtures in a 1.026-inch pipe. Flow regions: AM = annular mist, EB = elongated bubble, S = slug, ST = stratified, W = wave. Replotted from Govier and Omer (1962).

then to decrease to values below unity as the oil rate is increased. As discussed in detail in the original works and as shown in Figures 10.1 and 10.2, the reason for this is that the continuous phase changes from water to oil.

The data of Govier and Omer for the air-water system are given in Figure 10.6. Here much higher holdup ratios are observed and the similarity with the observations for air-water flow in vertical tubes (see Figure 8.3) is apparent. Each of the curves shows a clear maximum and minimum. (The flow pattern transitions shown in Figure 10.6 are discussed later.)

2.3 Pressure Gradient

Typical pressure gradient data for the horizontal flow of liquid-liquid and gas-liquid systems are presented in Figures 10.7, 10.8, and 10.9. Figure 10.7 gives data of Charles *et al.* for the equidensity oil-water system. The pressure gradient is observed to increase smoothly from the single-phase water value. Data

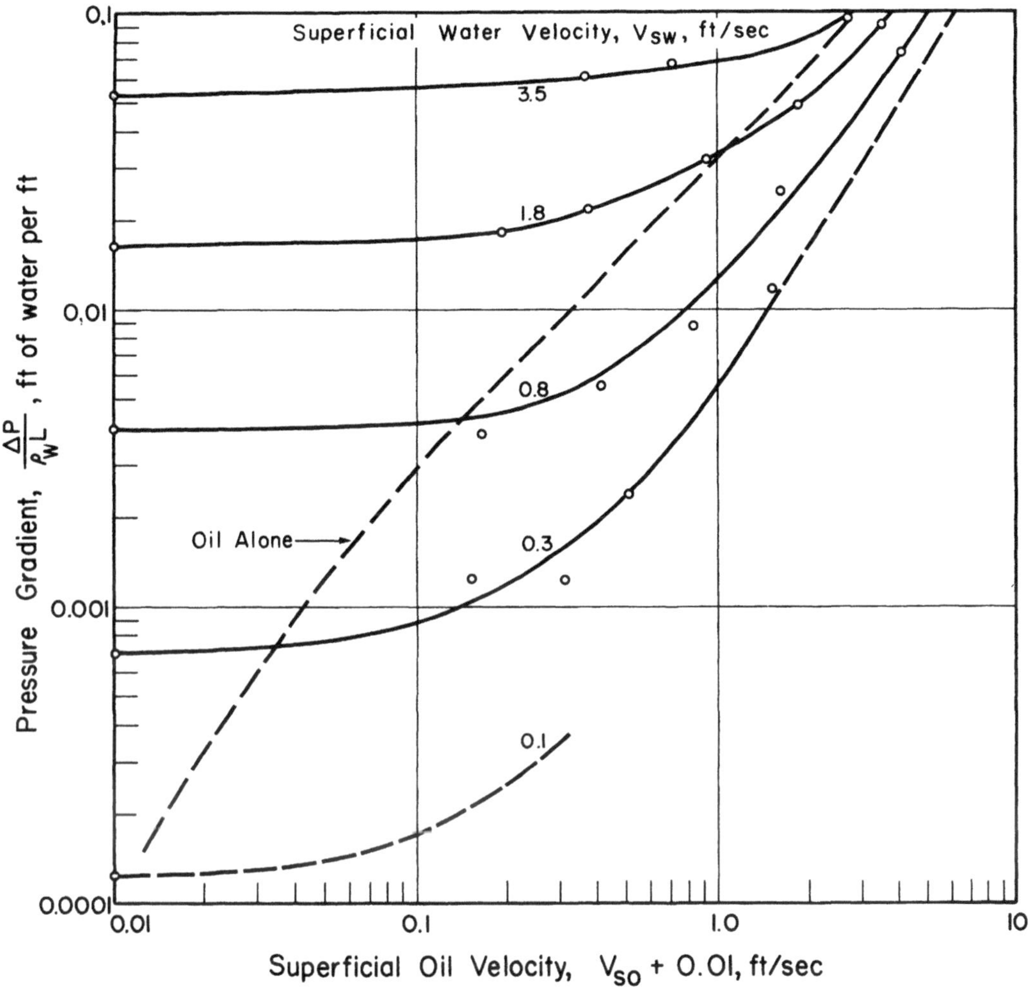

Figure 10.7. Pressure gradient data for 16.8-centipoise, 0.998-specific gravity oil and water mixtures in a 1.04-inch pipe. Replotted from Charles, Govier, and Hodgson (1961).

of Russell *et al.* on their oil-water system are given in Figure 10.8. The behavior is generally similar to that shown in Figure 10.7 but the curves are distorted, probably reflecting the greater complexity and asymmetry of the flow patterns. Figure 10.9 shows Govier and Omer's air-water data, indicating behavior similar to that shown in Figure 10.7. (The flow pattern transition and the locus lines shown in Figure 10.7 are discussed later.) In general, there are no significant inflections in the pressure gradient relationship due to changes in flow pattern reflecting the more gradual changes that occur in the horizontal system as compared with the vertical.

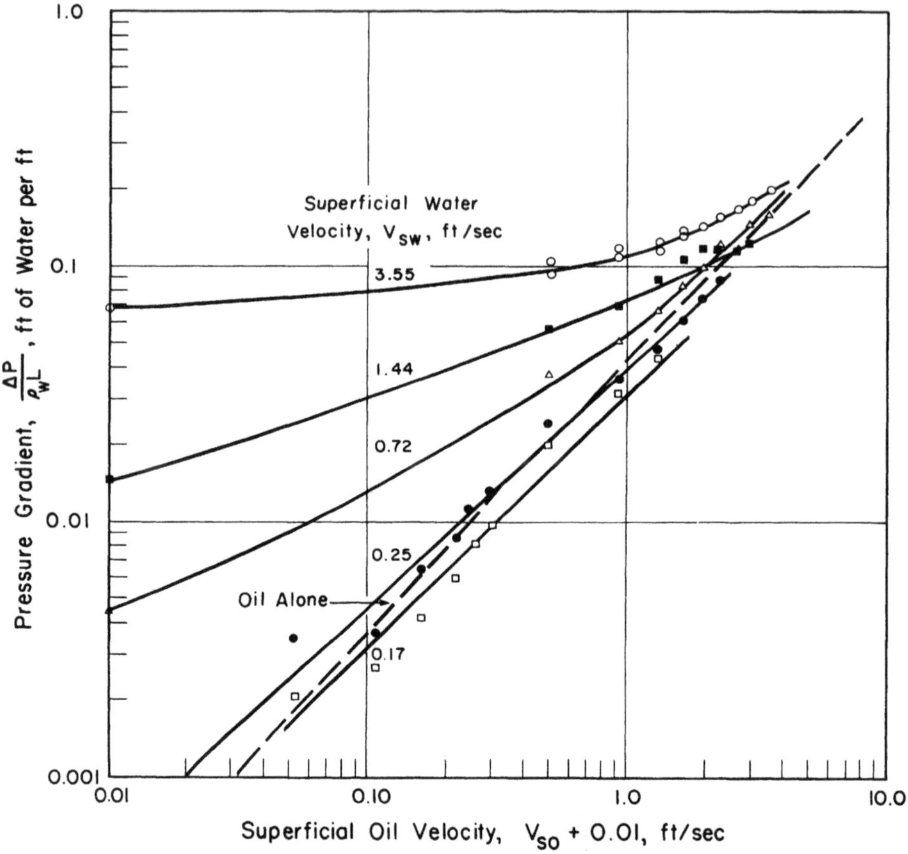

Figure 10.8. Pressure gradient data for 18.0-centipoise, 0.834-specific gravity oil and water mixtures in a 0.806-inch pipe. Replotted from Russell, Hodgson and Govier (1959).

3 EMPIRICAL OVERALL CORRELATIONS

No completely satisfactory general correlations for flow pattern or for holdup or pressure drop have yet been developed and, because of the complexity of the problem, it is doubtful if this will ever be possible. Those restricted correlations that are available apply only to gas-liquid mixtures.

3.1 Flow Pattern

Several empirical flow pattern correlations have been developed, but there is little agreement among them. Most of the correlations are expressed as flow pattern maps. As with vertical flow a number of different coordinates have been proposed for the maps.

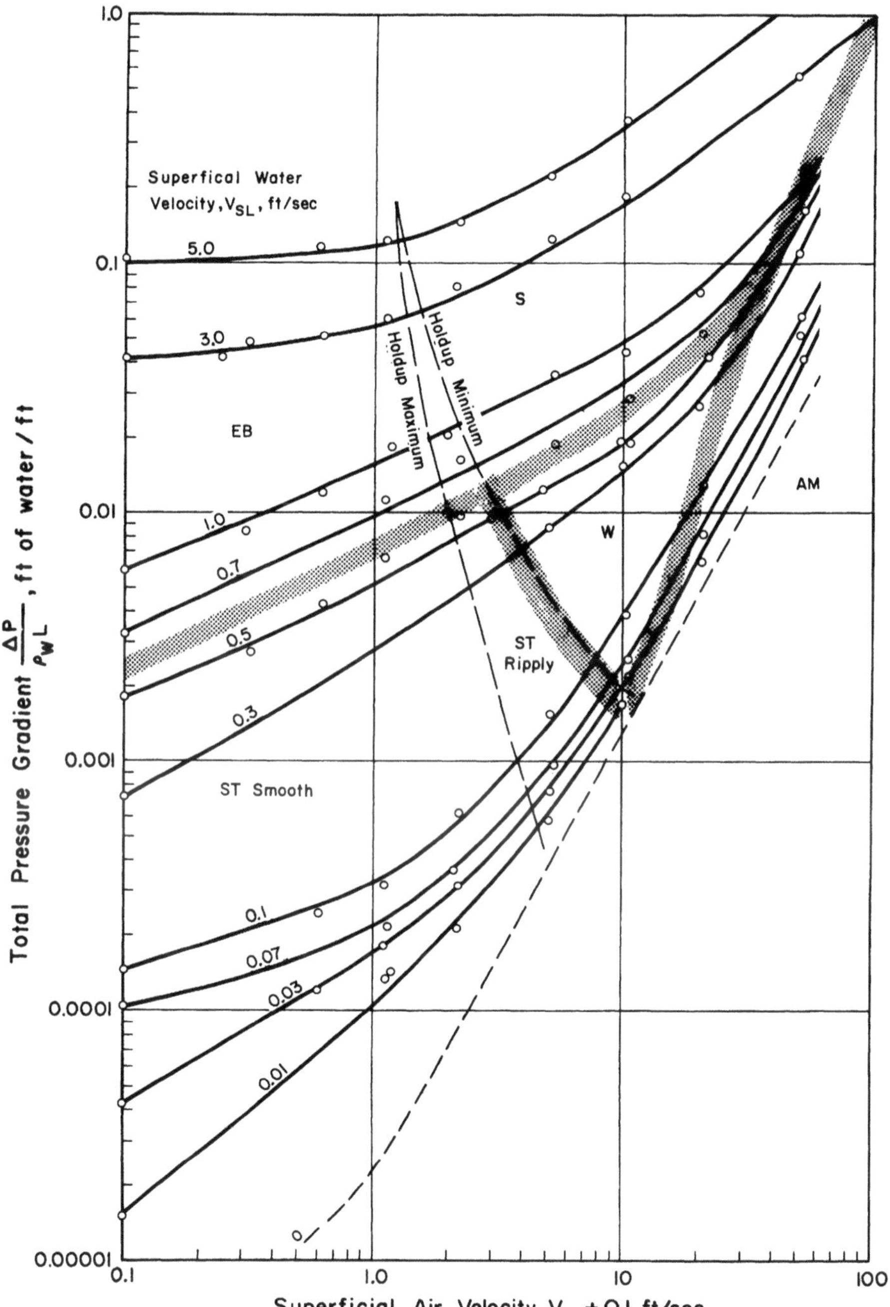

Figure 10.9. Pressure gradient data for air-water mixtures in a 1.026-inch pipe. Flow regions as defined in Figure 10.6. Replotted from Govier and Omer (1962).

Bergelin and Gazley (1949) introduced one of the first flow pattern maps for horizontal flow based upon limited data for the air-water system in a 1-inch pipe. The Bergelin and Gazley map was originally expressed in coordinates of mass flow rates for the liquid and gas, M_L and M_G. Their transition lines, transformed to coordinates of superficial liquid and gas velocities, V_{SL} and V_{SG}, are shown in Figure 10.10. Johnson and Abou-Sabe (1952) proposed a flow pattern map, similar to that of Bergelin and Gazley, based upon their air-water data.

In 1954 Alves studied the horizontal flow of air-water and air-oil (SAE 10) mixtures in a 1-inch pipe. He observed the full range of flow patterns and found that a single flow pattern map with coordinates V_{SL} and V_{SG} represented both systems. His flow pattern transition lines are also reproduced in Figure 10.10.

White and Huntington (1955) present a flow pattern map on coordinates of G_{SG} and G_{SL} based on their data in 1-, $1\frac{1}{2}$-, and 2-inch pipes with oil and natural gas, oil and air, and water and air systems. The correlation is limited to low liquid velocities.

Govier and Omer (1962) also developed a flow pattern map for their air-water data referred to earlier. Their transition lines transformed to V_{SG} and V_{SL} coordinates are also shown in Figure 10.10. The lack of agreement among the three correlations shown in Figure 10.10 reflects the problems of nomenclature and unequivocal identification.

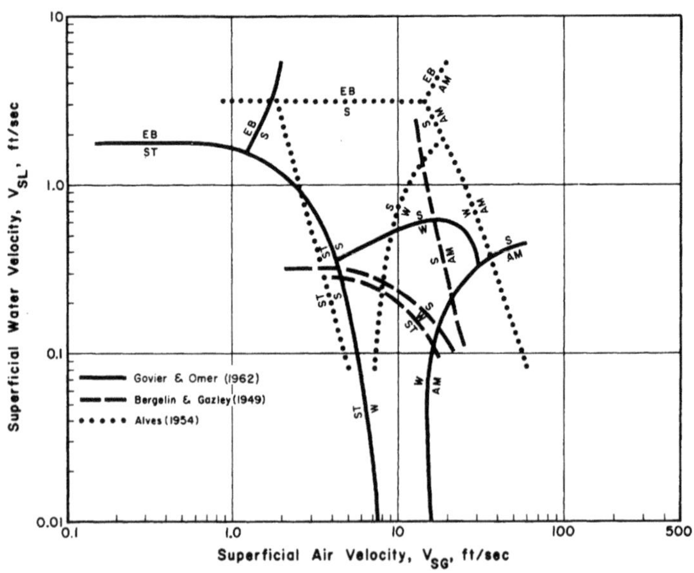

Figure 10.10. Comparison of flow pattern boundaries. Flow regions as defined in Figure10.6.

Baker (1954) proposed a flow pattern map based upon coordinates of G/λ and $L\psi\lambda/G$ which are proportional, respectively, to the mass velocity of the gas and the ratio of the mass velocity of the liquid to that of the gas. The quantities λ and ψ are defined as

$$\lambda = [(\rho_G/0.075)(\rho_L/62.3)]^{1/2} \tag{10.1}$$

$$\psi = \frac{73}{\sigma}\left[\left(\frac{\mu_L}{1}\right)\left(\frac{62.3}{\rho_L}\right)^2\right]^{1/3} \tag{10.2}$$

and are wholly empirical devices designed to bring the transition lines for systems other than air (of density 0.075 lb_m/ft^3) and water (of density 62.3 lb_m/ft^3, surface tension 73 dynes/cm, and viscosity 1 cp) into coincidence with the air-water system. Baker's coordinates reduce to G and L/G for the air-water system. Scott (1963) has modified Baker's correlation by making use of the more recent data of Hoogendoorn (1959) and of Govier and Omer (1962). Scott's modification of Baker's plot is reproduced as Figure 10.11. The Baker correlation is widely used in the petroleum industry.

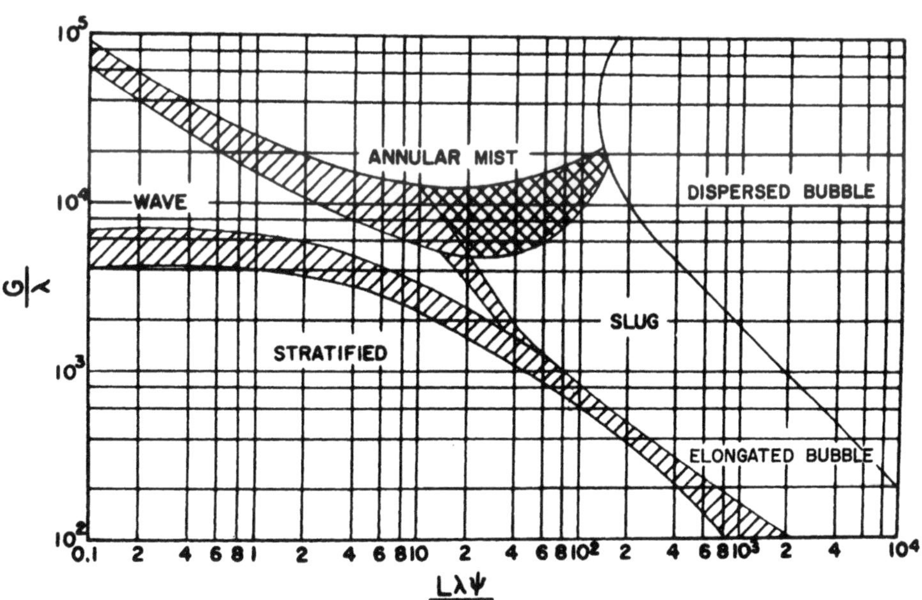

Figure 10.11. Baker's flow pattern map as modified by Scott. (From Scott, D. S., in *Advances in Chemical Engineering,* Vol. 4, p. 200, Academic Press, New York, 1963.)

Hoogendoorn (1959) and also Hoogendoorn and Buitelaar (1961) have used the coordinate system first proposed by Kosterin (1949) to describe the various flow regions in several different fluid systems. From their extensive data they found that the influence of diameter and of liquid and gas properties on the position of the transition lines was modest for air-oil systems except for high liquid viscosities ($\nu > 50$ cS). The air-water system showed nearly the same transitions except that the wave region occupied a larger area of the map. Hoogendoorn gives further discussion of the small effects of diameter and fluid properties in his paper. Figure 10.12 reproduces Hoogendoorn's recommended general flow pattern map. The Kosterin or Hoogendoorn coordinates result in crowding the important wave and annular-mist regions into a very small area on the map.

Bearing in mind the obvious importance of the relative volumes of the phases on the flow pattern and the importance of ease of visualization, it would appear, as discussed in Chapter 8, that the use of the coordinates V_{SG} and V_{SL} would be best for a simple flow pattern map.

The correlations of Baker and of Hoogendoorn, both transformed to V_{SG} and V_{SL} coordinates, and assuming the properties of atmospheric pressure

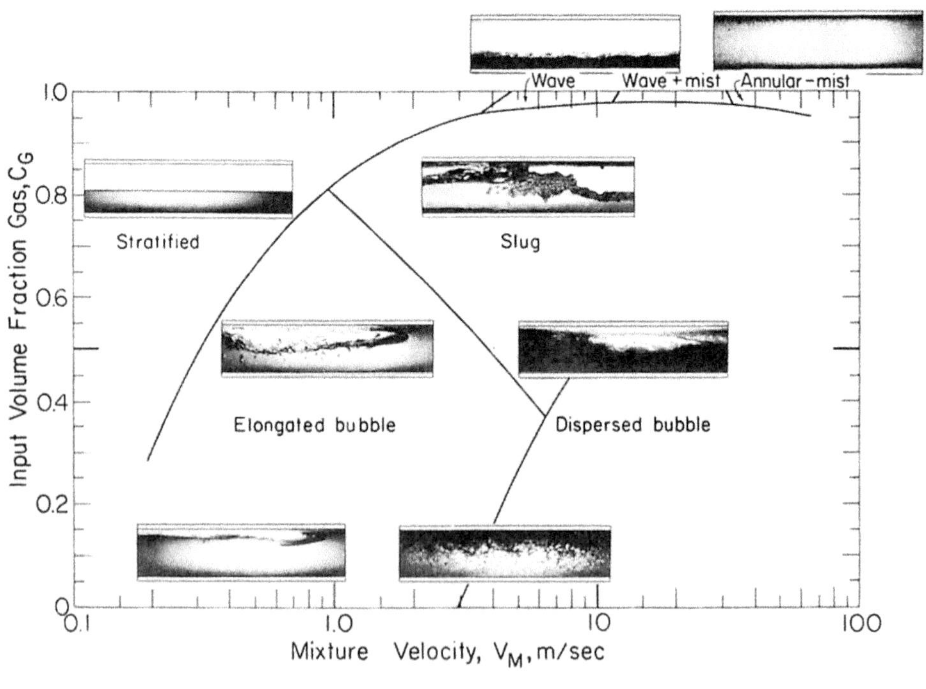

Figure 10.12. Hoogendoorn flow pattern map. (Retouched after Hoogendoorn, C. J., *Chem. Eng. Sci.*, **9**, 205, 1959.)

air and water in the case of the Baker correlation, are compared in Figure 10.13. Considering the amount of data these correlations are based upon, the extent of the disagreement disclosed is surprising and further confirms the difficulties inherent in the development of a sound correlation.

A careful reworking of the data of Govier and Omer, taking account both of the clearly defined inflection points in the holdup curves (Figure 10.6) and the visual observations that had been made at every data point, results in the revised Govier–Omer transition lines also shown in Figure 10.13. These lines, with the exception of those for slug to annular-mist flow, and for elongated to dispersed bubble flow that were not observed, are considered accurate for the air-water system in a 1-inch tube. This statement is based upon the direct correlation between the major transitions and the loci of the maximum and minimum points in the holdup curves.

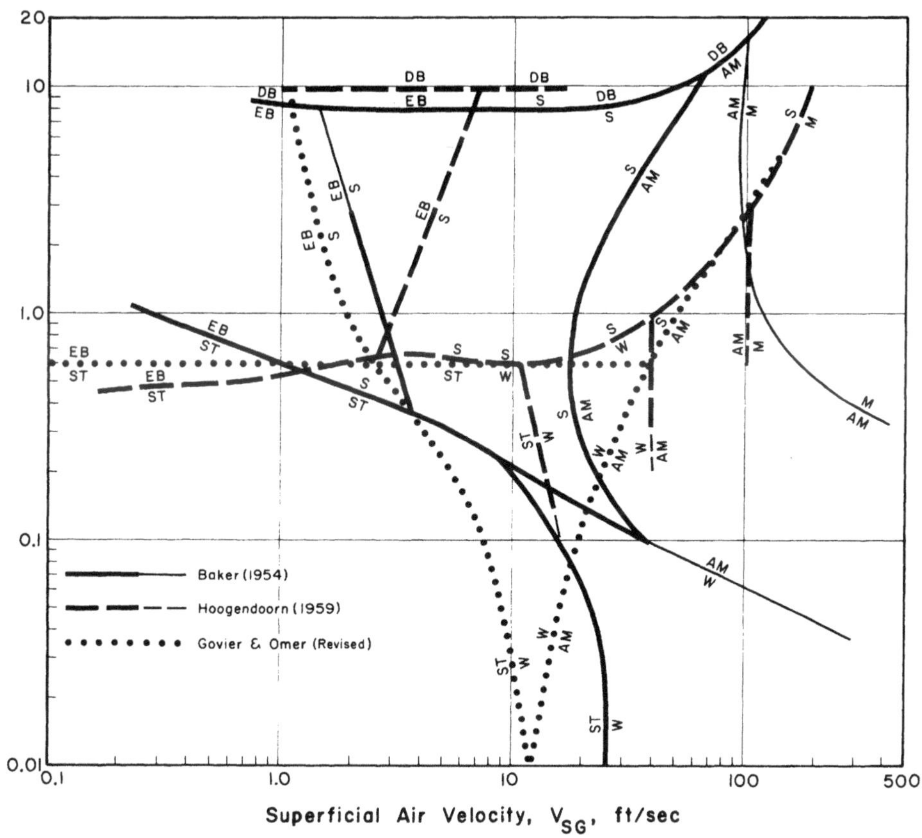

Figure 10.13. Comparison of flow pattern boundaries. Flow regions as defined in Figure 10.6.

Referring to Figure 10.6 the physical explanation is as follows: At V_{SG} values below those corresponding with the maximum locus, the steady increase in holdup, with V_{SG} at any constant V_{SL}, corresponds with smooth stratified flow at V_{SL} values below 0.6 ft/sec and with smooth-surfaced elongated bubbles at V_{SL} values above 0.6 ft/sec and to approximately $V_{SL} = 9$ ft/sec. From V_{SG} values corresponding to the maximum locus to values corresponding to the minimum locus, the holdup decreases. This reflects the lowering of the average slip velocity because of increased interfacial drag caused by the development first of a ripply and then of a pebbly interface. Above values of V_{SG} corresponding to the holdup minimum, the development of waves at V_{SL} values below 0.6 ft/sec and of slugs at V_{SL} values above 0.6 ft/sec causes an increase in holdup. The V_{SG} range over which waves are observed is very small (10 to 10.3 ft/sec) at a V_{SL} value of 0.01, but widens greatly to about 2.5 to 40 ft/sec as V_{SL} is increased to 0.6 ft/sec. This is as we would expect, having regard to the quantity of liquid present in the pipe. At V_{SL} values above 0.6, waves are replaced by slugs; and at $V_{SL} = 3.0$, the range of V_{SG} over which slugs are encountered extends to about 100 ft/sec. The revised Govier–Omer lines agree

a. for the elongated bubble to slug transition, with the Baker correlation;
b. for the elongated bubble to stratified transition, with the Hoogendoorn correlation;
c. for the slug to wave transition, with the Hoogendoorn correlation;
d. for the stratified to wave transition, roughly with the Baker correlation;
e. for the slug to annular mist (based on an extrapolation of the observed wave to annular mist) with the Hoogendoorn correlation from slug to mist.

Bearing in mind the observations of Alves and of Hoogendoorn concerning the modest effect of fluid properties and diameter, the revised Govier–Omer correlation is recommended for practical use. It is reproduced in working plot form in Figure 10.14 where the transitions are shown as bands of width equal to ± 10 percent of the flow rate. This figure incorporates the elongated bubble to dispersed bubble and slug to dispersed bubble transition at V_{SL} of approximately 9.0 ft/sec, which is consistent with both the Baker and the Hoogendoorn correlations. If we wish to take account of variations of fluid properties, we may use the Baker (or modified Baker) correlation directly in the regions where it checks with the revised Govier–Omer correlation. We may incorporate Baker's λ and ψ factors with the use of Figure 10.14, or we may incorporate the X and Y factors defined in Chapter 8 with the use of Figure 10.14. Neither of the latter procedures is tested.

Eaton, Andrews, Knowles, Silberberg, and Brown (1967) conducted an investigation dealing with the flow of natural gas–water, natural gas–crude oil,

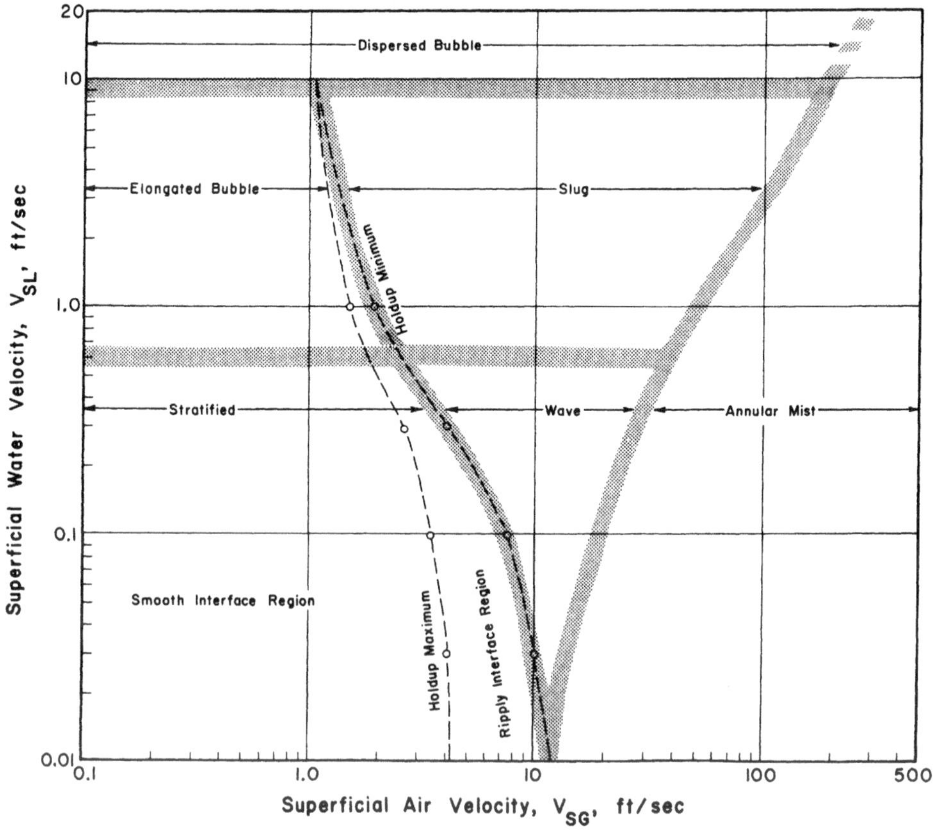

Figure 10.14. Revised Govier-Omer flow pattern map.

and natural gas–distilate systems in 2-, 4-, and 17-inch pipelines. They found all the existing flow pattern correlations to be inadequate and, instead of seeking a compromise of the previous correlations based solely on the fluid velocities, they developed a totally new correlation incorporating the fluid properties and the *in situ* volume fraction of the phases. They define a two-phase Reynolds number and a two-phase Weber number as follows:

$$\mathrm{Re}_{EA} = 1488\left(\frac{M_M E_L^2}{D\mu_M}\right) \tag{10.3}$$

$$\mathrm{We}_{EA} = 453D\left(\frac{\bar{\rho}_L V_L^2 \bar{E}_L^{1/2}}{\sigma} + \frac{\bar{\rho}_G S^2(1-\bar{E}_L)^{1/2}}{\sigma}\right) \tag{10.4}$$

where D = pipe diameter, ft,
$S = V_G - V_L$ = average slip velocity, ft/sec,
$\mu_M = \mu_L E_L + \mu_G(1 - E_L)$ = mixture viscosity, cp,
$M_M = M_L + M_G$ = total mass flow rate, lb_m/sec,
$\bar{\rho}$ = integrated average density, lb_m/ft^3,
σ = surface tension, dynes/cm,
V_G = actual average gas velocity, ft/sec,
V_L = actual average liquid velocity, ft/sec,
$\bar{E}_L$ = *in situ* average volume fraction liquid.

The constants in the defining equations for Re_{EA} and We_{EA} become unity if a consistent set of units is used. It is important to note that both Re_{EA} and We_{EA} incorporate the *in situ* volume fraction of liquid, a quantity not ordinarily known. Eaton *et al.* present a separate empirical correlation for E_L. This is discussed later. Their flow pattern map is reproduced in Figure 10.15. The region between the slug and froth flow labeled as "transition region" in this figure has been called "slug three" by Eaton *et al.* This correlation is supported by data for several systems in field-size pipelines.

Al-Sheikh, Saunders, and Brodkey (1970) have taken a different approach to the development of a correlation for the prediction of flow pattern. Their correlation is based on the AGA-API Data Bank referred to by Dukler, Wicks, and Cleveland (1964) covering a considerably greater range of data than has been used by others. Rather than attempt to define lines of separation between

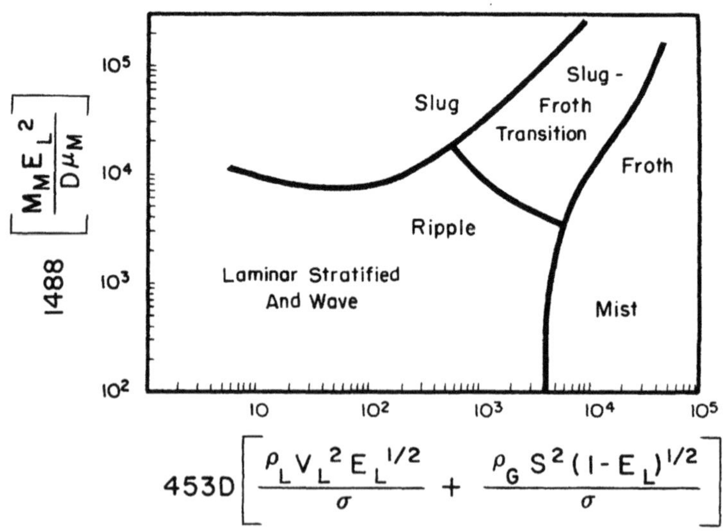

Figure 10.15. Eaton *et al.* flow pattern correlation. (From Eaton, B. A., D. E. Andrews, C. R. Knowles, I. H. Silberberg, and K. E. Brown, *J. Petrol. Technol.*, **19**, 815, 1967.)

flow patterns, Al-Sheikh *et al.* have chosen to consider entire areas enclosing all data corresponding to particular flow patterns. The correlation takes the form of a set of twelve figures on ten different coordinate systems. The original paper should be consulted for details.

Little information is available on the flow pattern transitions for liquid-liquid systems. The correlations developed by Charles *et al.* and Russell *et al.* for oil-water systems are reproduced in Figures 10.16 and 10.17. These correlations are not general but serve for orientation purposes.

Hasson, Mann, and Nir (1970) give a detailed qualitative description of the transitions of annular to stratified, or elongated bubble, flow by film breakup

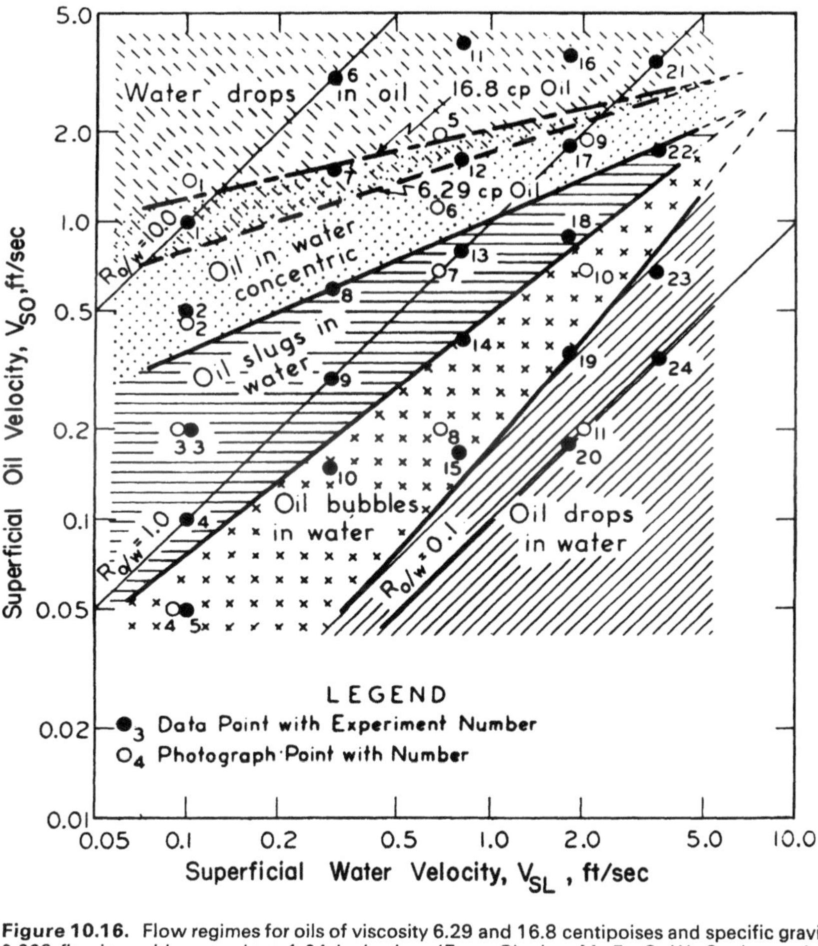

Figure 10.16. Flow regimes for oils of viscosity 6.29 and 16.8 centipoises and specific gravity of 0.998 flowing with water in a 1.04-inch pipe. (From Charles, M. E., G. W. Govier, and G. W. Hodgson, *Can. J. Chem. Eng.*, **39**, 27, 1961.)

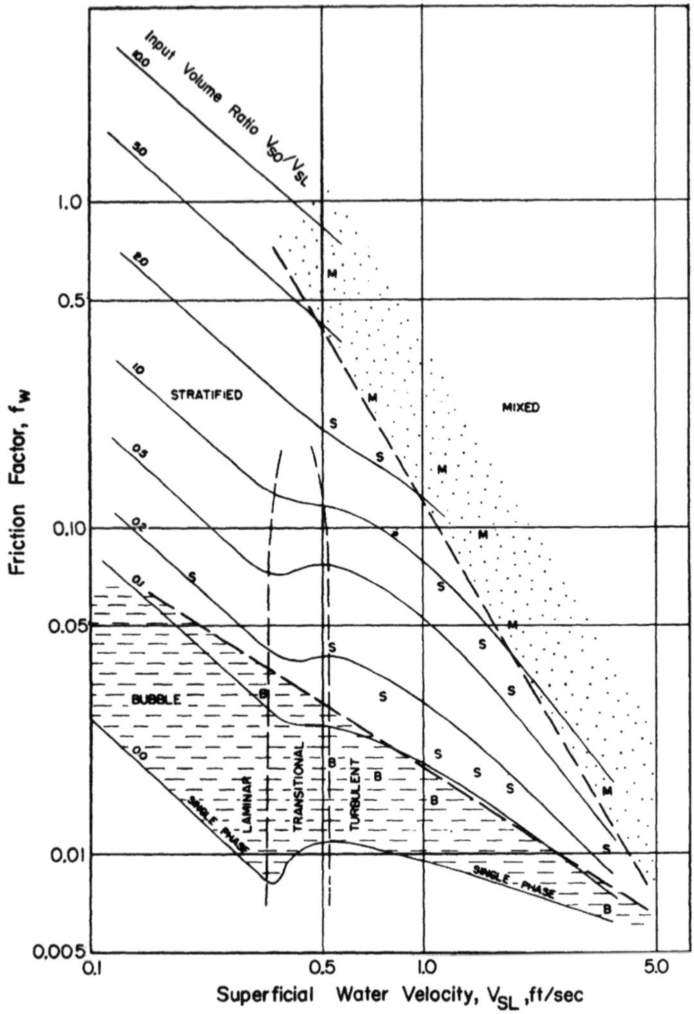

Figure 10.17. Friction factors and flow regimes for 18.0-centipoise, 0.834-specific gravity oil flowing with water in a 0.806-inch pipe. (From Russell, T. W. F., G. W. Hodgson, and G. W. Govier, *Can. J. Chem. Eng.*, **37**, 9, 1959.)

or by core breakup. Their observations indicate the importance of the condition of the pipe wall and the wetting properties of the fluids on the stability of annular flow. Hasson and Nir (1970) present a theoretical analysis of the trajectory of the core liquid for the annular flow of immiscible nonequidensity liquids. They confirm predicted trajectories for nonwavy interfaces with high-speed movie photographs.

3.2 Holdup

The various methods of expressing the holdup phenomenon were discussed in Chapters 7 and 8. In Chapter 8 we presented empirical methods for estimating the holdup for vertical flow. Here we will give a parallel treatment for horizontal flow.

Lockhart and Martinelli (1949) have presented an empirical correlation for the average *in situ* volume fraction of the liquid phase, E_L, as a function of a parameter X, which is defined by

$$X = \sqrt{\frac{(\Delta P/L)_{SL}}{(\Delta P/L)_{SG}}} \tag{10.5}$$

where $\left(\frac{\Delta P}{L}\right)_{SL}$ is the pressure gradient that would occur if the liquid were flowing alone at a velocity equal to its superficial velocity,

$\left(\frac{\Delta P}{L}\right)_{SG}$ is the pressure gradient that would occur if the gas were flowing alone at its superficial velocity.

From the definition of X and Equation 4.24 it may be noted that

$$X = \frac{V_{SL}}{V_{SG}}\sqrt{\frac{\rho_L f_{SL}}{\rho_G f_{SG}}} \tag{10.6}$$

where f_{SL} and f_{SG} are the appropriate single-phase friction factors. When turbulent flow would occur at both V_{SL} and V_{SG} and a Blasius-type relationship is assumed, Equation 10.6 reduces to

$$X = \left(\frac{V_{SL}}{V_{SG}}\right)^{7/8}\left(\frac{\rho_L}{\rho_G}\right)^{3/8}\left(\frac{\mu_L}{\mu_G}\right)^{1/8} \tag{10.7}$$

and X may be considered to be a property-modified velocity ratio. The Lockhart and Martinelli correlation, based upon data obtained in 1-inch pipe, is presented in Figure 10.18. Baker (1954) has compared his data for larger pipe sizes with Lockhart and Martinelli's correlation and found his holdup values to be higher. Baker's line is also shown in Figure 10.18.

Hoogendoorn and Buitelaar (1961) have developed an alternative empirical holdup correlation that incorporates V_{SG} as a variable in addition to the V_{SL}/V_{SG} ratio which is implicit in the Lockhart and Martinelli correlation. Hoogendoorn and Buitelaar's correlation for the air-oil system is given in Figure 10.19. It may also be expressed as

$$\frac{E_G}{1 - E_G} = 0.60\left\{V_{SG}\left[1 - \frac{E_G}{1 - E_G}\left(\frac{V_{SL}}{V_{SG}}\right)\right]\right\}^{0.85} \tag{10.8}$$

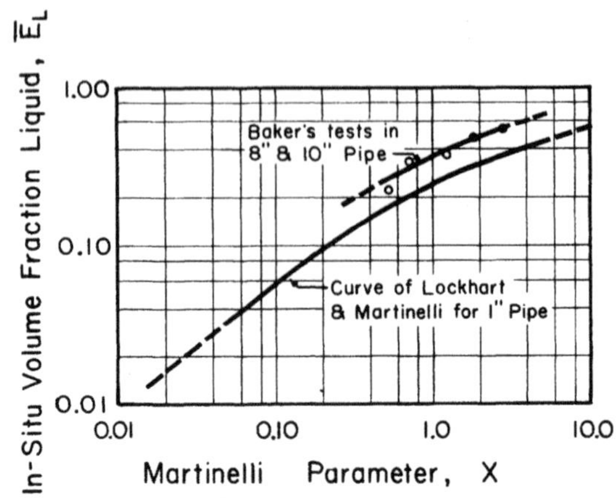

Figure 10.18. Lockhart and Martinelli, and Baker holdup correlations. (From Baker, O., *Oil and Gas J.*, **53** (12), 185, 1954.)

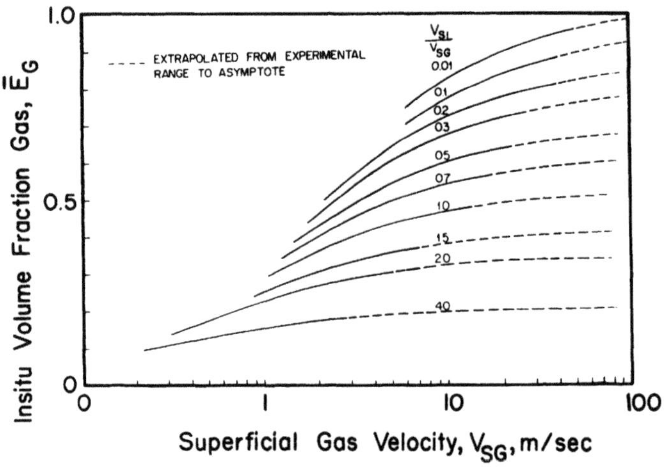

Figure 10.19. Hoogendoorn and Buitelaar holdup correlation. (From Hoogendoorn, C. J., and A. A. Buitelaar, *Chem, Eng. Sci.*, **16**, 208, 1961.)

It is to be noted that this correlation suggests that holdup is independent of pipe diameter and fluid properties.

Eaton *et al.* (1967) propose a further holdup correlation based upon their data on the flow of natural gas–water mixtures in 2- and 4-inch horizontal pipes. They incorporate three of the dimensionless ratios proposed by Ros (1961) in his work on vertical flow and a dimensionless pressure term to account for variations in the density of the natural gas with pressure. The dimensionless ratios used by, and the ranges of the data of, Eaton *et al.* are

$$N_{VL} = V_{SL}\sqrt[4]{\rho_L/g\sigma} \qquad 0.0697 \leq N_{VL} \leq 13.246 \qquad (10.9)$$

$$N_{VG} = V_{SG}\sqrt[4]{\rho_L/g\sigma} \qquad 1.5506 \leq N_{VG} \leq 140.537 \tag{10.10}$$

$$N_D = D\sqrt{\rho_L g/\sigma} \qquad 20.3395 \leq N_D \leq 39.6277 \tag{10.11}$$

$$\frac{P}{P_0} \qquad 5.0 \leq \frac{P}{P_0} \leq 65.0 \tag{10.12}$$

The Eaton *et al.* correlation also introduces a liquid viscosity influence term defined as N_L/N_{Lb}, where

$$N_L = \mu_L\sqrt[4]{g/\rho_L\sigma^3} \tag{10.13}$$

and N_{Lb} is the value of N_L for water at 60°F and 14.65 psia, i.e., 0.00226.

The correlation may be expressed as

$$E_L = \psi\left[\frac{N_{VL}^{0.575}}{N_{VG}N_D^{0.0277}}\left(\frac{P}{14.65}\right)^{0.05}\left(\frac{N_L}{0.00266}\right)^{0.10}\right] \tag{10.14}$$

and is reproduced in graphical form as Figure 10.20. The small exponents on the diameter and property terms confirm the suggestions of others that these influences are small. In fact, the work of Eaton *et al.* indicates that

$$E_L \simeq \psi_1\left[\left(\frac{V_{SL}}{V_{SG}}\right)^{0.575} V_{SG}^{-0.425}\right] \tag{10.15}$$

incorporating the same variables suggested by Hoogendoorn and Buitelaar.

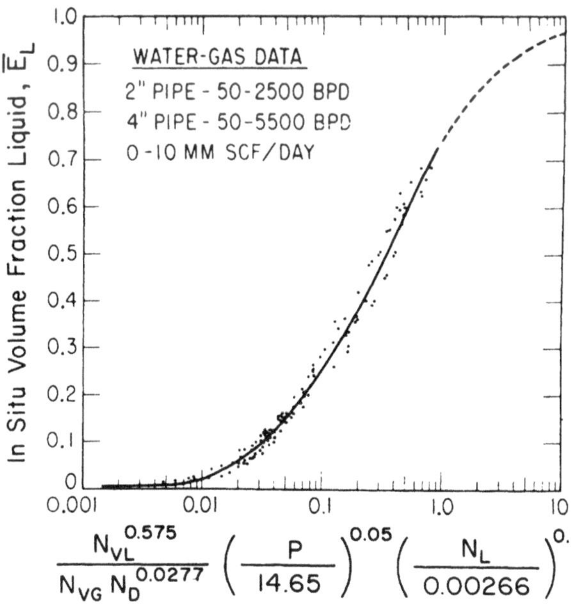

Figure 10.20. Eaton *et al.* holdup correlation. (From Eaton, B. A., D. E. Andrews, C. R. Knowles, I. H. Silberberg, and K. E. Brown, *J. Petrol. Technol.*, **19**, 815, 1967.)

Hughmark (1962) has shown that his correlation for holdup in vertical flow (discussed in Section 3.2 of Chapter 8) applies also to horizontal flow. This is confirmed by Dukler, Wicks, and Cleveland (1964).

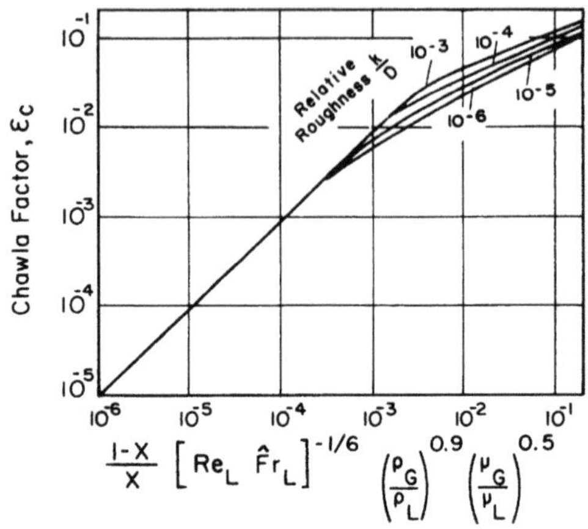

Figure 10.21. Chawla factor, ε_C. (From Chawla, J. M., *Forsch. Ing.-Wes.*, **34**, 53, 1968.)

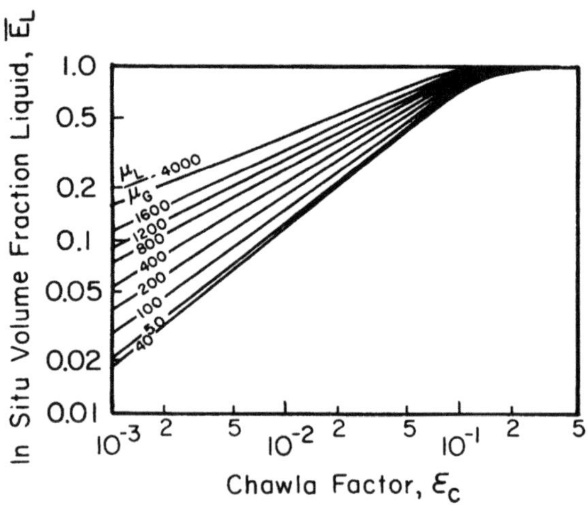

Figure 10.22. Chawla holdup correlation. (From Chawla, J. M., *Chem, Ing. Technik,* **41**, 328, 1969.)

Chawla (1969) has developed a correlation for the *in situ* volume fraction liquid in terms of liquid content in the pipe expressed through μ_L/μ_G and a parameter ε_C given by Figure 10.21 and expressed by the equation (for $\varepsilon_C < 2 \times 10^{-3}$)

$$\varepsilon_C = 9.1\,\frac{1-x}{x}\,(\mathrm{Re}_L\hat{\mathrm{Fr}}_L)^{-1/6}\left(\frac{\rho_G}{\rho_L}\right)^{0.9}\left(\frac{\mu_G}{\mu_L}\right)^{0.5} \tag{10.16}$$

where $x = M_G/M_M$,

$$\mathrm{Re}_L = \frac{M_M(1-x)D}{\mu_L},$$

$$\hat{\mathrm{F}}\mathrm{r}_L = \frac{M_M^2(1-x)^2}{\rho_F^2 g D}.$$

The correlation, given in Figure 10.22, is based on the previously mentioned data bank of Dukler *et al.* Chawla has compared his correlation with experimental data in smooth pipes ($k/D \leq 2.5 \times 10^{-3}$) and finds it to have an accuracy comparable to that of the experimental data. The correlation also applies to vertical flow.

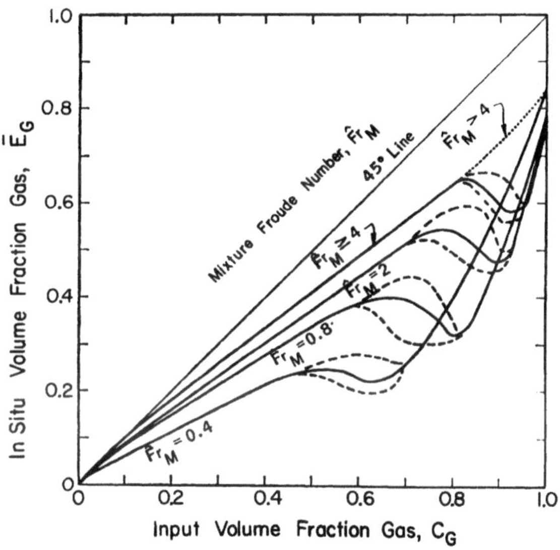

Figure 10.23. Guzhov *et al.*-Mamayev holdup correlation. (From Greskovich, E. J., A. L. Shrier, and R. H. Bonnecaze, *Ind. Eng. Chem. Fundam.*, **8**, 591, 1969.)

Mamayev (1965) proposed a simple correlation for calculating the *in situ* volume fraction gas from the input volume fraction and the mixture Froude number. The correlation as modified by Guzhov, Mamayev, and Odishariya (1967) and Greskovich, Shrier, and Bonnecaze (1969) is shown in Figure 10.23. The latter have discussed the correlation and compared it with some of their own data. The correlation is based primarily on data for the air-water system but apparently represents experimental data of other systems better than prior correlations. The dashed lines in the figure indicate the range of the scatter of data in the s-shaped region of the curves. This region seems to be associated with the transition from stratified through wave to slug flow and the scatter is probably due to the experimental problems associated with measurements in this region. For $\hat{\text{Fr}}_M > 4$ and at high values of the input gas fraction, the slug flow pattern predominates. For this region the correlation gives a good check with the Hughmark correlation

$$\bar{E}_G = 0.82\left(\frac{V_{SG}}{V_{SG} + V_{SL}}\right) \tag{10.17}$$

for slug flow at $\text{Re}_M > 400{,}000$. However, below this Reynolds number, Hughmark's correlation indicates a dependence of $\bar{E}_G$ on the Reynolds number.

Among others who have investigated holdup in horizontal pipes are Sobocinski and Huntington (1958), Hughmark (1962), Alves (1954), and Govier and Omer (1962). Sobocinski and Huntington obtained data in a 3-inch pipe with a three-phase system comprised of water, a gas-oil, and air. Some of these studies will be discussed later in connection with the individual flow patterns.

Holdup in liquid-liquid flow has not received as much treatment as the gas-liquid case. No generalized correlations are available, but the previously discussed data of Charles, Govier, and Hodgson (1961) and also of Russell, Hodgson, and Govier (1959) may be used as a general guide.

3.3 Pressure Gradient

In this section we will discuss some of the more important generalized pressure gradient correlations that apply in all flow patterns.

3.31 The Lockhart and Martinelli and Related Correlations

Lockhart and Martinelli (1949) were the first to present a general pressure drop correlation for two-phase flow in horizontal pipes. The method, although empirical, has some theoretical basis. It is simple to use but errors of up to 50 percent have been observed in its application. Many modifications of the correlation have been proposed. The original correlation and its modifications will be discussed in some detail here. We will present some theoretical justification of the method after presenting the development of the correlation in almost its original form.

The correlation is based upon the concept that the pressure drop for the liquid phase must equal the pressure drop for the gas phase regardless of the flow pattern, as long as an appreciable radial pressure difference does not exist. The pressure drop in the liquid phase may be expressed as

$$\left(\frac{\Delta P}{L}\right)_{tp} = 2f_L\left(\frac{\rho_L V_L^2}{D_L g}\right) \tag{10.18}$$

where D_L is the equivalent hydraulic diameter of the region of the pipe in which the liquid flows. Similarly, for the gas flow,

$$\left(\frac{\Delta P}{L}\right)_{tp} = 2f_G\left(\frac{\rho_G V_G^2}{D_G g}\right) \tag{10.19}$$

Clearly the two hydraulic diameters will each be less than the pipe diameter, D. The actual average velocities of the two phases may be related to the mass flow rate by

$$V_L = \frac{M_L}{\alpha \frac{\pi}{4} D_L^2 \rho_L} \tag{10.20}$$

$$V_G = \frac{M_G}{\beta \frac{\pi}{4} D_G^2 \rho_G} \tag{10.21}$$

The parameters α and β are introduced to account for the relative motion of the two fluids and the flow geometry. The two friction factors f_L and f_G may be related to Reynolds numbers using the generalized Blasius form:

$$f_L = \frac{B_L}{\mathrm{Re}_L^n} = \frac{B_L}{\left(\frac{4}{\pi}\frac{M_L}{\alpha D_L \mu_L}\right)^n} \tag{10.22}$$

$$f_G = \frac{B_G}{\mathrm{Re}_G^m} = \frac{B_G}{\left(\frac{4}{\pi}\frac{M_G}{\beta D_G \mu_G}\right)^m} \tag{10.23}$$

where B_L and B_G are constants.

Substituting Equations 10.20 and 10.22 into Equation 10.18, dividing by Equation 4.24, and assuming that B_L and n are the same for single-phase and two-phase flow, we obtain

$$\left(\frac{\Delta P}{L}\right)_{tp} = \left(\frac{\Delta P}{L}\right)_{SL} \alpha^{n-2}\left(\frac{D}{D_L}\right)^{5-n} \tag{10.24}$$

$$\sqrt{\frac{(\Delta P/L)_{tp}}{(\Delta P/L)_{SL}}} = \phi_L = \alpha^{(n-2)/2}\left(\frac{D}{D_L}\right)^{(5-n)/2} \tag{10.25}$$

Similar treatment of the equation for gas flow yields

$$\left(\frac{\Delta P}{L}\right)_{tp} = \left(\frac{\Delta P}{L}\right)_{SG} \beta^{m-2}\left(\frac{D}{D_G}\right)^{5-m} \tag{10.26}$$

and

$$\sqrt{\frac{(\Delta P/L)_{tp}}{(\Delta P/L)_{SG}}} = \phi_G = \beta^{(m-2)/2}\left(\frac{D}{D_G}\right)^{(5-m)/2} \tag{10.27}$$

The quantities ϕ_L and ϕ_G are defined by the left-hand sides of Equations 10.25 and 10.27. Lockhart and Martinelli postulated that ϕ_G and ϕ_L are both functions of X, the parameter defined by Equation 10.5. The actual relationship between ϕ_G and ϕ_L and X was determined by experiment. Since they expected the exact form of the friction factor in Equations 10.22 and 10.23 to depend upon the flow pattern, Lockhart and Martinelli correlated their data in four separate groupings as follows:

Laminar-laminar

$$\frac{DV_{SL}\rho_L}{\mu_L} \text{ and } \frac{DV_{SG}\rho_G}{\mu_G} \text{ both} < 1000$$

Laminar-turbulent

$$\frac{DV_{SL}\rho_L}{\mu_L} < 1000, \quad \frac{DV_{SG}\rho_G}{\mu_G} > 1000$$

Turbulent-laminar

$$\frac{DV_{SL}\rho_L}{\mu_L} > 1000, \quad \frac{DV_{SG}\rho_G}{\mu_G} < 1000$$

Turbulent-turbulent

$$\frac{DV_{SL}\rho_L}{\mu_L} \text{ and } \frac{DV_{SG}\rho_G}{\mu_G} \text{ both} > 1000$$

These four flow regions are indicated, respectively, by the subscripts *ll*, *lt*, *tl*, and *tt*.

The choice of Re = 1000, based on the superficial phase velocity, for the laminar-turbulent transition was based on the reasoning that the effective value of the Reynolds number for one phase would be increased by the presence of a second phase due to the resulting reduction in flow area. The empirical relations between ϕ_L and X and ϕ_G and X are reproduced in Figure 10.24.

Kriegel (1967) has studied the idealized case of annular flow assuming no entrainment and laminar flow of both phases and shows that theoretically

$$\phi_{Ltt}^2 = 1 + \frac{1}{X^2} \tag{10.28}$$

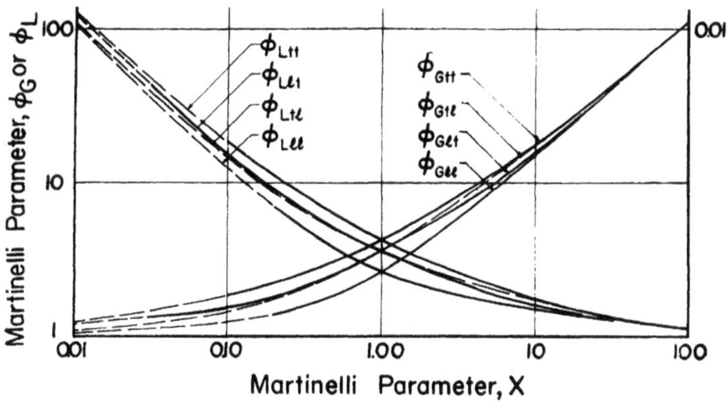

Figure 10.24. Lockhart and Martinelli relationships. (From Lockhart, R. W., and R. C. Martinelli, *Chem. Eng. Progress*, **45**, 39, 1949.)

and

$$\phi_{Gll}^2 = 1 + X^2 \tag{10.29}$$

Comparision of Equations 10.28 and 10.29 with experimental data indicates that the actual pressure drop is always higher than the values predicted by these equations.

Chisholm (1967) has attempted to give a theoretical basis for the Lockhart and Martinelli correlation. He used a force balance on each phase given by

$$A_L\left(\frac{\Delta P}{L}\right)_{tp} - \tau_{wL}\,p_L + S = 0 \tag{10.30}$$

and

$$A_G\left(\frac{\Delta P}{L}\right)_{tp} - \tau_{wG}\,p_G - S = 0 \tag{10.31}$$

where A_L and A_G are cross-sectional areas occupied by the liquid and gas phases, respectively,

τ_{wL} and τ_{wG} denote the wall shear stress, respectively, for the liquid and the gas phase,

p_L and p_G are the portions of the wall perimeter in contact with the liquid and the gas phase, respectively,

S is the interface force between the two phases per unit length.

These two equations are valid regardless of the type of flow as long as the two phases are continuous. In addition, we also have the relationship

$$A = A_G + A_L \tag{10.32}$$

Chisholm defines the quantities α, β, S_R, and Z as

$$\alpha = \frac{A_L}{p_L}\frac{4}{D} \tag{10.33}$$

$$\beta = \frac{A_G}{p_G}\frac{4}{D} \tag{10.34}$$

$$S_R = \frac{S}{A_G\left(\dfrac{\Delta P}{L}\right)_{tp}} \tag{10.35}$$

and

$$Z = \left(\frac{1 + S_R A_G/A_L}{1 - S_R}\right)^{0.5} \tag{10.36}$$

Using these relationships, Chisholm shows that

$$\frac{\Delta P_{tp}}{\Delta P_{SL}} = \frac{1}{\alpha}\frac{f_L}{f_L'}\left(1 + \frac{A_G}{A_L}\right)\left(\frac{A_G}{A_L Z^2} + 1\right) \tag{10.37}$$

where f_L' is a single-phase friction factor given by

$$f_L' = B_L\left(\frac{A\mu_L}{M_L D}\right)^n \tag{10.38}$$

and f_L is the individual phase friction factor during two-phase flow given by

$$f_L = B_L\left(\frac{A_L \mu_L}{\alpha M_L D}\right)^n \tag{10.39}$$

The friction factors for the gas phase may be defined in a similar fashion using a coefficient B_G and exponent m. Combination of these relations and some manipulation yields

$$\frac{1}{\beta} = \frac{A}{A_G} - \frac{A_L}{A_G \alpha} \tag{10.40}$$

$$\left(\frac{A_L}{A}\right)^{1-0.5n}\left(\frac{A}{A - A_L}\right)^{1-0.5m} = \frac{X\beta^{0.5(1+m)}}{Z\alpha^{0.5(1+n)}} \tag{10.41}$$

and

$$\phi_L^2 = \frac{1}{\alpha^{1+n}}\left(1 + \frac{A_G}{A_L}\right)^{1-n}\left(\frac{A_G}{A_L Z^2} + 1\right) \tag{10.42}$$

For a given value of A and A_L/A we can calculate ϕ_L from Equations 10.32, 10.40, 10.41, and 10.42. Chisholm has assumed Z to be a constant, and using

A_L/A data of Lockhart and Martinelli, found $Z = 14$ to give the best fit with the experimental data.

The equations discussed above are difficult to use and for design purposes Chisholm recommends

$$\phi_L^2 = 1 + \frac{C}{X} + \frac{1}{X^2} \tag{10.43}$$

where $C = Z + 1/Z$. Equation 10.43 may be derived from Equation 10.42 for the annular flow pattern case where both fluids are in turbulent flow. For the various flow conditions, Chisholm gives the following values of C:

turbulent-turbulent flow, $C = 20$
laminar-turbulent flow, $C = 12$
turbulent-laminar flow, $C = 10$
laminar-laminar flow, $C = 5$

Values predicted using Equation 10.43 are compared with the Lockhart and Martinelli correlation in Table 10.4.

The Lockhart and Martinelli correlation and its modification by Chisholm are based upon data for small diameter pipes and fluids flowing at pressures in the range of 16-52 psia. Modifications for higher pressures have been presented by Chisholm (1963) and others. Baker (1954, 1961) has modified the Lockhart and Martinelli correlation for application to large diameter pipes.

TABLE 10.4

Comparison of Values of ϕ_L Predicted by Equation 10.43 and the Lockhart–Martinelli Correlation[a]

	ϕ_L			
X	0.1	1.0	10	100
Turbulent-turbulent flow				
Lockhart–Martinelli	18.5	4.2	1.75	1.11
Equation 10.43 ($C = 20$)	17.3	4.7	1.73	1.10
Laminar-turbulent flow				
Lockhart–Martinelli	15.2	3.48	1.59	1.11
Equation 10.43 ($C = 12$)	14.9	3.75	1.49	1.06
Turbulent-laminar flow				
Lockhart–Martinelli	14.5	3.48	1.66	1.11
Equation 10.43 ($C = 10$)	14.1	3.47	1.42	1.05
Laminar-laminar flow				
Lockhart–Martinelli	12.4	2.61	1.50	1.11
Equation 10.43 ($C = 5$)	12.3	2.65	1.23	1.03

[a] D. Chisholm, *Int. J. Heat Mass Transf.*, **10**, 1767 (1967).

Separate curves for each flow pattern are required. These curves are reproduced in Figure 10.25. The Baker correlation may also be given in the form of separate equations for each flow pattern. These are

a. for stratified flow,

$$\phi_G = \frac{15400X}{G_{SL}^{0.8}} \tag{10.44}$$

b. for the elongated bubble flow pattern,

$$\phi_G = \frac{27.315X^{0.855}}{G_{SL}^{0.17}} \tag{10.45}$$

c. for the dispersed bubble flow pattern,

$$\phi_G = \frac{14.2X^{0.75}}{G_{SL}^{0.1}} \tag{10.46}$$

d. for slug flow,

$$\phi_G = \frac{1190X^{0.815}}{G_{SL}^{0.5}} \tag{10.47}$$

e. for the annular-mist flow pattern,

$$\phi_G = (4.8 - 0.3125D)X^{0.343-0.021D} \tag{10.48}$$

where G_{SL} is the superficial mass flow rate of the liquid in lb_m/ft^2 hr. No single equation is available for the wave flow pattern. These relations apply

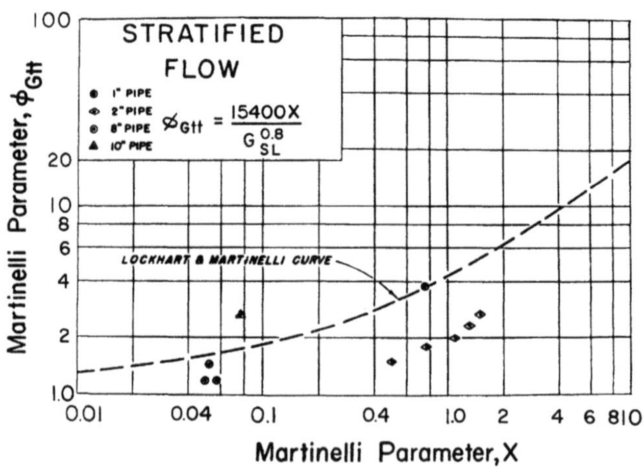

Figure 10.25. (a) Baker's modification of the Lockhart and Martinelli relationships. (From Baker, O., *Oil and Gas J.*, **53**(12), 185, 1954.)

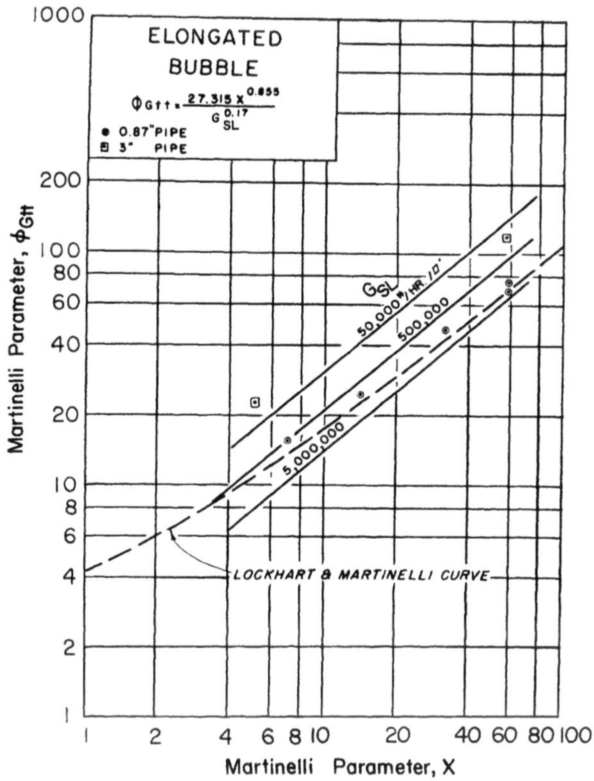

Fig. 25 (b).

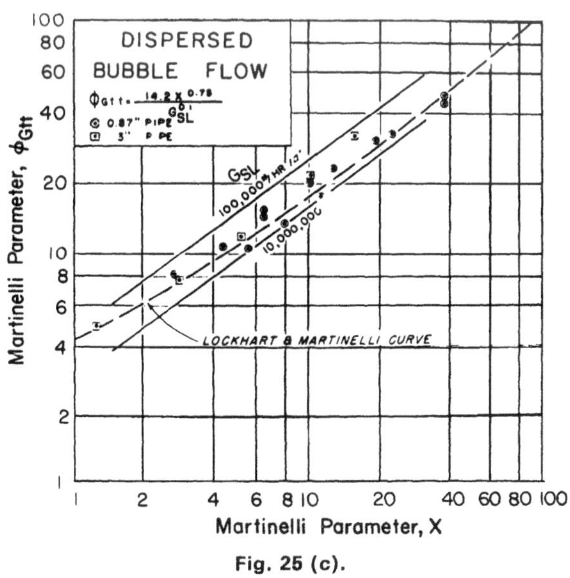

Fig. 25 (c).

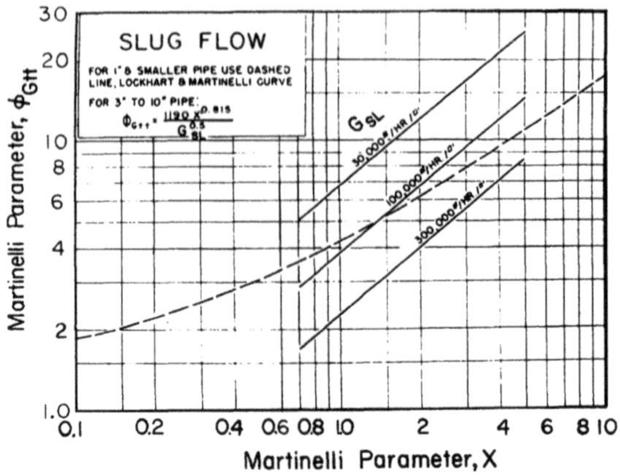

Fig. 25 (d)

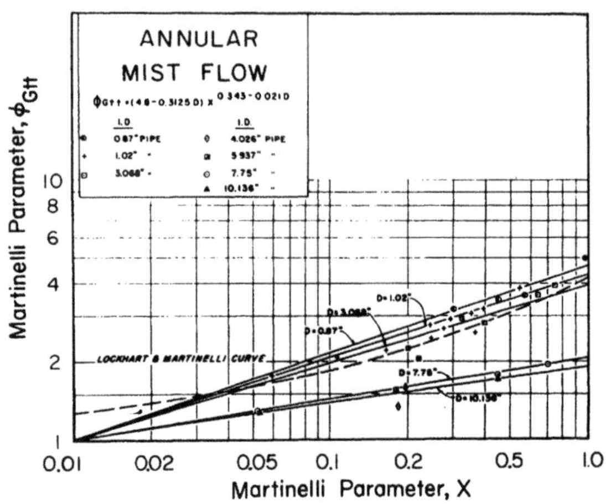

Fig. 25 (e).

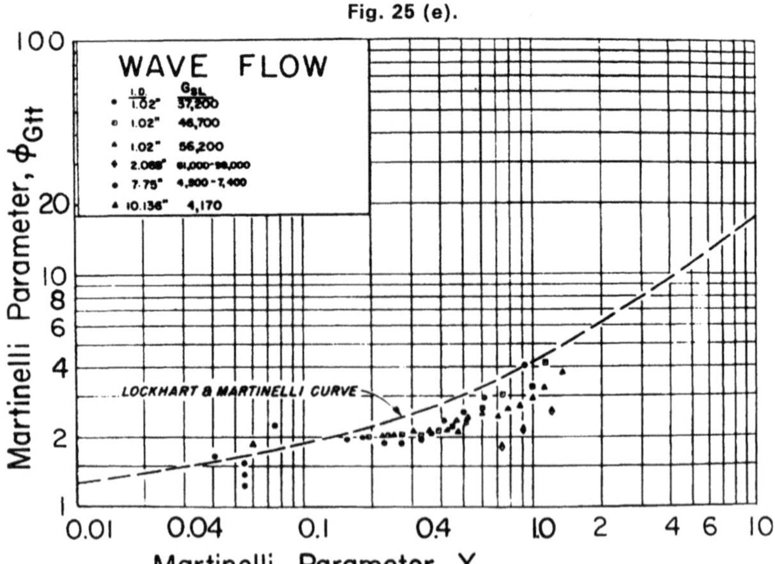

Fig. 25 (f).

only when the superficial Reynolds number for each phase exceeds 1000 (turbulent-turbulent case).

Chisholm and Laird (1958) developed a Lockhart and Martinelli type correlation for rough pipes but, because of the low overall accuracy of the generalized correlations, the modest effect due to pipe roughness may be neglected.

3.32 The Bertuzzi, Tek, and Poettmann Correlation

Bertuzzi, Tek, and Poettmann (1956) proposed a pressure drop correlation based upon the mechanical energy balance as applied to horizontal flow under conditions where the kinetic energy term is negligible and no work is done on or by the system. Equation 8.7 reduces to

$$\int_{P_1}^{P_2} \frac{dP}{\rho_M} + \frac{\Delta P_f}{\rho_M} = 0 \tag{10.49}$$

and a two-phase friction factor may be defined such that

$$\frac{\Delta P_f}{\rho_M} = \frac{2 f_{BTP} G_M^2 L}{g_c D \rho_M^2} \tag{10.50}$$

where ρ_M = superficial mixture density,
G_M = mixture mass flow rate per unit area.

Bertuzzi *et al.* defined a two-phase Reynolds number (Equation 8.15) as

$$\mathrm{Re}_{BTP} = (\mathrm{Re}_{SG})^{k/(1+k)} (\mathrm{Re}_{SL})^{1/e^{0.1k}}$$

where Re_{SG} = Reynolds number for the gas phase based on V_{SG},
Re_{SL} = Reynolds number for the liquid phase based on V_{SL},
k = mass ratio of gas to liquid, based on separator and stock-tank quantities.

As discussed in Chapter 8, Re_{BTP} reduces to Re_{SL} and Re_{SG}, respectively, as k reduces to zero or increases to infinity.

Figure 10.26 taken from Bertuzzi *et al.* gives their correlation of f_{BTP} versus Re_{BTP} for different values of the gas-liquid mass ratio, k. Confirmation of this method of correlating the two-phase pressure drop has been presented by Gallyamov and Goldzberg (1968), who find deviation of experimental data from the Bertuzzi *et al.* line shown in Figure 10.26 to be less than 30 percent.

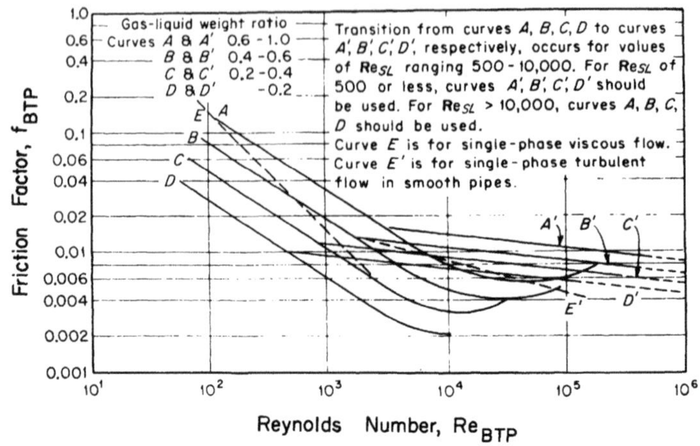

Figure 10.26. Bertuzzi *et al.* friction factor correlation. (From Bertuzzi, A. F., M. R. Tek, and F. H. Poettmann, *Petrol. Trans.*, **207**, 17, 1956.)

3.33 The Baxendell Correlation

Baxendell (1955) has also developed an empirical pressure drop correlation applicable to crude oil-gas mixtures based upon the mechanical energy balance in the single-phase form. He defines a friction factor, f_B, through the equation

$$144 \int_{P_1}^{P_2} \rho_M \, dP = \frac{4 f_B G_M^2 L}{g_c D} = \frac{64 f_B M_M^2 L}{\pi g_c D^5} \tag{10.51}$$

Note that $f_B = \frac{1}{2} f_{BTP}$ and that $M_M = (\pi D^2/4) G_M$. For the oil-gas mixtures which he considered, Baxendell expressed ρ_M and M_M in terms of pressure, gas-oil ratio, and oil flow rate through the relations

$$\rho_M = mP \tag{10.52}$$

and

$$M_M = Q_L / K \tag{10.53}$$

where ρ_M is in lb_m/ft^3, P is in psia, M is in lb_m/sec, Q_L is in m^3/day, and m and K are functions of gas-oil ratio determined from actual data for a particular crude oil. Figures 10.27 and 10.28 give these functions for the La Paz crude oil studied by Baxendell. Combining Equations 10.51, 10.52, and 10.53, he expressed the mechanical energy balance as

$$P_1^2 - P_2^2 = \frac{9.187 f_B Q_L^2 L_1}{m K^2 D^5} \tag{10.54}$$

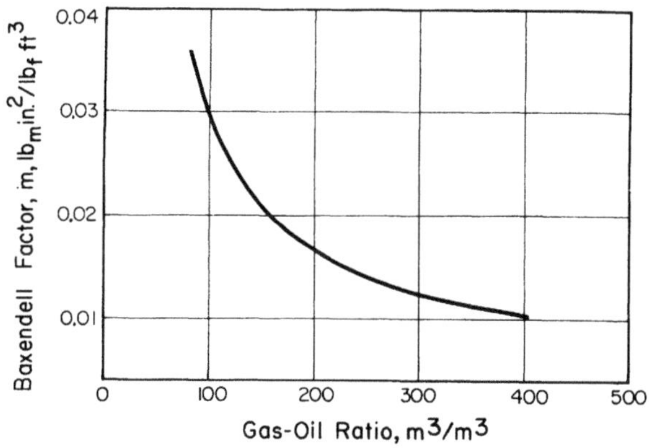

Figure 10.27. Baxendell factor, *m*. (From Baxendell, P. B., Fourth World Petroleum Congress, 1955.)

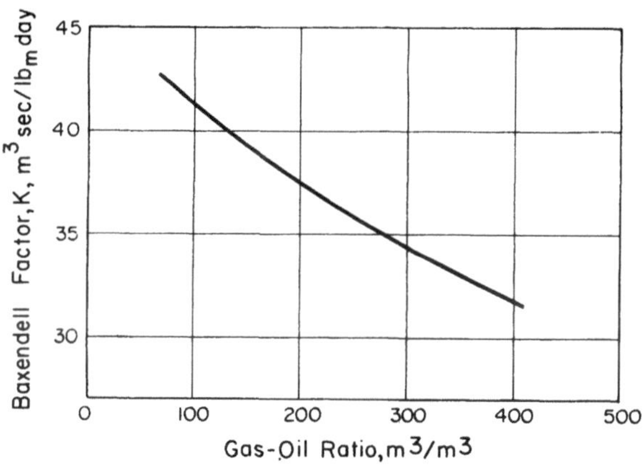

Figure 10.28. Baxendell factor, *K*. (From Baxendell, P. B., Fourth World Petroleum Congress, 1955.)

where D is in inches, and L_1 is in kilometers. Values of f_B calculated with this equation from his field data on 4 and 6-inch lines were correlated in terms of the quantity

$$DV\rho = \frac{1.2733Q_L}{KD}$$

as shown in Figure 10.29. Baxendell claims an accuracy of ± 5 percent for turbulent flow and ± 10 percent under other conditions. The correlation is restricted to the flow of oil-gas mixtures.

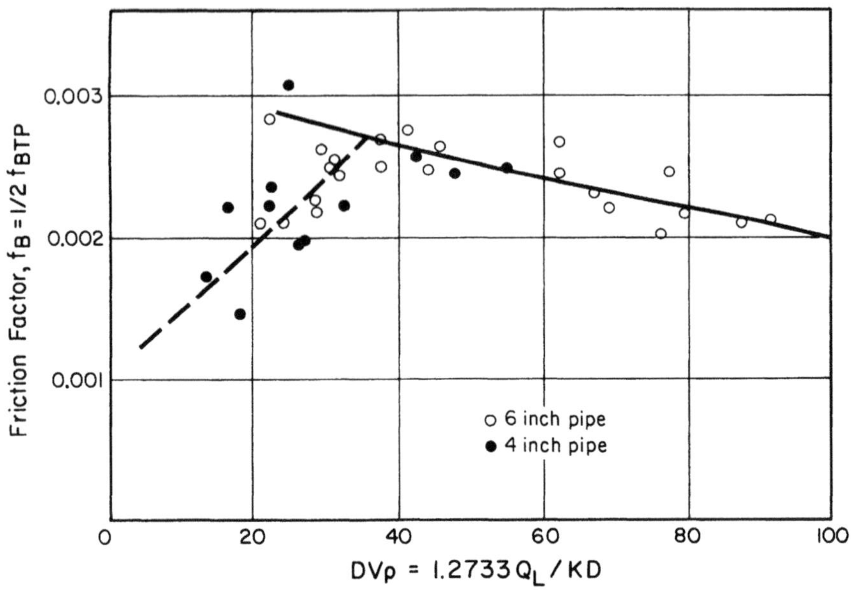

Figure 10.29. Baxendell friction factor correlation. (From Baxendell, P. B., Fourth World Petroleum Congress, 1955.)

3.34 The Hoogendoorn and Related Correlations

Hoogendoorn and Buitelaar (1961) have presented a correlation based on their own data and those of Hoogendoorn (1959) for the elongated bubble, slug, and froth flow patterns. Their correlation is given in Figure 10.30 in the form

$$\frac{f_{tpM}}{f_{LM}} = \phi\left(\frac{\rho_G}{\rho_L}, \frac{M_G}{M_M}\right) \tag{10.55}$$

where the two-phase friction factor is defined by

$$\frac{\Delta P}{L} = \frac{2 f_{tpM} G_M^2}{g_c D \rho_L} \tag{10.56}$$

and f_{LM} is the single-phase friction factor based on

$$\mathrm{Re}_{LM} = \frac{G_M D}{\mu_L} \tag{10.57}$$

For the stratified and wave flow patterns, Hoogendoorn (1959) found that the Lockhart and Martinelli correlation was inadequate. Instead he proposed, for $M_G/M_M < 0.8$,

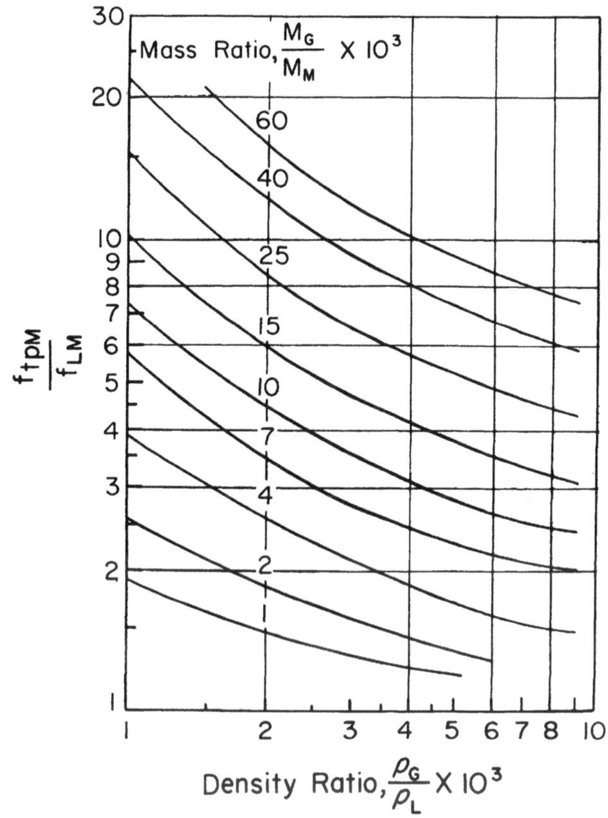

Figure 10.30. Hoogendoorn and Buitelaar friction factor correlation. (From Hoogendoorn, C. J., and A. A. Buitelaar, *Chem. Eng. Sci.*, **16**, 208, 1961.)

$$\frac{\Delta P}{L} = C\left(\frac{M_G}{M_M}\right)^{1.45} \frac{G_M^2}{2g_c D\rho_G} \tag{10.58}$$

The value of C depends on the pipe diameter as shown in Tables 10.5 (smooth pipe) and 10.6 (rough pipes). The influence of gas density was not investigated in the development of this correlation.

For the annular-mist flow pattern, Hoogendoorn proposed

$$\frac{\Delta P}{L} = \frac{2f_{tpG} G_{SG}^2}{g_c D\rho_G} \tag{10.59}$$

where

$$f_{tpG} = aG_{SG}^{-1/4} \tag{10.60}$$

and

$$a = 0.025 \text{ kg}^{1/4}\text{m}^{-1/2}\text{sec}^{-1/4}$$

TABLE 10.5

Influence of Pipe Diameter and Liquid Viscosity on C[a]

Diameter, m	Air–gas-oil	Air–spindle-oil
0.050	$C = 0.026$	$C = 0.028$
0.091	$C = 0\ 022$	
0.140	$C = 0.021$	$C = 0.022$

[a] C. J. Hoogendoorn, *Chem. Eng. Sci.*, **9**, 205 (1959).

TABLE 10.6

Influence of Pipe Roughness on C_{rough} for Air–Gas-Oil Flow in 50-mm Pipes[a]

Pipe Roughness, k/D	C_{rough}	C_{rough}/C
0.0012	0.026	1.0
0.0039	0.032	1.2
0.019	0.045	1.7
0.030	0.052	2.0

[a] C. J. Hoogendoorn, *Chem. Eng. Sci.*, **9**, 205 (1959).

This result is valid for $30 < G_{SL} < 200$ kg/m^{2}sec. The annular-mist flow pattern correlation, Equation 10.60, was found to be independent of gas density over the limited range $1.2 < \rho_G < 3$ kg/m^3. The influence of pipe roughness was also small for this flow pattern.

Equation 10.56 with Figure 10.30, and Equations 10.58 and 10.59 from the work of Hoogendoorn (1959) and Hoogendoorn and Buitelaar (1961) constitute a complete pressure drop correlation. The correlation is based on fairly extensive data on the flow of air with different oils and Freon 11 vapor, and water and flashing Freon 11 mixtures in 15-, 24-, 50-, 91, and 140-mm pipes. The equations are considered reliable.

3.35 The Dukler, Wicks, and Cleveland Correlation

Dukler, Wicks, and Cleveland (1964) have conducted a statistical study to compare the accuracy of correlations by Baker, Chenoweth and Martin, Lockhart and Martinelli, and two others (Bankoff, 1960; Yagi, 1954). The comparison is based upon a selected set of 2620 data points (the AGA-API Data Bank). For both one- and two-component systems, the Lockhart-

Martinelli correlation was found to be superior to all other correlations tested. Dukler *et al.* did not consider the Bertuzzi, Tek, and Poettmann correlation in their comparison because this correlation is based upon data on oil-gas systems and cannot be considered completely general.

Dukler *et al.* present the simple single-phase relationships which result if holdup is totally neglected and the phases are assumed to be thoroughly intermixed. Surprisingly they find predictions based upon this "no slip–homogeneous" model generally better than those based on the Lockhart and Martinelli method. DeGance and Atherton (1970a,b) discuss this further, pointing out that the no slip–homogeneous assumption must result in low predictions of the pressure gradient.

Dukler *et al.* also developed their own separate general pressure drop correlation through a similarity analysis. This correlation is based upon an assumption of finite but constant holdup. It has been discussed by Anderson and Russell (1965), Caughron (1966), and DeGance and Atherton. As a generally applicable correlation it appears statistically superior to all others tested by Dukler *et al.*

The pressure gradient due to friction is expressed as

$$\frac{\Delta P}{L} = \frac{2G_M^2 f_{tpM}}{g_c D \rho_M} \tag{10.61}$$

where

$$f_{tpM} = F\beta f$$

$$f = 0.00140 + 0.125\,\text{Re}_M^{-0.32} \tag{4.36}$$

$$\text{Re}_M = \frac{DG_M}{\mu_M}$$

$$\rho_M = \rho_L C_L + \rho_G(1 - C_L)$$
$$\mu_M = \mu_L C_L + \mu_G(1 - C_L)$$

$$F = 1 + \frac{\gamma}{1.281 - 0.478\gamma + 0.444\gamma^2 - 0.094\gamma^3 + 0.00843\gamma^4}$$

$$\gamma = -\ln C_L$$

$$\beta = \frac{\rho_L}{\rho_M}\frac{C_L^2}{\bar{E}_L} + \frac{\rho_G}{\rho_M}\frac{(1 - C_L)^2}{\bar{E}_G} \quad \text{(a dimensionless quantity)}$$

In Equation 10.61, F is a dimensionless function of C_L, the volume fraction liquid in the input stream. The function F is shown in Figure 10.31. The calculation of the factor β requires a knowledge of $\bar{E}_G$, the *in situ* gas fraction. To obtain this Dukler has used Hughmark's (1962) correlation discussed earlier.

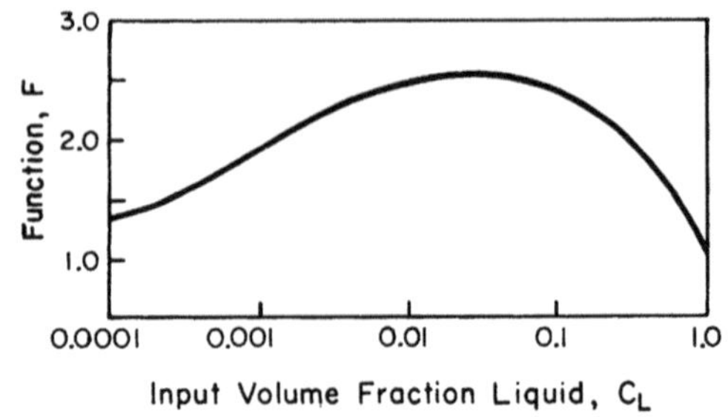

Figure 10.31. Dukler *et al.* function, *F*.(From Dukler, A. E., M. Wicks and R. G. Cleveland, *A.I.Ch.E. Journal,* **10**, 44, 1964.)

Dukler *et al.* also suggest the following expression for calculating the contribution to the overall pressure drop due to kinetic energy or acceleration effects

$$\left(\frac{\Delta P}{L}\right)_{KE} = \frac{1}{g_c A^2 L} M_G^2 \Delta\left(\frac{1}{\rho_G \bar{E}_G}\right) + \frac{M_L^2}{\rho_L} \Delta\left(\frac{1}{\bar{E}_L}\right) \tag{10.62}$$

The acceleration term is calculated by an iterative procedure. In many practical calculations where density variations are small, this contribution is negligible.

3.36 The Chawla Correlation

Chawla (1968) has presented a correlation based on extensive data including the AGA-API Data Bank of Dukler *et al.* (1964). His resulting equation for the frictional component of the pressure drop is

$$\frac{\Delta P}{L} = \frac{0.3164}{(M_M D \mu_G)^{1/4}} \frac{M_M^2 x^{7/4}}{2D\rho_G} \left(1 + \frac{(1-x)\rho_G}{x\varepsilon_C \rho_F}\right)^{19/8} \tag{10.63}$$

where $x = M_G/M_M$,

ε_C is the parameter given by Figure 10.22 and Equation 10.16.

The correlation is applicable to both smooth and rough pipes and checks with the experimental data to within about 30 percent.

3.37 Other Correlations

Several workers, Bergelin and Gazley (1949), White and Huntington (1955), Chenoweth and Martin (1956), Sobocinski and Huntington (1958), Govier and Omer (1962), and others have developed limited correlations in terms of two-phase friction factors defined by one of Equations 7.59, 7.60, 7.61, or a similar equation, or, in the case of Chenoweth and Martin, a two-phase pressure drop ratio. The White and Huntington correlation fits their previously mentioned data (Section 3.1) taken at low liquid velocities excluding those in the slug and stratified flow patterns.

The Chenoweth and Martin correlation represents their own data with an average deviation of about 20 percent, but Aziz and Govier (1962) and Isbin *et al.* (1959) found poor agreement with their data. The correlation of Sobocinski and Huntington is based upon their data on the three-phase water, gas-oil, air system in a 3-inch pipe. Govier and Omer's correlation is of f_{tpL} versus $DV_{SL}\,\rho_L/\mu_L$ and is shown, and compared with results of certain other works, in Figure 10.32. The correlation is for data on an air-water system in a 1-inch pipe, but it checks reasonably well with other data.

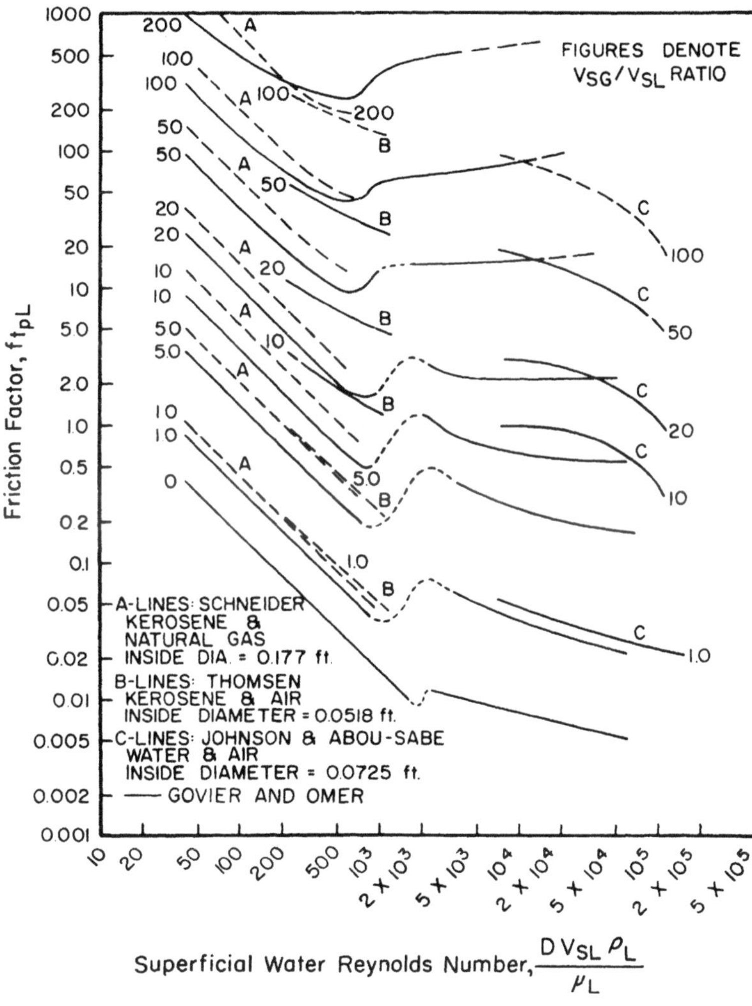

Figure 10.32. Govier and Omer friction factor correlation. (From Govier, G. W., and M. M. Omer, *Can. J. Chem. Eng.*, **40**, 93, 1962.)

3.4 Effect of Pipe Inclination

With two exceptions there are presently no satisfactory theoretical treatment available for flow in inclined pipes, and we must rely largely on our experience with vertical and horizontal flow and on the limited available data and empirical correlations, to deal with gas-liquid flow in inclined pipes. The exceptions are the analyses of Bonnecaze, Erskine, and Greskovitch (1969) and of Singh and Griffith (1970) both relating to slug flow.

It has been shown by several investigators (Brigham, Holstein, and Huntington, 1957; Baker, 1957; Flanigan, 1958; Katsuhara, 1959; Guzhov, Mamayev, and Odishariya, 1967; Vermeulen, 1968) that the effect of pipe inclination on pressure drop is high at relatively low gas flow rates but decreases with increasing gas flow rate. Baker (1957) has concluded that at the lower rates, uphill flow is predominantly in the slug flow regime, which gives rise to a relatively high hydrostatic head. As the gas flow rate is increased, the flow pattern changes to froth flow with a corresponding reduction in the hydrostatic head because of the reduced density of the frothy mixture.

In the downhill case the flow will usually be in the stratified or wave flow regime, and the total pressure drop is comparable to, or may even be slightly lower than, that encountered in horizontal flow. From a practical standpoint, this latter characteristic has an important consequence. In the usual design of pipelines for single-phase flow, one considers only the elevation difference at the ends of the line and the friction losses in calculating pressure requirements; this is possible because of the siphon effect on a downgrade—i.e., any gain in potential energy resulting from a decline in the pipe is a credit against a loss in potential energy from an incline. In the flow of gas-liquid mixtures, however, the siphon effect is normally absent because of the stratification of the phases on the downgrade. The result is that the pressure drop due to elevation changes is almost directly proportional to the sum of the uphill changes in elevation and is essentially independent of the elevation difference between the terminals. A consequence of this is that the pressure drop for two-phase flow may be very much higher than that for a single-phase system.

Brigham *et al.* (1957) have obtained data for the flow of air-oil and air-water mixtures in a 1.975-inch I.D. pipeline at 0°, 5.5°, and 12.4° inclinations to the horizontal. They concluded that a correlation proposed by White and Huntington (1955) adequately represented their data for all but the elongated bubble and slug flow patterns. However, since these latter patterns are likely the dominant ones in upward inclined flow, it would appear that the White and Huntington correlation is of limited practical use.

Probably the most useful generally applicable method of predicting pressure drop in a long-distance pipeline is one which has been proposed independently by Baker (1957) and Flanigan (1958). Each suggests the use of one of the general correlations for horizontal flow with a correction for the hydrostatic head resulting from the upward inclined sections. The overall pressure drop is therefore determined from the relation

$$\Delta P = \Delta P_f + \left(\sum \Delta z^+\right) \frac{g}{g_c} \rho_L E_H \tag{10.64}$$

or

$$\Delta P = \Delta P_f + \Delta P_{HH}$$

where ΔP_f = frictional contribution to the two-phase pressure drop calculated by one of the methods for horizontal flow for the total length of pipe,
$\sum \Delta z^+$ = summation of the uphill rises in the direction of flow,
E_H = "liquid head factor."

It is readily seen that the term on the right-hand side of Equation 10.64 is a height multiplied by an effective mixture density. Hence, the formula simply follows intuitive reasoning based on the summation of the uphill rises and neglects the actual angles of inclination involved.

Baker has determined empirically that the liquid head factor, E_H, is a function of the superficial gas velocity, V_{SG},

$$E_H = 1.61 V_{SG}^{-0.70} \tag{10.65}$$

for $V_{SG} > 2$ ft/sec.

This relation for E_H is based on limited data but covers pipe diameters from 2 inches to 16 inches, and V_{SG} from 2 to 90 ft/sec.

Flanigan (1958) proposes the identical form of the relationship, but gives his expression for E_H as a function of V_{SG} as shown in Figure 10.33. Baker's equation (Equation 10.65) is also shown in this figure for comparison. Using E_H from Figure 10.33, Flanigan reports that predicted pressure drops for five different sections of a 16-inch diameter pipeline, varying from 3 to 25 miles in length, with five different combinations of gas and liquid flow rates, were within ± 14 percent of the measured values.

Over long pipelines the apparent success of this method, which does not take into account the flow pattern, the actual angle of inclination, and which neglects downhill effects, likely represents an overall averaging of positive and negative errors. Care should be taken in using the relation for predicting the pressure drop effects in a relatively short inclined line such as may occur on a plant site. Also, although the method should work reasonably well for small angles of inclination to the horizontal, it is untested for cases when the pipe is

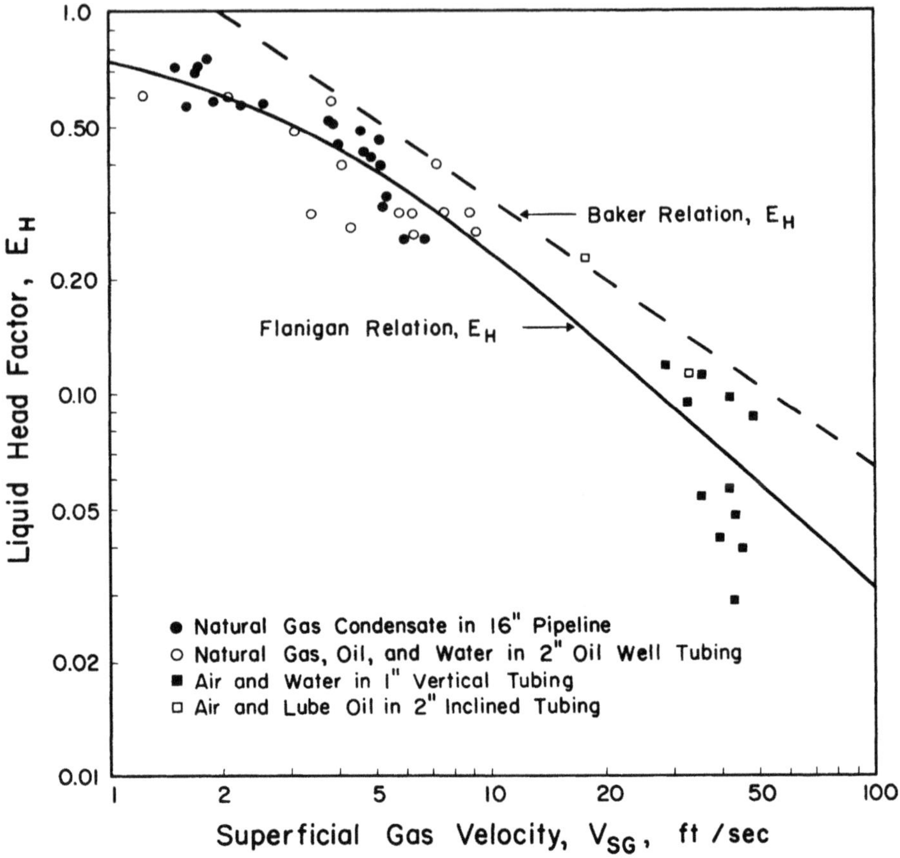

Figure 10.33. Liquid head factor, E_H, of Baker and of Flanigan. (From the data of Flanigan, O., *Oil and Gas J.*, **56**(10), 132, 1958.)

inclined at larger angles. The best present approach in such circumstances would seem to be either

a. to estimate the pressure drop for the same pipe for both vertical and horizontal configurations and use these estimates as a guide to the pressure drop at the appropriate inclination, or
b. to estimate the pressure drop assuming the line to be horizontal and to add to it a hydrostatic head correction based upon the sum of the uphill elevation changes and the *in situ* mixture density which would be expected in vertical flow as discussed in Chapter 8.

When the flow is known to be in the slug flow pattern, Bonnecaze *et al.* (1969) have proposed a simplified mechanistic analysis to calculate both the pressure drop and the holdup in an inclined pipe. By carrying out mass and approximate force balances around a simplified model of a slug in inclined flow, and neglecting the pressure drop contributions due to the liquid film and the gas bubble, they obtain

$$\frac{\Delta P}{L} = \frac{L_s}{L_s + L_b}\left(\frac{g}{g_c}\rho_L \sin\theta + \frac{2f_{LM}\rho_L V_M^2}{g_c D}\right) \tag{10.66}$$

where θ is the angle between the pipe axis and the horizontal, and f_{LM} is a friction factor. They correlate their own pressure drop data obtained in a $1\frac{1}{2}$-inch pipe inclined at various angles between $\pm 10°$ with Equation 10.66 and the following expression for f_{LM} :

$$f_{LM} = 0.0048 + 3980\mathrm{Re}_{LM}^{-1.285} \tag{10.67}$$

where $\mathrm{Re}_{LM} = DV_M\rho_L/\mu_L$.

In the derivation of their expression for holdup, they apply the equivalent of Equation 8.140 with

$$V_b = 1.20V_M + 0.35\delta\sqrt{gD\Delta\rho/\rho_L} \tag{10.68}^*$$

where $\delta = 0$ for the horizontal flow,
$= +1$ for uphill flow,
$= -1$ for downhill flow,

to obtain their holdup expression which is related to Equation 8.142, i.e.,

$$\bar{E}_G = \frac{C_G V_M}{1.20V_M + 0.35\delta\sqrt{gD\Delta\rho/\rho_L}} \tag{10.69}$$

This assumes that the bubbles in inclined flow will have a rise velocity relative to the slug which is the same as in vertical flow. The assumption is questionable but the equation is rather insensitive to the second term in the denominator in any case. Bonnecaze *et al.* approximate $L_S/(L_S + L_b)$ by $1 - \bar{E}_G$ for horizontal flow but derive a more complex relationship for estimating it in inclined flow. Their work should be consulted for details.

Bonnecaze *et al.* compare their predictions of $\bar{E}_G$ with actual measurements in the $1\frac{1}{2}$-inch line and find agreement within $+25.5$ and -8.6 percent. They compare predicted pressure gradients with field data over a 10,000-foot section of 6-inch line and find the maximum deviation to be about 5 percent.

*Actually Bonnecaze *et al.* write this with $\Delta\rho/\rho_L$ outside the root sign. We correct the apparent error here.

Singh and Griffith (1970) present essentially the same force balance analysis as Bonnecaze *et al.* which gives an equation that reduces to Equation 10.64 with $L_S/(L_S + L_b)$ approximated by the term $1 - \bar{E}_G$. They propose the evaluation of f_{LM} strictly from single-phase methods based upon $\text{Re} = DV_M\ \rho_L/\mu_L$. Singh and Griffith also base their relationship for holdup on Equation 8.140, but they suggest the empirical relation,

$$V_b = 0.95V_M + 1.15 \tag{10.70}$$

where V_b is independent of both tube diameter and inclination, as an alternate to Equation 10.68 for the bubble velocity. This results in

$$\bar{E}_G = \frac{C_G V_M}{0.95V_M + 1.15} \tag{10.71}$$

Singh and Griffith's work is supported by their own data on the air-water system in five tube sizes ranging from 0.626 to 1.600 inches in diameter and at inclinations of 5, 10, and 15 degrees from the horizontal and certain other data. They find the pressure drop data to be within ± 15 percent of predictions based upon Equations 10.70 and 10.71.

The analyses of Bonnecaze *et al.* and Singh and Griffith may be used, for the slug flow region, to complement empirical predictions based on the methods of Baker and of Flanigan.

4 THE BUBBLE FLOW PATTERNS

4.1 *Introduction*

The size, shape, and distribution of the bubbles depend both upon the gas and liquid rates and upon the manner in which the two phases are originally intermixed and the interfacial forces. At very low gas-liquid or liquid-liquid ratios the bubble shape is that of a slightly distorted sphere, but as the ratio is increased the bubble becomes larger both in cross section and in length.

With sufficiently high rates of the continuous phase, or sufficiently low density contrast, we may visualize symmetrically distributed, nearly spherical bubbles, the diameter of which may be small compared to the tube diameter or nearly equal to it (see Figures 10.1 and 10.3). Normally the bubble concentration will be greater in the center of the pipe. At somewhat lower velocities of the continuous phase, or with higher density contrast, the distribution becomes asymmetric, peaking near the top of the pipe. If the continuous phase velocity is reduced to the point where there is no turbulence, or insufficient turbulence exists to offset the rising tendency of the bubbles, and when coalescence among bubbles becomes significant, elongated bubbles that ride high in the tube are formed. These elongated bubbles are analogous to the Taylor bubbles discussed in Chapter 8 in connection with vertical flow. Figure 10.34 drawn from photographs of Zukoski (1966) shows how the Taylor bubble in vertical flow changes its shape as the angle of the tube axis with the horizontal is decreased. Still further reduction in the velocity of the continuous phase results in complete coalescence of the elongated bubbles to form the stratified flow pattern.

Increase in the gas velocity at constant liquid velocity results in the enlargement and elongation of the bubbles and the flow is transformed into the slug flow pattern.

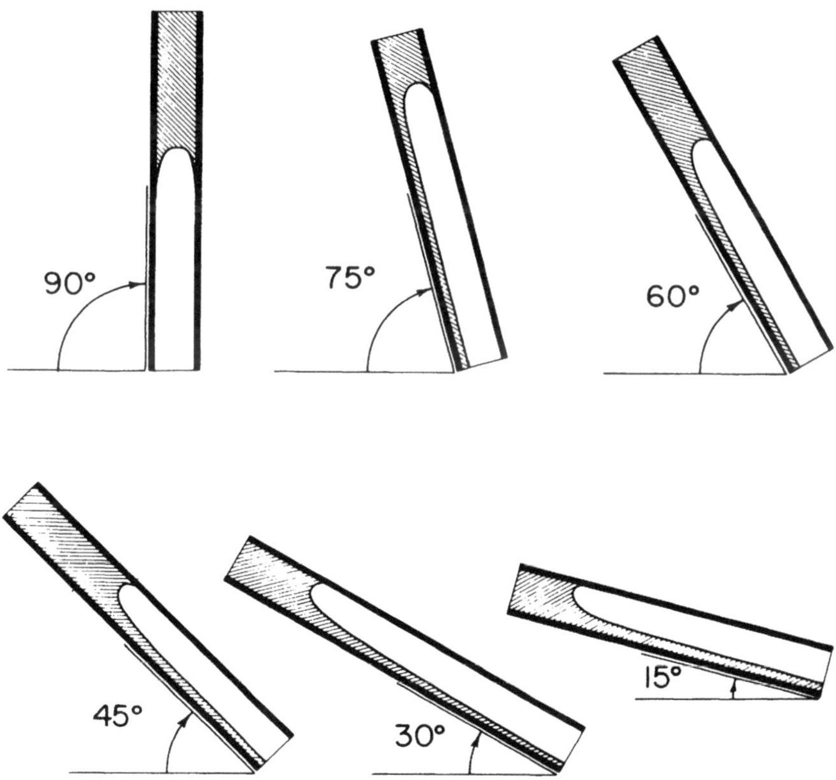

Figure 10.34. Taylor bubbles in inclined tubes. (After Zukoski, E. E., *J. Fluid Mech.*, **25**, 821, 1966.)

In a gas-liquid system the elongated bubble flow pattern normally occurs at modest and high liquid rates and at gas-liquid volume ratios of about 2 or less. The dispersed bubble region occurs only at the highest liquid velocities—normally above $V_{SL} = 10$ ft/sec. The same general situation is observed with liquid-liquid systems, although the superficial velocities of the continuous phase above which dispersed bubbles are obtained depends upon the density contrast.

For the small dispersed bubbles the local slip is probably insignificant, but, unless the liquid velocity and turbulence are high, the lack of a uniform concentration profile will result in some holdup. Theoretically this means that we should not treat the dispersed bubble region by methods appropriate for pseudohomogeneous fluids. Nonetheless, the homogeneous fluid model, with an approximate allowance for the holdup, is a reasonable one and we will consider it. Following this approximate treatment we will refer to the work of Huey and Bryant (1965, 1967), Huey (1966), and Brown and Kranich (1968) who treat the problem with regard to the nonuniform concentration profile. We have no specific way of dealing with the case of the large or elongated bubbles other than to consider it as the extreme of the asymmetrically distributed dispersed bubble flow pattern or in the limiting case of very long bubbles of the equivalent of stratified flow.

In the discussion in the remainder of Section 4 we address our attention primarily to gas-liquid mixtures, but the relations should also be applicable to liquid-liquid mixtures except where the density or the viscosity of the gas phase has been neglected.

4.2 Overall Continuity Considerations

Regardless of whether the bubbles are small and dispersed or large and elongated, the relationships between the velocities of the phases may be determined in a manner analogous to that we employed in Section 5.2 of Chapter 8. Continuity requires that

$$V_M A = (V_{SG} + V_{SL})A = Q_G + Q_L \tag{10.72}$$

$$V_M A = V_b A_b + V_L(A - A_b) \tag{10.73}$$

where $A = \pi D^2/4$ = pipe cross-sectional area,
A_b = total bubble cross-sectional area in a typical central vertical plane $= n(\pi d^2/4)$ for spherical bubbles,
V_M = average velocity of the liquid slug between the bubbles,
V_b = average velocity of the gas bubble,
V_L = average velocity of the liquid in a vertical plane through the center of the bubble.

Equations 10.72 and 10.73 lead to the overall volume balance

$$Q_G + Q_L = V_b A_b + V_L(A - A_b) \tag{10.74}$$

Now

$$Q_G = \omega v_b n \tag{10.75}$$

and

$$\omega = \frac{V_b}{S_b} \tag{10.76}$$

where v_b = the individual bubble volume,
n = the average number of bubbles across a cross section,
S_b = the center-to-center axial spacing of the bubbles.

Equations 10.75 and 10.76 lead to

$$Q_G = \left(\frac{v_b n}{S_b}\right) V_b = V_{SG} A \tag{10.77}$$

which is the gas continuity equation. Subtracting Equation 10.77 from Equation 10.74 we obtain the liquid volume balance as

$$Q_L = V_b A_b + V_L(A - A_b) - \frac{V_b v_b n}{S_b} \tag{10.78}$$

4.3 Approximate Treatment

4.31 Relative Velocity of the Bubbles

In horizontal flow after any initial period of acceleration, the bubbles would be expected to attain a velocity approximately equal to that of the immediately surrounding liquid, i.e., we assume no local slip. For small bubbles riding high in the tube, this may be much less than the average velocity of the liquid; for small bubbles at the axis, it would be above the average velocity; while for larger bubbles occupying, say, the upper three-quarters of the tube, it may be greater than the average velocity but never as great as the maximum liquid velocity at the center line. We may write

$$V_b = kV_M \tag{10.79}$$

where $k < 2$ for laminar flow of the continuous phase,
$k < 1.2$ for turbulent flow of the continuous phase.

For a known bubble size and position in the flow stream, k could, in principle, be approximated as

$$k = \frac{\bar{u}_{Ab}}{V_M} \tag{10.80}$$

where

$$\bar{u}_{Ab} = \frac{4\int_{A_b} u\,dA_b}{\pi d^2} = \text{the average fluid velocity in the region of the bubble,}$$

and

$$A_b = \frac{\pi d^2}{4} = \text{the portion of the pipe cross section occupied by the bubble.}$$

For this approximate treatment, however, we may take advantage of the observation of Nicklin, Wilkes, and Davidson (1962), as confirmed by Scott (1962), that in horizontal bubble flow (both dispersed and elongated), provided only that the superficial continuous phase Reynolds number exceeds 8000,

$$V_b \simeq 1.2 V_M \tag{10.81}$$

Bearing in mind the possible bubble distributions, Equation 10.81 cannot be exact but, as we will see later, calculations based upon it are confirmed by measurements.

4.32 Holdup

The *in situ* volume fraction of the bubble phase is

$$E_G = \frac{n v_b}{S_b A} \tag{10.82}$$

and from Equation 10.77

$$E_G = \frac{V_{SG}}{V_b} = \frac{V_{SG}}{k V_M} \tag{10.83}$$

We see that the *in situ* volume fraction gas may exceed or be less than the ratio V_{SG}/V_M, depending upon k, which in turn depends upon the bubble diameter and position, and whether the liquid flow is laminar or turbulent. If we use Equation 10.81 for V_b,

$$E_G = \frac{0.83 V_{SG}}{V_M} \tag{10.84}$$

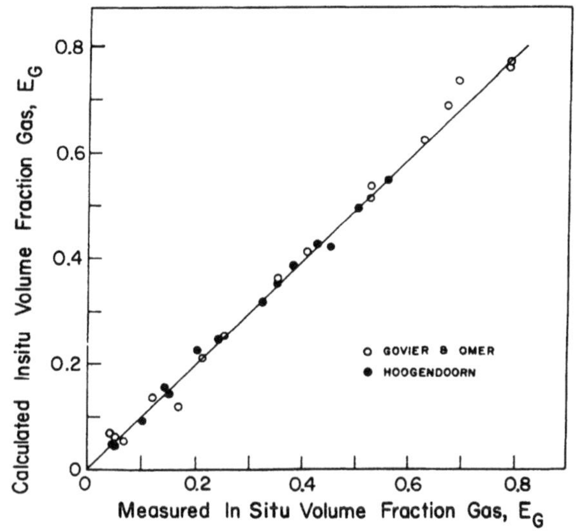

Figure 10.35. Comparison of calculated and measured values of E_G. (From Scott, D. S., *Can. J. Chem. Eng.*, **40**, 224, 1962.)

Scott (1962) has calculated values of E_G from Equation 10.84 to correspond with the dispersed bubble, elongated bubble, and slug air-water data of Govier and Omer (1962) and comparable air-oil data of Hoogendoorn (1959). His comparison is shown in Figure 10.35. The overall check is excellent, apparently confirming the suitability of Equations 10.81 and 10.84 for the entire bubble region.

4.33 Pressure Gradient

For all practical purposes we would expect the pressure gradient due to friction in the bubble region to be the same as that for a liquid phase of density equal to the average mixture flowing at an average velocity V_M. We therefore may apply

$$-\frac{dP}{dL} = \frac{2 f_{LM} \rho_M V_M^2}{g_c D} \tag{4.24}$$

where f_{LM} is evaluated from single-phase correlations at

$$\mathrm{Re} = \frac{D V_M \rho_L}{\mu_L}$$

and

$$\rho_M = (1 - E_G)\rho_L + E_G \rho_G \simeq (1 - E_G)\rho_L$$

This approach is similar to that of Dukler *et al.* (1964) except that they propose the use of μ_M defined as

$$\mu_M = (1 - E_G)\mu_L + E_G \mu_G$$

in the Reynolds number. The use of the viscosity of the continuous phase would seem more appropriate since it is the liquid phase which contacts the wall.

Huey and Bryant (1965), Huey (1966), and Huey and Bryant (1967) present an analysis of bubbly flow of gas-liquid systems based upon the assumption of pseudohomogeneity which, in this case, means not only no local slip but a uniform concentration profile and no holdup. They assume ideal gas behavior for the gas phase and develop the equation of state and relations for the thermodynamic properties of the assumed pseudohomogeneous fluid. From these, the continuity equation, the momentum equation, and an assumed constant friction factor, they develop the relations applicable to flow up to the acoustic velocity for the mixture. The emphasis of their studies is on the prediction of the critical flow rates and the distance along the pipe at which choking occurs. Consequently, the velocities considered (up to 80 ft/sec) and the pressure gradients encountered (up to 2 psi/ft in a 1-inch pipe) are well outside the range of interest in pipeline design. These references should be consulted for flow approaching the choking or acoustic velocity, but the approximate methods previously described and not involving the no holdup assumption are preferable for flow at normal velocities.

4.4 Treatment of Brown and Kranich for Dispersed Bubbles

Brown and Kranich (1968) have made a revealing and detailed study of the concentration and velocity profiles for horizontal dispersed bubble flow. Their study involved the application of the continuity, momentum, and energy equations in forms equivalent to those discussed in Chapter 7. They considered two-phase bubbly flow under adiabatic conditions in the absence of mass transfer, with negligible fluctuating and surface energy effects, and assuming no local slip in the axial direction. Instead of attempting the essentially impossible task of solution of the basic equations for the concentration and velocity profiles and the pressure gradient, they followed the route of separately estimating the profiles and, from them, the pressure gradient, and checking them all for consistency with the fundamental equations.

They presented a suite of velocity and concentration profiles, through the horizontal and the vertical diameters, covering a range of Reynolds and

Froude numbers. These showed a marked asymmetry in both profiles through the vertical diameter and a pronounced peaking of the concentration profile near the top of the pipe. The analysis of Brown and Kranich marks a significant advance in our understanding of bubbly flow, but in its present state it is too complex and time consuming for routine use.

4.5 Other Correlations

The Baker modification of the Lockhart and Martinelli correlation discussed in Section 3.31 and the Hoogendoorn and Buitelaar correlation presented in Section 3.34 both include separate treatments applicable to the bubble flow region.

4.6 Transition to Stratified or Slug Flow Pattern

With a decrease in liquid velocity the dispersed bubble flow pattern of gas-liquid systems transforms to the elongated bubble flow pattern at modest gas rates or to the slug flow pattern at high gas rates. The elongated bubble flow pattern transforms to the stratified flow pattern with further decrease in liquid velocity at low gas velocities and to the slug flow pattern with increase in gas velocity. A qualitative discussion of the transitions has been given in Section 2.1. This includes a discussion for liquid-liquid systems. At the present stage of our knowledge a mechanistic analysis of the transitions is not possible, and for a quantitative appraisal we must rely on the empirical flow pattern boundary correlations discussed in Section 3.1.

5 THE STRATIFIED FLOW PATTERN

5.1 Introduction

The stratified flow pattern occurs with gas-liquid systems at low liquid rates (below about $V_{SL} = 0.6$ ft/sec) and at superficial gas velocities below values ranging from 2 to 10 ft/sec. Depending upon the liquid rate, the liquid may be in either laminar or turbulent motion but the gas is normally in turbulent flow. At the lowest rates the interface is smooth, but with increase in either or both velocities small two-dimensional waves appear. Then the surface develops a pebbly appearance, and finally large roll waves develop on the interface marking the beginning of the vave flow pattern. Much the same sequence is observed with liquid-liquid systems, although here either phase may be in laminar or turbulent motion.

For the idealized stratified flow pattern where the interface is flat and smooth, it is possible to develop a reasonable theoretical explanation of the holdup and pressure gradient effects. This is especially applicable when both fluids flow in laminar motion. Although this situation is not of the greatest practical interest, it is instructive to review the results of theoretical analyses. The more practical case of turbulent motion in either or both the phases cannot be handled so well, but even here there are developments of interest.

Except where otherwise noted the discussion of this section applies equally to gas-liquid and liquid-liquid mixtures.

5.2 Overall Continuity Considerations

Referring to Figure 10.36 the simple relations between the velocities of the phases and the mixture under steady flow conditions are given by

$$Q_\alpha = V_\alpha A_a = V_{S\alpha} A \tag{10.85}$$

$$Q_\beta = V_\beta = A_\beta V_{S\beta} A \tag{10.86}$$

$$Q_\alpha + Q_\beta = V_\alpha A_\alpha + V_\beta A_\beta = (V_{S\alpha} + V_{S\beta})A = V_M A \tag{10.87}$$

where $A = \pi D^2/4$ = pipe cross-sectional area,
A_α = cross-sectional area occupied by the α phase = $\phi_1(h_\beta, D)$,
$A_\beta = (\pi D^2/4) - A_\alpha$ = cross-sectional area occupied by the β phase = $\phi_2(h_\beta, D)$,
V_α = average velocity of the α phase,
V_β = average velocity of the β phase.

We may also define the following quantities which we will require later:

p = perimeter of pipe = πD,
p_α = portion of pipe perimeter in contact with the α phase,
p_β = portion of pipe perimeter in contact with the β phase,
w_i = width of the interface,
h_β = depth of the β phase at the pipe center line,
γ = angle subtended by w_i, see Figure 10.36,
$D_{E\alpha}$ = equivalent diameter of the cross section occupied by the α phase = $4A_\alpha/(p_\alpha + w_i)$,

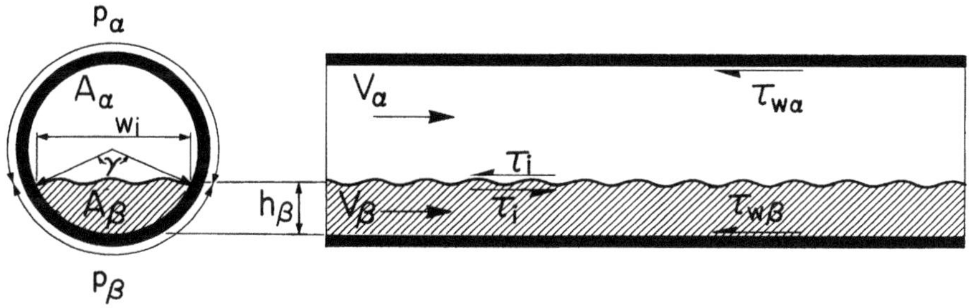

Figure 10.36. Schematic diagram of stratified flow.

$D_{E\beta}$ = equivalent diameter of the cross section occupied by the β phase $= 4A_\beta/(p_\beta + w_i)$.

The geometrical relationships among a number of these quantities are

$$\gamma = 2\cos^{-1}\left(1 - 2\frac{h_\beta}{D}\right) \tag{10.88}$$

$$\frac{w_i}{D} = 2\sqrt{\frac{h_\beta}{D} - \left(\frac{h_\beta}{D}\right)^2} \tag{10.89}$$

$$E_\beta = \frac{A_\beta}{A} = \frac{1}{2\pi}(\gamma - \sin\gamma) \tag{10.90}$$

$$E_\alpha = \frac{A_\alpha}{A} = 1 - \frac{A_\beta}{A} \tag{10.91}$$

$$\frac{p_\beta}{p} = \frac{\gamma}{2\pi} \tag{10.92}$$

$$\frac{p_\alpha}{p} = 1 - \frac{\gamma}{2\pi} \tag{10.93}$$

$$\frac{D_{E\beta}}{D} = \frac{\gamma - \sin\gamma}{\gamma + 2\sin\dfrac{\gamma}{2}} \tag{10.94}$$

$$\frac{D_{E\alpha}}{D} = \frac{2\pi - (\gamma - \sin\gamma)}{2\pi - \gamma + 2\sin\dfrac{\gamma}{2}} \tag{10.95}$$

where γ is in radians. The relationships are given in Figures 10.37 to 10.40

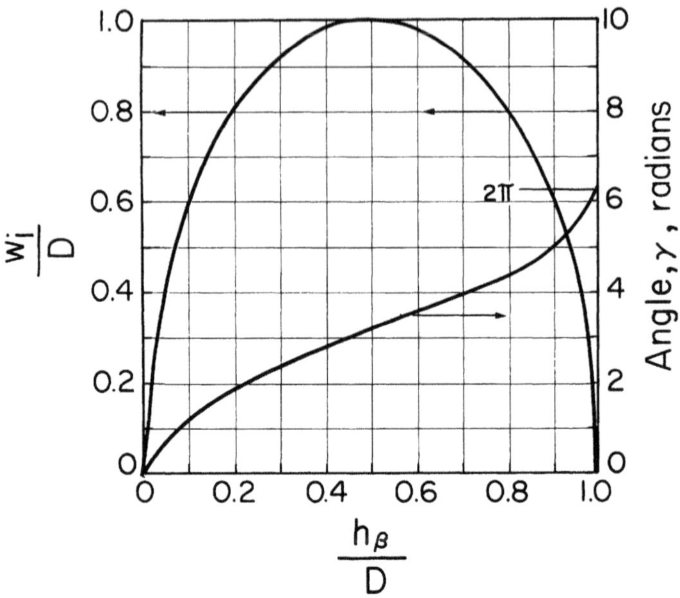

Figure 10.37. Geometrical relationships for stratified flow—w_i and γ.

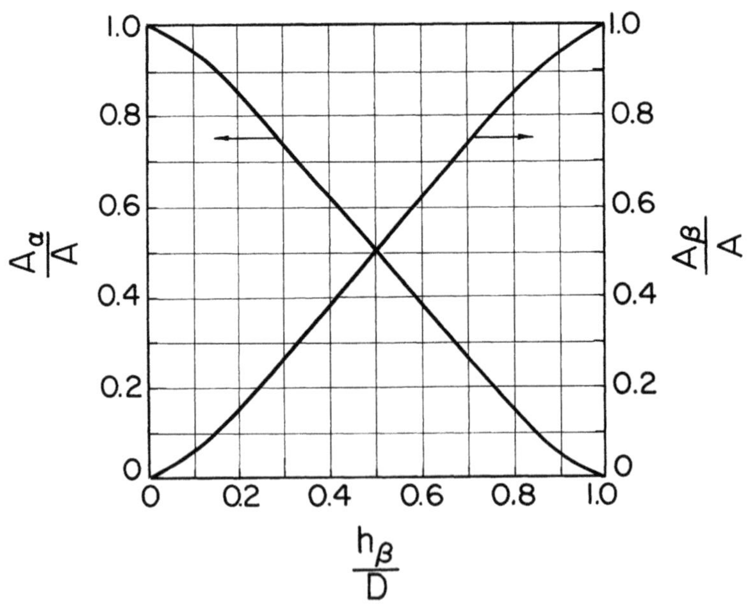

Figure 10.38. Geometrical relationships for stratified flow—A_α and A_β.

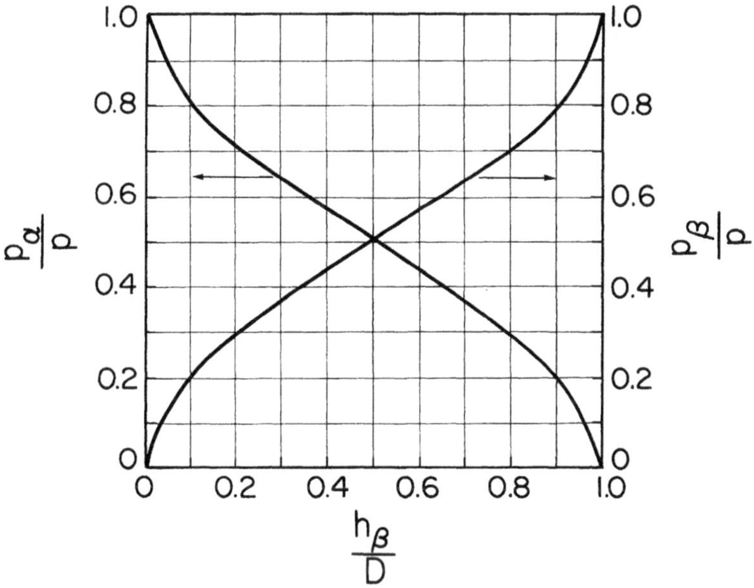

Figure 10.39. Geometrical relationships for stratified flow—p_α and p_β.

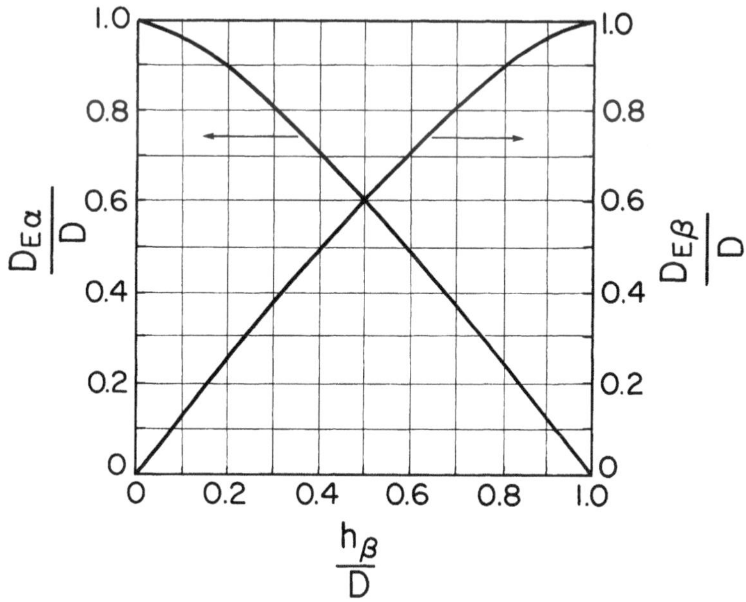

Figure 10.40. Geometrical relationships for stratified flow—$D_{E\alpha}$ and $D_{E\beta}$.

5.3 Momentum Considerations

The momentum relationships for stratified flow are dependent on whether both, one, or neither phase is in laminar flow.

Bergelin and Gazley (1949) have demonstrated that in stratified gas-liquid flow the gas phase becomes turbulent when the superficial gas Reynolds number, defined as $\mathrm{Re}_{SG} = DV_{SG}\rho_G/\mu_G$, exceeds 5000. The work of Govier and Omer (1962) shows that a superficial liquid Reynolds number, $\mathrm{Re}_{SL} = DV_{SL}\rho_L/\mu_L$, characterizes the transition from laminar to turbulent flow in the liquid layer. In this case, however, the end of the laminar region is given by Re_{SL} values which decrease from the single-phase value of 2100 with increasing gas rate to a value of the order of 500 at gas-liquid volume ratios of about 200.

Russell, Hodgson, and Govier (1959), in their studies of the flow of oil-water mixtures, show that laminar flow of the water phase persists to a superficial water Reynolds number of 2100 with the transition to turbulent flow extending to a superficial Reynolds number of 3000. They did not identify the circumstances when the oil phase became turbulent, but there would appear to be no reason why the critical Reynolds number for it would differ from the range for the water.

5.31 The Laminar-Laminar Case

For laminar-laminar stratified flow the momentum analysis is relatively simple since the Navier-Stokes equations apply and may be solved in each phase. For steady-state, incompressible, and unidirectional flow, we have

$$\frac{\partial^2 u}{\partial x^2} + \frac{\partial^2 u}{\partial y^2} = -\frac{g_c}{\mu}\frac{dP}{dL} \tag{10.96}$$

In addition to the above equation the boundary conditions at the pipe wall and the fluid-fluid interface are also required. These are

$$u = 0 \quad \text{at the pipe wall} \tag{10.97}$$

$$u_{i\alpha} = u_{i\beta} \tag{10.98}$$

$$\frac{\left(\dfrac{\partial u}{\partial y}\right)_\alpha}{\left(\dfrac{\partial u}{\partial y}\right)_\beta} = \frac{\mu_\beta}{\mu_\alpha} \quad \text{at the interface} \tag{10.99}$$

Equation 10.96 may be integrated to yield the velocity profiles in each phase. These velocity profiles, when integrated over the pipe cross section, yield the necessary relationship between flow rates and pressure drop.

Teletov (1946) solved the equations for flow between parallel plates and for the general case of flow inside a tube. Unfortunately the expressions for tube flow are too complex to be of any practical use except for the special case where the interface passes through the center of the pipe. Here the solution of the equations for the *in situ* volume fractions, the average velocities of the two phases, and the axial velocity, V_m, gives (Dumitrescu and Stanescu, 1959)

$$E_\alpha = E_\beta = 0.5 \tag{10.100}$$

$$V_\alpha = 2V_{S\alpha} \tag{10.101}$$

$$V_\beta = 2V_{S\beta} \tag{10.102}$$

$$V_m = \frac{g_c D^2}{8(\mu_\alpha + \mu_\beta)} \frac{\Delta P}{L} \tag{10.103}$$

Charles and Redberger (1962) have obtained a numerical solution of the problem. Their solution checks the analytical solution mentioned above for the case where the interface passes through the center of the pipe. Figures 10.41 and 10.42, taken from the work of Charles and Redberger, illustrate the velocity profiles in a 150-centipoise oil-water system. In these figures s and t are, respectively, the ratios of the vertical distance from the top of the pipe to the pipe diameter and the horizontal distance from the center of the pipe to the pipe radius.

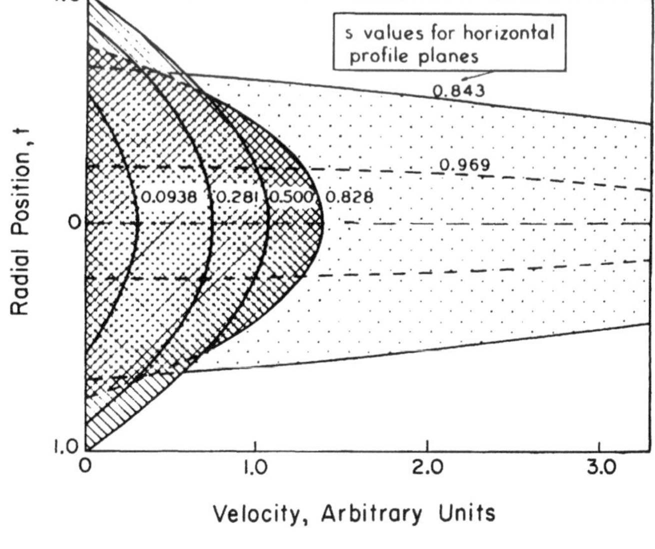

Figure 10.41. Plan views of velocity profiles. (From Charles, M. E., and P. J. Redberger, *Can. J. Chem. Eng.*, **40**, 70, 1962.)

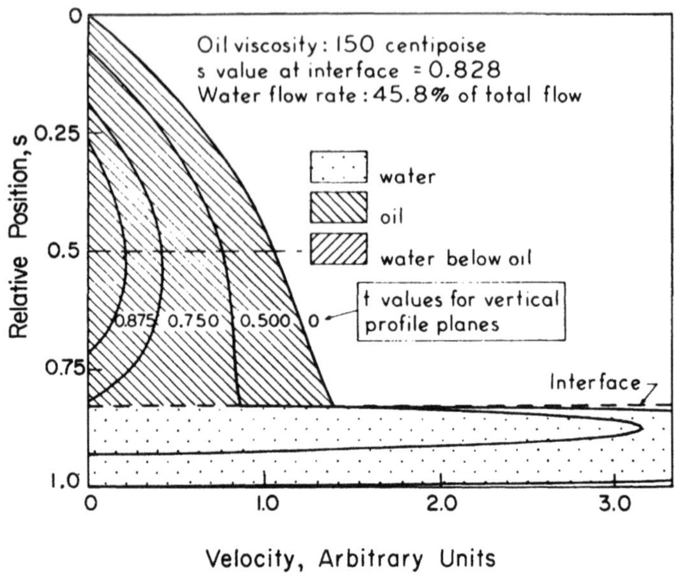

Figure 10.42. Elevation views of velocity profiles. (From Charles, M. E., and P. J. Redberger, *Can. J. Chem. Eng.*, **40**, 70, 1962.)

Gemmell and Epstein (1962) made an analysis essentially the same as that of Charles and Redberger. They developed velocity profiles for viscosity ratios of 1, 10, 100, and 1000 at eight positions of the horizontal interface and computed holdup ratios and pressure gradients from them. Their holdup ratio calculations confirm the conclusion of Russell *et al.* (1959) that for laminar flow of both phases the holdup ratio depends only upon the viscosity ratio and the input flow ratio. Figure 10.43 gives Gemmell and Epsteins's holdup results. Figure 10.44 expresses the pressure gradient results in terms of the ratio of the volume flow, Q_A, of fluid A, under the two-phase condition, to the volume flow, Q_{AF}, of fluid A flowing alone at the same pressure gradient. Because of the linearity between flow rate and pressure gradient in laminar flow, the ratio Q_A/Q_{AF} is identical with the ratio $(\Delta P/L)_A/(\Delta P/L)_{AF}$ between the pressure gradients required for the flow of equivalent volumes of fluid A.

Gemmell and Epstein compared their predictions as given in Figures 10.43 and 10.44 with the data of Russell *et al.* for fluids with a viscosity ratio of 20 and found almost exact agreement under conditions of laminar flow but, as would be expected, increasing deviations as the Reynolds number for the water phase increased. The analysis and Figures 10.43 and 10.44 may be considered confirmed and suitable for application in situations when both fluids flow in

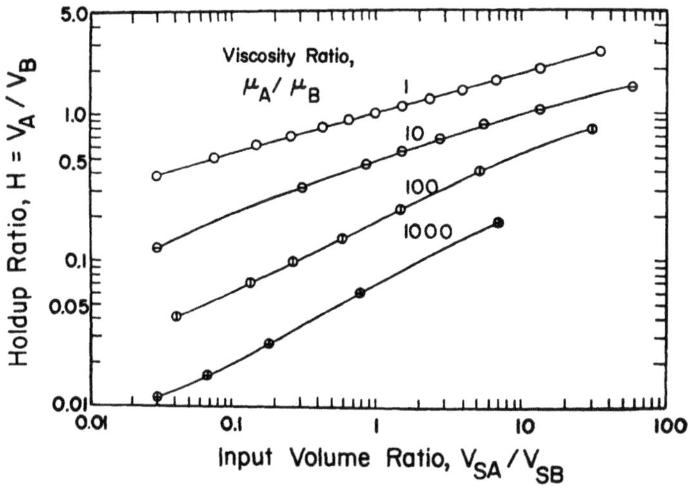

Figure 10.43. Calculated holdup relationships for laminar stratified flow. (From Gemmel, A. R., and N. Epstein, *Can. J. Chem. Eng.*, **40**, 215, 1962.)

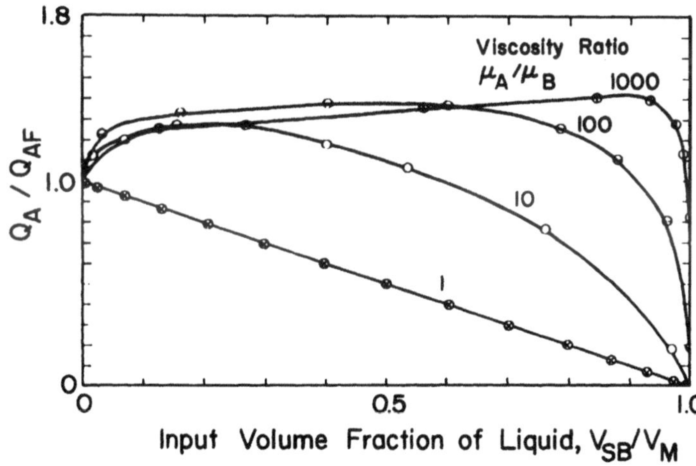

Figure 10.44. Calculated relative flow rates for laminar stratified flow. (From Gemmel, A. R., and N. Epstein, *Can. J. Chem. Eng.*, **40**, 215, 1962.)

laminar motion and the interface is relatively smooth. This may be taken to be the case provided the superficial Reynolds number for each phase is less than 1000 (Gemmell and Epstein, 1962; Lockhart and Martinelli, 1949). This condition is slightly more restrictive than the conditions suggested by Bergelin and Gazley, Govier and Omer, and Russell *et al.* for laminar flow.

5.32 The Laminar-Turbulent or Turbulent-Turbulent Case

For the relatively simple geometry of flat or slightly ripply or pebbled interface stratified flow, it should be possible to develop a reasonably sound theoretical analysis of the holdup and pressure drop effects using our present knowledge of turbulent flow. This has not yet been done and confirmed. Nonetheless, we may at least indicate the general approach and discuss such relevant data as are available.

Charles and Lilleleht (1966) have developed an empirical correlation based on the Lockhart and Martinelli parameters for the pressure gradient in the stratified flow of two immiscible liquids when one is in laminar and the other in turbulent flow. It does not provide information on the *in situ* volume fraction of the phases. Their correlation reproduced in Figure 10.45 is based upon three different liquid-liquid systems flowing in three different conduits, one rectangular and two circular in cross section. It includes data where the interface was slightly wavy. The correlation is in the Lockhart and Martinelli form with

$$\phi_M = \sqrt{\frac{\Delta P/L}{(\Delta P/L)_l}}$$

and

$$X = \sqrt{\frac{(\Delta P/L)_m}{(\Delta P/L)_l}}$$

where the subscripts m, l refer, respectively, to the more viscous and less viscous phases flowing alone at the total mixture velocity. For conditions

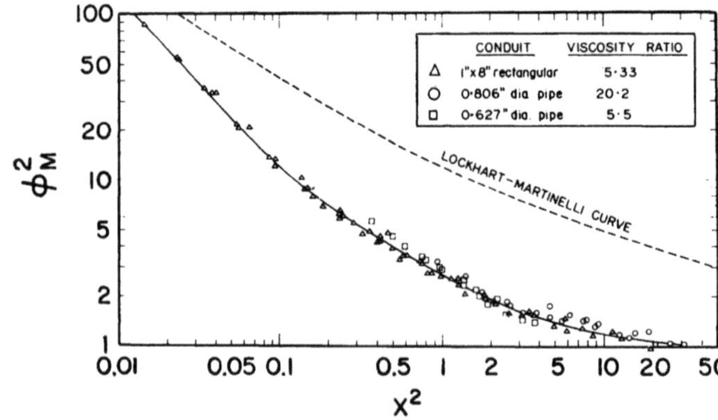

Figure 10.45. Charles and Lilleleht pressure gradient correlation for laminar-turbulent stratified flow. (From Charles, M. E., and L. U. Lilleleht, *Can. J. Chem. Eng.*, **44**, 47, 1966.)

similar to those covered by Charles and Lilleleht, this correlation is recommended.

Another approach, applicable particularly to the turbulent-turbulent case, is to consider each phase as though it were flowing through a duct of irregular shape defined by the part of the pipe wall which it contacts and by the moving interface. This is the rationale behind the Lockhart-Martinelli empirical correlation and the extension of it by Chisholm which we discussed in Section 3.31. Gazley (1949) also followed this approach.

Referring to Figure 10.36 and the nomenclature defined in Section 5.2, we may write force balances for the two phases as

$$A_\alpha \frac{dP}{dL} = \tau_{w\alpha} p_\alpha + \tau_i w_i \tag{10.104}$$

$$(A - A_\alpha) \frac{dP}{dL} = \tau_{w\beta}(\pi D - p_\alpha) - \tau_i w_i \tag{10.105}$$

where $\tau_{w\alpha}$, $\tau_{w\beta}$, and τ_i are, respectively, the shear stresses at the wall for fluid α, at the wall for fluid β, and at the interface. On addition of Equations 10.104 and 10.105, we have

$$A \frac{dP}{dL} = \tau_{w\alpha} p_\alpha + \tau_{w\beta}(\pi D - p_\alpha) \tag{10.106}$$

In these relations w_i, A_α, A_β, p_α, and p_β are all related geometrically to h_β and D as given in Figures 10.37, 10.38, and 10.39. The wall shear stresses, $\tau_{w\alpha}$ and $\tau_{w\beta}$, may be evaluated, at least approximately, by applying the single-phase methods, i.e.,

$$\tau_{w\alpha} = \frac{f_\alpha \rho_\alpha V_\alpha{}^2}{2g_c} \tag{10.107}$$

where

$$f_\alpha = \phi\left(\frac{D_{E\alpha} V_\alpha \rho_\alpha}{\mu_\alpha}\right)$$

$$\tau_{w\beta} = \frac{f_\beta \rho_\beta V_\beta{}^2}{2g_c} \tag{10.108}$$

with

$$f_\beta = \phi\left(\frac{D_{E\beta} V_\beta \rho_\beta}{\mu_\beta}\right)$$

Gazley (1949) reports data for the shear stresses on both the liquid and the gas side of the interface for the stratified flow of air-water mixtures in a tube 2.065 inches in diameter. His measurements cover gas-phase Reynolds numbers, Re'_G (based upon an equivalent diameter for the gas-filled section) from about 6000 to 13,000 and water rates from zero to some 2300 $\mathrm{lb}_m/\mathrm{hr}$. Expressing the shear stresses in terms of interfacial friction factors defined as

$$f_{iL} = \frac{2g_c\tau_{iL}}{\rho_G(V_G - V_L)^2}$$

and

$$f_{iG} = \frac{2g_c\tau_{iG}}{\rho_G(V_G - V_L)^2}$$

Gazley presents plots of f_{iL} and f_{iG} and the ratio f_{iL}/f_{iG} versus Re'_G with liquid flow rate as the parameter. For air flowing over stagnant water his data indicate both f_{iL} and f_{iG} to be close to the values which would be predicted by the Nikuradse equation at Re'_G as long as the interface is smooth. Upon the development of waves at Re'_G of about 10,000, f_{iG} was observed to increase rapidly. With flowing water, f_{iG} exceeded the Nikuradse value corresponding with Re'_G, whereas the values of f_{iL} scattered above and below the Nikuradse value.

Hanratty and Engen (1957), Ellis and Gay (1959), Smith and Tait (1966), and Davis (1969) all report data on the interfacial shear stresses encountered in two-phase flow. Ellis and Gay (1959) present data on velocity profile and interfacial shear stress observed with air flowing over a water surface in a rectangular channel. Their measurements covered a range of Re_G from 3000 to 7500 and Re_L from zero to 3200, with the bulk of the data taken at $1900 < \mathrm{Re}_L < 3200$ under conditions where a ripply surface was observed. In the range of Re_G covered the air flow was in the equivalent of PRW turbulence taken by Ellis and Gay to be within the region $0.1 < y_0 U_i^* \rho/\mu < 3.0$. Defining a friction factor, ϕ, through the expression $\tau = \phi^2 \rho V_G^2/g_c$ Ellis and Gay correlated their smooth wall shear stress data by a Blasius type equation:

$$\phi_{sw} = 0.804\ \mathrm{Re}_G^{-0.285}$$

This equation may be expected to be applicable to the prediction of the shear stress at a smooth gas-liquid interface.

Ellis and Gay correlated their shear stress data for the wavy interface by

$$\phi_i = 0.374\ \mathrm{Re}_G^{-0.111}(y_0/h_G)^{0.0862} \qquad (10.109)$$

An alternate to Equation 10.109 giving a better fit to the data is Equation 10.109a:

$$y_0/h_G = 2.84\,\mathrm{Re}_G^{-0.838}\left(\frac{y_0 U_i^* \rho}{\mu}\right)^{0.892} \tag{10.109a}$$

where y_0 = wave roughness height,
h_G = gas phase thickness,
$\mathrm{Re}_G = h_G V_G \rho_G/\mu_G$ = gas phase Reynolds number,
$U_i^* = \sqrt{\tau_i g_c/\rho}$ = friction velocity.

Conditions under which various surface effects are observed with the air-water system in a rectangular channel are indicated by the data of Hanratty and Engen (1957) replotted in Figure 10.46. These transitions have been generally confirmed by Smith and Tait (1966) and Davis (1969).

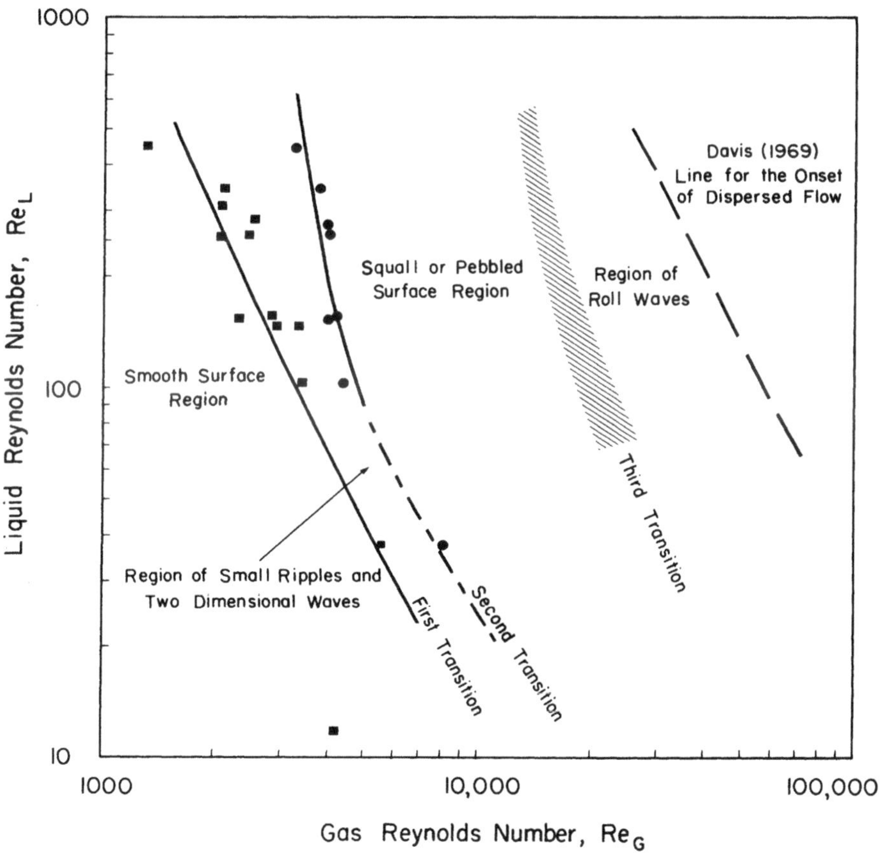

Figure 10.46. Interface structure for stratified flow of air-water mixtures. (From data of Hanratty, T. J., and J. M. Engen, *A.I.Ch.E. Journal*, **3**, 299, 1957.)

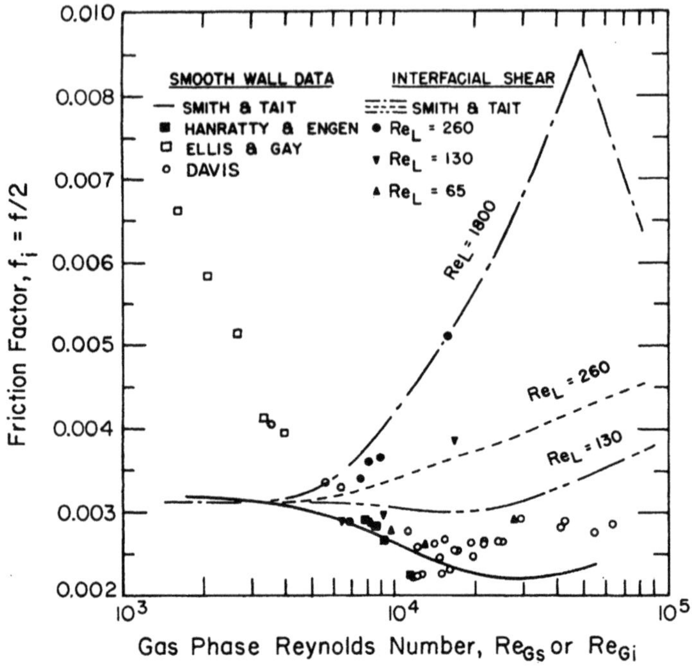

Figure 10.47. Davis' friction factor for interfacial shear. (From Davis, E. J., *Ind. Eng. Chem. Fundam.*, **8**, 153, 1969.)

Figure 10.47 taken from Davis (1969) summarizes and compares his data on the interfacial shear expressed as $f_i = f/2$ and most of that of others. In this figure the Reynolds numbers are defined as

$$\mathrm{Re}_{Gi} = \frac{h_{um} - \bar{h}}{\mu_G} V_G \rho_G \qquad \text{for the interfacial shear data}$$

$$\mathrm{Re}_{Gs} = \frac{h_T - h_{um}}{\mu_G} V_G \rho_G \qquad \text{for the smooth wall data}$$

$$\mathrm{Re}_L = \frac{Q_L \rho_L}{w \mu_L}$$

where h_{um} is the height above the bottom of the channel at which the maximum gas velocity is encountered,
$\bar{h}$ is the mean height of the liquid film,
h_T is the total height of the channel,
w is the width of the channel.

We may observe that where $\bar{h}$ is small, $\mathrm{Re}_{Gs} \simeq \mathrm{Re}_{Gi} \simeq \frac{1}{2}\mathrm{Re}_G$ where $\mathrm{Re}_G = h_T V_G \rho_G / \mu_G$, the normal single-phase gas Reynolds number based upon the total channel height. For the practical use of Figure 10.47, it is necessary to make an estimate of h_{um}.

All of these data to which we have referred have been taken in rectangular ducts of widths 8 to 10 inches and heights of the order of 1 inch where the flow was essentially one dimensional. There are certain unexplained inconsistencies among them. Their applicability to stratified flow in a pipe is somewhat questionable, but at the present state of knowledge there is little more that we can do. They should describe the phenomena in the pipe in an approximate way and at least in principle permit the prediction of the *in situ* volume fraction of the phases and the pressure gradient. Their application is discussed in the following section.

5.4 Flow Pattern, Holdup, and Pressure Gradient

With the stratified flow pattern it is more convenient to deal with the prediction of flow pattern, holdup, and pressure gradient in a single discussion than separately. In their empirical correlations which we discussed in Section 3.3, both Baker and Hoogendoorn gave methods separately applicable to the stratified flow pattern. These may be used either alone or in conjunction with the methods discussed below based upon the flow mechanics.

The starting point in flow pattern prediction is to determine whether one or both phases will be in laminar or turbulent motion. This may be done with the aid of the observations of Bergelin and Gazley, Govier and Omer, and Russell *et al.*

If both phases are in laminar motion we may estimate the holdup and the pressure gradient from Figures 10.43 and 10.44. From the relation

$$H = \frac{V_{S\alpha}}{V_{S\beta}} \frac{A_\beta}{A_\alpha} = \frac{V_{S\alpha}}{V_{S\beta}} \left(\frac{A_\beta}{A - A_\beta} \right)$$

we may obtain A_β and then, from Figures 10.37 and 10.38, w_i and h_β. For a liquid-liquid system where the superficial Reynolds numbers indicate that one or both of the phases are in turbulent motion, the Charles and Lilleleht modification of the Lockhart and Martinelli correlation may be employed. This, however, gives us only the pressure gradient, and recourse would then have to be made to the original Lockhart and Martinelli correlation, or the Chisholm modification of it, for a prediction of the *in situ* volume fraction.

For a gas-liquid system the continuity equations (Equation 10.85 and 10.86), the two force balance equations (Equations 10.104 and 10.105), and the appropriately determined friction factors discussed in Section 5.32, give a system of equations from which the holdup and the pressure gradient may be predicted approximately. To avoid extensive trial and error we may start by estimating the holdup through Lockhart and Martinelli's original correlation (Figure 10.18). From this and the appropriate geometrical relationships of Figures 10.38 and 10.39 we obtain A_α, A_β, w_i, p_α, p_β, and h_β. These permit calculation of V_α and V_β and, therefore, from the methods of Section 5.32, of f_α, f_β, and ϕ_i^2 or f_i, and finally of $\tau_{w\alpha}$, $\tau_{w\beta}$, and τ_i. Then we may solve for $\Delta P/L$ from each of Equations 10.104 and 10.105. If the two results are the same, the originally assumed holdup is consistent with the system of equations; otherwise an adjustment should be made and the calculations repeated.

We do not suggest that the prediction of the flow pattern and other details by this latter method will be exact, but it is sound in principle and will supplement computations based upon the strictly empirical techniques.

5.5 Transition to the Annular Mist and to the Wave Flow Patterns

As the gas or light phase velocity is increased beyond the point at which a pebbly interface occurs, the stratified flow pattern gives way either to annular-mist flow or wave flow. At low superficial liquid velocities, the modest depth of the liquid layer restricts the extent of development of large waves and the transition is to the annular-mist flow pattern; when the liquid rate and depth are greater, however, large waves develop and persist over a sufficient range of conditions to be identified as the separate wave flow pattern.

We have no theoretical basis for predicting which of these transitions will occur or the conditions of the occurrence. Apart from the wholly empirical generalizations of the flow pattern maps discussed in Section 3.2, there are only the data in rectangular channels of Hanratty and Engen (1957), Smith and Tait (1966), and Davis (1969) to guide us. Davis has suggested a transition from stratified to annular-mist flow which is equivalent to that indicated by the dashed line on Figure 10.46.

6 THE WAVE FLOW PATTERN

The wave flow pattern which is observed for the gas-liquid system is a transition pattern encountered at low and modest liquid rates ($V_{SL} < 0.6$ ft/sec) between the pebbly-surfaced stratified pattern and the annular-mist pattern. It is characterized by roll waves at the lowest gas rates where it is observed and by sharply cresting waves which touch the top of the pipe at higher gas rates. The analogous pattern is also encountered with liquid-liquid mixtures.

In neither case is the pattern amenable to theoretical or mechanistic analysis at this time.

From Figures 10.6 and 10.9 we may note, within the wave region for a gas-liquid mixture, that there are approximately linear logarithmic relationships at constant V_{SL} between H and V_{SG} and $\Delta P/\rho_L L$ and V_{SG}. This suggests a way in which we might estimate H and $\Delta P/\rho_L L$ from values determined from mechanistic considerations in the pebbly-surfaced stratified region and the lower range of annular-mist region. The procedure is as follows: For the V_{SL} of interest, estimate from Figure 10.6 the V_{SG} values corresponding with the end of pebbly stratified flow and the start of annular-mist flow. Using methods appropriate for each of these regimes as discussed elsewhere estimate the values of H and $\Delta P/\rho_L L$. Plot the values so calculated on logarithmic coordinates versus V_{SG} and join them with straight lines. The approximate values of H and $\Delta P/\rho_L L$ for the wave flow condition may then be read from the lines at the actual value of V_{SG}.

As discussed in Section 3.3 Baker and Hoogendorn both express their empirical correlations in separate forms applicable to the wave flow pattern.

Kordyban and Ranov (1970) have studied the transition from the wave to the slug flow pattern in a rectangular flow channel. They postulate that the formation of slugs, on increase in liquid and gas rate from those corresponding with the stratified or wave flow pattern, is due to the development of instability of the waves. The relationship which they derive expresses the transition to slug flow in terms of the gas velocity, the liquid height (or the *in situ* volume fraction liquid), the wave length, and the wave height. The latter two variables are shown by their experimental data not to be too important and they conclude that the transition is in fact controlled largely by the gas and the liquid rates (which together determine the holdup). A comparison of their predictions with the transition line of the Baker flow pattern map is satisfactory.

7 THE SLUG FLOW PATTERN

7.1 Introduction

As was described earlier, the important slug flow pattern observed with gas-liquid mixtures is characterized by a series of large distorted gas bubbles which nearly fill the cross section of the tube and which are separated by slugs of liquid which entirely fill the tube cross section. The gas bubbles may be many tube diameters in length; the liquid slugs are usually considerably shorter. The length of the gas bubbles depends upon the flow rates and the fluid properties, and for given values of these it may be influenced by the manner in which the fluids are introduced into the tube. However, if a sufficient length of pipe is available, a quasi-steady state will usually be established at a fixed point downstream, regardless of the method of introducing the two phases. Additionally, as in the case of vertical flow, the length of the gas bubble is influenced by the system pressure and therefore increases as the pressure declines in the direction of flow. The lack of axial symmetry makes the slug flow pattern in horizontal flow more difficult to analyze than its counterpart in vertical flow. The slug flow pattern is also observed with mixtures of immiscible liquids, and the preceding remarks apply to them also.

There are relatively few analyses of the slug flow pattern, and none of these is completely satisfactory for the prediction of pressure drop and holdup from flow variables and pipe size. The analyses available, however, do give some insight into the phenomenon and are useful for the correlation of experimental data. We will discuss some of the more important works and present one of these in some detail. Our discussion will be oriented to gas-liquid mixtures, but much of it will apply also to liquid-liquid mixtures.

7.2 Overall Continuity Considerations

The relationships resulting from the application of the continuity equation for horizontal slug flow are essentially the same as those we have already developed in Chapter 8 for vertical slug flow. In horizontal flow the liquid film surrounding the distorted gas bubble is ordinarily much thicker at the bottom than at the top of the pipe, but we may think of a mean film thickness, δ, such that

$$A_f = \pi \delta D$$

Thus we have

$$V_M = \frac{Q_G + Q_L}{A} = V_{SG} + V_{SL} \tag{8.88}$$

$$V_f = \frac{V_M - V_b(1 - 2\xi)}{2\xi} \tag{10.110}$$

$$V_b = \frac{V_M - 2\xi V_f}{1 - 2\xi} \tag{10.111}$$

where V_f and V_b are, respectively, the average actual velocities of the liquid in the film and the gas bubble and $\xi = 2\delta/D$. Equations 10.110 and 10.111 are identical with Equations 8.92a and 8.93a except that we now define V_f as positive in the direction of bulk flow.

When we may neglect the volume of the small gas bubbles in the liquid slug, Equations 8.31, 8.95, and 8.96 apply. Also if we neglect the volume of any liquid entrained in the large gas bubble, Equations 8.97, 8.98, and 8.99 are applicable. When the volume of small bubbles in the liquid slug is significant Equations 8.100, 8.101, and 8.102 are applicable.

Hubbard (1965) has also derived an equation for the liquid flow rate from continuity considerations. Hubbard's equation in our notation is

$$Q_L = [L_S A(1 - E_{GSL}) + L_b A_f]\omega' - Y \tag{10.112}$$

where E_{GSL} is the volume fraction of gas over the length L_S containing the liquid slug, Y is the volume rate at which the large bubble picks up liquid ahead of it, and ω' is a frequency defined by

$$\omega' = \left[V_M + \frac{Y}{A(1 - E_{GSL})}\right]\frac{1}{L_b + L_s} \tag{10.113}$$

We note that the quantity $V_M + Y/A(1 - E_{GSL})$ in Equation 10.113 is the slug velocity plus a component due to the buildup of the slug resulting from flow to it from the liquid film ahead. The total quantity is the bubble velocity, and Equation 10.113 is therefore an alternate form of Equation 8.95. We discuss this later.

The frequency used in Hubbard's work, ω', is the same as the frequency ω, defined by Equation 8.31. Hubbard's model will be described in more detail in the next section. In order to gain some insight into the significance of Y, we solve Equations 8.102, 10.112, and 10.113 for the liquid pickup rate to obtain

$$Y = \frac{A_f(V_M - V_f)}{1 - \dfrac{A_f}{A(1 - E_{GSL})}} \tag{10.114}$$

If we assume E_{GSL} to be small compared to 1 and A_f to be small compared to A, we have

$$Y = \frac{A_f A}{A - A_f}(V_M - V_f) \simeq A_f(V_M - V_f) \tag{10.115}$$

Equation 10.115 reveals that Hubbard's liquid pickup rate, Y, is approximately equal to a relative velocity of the slug multiplied by the cross-sectional area of the film.

7.3 Overall Momentum Considerations

7.31 The Kordyban Model

Kordyban (1961) was the first to propose a model for slug flow in horizontal tubes. The model is shown schematically in Figure 10.48. He assumed that part of the liquid flows in the lower part of the tube obeying the laws of open channel flow, whereas the remainder of the liquid flows in truncated disk-shaped slugs which alternate with the gas bubbles in the upper part of the tube. In this model, the liquid slug must move at the average velocity of the gas bubble and "skate" over the top of the more slowly moving liquid below it. Based upon this concept, a pressure drop expression containing a frictional and an acceleration term was developed. The frictional component predicted by this method is found to be less satisfactory than the prediction from the empirical Lockhart-Martinelli correlation. In a later paper Kordyban and Ranov (1963) present new data for slug flow and indicated that Kordyban's early model is oversimplified and inadequate.

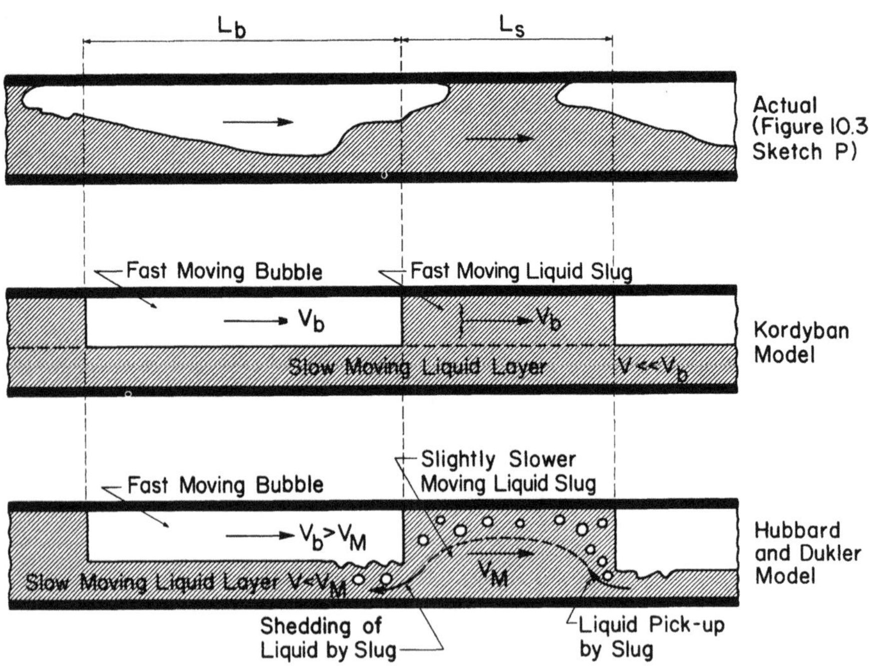

Figure 10.48. Schematic representations of slug flow.

7.32 The Hubbard and Dukler Model

Hubbard (1965) and Hubbard and Dukler (1968) have presented another model of the slug flow pattern based upon their visual and quantitative studies. The model is illustrated in Figure 10.48. Using essentially their words we may describe the sequence of events.

"*1.* Liquid and gas flow concurrently into a pipe. Near the entrance the liquid flows as a stratified phase with the gas passing above at a higher velocity. At gas and liquid velocities where slug flow takes place, the gas velocity is sufficient to generate waves on the surface of the liquid. These waves grow in amplitude as they travel along the pipe and in a short distance they bridge the pipe, blocking the gas flow.

2. As soon as the bridging occurs, the liquid in the bridge is accelerated to the gas velocity. The liquid appears to be accelerated uniformly across its cross section, thereby acting as a scoop, picking up all the slow moving liquid in the film ahead of it and accelerating it to the slug velocity. By this mechanism, the wave grows and becomes a liquid slug.

3. As the slug is formed and moves down the pipe, liquid is shed from its trailing edge and falls, under the influence of gravity, to the bottom of the pipe reforming the stratified film.

4. Shortly after a slug is formed it sweeps up all the excess liquid that had entered the pipe since the last slug was formed. At this time, the liquid film which is about to be picked up by the slug is only that liquid which has been shed from the preceding slug. Since the slug is picking up liquid at the same rate that it is shedding it, the length of the slug stabilizes.

5. Since the slug has a higher kinetic energy than the liquid film, the film penetrates a distance into the slug before it is finally assimilated at the slug velocity. This overrunning phenomenon creates an eddy at the front of the slug which is essentially a mixing vortex and, therefore, exists without a pressure gradient. The distance of penetration constitutes the length of the mixing eddy. Gas is entrapped in this mixing zone. This gas migrates to the top of the pipe due to gravity forces and is subsequently trailed off the end of the slug with its neighboring liquid.

6. As the gas rate and, consequently, the slug velocity increase, the degree of aeration of the slug increases. Ultimately the gas forms a continuous phase along the top of the slug. When this occurs the slug begins bypassing some of the gas. At this point, the slug no longer maintains a competent bridge to block the gas flow and the character of the flow changes. This point is referred to as the beginning of "blow-through." In the blow-through regime the slug velocity begins to fall below that of the gas phase due to the bypassing of gas through the slug.

7. As blow-through becomes more pronounced with increasing gas rate, the slugs begin to dissipate due to the entrainment of their liquid in the high velocity gas phase. This entrainment process begins the transition to annular-mist flow."

The significant differences between Hubbard's and Kordyban's visualization of the slug flow pattern are, first, that Hubbard assumes that the fast moving liquid slug extends over the full area of the pipe, picking up and shedding liquid from and to the liquid film and, secondly, that Hubbard provides for a gas saturation in the liquid slug.

Hubbard and Dukler's notation and their manipulation of equations are rather complex. We present in the following a simplified version of their development in our terminology. Figure 10.48 shows an average slug unit consisting of an average liquid slug plus the associated gas bubbles and liquid film. The pressure drop across an average liquid slug may be considered to be composed of two parts

$$\Delta P_s = \Delta P_a + \Delta P_f \tag{10.116}$$

The first part, ΔP_a, is the pressure drop that results from the acceleration of the slow moving liquid film to slug velocity. The second part, ΔP_f, is the pressure drop required to overcome wall shear associated with the slug. The acceleration component of pressure drop depends upon the mass of liquid picked up by the slug from the liquid film and accelerated to the slug velocity. The acceleration component, ΔP_a, may be expressed as

$$\Delta P_a = \frac{Y\rho_L}{Ag_c}(V_M - V_f) \tag{10.117}$$

The friction component of the pressure drop, ΔP_f, may be expressed in terms of the slug velocity and length through a modified form of the Fanning equation

$$\Delta P_f = \frac{2f_{LM}\rho_M V_M^2 L_s}{g_c D} \tag{10.118}$$

where $f_{LM} = \phi(\mathrm{Re}_M)$,

$$\rho_M = \rho_G E_{GSL} + \rho_L(1 - E_{GSL}) \simeq \rho_L(1 - E_{GSL}) \tag{10.119}$$

$$\mathrm{Re}_M = \frac{DV_M\rho_M}{\mu_M}$$

$$\mu_M = \mu_G E_{GSL} + \mu_L(1 - E_{GSL}) \simeq \mu_L(1 - E_{GSL}) \tag{10.120}$$

Hubbard and Dukler assumed that f_{LM} and Re_M, so defined, would be related as are f and Re in single-phase turbulent flow. They chose the relationship (Drew, Koo, and McAdams, 1932)

$$f = 0.00140 + 0.125\ \mathrm{Re}_M^{-0.32} \tag{4.36}$$

To calculate the two components of the pressure drop over a typical slug, we therefore require, in addition to the quantities which would normally be specified (V_{SL}, V_{SG}, ρ_L, μ_L, ρ_G, μ_G and D), values or estimates of V_M, Y, V_f, E_{GSL}, and L_s. The quantity V_M may immediately be identified, by an overall volume balance through a typical slug, as

$$V_M = V_{SL} + V_{SG} \tag{8.88}$$

As we saw in Equation 10.113, the liquid pickup, Y, is related to the average slug velocity, V_M, and the average velocity of the large bubble, V_b, by

$$V_b = V_M + \frac{Y}{A(1 - E_{GSL})} \tag{10.121}$$

We may also write a second volume balance for either the liquid

$$V_{SL} = [L_s(1 - E_{GSL}) + L_b(1 - E_{Gb})]\omega - \frac{Y}{A} \tag{10.122}$$

or the gas

$$V_{SG} = (L_s E_{GSL} + L_b E_{Gb})\omega + \frac{Y}{A}\frac{E_{GSL}}{(1 - E_{GSL})} \tag{10.123}$$

where E_{Gb} = the volume fraction gas over the length L_b containing the large bubble,

$$\omega = \frac{V_b}{L_s + L_b} \tag{8.95}$$

i.e., the frequency of passage of slugs or bubbles as discussed earlier.

Note that E_{Gb} is not the same as E_{GB} defined following Equation 8.120 in Chapter 8.

Hubbard has derived the following empirical equation relating Y, E_{GSL}, and V_M

$$Y = CA(1 - E_{GSL})V_M \tag{10.124}$$

where C is a function of Re_M, but for the case studied by Hubbard it is about 0.25. Gregory and Scott (1969) find that a better fit of data is obtained with

$C = 0.35$. Combining Equation 10.124 with Equation 10.121, we see that the two equations are equivalent to

$$V_b = (1 + C)V_M \simeq 1.25V_M \tag{10.125}$$

It is to be noted that the term $(1 + C)$ used by Hubbard, and by Gregory and Scott is really the same C_0 as we used it in Section 7.41 of Chapter 7 and also that the value of $(1 + C) \simeq 1.25$ is close to the value of C_0 for a Taylor bubble.

No general correlation is available for the volume fraction gas in the slug, E_{GSL}. Measured values of E_{GSL} by Hubbard using an impact probe are shown in Figure 10.49. This quantity is hard to measure and the absence of a correlation is a serious drawback of Hubbard's model. Approximate values for E_{GSL} may be estimated using Figure 10.49.

The data of Hubbard and Dukler also provide empirical relationships for V_f and L_s. These are

$$V_f = V_M\left(\frac{1}{1 + B\dfrac{V_M}{\omega}}\right) \tag{10.126}$$

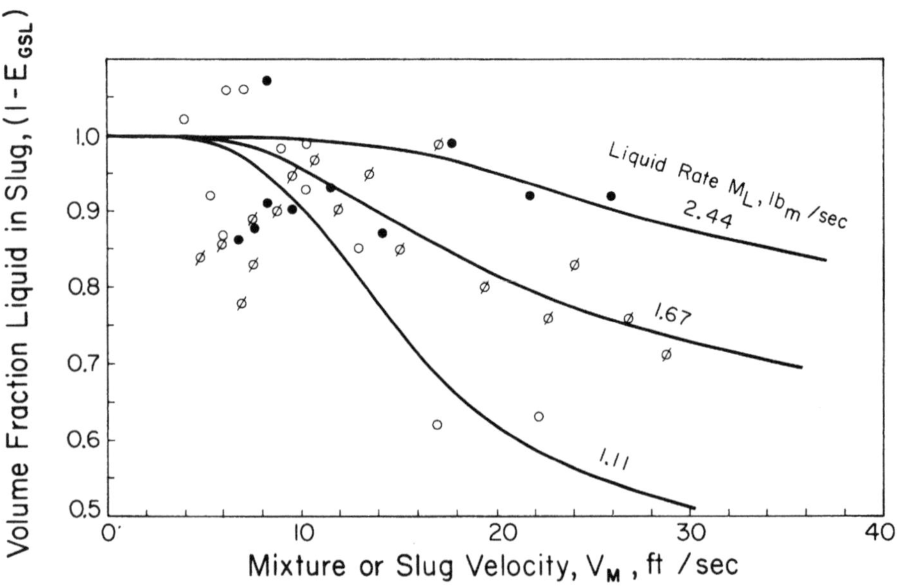

Figure 10.49. Volume fraction liquid in the liquid slug. (From Hubbard, M. G., Doctoral Dissertation, University of Houston, 1965.)

where B has the value 0.2, and

$$L_s = \frac{1 + C}{(1 - E_{GSL})\omega}\left(\frac{\dfrac{V_{SL}}{\omega} - \beta}{1 - \dfrac{\beta\omega}{V_M}}\right) \tag{10.127}$$

where β is defined as

$$\beta = \frac{(1 - E_{Gb})}{B} \ln\left(1 + B\frac{V_M}{\omega}\right) \tag{10.128}$$

Although not a part of the original Hubbard and Dukler correlation, Gregory and Scott (1969) have proposed a tentative empirical correlation for the frequency ω based upon their data and that of Hubbard (1965) in 0.75- and 1.5-inch pipes. The correlation is given by

$$\omega = 0.0226\left[\frac{V_{SL}}{gD}\left(\frac{19.75}{V_M} + V_M\right)\right]^{1.2} \tag{10.129}$$

Equations 8.88, 8.95, and 10.121 through 10.129, together with Figure 10.49, provide a system of relations from which estimates of L_s, L_b, V_f, V_b, Y, E_{GSL}, E_{Gb} and ω may be made. Hubbard and Dukler substitute Equations 10.124 and 10.126 into Equation 10.117 to get an expression for ΔP_a

$$\Delta P_a = \frac{\rho_L(1 - E_{GSL})V_M^2}{g_c} C\left(1 - \frac{1}{1 + \dfrac{BV_M}{\omega}}\right) \tag{10.130}$$

To obtain the frictional part of the pressure drop, ΔP_f, they substitute in Equation 10.118 for L_s from Equation 10.127 and for ρ_M from Equation 10.119. The resulting equation for ΔP_f is

$$\Delta P_f = \frac{2f_{LM}\rho_L V_M^2(1 + C)}{g_c D\omega}\left(\frac{\dfrac{V_{SL}}{\omega} - \beta}{1 - \dfrac{\beta\omega}{V_M}}\right) \tag{10.131}$$

We may add ΔP_a and ΔP_f to obtain the pressure drop over a single slug as

$$\Delta P_s = \frac{\rho_L V_M^2}{g_c}\left\{\frac{2f_{LM}(1 + C)}{D\omega}\left[\frac{\dfrac{V_{SL}}{\omega} - \beta}{1 - \dfrac{\beta\omega}{V_M}}\right] + \frac{(1 - E_{GSL})CBV_M}{\omega + BV_M}\right\} \tag{10.132}$$

Hubbard and Dukler now assume the pressure drop across the length, L_b, of the large bubbles to be negligible and, in effect, write

$$\frac{\Delta P}{L} = \frac{\Delta P_s}{L_s + L_b} \tag{10.133}$$

In spite of its shortcomings this is the only mechanistic model available for horizontal slug flow. Further work is required before this or any other model for slug flow can be applied to practical problems.

Hubbard and Dukler did not propose a method for calculating the overall holdup for the system, but this may be done by substituting for v_b in Equation 8.139 to obtain

$$E_G = \frac{1}{L_s + L_b}(L_s E_{GSL} + L_b E_{Gb}) \tag{10.134}$$

It is worth noting that this overall treatment of horizontal slug flow is similar to that outlined for vertical slug flow in Chapter 8. The differences are merely in the manner of estimating V_b and L_s.

7.4 Other Correlations

Baker's modification of the Lockhart and Martinelli correlation discussed in Section 3.31 includes a separate empirical correlation for slug flow. This may be used to estimate the pressure gradient. Also the correlation of Hoogendoorn and Buitelaar discussed in Section 3.34 gives relations specifically applicable to the slug region.

Hughmark (1965) has devised a simple empirical correlation for determining the overall volume fraction gas and the pressure gradient in the slug flow region. From Equations 8.67 and 8.141 derived for vertical flow but equally applicable to horizontal flow, we have

$$\bar{E}_G = \frac{V_{SG}}{V_b} = \frac{V_{SG}}{C_0 V_M + V_{bs}} \tag{8.141a}$$

For vertical flow the velocity of the bubbles was affected both by the rise velocity in a stagnant liquid, V_{bs}, and the position of the bubble in the cross section. In horizontal flow the former effect is not present but the latter is to be expected. Hughmark correlates data of Alves (1954), Baker (1954), Hoogendoorn (1959) and Brigham (1962) covering pipe sizes from 1 to 7.75 inches and air-water, air and gas-oil, and air-glycol systems in terms of the velocity of the bubble and that of the liquid in the slug, i.e.,

$$V_b = (1 + K)V_M = C_0 V_M \tag{10.135}$$

where K is determined by an empirical relation with the Reynolds number for the liquid slug, $DV_M\rho_L/\mu_L$. This relationship is given in Figure 10.50.

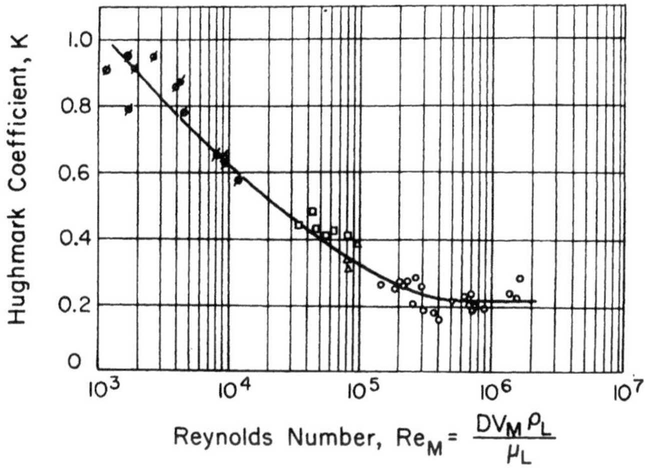

Figure 10.50. Hughmark coefficient, K. (From Hughmark, G. A., *Chem. Eng. Sci.*, **20**, 1007, 1965.)

Combining Equations 10.135 and 8.141a we have

$$\bar{E}_G = \frac{V_{SG}}{(1+K)V_M} \tag{10.136}$$

It is interesting to note that at liquid slug Reynolds numbers suggesting laminar flow, the constant K approaches unity (C_0 approaches 2.0) indicating that the absolute bubble velocity is approaching double the average liquid velocity or that the bubble tends to travel at the velocity of the liquid in the center of the tube. Similarly, at higher Reynolds numbers, the ratio of the absolute velocity of the bubble to the average velocity of the slug approaches the ratio of u_m/V for turbulent flow.

Hughmark proposes that the pressure gradient be calculated simply as though the pipe were carrying only the liquid at the average actual liquid velocity (not the liquid slug velocity), i.e.

$$\frac{\Delta P}{L} = \frac{2f\rho_L V_L^2}{g_c D}$$

where f = the single-phase friction factor based on Re = $DV_L\rho_L/\mu_L$,
$V_L = V_{SL}/(1 - \bar{E}_G)$.

The comparison of his calculations and experimental data for air and gas with a variety of liquids in pipes from 0.25 to 7.75 inches in diameter is shown in

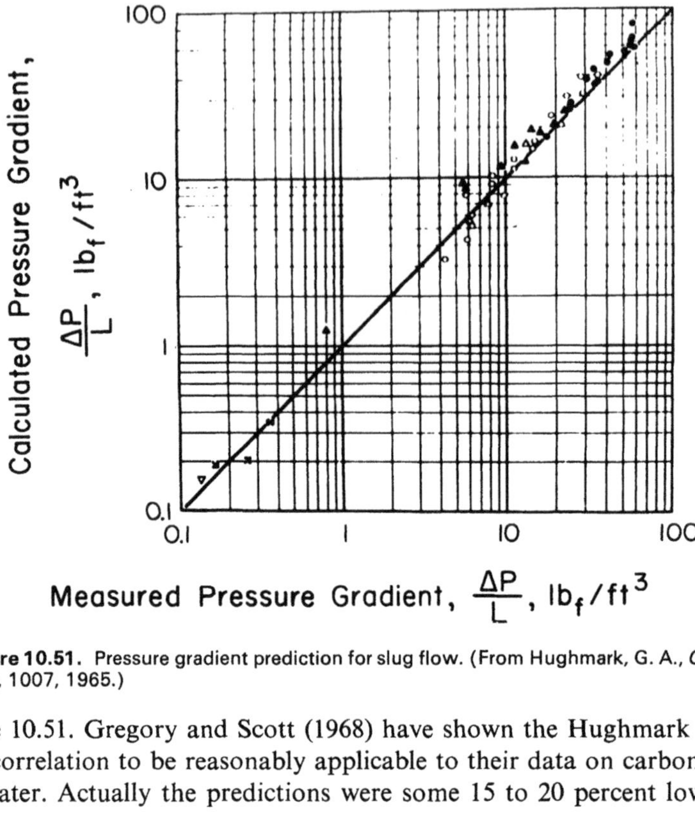

Figure 10.51. Pressure gradient prediction for slug flow. (From Hughmark, G. A., *Chem. Eng. Sci.*, **20**, 1007, 1965.)

Figure 10.51. Gregory and Scott (1968) have shown the Hughmark pressure drop correlation to be reasonably applicable to their data on carbon dioxide and water. Actually the predictions were some 15 to 20 percent low.

7.5 Flow Pattern, Holdup, and Pressure Gradient

The details of the flow pattern for horizontal slug flow cannot be predicted exactly, but the model of Hubbard and Dukler and some of their empirical relations are helpful in a broad definition of the pattern. From this work as discussed in Section 7.32 and the overall continuity equations, we may make estimates of L_s, L_b, V_b, V_f, A_b, A_f, ω, $\bar{E}_{GSL}$, and $\bar{E}_{Gb}$. These describe the flow pattern.

The overall volume fraction of gas $\bar{E}_G$ may be estimated from the quantities discussed above and Equation 10.134 or by Equation 10.136 using Hughmark's correlation.

For the pressure gradient we have the three empirical methods of Baker, Hoogendoorn and Buitelaar, and Hughmark discussed in Section 7.4. These methods would appear to be of comparable reliability. The procedure of Hubbard and Dukler for calculating the pressure gradient may also be applied but it is not fully confirmed.

7.6 Transition to Annular-Mist Flow

The transition from the slug to the annular-mist flow pattern occurs, as we discussed in Section 7.1 and as described in Section 7.32 relating to the Hubbard and Dukler model, when the gas rate and the gas-liquid ratio become sufficiently high that the liquid slugs are absorbed into an annular film and entrained droplets in the bubbles and the bubbles in turn grow to become a continuous phase.

We have no method based upon the mechanics of flow for calculating the conditions of the transition. The empirical transition lines and flow patterns discussed in Section 3.1 may be applied.

8 THE ANNULAR-MIST FLOW PATTERN

8.1 Introduction

In the gas-liquid system—at superficial gas velocities of the order of 20 ft/sec at very low, ranging to about 100 ft/sec at modest and high liquid velocities—the wave or slug flow pattern gives way to the annular-mist flow pattern. This pattern is very similar to that encountered in vertical flow. The differences are those one would expect having in mind the effects of gravity. Here at modest liquid and gas rates, all the liquid flows in an "annular" film on the pipe wall. The film is not uniform in thickness, the gravitational effects causing an increase in thickness in the low parts of the pipe. Theoretically the film could be in laminar motion, but this occurs only if the liquid is of abnormally high viscosity. Normally the film consists of a laminar sublayer, a transition zone, and a region in contact with the gas in which turbulence is present. At low liquid rates the film is relatively smooth with small ripples, but as the liquid rate is increased the surface of the film becomes more distorted with the formation of "disturbance waves." Under conditions of high gas rates, droplets of liquid are torn from the film and carried suspended in the gas. Under any given conditions within this range, a dynamic equilibrium is established between the rate of droplet formation from the film and the rate of droplet condensation by impingement on the film (see Anderson and Russell 1970a,b). This phenomenon of droplet interchange is of considerable importance in multiphase heat and mass transfer studies. Figure 8.53 depicting, in vertical annular flow, the range of velocities of the two phases for which various regimes of annular flow are encountered is also qualitatively valid for horizontal flow.

We might expect that at sufficiently high gas velocities the film would be entirely eroded and all of the liquid would be carried in the form of suspended droplets in the gas. Actually, data of Aziz and Govier (1962) suggest that this is not the case and that some equilibrium film thickness is eventually reached. These data for the flow of natural gas–water mixtures in a 1-inch pipe are shown in Figure 10.52. It may be seen that as the gas rate is increased the pressure drop ratio $\left(\frac{\Delta P}{L}\right)_{tp} \Big/ \left(\frac{\Delta P}{L}\right)_{SG}$ approaches a fixed value of approximately 2, regardless of the liquid rate. Similar behavior was observed in a 2-inch pipe, although it was not possible in this case to achieve velocities sufficiently high to demonstrate completely the attainment of the constant ratio. The ex-

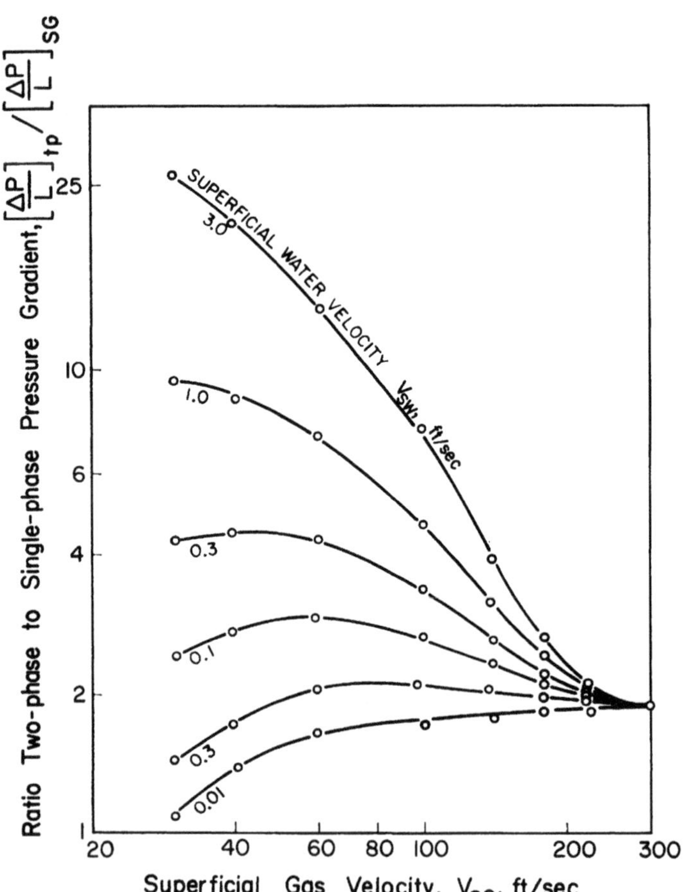

Figure 10.52. Pressure gradient ratio for the flow of natural gas–water mixtures in a 1.049-inch pipe. (From Aziz, K., and G. W. Govier, *Can. J. Chem. Eng.*, **40**, 51, 1962.)

planation for the constant pressure gradient ratio, notwithstanding substantial variation in the input mixture density, would appear to be that the increased pressure gradient over that for the gas alone is caused largely by a moving film of constant properties. Figure 10.52 clearly shows that even at a relatively high gas rate the assumption of homogeneous flow could cause large errors in pressure drop calculations.

The equivalent of the annular-mist flow pattern is also encountered with liquid-liquid systems where, in the limit where there is no film, it may also be considered the same as the dispersed bubble flow pattern with an inversion of the phases. Our discussion in this section is restricted to the gas-liquid case.

8.2 Overall Continuity Considerations

The main difference between the horizontal and vertical annular flow is the asymmetry caused by the gravitational force. A completely rigorous analysis must take this into account. For the present we will rely on a simplified treatment based upon a mean film thickness. Once this assumption is made all the equations derived for the vertical flow case in Section 7.2 of Chapter 8 apply.

8.21 Pure Annular Flow—No Entrainment

Assuming a film of mean thickness δ we have from Equations 8.151, 8.152, 8.154, and 8.155

$$V_{SG} = V_G(1-\xi)^2 \simeq V_G(1-2\xi) \tag{10.137}$$

$$V_{SL} = V_f(2\xi - \xi^2) \simeq V_f 2\xi \tag{10.138}$$

Some additional useful relations for mist free annular flow are

$$E_G = (1-\xi)^2 \tag{10.139}$$

$$\xi = 1 - \sqrt{E_G} \tag{10.140}$$

$$r_i = R\sqrt{E_G} \tag{10.141}$$

where r_i is the interface radius.

8.22 Annular-Mist Flow

Denoting the volume fraction of the droplets in the gas core by e, from Equations 8.158 and 8.157 we have

$$V_{SG} = V_c(1-\xi)^2(1-e) \simeq V_c(1-2\xi)(1-e) \tag{10.142}$$

and

$$V_{SL} = V_f(2\xi - \xi^2) + V_e e(1 - \xi)^2 \simeq V_f 2\xi + V_e e(1 - 2\xi) \qquad (10.143)$$

where V_c is the average velocity of the mixture in the core, and V_e is the average velocity of the entrained drops. Equation 10.143 may be further simplified if we assume no slip betweeen the droplets and the gas in the core, i.e., $V_c = V_e$. This is often a good approximation in horizontal flow.

8.3 Momentum Considerations

8.31 Concentric Film Flow—No Entrainment

In Chapter 8 (Section 7.31) we considered the case of vertical annular flow with no entrainment and derived the velocity profile, flow rate, and film thickness equations assuming laminar flow in the film and presented the results of some approximate analyses for the laminar-turbulent film. These equations may be simplified for horizontal flow in the limiting case of a concentric film by setting the gravitational force terms equal to zero.

The Combined Force Balance For concentric film, horizontal flow, the combined force balance equation (Equation 8.169) simplifies without assumption to

$$\tau = -\frac{dP}{dL}\frac{r}{2} \qquad (10.144)$$

where y has been replaced by its equivalent $r = R - y$ and the film is in the range $R - \delta \leq r \leq R$. Equation 10.144 is identical to the equation for the shear stress distribution in horizontal single-phase flow and is applicable over the full pipe cross section in the flow of two (or more) phases.

The Velocity Profile, Total Flow, and Mean Thickness of the Liquid Film; the interfacial Shear Stress The analyses given under the corresponding headings in Section 7.31 of Chapter 8 also apply, with the gravitational force term equated to zero, to the concentric film horizontal case. In addition, we have the analyses of Levy (1952) and of Abramson (1952), both of which are specific for horizontal flow, and that of Turner (1966) which is generalized for flow in a pipe of any inclination.

Levy (1952) was the first to present a rather complete analysis of the horizontal annular flow problem. He considered three different cases: (1) laminar flow in the film and laminar flow in the gas core (laminar-laminar), (2) laminar flow in the film and turbulent flow in the gas core (laminar-turbulent), and (3) turbulent flow in both the film and the gas core (turbulent-turbulent). He assumed the 1/7 power law velocity profile for turbulent flow. In the following we give a brief derivation of Levy's equation for the laminar-turbulent case and present his final equations for the other cases.

For laminar flow in the film we substitute the constitutive relationship for a Newtonian fluid in Equation 10.144 and integrate with respect to r with the condition that $u = 0$ at $r = R$. The resulting parabolic velocity profile is

$$u_L = -\frac{g_c}{4\mu_L}\frac{dP}{dL}(r^2 - R^2) \tag{10.145}$$

The flow rate in the film is given by

$$Q_L = -\frac{\pi g_c}{8\mu L}\frac{dP}{dL}R^4(2\xi - \xi^2)^2 \tag{10.146}$$

The velocity profile in the gas core may be written as

$$u_G = u_L + u' \tag{10.147}$$

For the case of laminar flow in the core, u' may be obtained by integrating

$$\frac{du'}{dr} = -\frac{g_c}{\mu_G}\left(\frac{dP}{dL}\right)\frac{r}{2} \tag{10.148}$$

with the boundary conditions

$$u' = 0 \qquad \text{at } r = r_i = R - \delta = R\sqrt{E_G}$$

and

$$\frac{du'}{dr} = 0 \qquad \text{at } r = 0$$

For the turbulent core the 1/7 power law velocity profile is

$$u' = 8.74\left[\frac{r_i}{2\rho_G}\left(-\frac{dP}{dL}\right)\right]^{4/7}\left(\frac{\mu_G}{\rho_G}\right)^{1/7}(r_i - r)^{1/7} \tag{10.149}$$

Equation 10.149 is the usual Blasius profile for flow in a circular pipe of radius r_i. Equations 10.147 and 10.149 may be combined to give the velocity profile in the core. The interface velocity, u_i, may be obtained from Equation 10.145,

$$u_i = -\frac{g_c}{4\mu_L}\frac{dP}{dL}(r_i^2 - R^2) \tag{10.150}$$

The gas flow rate is obtained by integrating the gas velocity equation, i.e.,

$$Q_G = 2\pi \int_0^{r_i} u_G \, r dr$$

$$= 2\pi \left[\frac{1}{\mu_L} \frac{dP}{dL} \frac{r_i^2(r_i^2 - R^2)}{2} + 3.57 \frac{U^{*8/7} r_i^{15/7}}{\left(\frac{\mu_G}{\rho_G}\right)^{1/7}} \right] \tag{10.151}$$

where

$$U^* = \sqrt{\frac{\tau_w g_c}{\rho_G}} = \sqrt{\frac{R}{2} \frac{g_c}{\rho_G} \left(-\frac{dP}{dL}\right)}$$

Equations 10.146 and 10.151 are two implicit equations which may be solved for the unknowns, dP/dL and δ. From D and δ we may calculate the *in situ* volume fraction gas, E_G.

Levy has expressed his final equations in terms of the Lockhart and Martinelli parameters. These equations are given below.

a. For the laminar-laminar case, we have

$$\phi_{Lll} = \phi_{Llt} = \frac{1}{1 - E_G} \tag{10.152}$$

$$\frac{\phi_{Lll}}{\phi_{Lll} - 1} = \frac{1}{E_G} = 1 + X^2 \frac{\mu_G}{\mu_L} + X\sqrt{1 + \left(X \frac{\mu_G}{\mu_L}\right)^2} \tag{10.153}$$

b. For the laminar-turbulent case, Equation 10.152 applies, and in addition we have

$$\frac{1 - E_G}{E_G} = \frac{X^{8/7} E_G^{5/14}}{(1 - E_G)^{1/7}} + 0.016 X^2 \frac{\mu_G}{\mu_L} \left(\frac{\rho_G V_{SG} D}{\mu_G}\right)^{3/4} \tag{10.154}$$

c. For the turbulent-turbulent case, the equations are

$$\phi_{Ltt} = \frac{1}{(1 - \sqrt{E_G})^{1/2} (1 + \frac{8}{7}\sqrt{E_G})^{7/8}} \tag{10.155}$$

$$\frac{(1 - \sqrt{E_G})(1 + \frac{8}{7}\sqrt{E_G})}{E_G} = X^{8/7} \left[\frac{60}{49} \left(\frac{\rho_G}{\rho_L}\right)^{3/7} \left(\frac{\mu_G}{\mu_L}\right)^{1/7} + \frac{E_G^{5/14}}{(1 - \sqrt{E_G})^{1/7}} \right] \tag{10.156}$$

The theoretical curves of Levy for ϕ vs. X are compared with the empirical correlations of Lockhart and Martinelli in Figures 10.53, 10.54, and 10.55. The agreement is quite good. There are three additional parameters involving

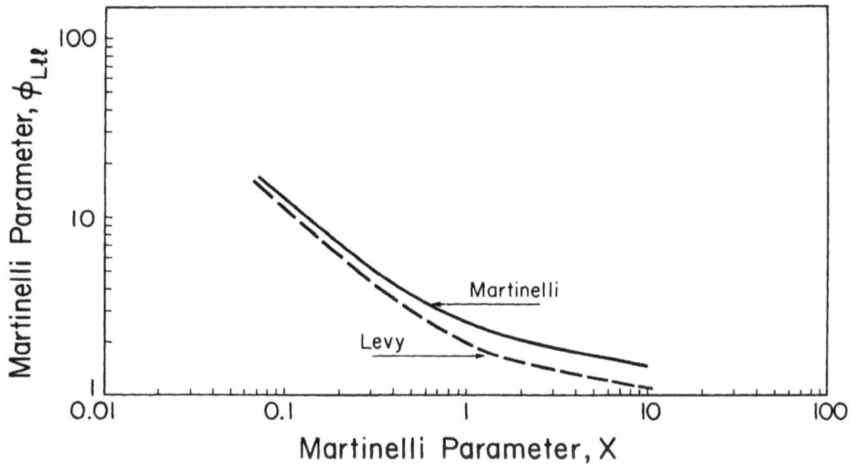

Figure 10.53. Comparison of the Levy and Martinelli relationships for ϕ_{Ltt}. (From Turner J. M., Doctoral Dissertation, Dartmouth College, Hanover, New Hampshire, 1966.)

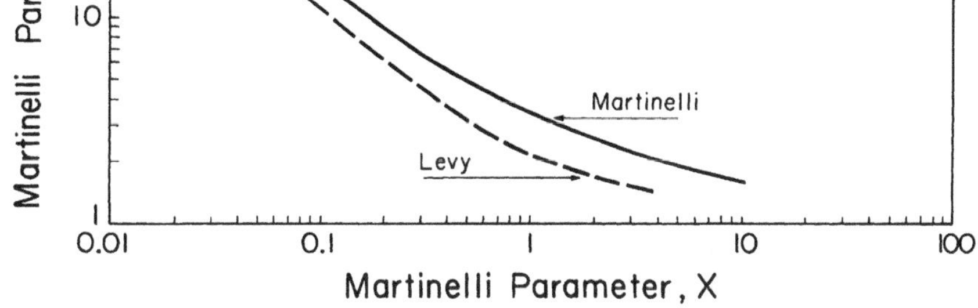

Figure 10.54. Comparison of the Levy and Martinelli relationships for ϕ_{Ltt}. (From Turner, J. M., Doctoral Dissertation, Dartmouth College, Hanover, New Hampshire, 1966.)

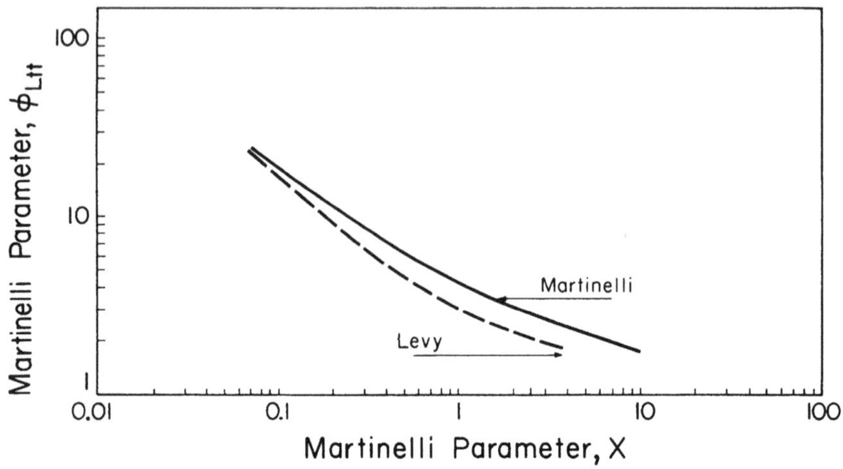

Figure 10.55. Comparison of the Levy and Martinelli relationships for ϕ_{Ltt}. (From Turner, J. M., Doctoral Dissertation, Dartmouth College, Hanover, New Hampshire, 1966.)

the ratio μ_G/μ_L in Levy's model discussed above. For most practical cases, terms involving this ratio can be neglected, and this has been done in Figures 10.53 to 10.55.

Abramson (1952) has presented another analysis of annular two-phase flow using the universal velocity profile in place of the 1/7 power law velocity profile employed by Levy. This analysis is difficult to use and will not be presented here.

Turner (1966) has derived general equations for annular flow in pipe inclined at any angle to the horizontal. Turner's equations are equally valid for upward and downward flow. His treatment is based upon the general form of the Navier–Stokes equations discussed in Chapter 7. He uses Prandtl's mixing length concept to relate shear stress to the velocity profile. He presents the exact solution for the laminar-laminar inclined flow case, but exact integration was not possible for inclined flow in the presence of turbulence. He was, however, able to obtain a solution to the horizontal turbulent flow case. We will summarize Turner's results here.

Laminar Flow in the Gas Core and the Liquid Film In the gas core

$$V_{SG}^* = \frac{N_G \Delta P^* E_G^2}{32} + \frac{E_G N_L}{16}\sqrt{\frac{\rho_G}{\rho_L}}\,[(\Delta P^* - \sin\theta)(1 - E_G) - E_G \sin\theta \ln E_G] \tag{10.157}$$

where $V_{SG}^* = V_{SG}\sqrt{\dfrac{\rho_G}{Dg(\rho_L - \rho_G)}}$, a dimensionless gas velocity,

$$\Delta P^* = \frac{-\frac{dP}{dL} - \rho_G \frac{g}{g_c} \sin\theta}{g(\rho_L - \rho_G)}, \text{ the dimensionless two-phase pressure drop,}$$

θ = angle of pipe to the horizontal

$$N_G = \frac{D\rho_G^{1/2}}{\mu_G}\sqrt{Dg(\rho_L - \rho_G)}$$

$$N_L = \frac{D\rho_L^{1/2}}{\mu_L}\sqrt{Dg(\rho_L - \rho_G)}$$

Equation 10.157 may also be written as

$$\Delta P_G^* = E_G^2 \Delta P^* + 2E_G \frac{N_L}{N_G}\sqrt{\frac{\rho_G}{\rho_L}}\,[(\Delta P^* - \sin\theta)(1 - E_G) - E_G \sin\theta \ln E_G] \tag{10.158}$$

where

$$\Delta P_G^* = \frac{\left(-\frac{dP}{dL}\right)_G}{g(\rho_L - \rho_G)} = \frac{32V_{SG}^*}{N_G} \tag{10.159}$$

$(-dP/dL)_G$ is the pressure drop in the same tube in the horizontal position when gas flows alone at a velocity equal to its superficial velocity. For $\rho_G \ll \rho_L$, Equation 10.158 may be simplified to

$$\Delta P_G^* \simeq E_G^2 \Delta P^* \tag{10.160}$$

In the liquid film

$$\Delta P_L^* = (\Delta P^* - \sin\theta)E_L^2 + 4\sin\theta\, E_G^2 \ln\sqrt{E_G} + 2\sin\theta\, E_G(1 - E_G) \tag{10.161}$$

where

$$\Delta P_L^* = \frac{\left(-\frac{dP}{dL}\right)_L}{g(\rho_L - \rho_G)} = \frac{32V_{SL}^*}{N_L} \tag{10.162}$$

Turner shows that for the range of E_G encountered in annular flow, we may approximate Equation 10.161 by

$$\Delta P_L^* = E_L^2 \Delta P^* - 0.684 \sin\theta\, E_L^3 \tag{10.163}$$

It should be noted that the equations in this section apply to vertical flow when $\sin\theta = 1$ and to horizontal flow when $\sin\theta = 0$. Furthermore these equations are valid only when both phases are in laminar motion. For these conditions simultaneous solution of Equations 10.160 and 10.163 yield values for the pressure gradient and the volume fraction gas.

Turbulent Flow in the Gas Core and the Liquid Film Turner was unable to obtain general expressions of the form of Equations 10.158 and 10.161 for turbulent flow. He was, however, able to treat the horizontal case of turbulent flow in a general way. He assumed the flow to be turbulent everywhere except close to the pipe wall in the laminar sublayer of the liquid film. This sublayer extends from the pipe wall a distance δ_l into the pipe. The following mixing length profile is assumed

$$l^* = l_G^* \qquad \text{for } 0 \le \frac{r}{R} \le \sqrt{E_G} \qquad \text{(gas core)} \tag{10.164a}$$

$$l^* = l_L^*\left(1 - \frac{r}{R}\right) \qquad \text{for } \sqrt{E_G} \le \frac{R}{r} \le \sqrt{\gamma} \qquad \text{(turbulent film)} \tag{10.164b}$$

$$l^* = 0 \qquad \text{for } \sqrt{\gamma} \le \frac{r}{R} \le 1 \qquad \text{(laminar sublayer)} \tag{10.164c}$$

where l_G^* and l_L^* are dimensionless constants and $\gamma = [1 - (\delta_l/R)]^2$. Following are his results for the gas core and the liquid film.

In the gas core

$$V_{SG}^* = \int_0^{\sqrt{E_G}} \left\{ \frac{\Delta P^{*1/2}(E_G^{3/4} - r^{*3/2})}{3 l_G^*} + \frac{1}{2 l_L^*} \sqrt{\frac{\Delta P^* \rho_G}{\rho_L}} \left\{ \ln\left[\frac{1 + \gamma^{1/4}}{1 - \gamma^{1/4}} \left(\frac{1 - E_G^{1/4}}{1 + E_G^{1/4}}\right)\right] - 2(\gamma^{1/4} - E_G^{1/4}) \right\} + \frac{N_L \Delta P^*(1 - \gamma)}{16} \sqrt{\frac{\rho_G}{\rho_L}} \right\} dr^{*2} \tag{10.165}$$

where $r^* = r/R$.

The first term on the right-hand side of Equation 10.165 shows that a constant mixing length approach leads to a 3/2 power low velocity profile in the gas core. The second two terms in this equation are dimensionless forms of the interface velocity, i.e.,

$$u_{Gi}^* = u_i \sqrt{\frac{\rho_G}{Dg(\rho_L - \rho_G)}} = \frac{1}{2 l_L} \sqrt{\frac{\Delta P^* \rho_G}{\rho_L}} \left\{ \ln\left[\frac{1 + \gamma^{1/4}}{1 - \gamma^{1/4}} \left(\frac{1 - E_G^{1/4}}{1 + E_G^{1/4}}\right)\right] - 2(\gamma^{1/4} - E^{1/4}) \right\} + \frac{N_L \Delta P^*(1 - \gamma)}{16} \sqrt{\frac{\rho_G}{\rho_L}} \tag{10.166}$$

Substituting Equation 10.166 into Equation 10.165 and integrating, we obtain

$$V_{SG}^* = E_G u_{Gi}^* + \frac{\Delta P^{*1/2} E_G^{7/4}}{7 l_G^*} \tag{10.167}$$

The expressions for the liquid film are based on the mixing length profile given by Equation 10.166. The equation for flow in the liquid film is

$$\begin{aligned} V_{SL}^* = \frac{\Delta P^{*1/2}}{2 l_L^*} \Bigg\{ &(1 - E_G) \ln \left[\frac{1 + \gamma^{1/4}}{1 - \gamma^{1/4}} \left(\frac{1 - E_G^{1/4}}{1 + E_G^{1/4}} \right) \right] - 2\gamma^{1/4} - 2E_G^{1/4} - \frac{2}{3}\gamma^{3/4} \\ &+ \frac{2}{3} E_G^{3/4} + \frac{8}{5}\gamma^{5/4} - \frac{8}{5} E_G^{5/4} - 2\gamma^{1/4}(\gamma - E_G) \Bigg\} \\ &+ \frac{N_L \Delta P^*(1 - \gamma)(\gamma - E_G)}{16} + \frac{N_L \Delta P^*(1 - \gamma)^2}{32} \end{aligned} \tag{10.168}$$

Substituting Equation 10.166 into Equation 10.168, we have

$$V_{SL}^* = (1 - E_G) u_{Li}^* + \frac{\Delta P^{*1/2}}{l_L^*} \left(\frac{E_G^{5/4}}{5} - \frac{\gamma^{5/4}}{5} + \frac{E_G^{3/4}}{3} - \frac{\gamma^{3/4}}{3} \right) - \frac{N_L \Delta P^*(1 - \gamma)^2}{32} \tag{10.169}$$

where $u_{Li}^* = u_{Gi}^* \sqrt{\rho_L / \rho_G}$.

The thickness of the laminar sublayer is very small and γ approaches unity. This assumption leads to the following simpler form of Equation 10.169:

$$V_{SL}^* = (1 - E_G) u_{Li}^* - \frac{\Delta P^{*1/2}}{15 l_L^*} (8 - 3E_G^{5/4} - 5E_G^{3/4}) \tag{10.170}$$

(It should be pointed out that γ may not be set equal to one in Equation 10.166 for the interface velocity. In that equation the term $(1 - \gamma)$ is important and γ must be determined experimentally or estimated.) Turner suggests that $l_L^* = 0.4$ and $l_G^* = 1/7$ may be used. The resulting final forms of Equations 10.167 and 10.168 are

$$V_{SG}^* = E_G u_{Gi}^* + \Delta P^{*1/2} E_G^{7/4} \tag{10.171}$$

$$V_{SL}^* = (1 - E_G) u_{Li}^* - \frac{\Delta P^{*1/2}}{6} (8 - 3E_G^{5/4} - 5E_G^{3/4}) \tag{10.172}$$

Equations 10.171 and 10.172, together with Equation 10.166 defining u_{Gi}^* and the definition of u_{Li}^*, permit the prediction of $\Delta P/L$ and E_G for specified conditions of pipe diameter, fluid flow rates, and fluid properties. The reliability of the prediction has not been thoroughly tested and is dependent upon Turner's assumptions regarding l_L^*, l_G^* and γ as discussed above. The treatment suggests that experimental data on ΔP^* and E_G could be correlated as functions of V_{SG}^* and V_{SL}^*. Such a correlation has not yet been developed for horizontal flow.

8.32 Circumferential Variation of Film Thickness

McManus (1957, 1959, 1960, 1961) provides us with data indicating that the liquid film reaches an equilibrium distribution and structure after some 23 diameters of travel from an entrance where the liquid is introduced as a concentric film. His data (McManus, 1960) on mean and minimum film thicknesses as functions of circumferential position (measured in degrees from the bottom of the pipe) and water rates at a gas phase Reynolds number of 178,000 are illustrated in Figures 10.56 and 10.57. Similar data were obtained at other gas velocities. McManus also made observations of the maximum wave height. His data, shown in Figure 10.58, indicate the extreme sensitivity of wave height to mean film thickness. These data include measurements taken at all circumferential positions.

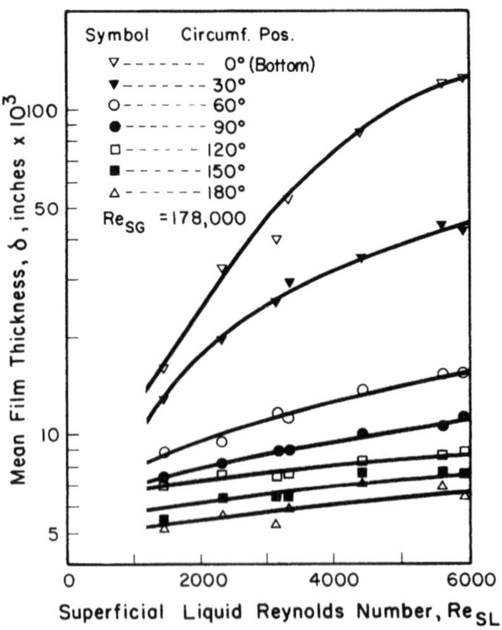

Figure 10.56. Circumferential variation of mean film thickness. (From McManus, H. N. Jr., A.S.M.E. Hydraulic Div., Paper No. 61-HYD-20, 1960.)

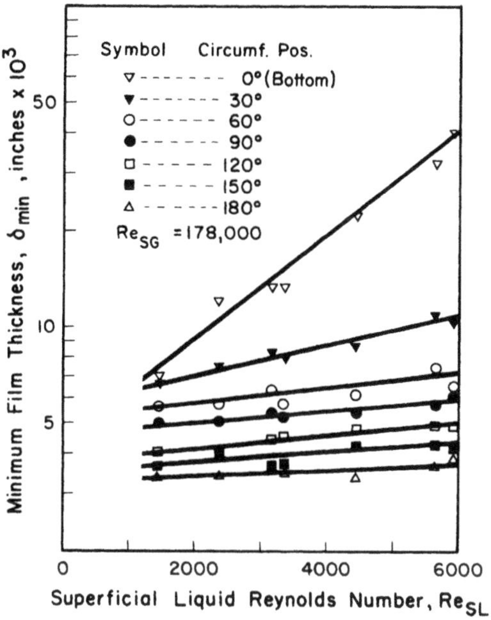

Figure 10.57. Circumferential variation of minimum film thickness. (From McManus, H. N. Jr., A.S.M.E. Hydraulic Div., Paper No. 61-HYD-20, 1960.)

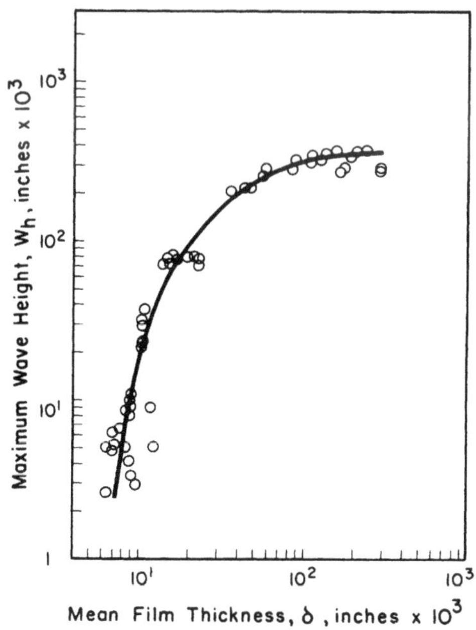

Figure 10.58. Maximum wave height in the liquid film. (From McManus, H. N. Jr., A.S.M.E. Hydraulic Div., Paper No. 61-HYD-20, 1960.)

TABLE 10.7

Values of Constant and Exponents for Correlation of Liquid Area Fraction and Mean Film Depth at Various Circumferential Positions for $Re_{SL} < 2100$[a]

		K	a	b	c	d
Re_{SL}		$e^{7.64}$	0.24	0.90	0.66	0.48
δ/D	0° (bottom)	$e^{-7.23}$	0.47	0.79	−0.78	0.84
	30°	$e^{-1.19}$	0.40	0.82	−0.06	0.70
	60°	$e^{10.73}$	0.23	0.97	1.15	0.42
	90°	$e^{14.23}$	0.09	0.97	1.45	0.24
	120°	$e^{14.38}$	0.03	0.90	1.49	0.14
	150°	$e^{14.48}$	0.02	0.85	1.57	0.10
	180°	$e^{15.20}$	0.03	0.83	1.73	0.12

[a] H. N. McManus, Jr., A.S.M.E. Paper No. 61-HYD-20 (1960).

McManus (1961) offers an empirical correlation of the mean film thickness in terms of δ/D for mist-free flow in the form

$$\frac{\delta}{D} = K \frac{Re_{SL}^a}{Re_{SG}^b} \left(\frac{\rho_G}{\rho_L}\right)^c \left(\frac{\mu_L}{\mu_G}\right)^d \tag{10.173}$$

where Re_{SL} and Re_{SG} are the liquid and gas Reynolds numbers based upon the superficial velocities. Note that the liquid Reynolds number based on the superficial liquid velocity is the same as a Reynolds number based upon the

TABLE 10.8

Values of Constant and Exponents for Correlation of Liquid Area Fraction and Mean Film Depth at Various Circumferential Positions for $Re_{SL} > 1800$[a]

		K	a	b	c	d
Re_{SL}		$e^{2.71}$	0.71	0.83	0.56	0.46
δ/D	0° (bottom)	$e^{-19.73}$	1.45	0.62	−1.19	0.90
	30°	$e^{-2.89}$	1.09	0.96	0.28	0.71
	60°	$e^{17.51}$	0.35	1.22	1.75	0.13
	90°	$e^{18.38}$	0.13	1.07	1.88	0.08
	120°	$e^{17.70}$	0.02	0.95	1.80	−0.06
	150°	$e^{19.16}$	−0.06	0.97	1.90	−0.06
	180°	$e^{20.26}$	−0.15	0.96	1.97	−0.05

[a] H. N. McManus, Jr., A.S.M.E. Paper No. 61–HYD–20 (1960).

average film velocity, and an equivalent diameter equal to four times the film cross-sectional area divided by the wetted perimeter, i.e.,

$$\mathrm{Re}_{SL} = \left(\frac{4\pi D\delta}{\pi D}\right) V_f \frac{\rho_L}{\mu_L} = D\left(\frac{\pi D\delta V_f}{\pi D^2/4}\right)\frac{\rho_L}{\mu_L} = \frac{DV_{SL}\rho_L}{\mu_L}$$

McManus' data on air-water and air-glycerine solutions in 1, 2, and 3-inch tubes enabled him to cover a wide range of variables. He confirms that when $\mathrm{Re}_{SL} < 2000$ the film appears to flow in laminar motion, and when $\mathrm{Re}_{SL} > 2000$ transitional and turbulent flow occur in the film. His coefficients for Equation 10.173 are given in Tables 10.7 and 10.8.

8.33 Film Waves

With the knowledge we now have of the structure of the film we may consider the more complex problem of annular-mist flow. We still find it necessary to neglect the circumferential variation in film thickness and to deal with an equivalent film of mean thickness δ but one which may have ripples or waves on its surface.

McManus (1961) characterizes the surface roughness of the film by the quantities m, n, $W_h = m + n$, and $\delta_p = \delta - n$, as shown in Figure 10.59. He defines a dimensionless mean film thickness, δ^+, in accordance with Equation 8.181. He assumes that the waves could be asymmetric about the mean depth line in the ratio 3 : 2 and that the dimensionless minimum film thickness, δ_p^+,

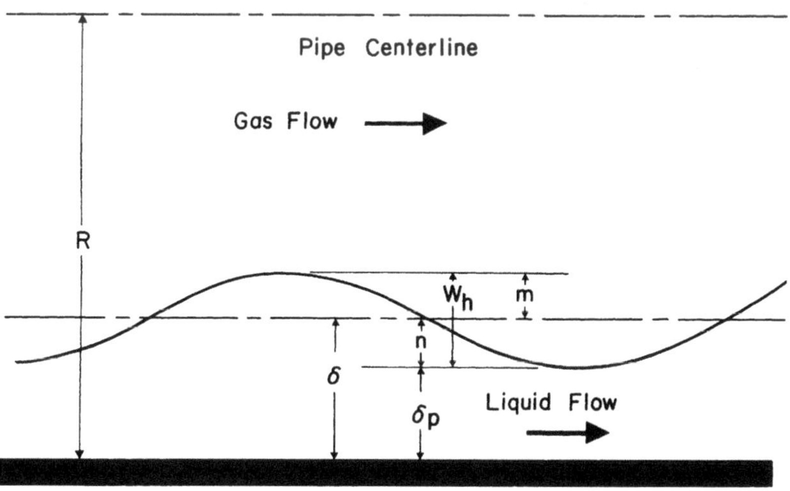

Figure 10.59. Schematic representations of film waves. (From McManus, H. N. Jr., Interim Report No. 3, Cornell Univ., Ithaca, N.Y., 1961.)

would be 10, i.e., twice the value of y^+ for the classical laminar sublayer. This leads to the following relationship for the wave height, W_h:

$$\frac{W_h}{\delta} = \frac{5}{2}\left(1 - \frac{10}{\delta^+}\right) \tag{10.174}$$

Where the wave height is less than the thickness of the laminar sublayer in the gas core, the gas will flow in "smooth-wall" or SW turbulence. The criterion for this condition is

$$W_h/\delta_{lG} \leq 1$$

where δ_{lG} is the thickness of the laminar sublayer in the gas phase determinable from

$$\frac{\delta_{lG}\, U_G^* \rho_G}{\mu_G} = 5$$

and

$$U_G^* = \sqrt{\tau_w g_c/\rho_G}$$

For the case of complete penetration of the liquid waves into the turbulent gas core McManus evaluates the dimensionless mean film thickness as

$$\delta_{cp}^+ = 10 + 28\left(\frac{\rho_G}{\rho_L}\right)^{1/2}\left(\frac{\nu_G}{\nu_L}\right) \tag{10.175}$$

Treating this case as FRW turbulence (see Chapter 4) with the wave height analogous to the roughness, introducing a relation for f in the fully rough region, and assuming the actual interface velocity to be $3V_f/2$, McManus obtains

$$V_G - u_i = V_{SG}\left[1 + \frac{2\delta}{R} - 3\left(\frac{R}{\delta}\frac{\nu_L}{\nu_G}\right)\right]\frac{Q_L^+}{\mathrm{Re}_{SG}} \tag{10.176}$$

where

$$Q_L^+ = \frac{DV_{Sf}\rho_L}{4\mu_L}$$ = the dimensionless liquid film flow rate defined by Equation 8.210,

and

$$\frac{\tau_i}{\rho_G} = \frac{\tau_w}{\rho_G} = \frac{V_{SG}^2\left[1 + \dfrac{2\delta}{R} - 3\left(\dfrac{R}{\delta}\right)\left(\dfrac{\nu_L}{\nu_G}\right)\dfrac{Q_L^+}{\mathrm{Re}_{SG}}\right]^2}{6g_c \ln\left[\dfrac{2.95}{1 - \dfrac{10}{\delta^+}}\left(\dfrac{R-\delta}{\delta}\right)\right]^2} \tag{10.177}$$

From the universal velocity profile and the definition of Q_L^+ he obtains

$$Q_L^+ = -64 + 3.0\delta^+ + 2.58\delta^+ \ln \delta^+ \qquad \delta^+ \geq 30 \tag{10.178}$$

$$Q_L^+ = 12.5 - 8.05\delta^+ + 5\delta^+ \ln \delta^+ \qquad \delta^+ \leq 30 \tag{10.179}$$

or approximately

$$\left(\frac{Q_L^+}{A} + 1\right)^2 - \left(\frac{\delta^+}{B}\right)^2 = 1 \tag{10.180}$$

where $A = 240$ and $B = 14.7$. Equations 10.178 and 10.179 are identical with those presented by Levy for vertical flow (see Equations 8.211 and 8.212). Combining Equations 10.176, 10.177 and 10.180, we have the following implicit relation for δ/R.

$$\frac{\dfrac{\delta}{R}\left[1 + \dfrac{2\delta}{R} - 3\left(\dfrac{R}{\delta}\right)\left(\dfrac{\nu_L}{\nu_G}\right)\dfrac{Q_L^+}{\mathrm{Re}_{SG}}\right]}{\ln\left[\dfrac{\dfrac{2.95}{10}\left(\dfrac{R}{\delta} - 1\right)}{1 - \dfrac{10}{\delta^+}}\right]} = \frac{\sqrt{24}}{\mathrm{Re}_{SG}}\left(\frac{\nu_L}{\nu_G}\right)\left(\frac{\rho_L}{\rho_G}\right)^{1/2}\left(\frac{B}{A}\right)(Q_L^{+2} + 2AQ_L^+)^{1/2} \tag{10.181}$$

Equation 10.181 is an expression for δ/R in terms of δ^+, Q_L^+, Re_{SG}, and the gas and liquid properties. Figure 10.60 presents McManus' solution of this equation for an air-water system. From this expression for δ/R McManus determines the following relationship for the interface friction factor, f_i,

$$f_i = -\frac{2}{3}\ln\left[\frac{\dfrac{7.4}{2.5}\left(\dfrac{R}{\delta} - 1\right)}{1 - \dfrac{10}{\delta^+}}\right] \tag{10.182}$$

and this may be converted to a superficial single-phase friction factor, f_{SG}, through the relation

$$f_{SG} = \left[1 + \frac{2\delta}{R} - \frac{3R}{\delta}\left(\frac{\nu_L}{\nu_G}\right)\frac{Q_L^+}{\mathrm{Re}_{SG}}\right]^2 f_i \tag{10.183}$$

McManus presents Figure 10.61, based upon Equations 10.181, 10.182, and 10.183 and the properties of air and water, for the estimation of the friction factor, f_{SG}. This figure may be used, along with Equation 4.24, to calculate the two-phase pressure drop from a knowledge of the film thickness as estimated from Figure 10.60.

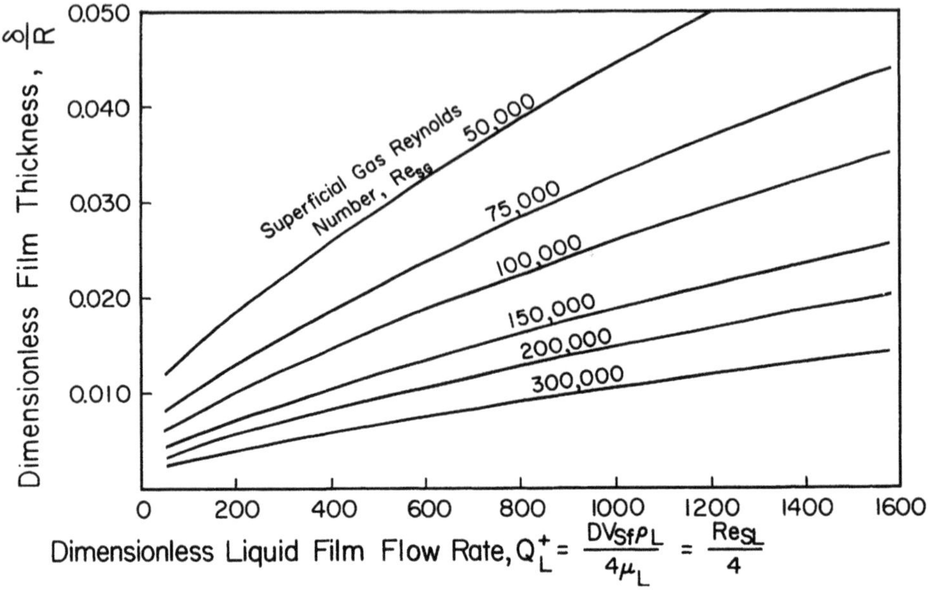

Figure 10.60. Dimensionless film thickness for air-water mixtures in annular-mist flow. (From McManus, H. N Jr., Interim Report No. 3, Cornell Univ., Ithaca, N.Y., 1961.)

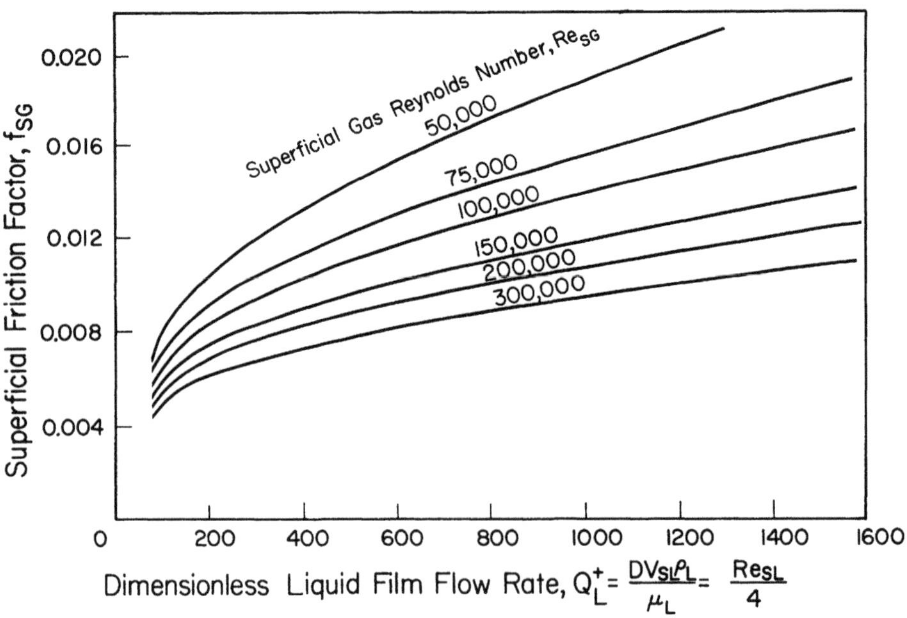

Figure 10.61. Superficial friction factor for horizontal, wavy annular flow of air-water mixtures. (From McManus, H. N. Jr., Interim Report No. 3, Cornell Univ., Ithaca, N.Y., 1961.)

McManus' analysis and the correlations given in Figures 10.60 and 10.61 have not been extensively proven. They are presented here to indicate a method of approach to the analysis of wavy annular flow. The correlations may be expected to give order of magnitude accuracy.

8.34 Effect of Entrainment

In the previous subsections we have dealt with the case of annular flow where there is no liquid entrainment in the gas core. We have seen that even for this relatively simple situation the equations may be solved only for some special cases. The general circumstances when entrainment is encountered and the complications which it introduces in the momentum analysis have been discussed in Section 7.32 of Chapter 8. This discussion and the corelations presented there are also applicable in a general way to horizontal flow under conditions when the film is more or less concentric (or of uniform thickness). Levy's analysis discussed in Section 7.32 of Chapter 8 may be applied to concentric annular-mist horizontal flow by setting $n = 0$ and neglecting all gravitational effects. In this section we will discuss briefly some data and two related correlations which are specifically applicable to horizontal flow with a liquid film which may vary in thickness.

Wicks and Dukler (1960) present entrainment and pressure drop measurements for the air-water system in 1- and 3-inch horizontal pipes. They examined several approaches and found that their data for different entrance conditions, and the data of Fritzlen (1951) for another type of entrance could be correlated through Martinelli's parameter, X (see Section 3.2), and their entrainment parameter $\mathscr{R}_E$ which we discussed in Section 7.32 of Chapter 8. They found that their own data and those of Fritzlen could be correlated provided values of We_c were taken as follows:

$\mathrm{We}_c = 13$ when the gas is introduced abruptly as with a tee entry,
$\mathrm{We}_c = 16$ when the liquid is introduced abruptly as with a tee entry,
$\mathrm{We}_c = 22$ when the liquid is introduced smoothly as with an annular entrance device.

Figure 10.62 gives the correlation of Wicks and Dukler.

In later work Magiros and Dukler (1961) attempted to generalize this correlation for various liquids. Their data for three different liquids are shown in Figure 10.63. Figure 10.63 cannot be considered a general correlation, but it does give some guide to the effect of liquid viscosity and interfacial tension.

If the amount of liquid in the gas core is determined from the previously described conditions, the liquid flowing in the film may be determined by simple difference. The gas core mixture may be assumed homogeneous, and the liquid in the film may be treated by the methods discussed previously.

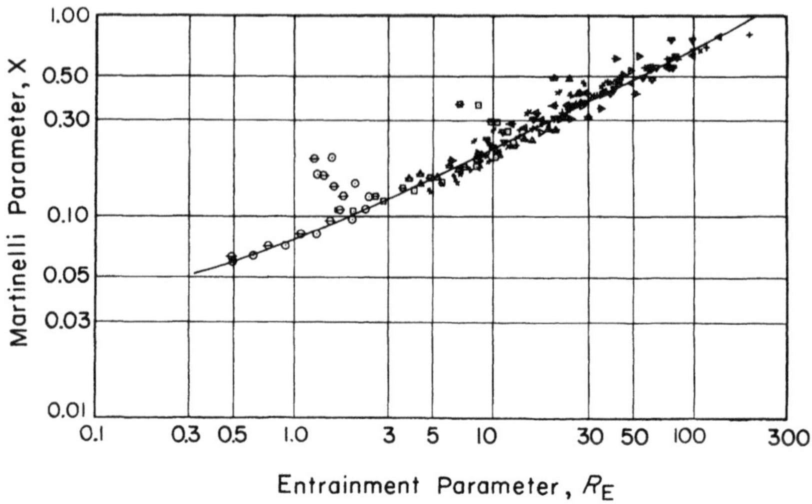

Figure 10.62. Wicks and Dukler entrainment parameter $\mathscr{R}_E$ for air-water mixtures. (From Wicks, M., and A. E. Dukler, *A.I.Ch.E., Journal*, **6**, 463, 1960.)

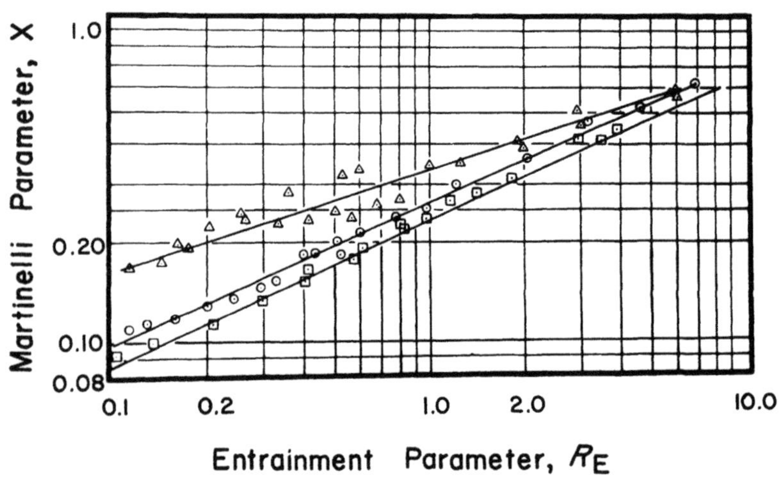

Figure 10.63. Magiros and Dukler correlation of the entrainment Parameter, $\mathscr{R}_E$.

○ $\mu = 4$ centipoise, $\sigma = 40$ dynes/cm
△ $\mu = 10$ centipoise, $\sigma = 40$ dynes/cm
□ Air-water data

(From Magiros, P. G., and A. E. Dukler, *Developments in Mechanics*, **1**, 532, Plenum Press, New York, 1961.)

8.4 Other Correlations

Of the general correlations presented in Section 3.3 those of Baker and of Hoogendoorn provide specific methods for the prediction of the pressure gradient for annular-mist flow. These may be used as alternates or to supplement the procedures discussed in Section 8.3.

8.5 Flow Pattern Details

We have already discussed, implicitly, various methods of determining the details of the flow pattern for annular-mist flow. As pointed out earlier, none of these methods is completely reliable, but taken together they enable us to obtain an approximate definition of the flow pattern. The following procedure may be used:

1. Estimate the amount of liquid flowing as entrained drops, M_e, from Figure 10.62 or Figure 10.63.
2. Determine, by difference, the amount of liquid flowing in the film.
3. From the amount of liquid flowing in the film calculate the approximate value of the film thickness by the method of Levy, or either of the methods of McManus (Equation 10.173 or Figure 10.60). All of these methods are discussed in Section 8.31. The choice of method will depend upon the relative flow rates in each phase and other system variables. This step will also yield a value of the pressure gradient if the method of Levy or the second method of McManus is used.
4. The interface velocity may be estimated by Levy's method (Equation 10.150) if the flow in the film is laminar. Turner's method involving Equation 10.166 may be used with an independent estimate of the film thickness if the flow in the film is turbulent. McManus' approximation of $u_i = 3V_f/2$ is also useful.
5. The circumferential variation of the film thickness may be roughly estimated through the correlation of McManus.

8.6 Holdup

Once an estimate is available of the amount of liquid in the gas core and of the mean film thickness, the calculation of the *in situ* gas fraction, E_G, is a simple matter. The methods discussed for vertical pipe flow apply directly.

8.7 Pressure Gradient

There are no completely satisfactory methods for the accurate prediction of the pressure drop for annular-mist flow. Of the methods based upon the flow mechanism, Levy's method usually results in low predictions. Turner's method depends upon an independent estimate of film thickness or interface velocity. The method of McManus involving Figure 10.61 has not been fully tested and its use cannot be generally recommended. Among the wholly empirical methods, as pointed out by Aziz and Govier (1962), the Lockhart and Martinelli annular-mist correlation predicts pressure drops only to within about 65 percent. The Baker modification of the Lockhart and Martinelli correlation for annular-mist flow (Figure 10.25e) is probably superior. The empirical method of Hoogendoorn, being based on data for a variety of fluids and pipe sizes, is probably as reliable as any.

In the annular-mist flow pattern the pressure drop due to acceleration or kinetic energy effects is often significant. It may be calculated from the appropriate terms in the mechanical energy equation (Equation 7.57).

9 RECOMMENDED DESIGN METHODS

9.1 Introduction

The general comments made in Section 8.1 of Chapter 8 for vertical flow are applicable and the design methods for horizontal flow closely parallel those given in that chapter. Most of the methods assume an incompressible fluid. Where the pressure drop is significant, the calculations may be performed in a stepwise fashion over sections of the pipe. As in the case of Chapter 8, certain of the relationships of this chapter are in integral form (Bertuzzi *et al.* and Baxendell) and may be applied over the entire pipe length. Where alternate methods are available the selection of the most suitable method is again a matter of judgment and experience.

9.2 Entrance Effects

The procedures outlined in Section 8.2 of Chapter 8 may be followed.

9.3 Effect of Bends and Fittings

The procedures outlined in Section 8.3 of Chapter 8 may be followed.

9.4 *Potential Energy or Hydrostatic Head Effects*

1. If the pipe is completely horizontal $\Delta P_{PE} = \Delta P_{HH} = 0$.
2. If the pipe is inclined the hydrostatic head effect, ΔP_{HH}, may be calculated from the second term of Equation 10.74 with E_H determined from Figure 10.33. Alternately the other procedures discussed in Section 3.4 may be used.

9.5 *Acceleration or Kinetic Energy Effects*

Acceleration effects are usually negligible with liquid-liquid systems but may be significant with gas-liquid mixtures particularly in the slug and annular-mist flow patterns. Alternate ways to calculate ΔP_{KE} are:

1. The quantity may be estimated from the appropriate term of the mechanical energy balance (Equation 7.60), i.e.,

$$\Delta P_{KE} = \frac{1}{Q_G + Q_L}\left[\frac{M_G(V_{G2}^2 - V_{G1}^2)}{2\alpha_G g_c} + \frac{M_L(V_{L2}^2 - V_{L1}^2)}{2\alpha_L g_c}\right]$$

with the α terms taken as 0.8.
2. It may be calculated from the simplified form given as Equation 10.62.

9.6 *Flow Pattern, Holdup, and Pressure Gradient for the Steady Flow of Incompressible Mixtures (Liquid-Gas Mixtures when $\Delta P < 0.10P_1$)*

9.61 Prediction by General Methods

1. a. The modified Govier–Omer Figure (Figure 10.14) may be used to predict the flow pattern for air-water or similar systems flowing in relatively small size pipes. This correlation may also be used with other systems for orientation purposes.
b. For gas-oil flow in commercial size pipes, the flow pattern correlations of Baker, Hoogendoorn, Eaton *et al.* or Al-Sheikh *et al.* may be used.

2. a. The holdup or E_G may be estimated by the Lockhart and Martinelli correlation for small size pipes and by Baker's modification of it for commercial size pipes. Both of these are given in Figure 10.18.
b. For air-water and similar systems in small pipes, Figure 10.6 based on the data of Govier and Omer is recommended.

c. The calculations of *2*a or *2*b may be checked by the methods of Hoogendoorn and Buitelaar, Eaton *et al.* or Hughmark. These are discussed in Section 3.2.

3. a. The pressure drop may be estimated from the Lockhart and Martinelli correlation (Figure 10.24 or Equation 10.43). For large commercial size pipes Baker's modification of this correlation given as Equations 10.44 to 10.48 or Figure 10.25 may be used.

b. The calculations of *3*a may be checked using the correlations of Hoogendoorn (Equations 10.56 to 10.60) or of Dukler *et al.* (Equation 10.61) or both.

9.62 Confirmation by Specific Flow Pattern Methods

(A) Bubble Flow Patterns

1. There are no reliable methods for estimating the details of the bubble flow pattern. The experimental observations reported in Sections 2.1 and 4.1 and the continuity relations discussed in Section 4.2 may serve as a guide.

2. The *in situ* volume fraction gas, E_G, may be predicted using Equation 10.82. Other measures of holdup are related to E_G through the relations discussed in Section 7 of Chapter 7.

3. The procedure discussed in Section 4.33 may be used for predicting the pressure drop in the bubble flow pattern based upon the single-phase friction factor.

(B) Stratified Flow Pattern

The procedure for predicting the details of the stratified flow pattern and the holdup and pressure drop in this flow pattern are discussed in Section 5.4. The principal steps are as follows.

1. Determine if each phase is in laminar or in turbulent motion (Section 5.3).

2. For laminar-laminar flow (Re_{SG}, $Re_{SL} < 1000$), the holdup may be predicted from Figure 10.43 and the pressure drop from Figure 10.44.

3. For the liquid laminar, gas turbulent flow case there appears to be no better method for estimating the *in situ* volume fractions than those of Lockhart and Martinelli, and Baker, as discussed in Section 9.61 item *2*a. The pressure drop may be predicted using Figure 10.45 based on the work of Charles and Lilleleht.

4. For the turbulent-turbulent flow case also there is no preferred alternative for either the holdup or the pressure drop prediction to the methods of Lockhart and Martinelli and of Baker discussed under items *2*a, *3*a, and *3*b in Section 9.61.

(C) Wave Flow Pattern

No methods of calculation specific to the wave flow pattern are available. The reader is referred to Section 6 for a discussion of an interpolation procedure which may be used to check the results of predictions discussed in Section 9.61.

(D) Slug Flow Pattern

1. It is not yet possible to predict the details of the slug flow pattern with any certainty but they may be determined approximately by estimating V_b from Equation 10.125, ω from Equation 10.129, L_s from Equation 10.127, and L_b from Equation 8.95. Section 7.2 gives relations for other details of the flow pattern.

2. The *in situ* volume fraction gas, E_G, may be estimated from Equation 10.134, with E_{Gb} obtained from Equation 10.122 or 10.123.

3. The pressure gradient may be estimated from Equation 10.133. Other methods of estimating E_G and ΔP are discussed in Section 7.4.

(E) Annular-Mist Flow Pattern

The flow pattern details may be estimated from the procedure outlined in steps *1* through *5* of Section 8.5. The holdup calculations based on the general methods of Section 9.61 may be further confirmed by Equation 8.221. The pressure drop calculation of Section 9.61 should be confirmed by the special annular-mist flow pattern correlation of Baker and that of Hoogendoorn discussed in Section 8.3.

Bibliography

Abramson, A. E., *J. Appl. Mech.*, **19**, 267 (1952).

Al-Sheikh, J. N., D. E. Saunders, and R. S. Brodkey, *Can. J. Chem. Eng.*, **48**, 21 (1970).

Alves, G. E., *Chem. Eng. Progress*, **50**, 449 (1954).

Anderson, R. J., and T. W. F. Russell, *Chem. Eng.*, **75**, 99 Dec. 20. (1965).

Anderson, R. J., and T. W. F. Russell, Film Formation in Two-Phase Annular Flow, Paper Presented at the 65th National Meeting of the American Institute of Chemical Engineers, Tampa, Fla., May 1968. Published in *A.I.Ch.E. Journal*, **16**, 626 (1970).

Aziz, K., and G. W. Govier, *Can. J. Chem. Eng.*, **40**, 51 (1962).

Baker, O., *Oil and Gas J.*, **53**(12), 185 (1954); **55**(45), 150 (1957).

Baker, O., *Can. Oil and Gas Industries*, March 1961, p. 43.

Bankoff, S. G., *J. Heat Transfer, Trans. A.S.M.E., Ser. C.*, **82**, 265 (1960).

Baxendell, P. B., Fourth World Petroleum Congress, Section 2/E, Paper No, 4, Rome, 1955.

Bergelin, O. P., and C. Gazley, Jr., Heat Transfer and Fluid Mechanics Inst., *Proc.* May 1949, p. 5.

Bertuzzi, A. F., M. R. Tek, and F. H. Poettmann, *Petrol. Trans.*, **207**, 17 (1956).

Bonnecaze, R. H., W. Erskin, Jr., and E. J. Greskovich, Holdup and Pressure Drop for Two-Phase Slug Flow in Inclined Pipelines. Paper presented at the Sixty-second Annual Meeting of the American Institute of Chemical Engineers, Washington, D. C., November 1969.

Brigham, W. E., Doctoral Dissertation, Univ. of Oklahoma, Norman, Oklahoma, 1962.

Brigham, W. E., E. D. Holstein, and R. L. Huntington, *Oil and Gas J.*, **55**(45), 145 (1957).

Brown, F. C., and W. L. Kranich, *A.I.Ch.E. Journal*, **14**, 750 (1968).

Caughron, R. D., Two-Phase Pressure Drop—A Literature Survey and Correlation Analysis, Report No. U.C.R.L-50123, University of California, September 1966.

Charles, M. E., G. W. Govier, and G. W. Hodgson, *Can. J. Chem. Eng.*, **39**, 27, (1961).

Charles, M. E., and L. U. Lilleleht, *Can. J. Chem. Eng.*, **44**, 47 (1966).

Charles, M. E., and P. J. Redberger, *Can. J. Chem. Eng.*, **40**, 70 (1962).

Chawla, J. M., *Forsch. Ing. Wes.*, **34**, 53 (1968).

Chawla, J. M., *Chemie-Ing. Technik*, **41**, 328 (1969).

Chenoweth, J. M., and M. W. Martin, *The Petroleum Engineer*, April (1956), p. C42.

Chisholm, D., *Engineering and Boiler House Review*, Aug. (1963) p. 287

Chisholm, D., *Int. J. Heat and Mass Transfer*. **10**, 1767 (1967).

Chisholm, D., and A. D. K. Laird, *Trans. A.S.M.E.*, **80**, 276 (1958).

Davis, E. J., *Ind. Eng. Chem. Fundam.*, **8**, 153 (1969).

DeGance, A. E. and R. W. Atherton, *Chem. Eng.:*
a. **77**(6), 135 (1970);
b. **77**(15), 95 (1970).

Drew, T. B., E. C. Koo, and W. H. McAdams, *Trans. A.I.Ch.E.*, **28**, 56 (1932).

Dukler, A. E., M. Wicks, and R. G. Cleveland, *A.I.Ch.E. Journal*, **10**, 44 (1964).

Dumitrescu, L., and C. Stanescu, Academy of the Democratic Rumanian Republic, *Appl. Mech. Review*, **2**, 79 (1959).

Eaton, B. A., D. E. Andrews, C. R. Knowles, I. H. Silberberg, and K. E. Brown *J. Petrol. Technol.*, **19**, 815 (1967).

Ellis, S. R. M., and B. Gay, *Trans. Inst. Chem. Engrs.*, **37**, 206 (1959).

Flanigan, O., *Oil and Gas J.*, **56**(10), 132, (1958).

Fritzlen, A. F., M. Ch. E. Thesis, University of Delaware, Newark, Del., 1951.

Gallyamov, A. K., and V. L. Goldzberg, *Izv. Vyssh. Ucheb. Zavedenii, Neft Gaz*, No. 3, p. 28 (1968).

Gazley, C., Jr., Heat Transfer and Fluid Mechanics Inst., *Proc.* May 1949, p. 29.

Gemmel, A. R., and N. Epstein, *Can. J. Chem. Eng.*, **40**, 215 (1962).

Govier, G. W., and M. M. Omer, *Can. J. Chem. Eng.*, **40**, 93, (1962).

Gregory, G. A., and D. S. Scott, Int. Symp. on Research in Cocurrent Gas-Liquid Flow, Univ. of Waterloo, Sept. 1968, *Proc.* p. 633, Plenum Press, New York, 1969.

Gregory, G. A., and D. S. Scott, *A. I. Ch. E. Journal*, **15**, 933 (1969).

Greskovich, E. J., A. L. Shrier, and R. H. Bonnecaze, *Ind. Eng. Chem. Fundam.*, **8**, 591 (1969).

Guzhov, A. I., V. A. Mamayev, and G. E. Odishariya, A Study of Transportation in Gas-Liquid Systems, Paper Presented at the Tenth International Gas Conference, Hamburg, 1967.

Hanratty, T. J., and J. M. Engen, *A.I.Ch.E. Journal*, **3**, 299 (1957).

Hasson, D., U. Mann, and A. Nir, *Can. J. Chem. Eng.*, **48**, 514 (1970).

Hasson, D., and A. Nir, *Can. J. Chem. Eng.*, **48**, 521 (1970).

Hoogendoorn, C. J., *Chem. Eng. Sci.*, **9**, 205, (1959).

Hoogendoorn, C. J., and A. A. Buitelaar, *Chem. Eng. Sci.*, **16**, 208 (1961).

Hubbard, M. G., and A. E. Dukler, Flow Model for Horizontal Gas-Liquid Slug Flow, Paper Presented at the 65th National Meeting of the American Institute of Chemical Engineers, Tampa, Fla., May 1968.

Hubbard, M. G., Doctoral Dissertation, University of Houston, 1965.

Huey, C. T., *Can. J. Chem. Eng.*, **44**, 313 (1966).

Huey, C. T., and R. A. A. Bryant, A Theory for Froth Flow in Horizontal Pipes, A.S.M.E., Paper No. 65-WA/FE-5, New York, 1965.

Huey, C. T., and R. A. A. Bryant, *A.I.Ch.E. Journal*, **13**, 70 (1967).

Hughmark, G. A., *Chem. Eng. Progress*, **58**(4), 62 (1962).

Hughmark, G. A., *Chem. Eng. Sci.*, **20**, 1007 (1965).

Isbin, H. S., R. H. Moen, R. O. Wickey, D. R. Mosher, and H. C. Larson *Chem. Eng. Progress Symp. Ser.* No. 23, **55**, 75 (1959).

Johnson, H. A., and Abou-Sabe, A. H., *Trans. A.S.M.E.*, **74**, 977 (1952).

Katsuhara, T., *Bulletin J. S.M.E.*, **2**, 439 (1959).

Kordyban, E. S., *J. Basic Eng.*, *Trans. A.S.M.E.*, *Ser. D*, **83**, 613 (1961).

Kordyban, E. S., and T. Ranov, Multiphase Flow Symposium, A.S.M.E. Winter Annual Meeting, Philadelphia Pa., Nov. 1963, *Proc.* p. 1, Amercian Society of Mechanical Engineers, New York.

Kordyban, E. S., and T. Ranov, *J. Basic Eng.*, *A.S.M.E. Trans.*, *Ser. D*, **92**, 857 (1970).

Kosterin, S. I., *Izv. Akad. Nauk. S.S.S.R.*, *Otd. Tekh. Nauk.*, No. 12, p. 1824 (1949).

Kriegel, E., *Chemie-Ing. Technik*, **39**, 1267 (1967).

Levy, S., Second Midwestern Conference on Fluid Mechanics, Ohio State University, Columbus, Ohio, 1952, *Proc.* p. 337.

Lockhart, R. W., and R. C. Martinelli, *Chem. Eng. Progress*, **45**, 39 (1949).

Magiros, P. G., and A. E. Dukler, *Developments in Mechanics*, **1**, 532 Plenum Press, New York, 1961.

Mamayev, V., *Int. Chem. Eng.*, **5**, 318 (1965).

McManus, H. N., Jr., Experimental Investigation of Film Characteristics in Horizontal Annular Two-Phase Flow, A.S.M.E. Paper, 57-A-144, American Society of Mechanical Engineers, Annual Meeting, New York, 1957.

McManus, H. N., Jr., Film Characteristics and Dimensions in Annular Two-Phase Flow, Sixth Annual Conference on Fluid Mechanics, Univ. of Texas, Austin, Texas, Sept. 1959, *Proc.* p. 292.

McManus, H. N., Jr., Local Liquid Distribution and Pressure Drops in Annular Two-Phase Flow, A.S.M.E. Hydraulic Div., Paper No. 61-HYD-20, New York, 1960.

McManus, H. N., Jr., Experimental Investigation of Liquid Distribution and Surface Character in Horizontal Annular Two-Phase Flow, Interim Report No. 3, Sibley School of Mechanical Engineering, Cornell Univ. Ithaca, N.Y., October 1961.

Nicklin, D. J., J. O. Wilkes, and J. F. Davidson, *Trans. Inst. Chem. Engrs.*, **40**, 61 (1962).

Ros, N. C. J., *J. Petrol. Technol.* **13**, 1037 (1961).

Russell, T. W. F., G. W. Hodgson, and G. W. Govier, *Can. J. Chem. Eng.* **37**, 9 (1959).

Schneider, P. N., M. Pet. Eng. Thesis, Univ. of Oklahoma, Norman, Oklahoma 1953.

Scott D. S. *Can. J. Chem. Eng.*, **40**, 224 (1962).

Scott, D. S. in *Advances in Chemical Engineering*, Vol. 4, p. 200, Academic Press, New York, 1963.

Singh, G., and P. Griffith, *J. Engineering for Industry*, *Trans. A.S.M.E.*, *Ser. B*, **92**, 717 (1970).

Smith, T. N., and R. W. F. Tait, *Chem. Eng. Sci.*, **21**, 63 (1966).

Sobocinski, D. P., and R. L. Huntington, *A.S.M.E. Trans.*, **80**, 252 (1958).

Teletov, S. G., *Compt. Rend. Acad. Sci. U.R.S.S.*, **51**, 579 (1946).

Turner, J. M., Doctoral Dissertation, Dartmouth College, Hanover, New Hampshire, 1966.

Vermeulen, L. R., M.Sc. Thesis, The University of Alberta, Edmonton, Alberta, 1968.

White, P. D., and R. L. Huntington, *The Petroleum Engineer*, **27**(9), D40 August (1955).

Wicks, M., and A. E. Dukler, *A.I.Ch.E. Journal*, **6**, 463 (1960).

Yagi, S., *Chem. Eng. Japan*, **18**, 2 (1954).

Zukoski, E. E., *J. Fluid Mech.*, **25**, 821 (1966).

The Horizontal Flow of Gas-Solid and Liquid-Solid Mixtures in Pipes

11

1 INTRODUCTION

The horizontal transport of solids in liquids by pipeline has been known and practiced for many years. Relatively short pipelines have long been used for the hydraulic transport of fine metallurgical wastes in and from process plants, of sand and gravel in dredging and similar operations, and of ores in mining operations. In recent years the long-distance hydraulic transport of solids has received increasing attention (Berkowitz and Jensen, 1963, Berkowitz, Moreland, and Round, 1963), and several major commercial installations have been made. The Colorado School of Mines (1963) and Gaessler (1967) give details of a number of these actual applications of hydraulic transport of solids. Wasp, Thompson, and Aude (1970) give a summary of the main features of selected commercial slurry pipelines. This is reproduced in Table 11.1.

The 108-mile 10-inch pipeline of Consolidation Coal Company in Ohio referred to in Table 11.1 is discussed in some detail by Wasp, Regan, Withers, Cook, and Clancey (1963). After six years (1956–1963) of successful operation during which the capital investment was recovered, substantial reductions in rail freight rates were achieved and the line was shut down. The so-called Black Mesa line in Arizona, also referred to in Table 11.1, is the largest slurry pipeline ever constructed and is designed for the transport of some 6.0 million tons per year of coal from the Black Mesa coal field of Arizona to Mohave in Nevada.

Dilute water suspensions of wood-pulp fibers have been transported in

TABLE 11.1

Summary of Selected Commercial Slurry Pipelines[a]

Location	Material	Length, miles	Diameter, inches	Million Tons/Year
Ohio	coal	108	10	1.25
Utah	gilsonite	72	6	0.38
England	limestone	57	10	1.70
Colombia	limestone	9	5	0.35
Trinidad	limestone	6	8	0.57
California[b]	limestone	17	7	2.0
South Africa	gold tailings	22	6 & 9	1.05
Tasmania	iron concentrates	53	9	2.25
Arizona[b]	coal	273	18	6.0
Japan	copper tailings	40	8	1.0
Canada[c]	sulfur/hydrocarbon	800	12–16	—
New Guinea[d]	copper concentrates	17	6	1.0
Africa[d]	phosphate	3	12	5.0

[a] E. J. Wasp, T. L. Thompson, and T. C. Aude, paper presented at C.I.M. Meeting, Toronto, 1970.
[b] Under construction. [c] In planning phase. [d] In design phase.

flumes and pipes in connection with paper making operations for years. We discussed the nature of such flow briefly in Chapter 5. Interest also persists in the possibility of hydraulic transport of wood chips in water (Faddick, 1963). Another possibility, which has attracted much interest in recent years, is that of the transport of sulfur in oil or water suspension. Wasp *et al.* (1970) make reference to this in Table 11.1.

Pneumatic conveying of grain has been common for many years and similar techniques are being used for the movement of dry chemicals, fertilizers, and other finely divided solids, particularly within processing plants.

We will see that solids may be transported in fluids both under conditions where they are totally suspended and where they are partially or wholly settled out and move as a bed. Most practical applications, however, are under conditions of total suspension—or, as commonly referred to, are of "nonsettling solids."

The general behavior of mixtures of solids and liquids and of solids and gases flowing in horizontal pipelines has been subjected to investigation since the turn of the century, but it is only within the last two decades that our understanding has reached a satisfactory level. Even now much less is known about the *in situ* conditions than is desirable. The relationship between hydraulic and pneumatic conveying is still not entirely clear, although it is possible to treat the flow of both liquid-solid and gas-solid systems together and by the same general approach. Our knowledge of the theory of the pipe flow of these mixtures has also developed and, although all problems are not amenable to theoretical analysis, we understand and can predict the major phenomena.

In our treatment of the topic we will follow, to the extent practical, the pattern established in earlier chapters. We will deal first with typical flow patterns, and holdup and pressure gradient data for actual systems of liquid-solid and gas-solid mixtures. Here we will find a considerable difference in behavior depending upon the shape, size, and density of the solid particles. Our treatment will then cover the more important empirical correlations which are applicable to a range of flow patterns. Following this we will deal, in the detail permitted by present knowledge, with the individual flow patterns and finally recommendations will be given for design methods.

2 TYPICAL FLOW PATTERNS, HOLDUP, AND PRESSURE GRADIENT

2.1 Introduction

The flow behavior of a fluid-solid system depends not only on the basic physical properties and relative amounts of the solid and the fluid, but also, in a most important way, on the size and shape of the solid particles and the range in size and shape in the common case where these are not uniform.

Suspensions of fibers in water behave in a manner much different from those of granular or similar solids. We treated the flow of wood pulp and similar fiber suspensions in Chapter 5.

Durand and Condolios (1952) were probably the first to attempt a classification of flow behavior of rigid particles in terms of their size. Modifying this classification slightly in the light of current knowledge and considering the full spectrum of particle sizes as indicated in Figure 11.1, we may generalize as follows:

1. Ultrafine particles, say those of 10 microns or smaller, are almost always carried in a fully suspended state and gravitational effects are negligible.
2. Fine particles, say those from about 10 to 100 microns, are usually carried in a fully suspended state, but concentration gradients due to gravitational effects may be encountered.
3. Medium-sized particles, say those from about 100 to 1000 microns, may be fully suspended at sufficiently high velocities but often will form a

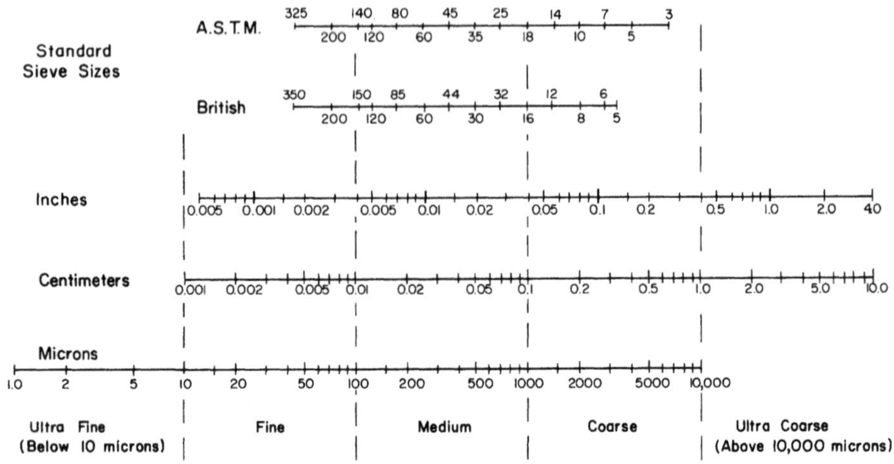

Figure 11.1. Range of particle sizes.

moving deposit on the bottom of the pipe; in all cases there is a decided concentration gradient.

4. Coarse particles in the range of 1000 to 10,000 microns (0.04 to 0.4 inch) are seldom fully suspended, and usually form a moving deposit on the bottom of the pipe.
5. Ultracoarse particles larger than 10,000 microns or 0.4 inch are not suspended at any normal velocities unless they are unusually light and may be transported in a fluid only as a moving bed on the bottom of the pipe.

As we shall see, this classification is far from completely satisfactory—the size divisions are entirely arbitrary, the behavior is influenced by particle density and shape, by the fluid density and viscosity, and by the diameter of the pipe. Furthermore, most suspensions of practical interest include particles covering a considerable range of sizes. Still the classification is useful in a general way and the behavior of particles will at least be within that described for the adjacent size ranges.

Govier and Charles (1961) in a general way; Newitt, Richardson, and Shook (1962), and others in more detail, have shown that the terminal settling velocity of the particles is more significant than the particle size in the determination of the flow behavior of a fluid-solid system. We shall explore this further as we proceed. First, however, it is useful to consider the actual behavior of a few fluid-solid systems.

As we discussed in Chapter 9, a serious problem with obtaining and interpreting data relating to the flow of gas-solid systems is caused by the acceleration of the particles at the entrance of the pipe. Clark, Charles, Richardson, and Newitt (1952) found that with pneumatic transport of even low density seeds (cress, rye, rape, etc.) the particles were not fully accelerated after a travel distance of 16 feet in a 1-inch pipe. It is important, therefore, either to account fully for the acceleration effects or to make observations under conditions where the flow has attained a quasi-steady state.

A further complication in connection with gas-solid flow, also discussed in Chapter 9, is that electrostatic effects may influence the concentration profile, and therefore the flow pattern, holdup, and pressure gradient. Unless specifically indicated to the contrary we will refer to data and correlations where electrostatic effects are negligible.

2.2 Flow Patterns

Meaningful photographic evidence of the flow patterns observed with fine and medium particles suspended in liquid or gas is not available, but from descriptions in the literature (Howard, 1939; Wilson, 1942; Bruce, Hodgson and Clark, 1952; Durand, 1953; Newitt, Richardson, Abbott and Turtle, 1955; Ellis and Round, 1963; Thomas, 1964; Gaessler, 1967; and Wicks, 1968) we may follow the sequence of flow patterns encountered, at any given concentration of solids, as the mixture velocity is decreased from high to low. Four distinct flow patterns are observed.

1. For liquid-solid systems, at the highest mixture velocities (say 10 ft/sec or more) fine or medium-sized solid particles are fully suspended and they are symmetrically, although not necessarily uniformly, distributed about the horizontal axis of the pipe. A symmetric concentration profile can be achieved even at modest velocities of 3 to 5 ft/sec if the flow is turbulent and the solids have a sufficiently low settling velocity. Under certain circumstances the concentration profile tends to approach the velocity profile (Soo and Regalbuto, 1960). This is the symmetric suspension pattern and it persists at all higher velocities and with decreasing velocity to the point where the turbulent and other lifting forces are no longer sufficient fully to maintain the suspension and prevent the particles from striking the wall of the pipe.

2. As the mixture velocity and the intensity of turbulence and lift forces are decreased, the resultant of the weakened forces tending to suspend the particles and the settling tendency of the particles causes a distortion of the concentration profile with more of the solids and particularly the larger particles being contained in the lower than in the upper half of the pipe—a consequence of the particles striking the lower regions of the pipe wall and rebounding into the fluid stream. This is the asymmetric suspension flow pattern and it persists as the velocity is further lowered until some of the particles remain on the bottom of the pipe.

3. Below the velocity at which all particles that strike the wall bounce back into the flow stream, particles tend to accumulate on the bottom of the pipe, first in the form of separated "dunes" (Sinclair, 1962) and then as a continuous moving bed. The dunes or the bed move along the pipe at a lower velocity than the liquid and the particles on the top of the dunes or bed move more rapidly than those beneath and are caused to roll and tumble by the shear force exerted on them by the liquid. For closely sized particles, essentially all of the particles are in the dunes or the moving bed and the liquid above is clear. For systems with particles of mixed size and a wide range of settling velocities, the bed is composed of the fastest settling particles, particles with moderate velocities are in asymmetric suspension, and the lowest settling velocity particles are in symmetric suspension. This is the so-called moving bed flow pattern.
4. As the mixture velocity is further reduced, the lowermost particles of the bed become nearly stationary, the bed thickens, and bed motion occurs caused by the uppermost particles tumbling over one another (saltation). As a consequence of the reduction in the cross section available for liquid flow the actual liquid velocity tends to be maintained. With particles covering a range of sizes, the smaller particles remain in asymmetric suspension above the bed. This flow pattern may be described as that of a stationary bed with saltation and asymmetric suspension. Eventually, with continued reduction in the mixture velocity and buildup of the bed, the pressure gradient to maintain the flow increases very rapidly and, in the absence of an abnormally high applied pressure, blockage of the pipe occurs.

Figure 11.2 taken from Newitt *et al.* (1955) indicates the range of velocity and quantity of solids at which the various flow patterns were obtained with

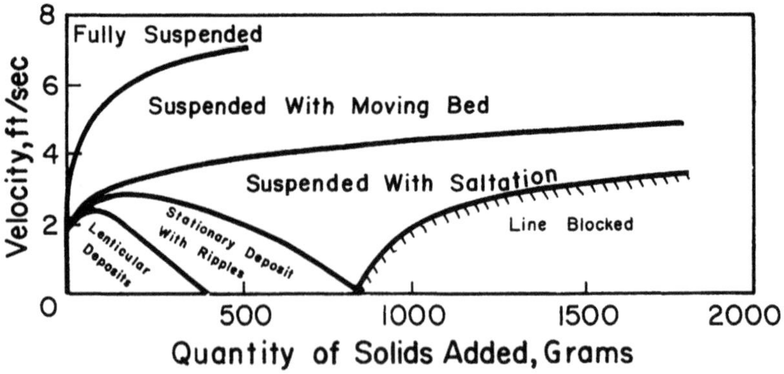

Figure 11.2. Illustrative flow pattern map—water suspension of 60-micron glass particles in $\frac{3}{4}$-inch pipe. (Redrawn from Newitt, D. M., J. F. Richardson, M. Abbott, and R. B. Turtle, *Trans. Inst. Chem. Engrs.*, **33**, 93, 1955.)

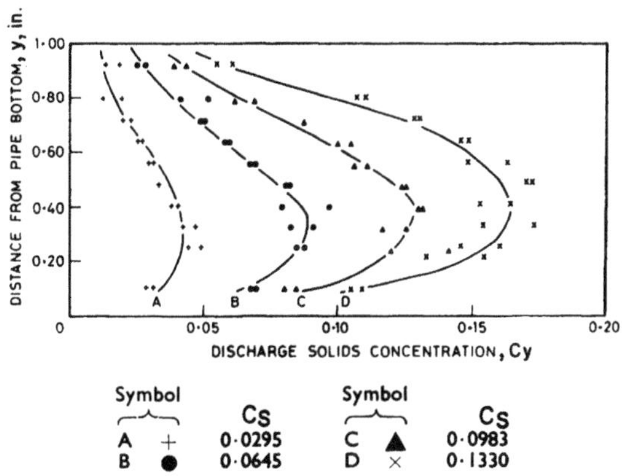

Figure 11.3. Concentration profiles measured for perspex in water, $i = 0.400$ ft of water/ft. (From Newitt, D. M., J. F. Richardson and C. A. Shook, Symp. Interaction Between Fluids and Particles, London, 1962, Proc., p. 87.)

the flow of 60-micron glass beads in water in a 3/4-inch pipe. Tests with other solids indicated the same flow patterns, although at different transition conditions.

Data of Newitt *et al.* (1962) on the concentration profiles observed with the flow of perspex particles (SG 1.17, mean diameter 0.073 in.), medium fine silica sand (SG 2.65, mean diameter 0.0072 in.), zircon sand (SG 4.60, mean diameter 0.0046 in.), and fine gravel (SG 2.67, mean diameter 0.061 in.) are given in Figures 11.3, 11.4, 11.5, and 11.6 taken from their work. All

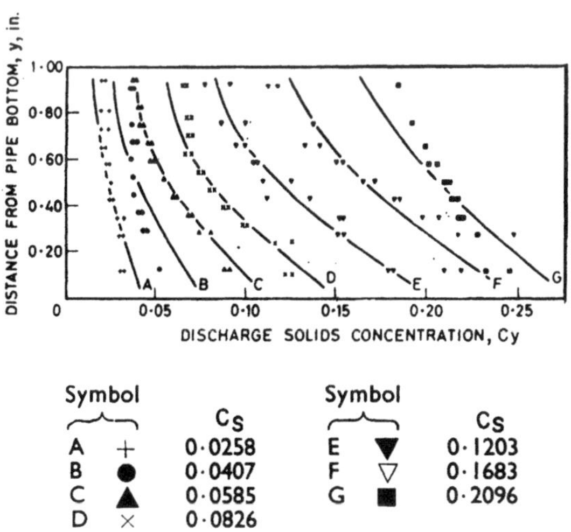

Figure 11.4. Concentration profiles measured for medium-fine sand in water, $i = 0.280$ ft of water/ft. (From Newitt, D. M., J. F. Richardson, and C. A. Shook, Symp. Interaction Between Fluids and Particles, London, 1962, Proc., p. 87.)

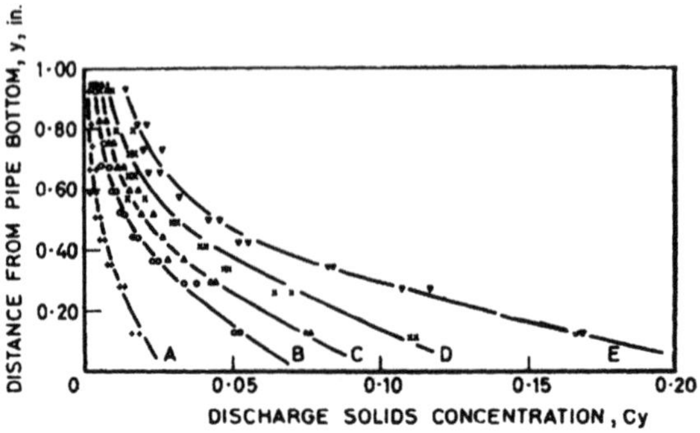

Symbol		C_S	Symbol		C_S
A	+	0·0063	D	×	0·0380
B	●	0·0179	E	▼	0·0571
C	▲	0·0251			

Figure 11.5. Concentration profiles measured for zircon sand in water, $i = 0.280$ ft of water/ft. (From Newitt, D. M., J. F. Richardson, and C. A. Shook, Symp. Interaction Between Fluids and Particles, London, 1962, Proc., p. 87.)

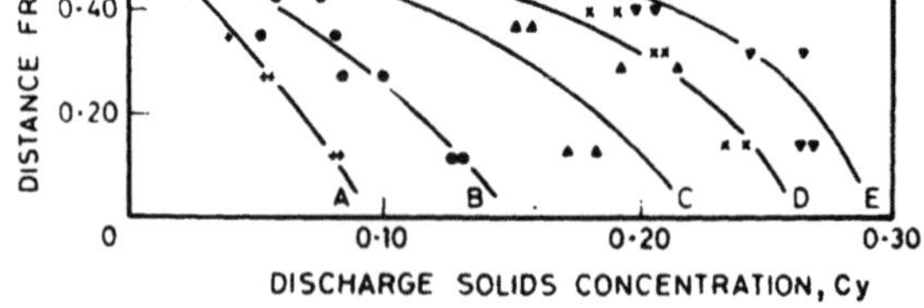

Symbol		C_S	Symbol		C_S
A	+	0·0205	D	×	0·0911
B	●	0·0377	E	▼	0·1155
C	▲	0·0677			

Figure 11.6. Concentration profiles measured for fine gravel in water, $i = 0.280$ ft of water/ft. (From Newitt, D. M., J. F. Richardson, and C. A. Shook, Symp. Interaction Between Fluids and Particles, London, 1962, Proc., p. 87.)

figures relate to the flow with water in a 1-inch pipe; the last three figures refer to a constant pressure gradient, *i*, of 0.280 ft of water per foot, whereas the figure for perspex is for a pressure gradient of 0.400 ft/ft.

The concentration profiles for the light perspex particles flowing under the 0.400 ft/ft pressure gradient (mixture velocity, V_M, approximately 11 ft/sec) are not far from being symmetric and represent about the point of change from the symmetric suspension to the asymmetric suspension flow pattern. In the case of the medium-fine silica sand and especially in the case of the zircon sand and the fine gravel, the concentration profiles are highly asymmetric, reflecting the asymmetric suspension and a mixed asymmetric suspension and sliding bed flow pattern.

Newitt *et al.* (1962) also measured velocity profiles of the water. These indicated an upward shift from the center of the pipe of the point of maximum velocity, a normal 1/7 power law profile above the point of maximum velocity, and a steepened profile below the maximum point.

The concentration and velocity profile data of Durand (1953) for the flow of water-sand mixtures in a 150-mm pipe illustrate further the difference between the asymmetric suspension and the moving bed flow pattern. Figures 11.7 and 11.8 depict Durand's data for fine sand (0.18 mm in diameter) indicating nearly symmetric velocity profiles but decidedly asymmetric sand concentration profiles which vary with the mean flow velocity. Figures

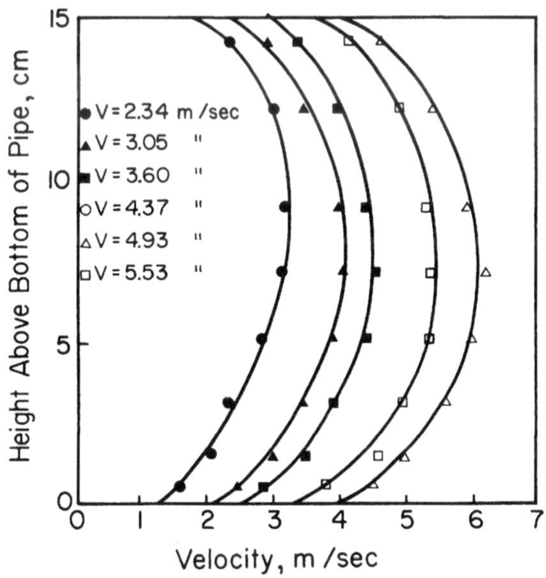

Figure 11.7. Velocity profiles—fine sand ($d_{av} = 0.18$ mm) in water in 150-mm pipe. (Redrawn from Durand, R., Proc. Minnesota Hydraulic Convention, 1953.)

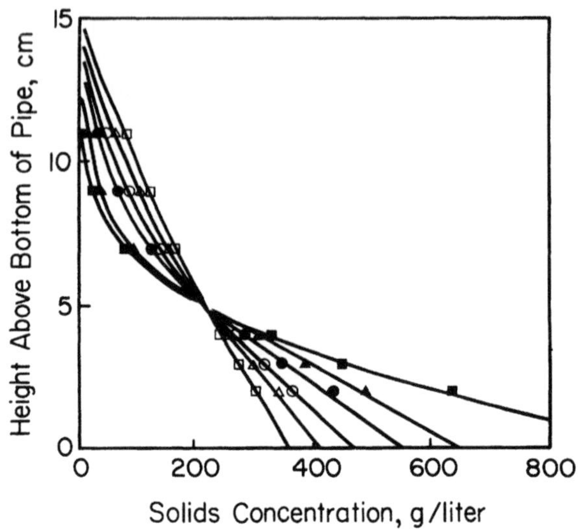

Figure 11.8. Concentration profiles—fine sand ($d_{av} = 0.18$ mm) in water in 150-mm pipe (Redrawn from Durand, R., Proc. Minnesota Hydraulic Convention, 1953.)

11.9 and 11.10, representing the moving bed pattern encountered with a sand 2.04 mm in diameter, show decided asymmetry in both the velocity profile and the concentration profile, with the latter being insensitive to the stream velocity.

The above description is for reasonably regularly shaped, medium (say 100 to 1000 microns), and closely sized solid particles in a liquid such as water. Sufficiently small particles (say 30 microns or less) remain in the symmetric suspension flow pattern to liquid velocities down to the minimum velocity for turbulence. Below that they may form a quiescent moving bed. The asymmetric suspension with bouncing of the particles is not observed since there are no turbulent forces present.

Larger particles, say 10 mesh or larger, are carried in symmetric suspension only at velocities above the normal range and the flow patterns normally encountered are those of the asymmetric suspension and the moving bed. Figure 11.11 shows sketches based on the photographs and concentration data of Worster (1953 and 1954) of 1/2-inch nominal size coal particles traveling in water in a $1\frac{1}{2}$-inch pipe. Table 11.2 gives details of the sketches shown in Figure 11.11. The asymmetric suspension pattern is observed at the highest velocity but at all other velocities the coal is in the moving bed flow pattern.

In most practical applications the solid particles are not uniformly or even closely sized, and the ratio of the dimensions of the largest to the smallest particle may be 10, 50, or more. This causes a variation in the flow

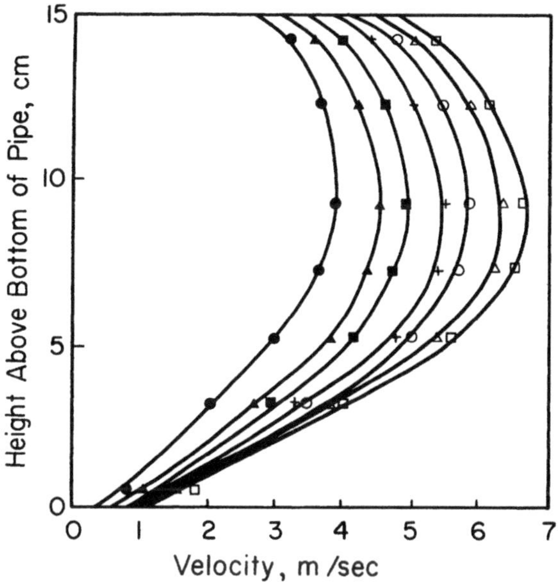

Figure 11.9. Velocity profiles—coarse sand ($d_{av} = 2.04$ mm) in water in 150-mm pipe. (Redrawn from Durand, R., Proc. Minnesota Hydraulic Convention, 1953.)

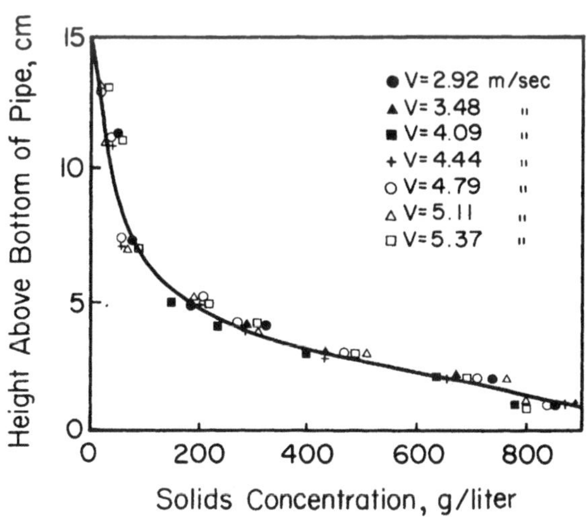

Figure 11.10. Concentration profiles—coarse sand ($d_{av} = 2.04$ mm) in water in 150-mm pipe. (Redrawn from Durand, R., Proc. Minnesota Hydraulic Convention, 1953.)

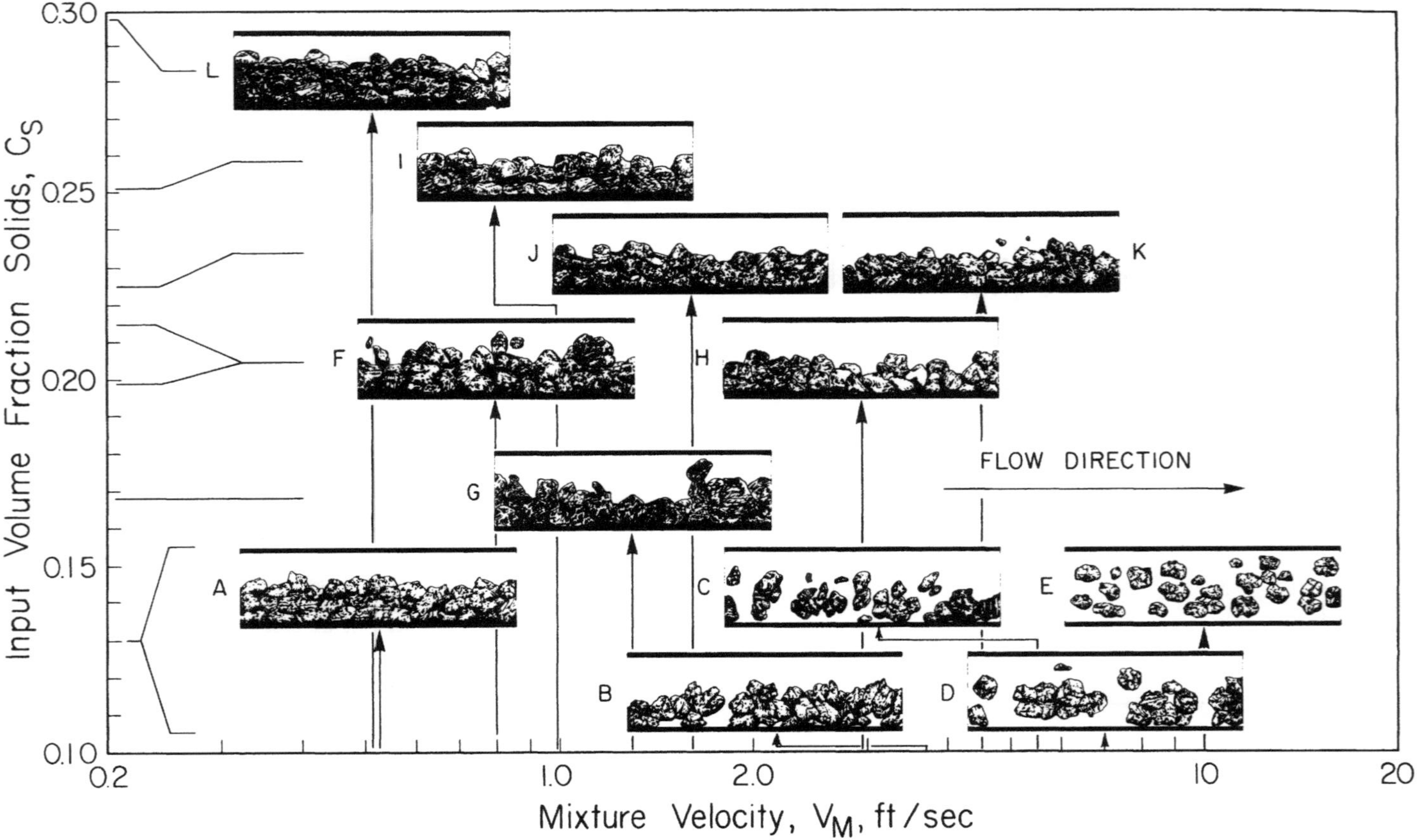

Figure 11.11. Illustrative flow patterns—$\frac{1}{2}$-inch coal particles in water in $1\frac{1}{2}$-inch pipe. (Extended from Worster, R. C., British Hydromechanics Research Association, S.P. 464, Sept. 1953, S.P. 477, Jan. 1954.)

TABLE 11.2

Flow Pattern Details for the Flow of 1/2 Inch Coal with Water in $1\frac{1}{2}$-Inch I.D. Horizontal Pipe (See Figure 11.11)

				Superficial Velocity					Approximate Only	
Sketch Identification	Flow Pattern	Solids Concentration C_S	Mixture Velocity V_M, ft/sec	Solids V_{SS}, ft/sec	Liquid V_{SL}, ft/sec	*In situ* Volume Fraction Solids E_S	Holdup Ratio H	Slip Velocity S, ft/sec	Number of Particles n_P[a]	Bed Depth h_{av}, in.
A	moving bed	0.13	0.525	0.068	0.457	0.33	3.3	0.48	110	0.55
B	moving bed with asymmetric suspension	0.131	3.7	0.48	3.22	0.27	2.5	2.6	90	0.50
C	asymmetric suspension	0.131	5.5	0.72	4.78	0.17	1.4	1.7	55	—
D	asymmetric suspension	0.13	7.0	0.91	6.09	0.17	1.37	2.0	55	—
E	asymmetric suspension	0.13	10.0	1.30	8.70	0.15	1.2	1.7	50	—
F	moving bed	0.212	0.8	0.17	0.63	0.30	1.6	0.34	110	0.55
G	moving bed with asymmetric suspension	0.168	1.3	0.22	1.08	0.36	2.8	1.1	120	0.60
H	moving bed with asymmetric suspension	0.199	2.94	0.59	2.35	0.27	1.5	1.1	90	0.65
I	moving bed	0.251	0.99	0.25	0.74	0.39	1.9	0.58	125	0.85
J	moving bed	0.225	1.61	0.36	1.25	0.33	1.7	0.77	110	0.75
K	moving bed with asymmetric suspension	0.218	4.5	0.98	3.52	0.29	1.5	1.7	100	0.55
L	moving bed	0.298	0.513	0.153	0.36	0.48	2.2	0.38	150	1.0

[a] Per foot of 1.5-inch pipe.

pattern in that the larger particles may flow as a moving bed and the smaller may be carried as a nearly symmetric suspension above it. In the extreme case of a mixture containing a substantial proportion of very fine particles, the larger particles present will behave as though immersed in a fluid whose properties are those of a suspension of the fine particles in the carrier liquid or gas. In such cases, with liquid carriers, non-Newtonian properties are usually encountered.

Similar flow patterns are encountered with solid particles carried in gaseous streams, although the gas velocities are approximately ten times higher and only relatively fine particles are carried in symmetric suspension. Table 11.3, taken from McCarthy and Olson (1968), indicates the range of particle size, the input mass ratio of solids to gas, and gas phase Reynolds numbers over which the symmetric suspension flow pattern has been studied. This flow pattern with gas-solid systems is often referred to as dilute phase flow and is generally thought of as extending to solid-gas mass ratios (or "loadings") of about 15. Beyond loadings of 15 or so and up to values of 1000, the flow pattern changes to the asymmetric suspension and ultimately gives way to the moving bed pattern with saltation. The asymmetric suspension flow pattern persists over a greater velocity range than in liquid suspensions, since particles striking the pipe wall tend to hit at a higher velocity and to rebound further in the gas of lower resistance than in a liquid. These latter two patterns are referred to as dense phase flow.

The data of Soo and Regalbuto (1960) shown in Figure 9.6, which were actually taken in a horizontal flow section, further illustrate the concentration profiles in the symmetric concentration flow pattern.

TABLE 11.3

Range of Variables over which the Symmetric Suspension Flow Pattern has been Studied[a]

Investigator	Reynolds Number, Re_{SG}	Mass Ratio, $\mathscr{R}$	Particle Size, microns	Flow Pattern
Soo *et al.* (1964)	$>10^5$	10–30	50	~10% density variations increase near wall
	$>10^5$	0.75–2.75	35	symmetric suspension
Peskin and Dwyer (1965)	$>10^5$	2.0	50	symmetric suspension
Thomas (1964)	$>10^5$	—	200	symmetric suspension
Goto and Iinoya (1964)	1.2×10^5	0.5	7200	symmetric suspension

[a] H. E. McCarthy and J. H. Olson, *Ind. Eng. Chem. Fundam.*, 7, 471 (1968).

2.3 Holdup

The holdup phenomenon has been clearly observed, especially in the moving bed flow pattern, for the flow of both liquid-solid and gas-solid mixtures, but relatively few reliable quantitative measurements of it have been made.

Figure 11.12 reproduces data of Bonnington (1959) on the flow of water and sand particles having a mean size of 0.048 inch (1220 microns) in a $1\frac{1}{2}$-inch pipe. The marked increase in holdup, as the liquid velocity is decreased and as the moving bed flow pattern is reached, is evident.

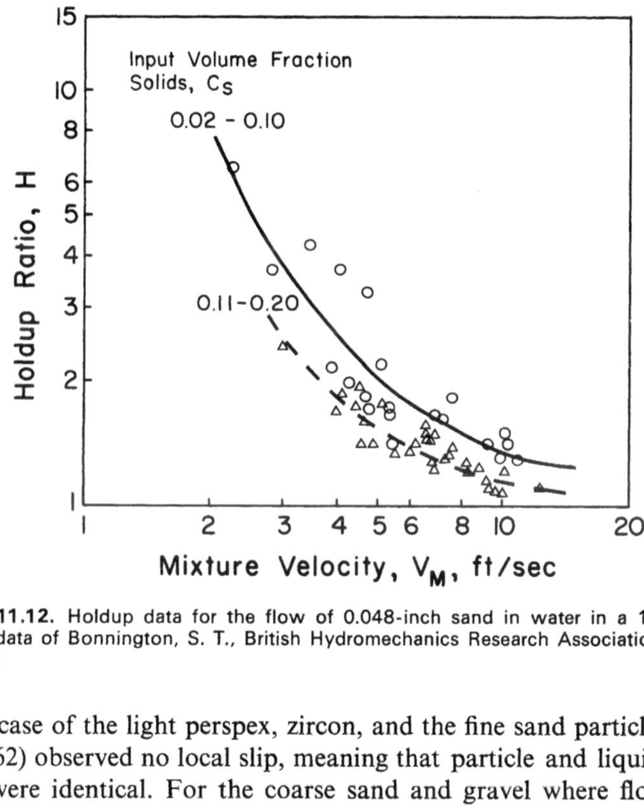

Figure 11.12. Holdup data for the flow of 0.048-inch sand in water in a $1\frac{1}{2}$-inch pipe. (From the data of Bonnington, S. T., British Hydromechanics Research Association, R.R. 637, Oct. 1959.)

In the case of the light perspex, zircon, and the fine sand particles, Newitt *et al.* (1962) observed no local slip, meaning that particle and liquid velocity profiles were identical. For the coarse sand and gravel where flow was by asymmetric suspension and sliding bed, the *in situ* and discharge concentrations were substantially the same in the upper reaches of the pipe but markedly different toward the bottom. This means that a local slip velocity exists and increases from zero in the upper 1/3 or so of the pipe to a maximum value at the bottom of the pipe. The situation is well illustrated in Figure 11.13 taken from Newitt *et al.* (1962). The figure shows the measured water velocity profile, u_W, the measured mean velocities of the solids, water, and the mixture, V_S, V_W, and V_M. The water velocity profile, u_W, is that across the

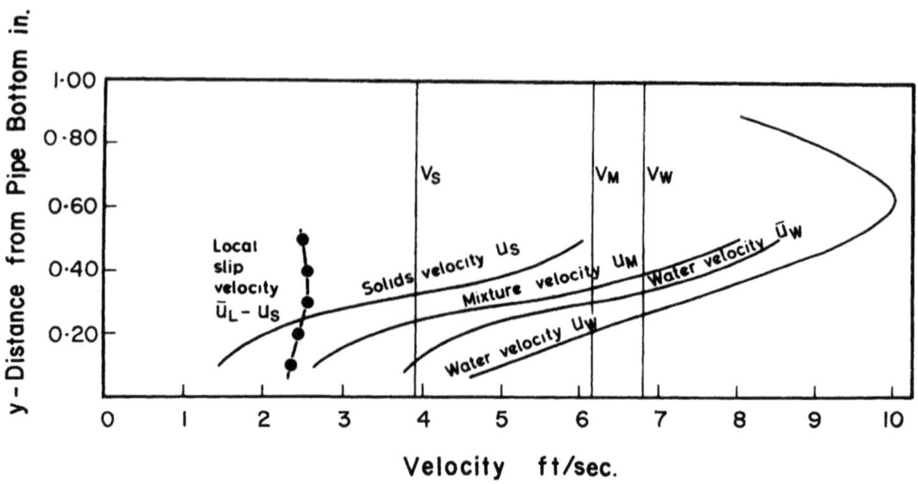

Figure 11.13. Water, mixture, and solids velocity profiles. Gravel, $i = 0.280$ ft of water/ft, $C_S = 0.10$. (From Newitt, D. M., J. F. Richardson, and C. A. Shook, Symp. Interaction Between Fluids and Particles, London, 1962, Proc., p. 87.)

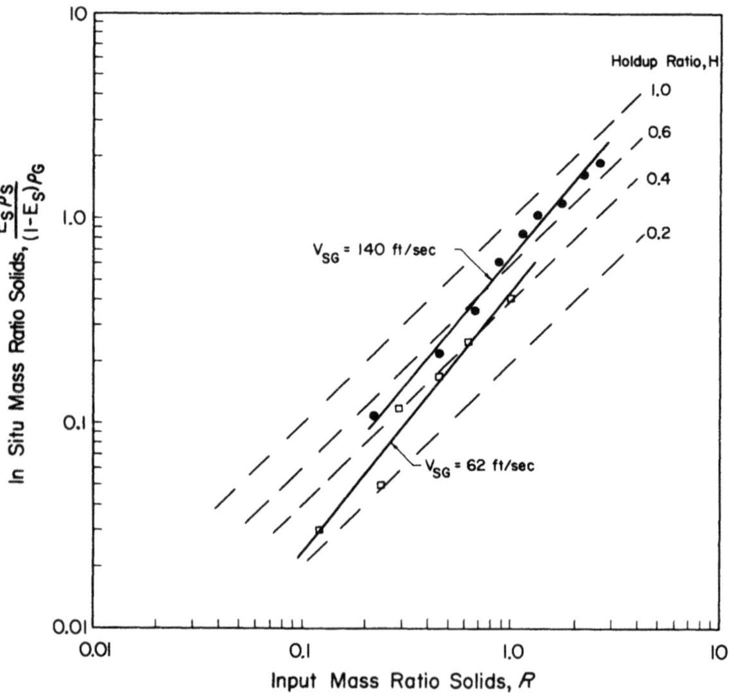

Figure 11.14. Holdup in the flow of dilute phase gas-solid mixtures—30-micron magnesium oxide particles in air. (From the data of Soo, S. L., and G. J. Trezek, *Ind. Eng. Chem. Fundam.*, **5**, 388, 1966.)

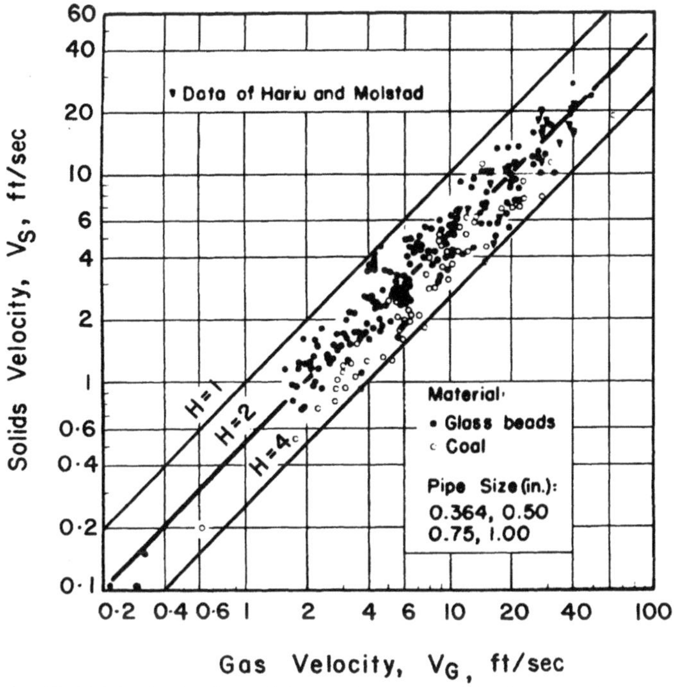

Figure 11.15. Holdup in the flow of dense phase gas-solid mixtures—glass beads and coal in air. (From Wen, C.-Y., and H. P. Simons, *A.I.Ch.E. Journal,* **5**, 263, 1959.)

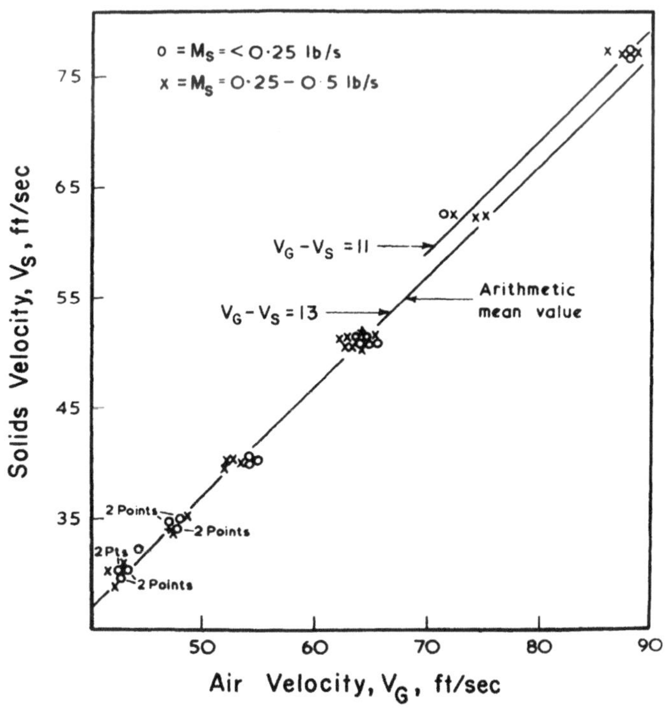

Figure 11.16. Holdup in the flow of dense phase gas-solid mixtures—coal in air. (From Richardson, J. F. and M. McLeman, *Trans. Inst. Chem. Engrs.,* **38**, 257, 1960.)

vertical diameter, and the profile $\bar{u}_W$ is the estimated average for the full width of the pipe. The velocity profiles of the solids and the mixture and the slip velocity are obtained from the estimated water velocity profile, $\bar{u}_W$, and the measured *in situ* concentrations.

Soo and Trezek (1966) report holdup data for the dilute phase transport of 30-micron magnesia particles in air at loadings below a mass ratio of three. Their data, expressed as input and *in situ* mass ratio of solid to gas, are given in Figure 11.14. The holdup ratio is less than unity because the dispersed solid particles tend to be more concentrated near the center of the pipe and their average velocity exceeds that of the carrier gas. The situation is analogous to the flow of dispersed bubbles of a gas in a liquid, but, with H defined as the holdup ratio for the heavier phase, H is less than one for the gas-solid system.

Data of Wen and Simons (1959) show the holdup effect in the flow of dense gas-solid mixtures (solid to gas ratios of 80 to 750 lb/lb). Their data for glass beads and coal powders of sizes from 0.0028 to 0.0297 inch, in 1/2-, 3/4-, and 1-inch pipes, are shown in Figure 11.15 in the form of average particle velocity, V_S, versus average gas velocity, V_G. The ratio V_G/V_S is the holdup ratio and we see that their data are bounded by lines corresponding to $H = 1$ and $H = 4$ averaging $H = 2$.

Richardson and McLeman (1960) also give data reflecting the holdup in dense phase gas-solid flow. Their data for coal particles (0.04 in.) flowing with air in a 1-inch pipe are shown in Figure 11.16. The data reflect a nearly constant slip velocity, S, of 11 to 13 ft/sec.

2.4 Pressure Gradient

We may obtain a good idea of the typical pressure gradient behavior of fluid-solid systems in horizontal flow by examining typical data for the systems for which we considered the flow patterns and the holdup. Some of the Newitt *et al.* (1955) pressure gradient data on the flow of closely graded fine sand (97 microns) in water in a 1-inch pipe are reproduced in Figure 11.17.

For this sand Newitt *et al.* reported that the sand traveled in "homogeneous" suspension. Whether the solids were homogeneously or uniformly suspended was probably not known, but it may be assumed that these particles were at least nearly symmetrically distributed. The data show a divergence from the water pressure gradient curve which increases with solids concentration, but, on logarithmic coordinates, results in a series of nearly parallel straight lines. This behavior is characteristic of the symmetric suspension flow pattern.

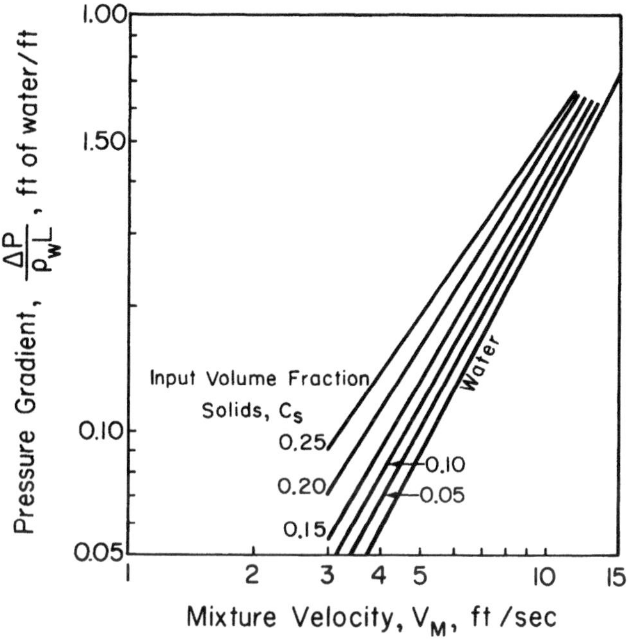

Figure 11.17. Pressure gradient—water suspensions of 97-micron sand. (Replotted from Newitt, D. M., J. F. Richardson, M. Abbott, and R. B. Turtle, *Trans. Inst. Chem. Engrs.*, **33**, 93, 1955.)

The data of Newitt *et al.* for the coarser sand (208 microns) and fine gravel are given in Figures 11.18 and 11.19. These show behavior different from the fine sand reflecting all three flow patterns. At the higher velocities and lower concentrations, the data compare with those for the finer sand indicating the symmetric suspension flow pattern. At intermediate velocities and the higher concentrations, the divergence of the curves from the water curve increases with decreasing flow velocity. This reflects the asymmetric suspension flow pattern and the increasing asymmetry with decreasing velocity. At velocities below 2.5 to 4.5 ft/sec, the divergence of the curves from the water curve becomes such as to result in an increase in pressure gradient with decrease in velocity. This is characteristic of the moving bed flow pattern.

Howard (1939), Durand (1953), Worster and Denny (1955), Gaessler (1967), and others have obtained data qualitatively similar to the data of Figure 11.18 for medium and coarse particles. Generally the coarser or the denser the particles, the higher the velocity required for symmetric suspension, and in many instances this was not reached at all, with the pressure gradient curves diverging from the water curve with decreasing water velocity at all water velocities.

Typical data reported by Gaessler (1967) on the flow in water of 0- to 10-mm coal particles in a 160-mm pipe are shown in Figure 11.20, at constant

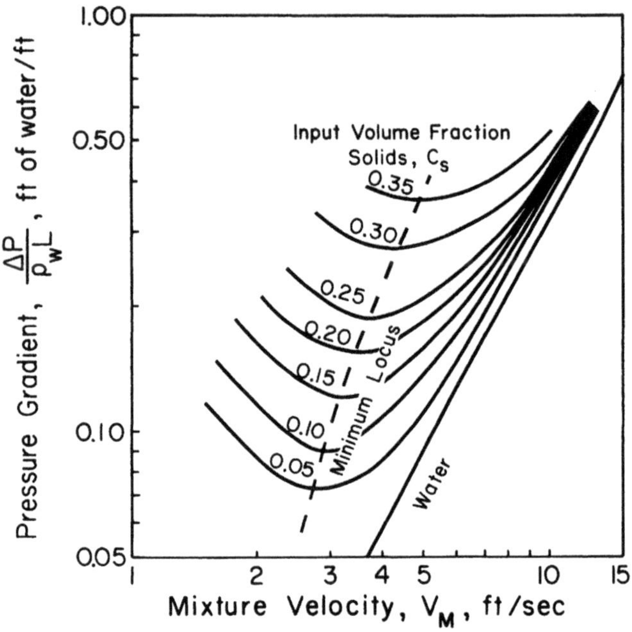

Figure 11.18. Pressure gradient—water suspensions of 208-micron sand. (Replotted from Newitt, D. M., J. F. Richardson, M. Abbott, and R. B. Turtle, *Trans. Inst. Chem. Engrs.*, **33**, 93, 1955.)

mass flow of solids, and in Figure 11.21, at constant input volume fraction, C_S. The similarity with the data of Newitt *et al.* on the coarse sand and gravel is apparent. Figure 11.22 presents data from Gaessler on the flow of 0- to 3-mm coal in a 46-mm pipe. These finer particles appear, at certain concentrations, to act as turbulence suppressors, causing a reduction of the pressure gradient to a level below that for water alone. (In all three of these figures the solid lines represent predictions based upon Gaessler's analysis which we will discuss in Section 3.42, but the lines closely reproduce the trend of actual data.)

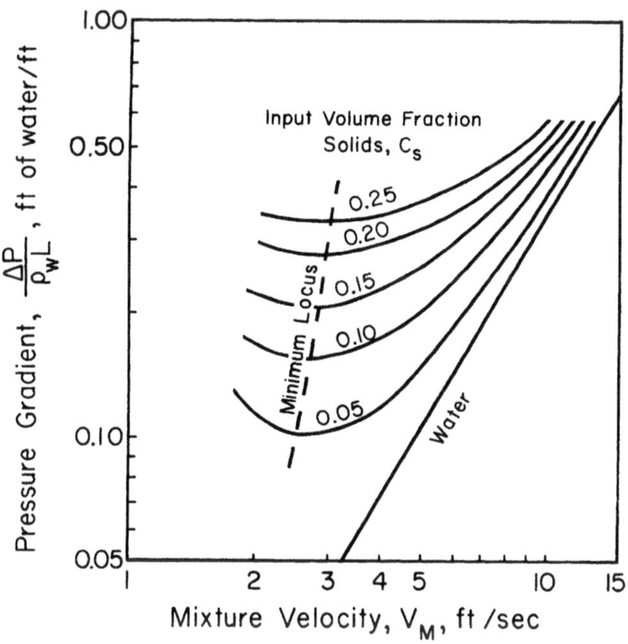

Figure 11.19. Pressure gradient—water suspensions of 0.125-inch gravel. (Replotted from Newitt, D. M., J. F. Richardson, M. Abbott, and R. B. Turtle, *Trans. Inst. Chem. Engrs.*, **33**, 93, 1955.)

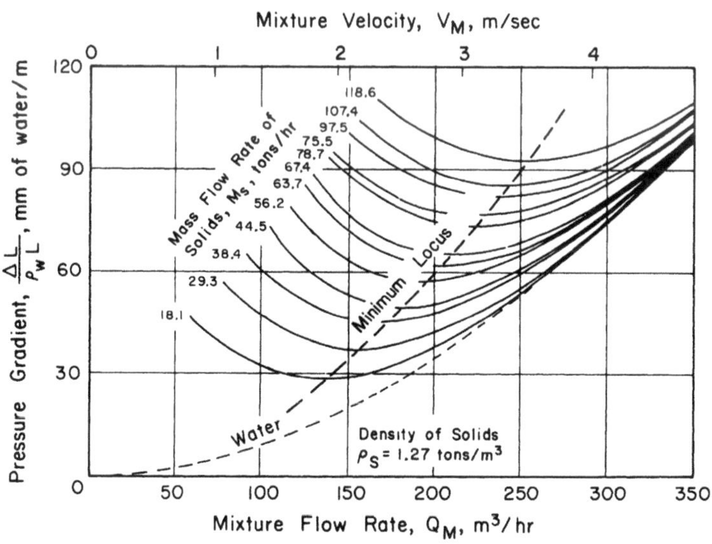

Figure 11.20. Pressure gradient—water suspensions of 0- to 10-mm coal particles in a 160-mm pipe. (From Gaessler, H., Doctoral Dissertation, Technische Hochschule, Karlsruhe, Germany, 1967.)

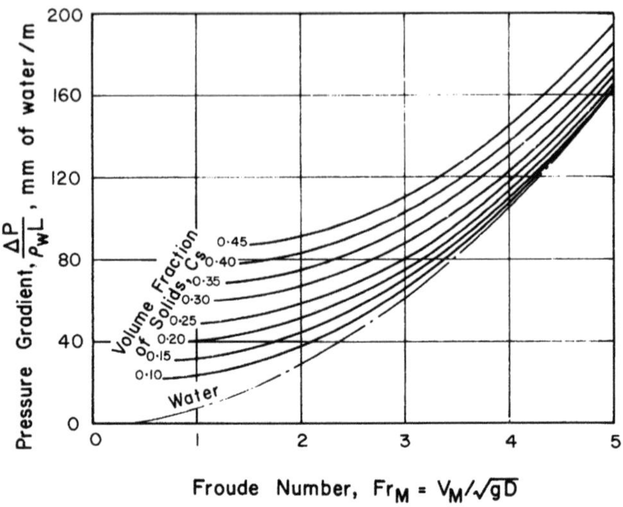

Figure 11.21. Pressure gradient—water suspensions of 0- to 10-mm coal particles in a 160-mm pipe. (From Gaessler, H., Doctoral Dissertation, Technische Hochschule, Karlsruhe, Germany, 1967.)

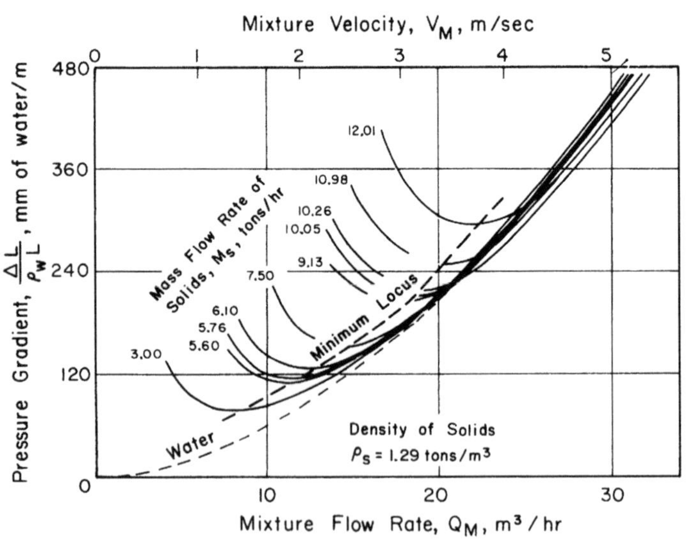

Figure 11.22. Pressure gradient—water suspensions of 0- to 3-mm coal particles in a 46-mm pipe. (From Gaessler, H., Doctoral Dissertation, Technische Hochschule, Karlsruhe, Germany, 1967.)

Data of Faddick (1963) on the flow of water-saturated $\frac{3}{4}$ inch nominal size wood chips are presented in Figure 11.23. These data also show the same trends exhibited by the coarse sand, gravel, and coal suspensions.

In the case of granular solids flowing in a gas stream, the data of many workers show qualitative agreement with the forms of Figures 11.17 and 11.18. To illustrate, Figure 11.24 shows the data of Mehta *et al.* (1957) for the horizontal flow of 97-micron glass beads in air at low and modest loadings and in the symmetric and asymmetric suspension flow patterns. The similarity with Figures 11.17 and 11.18 is apparent. Figure 11.25 shows curves based on the data of Worster (1953) for the flow of 1/30-inch coal particles in air in a 1-inch pipe. These data also reflect both the symmetric and the asymmetric suspension flow patterns. Data of Clark *et al.* (1952) on the pneumatic conveying of cress seed (91.9 percent, 16 BSS, specific gravity 1.17) in a 1-inch pipe are shown in Figure 11.26.

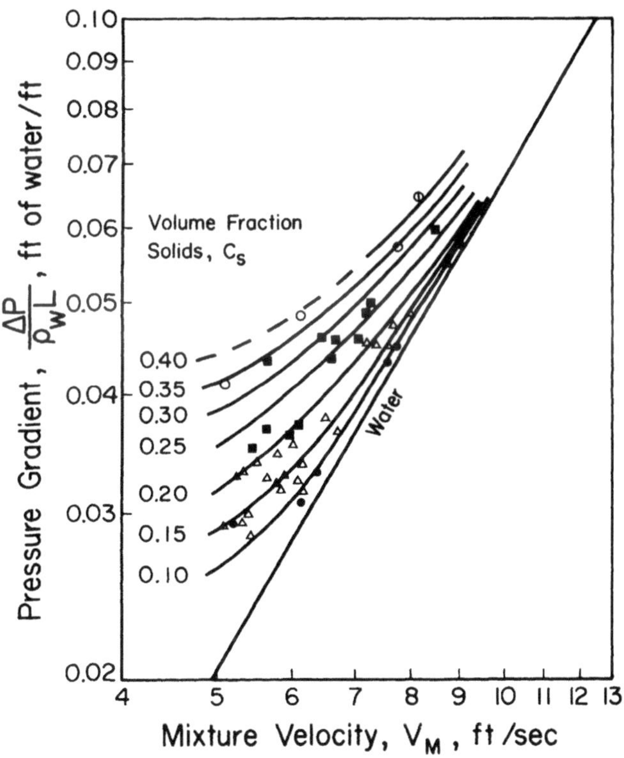

Figure 11.23. Pressure gradient—water suspensions of wood chips in 4-inch pipe. (Redrawn from Faddick, R. R., M.Sc. Thesis, Queen's University, Kingston, Ontario, 1963.)

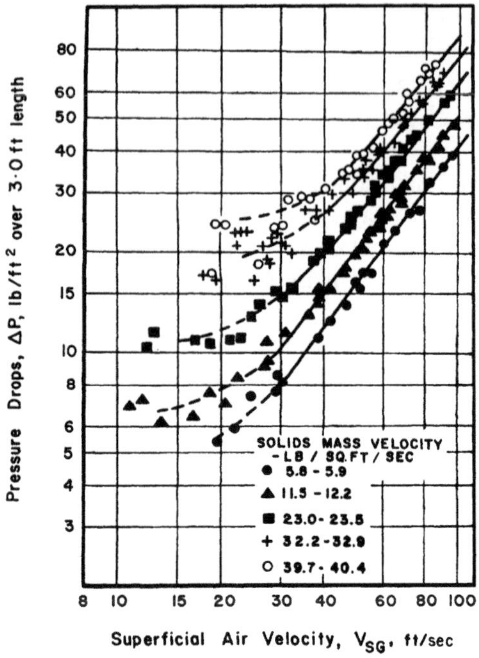

Figure 11.24. Pressure drop; 97-micron glass beads in air in a ½-inch pipe. (From Mehta N. C., J. M. Smith, and E. W. Comings, *Ind. Eng. Chem.*, **49**, 986, 1957.)

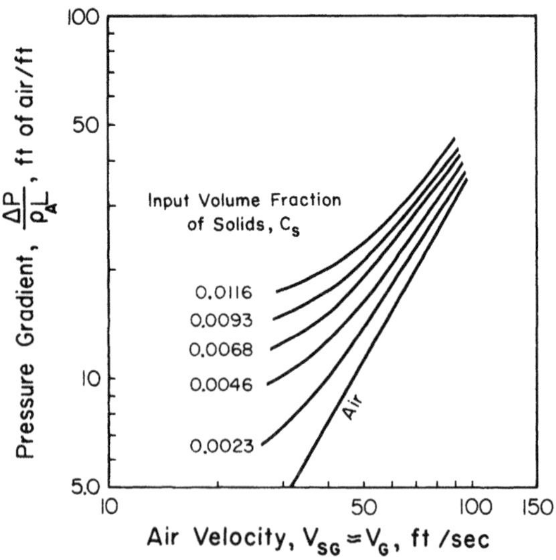

Figure 11.25. Pressure gradient; flow of 1/30-inch coal particles in air in a 1-inch pipe. (Redrawn from Worster, R. C., British Hydromechanics Research Association, S.P. 464, Sept. 1953.)

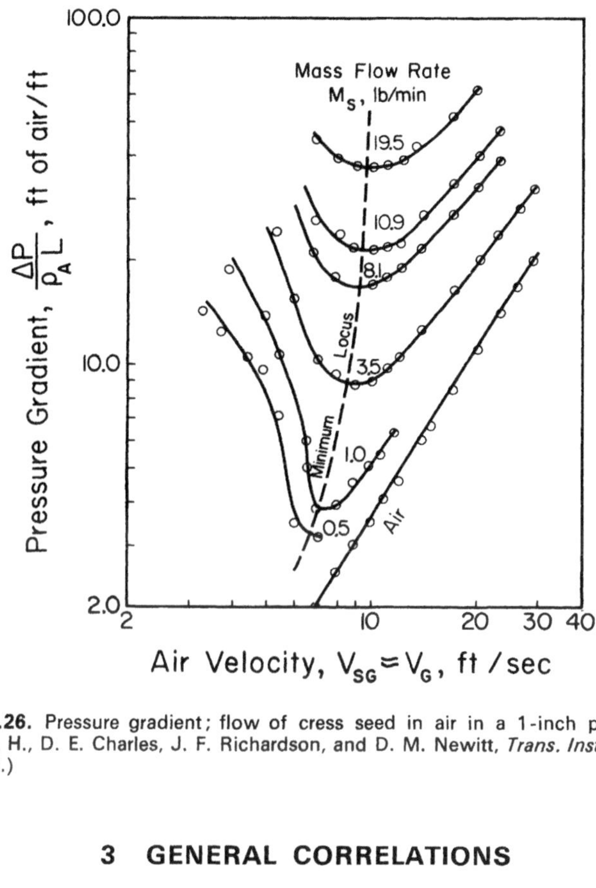

Figure 11.26. Pressure gradient; flow of cress seed in air in a 1-inch pipe. (Replotted from Clark, R. H., D. E. Charles, J. F. Richardson, and D. M. Newitt, *Trans. Inst. Chem. Engrs.*, **30**, 209, 1952.)

3 GENERAL CORRELATIONS

3.1 Introduction

As might be expected from the great variety and complexity of the flow behavior of liquid-solid and gas-solid systems, a host of empirical correlations relating to the flow patterns and the pressure gradient have been developed. Also some correlations of restricted applicability related to the holdup phenomenon have been proposed. None of these correlations is generally applicable to all flow patterns and to both gas and liquid carriers. Most of the correlations are strictly empirical, although there have been a few attempts to develop relationships based upon assumptions concerning the flow mechanism.

In this section we will deal with the more important of the available correlations, including those few having some theoretical basis. With one exception, separate correlations will be given for liquid and gas systems, although where possible these will be interrelated. We will treat the flow pattern and holdup correlations in single sections, but in our consideration of pressure gradient we will deal separately with correlations of general applicability and with those restricted to one flow pattern.

3.2 Flow Patterns

3.21 Introduction

Many different terms are in use to describe the various flow patterns encountered in the horizontal flow of liquid-solid and gas-solid mixtures. Although flow pattern maps of the type discussed with gas-liquid systems have not been developed extensively, there are many empirical equations relating the mixture transition velocity from one pattern to the next. The more important of these will be discussed and compared.

For our discussion related to the flow in either liquids or gases of granular and similar solids (excluding fiber-like materials) we will follow the terminology of Govier and Charles (1961) and define the average mixture velocities at which, in any given system, the transition takes place from one flow pattern to another, as follows:

V_{M1} = the velocity at and above which the mixture flows in the symmetric suspension pattern;

V_{M2} = the velocity at and above which (to V_{M1}) the mixture flows in the asymmetric suspension pattern; the velocity below which solids form a deposit on the bottom of the pipe;

V_{M3} = the velocity at and above which (to V_{M2}) a moving bed of particles exists on the bottom of the pipe and, with mixed-size particles, some particles move by saltation and some are in suspension; the velocity below which the part of the bed in contact with the pipe wall becomes stationary;

V_{M4} = the velocity at and above which (to V_{M3}) a bed exists the lower part of which is stationary and in which the particles in the upper part move by saltation and, with mixed-size particles, some particles are in suspension; the velocity below which pipe blockage occurs.

These transition velocities may be related to the pressure gradient–velocity curve as shown in the schematic diagram, Figure 11.27. Above V_{M1} the suspension behaves much as a true fluid, although nonuniform (even though symmetric) distribution of the particles over the cross section results in

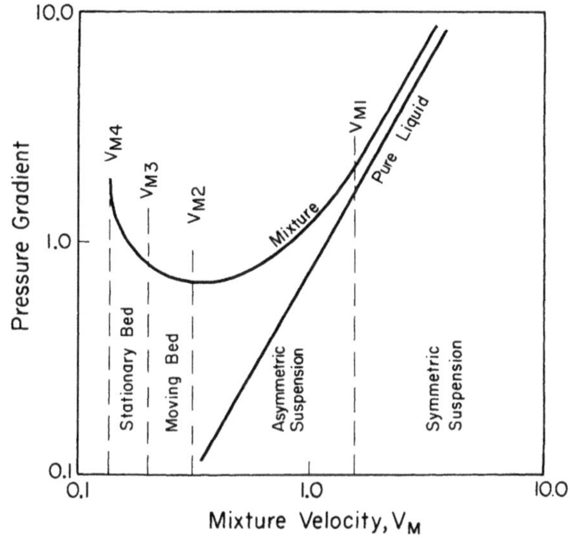

Figure 11.27. Schematic representation of pressure gradient curves.

holdup and a flowing mixture whose density is not uniform and exceeds that of the input mixture. Notwithstanding these effects, the pressure gradient line nearly parallels that for the carrier fluid and may often be approximated as for a pure fluid having the properties of the input mixture.

The concept of the transition velocities, V_{M1} to V_{M4}, is directly applicable only to suspensions of closely sized particles. For suspensions of wide particle size range it may be applied to arbitrarily selected narrow size fractions. In using the correlations to follow we should not use an average diameter for suspensions of wide particle size range.

Although there has been confusion in notation and some variation in interpretation, it now seems quite well established that the velocity V_{M2} coincides with that at which the minimum occurs in the pressure gradient–mixture velocity relationship (Durand, 1953; Thomas 1962, 1964; Condolios and Chapus, 1963). The transition velocity V_{M3} is not clearly reflected in the pressure gradient curve, and V_{M4} is seldom reached in actual measurements.

3.22 Transition Velocities

All Transition Velocities, Liquid-Solid Suspensions The work of Gaessler (1967), which we will discuss in some detail in Section 3.42, is the only known correlation that provides a method for determining the equivalent of all the transition velocities, V_{M1} to V_{M4}. The method applies to liquid-solid systems.

Actually Gaessler properly points out that with mixed-size particles the transitions are gradual, each flow pattern grading into the next.

Gaessler expresses his results in terms of C_S, C_w, C_{w1}, and C_{w2} defined in the following:

$$C_S = \text{input volume fraction of solids}$$

$$= \frac{Q_S}{Q_S + Q_L} = \frac{Q_S}{Q_M} = \frac{V_{SS}}{V_M}$$

$$C_w = \text{input mass fraction of solids}$$

$$= \frac{M_S}{M_M} = \frac{Q_S \rho_S}{Q_S \rho_S + Q_L \rho_L} = \frac{Q_S \rho_S}{Q_M \rho_M}$$

and two mass fractions of the solids

$$C_{w1} = \frac{M_{S1}}{M_M} = \frac{Q_{S1} \rho_S}{Q_M \rho_M}$$

$$C_{w2} = \frac{M_{S2}}{M_M} = \frac{Q_{S2} \rho_S}{Q_M \rho_M}$$

where M = mass rate of flow,
Q = volume rate of flow,
ρ = density,
V_{SS} = superficial velocity of solids,
V_M = average velocity of the mixture,

and the subscripts S, L, and M refer, respectively, to the solids, liquid, and the mixture; the subscripts 1 and 2 refer, respectively, to the solids flowing in suspension and to those in the bed.

The quantities C_S and C_w are interrelated by

$$C_w = \frac{sC_S(Q_S + Q_L)}{sQ_S + Q_L} \tag{11.1}$$

where $s = \rho_S/\rho_L$.

A mass balance on the flowing solids yields

$$C_w = C_{w1} + C_{w2} \tag{11.2}$$

We may also write a volume balance on the flowing solids:

$$V_S A \frac{Q_S}{Q_M} = V_{S1} A \frac{Q_{S1}}{Q_M} + V_{S2} A \frac{Q_{S2}}{Q_M} \tag{11.3}$$

where A is the cross-sectional area of the pipe, and V_S is the average velocity of the solids. If we multiply each term by $\rho_S/\rho_M A$, we obtain

$$V_S C_w = V_{S1} C_{w1} + V_{S2} C_{w2} \tag{11.4}$$

Gaessler devises an empirical method for estimating the mass fractions, C_{w1} and C_{w2}, from the input flow rates, the particle size, and the properties of the solids and the fluid. Then, in our notation, he proposes the flow pattern criteria as given in Table 11.4. The correlation suggests, for example, that bed formation or the first deposition of solids occurs at a value of the C_{w1}/C_w ratio of 0.70. This corresponds with V_{M2}. We will present the method of estimating C_{w1} in Section 3.42. Gaessler's flow pattern correlation is confirmed, somewhat indirectly, by the agreement between his prediction and a variety of observed pressure gradient data for water suspensions of a number of different types and sizes of solid particles. It may be considered reliable for medium and coarse particles suspended in water.

Transition Velocity V_{M1}, Liquid-Solid Suspensions The transition from the symmetric to the asymmetric flow pattern occurs gradually with decreasing fluid velocity and the consequent increasing relative significance of gravitational forces.

Spells (1955), using an empirical approach based on dimensional analysis, presented correlations for low concentration liquid suspensions of 80- to

TABLE 11.4

Flow Pattern Criteria of Gaessler

Transition Mixture Velocity	Flow Pattern	C_{w1}/C_w
	symmetric suspension	~ 1.0
V_{M1}		
	asymmetric suspension	$0.7 - \sim 1.0$
V_{M2}		
	moving bed with asymmetric suspension	$0.2 - 0.7$
V_{M3}		
	stationary bed with asymmetric suspension	$\sim 0 - 0.2$
V_{M4}		
	pipe blocked	~ 0

800-micron particles for what he called the standard velocity and the minimum velocity. His standard velocity appears to correspond with the transition velocity V_{M1}, whereas his minimum velocity corresponds with V_{M2}. Spells' equation for the standard velocity is

$$V_{M1}^2 = 0.075\left(\frac{DV_{M1}\rho_M}{\mu}\right)^{0.775} gd(s-1) \tag{11.5}$$

This may be rearranged to the form of Equation 1.35, and combined with Newton's settling equation (Equation 1.11) to give

$$V_{M1} = 134C_D^{0.816}D^{0.633}V_0^{1.63} \tag{11.6}$$

where D is in feet, and V_0 is the terminal settling velocity in ft/sec. This relation assumes that the kinematic viscosity of the mixture is the same as for water.

Additionally Govier and Charles (1961) have shown that Durand's (1953) equation for the pressure gradient in the asymmetric suspension region (to be discussed later) may be solved with the appropriate equation for the symmetric region to yield the following relation for V_{M1}:

$$V_{M1} = 11.9(V_0 D)^{0.5}d^{-0.25} \tag{11.7}$$

This relationship cannot be expected to give more than order of magnitude results.

Newitt *et al.* (1955), by semitheoretical analysis, also developed relations for V_{M1}. The analysis will be referred to later. His transition relations are

$$V_{M1} = 38.7(DV_0)^{1/3} \tag{11.8}$$

If we substitute for V_0 from Newton's law, we obtain

$$V_{M1} = 38.7D^{1/3}\,\frac{4gd}{3C_D}\,(s-1)^{1/6} \tag{11.9}$$

The concentration profile relationship of Ismail (1951) discussed in Section 3.23 is also relevant. The transition from symmetric to asymmetric flow may, for practical purposes, be taken as occurring when the ratio of the concentration near the bottom of the pipe to that near the top of the pipe significantly exceeds unity. Following Wasp *et al.* (1963), we may write

$$\ln\left(\frac{\varepsilon_{S,\,0.08}}{\varepsilon_{S,\,0.92}}\right) = 3.6\,\frac{V_c}{\beta\mathscr{K}V_M\sqrt{f_M/2}} \tag{11.10}$$

where V_c is the terminal settling velocity of the particles in the concentration present in the pipe,

$\varepsilon_{S,0.08}$ and $\varepsilon_{S,0.92}$ are the *in situ* volume fractions of the solids at elevations $0.08D$ and $0.92D$ above the bottom of the pipe,

β = constant of proportionality defined subsequently by Equation 11.39,
$\mathscr{K}$ = von Kármán constant
f_M = single-phase friction factor based upon the mixture velocity and properties.

Substituting

$$\beta = 1$$

$$\mathscr{K} = 0.35$$

$$f_M = 0.0225\left(\frac{DV_M\rho_M}{\mu_M}\right)^{-0.25}$$

we obtain

$$\ln \frac{\varepsilon_{S,\,0.08}}{\varepsilon_{S,\,0.92}} \simeq 100\,\frac{V_c D^{1/8}\rho_M^{1/8}}{V_M^{7/8}\mu_M^{1/8}} \tag{11.11}$$

Taking the transition from the symmetric to the asymmetric concentration profile arbitrarily at

$$\frac{\varepsilon_{S,0.08}}{\varepsilon_{S,0.92}} = 2$$

we obtain from Equation 11.11

$$V_{M1} = 294V_c\left(\frac{V_c D\rho}{\mu}\right)^{1/7} \tag{11.12}$$

The hindered settling velocity of the particles may be estimated from the methods discussed in Chapter 1. For the case of spherical particles in low concentration and settling in a turbulent field, Newton's law is applicable and

$$V_c = V_0 = 1.74\sqrt{gd(s-1)} \tag{1.11}$$

we then have, after rearranging,

$$V_{M1} = 554[gd(s-1)]^{0.571}\left(\frac{D\rho_M}{\mu_M}\right)^{0.143} \tag{11.13}$$

Equation 11.13 illustrates, for regions where Newton's law applies, the major roles of the particle size and the density ratio, and the lesser roles of the pipe diameter, the mixture density, and the mixture viscosity in determining the transition velocity. The similarity in form of this equation to Equations 11.5 and 1.35 of Spells is to be noted. In regions where Stokes' law is applicable, the transition velocity is even more sensitive to particle size and density ratio, and the mixture density and viscosity also have major effects.

Transition Velocity V_{M1}, ***Gas-Solid Suspensions*** Less attention has been given to the determination of V_{M1} for gas-solid suspensions. The work of Soo and Regalbuto (1960), although not providing definite criteria for the transition, indicates some of the factors involved and confirms that asymmetry in the concentration profile develops when an acceleration force related to the turbulent transport of particles across the flow stream becomes much less than the downward force due to the acceleration of gravity. This work, however, does not define the factors influencing the turbulent accelerating force other than to show that it probably increases more than linearly with mean stream velocity and decreases more than linearly with particle diameter.

Transition Velocity V_{M2}, ***Liquid-Solid Suspensions*** Durand (1953) was the first to attempt to define the conditions for liquid-solid systems beyond which a deposit on the pipe wall would be encountered. For the "limit deposit velocity," i.e., our V_{M2}, he proposed the empirical relation,

$$V_{M2} = F_L \, [2gD(s-1)]^{0.5} \tag{11.14}$$

where F_L is a dimensionless factor influenced by particle size and concentration, as given in Figure 11.28. Durand's correlation was based on data covering pipe diameters from 40 to 700 mm, for sand and coal particles in

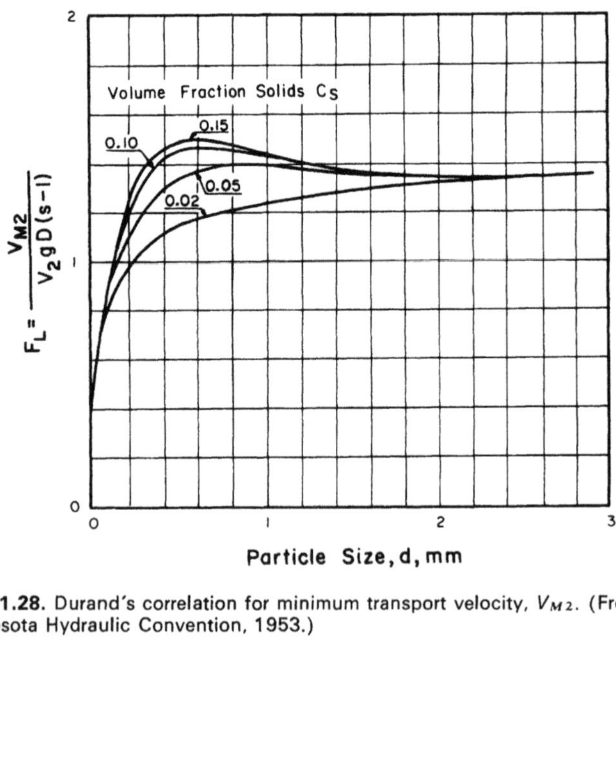

Figure 11.28. Durand's correlation for minimum transport velocity, V_{M2}. (From Durand, R., Proc. Minnesota Hydraulic Convention, 1953.)

concentrations to about 15 volume percent. It is to be noted that this correlation indicates no effect of particle size or concentration for particles 2 mm in diameter or larger.

The semitheoretical analysis of Newitt *et al.* (1955) resulted in the following equation for the transition to the moving bed flow pattern:

$$V_{M2} = 17V_0 \tag{11.15}$$

which, if we substitute for V_0 from Newton's law, leads to

$$V_{M2} = 17\left[\frac{4gd}{3C_D}(s-1)\right]^{1/2} \tag{11.16}$$

A flow pattern map has been developed by Newitt *et al.* (1955) (based on their equations for V_{M1} and V_{M2}, and Durand's equation for V_{M2} which they interpret as V_{M3}). The flow pattern map for mixtures of sand or gravel in water in 1- and 6-inch pipes is given in Figure 11.29. Because of the confusion with respect to V_{M2} and V_{M3}, the map should be used with caution.

Spells' minimum velocity is also equivalent to V_{M2}. His equation is

$$V_{M2}^2 = 0.0251\left(\frac{DV_{M2}\rho_M}{\mu}\right)^{0.775} gd(s-1) \tag{11.17}$$

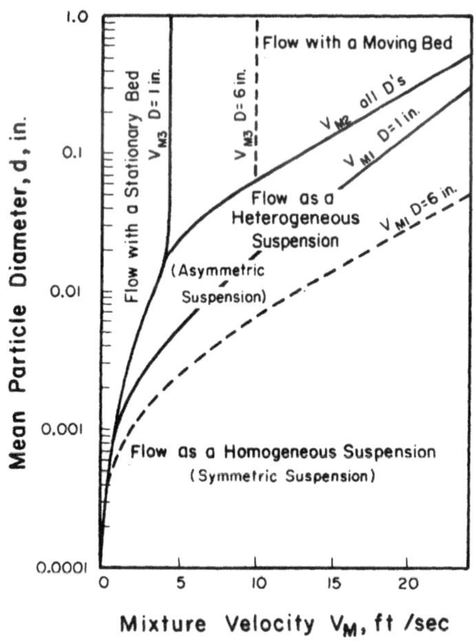

Figure 11.29. Newitt *et al.* flow pattern map. (Redrawn from Newitt, D. M., J. F. Richardson, M. Abbott, and R. B. Turtle, *Trans. Inst. Chem. Engrs.*, **33**, 93, 1955.)

which may be rearranged and combined with Newton's equation to give

$$V_{M2} = 54.4 C_D^{0.815} D^{0.633} V_0^{1.63} \tag{11.18}$$

Condolios and Chapus (1963) propose a further empirical correlation for V_{M2} based upon their own data for graded and mixed sands 0.44 mm in diameter and larger. Their relationship, in the form of $V_{M2}/\sqrt{gD}$ as a function of volume fraction solids, is given in Figure 11.30. The correlation has been confirmed in 23- and 28-inch pipes for solids concentrations of 15 to 20 percent.

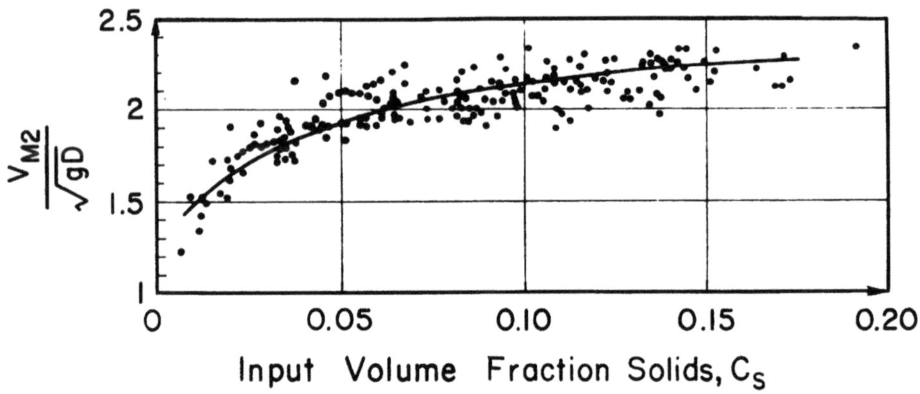

Figure 11.30. Critical deposit velocity, V_{M2}, for graded and mixed sands of more than 0.44 mm. (From Condolios, E., and E. E. Chapus, *Chem. Eng.*, **70** (July 8), 137, 1963. Reprinted by special permission of McGraw-Hill, Inc., New York.)

Hughmark (1961) has proposed a correlation for V_{M2} in the form

$$\frac{V_{M2}}{\sqrt{gD}} = \phi[C_S(s-1)F_D] \tag{11.19}$$

where F_D is a correlating factor given by Figure 11.31. Hayden and Stelson (1968) have followed this approach in presenting their data and those of others on this transition velocity. Figure 11.32 gives these results and may be taken as indicative of the scatter to be expected.

Zandi and Govatos (1967), after appraising some 1452 data points taken from the work of eleven references, proposed a flow pattern index

$$N_1 = \frac{V_M^2 \sqrt{C_D}}{C_S D g(s-1)}$$

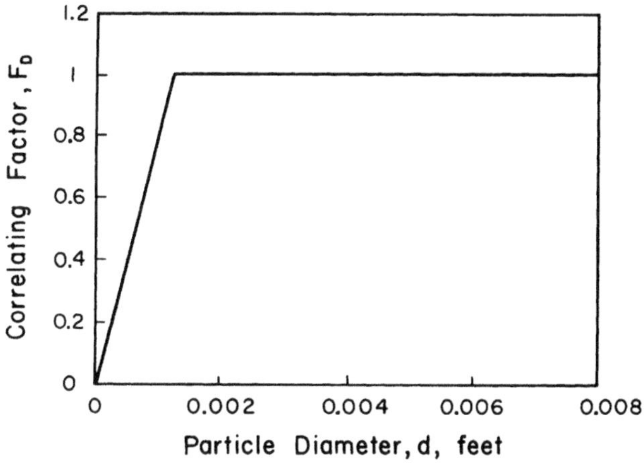

Figure 11.31. Hughmark's correlating factor, F_D. (From Hughmark, G. A., *Ind. Eng. Chem.*, **53**, 389, 1961.)

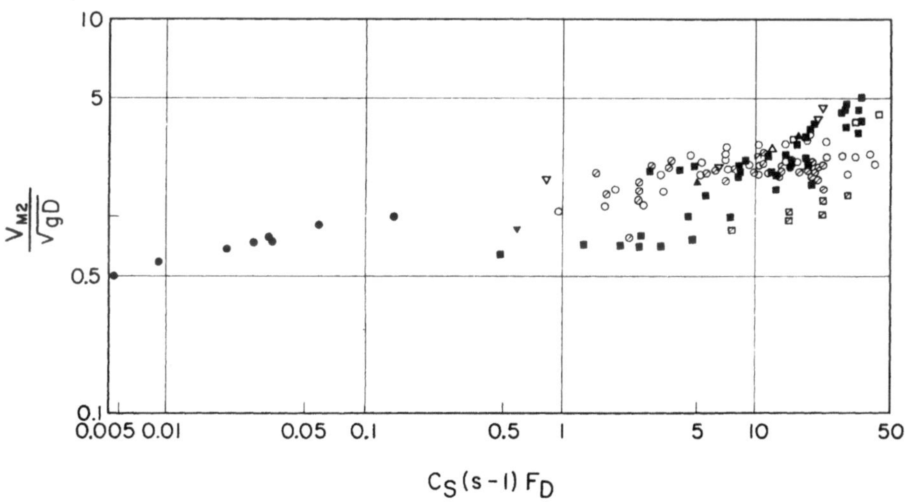

Figure 11.32. Hughmark's minimum transport velocity correlation. (From Hayden, J. W. and T. E. Stelson, paper presented at Int. Symp. on Solid-Liquid Flow in Pipes, Philadelphia, Penn., March, 1968.)

and indicated that when $N_1 \leq 40$, the flow would be by saltation. This is equivalent to

$$V_{M2} = \left[\frac{40C_S Dg(s-1)}{C_D^{0.5}}\right]^{1/2} \tag{11.20}$$

Babcock (1968) has correlated carefully taken pressure gradient data for a limited range of 20- to 30-mesh sand in water suspensions flowing in a 1-inch pipe by the equation

$$\frac{\Delta P}{\rho_W L} = \frac{\Delta P_W}{\rho_W L}\left[1 + 70\,\frac{C_S g D(s-1)}{V_M^2\sqrt{C_D}}\right] \tag{11.21}$$

which we will discuss later. This equation predicts minima in the relation of $\Delta P/\rho_W L$ versus V_M corresponding with our V_{M2}. Babcock differentiates the equation to find the locus of the minima as

$$N_1 = \frac{V_{M2}^2\sqrt{C_D}}{C_S g D(s-1)} = 10 \tag{11.22}$$

This is of the same form as Zandi's flow pattern index but suggests a transition value of 10 rather than 40. It would appear that N_1 is a good criterion of V_{M2} for suspensions of a single solid in a single pipe, but that other factors are required to accommodate a variety of solids, particle sizes, and pipe diameters.

Sinclair (1962) has carried out what is probably the most comprehensive experimental investigation of the limit-deposit velocity, V_{M2}, for liquid-solid mixtures in small pipes (1/2, 3/4, and 1 inch). He tested closely sized coal, sand, and iron particles in the size range from about 30 to 2000 microns. Sinclair confirmed observations of others that V_{M2} is dependent upon concentration and that a maximum value of V_{M2} occurs at intermediate concentrations in the range of 0.05 to 0.20. Average relationships for the trend in V_{M2} with concentration are given in Figure 11.33 taken from his work, where V_{M2} is normalized with respect to its maximum value, $(V_{M2})_{max}$. Sinclair elected to correlate $(V_{M2})_{max}$ in terms of a modified Froude number

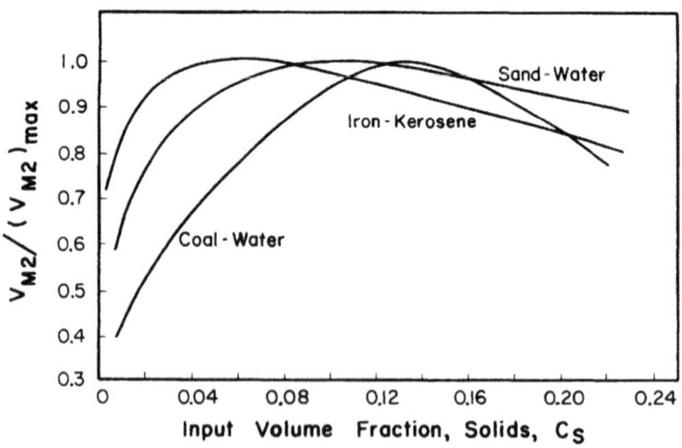

Figure 11.33. Sinclair's dimensionless minimum transport velocity relationships. (From Sinclair, C. G., Symp. Interaction Between Fluids and Particles, London, 1962, Proc. p. 78.)

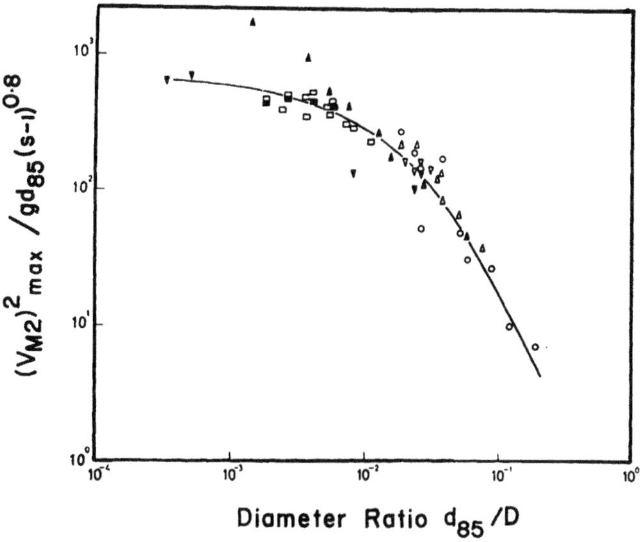

Figure 11.34. Sinclair's minimum transport velocity correlation. (From Sinclair, C. G., Symp. Interaction Between Fluids and Particles, London, 1962, Proc. p. 78.)

and the diameter ratio d_{85}/D, where d_{85} is the particle diameter such that 85 percent by weight of the particles are less than d_{85}. His correlation is given by Figure 11.34.

Sinclair shows that for d_{85}/D less than 0.001, where the limit velocity becomes independent of d/D, and

$$\frac{(V_{M2})^2_{\max}}{gd_{85}(s-1)^{0.80}} \simeq 650 \tag{11.23}$$

settled particles would be expected to penetrate about halfway through the buffer zone of the boundary layer. He reasons that increasing the particle size would lead to temporarily settled particles being increasingly exposed to the forces of the turbulent core, with a consequent reduction in the velocity required for their resuspension.

Wicks (1968) compares the predictions of Durand and Condolios (1952), Spells (1955), Hughmark (1961), Sinclair (1962), and Condolios and Chapus (1963) to calculate V_{M2} as a function of diameter for a 1 percent concentration of 0.25-mm sand in water. His results (see Figure 11.35) show the wide disparity in the predictions.

Thomas (1964) has developed separate empirical correlations for V_{M2} for particle sizes below and above the thickness of the laminar sublayer which he takes as

$$\delta = \frac{5\mu}{\rho U_M^*}$$

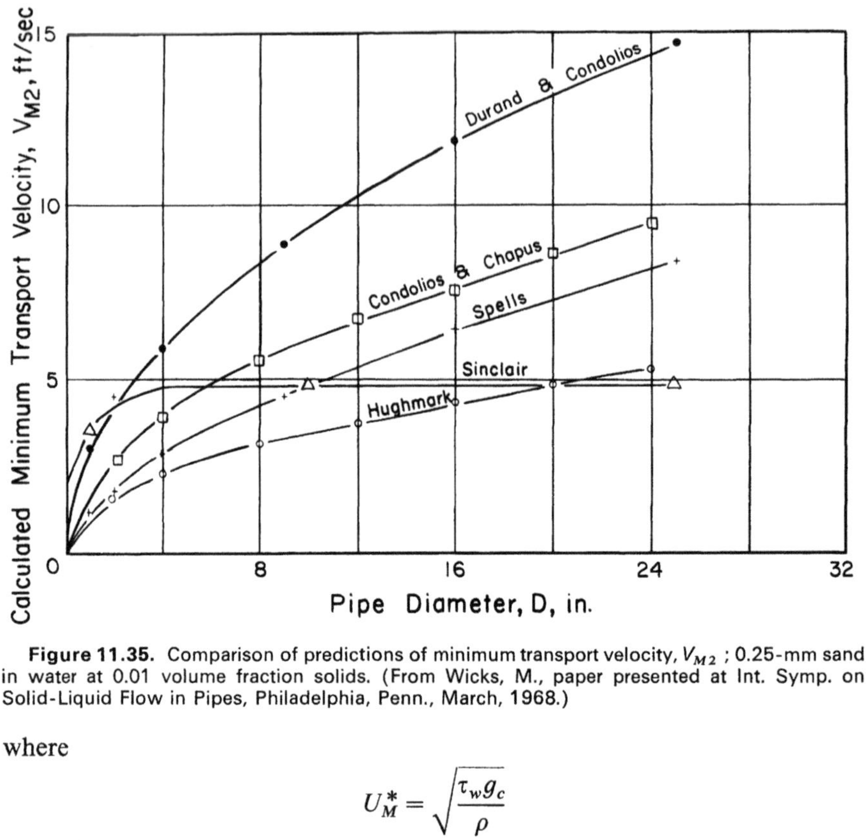

Figure 11.35. Comparison of predictions of minimum transport velocity, V_{M2} ; 0.25-mm sand in water at 0.01 volume fraction solids. (From Wicks, M., paper presented at Int. Symp. on Solid-Liquid Flow in Pipes, Philadelphia, Penn., March, 1968.)

where

$$U_M^* = \sqrt{\frac{\tau_w g_c}{\rho}}$$

For $d/\delta < 1$ he gives

$$\frac{V_0}{(U_{M2}^*)_0} = 0.010\left[\frac{d(U_{M2}^*)_0\rho}{\mu}\right]^{2.71} \tag{1.33}$$

and for $d/\delta > 1$

$$\frac{V_0}{(U_{M2}^*)_0} = 4.90\left[\frac{d(U_{M2}^*)_0\rho}{\mu}\right]\left[\frac{\mu}{D(U_{M2}^*)_0\rho}\right]^{0.60}(s-1)^{0.23} \tag{11.24}$$

where $(U_{M2}^*)_0$ is the friction velocity corresponding with flow at V_{M2}, assuming a zero concentration of particles. He also developed an empirical correlation based on his own data and those of Shields (1936), Liu (1956), and Zenz (1964) for the region below V_{M2} where, at the beginning of bed formation, transport of particles changes from moving longitudinal to moving transverse waves. His relationship for this condition takes the form

$$\frac{V_0}{U_{0w}^*} = \left(\frac{1}{s-1}\right)^{0.33}\phi\left(\frac{dU_{0w}^*\rho_F}{\mu}\right) \tag{11.25}$$

where U_{0w}^* is the friction velocity corresponding with flow at the velocity where longitudinal waves give way to transverse waves,

ϕ denotes a "function of."

Thomas has developed a sophisticated flow pattern map based upon Equations 1.33, 11.24, and 11.25. This is reproduced as Figure 11.36.

Rose and Duckworth (1969), in connection with their detailed study of the pressure gradient for the flow of solids suspended in liquids or gases, present an empirical relationship for the minimum transport velocity. They define this velocity as that "at which the particles settle out of suspension and thereby tend to block the pipe." We interpret this as our V_{M2}. Their equation is

$$V_{M2} = 3.2 V_0 \mathscr{R}^{0.2} \left(\frac{D}{d}\right)^{0.6} s^{-0.7} \left(\frac{V_{M2}^2}{gD}\right)^{0.25} \tag{11.26}$$

Equation 11.26 may be rearranged and combined with Newton's law to give

$$V_{M2} = 13.6 \left[\frac{gd}{C_D}(s-1)\right] \left(\frac{1}{gD}\right)^{1/2} \frac{1}{s^{1.7}} \mathscr{R}^{0.4} \left(\frac{D}{d}\right)^{1.2} \tag{11.27}$$

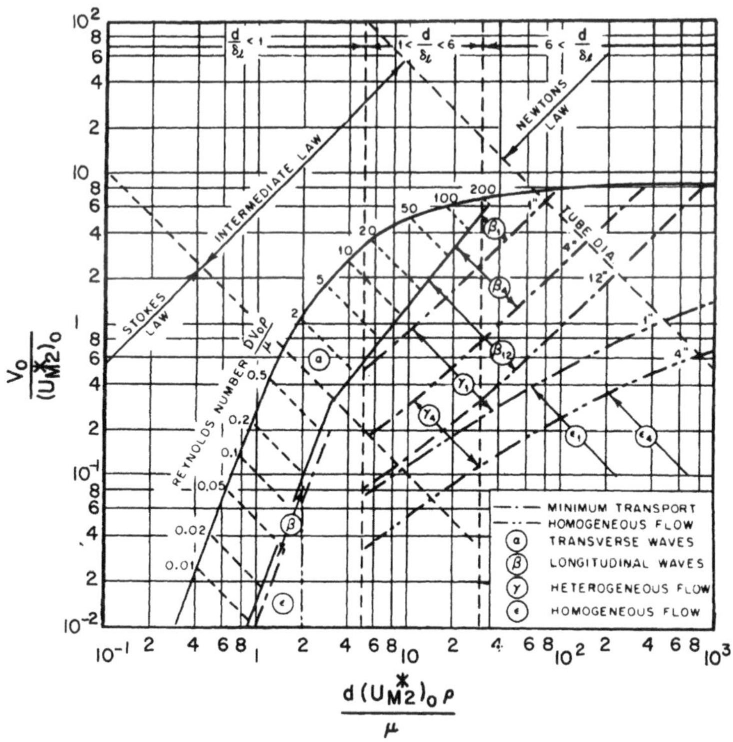

Figure 11.36. Thomas flow pattern map; $(s-1) = 1.65$. (From Thomas, D. G., *A.I.Ch.E. Journal*, **10**, 303, 1964.)

Rose and Duckworth state that the correlation is based upon their own and other data but do not define its reliability.

Shook (1969) points out that Durand's equation for the pressure gradient for the asymmetric suspension flow pattern (which we will discuss later) may be differentiated and solved for the condition where C_S is a maximum at a fixed pressure gradient and for a given system. This condition should correspond to V_{M2}, although to accept the correspondence is to assume a rather precise applicability of the Durand equation at least somewhat into the moving bed flow pattern. Shook obtains

$$V_{M2} = \frac{2.43 C_S^{1/3}[2gD(s-1)]^{1/2}}{C_D^{1/4}} \tag{11.28}$$

from which he writes for Durand's F_L

$$\frac{F_L}{C_S^{1/3}} = \frac{2.43}{C_D^{1/4}} \tag{11.29}$$

Shook also expresses Zandi's criterion for V_{M2} in terms of F_L as follows:

$$\frac{F_L}{C_S^{1/3}} = \frac{4.47 C_S^{0.167}}{C_D^{1/4}} \tag{11.30}$$

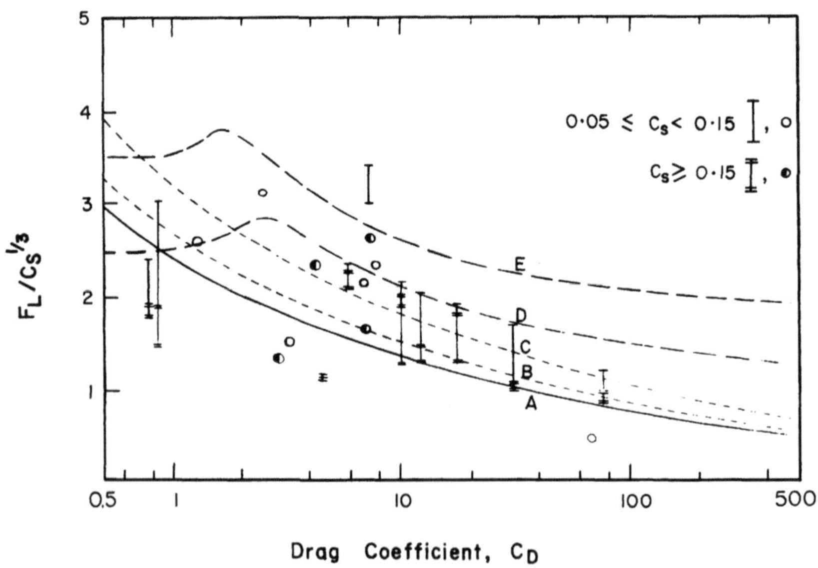

Figure 11.37. Comparison of predicted and observed values of $F_L/C_S^{1/3}$. $D \geq 1$in., $C_S \geq 0.05$. (From Shook, C. A., paper presented at the C.S.Ch.E. Symposium on Pipeline Transport of Solids, Toronto, Nov. 1969.)

Figure 11.37 taken from Shook shows the following lines:

A—Equation 11.29 derived from Durand's pressure gradient equation,
B—Equation 11.30 derived from Zandi's criterion with $C_S = 0.05$,
C—Equation 11.30 derived from Zandi's criterion with $C_S = 0.15$,
D—Durand's F_L line from Figure 11.28 for $C_S = 0.15$,
E—Durand's F_L line from Figure 11.28 for $C_S = 0.05$.

The figure also indicates the scatter of selected data for hydraulic suspension in pipes of diameters ranging from 1 to 14 inches, with particles ranging from 0.001 to 0.15 inch.

Shook's conclusions resemble those of Wicks, i.e., that much work remains to be done before we can predict V_{M2} with good accuracy. Still, from the various correlations and particularly if care is taken to select those based on data similar to a proposed application, a reasonable estimate of V_{M2} should be possible.

Transition Velocity V_{M2}, Gas-Solid Suspensions In contrast to the situation for V_{M1}, there has been a fair amount of work directed to correlating V_{M2} for gas-solid suspensions. The semitheoretical expressions of Thomas discussed in the preceding section are applicable and, in fact, have been tested with data for gas-solid as well as liquid-solid mixtures. In addition, the Rose and Duckworth correlation is said to be applicable to gas-solid suspensions.

The minimum velocity required to keep particles from depositing on the bottom of the pipe (our V_{M2}) is referred to as the saltation velocity by Zenz (1964); he first develops a correlation for $(V_{M2})_0$ which is the minimum velocity required to transport single particles. The correlation is based upon an analogy with the terminal settling velocity of particles and introduces the drag coefficients, C_D, and the particle Reynolds number at $(V_{M2})_0$, i.e., $d(V_{M2})_0\rho/\mu = \mathrm{Re}_P$. By some manipulation he devises the parameters d/Δ and $(V_{M2})_0/\omega$, where Δ and ω are functions of the fluid and particle properties as follows:

$$\Delta = [3\mu^2/4g\rho(\rho_S - \rho)]^{1/3}$$
$$\omega = [4g\mu(\rho_S - \rho)/3\rho^2]^{1/3}$$

The correlation is presented in Figures 11.38 and 11.39 for $1\frac{1}{4}$- and $2\frac{1}{2}$-inch pipes, respectively. The terminal settling velocity of spherical particles is also shown in these figures. He has compared these correlations with the data of several investigators and considering the complex nature of the problem finds satisfactory agreement. It is interesting to observe that there is a mini-

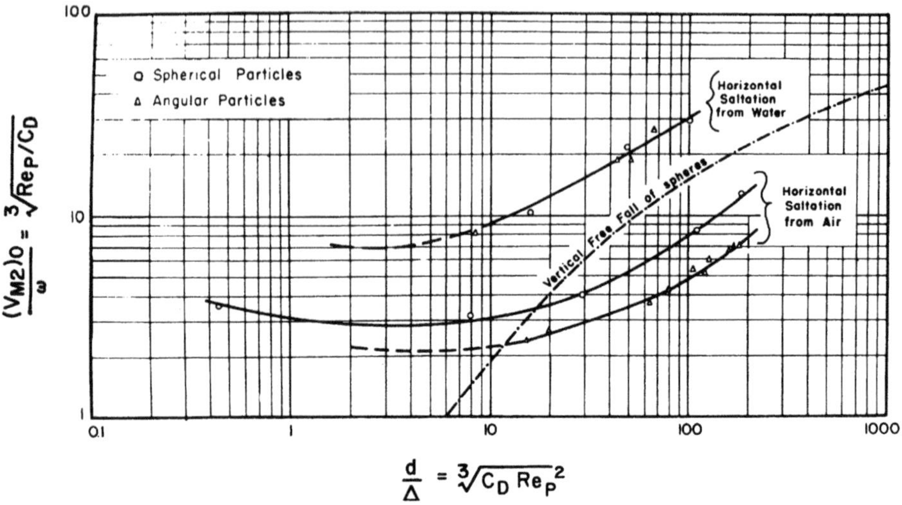

Figure 11.38. Zenz correlation for single-particle minimum transport velocity, $(V_{M2})_0$; $1\frac{1}{4}$-inch pipe. (From Zenz, F. A., *Ind. Eng. Chem. Fundam.*, **3**, 65, 1964.)

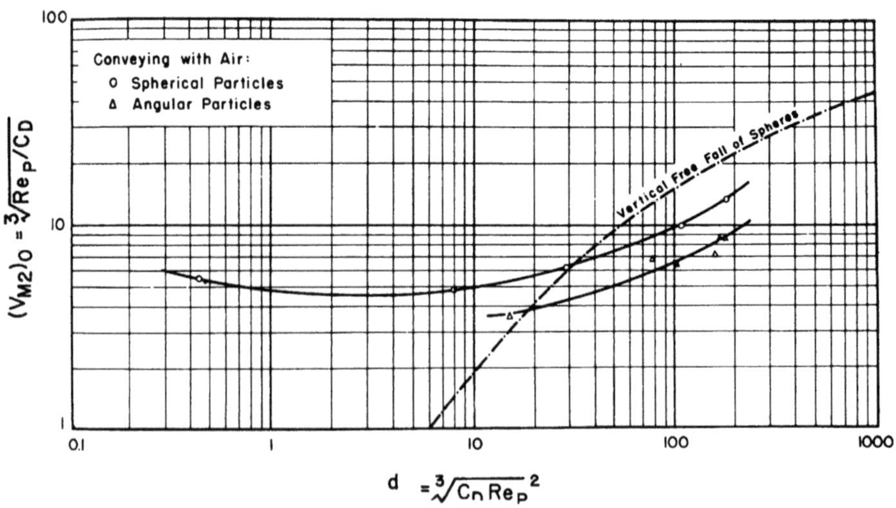

Figure 11.39. Zenz correlation for single-particle minimum transport velocity, $(V_{M2})_0$; $2\frac{1}{2}$-inch pipe. (From Zenz, F. A., *Ind. Eng. Chem. Fundam.*, **3**, 65, 1964.)

mum in the $(V_{M2})_0$ curves shown in these figures. Zenz (1964) explains this behavior as follows:

> ... If a particle is sufficiently small to sink into the viscid boundary layer at the tube wall, it will be out of reach of the influence of the eddies of turbulence which could pick it up. Thus the smaller the particle the deeper it is engulfed in the boundary layer and therefore the higher the velocity required to thin this layer and reach the particle.

Zenz has correlated his observations of V_{M2} for a variety of air suspensions of materials ranging in size from 0.004 to 0.25 inch in terms of $(V_{M2})_0$ and the mass velocity of the solids. Figure 11.40 gives this correlation with separate curves labeled by the parameter S_Δ which characterizes the slope of the single particle minimum transport velocity as illustrated by Figure 11.41. He also expresses the correlation of Figure 11.40 by an analytical expression as

$$V_{SS} = \frac{M_S}{A\rho_S} = 0.7S_\Delta^{1.5}\,\frac{V_{M2} - (V_{M2})_0}{(V_{M2})_0} \tag{11.31}$$

Zenz has compared his correlations with published data on the transport of fine and coarse coal particles in 8- and 12-inch pipes and found good agreement.

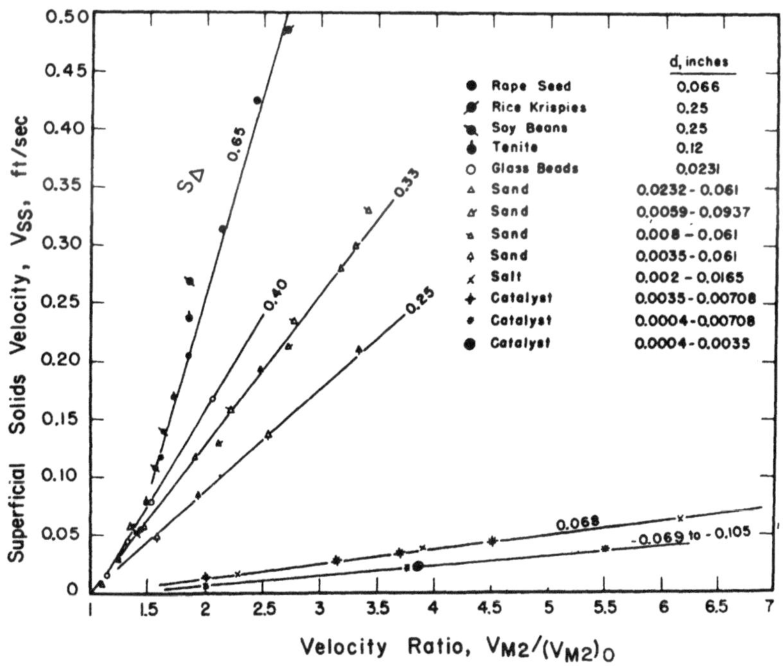

Figure 11.40. Zenz correlation for minimum transport velocity, V_{M2}. (From Zenz, F. A., *Ind. Eng. Chem. Fundam.*, **3**, 65, 1964.)

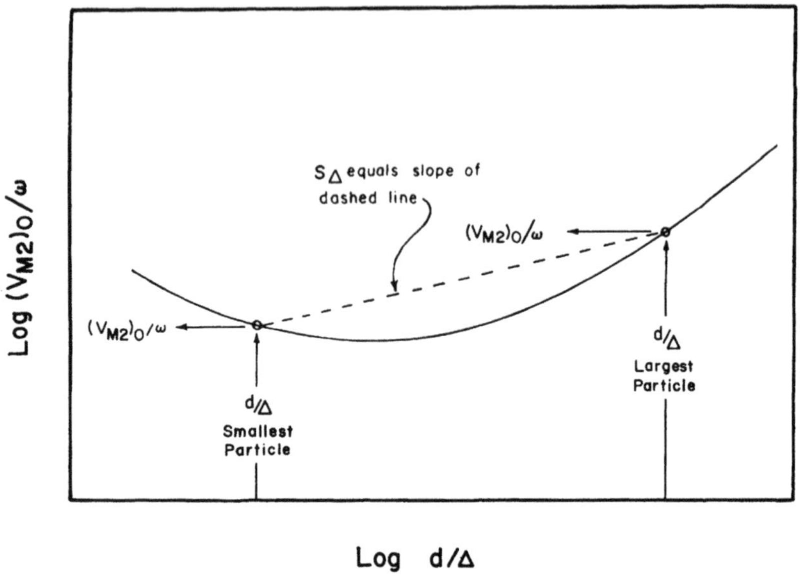

Figure 11.41. Definition of S_Δ as determined graphically from a single-particle saltation velocity curve. (From Zenz, F. A., *Ind. Eng. Chem. Fundam.*, **3**, 65, 1964.)

Doig and Roper (1963) have reviewed the available literature on the minimum gas rate required for the transport of solids without settling in horizontal pipes and without choking in vertical pipes. Their minimum transport velocity V_{SG2} seems to correspond to our V_{M2}. It is given by

$$\log \mathrm{Fr}_2 = \frac{V_0 - 2}{28} + 0.25 \log \mathscr{R} \qquad \text{for } 10 \le V_0 \le 40 \text{ ft/sec} \qquad (11.32)$$

and

$$\log \mathrm{Fr}_2 = 0.03 V_0 + 0.25 \log \mathscr{R} \qquad \text{for } V_0 < 10 \text{ ft/sec} \qquad (11.32a)$$

where V_0 = the terminal settling velocity of the particles,
Fr_2 is based on the superficial gas velocity, V_{SG2} corresponding with V_{M2}, i.e., $\mathrm{Fr}_2 = V_{SG2}/\sqrt{gD}$,
$V_{M2} = V_{SS2} + V_{SG2}$.

The correlation is based on extensive data of several investigators covering a variety of solids from 0.0016 to 0.25 inch in diameter and pipes up to 16.52 inches in diameter. It represents a reasonably close fit to much of the data, but deviations of up to 80 percent occur with the data of Zenz and Othmer (1960).

Transition Velocities, V_{M3} and V_{M4} No direct methods for the prediction of the transition velocities V_{M3} or V_{M4} are known. The analyses of Condolios and Chapus (1963) and of Wicks (1968) for the moving bed flow pattern which we discuss later may be used indirectly to obtain some idea of V_{M3} for liquid-solid suspensions if this be taken to correspond with the condition where a bed reaches a height equal to a certain fraction of the pipe diameter.

Wilson (1970) has made a force balance analysis for the condition of incipient motion of a bed of granular solids in a pipe, corresponding with V_{M3}. He assumed a hydrostatic distribution of the pressure of the particles on the pipe wall. His equation for the pressure gradient at the start of bed motion is

$$\frac{1}{\rho_L}\left(\frac{\Delta P}{L}\right)_3\left[\frac{\theta - \sin\theta\cos\theta}{4} + \frac{R'}{D}\left(\sin\theta - \frac{\theta\eta_s}{\tan\phi}\right)\right] = \frac{\eta_s(s-1)E_{S2}(\sin\theta - \theta\cos\theta)}{2}\left(\frac{g}{g_c}\right) \quad (11.33)$$

where $(\Delta P/L)_3$ = pressure gradient at velocity V_{M3},
θ = one-half the angle subtended at the pipe center by the surface of the bed of solids, equivalent to $\gamma/2$ of Figure 10.36, radians,
η_s = coefficient of static friction for the solids against the pipe wall,
R' = cross-sectional area of the bed divided by the bed width,
ϕ = angle of repose of solid particles,
s = ratio of density of solids to density of liquid, ρ_S/ρ_L,
E_{S2} = volume fraction of solids in the bed.

From tests with 0.7-mm sand with water flowing in a 3.485-inch perspex pipe, Wilson confirmed the constancy of η_s (at about 0.35) and from that concluded that his force analysis and the assumption of a hydrostatic distribution of the granular pressure were correct.

Wilson's analysis, with the appropriate measured or estimated value of η_s in each case, should be generally applicable to the prediction of the pressure gradient required to initiate the movement of a bed of solids of known height (or bed angle). In principle at least, one should be able to predict the corresponding velocity, V_{M3}, from the pressure gradient–velocity relationship for the mixture flowing above the bed as discussed in Section 3.45.

We recall that the general method of Gaessler (1967) presented earlier is suggested as suitable for predicting the equivalent of both V_{M3} and V_{M4} for hydraulic transport. The correlation of Acaroglu and Graf (1968), which we discuss in Section 3.45, permits an estimation of V_{M2}, V_{M3}, and V_{M4} for liquid-solid suspensions provided an estimate of the pressure gradient is available.

The lack of proven methods for estimating V_{M4} is not too serious from a practical point of view because only rarely would we wish to consider flow in the range below V_{M2} and even then not below V_{M3}.

3.23 Velocity and Concentration Profiles

Related to the description and correlations of the flow patterns are the velocity and concentration profiles of the flowing mixture. No correlations for predicting the velocity profile of the solids are known, but methods of limited applicability have been proposed for the prediction of the water velocity and the concentration profiles in water-solid suspensions.

Newitt *et al.* (1962) have developed a method for predicting the velocity profile for water mixtures of coarse sand and gravel in the asymmetric suspension pattern. They correlated the displacement of the maximum velocity point with the equation:

$$y' - \frac{D}{2} = 15.2\left(C_S\sqrt{\frac{\tau_w g_c}{\rho_W}}\right)^{1.81} \qquad (11.34)$$

where y' = distance from the bottom of the pipe to the point of the maximum fluid velocity. The velocity profile below the maximum point is correlated by

$$(1 - R^*)\frac{\tau_w g_c}{\rho_W} = (0.1176 - 0.0538 \log C_S)\left(1 - \frac{y}{y'}\right) \qquad (11.35)$$

where R^* = ratio of the velocity at a point to the maximum velocity in the presence of solids, divided by the corresponding ratio for water alone at the same hydraulic gradient,

y = distance from the bottom of pipe.

Above the point of the maximum velocity they found the velocity profile to be described by the simple 1/7 power law.

Newitt *et al.* (1962) also derive empirical relations to describe their observed concentration profiles with coarse sand and gravel and with fine sand flowing with water. For the coarse materials, as illustrated by Figure 11.6, their data in the upper part of the pipe are described by

$$\frac{1}{\varepsilon_S}\frac{d\varepsilon_S}{dy} = -1.66\frac{V_0}{U^*}\left(\frac{C_D}{d}\right)^{1/2} \qquad (11.36)$$

where $U^* = \sqrt{\tau_w g_c/\rho}$

ε_S = local *in situ* volume fraction solids,

y = distance from the pipe bottom in inches,

d = particle diameter in inches.

For the finer materials, such as those shown in Figures 11.4 and 11.5 they correlate a ratio, P_y/P_0, with position, settling velocity, and friction velocity. Here

$$P = \frac{\varepsilon_S}{1 - \varepsilon_S} \exp \frac{1}{1 - \varepsilon_S} \qquad (11.37)$$

and the subscripts y and 0 refer to evaluations at a distance y or zero from the bottom of the pipe. The correlation is reproduced in Figure 11.42.

Ismail (1951) has made a careful study of the problem of concentration distribution in water-sand mixtures in closed rectangular channels. His analysis is based upon an assumed balance of the rate of settling of particles and the rate of their upward transport by turbulent forces. Following O'Brien (1933) he writes

$$e_s \frac{d\varepsilon_S}{dy} + V_0 \varepsilon_S = 0 \tag{11.38}$$

where e_s is a sediment transfer coefficient assumed to be related to the momentum transfer coefficient through a linear relationship,

$$e_s = \beta e_m \tag{11.39}$$

where β is a constant, and for turbulent flow e_m is defined by

$$\tau g_c = \rho e_m \frac{du}{dy} \tag{11.40}$$

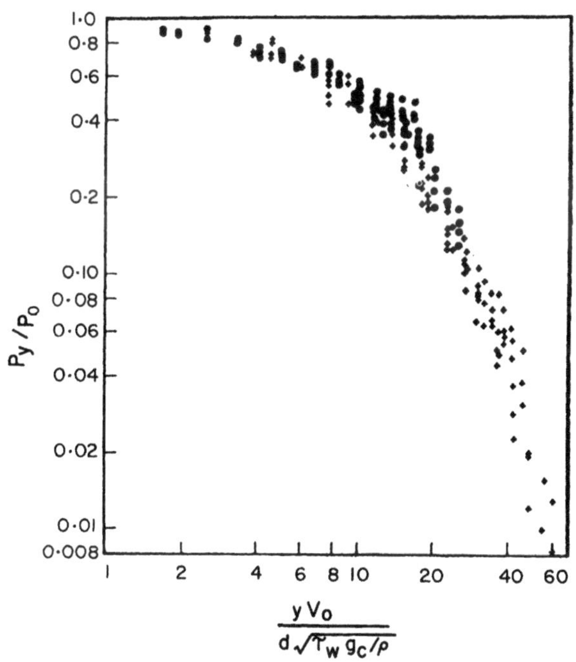

Figure 11.42. Newitt *et al.* correlation of concentration profiles for fine materials. (From Newitt, D. M., J. F. Richardson, and C. A. Shook, Symp. Interaction Between Fluids and Particles, London, 1962, Proc., p. 87.)

For two-dimensional flow in a rectangular duct,

$$\tau = \tau_w\left(1 - \frac{y}{y_m}\right) \tag{11.41}$$

where y_m is the distance at which $u = u_{\max}$ and $\tau = 0$. Using Equation 11.41 we can solve for e_m from Equation 11.40

$$e_m = \frac{\tau_w g_c(1 - y/y_m)}{\rho \dfrac{du}{dy}} \tag{11.42}$$

Ismail evaluates the derivative of the velocity from von Kármán's universal velocity profile equation,

$$\frac{u - u_{\max}}{U^*} = \frac{1}{\mathscr{K}} \ln \frac{y}{y_m} \tag{11.43}$$

and combining Equations 11.42 and 11.43, he obtains

$$e_m = \mathscr{K} U^* y\left(1 - \frac{y}{y_m}\right) \tag{11.44}$$

Integration of Equation 11.38, with the value of e_s obtained from Equations 11.39 and 11.44, yields

$$\frac{\varepsilon_S}{\varepsilon_{Sa}} = \left[\frac{y_a(y_m - y)}{y(y_m - y_a)}\right]^z \tag{11.45}$$

where

$$z = \frac{V_0}{\beta \mathscr{K} U^*}$$

and the subscript a refers to some reference position, y_a. Equation 11.45 only permits a prediction of the ratio $\varepsilon_S/\varepsilon_{Sa}$ and not the absolute value of ε_S. The equation has been confirmed by Vanoni (1946) for open channel flow and by Ismail for a closed rectangular channel. Both of these authors have found that $\mathscr{K}$ is a function of solids loading and decreases with increasing loading due to the damping effect of the solids on turbulent eddies.

Wasp *et al.* (1963) have applied the equation to a 12-inch-diameter pipe transporting coal slurries. They assumed $\mathscr{K} = 0.35$, a value lower than the classical value of 0.38. In applying Equation 11.45 Wasp *et al.* have assumed that the point of maximum velocity is at the center line of the pipe and that the *in situ* volume fraction of solids at the center line may be taken as the input volume fraction C_S. They found excellent agreement between calculated and measured values of ε for positions near the top and near the bottom of the 12-inch pipe.

Wasp's method for estimating the *in situ* concentrations near the top and the bottom of the pipe may be considered established and satisfactory for pipeline design for particles in the size range from 48 mesh to 1/4 inch. We will refer to it again in connection with the prediction of pressure gradient for the flow of mixed-size particles in asymmetric suspension.

Shook and Daniel (1965) and Shook, Daniel, Scott, and Holgate (1968) explain deviations between observed concentration profiles and those calculated by Equation 11.38 by postulating that a normal dispersive force, P_B, is associated with the shear stress. This is the so-called Bagnold force (Bagnold 1954, 1956) described (Shook and Daniel, 1965) by

$$\tau = P_B \tan \alpha \tag{11.46}$$

where $\tan \alpha$ = a coefficient of dynamic friction,
= 0.32 for $G^2 > 3700$,
= 0.75 for $G^2 < 28$,
= $\phi(G^2)$ for $28 < G^2 < 3700$,
σ = average distance between particle surfaces,
G^2 = dimensionless group = $\rho_S d P_B g_c \sigma / \mu^2$.

This relationship is assumed valid for conditions where the volume fraction of solids exceeds about 0.15.

Shook *et al.* (1968) present detailed concentration profile studies for the flow of water suspensions of a variety of sands and of nickel powders in a rectangular channel at concentrations higher than those studied by Ismail (1951). Their studies confirm that the suspension of the particles cannot be fully explained by turbulence forces and lend support to the role of the Bagnold stresses. We will make further references to this in Sections 3.43 and 3.44. Notwithstanding this conclusion they found that Equation 11.38 may be used for a crude estimate of concentration distribution in the upper half of a pipe by setting $2e_s/DU^* \simeq 0.07$ or 0.08.

3.3 Holdup

There are no generally applicable methods for the prediction of holdup for all flow patterns and for both liquid-solid and gas-solid systems.

In the symmetric suspension flow regime for both liquid-solid and gas-solid mixtures it is common to assume that holdup is negligible and that the *in situ* and input concentrations are identical. This is not entirely correct but is a suitable approximation in many cases.

Referring to the fundamental relations, Equation 7.104, we might reasonably assume that the weighted average drift velocity, $V_{\alpha j}$, would be zero for horizontal flow of any fluid-solid mixture where the solids are fully suspended, and this leads us to the relation

$$V_S = C_0 V_M \tag{11.47}$$

For symmetric suspensions at low velocities (see Toda *et al.*, 1969, and Newitt *et al.*, 1962) as for vertical flow we would expect a nearly uniform concentration profile and C_0 will approach unity. At higher velocities the concentration profile peaks at the pipe axis and C_0 will approach 1.2 or may even be higher (see Figure 7.5). Given an estimate of C_0 we may calculate H and E_S from Equation 11.47 and the relations of Table 7.1.

As we have seen from the typical data we discussed, the holdup may be quite significant in the asymmetric suspension and bed flow regimes.

The correlation of Rose and Duckworth discussed in detail in Chapter 9 is reported to be applicable for both liquid-solid and gas-solid systems in both the symmetric and asymmetric flow patterns. For the former flow pattern it provides an alternate to the method based upon an estimate of C_0. For the latter it appears to be the only generally applicable method. The Rose and Duckworth correlation yields estimates of V_S and V_F. The holdup ratio H and the *in situ* volume fraction solids, E_S, may be determined from these quantities through the relations discussed in Section 7 of Chapter 7.

Gaessler's pressure gradient correlation (discussed subsequently in Section 3.42) incorporates a relationship for the prediction of V_S/V_M in terms of Froude numbers based upon the mixture velocity and the particle settling velocity. We may note from Equation 11.47 that the ratio V_S/V_M is our concentration distribution coefficient C_0 and that H or E are readily determined from it. Some lines from his final correlation and actual data are shown in Figures 11.43 and 11.44 for various size ranges of coal particles.

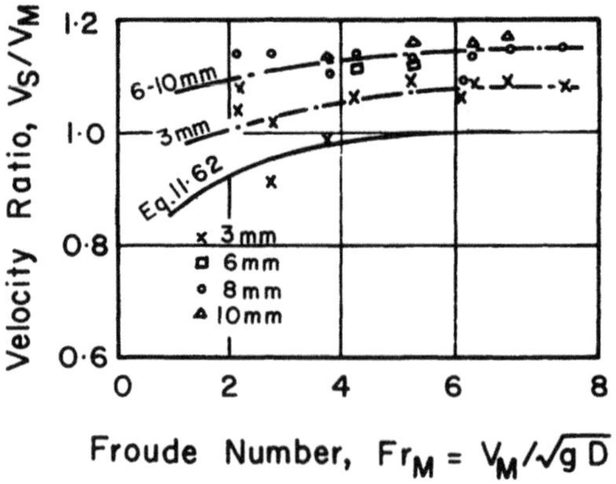

Figure 11.43. Velocity ratio data for water suspensions of 0- to 10-mm coal particles in a 46-mm pipe. (From Gaessler, H., Doctoral Dissertation, Technische Hochschule, Karlsruhe, Germany, 1967.)

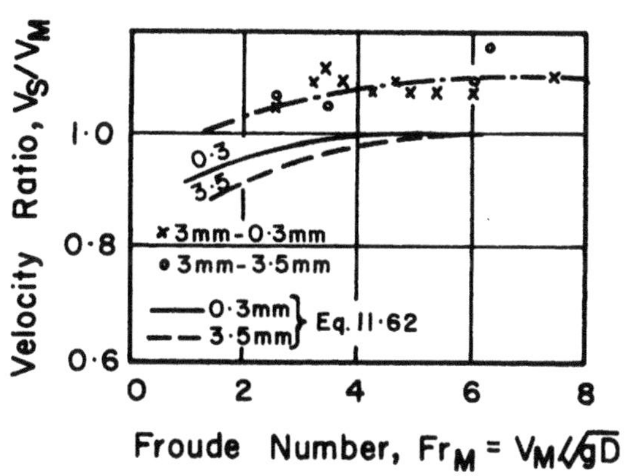

Figure 11.44. Velocity ratio data for water suspensions of 0- to 3-mm and 3- to 5-mm coal particles in a 46-mm pipe. (From Gaessler, H., Doctoral Dissertation, Technische Hochschule, Karlsruhe, Germany, 1967.)

Gaessler's recommended correlation appears to be based on an assumed maximum value of 1.0 for the velocity ratio V_S/V_M. His data, however, confirm our expectation that this ratio could be as high as 1.2 for the symmetric suspension flow pattern. We therefore recommend the use of the lines through the data points in Figures 11.43 and 11.44. The theoretical equations upon which Gaessler's correlation is based are discussed in Section 3.43. The correlation applies only to liquid-solid systems but is applicable to all flow patterns.

Newitt *et al.* (1962) offer a correlation for the mean slip velocity of coarse sand and gravel in water suspensions in a 1-inch pipe. The correlation is

$$S = 3.70 - 0.0397(1 - C_S)\frac{V_M}{V_0} \tag{11.48}$$

Their data cover slip velocities in the range of 2 to 4 ft/sec. The scatter is considerable. The other measures of holdup may be calculated from the relations of Section 7 of Chapter 7.

3.4 Pressure Gradient

3.41 Introduction

No overall pressure gradient correlation applicable to all flow patterns for both gas-solid and liquid-solid systems is available.

The correlation of Rose and Duckworth discussed in Chapter 9 applies to horizontal flow for both the symmetric and the asymmetric flow pattern for both liquid-solid and gas-solid mixtures. A correlation due to Gaessler (1967) and a related method suggested by Toda *et al.* (1969) apply to all flow patterns for liquid-solid systems. A computation method due to Wasp,

Regan, Withers, Cook, and Clancey (1963) and based upon empirical correlations of others is applicable to the symmetric and the asymmetric suspension patterns for liquid-solid systems. Only the symmetric suspension flow pattern has yet yielded to any kind of theoretical analysis. A number of empirical correlations have been developed for the individual flow regions and for one or the other of liquid or gas as the carrier fluid.

Most of the correlations are based upon the mechanical energy balance in its single-phase form (i.e., Equation 3.29) and they are therefore analogous to those we discussed in Chapter 9, Section 3.31. One important correlation is based on the two-phase form of the mechanical energy balance (Equation 7.57).

Many of the correlations are expressed in terms of the dimensionless excess pressure gradient

$$\frac{\Delta P/L - (\Delta P/L)_F}{(\Delta P/L)_F} = \frac{i - i_F}{i_F}$$

where the subscript F refers to the pressure gradient that would be observed in the same pipe if the carrier fluid were flowing alone at the mixture velocity, V_M.

We will deal first with the two most generally applicable correlations, with the Wasp *et al.* computation method, and then with the other correlations under the headings of the individual flow patterns.

3.42 All Flow Patterns

The Rose and Duckworth Correlation The Rose and Duckworth correlation is discussed in detail in Chapter 9. It is directly applicable to horizontal flow of both liquid-solid and gas-solid mixtures provided all the solids are suspended. The angle $\theta = 0$ in Equation 9.66, and in the determination of the functions ϕ_5 and ψ_5. The correlation has been tested for $1\frac{1}{4}$- to 16-inch diameter pipes, conveying solids of specific gravities from 1.5 to 11.1. The correlation may also be applied to inclined flow with the appropriate value assigned to θ.

The Gaessler Correlation Gaessler (1967), following the approach of Barth (1958), has conducted a detailed analysis of the flow of water-solid suspensions in horizontal pipes. His work is based on extensive experimental data obtained by the Association of Hydraulic Transport of Germany in pipes of 160, 125, and 46 mm inside diameter. The two large-size pipes contained two 30-meter test sections each. The 46-mm pipe contained two test sections of 10 meters each. Three size ranges of coal were used in the tests, 0–10 mm, 0–3 mm, and 3–5 mm. The main properties of the coal particles are given in Table 11.5. Typical pressure drop data reported by Gaessler are shown in Figures 11.20, 11.21, and 11.22 discussed earlier. The solid lines shown on these figures are calculated from the correlation developed by Gaessler.

Gaessler divides the total pressure drop into three net force components, i.e.,

$$\frac{\pi D^2}{4}\Delta P = F_1 + F_2 + F_3 \tag{11.49}$$

where F_1, F_2, and F_3 are, respectively, the frictional force due to the sliding of any bed of solids which may be present, the wall shear force attributable to the impact of any suspended solids on the pipe wall, and the wall shear force due to the fluid itself. Through an analysis based on a logical explanation of the mechanism of the flow, he develops the following expressions:

$$F_1 = \beta W_S(s-1)\frac{g}{g_c} \tag{11.50}$$

$$F_2 = \pi D L \tau_{wS} \tag{11.51}$$

$$F_3 = \frac{f_W A L(1 - E_S)\rho_W V_W^2}{g_c D} \tag{11.52}$$

where β = factor of proportionality dependent upon the fraction of the flowing solids in the form of a bed, the velocity of any such bed, and the coefficient of sliding friction,

TABLE 11.5

Properties of Coal Particles in Gaessler's Tests[a]

Mixture	Size Range, mm	Specific Gravity	Mean Diameter, mm	Weight Percent Size Distribution, mm						Mean Settling Velocity, m/sec
				+10 mm	10–5 mm	5–3 mm	3–1 mm	−1 mm		
I	0–10	1.27	5.2	4.5	38.0	32.0	21.5	4.0		0.168
II	0–10	1.29	4.2	6.0	26.0	21.0	31.0	16.0		0.135
				+5 mm	5–3 mm	3–2 mm	2–1 mm	1–0.5 mm	−0.5 mm	
III	0–3	1.29	1.75		13	20	35	20	12	0.078
IV	3–5	1.29	3.2	2	38	39	18	3		0.141

[a] H. Gaessler, Doctoral Dissertation, Technische Hochschule, Karlsruhe, Germany, 1967.

W_S = mass of solids in a length of pipe, L,
V_S, V_W, V_M = average actual velocities in the pipe of the solids, the water, and the mixture,
$V_{SS} = E_S V_S$ = superficial velocity of the solids,
$V_{SW} = (1 - E_S)V_W$ = superficial velocity of the water,
$M_S = Q_S \rho_S = V_{SS}\rho_S(\pi D^2/4)$ = mass flow rate of solids,
τ_{wS} = wall shear stress due to suspended solids, defined as

$$\tau_{wS} = \frac{4W_S V_S^2}{g_c \pi D^2 L} f_S^* \tag{11.53}$$

f_S^* = friction factor for suspended solids defined by Equation 11.53.

Combination and rearrangement of Equations 11.49 to 11.53 results in the following relation:

$$\frac{\Delta P}{L} = \frac{g}{g_c}\left(\frac{V_{SS}}{V_S}\right)\left[\rho_W(s-1)\beta + \rho_S\left(\frac{V_S^2}{2gD}\right)f_S^* + \frac{\rho_W V_S^2(V_M - V_{SS})^2 f_W}{(V_S - V_{SS})V_{SS}\,2gD}\right] \tag{11.54}$$

In this equation the quantities β, f_S^*, and V_S remain to be evaluated; f_W is the conventional single-phase friction factor taken at $\mathrm{Re} = DV_M\rho_W/\mu_W$, and the other quantities are normally specified.

Evaluation of β Gaessler develops the relationship

$$\beta = \beta^*\left(\frac{C_{w2}}{C_w}\right)\frac{V_S}{V_{S2}} \tag{11.55}$$

which leads to

$$\beta = \beta^*\left(\frac{C_{w2}}{C_w}\right)^2\left(\frac{C_{w1}}{C_{w2}}\frac{V_{S1}}{V_{S2}} + 1\right) \tag{11.56}$$

The quantities C_w, C_{w1}, C_{w2}, V_{S1} and V_{S2} are those discussed and defined in Section 3.22.

Gaessler has found that β^* is in fact essentially constant for any solid material, liquid and pipe-wall material combination. Gaessler's tests yield the following values for β^*:

$\beta^* = 0.25 - 0.28$ for coal
$\beta^* = 0.34 - 0.38$ for sand and gravel
$\beta^* = 0.40 - 0.45$ for ore

Values of C_{w2}/C_w and V_{S2}/V_S are required for the evaluation of β in Equation 11.56. For the ratio C_{w2}/C_w Gaessler has found that his data are well correlated by

$$\frac{C_{w2}}{C_w} = \frac{\mathrm{Fr}_0}{0.1\phi_S}\sqrt{\frac{3}{4}C_D}\left(\frac{C_S}{\phi_S}\right)^a \tag{11.57}$$

where $a = \dfrac{1}{\phi_S}\sqrt{\dfrac{3}{4}\dfrac{\rho_T}{(\rho_S - \rho_T)}}\left(\dfrac{\mathrm{Fr}_M}{3.7}\right)^b$

$b = \mathrm{Fr}_0^{-1/3}$

ϕ_S = the volume fraction occupied by the solids if packed in a tube,

$\rho_T = C_{S1}(\rho_S - \rho_W) + \rho_W$ = the bulk fluid density,

C_{S1} = the volume fraction of fines which do not settle,

$\mathrm{Fr}_0 = V_0/\sqrt{gD}$ = Froude number based on the settling velocity, V_0, of the average size particle,

$\mathrm{Fr}_M = V_M/\sqrt{gD}$ = Froude number based on the mixture velocity, V_M,

d_m = diameter of average-size particles.

Gaessler recommends the use of an average particle diameter, d_m, defined as the mass-weighted mean value, i.e.,

$$d_m = \sum_i d_i m_i$$

where d_i is the average screen opening for a small size interval i, and m_i is the mass fraction of solids within the size interval i.

The value of ϕ_S depends upon the mode of packing of the particles. For random packing of random shaped particles, Gaessler's tests indicate a value of $\phi_S = 0.65$. Other values from the literature for different solids, as suggested by Gaessler, are

$$\phi_S = 0.60 - 0.65 \text{ for coal}$$
$$\phi_S = 0.55 - 0.60 \text{ for sand}$$
$$\phi_S = 0.45 - 0.50 \text{ for ore}$$

In order to simplify calculations Gaessler has prepared a nomograph for the solution of Equation 11.57 with $\phi_S = 0.65$ and $C_D = 0.44$. The nomograph is reproduced here as Figure 11.45. If the value of C_D deviates from 0.44, then the solution may be approximated by multiplying the value obtained from Figure 11.45 by $\sqrt{C_D/0.44}$.

The velocity ratio V_{S2}/V_S is correlated by Equation 11.58,

$$\frac{V_{S2}}{V_S} = \frac{C_{w2}}{C_S}\frac{0.1\phi_S}{\mathrm{Fr}_0\sqrt{\frac{3}{4}C_D}}\,\phi\left(\frac{C_S}{\phi_S}\right)^{-a} \tag{11.58}$$

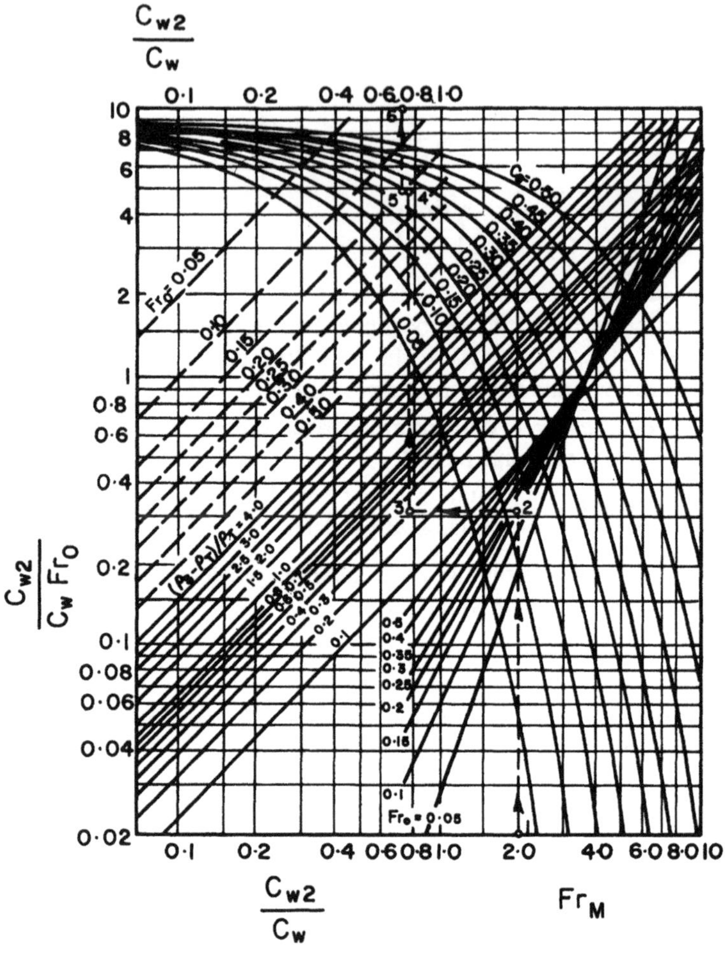

Figure 11.45. Gaessler nomogram for Equation 11.57; $\phi_S = 0.65$, $C_D = 0.44$. (From Gaessler, H., Doctoral Dissertation, Technische Hochschule, Karlsruhe, Germany, 1967.)

where

$$a = \frac{\phi}{\phi_S}\sqrt{\frac{3}{4}\frac{\rho_T}{(\rho_S - \rho_T)}}\left(\frac{\mathrm{Fr}_M}{3.7}\right)^b - \frac{1}{\phi_S}$$

$$b = 2\mathrm{Fr}_0^{-1/3} - \frac{1}{\phi_S}$$

$$\phi = \frac{1 - \phi_S}{\phi_S}$$

Substituting Equations 11.57 and 11.58 into Equation 11.56, we obtain

$$\frac{\beta}{\beta^*} = \frac{\mathrm{Fr}_0}{0.1(1 - \phi_S)}\sqrt{\frac{3}{4}\,C_D\left(\frac{C_S}{\phi_S}\right)^a} \tag{11.59}$$

Gaessler has also presented a nomograph for the solution of Equation 11.59 which is reproduced here as Figure 11.46. Like Figure 11.45 this nomograph was prepared for $\phi_S = 0.65$ and $C_D = 0.44$. For other values of C_D the solution may be approximated by multiplying the value obtained from the figure by $\sqrt{C_D/0.44}$.

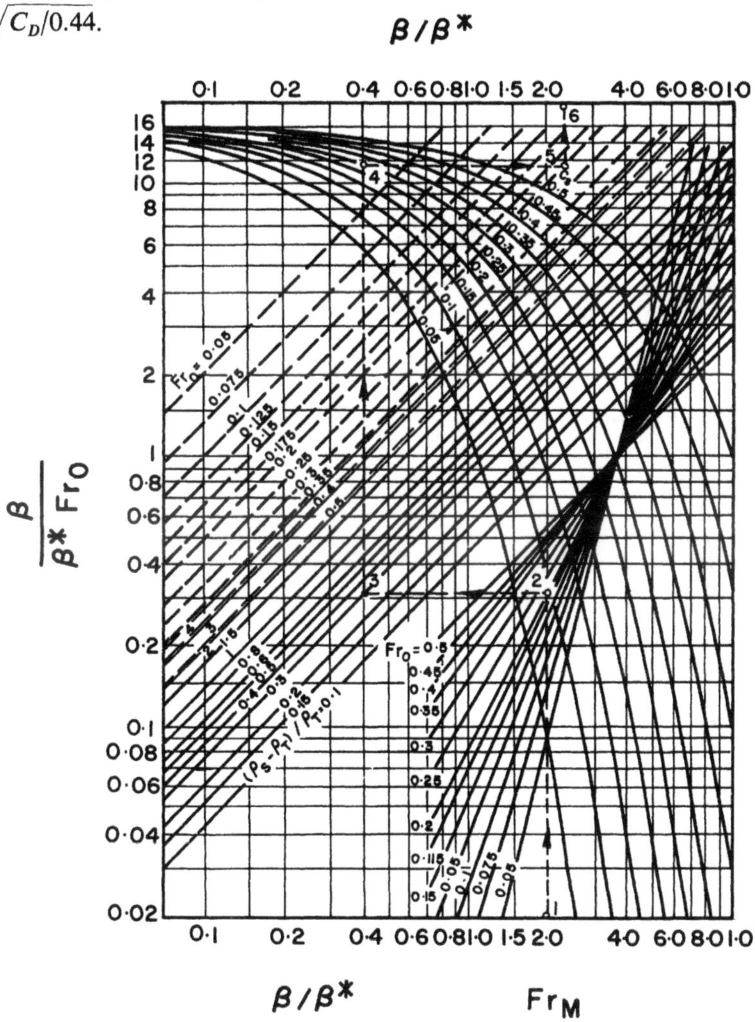

Figure 11.46. Gaessler nomogram for Equation 11.59; $\phi_S = 0.65$, $C_D = 0.44$. (From Gaessler, H., Doctoral Dissertation, Technische Hochschule, Karlsruhe, Germany, 1967.)

Evaluation of f_S^* Gaessler develops the following relationships for the evaluation of f_S^*:

$$\frac{f_S^*}{f_S^0} = 1 - \left(\frac{C_{w2}}{C_w}\right)\left(\frac{V_{S2}}{V_S}\right) \tag{11.60}$$

and

$$\frac{f_S^*}{f_S^0} = 1 - \left(\frac{C_{w2}}{C_w}\right)\left(\frac{\beta^*}{\beta}\right) \tag{11.61}$$

where f_S^0 is a material constant that takes into account the properties of the solid particles and the pipe surface. Values for f_S^0 determined by Gaessler are as follows:

coal (3 to 5 mm)	in hardened steel pipe,	$f_S^0 = 0.0046$
	in non-hardened steel pipe,	$f_S^0 = 0.0038$
quartz sand (3 to 5 mm)	in hardened steel pipe,	$f_S^0 = 0.012$
	in non-hardened steel pipe,	$f_S^0 = 0.0144$

The ratio $f_S^*/f_S^0 = 1$ when there is no tendency for sedimentation.

Evaluation of V_S From a momentum analysis Gaessler develops the following relationship for the prediction of the average solids velocity, V_S,

$$\left(1 - \frac{V_S}{V_M}\right)^2 = \left(\frac{\mathrm{Fr}_0}{\mathrm{Fr}_M}\right)^2\left(1 - \frac{V_M}{V_S}\,C_S\right)$$
$$\times\left\{\beta + \left(\frac{V_S}{V_M}\right)^2 \frac{\mathrm{Fr}_M}{2}\left[f_S^* \frac{\rho_S}{\rho_S - \rho_W} - f_W \frac{\rho_W}{(\rho_S - \rho_W)} \frac{(1 - C_S)^2}{(V_S/V_M) - C_S}\right]\right\} \tag{11.62}$$

Equation 11.62 is implicit in V_S/V_M and requires an iterative solution. Figure 11.47, taken from Gaessler, shows V_S/V_M as function of Fr_M and Fr_0

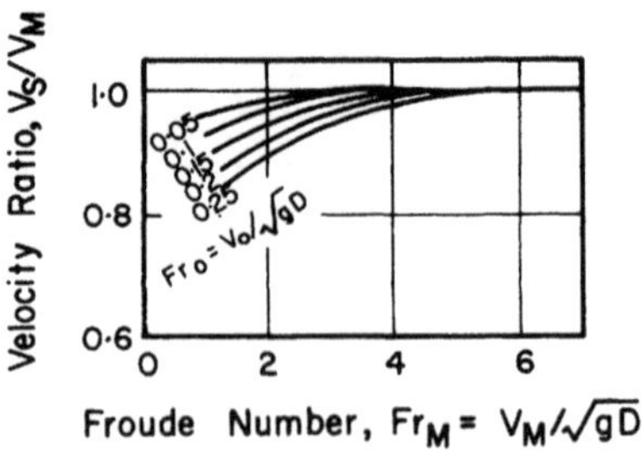

Figure 11.47. Gaessler velocity ratio correlation. (From Gaessler, H., Doctoral Dissertation, Technische Hochschule, Karlsruhe, Germany, 1967.)

which is approximately valid for all values of C_S and $\rho_W/(\rho_S - \rho_W)$, i.e., V_S/V_M is not sensitive to the latter two quantities.

Equation 11.62 or Figure 11.47 is in effect a holdup correlation since any of the measures of holdup may be calculated from V_S, V_{SS}, and V_{SW}. Some limitations of this correlation have been discussed in Section 3.3.

Overall Correlation. With the methods discussed for the prediction of β, f_S^*, and V_S, Equation 11.53 may be used to predict the pressure gradient for the flow of any suspension of granular solids in water in any size pipe.

Gaessler has compared his correlation (Equation 11.54) with his own data and the data of several other investigators. The prediction of pressure drop when compared with his own data are within 5 percent.

Alternate Form of the Correlation Recognizing the complexity of the form of the correlation, Gaessler expresses the pressure gradient predictions from it in a simpler and more familiar form representing the dimensionless excess pressure gradient. Expressing

$$\Delta P = \Delta P_W + \Delta P_S \tag{11.63}$$

$$\Delta P_W = \frac{2 f_W \rho_W V_M^2 L}{g_c D} \tag{11.64}$$

$$\Delta P_S = \frac{2 f_S (\rho_S - \rho_W) V_M^2 L}{g_c D} \tag{11.65}$$

Gaessler derives

$$\frac{\Delta P - \Delta P_W}{\Delta P_W} = \frac{i - i_W}{i_W} = \frac{f_S}{f_W} C_S (s - 1) \tag{11.66}$$

and calculates the values of f_S corresponding with his predictions of ΔP. He then compared his predictions of the effect of the variables on f_S with published data of several other investigations and found excellent agreement providing strong support for the correlation. Figure 11.48, where $f_S(\rho_S - \rho_T)/\rho_T$ is plotted against $\mathrm{Fr}_M\, [\rho_T/(\rho_S - \rho_T)]^{1/2}$ for three different values of Fr_0 is the alternate form of Gaessler's correlation for $C_S = 0.1$. To facilitate use of the correlation, Gaessler has also developed Figures 11.49 and 11.50 as counterparts of Figure 11.48 for C_S values of 0.2 and 0.4. Linear interpolation should be permissible between pairs of the figures.

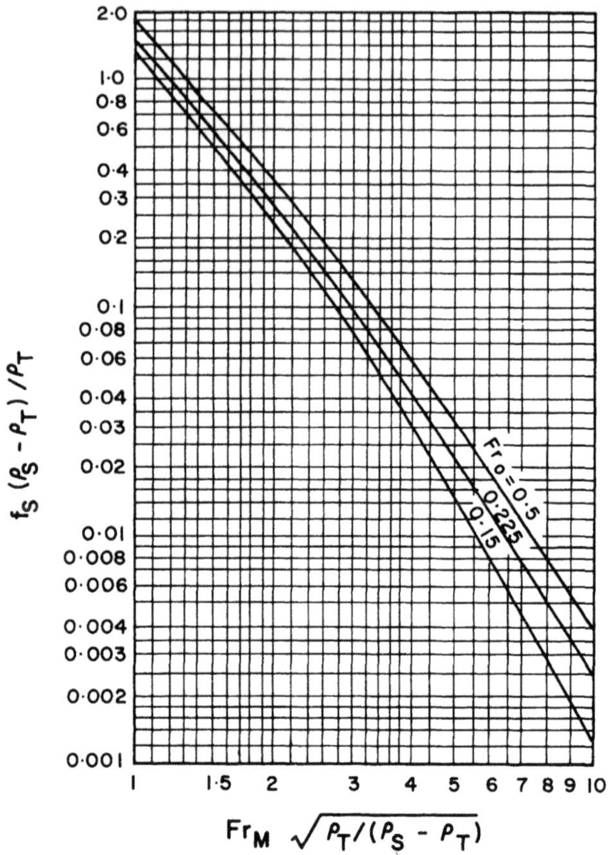

Figure 11.48. The Gaessler correlation for the solids friction factor, f_S. $C_S = 0.1$, $C_D = 0.44$, sphericity, $\psi = 0.54$. (From Gaessler, H., Doctoral Dissertation, Technische Hochschule, Karlsruhe, Germany, 1967.)

The Wasp et al. Calculation Method Wasp *et al.* (1963, 1968) have pointed out not only that most practical applications of hydraulic transport involve mixed sizes of particles, but that long distance transport is facilitated by a properly designed mix of particle sizes. With a reasonable range of particle sizes the smallest particles will normally be in the symmetric concentration flow pattern, the intermediate and large in the asymmetric pattern, and the largest may slide on the bottom of the pipe. Wasp *et al.* describe an empirical method for calculating the pressure gradient as the sum of the gradients due to the (more or less) symmetrically suspended material and to the asymmetrically suspended and sliding material. The method provides that if all the solids were uniformly suspended (as predicted by the method) the

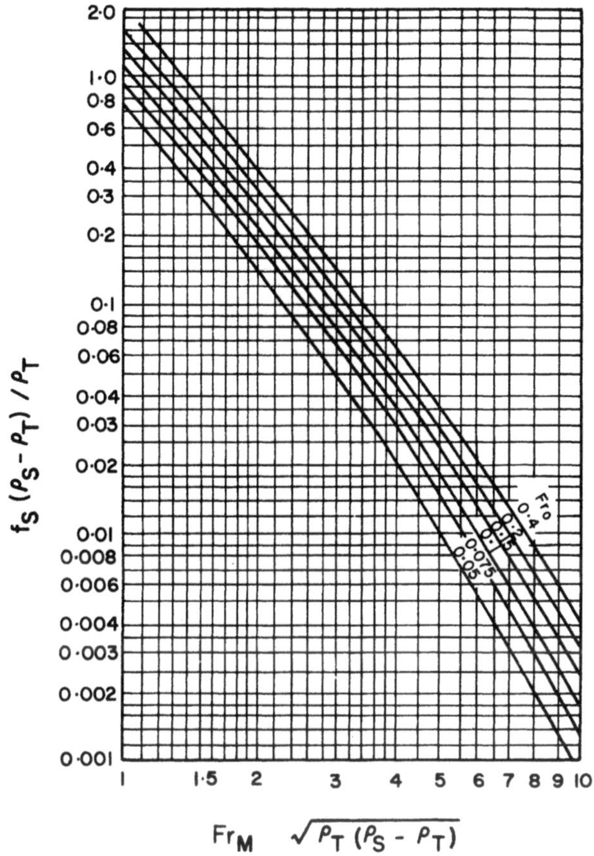

Figure 11.49. The Gaessler correlation for the solids friction factor, f_S. $C_S = 0.2$, $C_D = 0.44$, sphericity, $\psi = 0.54$. (From Gaessler, H., Doctoral Dissertation, Technische Hochschule, Karlsruhe, Germany, 1967.)

calculated pressure gradient would be the same as that for a homogeneous fluid having the density of the mixture.

On the other hand, if the material were so coarse or settling so fast that essentially all of it would be in asymmetric suspension and sliding bed (the method also predicts this condition), the calculated pressure gradient would be identical with that determined by the Durand correlation. In effect, this method provides a systematic means of interpolating between the two flow pattern extremes for the usual mixed-size systems. It has definite possibilities for refinements as our understanding of the flow patterns improves. The work of Wasp *et al.* is of great practical significance since it is based on, and confirmed by, extensive tests on large diameter commercial pipelines transporting coal-water slurries.

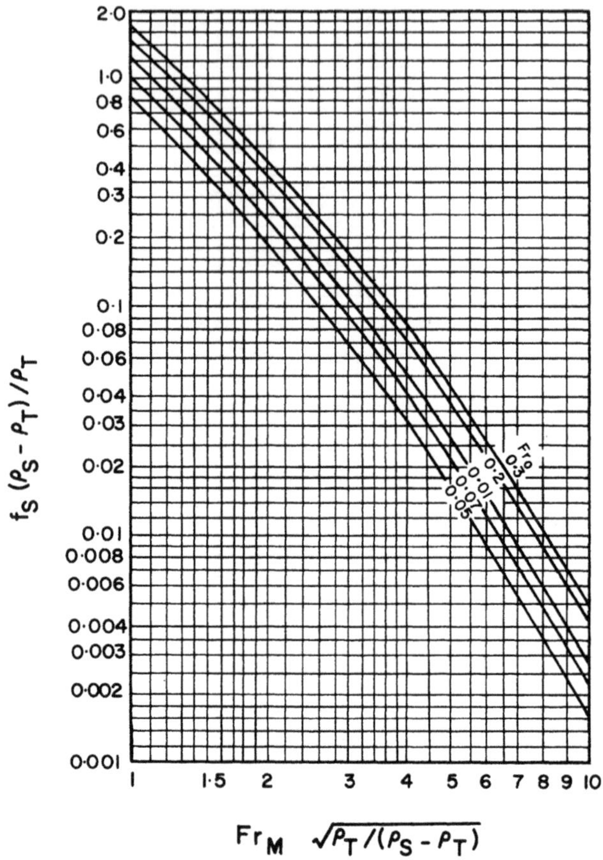

Figure 11.50. The Gaessler correlation for the solids friction factor, f_S. $C_S = 0.4$, $C_D = 0.44$ sphericity, $\psi = 0.54$. (From Gaessler, H., Doctoral Dissertation, Technische Hochschule, Karlsruhe, Germany, 1967.)

The procedure involves five separate steps as follows:

1. The approximate average volume fraction of the solids in symmetric suspension is determined by the method based on Ismail's (1951) work as the volume fraction calculated from Equation 11.45 at an elevation $0.80D$ above the bottom of the pipe and assuming that $\varepsilon_{S,0.30} = C_S$, i.e.,

$$\log \frac{C_{i,S}}{\varepsilon_{i,S,0.80}} = 1.8 \frac{V_c}{0.35 V_M \sqrt{f/2}} \tag{11.67}$$

where $C_{i,S}$ and $\varepsilon_{i,S}$ are the values of C_S and ε_S for particles in the size range i. Equation 11.67 is applied successively to each small size-range interval, i, following which the total volume fraction of solids at elevation $0.80D$ is determined as

$$\varepsilon_{S,\,0.80} = \sum_i \varepsilon_{i,\,S,\,0.80}$$

2. The pressure gradient attributable to the symmetrically (or nearly so) suspended solids, $(\Delta P/L)_{\text{sym}}$, is then determined by Wasp *et al.* by single-phase methods assuming Newtonian behavior and a velocity V_M. He uses a density $\rho_{M,0.80}$ and a viscosity $\mu_{M,0.80}$, either measured or estimated (see Chapter 2), for a homogeneous suspension of size consist and concentration corresponding with $\varepsilon_{S,0.80}$. The procedure could be made more general by taking account of any non-Newtonian properties. Additionally our discussion in Section 3.43 is applicable.
3. The proportion of the total solids in each size range flowing in asymmetric suspension (possibly including some particles sliding on the bottom of the pipe) is determined by a simple volume balance, i.e.,

$$\varepsilon_{i,\,S,\,\text{asym}} = C_{i,\,S} - \varepsilon_{i,\,S,\,0.80} \tag{11.68}$$

4. The pressure gradient attributable to the solids in asymmetric suspension or bed flow, $(\Delta P/L)_{\text{asym}}$, is estimated by a modification of the empirical correlation of Durand and Condolios (1952) which we discuss in Section 3.44. Wasp *et al.* recommend against the use of an overall weighted average drag coefficient as proposed by Condolios and Chapus (1963) and advise the treatment of each size fraction as a separate entity, i.e., the evaluation of

$$\left(\frac{\Delta P}{L}\right)_{\text{asym}} = \sum_i \left(\frac{\Delta P}{L}\right)_{i,\,\text{asym}}$$

Further we should note that for each size fraction the pressure gradient, $(\Delta P/L)_{i,\,\text{asym}}$, is evaluated as the gradient in excess of that for water alone, $(\Delta P/L)_W$, at the total mixture velocity, V_M, i.e.,

$$\left(\frac{\Delta P}{L}\right)_{i,\,\text{asym}} = \left(\frac{\Delta P}{L}\right)_{i,\,D} - \left(\frac{\Delta P}{L}\right)_W$$

where $(\Delta P/L)_{i,D}$ is the total pressure gradient determined by the Durand correlation for the size fraction i in the concentration $\varepsilon_{i,S,\text{asym}}$.
5. The total pressure gradient, $\Delta P/L$, is taken as the sum of $(\Delta P/L)_{\text{sym}}$ and $(\Delta P/L)_{\text{asym}}$.

Figure 11.51 taken from Wasp *et al.* (1963) compares predicted and actual wall shear stresses for coal-water slurries in 10- and 12-inch pipes. The check is good for all velocities resulting in turbulent flow (taken as when $f < 0.0072$).

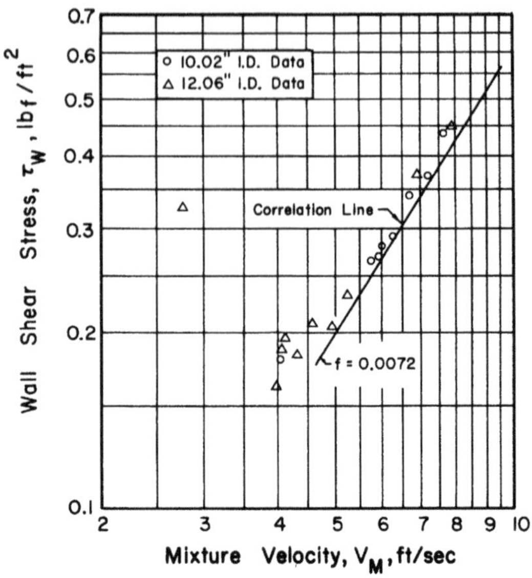

Figure 11.51. Predicted and measured wall shear stress in 10- and 12-inch pipes—water suspensions of 54–55 weight percent coal. (From Wasp, E. J., T. J. Regan, J. Withers, P. A. C. Cook, and J. T. Clancey, *Pipeline News*, **35**, 20, 1963.)

3.43 Symmetric Concentration Flow Pattern ($V_M > V_{M1}$)

Introduction Systems in which the particles are fully suspended and for which the concentration profile is symmetric, or in the limiting case, uniform, have traditionally been treated as pseudohomogeneous suspensions. The assumption is also usually made that there is no holdup effect or that the average *in situ* volume fraction solids is the same as the input fraction. As discussed in Section 3.3 this assumption is often quite reasonable.

If the solids concentration is high and particularly if the particles are fine (say under 100 microns) the pseudohomogeneous mixtures may exhibit non-Newtonian behavior. If so, the pressure gradient should be calculated by the methods of Chapter 5 or 6. On the other hand, more dilute suspensions ($C_S <$ say 0.25) and suspensions of coarser particles exhibit the behavior of two-phase systems. We deal here with such systems.

There have been two independent approaches assuming pseudohomogeneity, one based upon a direct analogy with the Newtonian fluid case and involving the concept of dimensionless excess pressure gradient; the other, developed by Julian and Dukler (1965) and discussed in Chapter 9, based upon an assumed mechanism postulating an increase in turbulence. A third approach, that of Shook and Daniel (1969) recognized the nonuniform distribution of the solids, and accounts for it by treating the mixture as an equivalent fluid of variable density. We will discuss all of these and indicate the extent to which they apply to liquid-solid and gas-solid systems.

Dimensionless Excess Pressure Gradient With the assumptions of pseudo-homogeneity and negligible holdup, and from the Fanning equation, we may express the dimensionless excess pressure gradient due to frictions as

$$\frac{i - i_F}{i_F} = \frac{\dfrac{2f_M \rho_M V_M^2}{g_c D} - \dfrac{2f_F \rho V_M^2}{g_c D}}{\dfrac{2f_F \rho V_M^2}{g_c D}} \tag{11.69}$$

or

$$\frac{i - i_F}{i_F} = \frac{f_M \rho_M - f_F \rho}{f_F \rho} \tag{11.70}$$

where f_M and f_F are, respectively, the friction factors for the mixture and the carrier fluid, the former evaluated for a fluid having the properties of the mixture, and both evaluated at V_M.

Both the density and the viscosity of the mixture will normally exceed those of the fluid. The mixture density, as discussed in Chapter 9, is

$$\rho_M = C_S \rho_S + (1 - C_S)\rho \tag{9.30}$$

or, substituting $s = \rho_S/\rho$,

$$\rho_M = [1 + (s - 1)C_S]\rho$$

The viscosity of the mixture, as discussed in Chapter 2, is

$$\mu_M = \mu(1 + 2.5C_S + 14.1C_S^2) \tag{2.62}$$

or

$$\mu_M = \mu[1 + 2.5C_S + 10.05C_S^2 + 0.00273 \exp(16.6C_S)] \tag{2.63}$$

Having regard for Equations 9.30 and 2.62, the Reynolds number ratio for the mixture and the carrier fluid may be approximated by

$$\frac{\mathrm{Re}_M}{\mathrm{Re}_F} = \frac{1 + (s - 1)C_S}{1 + 2.5C_S + \cdots} \tag{11.71}$$

Unless the presence of the particles should affect the development and the scale of the turbulence we would expect a Blasius type equation to apply to the mixture as well as to the single-phase fluid and

$$\frac{f_M}{f_F} = \left[\frac{1 + 2.5C_S + \cdots}{1 + (s - 1)C_S}\right]^{0.25} \tag{11.72}$$

Depending on s and C_S, f_M may be slightly more or slightly less than f_F.

Actually, with flow of fine particles in a gas and of flexible fibers in a liquid, the particles do affect and significantly repress the development of turbulence, with the result that for such particles $f_M < f_F$. We have discussed this in Chapter 9. If the friction factors f_M and f_F are taken as equal, an assumption valid if the mixture remains Newtonian and for modest concentrations of granular solids, we obtain

$$\frac{i - i_F}{i_F} = \frac{\rho_M}{\rho} - 1 \tag{11.73}$$

Substituting for ρ_M from Equation 9.30, we have

$$\frac{i - i_F}{C_S i_F} = s - 1 \tag{11.74}$$

In practice we find Equation 11.74 a good general approximation, but we also observe significant deviations.

First, at relatively high velocities and with low and modest concentration of solids where the solids tend to concentrate near the axis of the pipe, the region near the pipe wall is occupied by the carrier fluid only and we find that

$$i - i_F = 0$$

as in the corresponding case for ΔP_f for vertical flow.

Secondly, there are cases such as those reported by Boothroyd (1966) where fine particles in low concentration appear to suppress turbulence and consequently i becomes less than i_F. Related to this situation is that encountered with liquid suspensions of flexible fibers where again, because of suppression of turbulence, we find that $i < i_F$. There are no general correlations for these situations, but some guide to design is provided by the observations with respect to fine particles of Boothroyd (1966) discussed in Chapter 9, and of Gaessler (1967) discussed in Section 2.4 of this chapter.

Graf and Acaroglu (1967) studied the symmetric concentration flow of 2.85-mm sand particles with water in smooth 3-inch horizontal, inclined and vertical pipes. Assuming no holdup at the velocities of 16.1 to 20.5 ft/sec and ρ_M calculated from Equation 9.28, they found agreement within ± 5 percent of values predicted from single phase methods. For horizontal flow this corresponds with Equation 11.73 or 11.74.

Data of Newitt *et al.* (1962) for water-solid suspensions suggest that

$$\frac{i - i_W}{C_S i_W} = K(s - 1) \tag{11.75}$$

with K having a value of about 0.6. The reduction in the value of K from unity is apparently due to some turbulence suppression by the solids. Equation 11.75 with a value of K in the range of 0.6 to 1.0, depending on the nature of the solids, is a good empirical basis for estimating the pressure gradient in the symmetric concentration profile region.

The Julian and Dukler Analysis The analysis of Julian and Dukler (1965), which was discussed in Chapter 9, applies in its entirety to horizontal flow of dilute phase gas-solid suspensions, provided that the concentration distribution is symmetric. Furthermore, the analysis should also be applicable to liquid-solid systems.

The Shook and Daniel Analysis An analysis similar to the one by Julian and Dukler is that of Shook and Daniel (1969). They base their analysis on the concept that a nonuniform suspension of solids in a fluid may be represented as a fluid of density which varies with radial position. The method is applicable to either the symmetric or the asymmetric concentration profile, but depends upon some independent procedure for estimating the ratio of the density of the mixture at the center line of the pipe to the density of the average flowing stream. Shook and Daniel start with the shear stress distribution equation (Equation 3.47) reduced to the form

$$\tau = -\frac{1}{g_c} v_e \frac{d}{dy}(\rho u) \tag{11.76}$$

which neglects the viscous component and therefore is not applicable at and close to the pipe wall. As an alternate to the normal boundary condition of $u = 0$ at $y = 0$, Shook and Daniel state:

> ... the boundary condition is obtained intuitively by assuming that at the axis of the pipe the mixture of fluid and solids has a point velocity u_0 equal to that which a Newtonian fluid of density ρ'_M and viscosity μ'_M would have if it were flowing in the same pipe at the same pressure drop.

The properties ρ'_M and μ'_M are those of the *in situ* mixture. This enables them to write

$$\frac{u_0}{U^*} = 2.5 \ln \frac{DU^*\rho'_M}{2\mu'_M} + 5.5 \tag{11.77}$$

Shook and Daniel assumed that the variation of $2v_e/DU^*$ with y would be the same for the two-phase as for a single-phase system and chose the relationship

$$\frac{2v_e}{DU^*} = 0.40\left(\frac{2y}{D}\right)\left(1 - \frac{2y}{D}\right) \tag{11.78}$$

Substitution of Equation 11.78 into Equation 11.77 and integration gives us

$$\frac{\rho u}{\rho'_M U^*} = 2.5 \ln \frac{2y}{D} + 2.5 \frac{\rho_0}{\rho'_M} \ln \frac{DU^* \rho'_M}{2\mu'_M} + 5.5 \frac{\rho_0}{\rho'_M} \tag{11.79}$$

Integrating Equation 11.79 over the area of the pipe results in

$$\frac{1}{\sqrt{f}} = \frac{2.5}{\sqrt{2}} \frac{\rho_0}{\rho_M} \ln \frac{DU^* \rho'_M}{2\mu'_M} + \frac{5.5}{\sqrt{2}} \frac{\rho_0}{\rho_M} - \frac{7.5}{2\sqrt{2}} \frac{\rho'_M}{\rho_M} \tag{11.80}$$

where $\rho_M = [1 + (s - 1)C_S]\rho$,
ρ_0 = density of the *in situ* mixture at the center line of the pipe,
$f = \left(\frac{\Delta P}{L}\right)_M \frac{g_c D}{2\rho'_M V_M^2}$.

Shook and Daniel define a reference friction factor as for a Newtonian fluid, i.e.,

$$f_N = \left(\frac{\Delta P}{L}\right)_M \frac{g_c D}{2\rho_M V^2}$$

where

$$f_N = \phi\left(\frac{DV\rho'_M}{\mu'_M}, \frac{k}{D}\right)$$

and V is the velocity of a Newtonian fluid of density ρ'_M and viscosity μ'_M flowing in the same pipe at the same pressure gradient as the suspension.

They then write for this equivalent Newtonian fluid

$$\frac{1}{\sqrt{f_N}} = \frac{2.5}{\sqrt{2}} \ln \frac{DU^* \rho'_M}{2\mu'_M} + \frac{3.5}{2\sqrt{2}} \tag{11.81}$$

and combine Equations 11.75 and 11.76 to obtain, after assuming $\rho_M = \rho'_M$,

$$\frac{1}{\sqrt{f_N}} - \frac{1}{\sqrt{f}} = \left(1 - \frac{\rho_0}{\rho_M}\right)\left(\frac{1}{\sqrt{f_N}} + \frac{15}{4\sqrt{2}}\right) \tag{11.82}$$

From Equation 11.82 and the conventional relationship for f_N with the Reynolds number, Shook and Daniel have developed Figure 11.52. Use of this figure requires an independent estimate of ρ_0/ρ_M.

No general method of predicting this density ratio is known for either the symmetric or the asymmetric concentration profile cases. Shook, Daniel, Scott, and Holgate (1968), however, do propose an approximate procedure for the asymmetric profile. We will discuss this later.

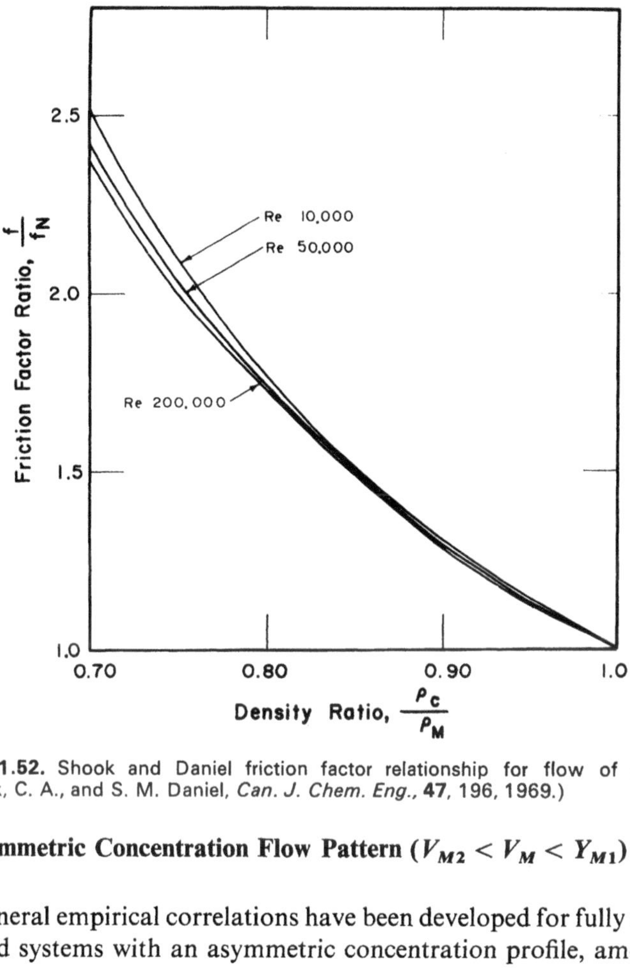

Figure 11.52. Shook and Daniel friction factor relationship for flow of suspensions. (From Shook, C. A., and S. M. Daniel, *Can. J. Chem. Eng.*, **47**, 196, 1969.)

3.44 Asymmetric Concentration Flow Pattern ($V_{M2} < V_M < Y_{M1}$)

Several general empirical correlations have been developed for fully suspended liquid-solid systems with an asymmetric concentration profile, among which that of Durand and Condolios (1952) and variations of it due to Zandi and Govatos (1967) and Hayden and Stelson (1968) are probably the most important. A correlation developed by Worster (1952) has been shown by Zandi and Govatos not to be generally applicable. The correlation that is due to Newitt, Richardson, Abbott, and Turtle (1955) gives results comparable to the Durand type correlations, has some theoretical basis, and therefore is also of interest. Yen and Zandi (1969) have made a detailed, theoretically based analysis for flow in a rectangular duct.

The Durand and Condolios and Related Correlations Durand and Condolios obtained extensive data on water slurries of a variety of sands and gravels in horizontal pipes up to 22 inches in diameter with solids up to 1 inch size and to concentrations of 22 percent. For flow conditions where the solids were fully suspended ($V_M > V_{M2}$ and including some data where $V_M > V_{M1}$), they found a satisfactory correlation of the form

$$\frac{i - i_W}{C_S i_W} = K\left[\left(\frac{gD}{V_M^2}\right)\left(\frac{V_0}{\sqrt{gd}}\right)\right]^{1.5} \tag{11.83}$$

Durand and Condolios did not give a value of K, but this may be inferred from their data and is 176. Worster (1952) and Durand (1953) show that Equation 11.83 may be generalized to apply to solids of different densities to

$$\frac{i - i_W}{C_S i_W} = 121\left[\left(\frac{gD(s-1)}{V_M^2}\right)\left(\frac{V_0}{\sqrt{gd(s-1)}}\right)\right]^{1.5} \tag{11.84}$$

Durand (1953) also gives Equation 11.83 in an alternate form with V_0 replaced through Equations 1.1 and 1.2, i.e.,

$$V_0 = \left[\frac{4}{3}\frac{(s-1)gd}{C_D}\right]^{1/2}$$

where C_D is the drag coefficient of the particles. We may replace V_0 in Equation 11.84 to obtain

$$\frac{i - i_W}{C_S i_W} = 150\left[\left(\frac{gD(s-1)}{V_M^2}\right)\left(\frac{1}{\sqrt{C_D}}\right)\right]^{1.5} \tag{11.84a}$$

We should note that the abscissa of Figure 10 in the paper by Durand (1953) appears to be mislabeled, and Babcock (1964) and others appear to have erroneously determined from it the value of the constants in Equations 11.84 and 11.84a as 65 and 81, respectively.

Zandi and Govatos (1967) collected and correlated most of the available data (2549 data points) for the horizontal flow of water-solid systems, excluded those for which they thought a moving bed had been present and those for concentrations less than 5 percent, and developed two equations to describe the remaining data (some 1452 data points). Their exclusion of suspected moving bed data was based on their criterion for V_{M2}, discussed in Section 3.22, i.e.,

$$N_1 = \frac{V_M^2\sqrt{C_D}}{C_S Dg(s-1)} < 40$$

The equations of Zandi and Govatos are

$$\frac{i - i_W}{C_S i_W} = 6.3(C_S N_1)^{-0.354} \qquad \text{for } C_S N_1 > 10 \tag{11.85a}$$

$$\frac{i - i_W}{C_S i_W} = 280(C_S N_1)^{-1.93} \qquad \text{for } C_S N_1 < 10 \tag{11.85b}$$

Hayden and Stelson (1968) propose a modification of Durands' equation to the form

$$\frac{i - i_W}{C_S i_W} = 100\left[\frac{gD(s-1)V_0}{V_M^2\sqrt{gd(s-1)}}\right]^{1.3} \tag{11.86}$$

and support this form with their data on fine and coarse sand and fine gravel in 1- and 2-inch pipes. The generalized Durand and Condolios equation (Equation 11.84), Hayden and Stelson's modified equation, and the two equations of Zandi and Govatos are shown in Figure 11.53, modified after Hayden and Stelson, along with Hayden and Stelson's data. These data do not particularly support the modified equations of Zandi and Govatos and suggest that the original Durand and Condolios equation or the Hayden and Stelson relationship are about as good as may be expected.

Babcock (1968) is properly critical of all equations of the Durand and Condolios form and illustrates, for example, that data that have been carefully tested for internal consistency and that correlate perfectly on a simple

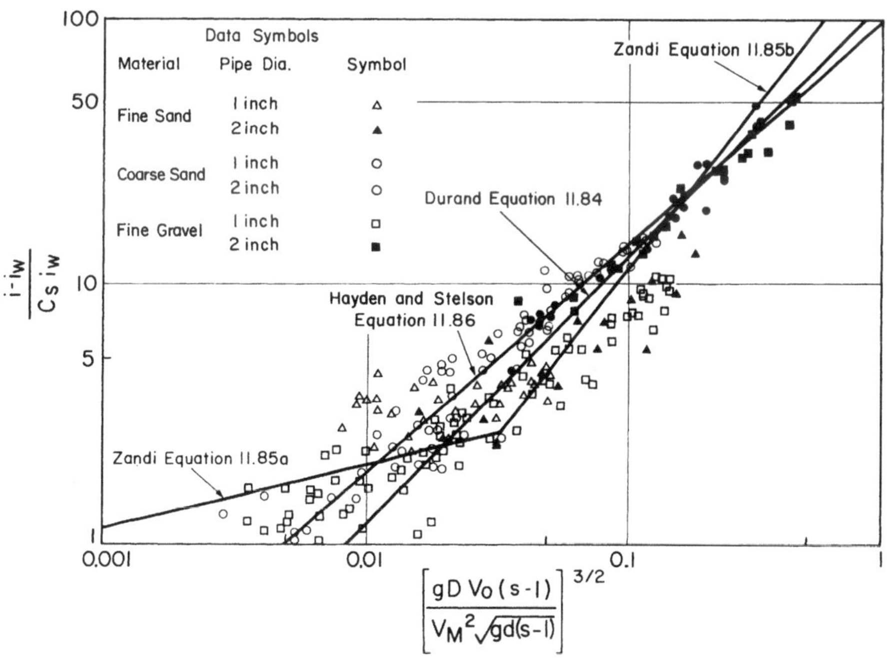

Figure 11.53. Comparison of predictions for the pressure gradient in the flow of asymmetric suspensions. (From Hayden, J. W., and T. E. Stelson, Paper presented at Int. Symp. on Solid-Liquid Flow in Pipes, Philadelphia, Penn., March, 1968.)

plot of $(i - i_W)/C_S i_W$ versus $V_M^2/gD(s-1)$ are actually separated on a plot of $(i - i_W)/C_S i_W$ versus $V_M^2\sqrt{C_D}/gD(s-1)$. This does raise the question of whether the abscissa originally suggested by Durand is in fact the most suitable one. Nonetheless the Durand and Condolios type of correlation remains one of our best overall empirical tools. We must recognize, however, that uncertainties of the order of ± 40 percent may be present in predicted values of $(i - i_W)/C_S i_W$.

Condolios and Chapus (1963) reported that the Durand and Condolios correlation, in the form of Equation 11.84a, applies to materials having a wide range of particle size provided a weighted drag coefficient is used. They define such a coefficient as

$$\sqrt{C_{DM}} = \sum_i m_i \sqrt{C_{Di}}$$

where m_i = mass fraction of solids having a drag coefficient C_{Di}.

The Newitt et al. Correlation The correlation of Newitt *et al.* (1955) is based on their own data on the flow in a 1-inch pipe of various sizes of perspex, fine coal, and sands all traveling in water suspension, with $V_{M2} < V_M < V_{M1}$. Assuming no local slip, they reasoned that the work done in maintaining particles of settling velocity V_0 in suspension would be proportional to their effective weight and their settling velocity. They also assumed that this work would equal the energy dissipation due to the presence of the solids. Thus in a volume Q' of a suspension moving at a velocity V_M, the work performed in the time $1/V_M$ to travel a unit distance would be, $K_S C_S Q'(\rho_S - \rho_W)V_0 g/V_M$, and this is equal to $i_S Q' \rho_W g$, the energy dissipated due to the pressure gradient. Equating these terms and rearranging, Newitt *et al.* obtained

$$\frac{i - i_W}{C_S i_W} = K_1 \frac{gDV_0}{V_M^3}(s-1) \tag{11.87}$$

Their data and correlation are illustrated in Figure 11.54. The final equation is

$$\frac{i - i_W}{C_S i_W} = 1100 \frac{gDV_0}{V_M^3}(s-1) \tag{11.88}$$

Newitt *et al.* observe that where the solids cover a range of particle sizes, the finer particles tend to increase the carrying capacity of the mixture for larger particles. They present no analytical method of using their correlation for mixed sizes but say that if a weighted mean diameter,

$$d_m = \sum_i d_i m_i$$

where m_i = mass fraction of solids of diameter d_i, is used, conservative results are obtained.

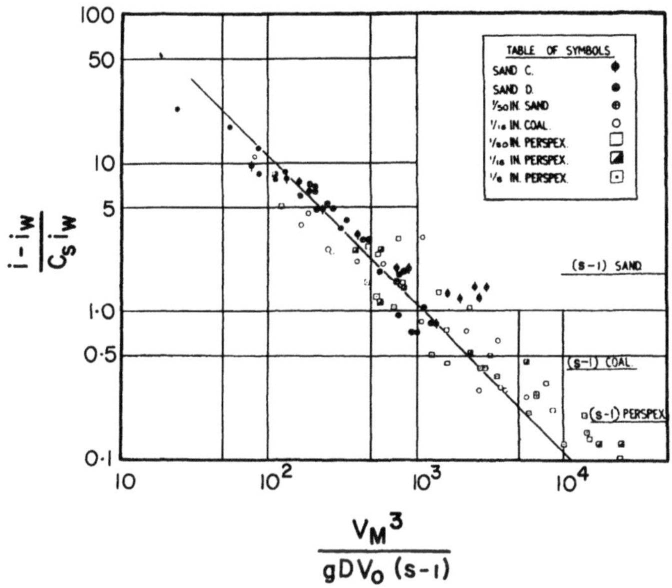

Figure 11.54. Test of the Newitt *et al.* equation (Eq. 11.88) for the flow of asymmetric suspensions. (From Newitt, D. M., J. F. Richardson, M. Abbott, and R. B. Turtle, *Trans. Inst. Chem. Engrs.*, **33**, 93, 1955.)

Notwithstanding that Equation 11.88 was based on data taken only in small pipes, it has been found (Zandi and Govatos, 1967) to be comparable in reliability to the Durand and Condolios type relations.

We mentioned in Section 3.43 that the Shook and Daniel (1969) analysis based upon the concept of a variable fluid density is also applicable to the asymmetric concentration flow pattern. They have applied Equation 11.82 with ρ_0 estimated, as proposed by Shook *et al.* (1968), with the aid of the equivalent of Equation 11.38, and assuming that $2e_s/DU^* = 0.07$. Shook and Daniel compare their calculated and observed velocities under various pressure gradients for the flow in a 1-inch pipe of 40 to 60 mesh sand and water. They report discrepancies generally under 5 percent and less than those obtained from the empirical Durand equation. The Shook and Daniel analysis is not sufficiently confirmed to serve as a general prediction method and its reliance on a separate correlation for ρ_0/ρ_M is clearly a disadvantage. Nonetheless it is valuable as an aid to our general understanding of the flow of fluid-solid suspensions and may be of supplementary value in connection with design problems.

Yen and Zandi (1969) present a detailed analysis of the asymmetric concentration flow pattern in a rectangular duct. The analysis is based upon the continuity and momentum equations, O'Brien's equation (Equation

11.38) for the vertical distribution of solids, the drag force relation for spherical particles (Equation 1.2) and certain assumptions concerning the distribution of the turbulent shear stresses. Their predictions of the velocity and concentration profiles and, through them, the friction factor check well with experimental data of Ismail (1949). The approach shows promise for application to flow in pipes.

3.45 Moving and Stationary Bed Flow Patterns

Moving Bed, with Saltation and Suspension $V_{M3} < V_M < V_{M2}$; Stationary Bed, with Saltation and Suspension $V_{M4} < V_M < V_{M3}$ As discussed earlier at mixture velocities below V_{M2}, some or all of the solids deposit on the bottom of the pipe and move first in transverse waves, then in longitudinal waves, then in a sliding bed with the uppermost particles moving also by saltation, and finally a stationary bed forms with the uppermost particles moving by saltation. The empirical correlations applicable to flow rates below V_{M2}, while based mainly on data for $V_{M3} < V_M < V_{M2}$, overlap into the range of the stationary bed. Accordingly we will treat the two regions together.

As the mixture velocity is decreased below V_{M2} where a moving, and eventually a stationary, bed develops, the average velocity of the liquid tends to be maintained as the cross section available to it decreases with the buildup of the bed. This may be illustrated as follows. Assume a mixture of input concentration of C_S flowing at the critical mixture velocity V_{M2} with no bed. Now assume the mixture velocity is decreased to V_M, C_S being maintained constant. Under the new conditions a bed of height h_b will form and the bulk of the transport will be by asymmetric suspension flowing through the reduced cross section A_u above the bed. Then the actual average suspension velocity will be

$$V_{Mu} = V_M \frac{A}{A_u} \tag{11.89}$$

Condolios and Chapus (1963) report on the basis of actual measurements with water-solid suspensions that the true average velocity of the flowing mixture above the bed, V_{Mu}, adjusts itself so that

$$\frac{V_{Mu}}{\sqrt{gD_E}} = K \tag{11.90}$$

where D_E is the equivalent diameter of the conduit above the bed and K is a function only of the solids concentration and particle size. They state that

Equation 11.90 also applies at the condition of first deposition of solids, namely when $D_E = D$ and $V_{Mu} = V_{M2}$, indicating that

$$\frac{V_{Mu}}{\sqrt{D_E g}} = \frac{V_{M2}}{\sqrt{Dg}} \tag{11.91}$$

Actually Wicks (1968) finds that if V_{Mu} is calculated from Equation 11.89 with measured data for V_M and A_u, the ratio $V_{Mu}/\sqrt{D_E g}$ is not constant and varies by as much as 65 percent from the value corresponding with $V_{M2}/\sqrt{Dg}$, as reported by Condolios and Chapus for similar suspensions.

Wicks has made a semitheoretical analysis of some aspects of the flow of liquid-solid mixtures under conditions where a bed is present based upon the mechanics associated with the lifting and rolling of particles on the bed surface. He describes the forces acting on a particle at the moment of lifting, as shown in Figure 11.55, and states that

$$F_B x + F_L x + F_D y > F_G x$$

or

$$F_B + F_L + \frac{y}{x} F_D > F_G$$

where y/x is a shape factor. He identifies the various forces as

$$F_G = v_S \rho_S \frac{g}{g_c} \tag{11.92}$$

$$F_B = v_S \rho \frac{g}{g_c} \tag{11.93}$$

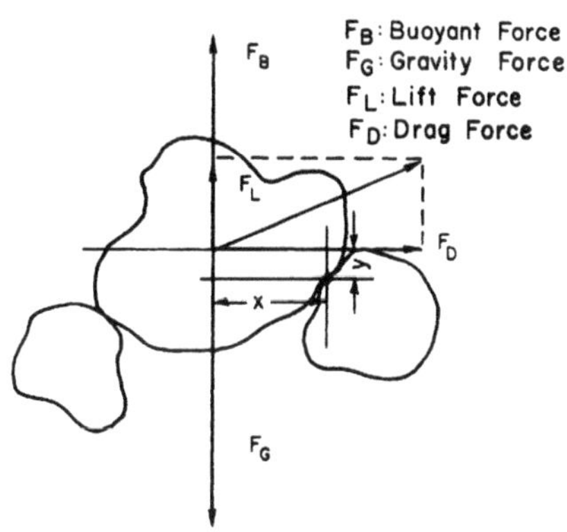

Figure 11.55. Forces acting on a typical particle at the bed surface. (From Wicks, M., Paper presented at Int. Symp. on Solid-Liquid Flow in Pipes, Philadelphia, Penn., March, 1968.)

$$F_L = C_L a \frac{\rho u^2}{2g_c} \tag{11.94}$$

$$F_D = C_D a \frac{\rho u^2}{2g_c} \tag{11.95}$$

where v_S = the average volume of the solid particles,
a = the average projected area of the solid particles,
u = local velocity of the fluid,
C_L and C_D are, respectively, lift and drag coefficients,

obtaining, for the condition of incipient motion,

$$\frac{(C_L + yC_D/x)a\rho u^2}{2v_S(\rho_S - \rho)} = 1 \tag{11.96}$$

On the basis of the expected dependence of u, C_L, and C_D on the fluid properties and geometry factors, he postulates that

$$u = \frac{g_c}{\mu} d \frac{\rho V_L^2}{2g_c} \phi\left(\frac{D_E V_{Mu} \rho}{\mu}\right) \tag{11.97}$$

$$C_L = f_1\left(\frac{d}{D}, \frac{D_E V_{Mu} \rho}{\mu}\right) \tag{11.98}$$

$$C_D = f_2\left(\frac{d}{D}, \frac{D_E V_{Mu} \rho}{\mu}\right) \tag{11.99}$$

and taking $y/x = 1.44$ converts Equation 11.96 to

$$\frac{f_1 + 1.44 f_2 \phi}{8}\left[\frac{\rho^3 d V_{Mu}^4}{(\rho_S - \rho) g \mu^2}\right] = 1 \tag{11.100}$$

This equation suggests that in the region of stable bed flow the bed height (expressed through D_E) may be correlated with the true average suspension velocity, V_{Mu}, by a relation of the form

$$\psi = \phi(S) \tag{11.101}$$

where

$$\psi = \frac{\rho^3 d V_{Mu}^4}{(\rho_S - \rho) g \mu^2}$$

and

$$S = \left(\frac{d}{D}\right)^{2/3}\left(\frac{D_E V_{Mu} \rho}{\mu}\right)$$

Wicks' correlation based on this approach is given in Figure 11.56. The extreme ends of the correlation may be represented by the following equations.

$$\psi = 0.1S^3 \qquad \text{for } S < 40 \tag{11.102}$$

and

$$\psi = 100S^{3/2} \qquad \text{for } S > 400 \tag{11.103}$$

A trial and error simultaneous solution of Equation 11.89 and Figure 11.56, together with the relationships of Figures 10.38 and 10.40, permits estimates of V_{Mu} and h_b or D_E for any mixture velocity $V_M < V_{M2}$. Although not extensively tested, this correlation provides an alternate and possibly superior approach to the strictly empirical one of Condolios and Chapus for determining

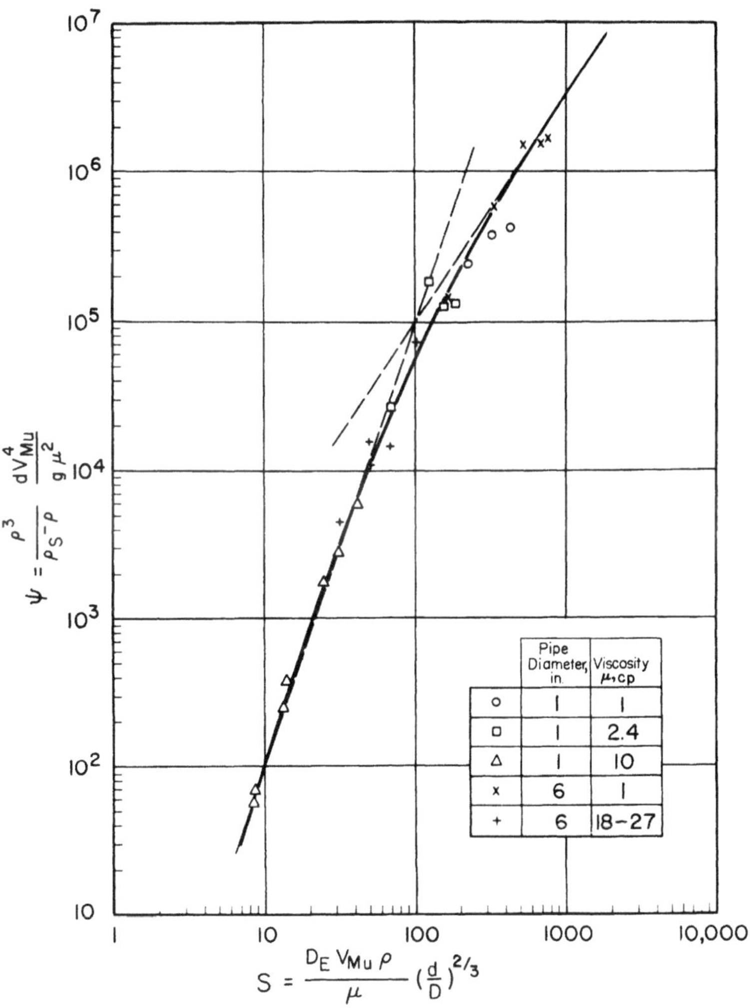

Figure 11.56. Wicks' relationship for the average velocity of a suspension above a bed (From Wicks, M., Paper presented at Int. Symp. on Solid-Liquid Flow in Pipes, Philadelphia Penn., March, 1968.)

the relationship between V_{Mu} and the bed height for liquid-solid suspensions.

Condolios and Chapus also report that, if V_{Mu} is substituted for V_M in the Durand equation, their data for flow in the presence of a bed are correlated by the so-modified equation. In confirming this, Condolios and Chapus estimated i_W using the relation

$$i_W = \frac{2f_W \rho_W V_L^2}{g_c D_E} \tag{11.104}$$

with f_W determined as for single-phase flow in a circular pipe using a Reynolds number $\text{Re} = D_E V_L \rho_W / \mu_W$. Thus, according to Condolios and Chapus, if we know V_{M2}, we have a way of estimating the true average mixture velocity, V_{Mu}, for any given bed height (which determines D_E); from this we may estimate i_W; using Durand's correlation (or one of the modifications) we may obtain $(i - i_W)/C_S i_W$ and the pressure gradient i.

Although Condolios and Chapus estimated V_{Mu} from Equation 11.90, we might expect a similar agreement with the Durand equation if V_{Mu} is estimated with the aid of Equation 11.89 and Wicks' correlation between V_{Mu} and the bed height.

Newitt *et al.* (1955) arrived at a relationship for the moving bed case again by equating the work done on the solid particles to the energy dissipation as a result of the excess pressure gradient. They expressed the work done on the sliding bed of particles per unit length of pipe as $\eta_m C_S Q'(\rho_S - \rho_W)g$, where Q' is the volume of a suspension occupying a unit length of pipe, and η_m is a coefficient of sliding friction. Equating this to the energy loss due to the solids, $i_S Q' \rho_W g$, and rearranging, they obtained

$$\frac{i_S}{C_S} = \frac{i - i_W}{C_S} = K_2(s - 1) \tag{11.105}$$

or, dividing by i_W,

$$\frac{i - i_W}{C_S i_W} = K_3(s - 1)\frac{gD}{V_M^2} \tag{11.106}$$

The data of Newitt *et al.* for materials traveling in water in a 1-inch pipe both by sliding bed and by stationary bed and saltation are given in Figure 11.57 with his best fit line represented by either

$$\frac{i - i_W}{C_S} = 0.8(s - 1) \tag{11.107}$$

or

$$\frac{i - i_W}{C_S i_W} = 66(s - 1)\frac{gD}{V_M^2} \tag{11.108}$$

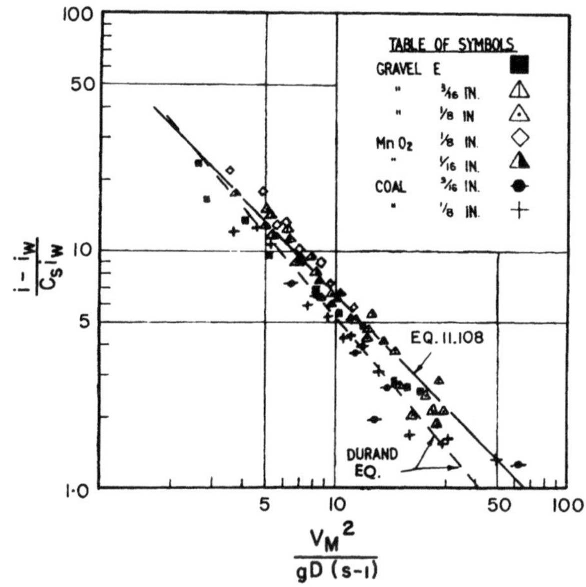

Figure 11.57. Test of the Newitt *et al.* equation (Eq. 11.108) for the bed flow of suspensions. (From Newitt, D. M., J. F. Richardson, M. Abbott, and R. B. Turtle, *Trans. Inst. Chem. Engrs.*, **33**, 93, 1955.)

Babcock (1968) also finds that data for the moving bed flow pattern may be correlated by Equation 11.106. For water suspensions of coarse sand and other particles, including steel shot, in a 1-inch pipe with $5 < V_M^2/gD < 40$, he obtains

$$K_3 = 60.6$$

which is close to the value of Newitt *et al.* On the other hand, Babcock reports

$$K_3 = 6.6$$

for the flow in a 6-inch pipe of water suspensions of taconite ore with $1 < V_M^2/gD \leqslant 12$. It is clear that Newitt's equation (Equation 11.108), although appropriate for pipe 1 inch in diameter, is not applicable to larger sizes and that K_3 is an inverse function of pipe diameter. As a crude approximation we may take

$$K_3 = 66D'^{-1.27}$$

where D' is the pipe diameter in inches.

Shook and Daniel (1965) devise a relationship for the stationary bed with saltation and suspension flow pattern in a rectangular duct. Their work is also relevant to flow in a pipe. It is applicable to conditions where the turbulent forces are not adequate to maintain particles in suspension, and where the previously mentioned Bagnold dispersive force is significant. They write a force balance in the form

$$-\frac{dP}{dL} = \frac{\tau_{wS} + \tau_b}{h_u} \qquad (11.109)$$

where τ_{wS} = shear stress caused by the suspended mixture on the upper surface of the channel,
τ_b = shear stress at the upper surface of the bed,
a = depth of channel,
h_b = depth of bed,
$h_u = a - h_b$.

The Bagnold or dispersive force acting at the surface of the bed, P_{Bb}, is expressed as

$$P_{Bb} = (\rho_S - \rho)\frac{g}{g_c}\int_0^{h_u} \varepsilon_S \, dy \qquad (11.110)$$

From Equations 11.46 and 11.110, we have

$$\tau_b = (\rho_S - \rho)\frac{g}{g_c}\tan\alpha \int_0^{h_u} \varepsilon_S \, dy \qquad (11.111)$$

Shook and Daniel express the shear stress on the wall due to the mixture in suspension as

$$\tau_{wS} = \frac{f\rho V_{Mu}^2}{2g_c} \qquad (11.112)$$

where V_{Mu} = average mixture velocity above the bed $= V_M a/h_u$.

From Equations 11.109, 11.111, and 11.112, they write

$$i = \frac{fV_{MS}}{2g_c h} + (s-1)\bar{\varepsilon}_{Sh}\frac{g}{g_c}\tan\alpha \qquad (11.113)$$

where $\bar{\varepsilon}_{Sh}$ is the mean value of the solids concentration in the moving bed. They compare Equation 11.113 with Equation 11.107 of Newitt *et al.* for pipes, which may be written as

$$i = \frac{2fV_M^2}{g_c D} + 0.8C_S(s-1) \qquad (11.114)$$

The similarity is apparent and the two equations become identical if $D = D_E = 2h_u$, $a = \sqrt{2}\, h_u$, and $\tan\alpha = 0.8C_S/\bar{\varepsilon}_{Sh}$.

Shook and Daniel indicate that the basic assumption of the work, i.e., that particle suspension is due to Bagnold dispersive stresses, is justified so long as

$$V_0 > U^*$$

where V_0 is the particle settling velocity,
$U^* = V_M\sqrt{f/2}$ is the friction velocity.

They compare values of tan α derived from their data in a rectangular channel with sand, lead, and nickel particles with Bagnold's results and find good order of magnitude agreement.

Equation 11.113 is not recommended for pipeline design purposes, but it lends support to Equation 11.114 of Newitt *et al.* and indicates that the factor, 0.8, of that equation (or the related factor, 66, of Equation 11.108) is not really a constant but is related to $\bar{\varepsilon}_{Sh}$, C_S, and tan α.

Carstens' Theory Carstens (1969) has developed a theory of hydraulic transport in the stationary bed and asymmetric suspension flow pattern based on the simplifying assumption that any deviation from the relation

$$\rho_M\left(\frac{\Delta P}{L}\right) - \rho_L\left(\frac{\Delta P}{L}\right)_L = 0$$

is due to the formation of a stationary bed of solids in the pipe which causes a reduction in the cross-sectional flow area. Defining A_u as the cross-sectional area of the pipe above the bed, and through which the suspension is flowing, $(p_u + w_i)$ its wetted perimeter, and D_E its equivalent diameter, Carstens expresses A_u and p_u in terms of θ_b as defined in Figure 11.58 as

$$A_u = \frac{R^2(\pi - 2\theta_b)}{2} - R^2 \sin\theta_b \cos\theta_b \tag{11.115}$$

$$p_u = R(\pi - 2\theta_b) + 2R\cos\theta_b \tag{11.116}$$

$$D_E = \frac{4A_u}{p_u + w_i} \tag{11.117}$$

The geometrical relationship between bed height, h_b, and the angle θ_b may be obtained from Figure 10.37 noting that $\theta_b = (\pi - \gamma)/2$ (see Figure 10.37). The relationships of w_i, h_b, A_u, and p_u are given in Figures 10.36 to 10.40. The velocity of the suspension flowing in the upper portion of the pipe is

$$V_{Mu} = \frac{Q_S + Q_F}{A_u} \tag{11.118}$$

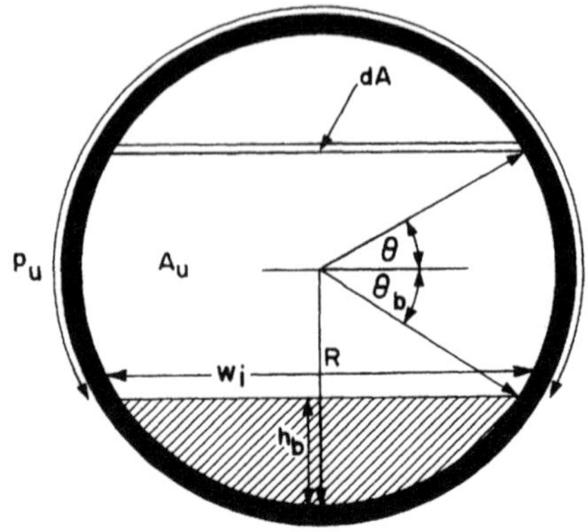

Figure 11.58. Terminology for Carstens' theory. (After Carstens, M. R., *A.S.C.E. Proc. Hydraulics Div.*, **95**, (HY1), 275, 1969.)

and

$$\frac{V_{Mu}}{V_M} = \frac{A}{A_u} \tag{11.119}$$

where $A = \dfrac{\pi D^2}{4}$

The pressure gradient for the suspension may be expressed as

$$\frac{\Delta P}{L} = \frac{2f_{MC}\rho V_{Mu}^2}{g_c D_E} \tag{11.120}$$

where f_{MC} is a friction factor defined by Equation 11.120. We may also write for the pressure gradient for the carrier liquid flowing alone in the pipe at a velocity V_M

$$\left(\frac{\Delta P}{L}\right)_L = \frac{2f\rho V_M^2}{g_c D} \tag{11.121}$$

Combining Equations 11.119, 11.120, and 11.121, we obtain

$$\frac{\Delta P/L}{(\Delta P/L)_L} = \frac{f_{MC}}{f}\frac{D}{D_E}\left(\frac{A}{A_u}\right)^2 \tag{11.122}$$

Carstens assumes that $f_{MC}/f = 1$ and introduces Equations 11.115, 11.116, and 11.117 into Equation 11.122, obtaining

$$\frac{\Delta P/L}{(\Delta P/L)_L} = \frac{4\pi^2(\pi - 2\theta_b + 2\cos\theta_b)}{(\pi - 2\theta_b - \sin 2\theta_b)^3} \tag{11.123}$$

The next step in the development is to relate the depth of bed deposit, or θ_b, to other flow variables. Carstens uses the equation

$$e_s \frac{d\varepsilon_S}{dy} + V_0 \varepsilon_S = 0 \tag{11.38}$$

discussed earlier. He replaces e_s by a constant value, $\bar{e}_m$, and integrates Equation 11.38 along a vertical line from the bed surface to obtain

$$\frac{\varepsilon_S}{\varepsilon_{Sb}} = \exp\left[-\frac{V_0 D}{2\bar{e}_m}(\sin\theta - \sin\theta_b)\right] \tag{11.124}$$

where ε_{Sb} is the solids concentration at the bed surface. Carstens neglects any transport due to saltation or moving bed and assumes that the solids flow rate may be approximated by

$$Q_S = V_{Mu} \int_{A_u} \varepsilon_S \, dA \tag{11.125}$$

From Figure 11.58 we have

$$dA = 2R^2 \cos^2\theta \, d\theta \tag{11.126}$$

Expressing Q_S in terms of the volume fraction solids of the input mixture, C_s, and combining Equations 11.124, 11.125, and 11.126, we obtain

$$\frac{C_S}{\varepsilon_{Sb}} = \frac{2A}{\pi A_u} \int_{\theta_b}^{\pi/2} \exp\left[\left(-\frac{V_0 D}{2\bar{e}_m}\right)(\sin\theta - \sin\theta_b)\right] \cos^2\theta \, d\theta \tag{11.127}$$

Equation 11.127 may be integrated either by the series expansion of the exponential term or by some numerical technique. Carstens has evaluated the integral for several values of $V_0 D/2\bar{e}_m$ and for θ_b in the range $-\pi/2$ to $\pi/2$. To obtain $\bar{e}_m$ he assumes

$$\frac{2\bar{e}_m}{U^* D} = 0.052 \tag{11.128}$$

where $U^* = V_M\sqrt{f/2}$ which yields, for pipe flow without bed deposit,

$$\bar{e}_m = 0.0184 V_M \sqrt{f} \, D \tag{11.129}$$

and with bed deposit

$$\bar{e}_m = 0.0184 V_{Mu} \sqrt{f_{MC}} \, D_E \tag{11.130}$$

Equations 11.119 and 11.130 lead to

$$\frac{V_M\sqrt{4f}}{V_0} = 54.4\sqrt{\frac{f}{f_{MC}} \frac{2\bar{e}_m}{V_0 D} \frac{A_u}{A} \frac{D}{D_E}} \tag{11.131}$$

for asymmetric suspension flow in the presence of a stationary bed. The left-hand side of Equation 11.131 may easily be evaluated from a knowledge of V_M, D, and V_0 or the fluid and particle properties.

The solution of Equations 11.123, 11.127, and 11.131 gives the final theoretical relationship between $\Delta P/L$ and the factors assumed to influence it: V_M, V_0, C_S, ρ_M, ε_{Sb}, θ_b, or h_b, and f. Carstens gives the solution in graphical form in Figures 11.59 and 11.60, being alternate representations of the relationship

$$\frac{\Delta P/L}{\rho_M V_0^2/2g_c D} = \phi\left(\frac{V_M\sqrt{4f}}{V_0}, \frac{C_S}{\bar{\varepsilon}_{Sb}}\right) \tag{11.132}$$

In Figure 11.59 he shows both a set of lines of constant concentration ratio, $C_S/\bar{\varepsilon}_{Sb}$, and lines of constant bed depth ratio, h_b/D. Figure 11.60 gives the same solution expressed through lines of constant dimensionless rate of solids flow, i.e.,

$$\frac{C_S}{\bar{\varepsilon}_{Sb}}\frac{V_M\sqrt{4f}}{V_0} = \frac{4Q_S\sqrt{4f}}{\pi D^2 \bar{\varepsilon}_{Sb} V_0} \tag{11.133}$$

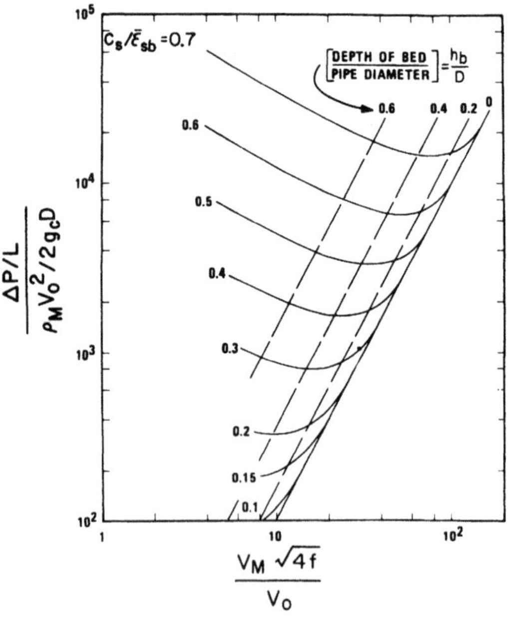

Figure 11.59. Carstens' pressure gradient prediction for suspension flow with a bed—concentration ratio parameter. (From Carstens, M. R., *A.S.C.E. Proc. Hydraulics Div.*, **95** (HY1), 275, 1969.)

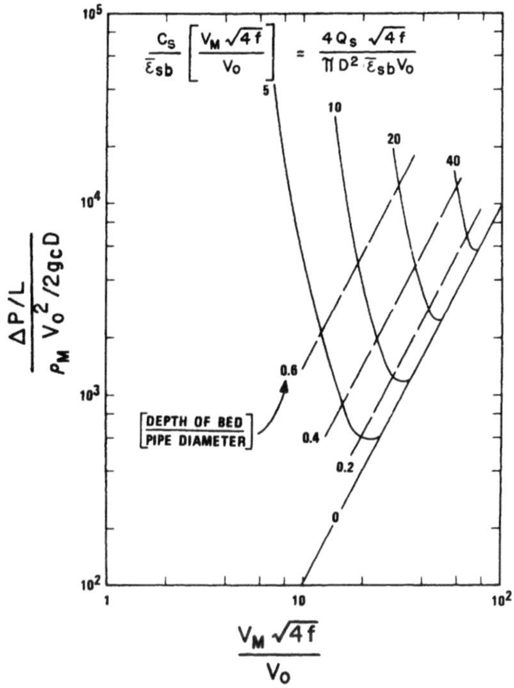

Figure 11.60. Carstens' pressure gradient prediction for suspension flow with a bed—solids flow rate parameter. (From Carstens, M. R., *A.S.C.E. Proc. Hydraulics Div.*, **95** (HY1), 275, 1969.)

It is interesting to note how closely Figures 11.59 and 11.60 conform, in a qualitative way, to the pressure gradient curves discussed in Section 2.4 and resulting from actual measurements. Carstens has not made quantitative comparisons of his predictions with actual data and he does not recommend the use of Figure 11.59 or 11.60 for design purposes. Despite this and the approximations of the development, Carstens' theory provides a useful basis for our understanding. We should note that, although the relations were developed for hydraulic transport, there is no reason why they should not be applicable to dense phase gas-solid flow.

Graf and Acaroglu (1968) have studied the hydraulic transport of solids in the presence of a bed in open channels, rectangular conduits, and pipes. By a consideration of the forces acting on the particles, they show that a "shear intensity parameter" Ψ must be a function of a "transport parameter" Φ, i.e.,

$$\Phi = \phi(\Psi) \tag{11.134}$$

where

$$\Phi = \frac{C_S V_M D'_E}{4\sqrt{(s-1)gd^3}} \tag{11.135}$$

D'_E = equivalent diameter of the entire channel = D for a pipe,

$$\Psi = \frac{4(s-1)d}{\dfrac{\Delta P}{\rho_M L} D'_E} = \frac{(s-1)d\rho_M}{\tau_w} \tag{11.136}$$

We may note that, where Newton's settling law applies, Ψ is proportional to V_0^2/τ_w. Using data obtained in closed conduits, flumes, and riverbeds by eight different investigators, they derive the following approximate relationship

$$\Phi = 10.39\Psi^{-2.52} \tag{11.137}$$

Figure 11.61 gives a comparison of the data with Equation 11.137. There is a large scatter of the data and the equation can be considered a crude approximation only.

Acroglu and Graf (1968) have made use of their shear intensity parameter, Ψ, to develop a detailed map of the various modes of solids transport in the presence of a bed. Their map is reproduced in simplified form in Figure 11.62

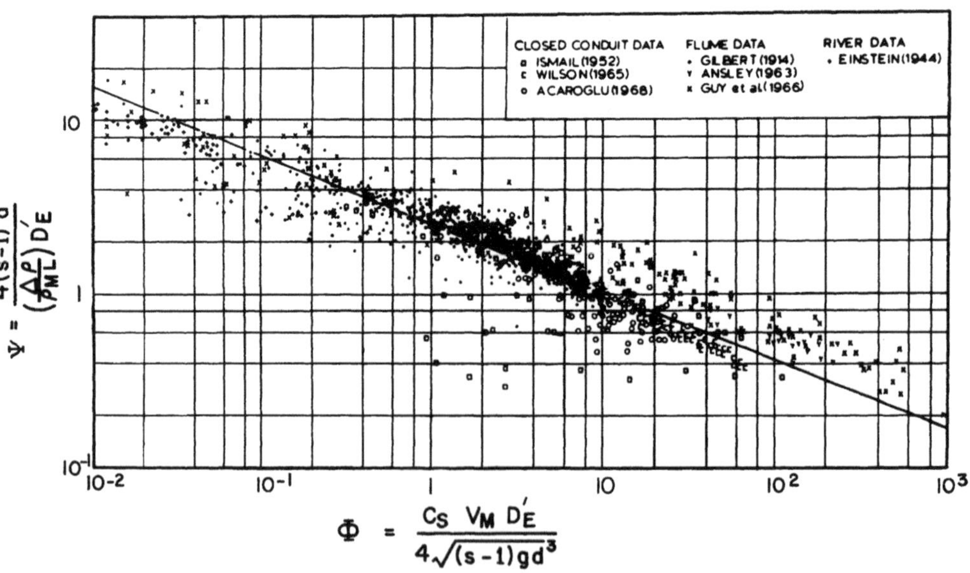

Figure 11.61. Relationship of the shear intensity parameter, Ψ, and the transport parameter, Φ. (From Graf, W. H., and E. R. Acaroglu, *Bull. Int. Assoc. Scientific Hydrology*, **13**, 20, 1968.)

on coordinates of Ψ and Re_S^*, where $\mathrm{Re}_S^* = dU^*\rho/\mu$ and $U^* = V_M\sqrt{f/2}$. Although based upon the sequence of behavior observed when a carrier

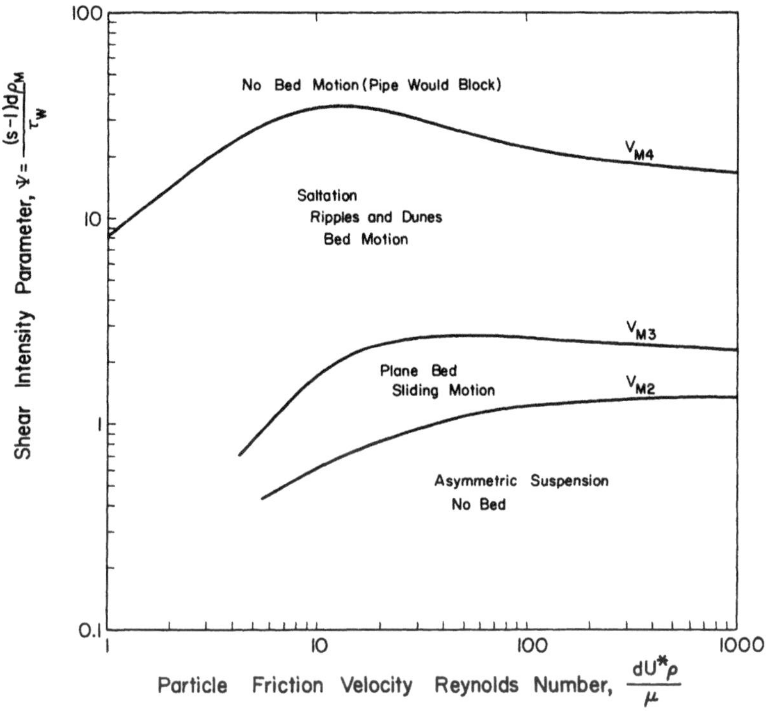

Figure 11.62. Modes of solids movement in the presence of a bed. (Modified after Acaroglu, E. R., and W. H. Graf, *Bull. Int. Assoc. Scientific Hydrology*, **13**, 123, 1968.)

liquid is passed at increasing velocity over an initially stagnant, smooth surfaced bed $h_b/D \simeq 3$, nonetheless the bed forms observed are relevant to the pipe flow of a liquid-solid mixture.

At sufficiently high values of Ψ (corresponding with low values of V_L or V_M), the bed is found to remain stagnant and its surface smooth. Considering the situation at an abscissa value of about 20 if the flow velocity is increased (Ψ is decreased) to the point where $\Psi \simeq 23.0$, a few surface particles start moving by saltation. This corresponds with our V_{M4}. With further increase in flow, moving dunes are formed on the bed surface and these increase in size as Ψ decreases to about 2.3. In this region, $2.3 < \Psi < 23.0$, an increasing proportion of the bed is in dune-like, saltation motion. Below $\Psi \simeq 2.3$ the dunes disappear, the bed surface becomes flat, some particles may become suspended above the bed, and much of the bed is in sliding motion.

Flow under conditions where $\Psi \simeq 2.3$ corresponds approximately with our V_{M3}. As the flow rate is increased still further the particles in the bed are carried more and more into suspension, and when $\Psi \simeq 1.35$ the bed is entirely eroded away and all of the solids are suspended. This corresponds with our V_{M2}.

Figure 11.62 is useful for estimating V_{M2}, V_{M3}, or V_{M4} and the details of the solids transport in cases where some estimates of the pressure gradient (required for the evaluation of Ψ) may be made. A crude estimate of Ψ and the pressure gradient may always be obtained from the correlation of Graf and Acaroglu (1968) given in Figure 11.61.

4 RECOMMENDED DESIGN METHODS

4.1 Introduction

As discussed in connection with Chapters 8, 9, and 10 the first step in designing for the pipe flow of liquid-solid or gas-solid mixtures is to determine, by the procedure outlined in Section 3.5 of Chapter 1, whether the mixture may be expected to behave as a pseudohomogeneous system. If it may, the design methods of Chapter 4, 5, or 6 will apply; otherwise the methods outlined here may be followed. Also, as discussed in Section 8.1 of Chapter 8 and 4.1 of Chapter 9, if the mixture is one of solid and liquid, or if one of solid and gas where the expected pressure drop is less than about 0.10 P_1, it may be considered incompressible and the methods discussed in the following may be applied in a single computation to the full length of the pipe. If otherwise, the equations may require integration or application in stepwise calculations to sections of the pipe.

For simplicity in presentation, we assume that the rate of flow of solids and fluid, the properties of the solids and fluid, and the diameter and length of the pipe are specified and that the design problem is to determine the flow pattern and holdup and to calculate the pressure drop. The approach to alternate problems should be apparent from what is discussed.

4.2 Entrance Effects

We noted in Section 4.2 of Chapter 9 that entrance effects of the type encountered with single-phase fluids and involving the growth of the boundary layer are not normally important in connection with the flow of fluid-solid mixtures. On the other hand, and particularly with gas-solid mixtures, the travel distance and energy required for the acceleration of the solids from essentially zero to the stabilized pipe velocity may be significant. The method of Rose and Duckworth discussed in Section 4.2 of Chapter 9 and involving Equations 9.64 and 9.65 is applicable for the estimation of these quantities.

4.3 Effect of Bends and Fittings

No methods are available for the accurate prediction of the effects of bends and fittings in the flow of fluid-solid mixtures. Approximations may be made based on the equivalent length technique as discussed in Section 8.3 of Chapter 8. (We are not dealing here with problems of local accumulation or erosion which may be encountered.)

4.4 Potential Energy or Hydrostatic Head Effects

1. If the pipe is completely horizontal

$$\Delta P_{PE} = \Delta P_{HH} = 0$$

2. If the pipe is inclined

$$\Delta P_{PE} \neq \Delta P_{HH} \neq 0$$

and an adjustment must be made to the pressure drop, ΔP_f, calculated for horizontal flow. The term ΔP_{HH}, which incorporates the effect of holdup, is the appropriate adjusting quantity, although when it is not possible to estimate it, ΔP_{PE} is a reasonable approximation. The potential energy change may be calculated directly from the appropriate term of Equation 7.60, i.e.,

$$\Delta P_{PE} = \frac{M_S + M_F}{Q_S + Q_F}\left(\frac{g}{g_c}\right) L \sin\theta$$

3. If the pipe is inclined, and if the solids are fully suspended, the hydrostatic head effect may be accounted for by using the Rose and Duckworth method (Equation 9.66 and related equations). The hydrostatic head component of the total pressure drop may be separately evaluated, if desired, by

$$\Delta P_{HH} = \frac{g}{g_c} L\rho_F \sin\theta\left[1 + \mathscr{R}\frac{V_{SF}}{V_S}\left(1 - \frac{1}{s}\right)\right]$$

In any case any inclined section of the pipe should be treated separately from the remainder of the pipe.

4. If the pipe is inclined and some or all of the solids are in the moving bed pattern,

a. the effect of an uphill incline may be estimated conservatively by

$$\Delta P_{HH} = \frac{g}{g_c} L\rho_M \sin\theta$$

where ρ_M is the *in situ* mixture density;

b. the effect of a downhill incline should be estimated neglecting the contribution of the solids to the density of the flowing mixture, i.e., by

$$\Delta P_{HH} = \frac{g}{g_c} L\rho_F \sin\theta$$

where ρ_F is the carrier fluid density. We note that for downhill flow, $\sin\theta$ is a negative quantity.

4.5 Kinetic Energy or Acceleration Effects

As mentioned in Section 4.2, kinetic energy effects may be important in the entrance section of the pipe and methods for estimating them have been referred to. For the remainder of the pipe, kinetic energy effects are small except in the case of gas-solid mixtures where the overall pressure drop is large and consequently the velocity change is significant. The method described in Section 4.5 of Chapter 9 is recommended for estimating this latter portion of the kinetic energy effect.

4.6 Flow Pattern, Holdup, and Pressure Gradient for the Steady Flow of Incompressible Mixtures (Liquid-Solid Mixtures; Gas-Solid Mixtures when $\Delta P < 0.10P_1$)

4.61 Predictions by General Methods

1. The flow pattern to be expected may be determined by estimating the appropriate of V_{M1}, V_{M2}, V_{M3}, and V_{M4} from methods discussed in Section 3.22. If it is specifically desired to design for sliding bed conditions, V_M should lie in the range $V_{M3} < V_M < V_{M2}$ but normally, and for flow in suspension, V_M should exceed V_{M2}.

2. For flow in suspension ($V_M > V_{M2}$), the pressure gradient and holdup, for both liquid-solid and gas-solid mixtures, may be estimated by the Rose and Duckworth method discussed in detail in Section 3.33 of Chapter 9.

3. If the mixture is one of liquid and granular solids, and regardless of the flow pattern, the holdup and pressure gradient may be estimated by the method of Gaessler discussed in detail in Section 3.42.

4. If the mixture is one of liquid and granular solid, and the solids are in suspension ($V_M > V_{M2}$), the design procedure of Wasp *et al.* discussed in Section 3.42 may be applied. This procedure is especially recommended for mixtures covering a wide range of particle sizes.

4.62 Confirmation by Specific Flow Pattern Methods

(A) Symmetric Suspension Flow Pattern ($V_M > V_{M1}$)

1. For either gas-solid or liquid-solid mixtures, the holdup, H, and the *in situ* volume fraction solids, E_S, may be estimated from Equations 11.47 and 7.87. The *in situ* mixture density ρ_M may be estimated from Equation 9.28.

2. For either gas-solid or liquid-solid mixtures, the pressure gradient may be estimated from Equation 11.75, with E_S replacing C_S if the holdup effect is significant and with $0.6 < K \leq 1$ depending upon the shape, concentration, and size of the solid particles as discussed in Section 3.43.

3. Alternately, for gas-solid suspensions, the method of Julian and Dukler discussed in Chapter 9 may be applied.

(B) Asymmetric Suspension Flow Pattern ($V_{M2} < V_M < V_{M1}$)

1. For liquid-solid systems where the range of particle size is narrow, the pressure gradient may be estimated from the Durand and Condolios correlation or one of its modifications. Equation 11.86 is recommended.

2. If the range of particle size is wide, the technique of Wasp *et al.* for separately treating the solids in pseudohomogeneous suspension and applying the Durand and Condolios equation separately to narrow size ranges is suggested. This is discussed in Section 3.42.

3. No method specific to the asymmetric flow pattern is known for gas-solid mixtures and the Rose and Duckworth general method (see Section 4.61) is suggested.

(C) Flow with a Bed of Solids ($V_{M3} < V_M < V_{M2}$)

This flow pattern is only important in the design of systems for liquid-solid mixtures and even here it is only with short pipe lines that design deliberately provides for the presence of a moving bed.

1. A rough estimate of the pressure drop may be made using the correlation of Graf and Acaroglu, Equation 11.137.

2. The type of bed to be expected may be determined from the flow pattern map, Figure 11.62.
3. The height of bed and the mixture velocity in the upper portion of the pipe, V_{Mu}, may be predicted using the method of Wicks involving Equation 11.89, and Figures 10.38, 10.40, and 11.56.
4. A further estimate of the pressure drop may be made, as discussed following Equation 11.103, by applying the Durand and Condolios correlation, or a modification of it, and substituting V_{Mu} for V_M.
5. Newitt's method, involving Equation 11.106, provides a still further alternative estimate of the pressure drop. Babcock's and the other comments of Section 3.45 with respect to the value of K_3 should be kept in mind.

Bibliography

Acaroglu, E. R., Doctoral Dissertation, Cornell University, Ithaca, N. Y., 1968.

Acaroglu, E. R., and W. H. Graf, *Bull. Int. Assoc. Scientific Hydrology*, **13**, 123 (1968).

Ansley, R. W., Doctoral Dissertation, University of Alberta, Edmonton, Alta., 1963.

Babcock, H. A., *Chem. Eng. Review*, **48**, 36 (1964).

Babcock, H. A., Heterogeneous Flow of Heterogeneous Solids, Paper presented at the Int. Symposium on Solid-Liquid Flow in Pipes, University of Pennsylvania, Philadelphia, Pa. March, 1968.

Bagnold, R. A., *Proc. Royal Soc. London*, **225**, 49 (1954).

Bagnold, R. A., *Royal Soc. London Phil. Trans. Ser. A*, **249**, 235 (1956).

Barth, W., *Chemie-Ing.-Techn.*, **30**, 171 (1958).

Berkowitz, N., and E. J. Jensen, *Trans. Can. Inst. Mining Met.*, **66**, 258 (1963).

Berkowitz, N., C. Moreland, and G. F. Round, *Can. J. Chem. Eng.*, **41**, 116 (1963).

Bonnington, S. T., Experiments on the Hydraulic Transport of Mixed-Size Solids, British Hydromechanics Research Association, R. R. 637, Cranfield, Bedford, October, 1959.

Boothroyd, R. G., *Trans. Inst. Chem. Engrs.*, **44**, 306 (1966).

Bruce, W. R., G. W. Hodgson, and K. A. Clark, *Trans. Can. Inst. Mining Met.*, **60**, 422 (1952).

Carstens, M. R., *A.S.C.E. Proc., Hydraulics Div.*, **95** (HY1), 275 (1969).

Clark, R. H., D. E. Charles, J. F. Richardson, and D. M. Newitt, *Trans. Inst. Chem. Engrs.*, **30**, 209 (1952).

Colorado School of Mines Research Foundation Inc., The Transportation of Solids in Steel Pipelines, Golden, Colorado, 1963.

Condolios, E., and E. E. Chapus, *Chem. Eng.*, **70** (July 8), 137 (1963).

Doig, I. D., and G. H. Roper, *Australian Chem. Eng.*, **4**, 9 (1963).

Durand, R., Minnesota Int. Hydraulics Convention, *Proc.*, p. 89, Int. Assoc. for Hydraulic Research, 1953.

Durand, R., and E. Condolios, Compte Rendue des Deuxième Journées de l'Hydraulique, Société Hydrotechnique de France, Paris, June, 1952, p. 29.

Einstein, H. A., Bed Load Transportation in Mountain Creek, Report SCS-TP-55, U.S. Dept. Agr. Soil Conservation Service, 1944.

Ellis, H. S., and G. F. Round, *Trans. Can. Inst. Mining Met.*, **66**, 345 (1963).

Faddick, R. R., M. Sc. Thesis, Queen's University, Kingston, Ontario, 1963.

Gaessler, H., Experimentelle und theoretische Untersuchungen über die Strömungsvorgänge beim Transport von Feststoffen in Flüssigkeiten durch horizontale Rohrleitungen, Doctoral Dissertation, Technische Hochschule, Karlsruhe, West Germany, 1967.

Gilbert, G. K., Transportation of Debris by Running Water, U.S. Geological Survey, Professional Paper No. 86, 1914.

Goto, K., and K. Iinoya, *Chem. Eng. Japan*, **2**, 144 (1964).

Govier, G. W., and M. E. Charles, *Engineering Journal*, **44**, 50, August (1961).

Graf, W. H., and E. R. Acaroglu, *A.S.C.E. Proc., Pipeline Div.*, **93** (PL2), 63 (1967).

Graf, W. H., and E. R. Acaroglu, *Bull. Int. Assoc. Scientific Hydrology*, **13**, 20 (1968).

Guy, H. P., D. B. Simons, and E. V. Richardson, Summary of Alluvial Channel Data From Flume Experiments, 1956-1961, U.S. Geological Survey, Professional Paper No. 462-1, 1966.

Hayden, J. W., and T. E. Stelson, Hydraulic Conveyance of Solids in Pipes, Paper presented at the Int. Symposium on Solid-Liquid Flow in Pipes, University of Pennsylvania, Philadelphia, Pa., March, 1968.

Howard, G. W., *Trans. A.S.C.E.*, **104**, 1334 (1939).

Hughmark, G. A., *Ind. Eng. Chem.*, **53**, 389 (1961).

Ismail, H. M., Doctoral Dissertation, California Institute of Technology, Pasadena, Calif., 1949.

Ismail, H. M., *A.S.C.E. Proc.*, Vol. **77**, Separate No. 56, Feb. 1951; see also *Trans. A.S.C.E.*, **117**, 409 (1952).

Julian, F. M., and A. E. Dukler, *A.I.Ch.E. Journal*, **11**, 853 (1965).

Liu, H. K., *Trans. A.S.C.E.*, **121**, 877 (1956).

McCarthy, H. E., and J. H. Olson, *Ind. Eng. Chem. Fundam.* **7**, 471 (1968).

Mehta, N. C., J. M. Smith, and E. W. Comings, *Ind. Eng. Chem.*, **49**, 986 (1957).

Newitt, D. M., J. F. Richardson, M. Abbott, and R. B. Turtle, *Trans. Inst. Chem. Engrs.*, **33**, 93 (1955).

Newitt, D. M., J. F. Richardson, and C. A. Shook, Symposium on Interaction Between Fluids and Particles, London, 1962, *Proc.*, p. 87. Published by the Institution of Chemical Engineers, London.

O'Brien, M. P., *Trans. Amer. Geophysical Union*, **14**, 487 (1933).

Peskin, R. L., and H. A. Dwyer, Study of the Mechanics of Turbulent Gas-Solid Shear Flows, Paper presented at the Annual Meeting of the American Society of Mechanical Engineers, Chicago, Ill., November, 1965.

Richardson, J. F., and M. McLeman, *Trans. Inst. Chem. Engrs.*, **38**, 257 (1960).

Rose, H. E., and R. A. Duckworth, *The Engineer*, **227** (5903), 392 (1969); **227** (5904), 430 (1969); **227** (5905), 478 (1969).

Shields, A., Mitt. Preuss. Versuchanstalt Wasserbau und Schiffbau, Berlin, 1936.

Shook, C. A., Pipelining Solids: The Design of Short-Distance Pipelines, Paper presented at the Symposium on Pipeline Transport of Solids, Canadian Society for Chemical Engineering, Toronto, November, 1969.

Shook, C. A., and S. M. Daniel, *Can. J. Chem. Eng.*, **43**, 56 (1965).

Shook, C. A., and S. M. Daniel, *Can. J. Chem. Eng.*, **47**, 196 (1969).

Shook, C. A., S. M. Daniel, J. A. Scott, and J. P. Holgate, *Can. J. Chem. Eng.*, **46**, 238 (1968).

Sinclair, C. G., Symp. Interaction Between Fluids and Particles, London, 1962, *Proc.*, p. 78.

Soo, S. L., and J. A. Regalbuto, *Can. J. Chem. Eng.*, **38**, 160 (1960).

Soo, S. L., and G. J. Trezek, *Ind. Eng. Chem. Fundam.*, **5**, 388 (1966).

Soo, S. L., G. J. Trezek, R. C. Dimick, and G. F. Hohnstreiter, *Ind. Eng. Chem. Fundam.*, **3**, 98 (1964).

Spells, K. E., *Trans. Inst. Chem. Engrs.*, **33**, 79 (1955).

Thomas, D. G., *A.I.Ch.E. Journal*, **8**, 373 (1962); **10**, 303 (1964).

Toda, M., H. Konno, S. Saito, and S. Maeda, *Int. Chem. Eng.*, **9**, 553 (1969).

Vanoni, V. A., *Trans. A.S.C.E.*, **111**, 67 (1946).

Wasp, E. J., T. C. Aude, R. H. Seiter, and T. L. Thompson, Hetero-Homogeneous Solids-Liquid Flow in the Turbulent Regime, Paper presented at the Int. Symposium on Solid-Liquid Flow in Pipes, University of Pennsylvania, Philadelphia, Pa., March, 1968.

Wasp, E. J., T. J. Regan, J. Withers, P. A. C. Cook, and J. T. Clancey, *Pipeline News*, **35**, 20 (1963).

Wasp, E. J., T. L. Thompson, and T. C. Aude, Process Basis and Economics of Pipeline Systems, Paper presented at the Annual Meeting of Canadian Institute of Mining and Metallurgy, Toronto, Ont., April, 1970.

Wen, C.-Y., and H. P. Simons, *A.I.Ch.E. Journal*, **5**, 263 (1959).

Wicks, M., Transport of Solids at Low Concentration in Horizontal Pipes, Paper presented at the Int. Symposium on Solid-Liquid Flow in Pipes, University of Pennsylvania, Philadelphia, Pa., March, 1968.

Wilson, K. C., Derivation of the Regime Equations from Relationships for Pressurized Flow by Use of the Principle of Minimum Energy Degradation Rate, Report No. 51, Dept. Civil Engineering, Queen's University, Kingston, Ont., 1965.

Wilson, K. C., *A.S.C.E. Proc., Hydraulics Div.*, **96** (HY1), 1 (1970).

Wilson, W. E., *Trans. A.S.C.E.*, **107**, 1576 (1942).

Worster, R. C., The Hydraulic Transport of Solids, *Proc.* Colloquium on Hydraulic Transport of Coal, National Coal Board, London, November, 1952.

Worster, R. C., Relationship Between Hydraulic and Pneumatic Conveying of Solids in Pipes, British Hydromechanics Research Association, SP 464, Cranfield, Bedford, September, 1953.

Worster, R. C., Measurements of the Distribution of Forces in the Horizontal Flow of Heavy Solids with Water in a Round Pipe, British Hydromechanics Research Association, Publication 477, Cranfield, Bedford, January, 1954.

Worster, R. C., and D. F. Denny, *Proc. Inst. Mech. Engrs.*, **169**, 563 (1955).

Yen, J. G., and I. Zandi, Transport of Slurries in Heterogeneous Regime, Soc. Mining Engrs, A.I.M.E., Preprint No. 69-B-70, 1969.

Zandi, I., and G. Govatos, *A.S.C.E. Proc., Hydraulics Div.*, **93** (HY3), 145 (1967).

Zenz, F. A., *Ind. Eng. Chem. Fundam.*, **3**, 65 (1964).

Zenz, F. A., and D. F. Othmer, *Fluidization and Fluid Particle Systems*, Reinhold Publishing Corp., New York, 1960.

The Flow of Capsules in Pipes

12

1 INTRODUCTION

The concept of the flow of capsules in pipelines has been proposed by workers at the Research Council of Alberta (Hodgson and Bolt, 1962; Hodgson and Charles, 1963, also Canadian Patent 673,280; Jensen and Berkowitz, 1965; Berkowitz, Brown, and Jensen, 1965; Berkowitz and Jensen, U.S. Patent 3,190,701) as a practical means for the large-scale transport of solids by pipeline. The idea developed out of observations of Charles, Govier, and Hodgson (1961), made during a study of two-phase flow of oil and water. The capsules may be cylindrical with square or shaped ends, and of a range of lengths and diameters, or they may be spherical. They may be containers of metal or plastic filled with any solid, or they may be formed out of the solid itself as in the case of castings of sulfur or formed shapes of pastes such as of coal and water (Berkowitz, Brown, and Jensen, 1965; Brown and Jensen, 1968). The capsule container may be rigid or, as in the case of a plastic container, somewhat flexible. The paste capsule would ordinarily be flexible.

The Research Council of Alberta has proposed that capsule pipelining may be practical for a wide variety of solids, including wheat and other grains, mineral solids, chemicals, sulfur, and coal. Although no commercial developments have yet been made, extensive laboratory work and theoretical analyses provide a basis for pilot development and large-scale application.

Advantages claimed for capsule pipelining as distinct from slurry pipeline flow are:

a. freedom from contamination and ease of separation of the solid,
b. lower power requirements per unit of solid transported.

Disadvantages include:

a. cost of encapsulating or forming into capsule shape,
b. mechanical complications associated with the introduction and removal of the capsules to and from the pipe and at pumping stations.

Capsule pipelining involves the flow in a pipe of a long train of capsules, the behavior of which is the resultant of the behavior of each individual capsule in the group and the hydrodynamic influence of one capsule on another. Before considering some of the actual test data that have been obtained, it is instructive to visualize what happens in the case of a single capsule, initially at rest on the lower wall in a pipeline, as the velocity of the carrier fluid is gradually increased.

With the capsule at rest the carrier fluid flowing from an upstream position is restricted to the crescent-shaped annular space between the pipe and the capsule and is caused to flow at a much greater velocity as it passes the capsule. Under these conditions, the pressure gradient over the length of the pipeline containing the capsule is significantly greater than that either upstream or downstream of the capsule. The capsule is subjected both to a pressure force, equal to the difference of the pressure forces exerted by the flowing fluid on the two ends of the capsule, and to a shear force due to the streaming of the fluid past it.

With increasing fluid velocity the capsule remains stationary until the sum of the net pressure force and the shear force acting exceeds the force due to static friction between the capsule and the pipe wall in the presence of the carrier liquid. Once this friction force is exceeded, the capsule begins to move. The liquid velocity in the free pipe, which is just sufficient to start motion of the capsule, is called the "threshold velocity." When the capsule is in motion, the friction force is reduced because the coefficient of sliding friction is lower than that of static friction. As a consequence, the capsule accelerates until a balance of forces results. The shear force exerted on the surface of the capsule is reduced because of the decreased relative velocity between the liquid and the capsule. In fact, over part of the circumferential area of the cylinder the shear force reverses in direction, since the capsule moves at a higher velocity than the liquid nearest the wall of the pipe.

For a stationary or a slowly moving capsule, the resultant of the forces acting on it is such that there is a net moment of force in a direction to lift the tail of the capsule from the bottom of the pipe. When the capsule velocity increases, however, another moment develops which is opposite in direction and which tends to lift the capsule nose. The development of this opposite moment is not fully understood but is believed to be connected with secondary or eddy flows around the capsule and with the flow in the thin liquid layer between the capsule and the pipe wall. When the capsule velocity increases sufficiently, this second moment of force becomes such that the nose of the capsule actually rises and the capsule rides on a wedge of liquid which is thicker under the nose than under the tail. The lowest velocity at which this is readily perceptible is the "lift-off velocity." At velocities above the lift-off velocity, we may think of the capsule as "planing" over the more slowly moving liquid beneath it. In the extreme, and especially with light capsules, the capsule may be lifted to an approximate concentric position.

In the following we will discuss typical flow patterns, holdup, and pressure gradient data; the mechanism of capsule flow and theoretically based relationships for capsule velocity and pressure gradients.

2 TYPICAL FLOW PATTERNS, HOLDUP, AND PRESSURE GRADIENT

2.1 *Flow Patterns*

In capsule and paste slug flow the flow patterns are established essentially by the size and shape of the capsule and the relative quantities of capsulated solid and liquid flowing in the line.

We will consider first the somewhat idealized case of capsules having the same overall density as the carrier liquid. Ellis (1964a) and Ellis and Bolt (1964) have reported fundamental work on the flow of single spherical capsules and single cylindrical capsules of various lengths and with square and ellipsoidal ends in a $1\frac{1}{4}$-inch plastic pipe. For both the spheres and the cylinders, diameter ratios from 0.39 to 0.89 were studied; the length to cylinder diameter ratios were varied from 4 to 14 for the smallest cylinder diameter, and from 1.8 to 6.25 for the largest. Water and two oils (kinematic viscosities of 6.90 and 45.0 centistokes) were used.

In nearly all instances the equidensity capsules traveled concentrically in the pipe. The only exceptions were the smallest diameter cylinders which tended to oscillate and to flow in a random nose-up or nose-down pattern. With the smallest spheres the flow pattern was extremely unstable and the spheres bounced from wall to wall, often spinning as they progressed down the pipe.

No work appears to have been published on trains of equidensity capsules, but from the observed behavior of trains of nonequal density capsules we would expect the trains to behave much the same as individual capsules.

Ellis (1964b,c) also reports results of the flow of capsules of density greater than that of the carrier fluid. Spheres of 1.19 and 2.84 specific gravity, and of the same diameter ratio mentioned above, were observed both to slide and to roll along the bottom of the pipe at modest liquid velocities, but the smallest lighter spheres became "water-borne" at velocities above about 4 ft/sec. It would appear that at sufficient velocities the spheres would tend to the axial position. Ellis' tests indicated that the less dense spheres were more prone to slide than the more dense and that the percentage of travel distance covered by sliding (as distinct from rolling) increased with liquid velocity and with decreasing ratio of sphere to pipe diameter.

The work on cylinders revealed only that the cylinders slid along the bottom of the pipe separated from it by a thin film of liquid. At sufficient liquid velocities, greater for greater density cylinders, the liquid film thickens to the point that the cylinder may be lifted from the bottom of the pipe. The noses of the capsules lifted at a water velocity of 6 to 8 ft/sec, and at higher velocities the entire capsule was lifted.

Liddle (1968) reports data on the behavior of single cylindrical capsules flowing with water in a 1.310-inch acrylic plastic pipe. He tested capsules having lengths of 3 and 6 inches; capsule to pipe diameter ratios, k, of 0.701, 0.802, and 0.902; and capsule to liquid density ratios, $s = \rho_c/\rho$, of 1.010, 1.020, 1.050, 1.100, and 1.50. Below a critical capsule or mixture velocity (the lift-off velocity), he observed a dimensionless clearance, c, between the underside of the capsule and the pipe wall of the order of 0.001 to 0.01. Above this velocity, however, he obtained reproducible photographic evidence of significant "lift-off" at the capsule nose. His data, shown in Figure 12.1, illustrate the roles of s–1, k, and L_c on the lift-off velocity, V_{ML}, for single, plain, cylindrical capsules.

Kruyer, Redberger, and Ellis (1967) have presented a theoretical analysis for the laminar flow of cylindrical capsules which we will discuss later. Their analysis permits the calculation, from the capsule velocity and the other variables, of the thickness of the laminar film, or the clearance, between the cylindrical capsule and the pipe wall. Because of the relationship of clearance to flow pattern we may refer to some of their results presented subsequently in Figure 12.20. This figure shows the calculated dimensionless clearances to be very small (from 10^{-5} to 10^{-2}), and to increase with capsule velocity and to decrease with capsule density.

Jensen and Bruce (1970) indicate that cylindrical capsules fitted with circumferential bands located near and upstream of the midpoint of their length tend to rise and plane more readily than simple cylinders. Thus, at equivalent capsule density, fluid velocity, and fluid properties, "banded"

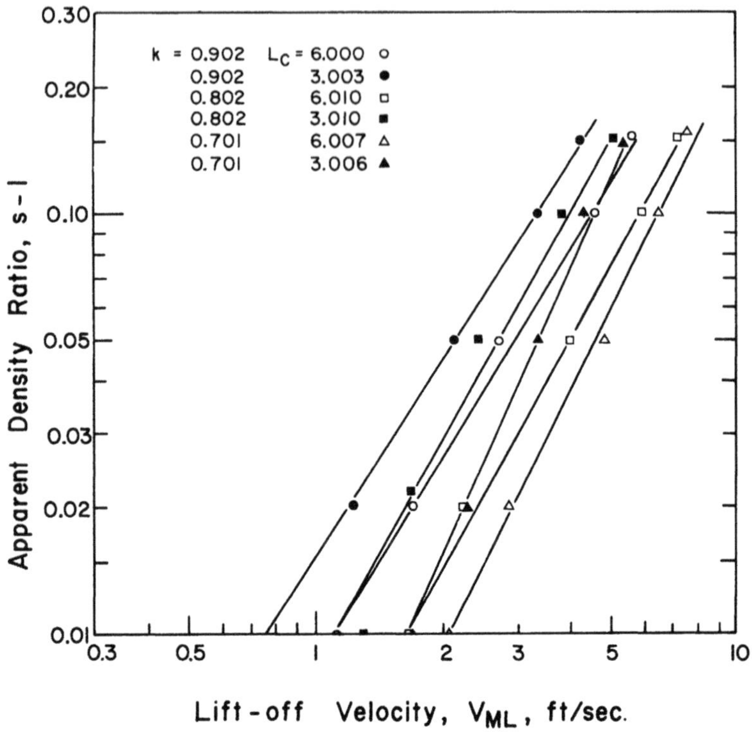

Figure 12.1. Lift-off velocity data for cylindrical capsules in water in a 1.310-inch plastic pipe. (Redrawn from Liddle, R. T., M.Sc. Thesis, Univ. Alberta, Edmonton, Alta., 1968.)

cylinders result in a slightly different flow pattern than simple cylinders. Capsules with radially extending fins (Shellene and Chu, Canadian Patent 836,381) tend to flow in a concentric position.

The above discussion relates to single capsules. Trains of capsules would be expected to exhibit much the same flow pattern behavior as relatively long individual capsules except that a train of short capsules should permit greater nose lift of the individual capsules than for a single capsule of the total length.

2.2 Holdup

2.21 Overall Continuity Considerations

Before proceeding further it is useful to define our nomenclature and consider certain overall continuity matters. Because of the greater importance of cylindrical capsules we will restrict detailed consideration to them. The extension to other shapes is straightforward. We define Q_c and Q_L, respectively, as the volume flow rate of capsules and liquid. From these we define the superficial velocities as

$$V_{SL} = \frac{4Q_L}{\pi D^2} \tag{12.1}$$

$$V_{Sc} = \frac{4Q_c}{\pi D^2} \tag{12.2}$$

$$V_M = \frac{4(Q_c + Q_L)}{\pi D^2} = V_{SL} + V_{Sc} \tag{12.3}$$

Under steady flow conditions with incompressible capsules and liquid, V_M is the constant average bulk velocity in the pipe.

Figure 12.2 illustrates a typical capsule train flowing in a pipe with the capsules moving at an actual velocity V_c and the liquid surrounding the capsule at an actual average velocity V_L. If L_c is the capsule length and ΔL the space between capsules, the volumes of capsule and liquid within the length $L_c + \Delta L$, are respectively,

$$\text{volume of capsule} = \frac{\pi}{4} d^2 L_c$$

$$\text{volume of liquid} = \frac{\pi}{4} D^2(L_c + \Delta L) - \frac{\pi}{4} d^2 L_c$$

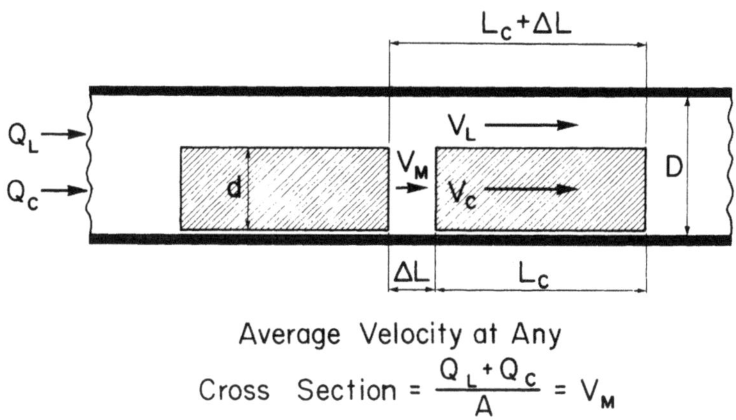

Figure 12.2. Schematic diagram of capsules flowing in a pipe.

Continuity considerations require that

$$V_c = \frac{Q_c(L_c + \Delta L)}{\left(\frac{\pi d^2}{4}\right) L_c} \tag{12.4}$$

$$V_L = \frac{Q_L(L_c + \Delta L)}{\frac{\pi D^2}{4}(L_c + \Delta L) - \frac{\pi}{4} d^2 L_c} \tag{12.5}$$

The *in situ* volume fraction of capsules is simply

$$\bar{E}_c = \left(\frac{d}{D}\right)^2 \frac{L_c}{L_c + \Delta L} = k^2 \frac{L_c}{L_c + \Delta L} \tag{12.6}$$

where

$$k = \frac{d}{D}$$

When $\Delta L = 0$, as for a continuous train of capsules

$$\bar{E}_c = k^2 \tag{12.7}$$

The holdup ratio, as defined in Chapter 7, is

$$H = \frac{\text{in situ volume ratio of liquid to capsule}}{\text{input volume ratio of liquid to capsule}}$$

$$= \frac{(L_c + \Delta L)D^2 - L_c d^2}{L_c d^2} \left(\frac{Q_c}{Q_L}\right) \tag{12.8}$$

Utilizing Equations 12.4 and 12.5 to define Q_L/Q_c in terms of V_L and V_c, we obtain

$$H = \frac{V_c}{V_L} \tag{12.9}$$

which is analogous to relations derived for the flow of gas-liquid and liquid-solid mixtures.

Employing Equations 12.3, 12.4, and 12.5, it is possible to express V_M in terms of V_c and V_L and to express V_L in terms of V_c and V_M. Thus

$$V_M = V_c k^2 \left(\frac{L_c}{L_c + \Delta L}\right) + V_L\left(1 - k^2 \frac{L_c}{L_c + \Delta L}\right) \tag{12.10}$$

and we may express Equation 12.9 in terms of V_M as

$$H = \frac{1 - k^2\left(\dfrac{L_c}{L_c + \Delta L}\right)}{\dfrac{V_M}{V_c} - k^2\left(\dfrac{L_c}{L_c + \Delta L}\right)} \tag{12.11}$$

The Research Council of Alberta workers have found it convenient to discuss the holdup phenomenon in terms of the simple velocity ratio $R_V = V_c/V_M$. From Equations 12.4, 12.3, and 12.6 we may show this ratio to be

$$R_V = \frac{Q_c}{Q_c + Q_L}\left(\frac{1}{\bar{E}_c}\right)$$

or, in other words, it is identical with the Bankoff K factor discussed in Chapter 7. Also from Equation 12.11 we have

$$R_V = \frac{(L_c + \Delta L)H}{L_c + \Delta L - k^2 L_c + k^2 L_c H} \tag{12.12}$$

or, with $\Delta L = 0$ for a continuous train of capsules,

$$R_V = \frac{H}{1 - k^2 + k^2 H} \tag{12.13}$$

2.22 Typical Holdup Data

Data of Ellis (1964a,b,c) and others on the flow of single, both equidensity and heavy spherical, and cylindrical capsules in water and in oil in a $1\frac{1}{4}$-inch pipe illustrate the holdup phenomenon. These workers observed that under many conditions the velocity of the capsule exceeded the bulk velocity of the mixture. Recognizing a possible economic significance in the observation, they studied in detail the effect of capsule size, shape and density, carrier liquid density and viscosity, and bulk velocity on the capsule velocity.

Figure 12.3 taken from Ellis shows the variation of R_V with V_M for equidensity flat-ended cylinders, 7 inches long, with water as the carrier liquid. Data for equidensity spheres are similar. Figure 12.4, also from Ellis, shows the effect of cylinder length on R_V at a fixed bulk velocity. Ellis observed relatively little effect of capsule end shape on the R_V ratio except at small k ratios.

Ellis and Bolt (1964) present similar data for equidensity spheres and cylinders in two oils having viscosities of 6.90 and 45.0 centistokes.

The data for spheres heavier than the carrier liquid exhibit different behavior at low liquid velocities but approach the behavior of equal density

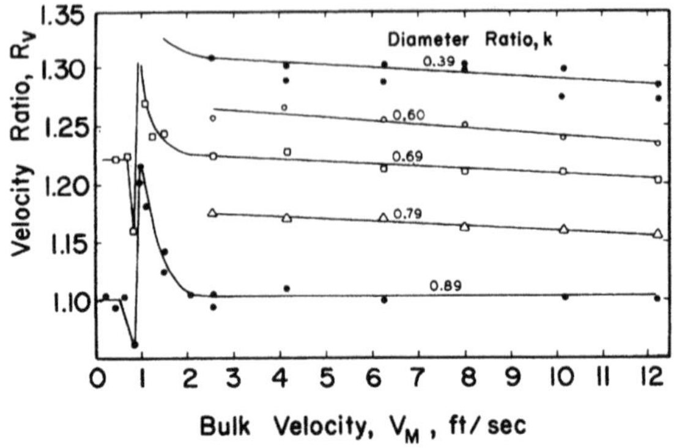

Figure 12.3. Velocity ratios for 7-inch cylindrical capsules in a $1\frac{1}{4}$-inch plastic pipe. (From Ellis, H. S., *Can. J. Chem. Eng.*, **42**, 1, 1964.)

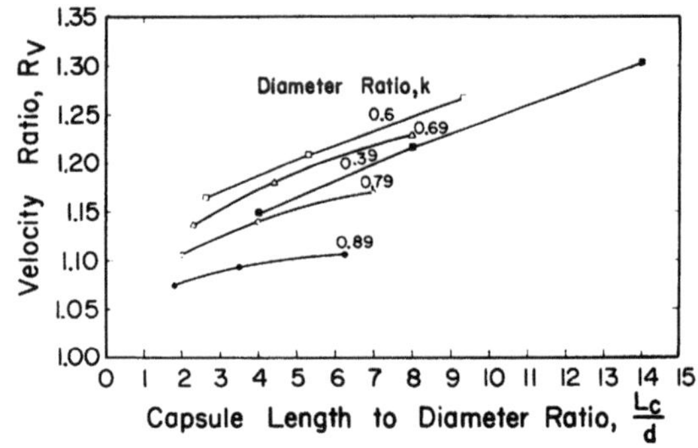

Figure 12.4. Effect of capsule length on velocity ratio. Plastic pipe, $D = 1\frac{1}{4}$ in., constant bulk velocity = 4.14 ft/sec. (From Ellis, H. S., *Can. J. Chem. Eng.*, **42**, 1, 1964.)

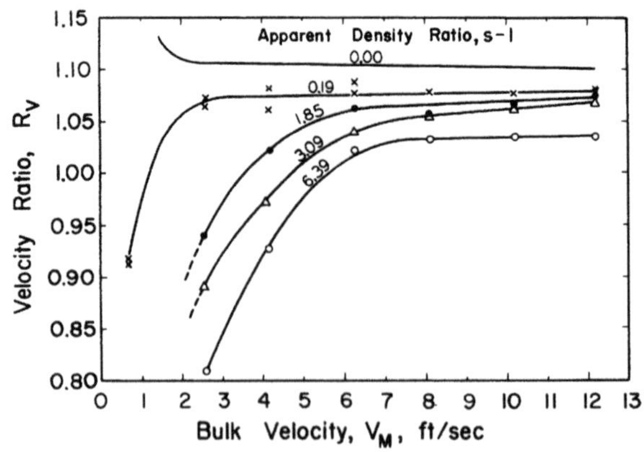

Figure 12.5. Effect of capsule density on velocity ratio. Plastic pipe, $D = 1\frac{1}{4}$ in., 7-inch cylindrical capsules; diameter ratio, $k = 0.89$. (From Ellis, H. S., *Can. J. Chem. Eng.*, **42**, 69, 1964.)

spheres at higher velocities. Figure 12.5 illustrates the effect of capsule density for cylindrical capsules.

Figures 12.6 and 12.7 summarize much of the data on spheres ($k = 0.90$) and cylinders ($k = 0.89$, $L_c = 7$ inches) of different densities carried in water and in oils of different densities and viscosities. These figures should not be

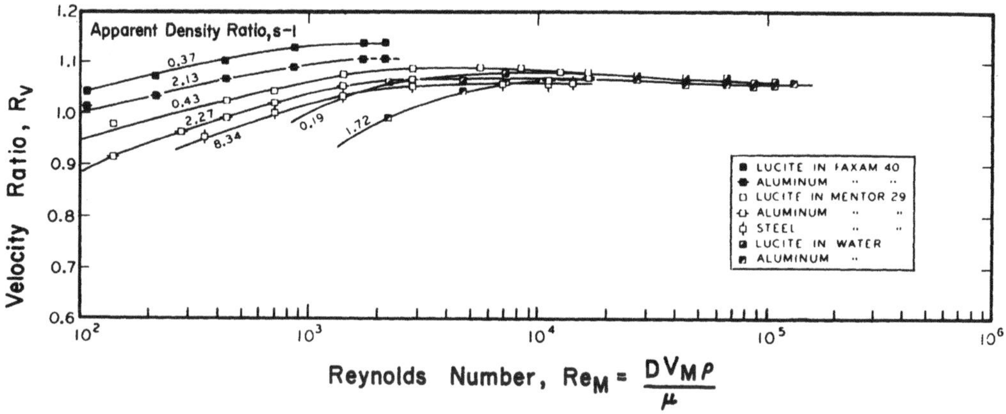

Figure 12.6. Velocity ratios for spherical capsules. Plastic pipe, $D = 1\frac{1}{4}$ in., diameter ratio $k = 0.90$.
Faxam 40, specific gravity = 0.865,
viscosity = 39.9 centipoises ;
Mentor 29, specificgravity = 0.828,
viscosity = 5.56 centipoises.
(From Round, G. F., and L. H. Bolt, *Can. J. Chem. Eng.*, **43**, 197, 1965.)

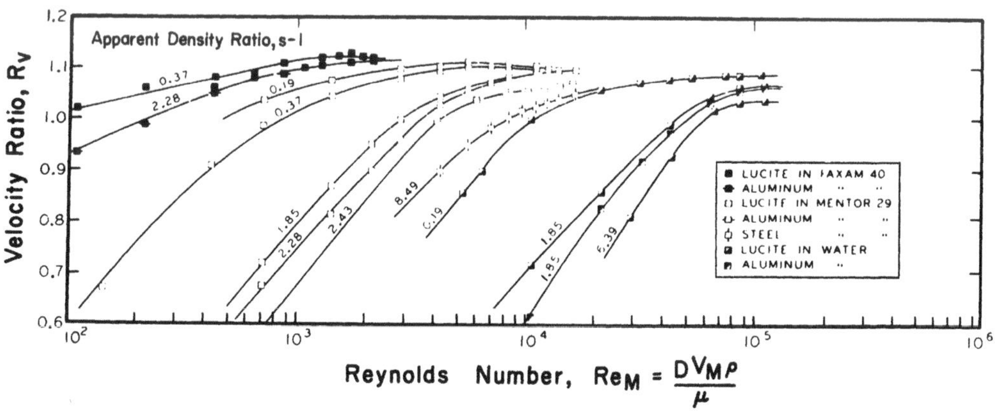

Figure 12.7. Velocity ratios for 7-inch cylindrical capsules. Plastic pipe, $D = 1\frac{1}{4}$ in., diameter ratio, $k = 0.89$.
Faxam 40, specific gravity = 0.865,
viscosity = 39.9 centipoises;
Mentor 29, specific gravity = 0.828,
viscosity = 5.56 centipoises.
(From Round, G. F., and L. H. Bolt, *Can. J. Chem. Eng.*, **43**, 197, 1965.)

interpreted as general correlations but merely as displays of typical data. In particular they apply only to capsule flowing in a plastic pipe 1.254 inches in diameter. The Reynolds number is not suggested as a correlating device to account for the influence of liquid viscosity or the role of pipe diameter.

Hodgson *et al.* (1967) report velocity ratio data for a cylindrical capsule of density 1.95 times that of the carrier fluid, $k = 0.83$, and $L_c/d = 3.17$ for a $1\frac{1}{4}$-inch, a 4-inch, and a 20-inch pipe. Their data are shown in Figure 12.8. They conclude from these curves that, at constant bulk or mixture velocity, the velocity ratio falls progressively as the pipeline diameter increases.

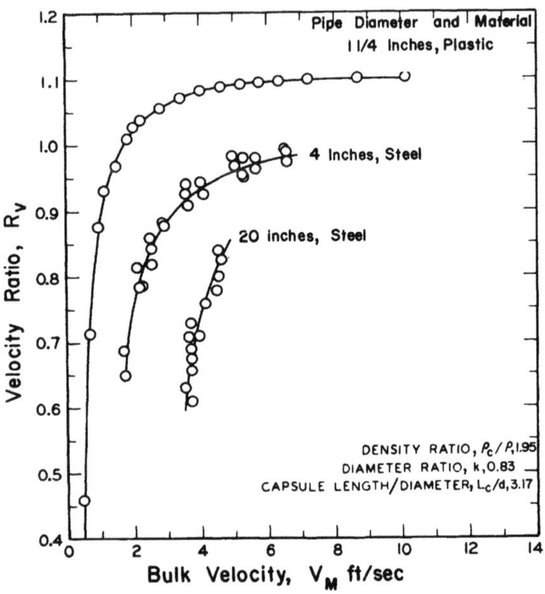

Figure 12.8. Velocity ratios for cylindrical capsules. (From Hodgson, G. W., J. Kruyer, A. A. Roehl, G. F. Round and H. S. Ellis, *J. Can. Petrol. Technol.*, **6**, 1, 1967.)

The Research Council of Alberta is known to have carried out extensive tests with long trains of $3\frac{5}{8}$ inches in diameter, 15 inch long capsules in a 4-inch commercial steel pipe (*Oilweek*, Sept. 18, 1967; Jensen and Bruce, 1970). The quantitative results of these tests have not yet (June 15, 1971) been made public. On the basis of other published data as summarized in Figures 12.6, 12.7, and 12.8, however, we would expect cylindrical capsules of practical diameter ratios ($0.80 < k < 0.90$) and normal specific gravity (say 1.1 to 2.5), to flow at much the same velocity as or slightly higher than the mixture provided only that the mixture velocity is not too far below the lift-off value.

2.3 Pressure Gradient

Although the early work of the Research Council of Alberta Concentrated on the velocity ratio effect, more recently its work has been extended to measurements of pressure gradient.

Hodgson and Charles (1963), referring to the work of Charles, Govier, and Hodgson (1961), reasoned that, for concentrically flowing equidensity capsules, the pressure gradient in turbulent flow would be expected to be somewhat less than that with the carrier liquid flowing alone. They plot the data of the latter group for bubble, slug or concentric flow in terms of a pressure gradient ratio, R_P = ratio of observed two-phase pressure drop to the pressure drop for water flowing alone, as in Figure 12.9.

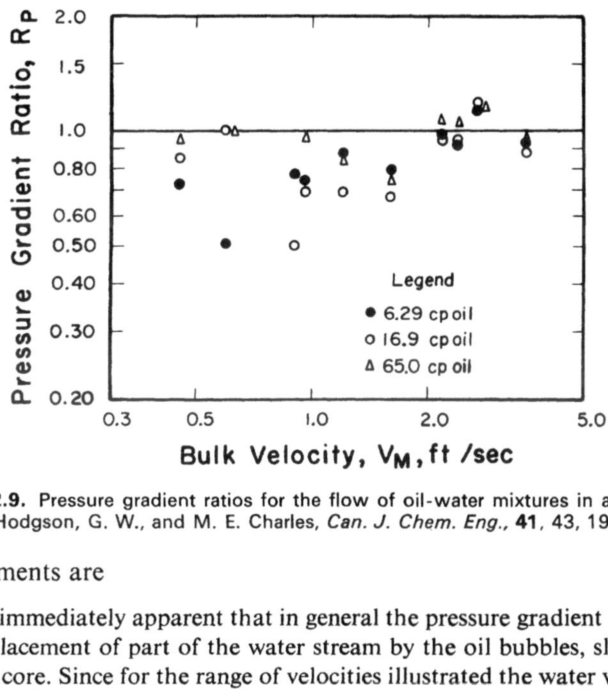

Figure 12.9. Pressure gradient ratios for the flow of oil-water mixtures in a 1-inch plastic pipe. (From Hodgson, G. W., and M. E. Charles, *Can. J. Chem. Eng.*, **41**, 43, 1963.)

Their comments are

> It is immediately apparent that in general the pressure gradient is reduced by the replacement of part of the water stream by the oil bubbles, slugs and concentric core. Since for the range of velocities illustrated the water was always in turbulent flow (a velocity of 1.0 ft/sec corresponds to a Reynolds number of approximately 10,000) it is reasonable to conclude that the presence of the semi-rigid oil bodies eliminates the turbulence within the portion of the stream occupied by the oil, in addition to suppressing the turbulence in the continuous water phase, at least to the extent of the formation of a laminar sublayer adjacent to the water-oil interface. It is further apparent that the greatest reduction in pressure gradient occurs for Reynolds numbers based on the overall velocity and the physical properties of the water, up to about 15,000 above

which the pressure gradient ratio is approximately unity. As would be expected, no definite effect of the magnitude of the oil viscosity on the pressure gradient is evident.

On the other hand, for conditions in which the basic water stream is in laminar flow (for which few reliable data were obtained) the replacement of part of the water stream by semi-rigid oil bodies would be expected to increase the pressure gradient in the water phase between the oil bodies and the pipe wall.

Kruyer *et al.* (1967) present some data on the pressure gradient for single cylindrical capsules in ½- to 4-inch pipes and for a variety of k and L_c/d ratios. The three curves in the lower left-hand side of Figure 12.18, presented later, give typical data for a ½-inch line for conditions where the liquid flow is laminar. The data were obtained from tests involving the flow of single capsules preceded and followed by the carrier liquid. All such results are presented as though for a continuous tight-packed train (i.e., $\Delta L = 0$) in the form of $(\Delta P/L)_c$ calculated as

$$\left(\frac{\Delta P}{L}\right)_c = \frac{\Delta P' - L'\left(\frac{\Delta P}{L}\right)_L}{L_c} + \left(\frac{\Delta P}{L}\right)_L \tag{12.14}$$

where L' is the length of pipeline over which the total pressure drop is $\Delta P'$, $\left(\frac{\Delta P}{L}\right)_L$ is the pressure gradient for the carrier liquid flowing at velocity V_M.

Further typical data are the data points and dashed line of Figure 12.19 also discussed later, for a 48-inch cylinder of aluminum in a 4.03-inch pipe. The general form of the data in the laminar region is the same as that for the ½-inch pipe.

Pressure drop data were also measured in the previously mentioned Research Council tests on trains of cylindrical capsules. These data are not yet generally available but on hydrodynamic grounds we would not anticipate the pressure gradient to be significantly dependent on train length. The single capsule data of Kruyer *et al.* may be considered at least generally representative of what would be obtained with long trains.

Under any capsule flow condition where any part of a capsule is in contact or very close proximity to the wall of the pipe we would expect an effect on the pressure gradient due to the surface condition of both the capsule and the inner wall of the pipe. Rough surfaces on either the capsule or the pipe would result in higher pressure gradients than would be observed with smooth surfaces. Jensen and Bruce (1970) report pressure gradients for steel and for plastic capsules of identical dimensions and weight flowing in water in a copper pipe 1.267 inches in diameter. They found that the pressure gradients for the steel capsules were 50 percent higher than for the plastic capsules. They also observed higher pressure gradients for identical steel capsules in rough steel pipe than in a smooth steel pipe. The exact effect will be dependent on the frictional characteristics of the particular surfaces and the normal force between them.

3 GENERAL ANALYSES AND CORRELATIONS

In this section we will present the relevant theoretical analyses of the flow of concentric and eccentric capsules and extend it in the direction suitable for practical application. We will deal with the holdup and pressure gradient phenomena together.

3.1 Spherical Capsules

Apart from the work of Ellis (1964a), and Ellis and Bolt (1964) already reported, no specific correlations are available for the prediction of holdup of equidensity, spherical capsules in liquids. Notwithstanding this, for concentrically flowing capsules, the analysis of Zuber and Findlay (1965) discussed in Section 7.42 of Chapter 7 is directly applicable. From Equation 7.104 with $V_{\alpha j} = 0$ we may write, as in Section 7.4 of Chapter 10,

$$V_c = C_0 V_M \tag{12.15}$$

The distribution coefficient C_0 will be dependent upon both the liquid velocity profile and the capsule diameter ratio, k. For normal velocities and for values of k in the normal high range of 0.80 to 0.90 analogy with the flow of Taylor bubbles in vertical pipes suggests that $C_0 = 1.2$. We will see that this gives a result very similar to that which may be deduced by a more detailed consideration of the velocity profile for the case of concentrically flowing cylindrical capsules.

Thinking in terms of the distribution coefficient C_0, for mixture velocities above the lift-off velocity, V_{ML}, we could apply Equation 12.15 recognizing that C_0 would be somewhat less than 1.2, but greater than 1.0 and probably in the range of 1.10 to 1.15 for k values of 0.80 to 0.90, when the capsule is eccentrically positioned. This would seem to be the practical approach to the estimation of V_c at the present time provided V_M is near or exceeds V_{ML}.

No analysis of the pressure gradient for the flow of either concentric or eccentric spherical capsules is known. In the former case, however, where there is no energy loss due to solid friction, it is reasonable to assume that the pressure gradient in the presence of a capsule train would be close to that which would occur with the flow of a single-phase liquid of mixture density and at the mixture velocity. Specifically we might expect a pressure gradient in a horizontal line

$$\left(\frac{\Delta P}{L}\right)_c = \frac{2f\rho_{Mc} V_M^2}{g_c D}$$

where f is the single-phase friction factor based upon $\mathrm{Re}_L = DV_M \rho_L/\mu_L$ and ρ_{Mc} is the density of the capsule-liquid mixture. It may readily be shown that capsule-liquid mixture densities are given by

$$\rho_{Mc} = \rho[k^2(s-1)+1] \qquad \text{for cylinders in contact} \tag{12.16}$$

$$\rho_{Ms} = \rho[\tfrac{2}{3}k^2(s-1)+1] \qquad \text{for spheres in contact} \tag{12.17}$$

This is analogous to our treatment of the symmetric concentration flow pattern in Chapter 11. It neglects the effects of distortion of the liquid velocity profile and possible suppression of large-scale turbulence. Even for the case of eccentric spherical capsules, if we assume that the motion is largely one of rolling or that the contribution of the solid-to-solid friction is small relative to the fluid friction, the above approach should be realistic. This should be so particularly for $V_M > V_{ML}$.

3.2 Concentric Cylindrical Capsules

For cylindrical capsules flowing in a concentric position, considerable progress has been made in the development of theoretical correlations both for laminar and for turbulent flow.

3.21 Velocity and Holdup Analysis—Laminar Flow

Charles (1963) has presented an analysis of the concentric flow of long cylindrical capsules carried in an equidensity liquid in laminar motion. The steady-state unidirectional form of the Navier–Stokes equation (Equation 3.11) is

$$\frac{\partial}{\partial r}\left(r\frac{\partial u}{\partial r}\right) = \frac{g_c}{\mu}\left(\frac{\Delta P}{L}\right)_c r \tag{12.18}$$

where the subscript c refers to the capsule system. The boundary conditions are

a. $u = 0$, when $r = R$
b. $u = V_c$, when $r = r_c$

and, as Epstein (1963) has shown,

$$\text{c.}\quad g_c\left(\frac{\Delta P}{L}\right)_c \pi r_c^2 L = -\mu\left(\frac{du}{dr}\right)_{r_c} 2\pi r_c L \tag{12.19}$$

Equation 12.19 imposes the limitation that the cylinder is free flowing, i.e., propelled by the fluid alone.

Integration of Equation 12.18 and incorporation of the boundary conditions lead to

$$u = \frac{g_c}{4\mu}\left(\frac{\Delta P}{L}\right)_c (R^2 - r^2) \qquad \text{for } r_c < r < R \tag{12.20}$$

$$V_c = \frac{g_c}{4\mu}\left(\frac{\Delta P}{L}\right)_c (R^2 - r^2) \qquad \text{for } r < r_c \tag{12.21}$$

The discharge terms are

$$Q_{\text{ann}} = 2\pi \int_{r_c}^{R} urdr = \frac{\pi}{2} V_c(R^2 - r_c^2) \tag{12.22}$$

$$Q_c = \pi r_c^2 V_c \tag{12.23}$$

from which we have

$$V_M = \frac{Q_{\text{ann}} + Q_c}{\pi R^2} \tag{12.24}$$

$$V_M = \frac{R^2 + r_c^2}{2R^2} V_c \tag{12.25}$$

or

$$R_V = \frac{V_C}{V_M} = \frac{2}{1 + k^2} \tag{12.26}$$

From Equation 12.26 we observe that $V_c = 2V_M$ when $k = 0$ as for normal laminar flow, and $V_c = V_M$ when $k = 1$.

We may note from Equations 12.20 and 12.21, and recognizing that $(\Delta P/L)_c > (\Delta P/L)_L$, that the velocities and the velocity gradient in the annular space are greater than in the same radial region for a liquid flowing at the average velocity, V_M. If we visualize the velocity profile obtained for the annular region (Equation 12.20) extended over the complete pipe cross section we obtain

$$u = \frac{g_c}{4\mu}\left(\frac{\Delta P}{L}\right)_c (R^2 - r^2) \qquad \text{for } 0 < r < R \tag{12.27}$$

The fictitious center line velocity obtained from this relationship is

$$u'_M = \frac{g_c}{4\mu}\left(\frac{\Delta P}{L}\right)_c R^2 = \frac{V_c}{1 - k^2} \tag{12.28}$$

where Equation 12.21 is employed to relate $(\Delta P/L)_c$ to V_c. The mean or bulk velocity corresponding with this profile is

$$V'_M = 0.5u'_M = \frac{0.5V_c}{1 - k^2} \tag{12.29}$$

and from Equation 12.26 we obtain

$$V'_M = \left(\frac{2V_M}{1 + k^2}\right)\frac{1}{2(1 - k^2)} = \frac{V_M}{1 - k^4} \tag{12.30}$$

Thus a liquid flowing in a pipe at a bulk velocity of V'_M would result in the same velocity profile (and consequently the same shear stresses) near the pipe wall as a liquid-capsule combination flowing at a bulk or average velocity of V_M. The relationship between V'_M and V_M is shown in Figure 12.10, taken from the work of Charles (1963).

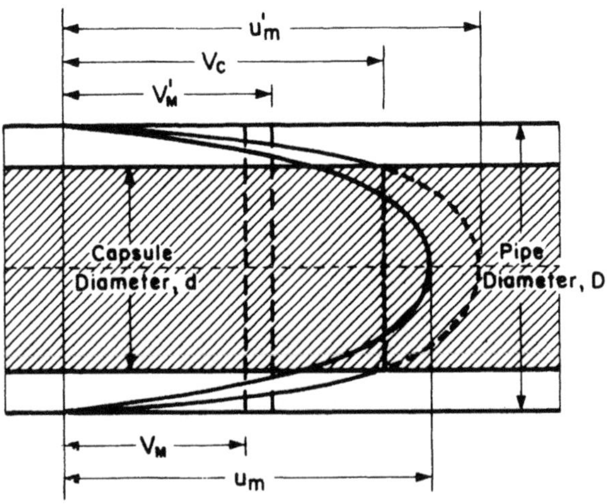

Figure 12.10. Schematic velocity profiles in the concentric capsule and the capsule-free systems. (From Charles, M. E., *Can. J. Chem. Eng.*, **41**, 46, 1963.)

3.22 Velocity and Holdup Analysis—Turbulent Flow

A similar analysis may be made when the fluid is in turbulent flow provided a velocity profile is assumed. Charles chose the power law profile

$$u = u'_M\left(1 - \frac{r}{R}\right)^{1/7} = \frac{V_c}{(1-k)^{1/7}}\left(1 - \frac{r}{R}\right)^{1/7} \tag{12.31}$$

Following the procedure described in Section 3.21 he obtained the following relationship between V_M and V_c:

$$V_M = V_c\left[\frac{4}{7}k(1-k) + \frac{49}{60}(1-k)^2 + k^2\right] \tag{12.32}$$

Kennedy (1966) has noted that the analysis of Charles (1963) did not account for the "slip," V_R, in the viscous sublayer next to the cylindrical capsule. Figure 12.11 shows in schematic fashion the probable range of the actual profiles which would be expected in the space between the capsule and the pipe wall and those assumed by Charles and by Kennedy. The uppermost sketch of the figure shows what would be expected, including the laminar sublayers at each boundary, if the average velocity in the liquid-filled space is greater than that of the capsule. The second sketch shows the converse case which would be representative for a concentrically flowing cylinder where $V_{\text{ann}} < V_c$. These sketches confirm Kennedy's contention

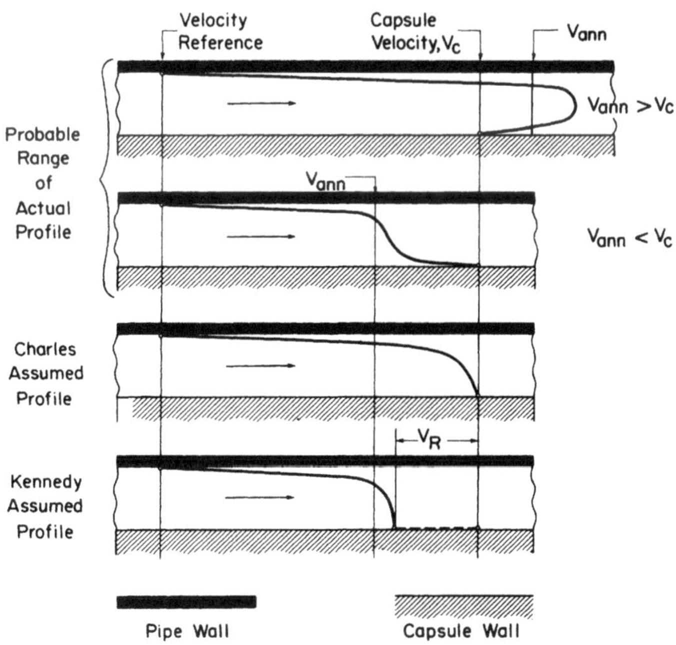

Figure 12.11. Schematic comparison of velocity profiles between pipe wall and capsule.

that an allowance for "slip" at the capsule wall is appropriate for the concentric case. (Nonetheless, as we will see, Charles' approximation is useful for eccentric capsules where the velocity profile varies circumferentially from extremes corresponding to the two upper sketches.) Here we will follow Kennedy's analysis noting that it reduces to that of Charles if V_R is put equal to zero. This leads to the following equation relating V_M and V_c:

$$V_M = (V_c - V_R)\frac{7}{4}k(1-k) + \frac{49}{60}(1-k)^2 + V_c k^2 \tag{12.33}$$

We may note that $V_M = 0.82(V_c - V_R)$ for $k = 0$ and $V_M = V_c$ for $k = 1.0$.

In order to solve for V_c we must first evaluate V_R. Kennedy has developed the equation

$$V_R = \frac{B\sqrt{\dfrac{dg_c}{4\rho}\left|\dfrac{dP}{dL}\right|}}{1.22(1-k)^{1/7}\left[\dfrac{7}{4}k(1-k) + \dfrac{49}{60}(1-k)^2 + k^2\right] + \dfrac{Bk^2}{V_M}\sqrt{\dfrac{dg_c}{4\rho}\left|\dfrac{dP}{dL}\right|}} \tag{12.34}$$

where B is a constant defined by Kaufmann (1963) as

$$V_R = B\sqrt{\frac{\tau_c g_c}{\rho}} \tag{12.35}$$

and has a value between 11 and 13. Kennedy assumes that $B = 12$.

Following the same procedure as in Section 3.21, we have

$$V'_M = 0.82u'_M = 0.82\,\frac{V_c - V_R}{(1-k)^{1/7}} \tag{12.36}$$

Utilizing Equation 12.33 to relate V_c to V_M, we obtain

$$V'_M = \frac{0.82(V_M - k^2 V_R)}{\left[\dfrac{7}{4}k(1-k) + \dfrac{49}{60}(1-k)^2 + k^2\right](1-k)^{1/7}} \tag{12.37}$$

Again this equation reduces to one proposed by Charles if V_R is taken zero.

3.23 Pressure Gradient Analysis—Laminar and Turbulent Flow

The equivalent liquid concept discussed in Section 3.1 would appear applicable to the flow of cylinders under conditions where they are lifted free of the pipe wall and approach concentricity. The analyses of Charles and of Kennedy provide alternates having a stronger theoretical basis.

Charles extended his analysis of the concentric, free flowing cylinder to predict the pressure gradients to be expected. He identified four possible flow cases as follows:

Model 1. Laminar flow exists in both the annulus between the capsule and the pipe and in the capsule-free system.

Model 2. Turbulent flow exists in both the annulus and in the capsule-free system.

Model 3. Laminar flow exists in the annulus, whereas turbulent flow prevails in the capsule-free system.

Model 4. Turbulent flow exists in the annulus and laminar flow prevails in the capsule-free system.

In Figure 12.12 Charles shows the ranges of the average Reynolds number and k corresponding with the four models. Model 1 applies at Reynolds numbers less than 2000 for all values of k, whereas Model 3 applies to higher average Reynolds numbers for high values of k such that the laminar sub-

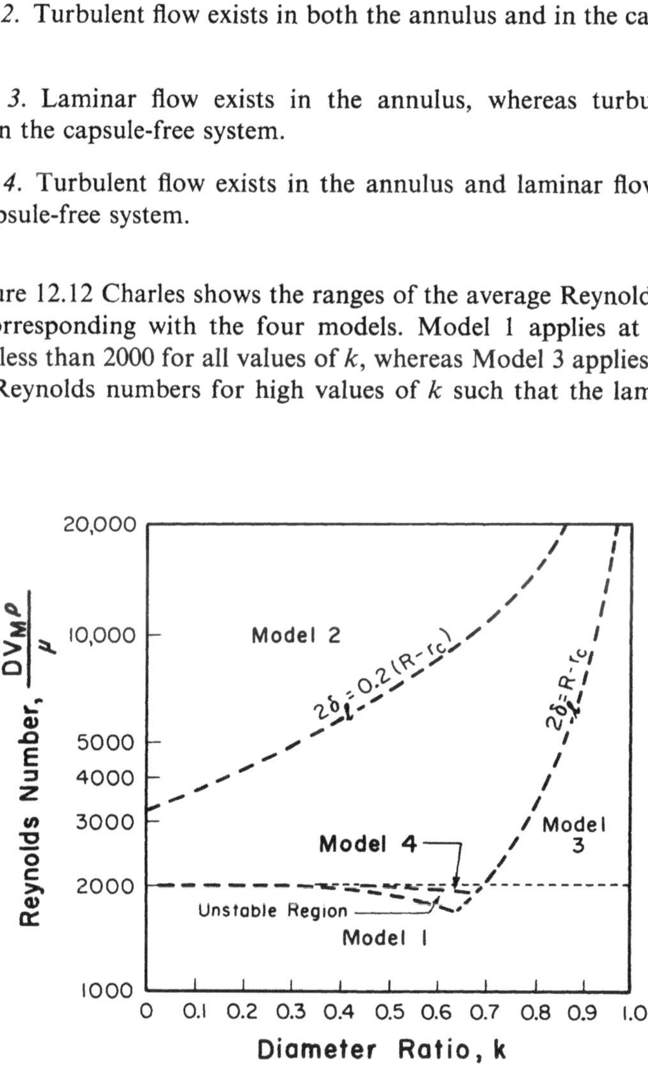

Figure 12.12. Regimes of applicability of the different models of Charles. (From Charles, M. E., *Can. J. Chem. Eng.*, **41**, 46, 1963.)

layers at the pipe and cylinder wall touch one another. The latter region is defined on the assumption that the two laminar sublayers are each of the thickness given by Coulson and Richardson (1954) as

$$\frac{\delta_l}{D} = 31\left(\frac{DV_M\rho}{\mu}\right)^{-7/8} \tag{12.38}$$

noting that $k = 1-(4\delta_l/D)$.

We see that Equation 12.38 presents essentially the same relationship as would be obtained from the universal velocity profile. The transition from laminar to turbulent flow is found experimentally to begin at $u^+ = y^+ = 5$ (see Chapter 4, Section 4.12) which may be rewritten as

$$\frac{\delta_l}{D} = \frac{5}{\text{Re}}\sqrt{\frac{2}{f}}$$

Using the Blasius equation (Equation 4.33) to define the friction factor, we obtain

$$\frac{\delta_L}{D} = 25.2\left(\frac{DV_M\rho}{\mu}\right)^{-7/8}$$

The line indicating the limit of this laminar flow region is labeled $2\delta_l = R - r_c$ in Figure 12.12. As a guide to the region where the bulk of the fluid in the annulus flows in turbulent motion, a line labeled $2\delta_l = 0.20(R - r_c)$ is indicated. For regions to the left of this line Model 2 is applicable. The region between the two lines may be taken as one where both laminar and turbulent motion is significant and something between Models 2 and 3 applies. Model 4 applies only to the small region so labeled in the figure to Reynolds numbers less than 2100 and where $0.4 < k < 0.7$. In this region the true bulk Reynolds number $DV'_M\rho/\mu$ may exceed 2100 even though $DV_M\rho/\mu$ is less than 2100. This region is not important and Model 4 will not be discussed further.

Considering Model 1 which covers the true laminar situation, the theoretical analysis of Section 3.21 applies. Recognizing the linearity between velocity and pressure gradient for horizontal laminar flow (Equation 4.21), we obtain from Equation 12.30

$$R_P = \frac{\left(\frac{\Delta P}{L}\right)_c}{\left(\frac{\Delta P}{L}\right)_L} = \frac{V'_M}{V_M} = \frac{1}{1-k^4} \tag{12.39}$$

This equation indicates that R_P remains close to 1 for k values below about 0.6 but increases to about 1.7 when $k = 0.8$ and 2.9 when $k = 0.9$.

In the case of Model 2 we may apply the relationship between velocity and pressure gradient suggested by the Blasius equation. Thus

$$R_P = \left(\frac{V'_M}{V_M}\right)^{1.75} = \left\{\frac{0.82(1-k^2)\dfrac{V_R}{V_M}}{\left[\dfrac{7}{4}k(1-k) + \dfrac{49}{60}(1-k)^2 + k^2\right](1-k)^{1/7}}\right\}^{1.75} \tag{12.40}$$

where Equation 12.37 due to Kennedy (1966) is employed to relate V_M to V'_M. Equation 12.40 predicts a decrease in R_P with increase in k, the reverse of the original prediction by Charles neglecting V_R. It shows (Kennedy, 1966) that R_P is less than 1 for normal values of k, being about 0.65 when $k = 0.8$, and 0.60 when $k = 0.9$.

For Model 3 it is assumed that the turbulence in the free liquid stream is so suppressed by the presence of the capsule that laminar flow exists in the annulus. For the flow of the liquid in the full cross section we may assume the Blasius equation and write

$$\left(\frac{\Delta P}{L}\right)_L = 2(0.079)\left(\frac{\mu}{DV_M\rho}\right)^{0.25}\frac{V_M^2\rho}{g_c D} \tag{12.41}$$

In the annulus with turbulence suppressed the pressure gradient is (from Equation 12.21)

$$\left(\frac{\Delta P}{L}\right)_c = \frac{4\mu}{g_c(R^2 - r_c^2)}V_c \tag{12.42}$$

or utilizing Equation 12.25 to express V_c in terms of V_M

$$\left(\frac{\Delta P}{L}\right)_c = \frac{8\mu}{g_c(1-k^4)R^2}V_M \tag{12.43}$$

Combining Equations 12.41 and 12.43 we obtain

$$R_P = \frac{202}{(1-k^4)\left(\dfrac{DV_M\rho}{\mu}\right)^{0.75}} \tag{12.44}$$

Figure 12.13 from Charles (1963) shows the variation of R_P with k and $DV_M\rho/\mu$ as predicted by Equation 12.44.

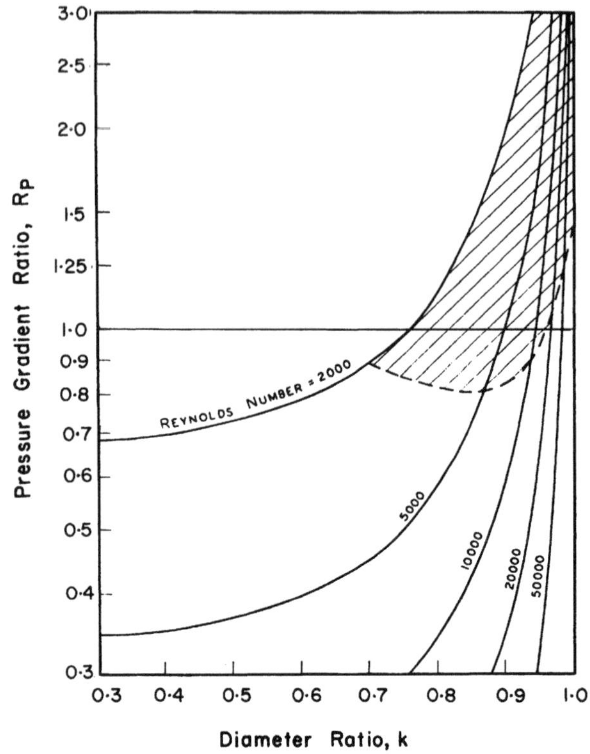

Figure 12.13. Pressure gradient ratios for Model 3. (In cross-hatched region $2\delta \geq R - r_c$.) (From Charles, M. E., *Can. J. Chem. Eng.*, **41**, 46, 1963.)

3.3 Nonconcentric Cylindrical Capsules

3.31 Laminar Flow—Analysis in Terms of Clearance

Kruyer, Redberger, and Ellis (1967) have made a revealing analysis of the free flow of nonconcentric cylindrical capsules carried by a liquid in laminar flow. They have drawn on the work of Bentwich, Kelly, and Epstein (1970) who derived the following expression for the free flow under laminar conditions of the annular liquid and an infinitely long nonconcentric cylindrical capsule:

$$Q_{\text{ann}} = B\left(\frac{\Delta P}{L}\right)_c \left\{(1 - k^2)^2 + 2k^2\varepsilon^2 - [8k \sinh(\alpha - \beta)] \times \left[\sum_{n=1}^{\infty} \coth[n(\alpha - \beta)]\left(\frac{2(n\varepsilon \sinh \beta - k^2)}{e^{2n\beta} + e^{2n\alpha}} + \frac{k^2}{e^{2n\alpha}}\right)\right]\right\} \quad (12.45)$$

where $\alpha = \cosh^{-1}\left(\frac{1-k^2-\varepsilon^2}{2k\varepsilon}\right)$ and $\beta = \cosh^{-1}\left(\frac{1-k^2+\varepsilon^2}{2\varepsilon}\right)$,

ε = eccentricity, distance between the axes of the pipe and the cylinder expressed as a fraction of the pipe radius,

n = integer (summation index),

$$B = -\frac{\pi g_c D^4}{128\mu}$$

For the movement of the cylinder, Bentwich *et al.* obtained

$$Q_c = B\left(\frac{\Delta P}{L}\right)_c [4k^3 \sinh(\alpha - \beta)] \tag{12.46}$$

The preceding equations (Equations 12.45 and 12.46) reduce to Equations 12.22 and 12.23 for the concentric case ($\varepsilon = 0$), but become indeterminate for the fully eccentric case ($\varepsilon = 1 - k$). Caldwell (1930) studied the fully eccentric case (stationary capsule touching the pipe wall) and determined analytically that

$$Q_{\text{ann}} = B\left(\frac{\Delta P}{L}\right)_c \left[1 - k^4 - 4k^2(1-k)^2 \sum_{n=0}^{\infty} \frac{1}{(1+n-nk)^2}\right] \tag{12.47}$$

$$Q_c = 0 \quad \text{(by assumption)} \tag{12.48}$$

From Equations 12.47 and 12.48, along with the equation relating $(\Delta P/L)_L$ to the liquid flow rate, the following equation may be derived for the stationary capsule or fully eccentric case:

$$R_P = \frac{1}{1 - k^4 - 4k^2(1-k)^2 \sum_{n=0}^{\infty} \frac{1}{(1+n-nk)}} \tag{12.49}$$

For k values greater than 0.4, the following simpler relationship matches Equation 12.49 to within 2 percent:

$$R_P = 0.645 + \frac{0.355}{(1-k)^{2.95}} \tag{12.50}$$

Equation 12.49 or 12.50 would permit the estimation of the threshold pressure gradient under laminar flow conditions provided the threshold velocity were known or could otherwise be estimated (see Section 3.32).

For the intermediate values of eccentricity, R_V and R_P may be obtained by the use of Equations 12.45 and 12.46. Kruyer *et al.* (1967) derived Figures 12.14 and 12.15 as the predicted relationship between R_V, c, and k, and Figure 12.16 as the predicted relationship between R_P, c, and k. The dimensionless clearance, c, is related by simple geometry to the eccentricity as

$$\varepsilon = 1 - k - 2c \tag{12.51}$$

and is defined as the ratio of the minimum distance between the pipe wall and the capsule to the pipe diameter.

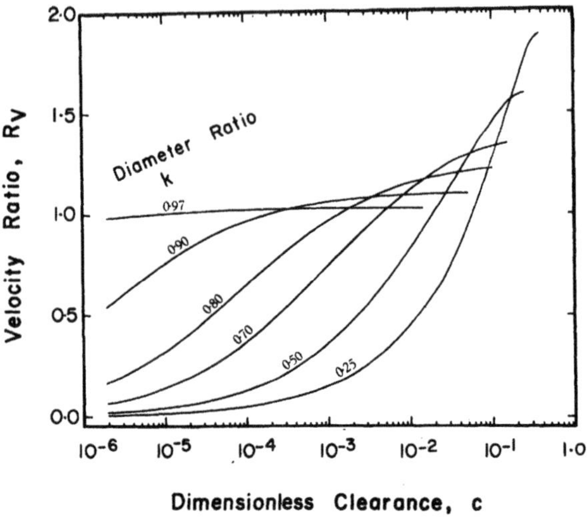

Figure 12.14. Predicted effect of clearance on the velocity ratio. (From Kruyer, J., P. J. Redberger, and H. S. Ellis, *J. Fluid Mech.*, **30**, 513, 1967.)

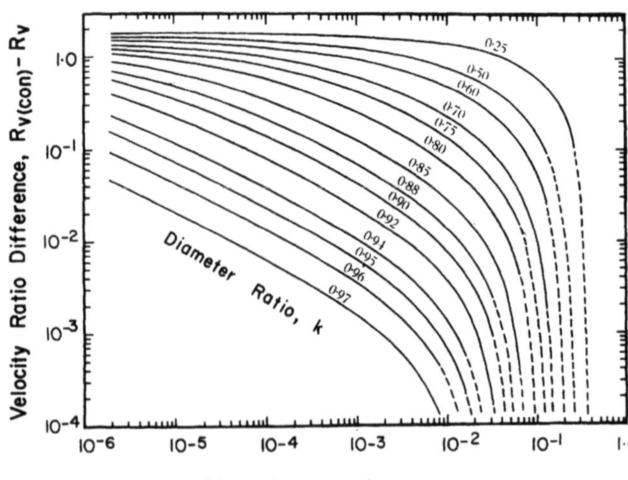

Figure 12.15. Predicted difference between the velocity ratios of a concentric and of an eccentric cylinder. (From Kruyer, J., P. J. Redberger, and H. S. Ellis, *J. Fluid Mech.*, **30**, 513, 1967.)

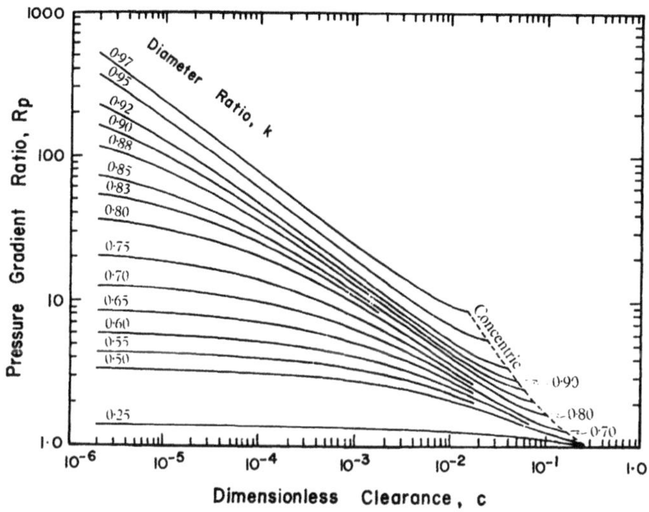

Figure 12.16. Predicted effect of diameter ratio and clearance on the pressure gradient ratio. (From Kruyer, J., P. J. Redberger, and H. S. Ellis), *J. Fluid Mech.*, **30**, 513, 1967.)

Kruyer *et al.* show that there is a nearly linear relationship between R_P and R_V for a fixed value of k and on this basis develop Figure 12.17 which relates R_P, R_V, k and a quantity N related to the dimensionless clearance by the equation

$$c = \frac{2^N}{1000} \tag{12.52}$$

The relations developed by Kruyer *et al.* have been tested with actual data covering the range of variables shown in Table 12.1.

TABLE 12.1

Characteristics of the Research Council of Alberta Experimental Capsule Pipelines[a]

Average internal pipe diameter in the test section, in.	0.532	1.254	4.03
Diameter tolerance, in.	0.001	0.01	0.02[b]
Pipe material	copper	acrylic plastic	steel
Test section length, ft	8.00	10.00	20.00
Upstream calming section, ft	12	21	110
Downstream calming section, ft	3	3	60

[a] J. Kruyer, P. J. Redberger, and H. S. Ellis, *J. Fluid Mech.*, **30**, 513 (1967).
[b] Estimated.

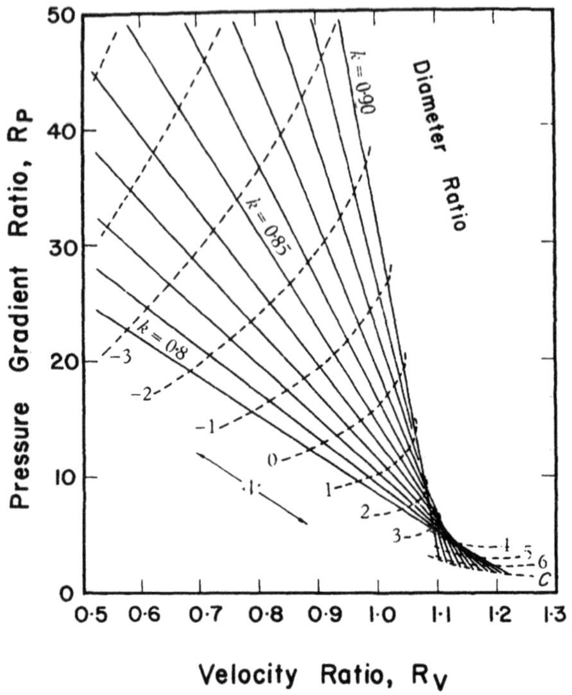

Figure 12.17. Predicted effect of diameter ratio, clearance, and velocity ratio on the pressure gradient. (From Kruyer, J., P. J. Redberger, and H. S. Ellis, *J. Fluid Mech.*, **30**, 513, 1967.)

The data for 24-inch-long cylinders of 0.824 diameter ratio flowing in a 0.532-inch pipe with a 36-centipoise, 0.86-specific gravity oil are presented in Figure 12.18. Similar data were obtained for 3-inch cylinders in the 0.532-inch pipe and for the flow of 18-inch and 24-inch cylinders in a $1\frac{1}{4}$-inch pipe. The data confirm the theoretical prediction of linearity between R_P and R_V for the laminar flow case.

Kruyer *et al.* also present data for the case where the liquid flows in laminar motion at low velocities but in turbulent motion at higher velocities. These are for a 48-inch, 0.90-diameter ratio capsule in a 4-inch commercial pipe. The data are given in Figure 12.19, where the crosses represent the actual R_P values and the circles are R_P values assuming $(\Delta P/L)_L$ to be that which would exist under laminar flow conditions but at the actual liquid velocity. The R_P versus R_V relationship based upon the actual data deviates, as might be expected, from the laminar-laminar prediction. On the other hand, the relationship between the R_P values based upon laminar flow and R_V is in accord with the theory and the concept of prolonged stability of laminar flow in the annulus.

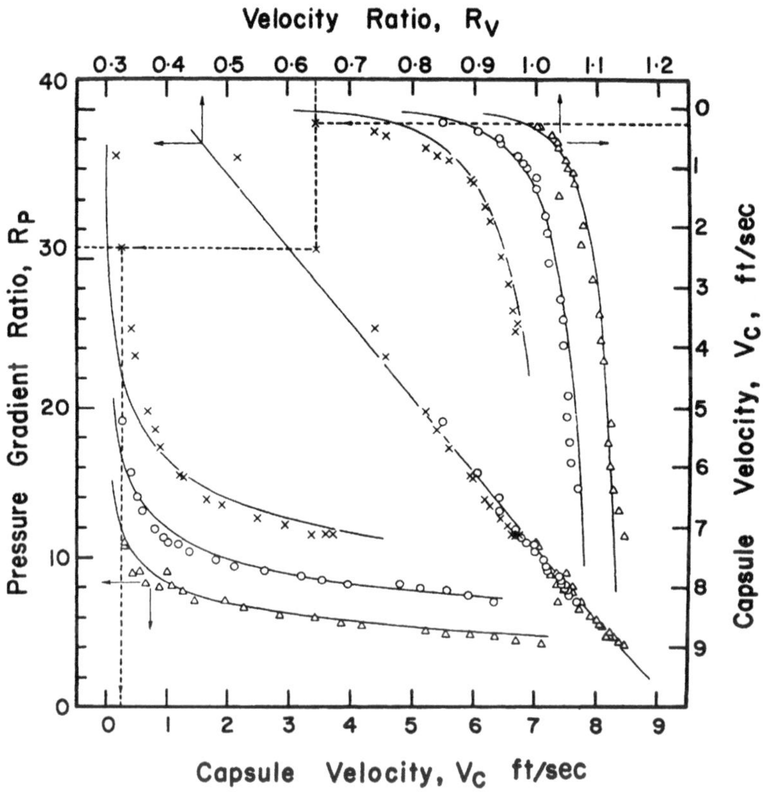

Figure 12.18. Experimental and predicted pressure gradient and velocity ratios—laminar data. A 24-inch long, 0.824 diameter ratio capsule flowing in lubricating oil (viscosity = 36 centipoises, specific gravity = 0.86) in a 0.532-inch pipe. The hollow cylindrical capsule was loaded to three specific gravities: Δ, 2.03; ○, 5.01; ×, 11.75. The straight line is the predicted R_P, R_V relationship for this diameter ratio and the curves are drawn by means of the correlation of Figure 12.20. (From Kruyer, J., P. J. Redberger, and H. S. Ellis, *J. Fluid Mech.*, **30**, 513, 1967.)

Kruyer *et al.* have applied this theory to their actual data on capsule flow to derive the theoretically indicated clearance, *c*. Typical results are given in Figure 12.20 for 24-inch long capsules in a 0.532-inch pipe. In this illustration a deviation from a linear trend occurs at a derived clearance of about 2×10^{-4} which is about the value of the relative roughness of "smooth" 1/2-inch copper pipes. It is probable that capsule or pipe wall roughness affected the capsule velocity at and below this indicated clearance.

Inasmuch as there is, as yet, no theoretical or general empirical method for predicting the clearance between the capsule and the pipe wall, the development of Kruyer *et al.* cannot be applied to the prediction of either R_V or R_P. On the other hand, if one of R_V or R_P is known or may otherwise

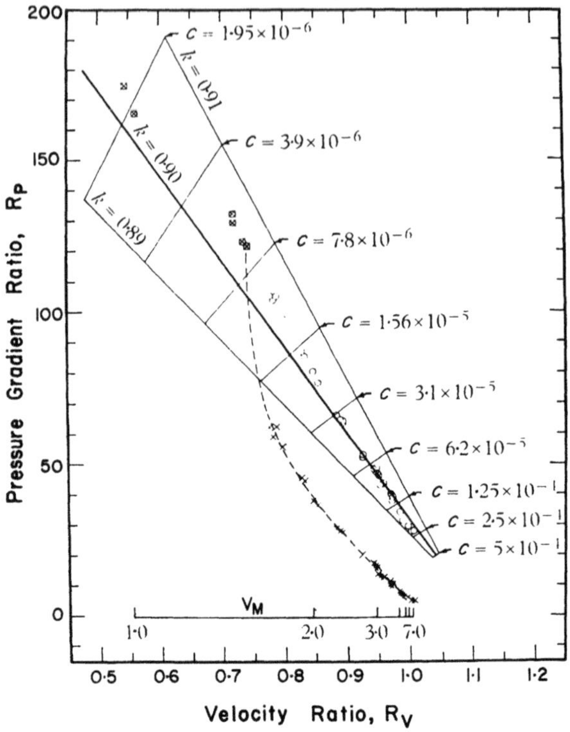

Figure 12.19. Experimental and predicted pressure gradient and velocity ratios—laminar-turbulent data. A 48-inch long, 0.90-diameter ratio aluminium (specific gravity = 2.71) cylindrical capsule flowing in transformer oil (viscosity = 16.7 centipoises, specific gravity = 0.853) in a 4.03-inch pipe. (From Kruyer, J., P. J. Redberger, and H. S. Ellis, *J. Fluid Mech.*, **30**, 513, 1967.)

be estimated, the relationships of Kruyer *et al.* will permit the estimation of c and of whichever of R_V or R_P is not known. The relationships are only valid in laminar flow or to what Charles referred to as Model 1 or 4, and to cases where the surfaces of the pipe and the capsule are sufficiently smooth that there is no solid-to-solid contact.

3.32 Laminar and Turbulent Flow—Approximate Force Balance Analyses

An alternative to the analysis just described, and one which is probably more useful in practical applications, is based upon simple overall force balance considerations. We will deal with two cases, the first representing threshold conditions for cylindrical capsules and the second representing steady flow conditions also for cylindrical capsules.

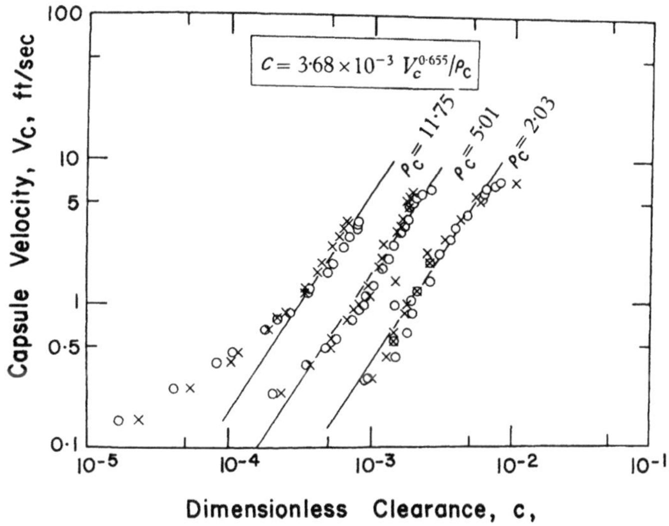

Figure 12.20. Measured capsule velocity as a function of theoretical clearance. The 24-inch long, 0.824-diameter ratio capsule of Figure 12.18 flowing in lubricating oil (viscosity = 36 centipoises, specific gravity = 0.86) in a 0.532-inch pipe. (From Kruyer, J., P. J. Redberger, and H. S. Ellis, *J. Fluid Mech.*, **30**, 513, 1967.)

Threshold Velocity and Pressure Gradient. We start with the fundamental relationships which determine the threshold velocity and the pressure gradient required to start the motion of an initially stationary cylindrical capsule, i.e., $(\Delta P/L)_{ct}$. Referring to Figure 12.21 we may write force balances over a length of pipe, L_c, containing a single capsule, as follows:

a. on the liquid flowing in the annulus between the pipe and the capsule

$$\frac{\pi}{4}(D^2 - d^2)\Delta P_c = \pi L_c(D\tau_{wa} + d\tau_{ca}) + W_L \sin\theta \tag{12.53}$$

b. on the capsule itself

$$\frac{\pi}{4}d^2\Delta P_c = -\pi d L_c \tau_{ca} + \eta W_B \cos\theta + W_c \sin\theta \tag{12.54}$$

where τ_{wa} and τ_{ca} are the average shear stresses in the fluid at the pipe wall and the capsule, respectively,

η = proportionality coefficient for the force parallel to the pipe wall caused by the interaction between the capsule and the pipe wall,

W_B = the buoyed weight of the capsule,

W_c = the weight of the capsule,

W_L = the weight of the fluid in the annulus.

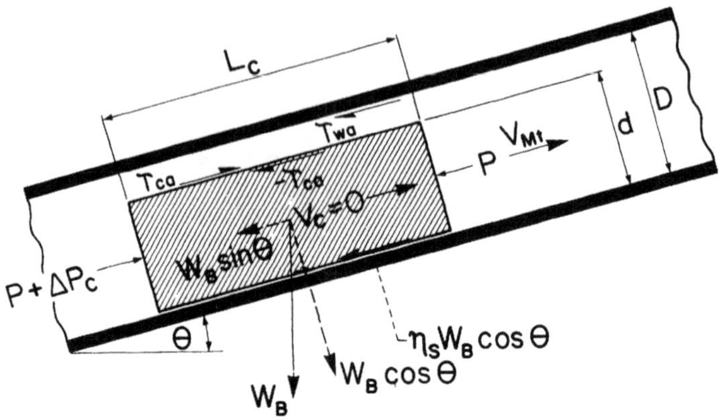

Figure 12.21. Forces acting on a capsule at threshold condition.

Thus

$$W_B = \frac{\pi d^2}{4}(\rho_c - \rho)L_c \frac{g}{g_c} \tag{12.55}$$

$$W_c = \frac{\pi d^2}{4}\rho_c L_c \frac{g}{g_c} \tag{12.56}$$

$$W_L = \frac{\pi}{4}(D^2 - d^2)\rho L_c \frac{g}{g_c} \tag{12.57}$$

Equations 12.53 and 12.54 are free of assumptions for the steady-state flow of a liquid. At a liquid velocity equal to the threshold velocity η represents the well known coefficient of static friction, η_s, and thus may be estimated.

Equation 12.53 may be rewritten to yield

$$\left(\frac{\Delta P}{L}\right)_c = \frac{4k(\tau_{wa} + k\tau_{ca})}{d(1-k^2)} + A \tag{12.57}$$

where

$$A = \frac{4W_L \sin\theta}{\pi L_c(D^2 - d^2)} = \frac{g\rho \sin\theta}{g_c}$$

For the condition where the liquid velocity, V_L, is equal to the threshold velocity, V_{Lt}, Equation 12.54 may be rewritten to yield

$$\left(\frac{\Delta P}{L}\right)_c = -\frac{4\tau_{ca}}{d} + B \tag{12.59}$$

where

$$B = \frac{4(W_B \eta_s \cos\theta + W_c \sin\theta)}{\pi L_c d^2} = [(\rho_c - \rho)\eta_s \cos\theta + \rho \sin\theta]\,\frac{g}{g_c}$$

Equating Equations 12.58 and 12.59 we obtain a relationship independent of the pressure gradient

$$\frac{4(k\tau_{wa} + \tau_{ca})}{d(1 - k^2)} + A - B = 0$$

that is,

$$\frac{4(k\tau_{wa} + \tau_{wc})}{d(1 - k^2)} = \frac{g}{g_c}(\rho_c - \rho)\eta_s \cos\theta \tag{12.60}$$

The shear stresses τ_{wa} and τ_{ca} cannot be predicted with any accuracy but they are clearly dependent upon the liquid velocity profile and μ, ρ, d, and D. With our present knowledge we must make assumptions in order to estimate one or both of τ_{wa} and τ_{ca}. An assumption upon which we could proceed is that the ratio of the average pipe wall shear stress to the average capsule wall shear stress is the same as the ratio obtained for laminar flow in a concentric annulus. The reasonability of this assumption is confirmed by experimental data of Jonsson and Sparrow (1966). Knudsen and Katz (1958) indicate the ratio for concentric annuli to be

$$\frac{\tau_{wa}}{\tau_{wc}} = \frac{k(2\ln k - k^2 + 1)}{(k^2 - 1 - 2k^2 \ln k)} = \phi_1(k) \tag{12.61}$$

If we now consider the friction component of the pressure gradient as given by Equation 12.58, we may express it in terms of a friction factor, f_{ann},

$$\left(\frac{\Delta P_f}{L}\right)_c = \frac{4k(\tau_{wa} + k\tau_{ca})}{d(1 - k^2)} = \frac{2f_{\text{ann}}\,\rho V_L^2}{D(1 - k)g_c} \tag{12.62}$$

where $(\Delta P_f/L)_c$ = pressure gradient in the fluid due to friction,
f_{ann} = friction factor defined by Equation 12.62.

We may utilize Equation 12.61 to express τ_{wa} in terms of τ_{ca} and Equation 12.62 may be rewritten to yield

$$\tau_{ca} = f_{\text{ann}}\left(\frac{\rho V_L^2}{2g_c}\right)\left(\frac{1 + k}{\phi_1(k) + k}\right) = f_{\text{ann}}\left(\frac{\rho V_L^2}{2g_c}\right)\phi_2(k) \tag{12.63}$$

where

$$\phi_2(k) = \frac{1 + k}{\phi_1(k) + k}$$

Equations 12.61 and 12.63 would apply up to and at the threshold velocity,

V_{Lt}, permitting us to substitute for τ_{ca} and τ_{wa} in Equation 12.60 to obtain

$$\frac{2(f_{\text{ann}})_t \rho V_{Lt}^2}{g_c d\phi_3(k)} = \frac{g}{g_c}(\rho_c - \rho)\eta_s \cos\theta \tag{12.64}$$

where $\phi_3(k) = \dfrac{1 - k^2}{\phi_2(k([k\phi_1(k) + 1]}$

$V_{Lt} = V_L$ at the threshold conditions,

$(f_{\text{ann}})_t = f_{\text{ann}}$ at the threshold conditions.

The functions $\phi_1(k)$, $\phi_2(k)$, and $\phi_3(k)$ are given in Figure 12.22. It is worth noting that both $\phi_1(k)$ and $\phi_2(k)$ are close to unity for k over about 0.80 and $\phi_3(k)$ approaches 0 as k approaches 1. The friction factor $(f_{\text{ann}})_t$ may be evaluated by noting the similarity between the threshold case considered here and the case of liquid flow in a fully eccentric annulus formed

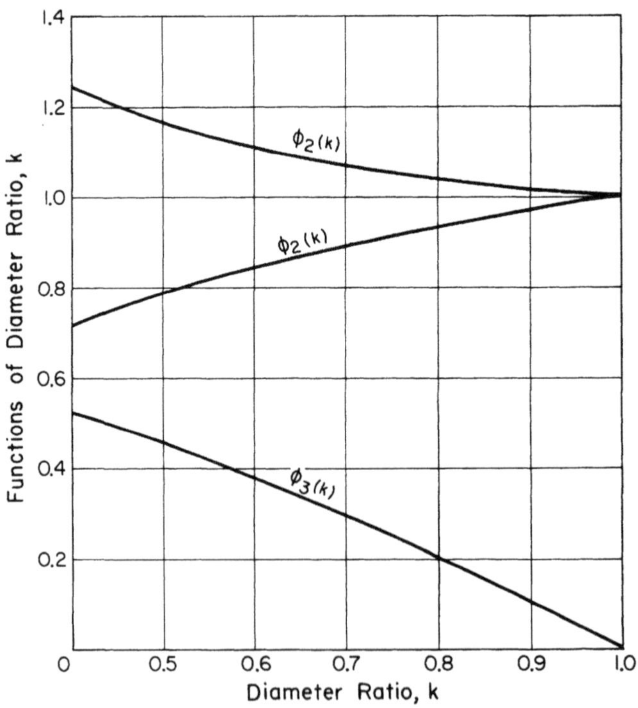

Figure 12.22. Functions of the diameter ratio, *k*.

by two stationary pipes. The latter case was studied experimentally by Bourne, Figueiredo, and Charles (1968). Also Tiedt (1968) has evaluated the theoretical expression of Caldwell (Equation 12.47) in terms of the friction factor f_{ann}. Bourne *et al.* find that the product (f_{ann} Re_{ann}) declines from 16.0 at $k = 0$ to a minimum of 12.0 at $k \simeq 0.75$, increasing to higher values as k increases toward unity. The theoretical solutions of Caldwell and of Tiedt show (f_{ann} Re_{ann}) declining from 16 at $k = 0$ to an essentially constant value of 9.6 for $k > 0.75$.

We have the choice, therefore, of taking the (f_{ann} Re_{ann}) product, within the k range of our interest, at the theoretical value of 9.6 or the experimental value in the range of 12. For our purposes we choose

$$f_{ann} = \frac{12}{\mathrm{Re}_{ann}} = \frac{12\mu}{(D-d)V_L\rho} \tag{12.65}$$

Substituting for f_{ann} in Equation 12.64 from Equation 12.65, we obtain a relationship for V_{Lt} applicable for laminar flow,

$$V_{Lt} = \frac{g}{g_c}(\rho_c - \rho)\eta_s \cos\theta \left[\frac{(D-d)\,dg_c\phi_3(k)}{24\mu}\right] \tag{12.66}$$

For turbulent flow ($\mathrm{Re}_{ann} > 1000$) Bourne *et al.* suggest the equation

$$f_{ann} = 0.06\,\mathrm{Re}_{ann}^{-0.25} = 0.06\left[\frac{\mu}{(D-d)V_L\rho}\right]^{1/4} \tag{12.67}$$

A relationship for V_{Lt} applicable for turbulent flow may now be obtained by introducing Equation 12.67 into Equation 12.64,

$$V_{Lt} = \left[\frac{g}{g_c}(\rho_c - \rho)\eta_s \cos\theta \left(\frac{D-d)\rho}{\mu}\right)\frac{g_c d\phi_3(k)}{0.12\rho}\right]^{4/7} \tag{12.68}$$

From the appropriate one of Equations 12.66 or 12.68 and a knowledge of η_s, we may estimate the threshold velocity, V_{Lt}. In the absence of actual measurements, η_s may be estimated having regard to its normal range as given in handbooks of mechanical engineering (e.g., Marks, 1967, Kent, 1958). For metal surfaces of normal pipe roughness and in the presence of water or oil, η_s will probably fall in the range of 0.15 to 0.30.

Given V_{Lt} we may now estimate τ_{cat}, τ_{wat}, and $(\Delta P/L)_{ct}$ from Equations 12.63, 12.61, and 12.58, respectively. Alternately in laminar flow we may estimate R_P and from it $(\Delta P/L)_{ct}$ from Equation 12.49 or 12.50 with $(\Delta P/L)_L$ evaluated at V_{Lt}. The threshold bulk velocity may be obtained by

$$V_{Mt} = (1 - k^2)V_{Lt} \tag{12.69}$$

This procedure for the estimation of V_{Lt} (or V_{Mt}) and $(\Delta P/L)_{ct}$ is based upon a rational analysis and reasonable assumptions. It should yield good order of magnitude results. Full confirmation of it by comparison with direct measurements covering a variety of pipe, capsule, and fluid properties is desirable but not yet possible from available published data. The threshold velocity and pressure gradient predicted for the test conditions of Figure 12.19 are compared with the steady flow data in Figure 12.25 given subsequently. The predictions, based upon an assumed value of $\eta_s = 0.30$, are compatible with the measured data.

Lift-off and Minimum Practical Transport Velocity Liddle (1968) has correlated his data on the mixture lift-off velocity, V_{ML}, in terms of a Froude number, $V_M/\sqrt{g_c D}$, as shown in Figure 12.23. The correlation covers a reasonable range of density ratios, s, and diameter ratios, k, but is based only on water as the carrier liquid and on measurements for plain cylindrical capsules in a 1.310-inch plastic pipe. The suggested role of the Froude number is reasonable but not proven.

A comparison of the lift-off velocities indicated by Figure 12.23 with the R_V data reported by Round and Bolt (1965) in Figure 12.7 and by Hodgson *et al.* (1967) in Figure 12.8 suggests that the lift-off velocity corresponds with the attainment of an R_V of 1.1 or higher. This is also what we would expect from Equation 12.15. Further, if we accept the role of the Froude number as suggested by Liddle, we may place in Figure 12.23 parallel lines reflecting other values of R_V. A line for $R_V = 1$ has been positioned on the figure having regard for the data of Figures 12.7 and 12.8. This line leads to predicted values compatible with the limited data of Figure 12.8 for the 4-inch and the 20-inch pipe thus giving some support to the choice of the Froude number for the correlation.

The significance of Liddle's correlation and the approximate line for $R_V = 1$ in Figure 12.23 is that they provide at least some basis for choosing a practical transport velocity. At velocities significantly below the lift-off velocity, the capsule-wall friction is relatively high, capsule abrasion could be serious, capsule throughput is low, and the pressure gradient is relatively high. Conditions resulting in $R_V \geq 1$ would seem to be necessary in any practical application and the $R_V = 1$ line of Figure 12.23 may be considered at least a rough criterion for a minimum practical transport velocity for plain cylindrical capsules. Clearly the minimum transport velocity and the velocity for $R_V = 1$ would be lowered by any capsule shape or device which would promote capsule lift—such as the circumferential bands referred to earlier.

Velocity and Pressure Gradient under Conditions of Steady Capsule Motion Under conditions of steady capsule motion at $V_M > V_{Mt}$, a number of factors

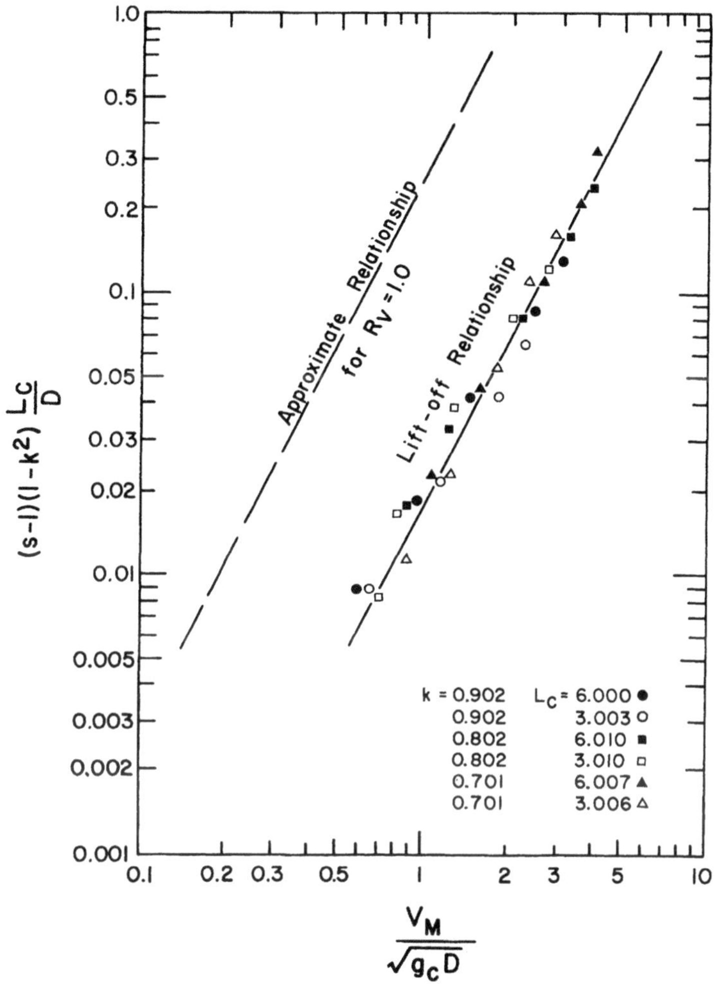

Figure 12.23. Liddle lift-off velocity correlation. (Modified from Liddle, R. T., M.Sc. Thesis, Univ. Alberta, Edmonton, Alta., 1968.)

differ from their counterparts before capsule motion. First, in place of a coefficient of static friction we must use a coefficient under sliding or moving conditions, η_m. Secondly, when the capsule is in steady motion its velocity will exceed that of the fluid nearest the pipe wall and may slightly exceed the velocity of the fluid at all points on the circumference of the capsule. In this case the absolute value of τ_{ca} is small and its direction is reversed. A third factor of importance is that at mixture velocities above V_{ML} the capsule takes on a slight nose-up orientation and tends to plane over the more slowly moving liquid beneath it. This results in a lift force which may be assumed

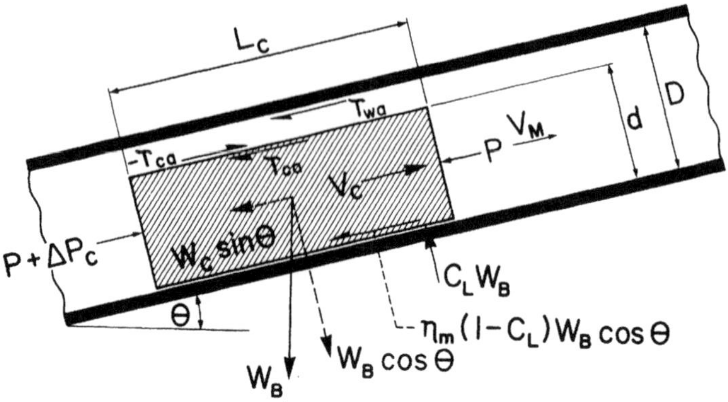

Figure 12.24. Forces acting on a capsule in motion

to be progressively larger as the capsule and liquid velocity increase. The situation is illustrated in Figure 12.24.

We may write the following equations:

a. for the total contents of the pipe,

$$\frac{\pi D^2}{4}\Delta P_c = L_c \pi D \tau_{wa} + \eta_m W_B \cos\theta(1 - C_L) + (W_c + W_F)\sin\theta \quad (12.70)$$

b. for the capsule itself,

$$\frac{\pi d^2}{4}\Delta P_c = L_c \pi d \tau_{ca} + \eta_m W_B \cos\theta(1 - C_L) + W_c \sin\theta \quad (12.71)$$

where η_m is the coefficient of moving friction and C_L is a lift coefficient. The other terms were defined earlier. Now the magnitude and center of action of the lift force, $C_L W_B$, will depend upon the capsule density and geometry, the pipe size, the fluid properties, and the fluid velocity. It will be zero at the threshold velocity and will increase as a function of velocity ultimately to a value of W_B when the capsule will behave effectively as one of density equal to that of the fluid. At the present state of knowledge we have no effective way of predicting C_L from theory alone.

As in the threshold case there is no accurate way of calculating either of the average shear forces τ_{wa} or τ_{ca} but τ_{wa} may be estimated at least roughly by comparison with the concentric case. Equation 12.70 may be rewritten as

$$\left(\frac{\Delta P}{L}\right)_c = \frac{4\tau_{wa}}{D} + \frac{4}{\pi D^2 L_c}[\eta_m W_B \cos\theta(1 - C_L) + (W_c + W_L)\sin\theta] \quad (12.72)$$

or, introducing the relations of Equations 12.55, 12.56, and 12.57, as

$$\left(\frac{\Delta P}{L}\right)_c = \frac{4\tau_{wa}}{D} + \eta_m k^2 \frac{g}{g_c}(\rho_c - \rho)(1 - C_L)\cos\theta + \frac{g}{g_c}[k^2(\rho_c - \rho) + \rho]\sin\theta$$

(12.73)

We now recognize that, for this eccentric case, the velocity in the liquid-filled space will be greater than the capsule velocity over part of the pipe circumference (as shown in the uppermost sketch of Figure 12.11) and that the converse will be the case over other regions (as shown in the second sketch of Figure 12.11). To a first approximation, we could assume that the average wall shear stress τ_{wa} would be the same as in the concentric capsule case at the same bulk velocity, V_M, but based upon the velocity profile assumed by Charles, i.e. Equation 12.32, or taking Kennedy's V_R to be zero. This represents a crude averaging of the profiles of the two upper sketches of Figure 12.11. With this assumption the term $4\tau_{wa}/D$ becomes identical with $(\Delta P/L)_c$ for the concentric capsule case, here designated as $(\Delta P/L)_c^*$. It may therefore be expressed as

$$\frac{4\tau_{wa}}{D} = \left(\frac{\Delta P}{L}\right)_c^* = R_P^*\left(\frac{\Delta P}{L}\right)_L \tag{12.74}$$

and may be evaluated using the relationships developed in Section 3.23 assuming $V_R = 0$.

Thus depending on the flow regime encountered, Equations 12.39, 12.40, and 12.44 may be employed to calculate R_P^*. If the free liquid stream is in turbulent flow, $(\Delta P/L)_L$ may be calculated at any value of V_M from Equation 12.41; otherwise it may be obtained using Equation 4.21. With $4\tau_{wa}/D$ calculated from Equation 12.74, $(\Delta P/L)_c$ may be evaluated from Equation 12.73 and a knowledge of η_m and C_L. If desired, the average capsule shear stress, τ_{ca}, may subsequently be determined from Equation 12.71.

The analysis may be checked against actual data of Kruyer *et al.*, but first we must have some way of estimating the lift coefficient, C_L. A form of relationship which would satisfy the condition that $C_L = 0$ when $V_M = V_{Mt}$ and also provide the expected increase of C_L with V_{Mt} is

$$C_L = 1 - \exp\left(\frac{V_{Mt} - V_M}{a}\right) \tag{12.75}$$

where a is a constant characteristic of the capsule. Analysis of some of the data of Kruyer *et al.* suggests that Equation 12.75 is of the right form and that the constant a has a value of the order of 50 for such capsules. Presumably a will be dependent on the ratio d/L_c and would be significantly affected by the presence of bands or any lift device on the capsules.

If we now use Equation 12.75 with $a = 50$ and values of η_m in its expected range, we may predict the relationship between $(\Delta P/L)_c$ and V_M by the method described above. Figure 12.25 presents the comparison of predicted and measured gradients for the horizontal flow of 48-inch, 0.90-diameter ratio capsules in a 4-inch steel pipe with a transformer oil carrier fluid. The coefficient of moving friction, η_m was taken at 0.15, 0.20 and 0.25 in an attempt

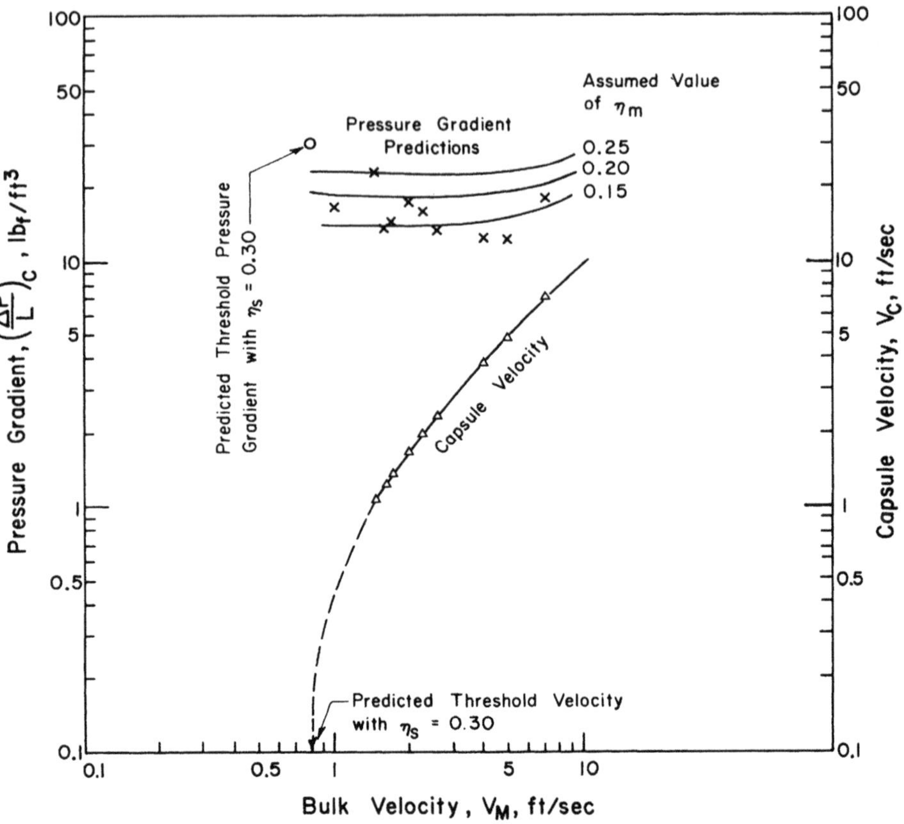

Figure 12.25. Comparison of predictions based upon force balance analysis with data of Kruyer *et al.* from Figure 12.19. (×, Δ represent actual data as read from the curves of Figure 12.19.)

to cover the probable range. The agreement indicated in Figure 12.25 is of good order of magnitude.

This check with the very limited actual data certainly does not confirm all aspects of the analysis, including in particular the major assumption that R_P will be the same as for the Charles analysis of the concentric case at the same V_M, but it does, however, provide some support for the general approach.

Ellis and Kruyer (1970) and Kruyer and Ellis (1970) have made an approximate force balance analysis somewhat similar to that discussed above. Their approach was to recognize that the distortion of the velocity profile in the eccentric annulus was related to the capsule velocity and to find experimentally an effective average velocity in the annulus such that the average shear stress over the two walls could be estimated from the conventional friction factor and Reynolds number. The relationship is shown by the force balance on the liquid

$$\frac{\Delta P\pi(D^2 - d^2)}{4} = (\tau_{wa}\pi D + \tau_c \pi d)L + W_L \sin\theta \qquad (12.76)$$

which may be rewritten in terms of τ_{av}, the average shear stress, to yield, after substituting for W_L from Equation 12.57,

$$\left(\frac{\Delta P}{L}\right)_c = \frac{4\tau_{av}}{D - d} + \frac{\rho g}{g_c}\sin\theta \qquad (12.77)$$

where

$$\tau_{av} = \frac{\tau_{wa} + k\tau_{ca}}{1 + k} \qquad (12.78)$$

Kruyer and Ellis express τ_{av} in terms of an effective velocity and friction factor as

$$\tau_{av} = \frac{f_{eff}\,\rho\, V_{eff}^2}{2g_c} \qquad (12.79)$$

where $V_{eff} = V_L - \bar{a}V_c$ = an effective average velocity of the liquid in the annulus,

$V_L = \dfrac{V_M - k^2 V_c}{1 - k^2}$ (from Equation 12.10),

= average velocity of the liquid in the annulus,

$_{eff} = \phi\,(\mathrm{Re}_{eff}) = \phi\left[\dfrac{(D - d)V_{eff}\,\rho}{\mu}\right]$

The coefficient $\bar{a}$ is dependent on the flow regime. Kruyer and Ellis suggest the following experimentally determined values:

$\bar{a} = 0.05 + 0.455k$ for laminar flow
$\bar{a} = 0.35$ for turbulent flow.

Using these values for the factor $\bar{a}$, Kruyer and Ellis computed f_{eff} and Re_{eff} from Research Council data. They found their data to be bounded by the lines $f\,\mathrm{Re} = 16$ and $f\,\mathrm{Re} = 9.6$ in the laminar flow region, and the lines $f\,\mathrm{Re}^{0.25} = 0.079$ and $f\,\mathrm{Re}^{0.25} = 0.05$ in the turbulent region. Figure 12.26 shows typical data for capsules 18 to 48 inches long, of diameter ratios 0.693 to 0.900 and of a range of densities and viscosities in 0.5- and 4-inch pipes.

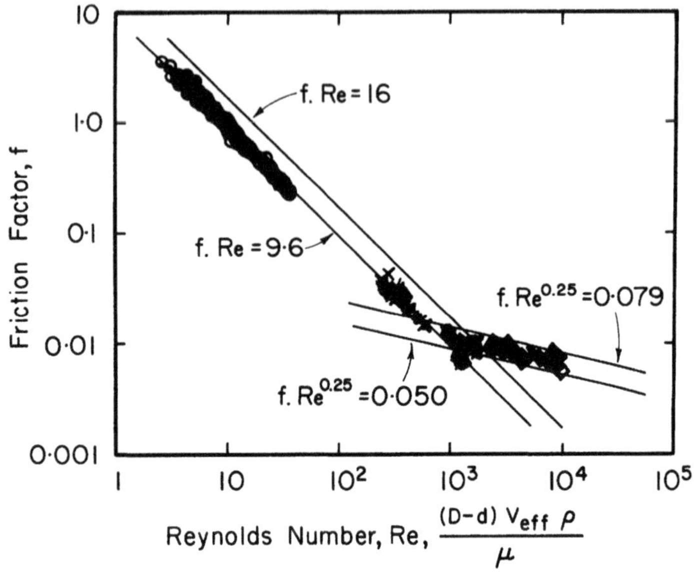

Figure 12.26. Kruyer and Ellis friction factor correlation. (From Kruyer, J., and H. S. Ellis, *B.H.R.A. Hydrotransport 1*, Sept. 1970.)

With the explicit relationship for τ_{av} given by Equation 12.79, Equation 12.77 may be solved for the pressure gradient if all the other variables are specified. It should be noted that, in order to obtain a solution, V_c, the capsule velocity, must be known or estimated.

Kruyer and Ellis consider this correlation a preliminary one and recognize some inconsistencies in their treatment of the effective average velocity in the annulus. The approach is interesting and may well lead to relations useful for design purposes.

4 RECOMMENDED DESIGN METHODS (PRELIMINARY ONLY)

4.1 Introduction

At the present stage of knowledge it is not possible to design for capsule pipeline flow with an accuracy comparable to that for single-phase or even many multiphase systems. No large-scale capsule pipeline should be designed without extensive pilot pipeline work closely simulating the anticipated operating conditions. Nonetheless, from the data available and the analyses we have discussed, preliminary estimates may be made of the velocity and pressure gradient ratios and these at least may serve as a rational basis for pilot pipeline design.

4.2 Entrance Effects

Entrance effects of the type encountered with the flow of single-phase fluids or mixtures of fluids are not significant in the flow of capsules, and only the kinetic energy effect discussed in Section 4.5 would normally be important.

4.3 Effect of Bends and Fittings

By its very nature a capsule pipeline requires full opening fittings and long radius bends which would be expected to have a negligible effect on capsule velocity or pressure gradient.

4.4 Potential Energy or Hydrostatic Head Effects

The effect of incline or decline of the pipe on the threshold velocity and pressure gradient is implicitly incorporated in Equations 12.66 and 12.67 for V_{Lt} and Equation 12.58 for $(\Delta P/L)_{ct}$. Similarly, for normal flow conditions it is incorporated in Equation 12.73. Alternately, for the normal flow case if the significance of the cos θ term is small, Equation 12.73 suggests that the effect of incline or decline may be estimated by adding or subtracting to the pressure gradient for horizontal flow a hydrostatic head component based upon the equivalent liquid concept, i.e., an amount

$$\left(\frac{\Delta P}{L}\right)_{HH} = \left[\frac{2}{3}\,k^2(\rho_c - \rho) + \rho\right]\frac{g}{g_c}\sin\theta$$

for spherical capsules in contact, and

$$\left(\frac{\Delta P}{L}\right)_{HH} = \left[k^2(\rho_c - \rho) + \rho\right]\frac{g}{g_c}\sin\theta$$

for cylindrical capsules in contact.

4.5 Acceleration or Kinetic Energy Effects

The energy required for the initial acceleration of the capsules to their pipe flow velocity, V_c, may be significant, and may be estimated from the equation

$$\Delta P_{KEC} = \frac{\pi}{12}\frac{\rho_c\, d^2 V_c^2}{g_c}$$

for spherical capsules in contact, and

$$\Delta P_{KEC} = \frac{\pi}{8} \frac{\rho_c d^2 V_c^2}{g_c}$$

for cylindrical capsules in contact. The kinetic energy effect associated with the carrier liquid is usually negligible.

4.6 *Pressure Gradient and Holdup for the Steady Flow of Capsules*

The procedure for making preliminary estimates of the velocity and pressure gradient relationships for a given capsule flow capacity, Q_c, is

(A) For Spherical Capsules in Contact

1. Assume a diameter D and a diameter ratio k in the range $0.80 < k < 0.90$ such that V_c as determined from the continuity relationship

$$Q_c = \frac{\pi d^3}{6}\left(\frac{V_c}{d}\right) = \frac{\pi}{6} d^2 V_c$$

is in the normal economic range of say $3.0 < V_c < 8.0$.

2. Assume $V_M = 0.90 V_c$, which is a close approximation.

3. Calculate Q_L from Equation 12.3 and the known values of Q_c and V_M.

4. Estimate the pressure gradient $(\Delta P/L)_c$ at the bulk velocity V_M from single-phase methods assuming an equivalent liquid of density ρ_{Ms} given by Equation 12.17.

5. The calculations and considerations of steps *1* to *4* may be repeated for other assumed values of k and D to evaluate the impact of these specified quantities on the velocities and the pressure gradient.

(B) For Cylindrical Capsules in Contact

1. Specify the capsule length L_c and the diameter ratio k within the normal range 0.80 to 0.90.

2. From the known capsule and fluid properties, with an assumed value of D, estimate the minimum practical transport velocity from the $R_V = 1$ line of Figure 12.23. Choose a value of V_M somewhat above $(V_M)_{min}$ and estimate R_V in the range $1.0 < R_V < 1.2$.

3. From the values of R_V, V_M, and k, and the continuity relation

$$Q_c = \frac{\pi d^2}{4} V_c$$

determine d, D, and V_c. Check step *2*. Make adjustments to d, D, V_M, and V_c as appropriate to suit standard pipe dimensions or other restraints.

4. Calculate Q_L from Equation 12.3 and the known values of Q_c and V_M.

5. Estimate the pressure gradient for the flow of the liquid phase alone at an average velocity, V_M, by conventional single-phase methods.

6. Provided a measured value for η_s is available, or a good estimate may be made, the threshold bulk velocity, V_{Mt}, and the pressure gradient $(\Delta P/L)_{ct}$ may be roughly estimated by the method discussed in the first part of Section 3.32.

7. Provided a measured value for η_m is available, or a good estimate may be made, the relationship between the bulk velocity, V_M, and the pressure gradient may be roughly estimated by the method discussed in the central part of Section 3.32, with C_L taken as

$$1 - \exp[0.02\ (V_{Mt} - V_M)]$$

for plain cylindrical capsules with L_c/d in the range of, say, 6 to 12. For other capsule shapes or proportions, a similar approach is applicable, but the value of the constant a in Equation 12.75 would have to be determined by pilot testing.

8. The calculations of step *7* may be compared with an estimate based upon the method of Kruyer and Ellis (1970) discussed in the last part of Section 3.32.

9. The calculations and considerations of steps *1* to *8* may be repeated for other assumed values of k and especially of V_M and D to evaluate the impact of these specified quantities on the velocities and the pressure gradient.

Bibliography

Bentwich, M., D. A. I. Kelly, and N. Epstein, *J. Basic Eng., Trans. A.S.M.E., Ser. D*, **92**, 32 (1970).

Berkowitz, N., R. A. S. Brown, and E. J. Jensen, *Can. J. Chem. Eng.*, **43**, 280 (1965).

Berkowitz, N., and E. J. Jensen, U.S. Patent 3, 190, 701.

Bourne, D. E., O. Figueiredo, and M. E. Charles, *Can. J. Chem. Eng.*, **46**, 289 (1968).

Brown, R. A. S., and E. J. Jensen, *Can. J. Chem. Eng.*, **46**, 157 (1968).

Caldwell, J., Royal Technical College, Glasgow, *Journal*, **2**, 203(1930).

Charles, M. E., *Can. J. Chem. Eng.*, **41**, 46 (1963).

Charles, M. E., G. W. Govier, and G. W. Hodgson, *Can. J. Chem. Eng.*, **39**, 27 (1961).

Coulson, J. M., and J. F. Richardson, *Chemical Engineering*, McGraw-Hill Book Co., New York, 1954.

Ellis, H. S., *Can. J. Chem. Eng.:*

a. **42**, 1 (1964);

b. **42**, 69 (1964);

c. **42**, 155 (1964).

Ellis, H. S., and L. H. Bolt, *Can. J. Chem. Eng.*, **42**, 201 (1964).

Ellis, H. S. and J. Kruyer, The Pipeline Flow of Capsules, Part 10, paper presented at the First International Conference on the Hydraulic Transport of Solids in Pipes, British Hydromechanics Research Association, Cranfield, Bedford, England, Sept. 1970.

Epstein, N., *Can. J. Chem. Eng.*, **41**, 181 (1963).

Hodgson, G. W., and L. H. Bolt, *Engineering Journal*, **45**, 25 (Dec. 1962).

Hodgson, G. W., and M. E. Charles, *Can. J. Chem. Eng.*, **41**, 43 (1963).

Hodgson, G. W., and M. E. Charles, Canadian Patent 673,280.

Hodgson, G. W., J. Kruyer, A. A. Roehl, G. F. Round, and H. S. Ellis, *J. Can. Petrol. Technol.*, **6**, 1 (1967).

Jensen, E. J., and N. Berkowitz, First Intermountain Symposium on Fossil Hydrocarbons, Salt Lake City, 1965, *Proc.* p. 141.

Jensen, E. J., and J. G. Bruce, Capsule Pipelining—A Feasibility Assessment, paper presented at the First International Conference on the Hydraulic Transport of Solids in Pipes, British Hydromechanics Research Association, Cranfield, Bedford, England, Sept. 1970.

Jonsson, V. K., and E. M. Sparrow, *J. Fluid Mech.*, **25**, 65 (1966).

Kaufmann, W., *Fluid Mechanics*, McGraw-Hill Book Co., New York, 1963, p. 92.

Kennedy, R. J., *Can. J. Chem. Eng.*, **44**, 354 (1966).

Kent's Mechanical Engineers' Handbook, C. Carmichael, Editor, John Wiley & Sons New York, Twelfth Edition, 1958.

Knudsen, J. G., and D. L. Katz, *Fluid Dynamics and Heat Transfer*, McGraw-Hill Book Co., New York, 1958.

Kruyer, J., and H. S. Ellis, Private Communication, 1970.

Kruyer, J., P. J. Redberger, and H. S. Ellis, *J. Fluid Mech.*, **30**, 513 (1967).

Liddle, R. T., M. Sc. Thesis, Univ. Alberta, Edmonton, Alta., 1968.

Marks, L. S. (Editor 1916 to 1951), Baumeister, T., Editor, *Standard Handbook for Mechanical Engineers*, McGraw-Hill Book Co., New York, Seventh Edition, 1967.

Round, G. F., and L. H. Bolt, *Can. J. Chem. Eng.*, **43**, 197 (1965).

Shellene, K. R., and P. T. Chu, Canadian Patent 836, 381.

Tiedt, W., Technischer Bericht No. 4, Institut für Hydraulik und Hydrologie der Technischen Hochschule, Darmstadt, 1968.

Zuber, N., and J. A. Findlay, *J. Heat Transfer, Trans. A.S.M.E. Ser. C*, **87**, 453 (1965).

Appendix

TABLE A-1 Constants and Physical Properties

Item	Substance	Formula	Molecular Weight	Boiling Point at 14.696 psia		Vapour Pressure at 100°F		Freezing Point at 14.696 psia		Specific Gravity Liquid (60°F/60°F, 14.696 psia)
				°R	°K	psia	atm	°R	°K	
			Ref. 1	Ref. 1		Ref. 1		Ref. 1		Ref. 1
	PARAFFINS									
1.	Methane	CH_4	16.042	201.32	111.7	—	—	163.54[b]	90.72	0.3[d]
2.	Ethane	C_2H_6	30.068	332.47	184.6	—	—	162.11[b]	89.93	0.3771[e]
3.	Propane	C_3H_8	44.094	416.27	231.1	190	12.9	154.16[b]	85.51	0.5077[e]
4.	*n*-Butane	C_4H_{10}	58.120	491.10	272.7	51.6	3.5	242.97	134.85	0.5844[e]
5.	2-Methyl-propane (isobutane)	C_4H_{10}	58.120	470.89	261.5	72.2	4.9	204.72	113.60	0.5631[e]
6.	*n*-Pentane	C_5H_{12}	72.146	556.93	309.3	15.570	1.06	258.50	143.48	0.6312
7.	2-Methyl-butane (isopentane)	C_5H_{12}	72.146	542.13	301.0	20.44	1.39	204.18	113.30	0.6248
8.	2,2-Dimethyl-propane (neopentane)	C_5H_{12}	72.146	509.10	282.7	35.9	2.4	462.21	256.65	0.5967[e]
9.	*n*-Hexane	C_6H_{14}	86.172	615.73	341.9	4.956	0.337	320.37	177.85	0.6640
10.	*n*-Heptane	C_7H_{16}	100.198	669.17	371.6	1.620	0.110	328.90	182.59	0·6882
11.	*n*-Octane	C_8H_{18}	114.224	718.20	398.9	0.537	0.036	389.77	216.41	0.7068
12.	*n*-Nonane	C_9H_{20}	128.250	763.44	424.0	0.179	0.012	395.67	219.68	0.7217
13.	*n*-Decane	$C_{10}H_{22}$	142.276	805.42	447.3	0.073	0.005	438.61	243.54	0.7341
	OLEFINS									
14.	Ethane (Ethylene)	C_2H_4	28.052	305.32	169.5	—	—	187.53[b]	104.05	0.35
15.	Propene	C_3H_6	42.078	406.14	225.5	226.4	15.4	158.55[b]	87.95	0.5220[e]
16.	1-Butene	C_4H_8	56.104	480.73	266.9	63.05	4.29	158.37[b]	87.85	0.6013[e]
17.	*cis*-2-Butene	C_4H_8	56.104	498.70	276.9	45.54	3.10	241.96	134.29	0.6271[e]
18.	*trans*-2-Butene	C_4H_8	56.104	493.58	274.1	49.80	3.39	302.01	167.65	0.6100[e]
19.	2-Methyl-propene (isobutene)	C_4H_8	56.104	479.58	266.3	63.40	4.31	239.37	132.85	0.6004[e]
20.	1-Pentene	C_5H_{10}	70.130	545.94	303.2	19.12	1.30	194.60	107.98	0.6457
21.	Ethyne (Acetylene)	C_2H_2	26.036	341.00[a]	189.3	—	—	346[b]	192	0.6150[f]
	AROMATICS									
22.	Benzene	C_6H_6	78.108	636.18	353.3	3.224	0.219	501.96	278.73	0.8845
23.	Toluene	C_7H_8	92.134	691.12	383.8	1.032	0.070	321.02	178.21	0.8719
24.	Ethylbenzene	C_8H_{10}	106.160	737.13	409.4	0.371	0.025	321.04	178.22	0.8717
	ALCOHOLS									
25.	Methyl alcohol	CH_4O	32.042	608.1	337.7	4.63	0.31	316.18	175.52	0.796
26.	Ethyl alcohol	C_2H_6O	46.069	632.92	351.5	2.3	0.16	286.6	159.09	0.794
27.	*n*-Propyl alcohol (ref. 2)	C_3H_8O	60.09	667	370.4	—	—	263.4	146.2[c]	0.804[g]
28.	*n*-Butyl alcohol (ref. 2)	$C_4H_{10}O$	74.12	703.7	390.8	—	—	348.2	193.3[c]	0.810[g]
	GASES									
29.	Carbon monoxide	CO	28.010	146.4	81.2	—	—	119.4	66.2	0.801[h]
30.	Carbon dioxide	CO_2	44.010	350.7	194.7	—	—	—	—	0.827[e]
31.	Sulfur dioxide	SO_2	64.060	474.0	263.2	88	6.0	356.1	197.7	1.397[e]
32.	Hydrogen chloride	HCl	36.465	339	188.2	925	62.9	286.4	159.0	0.8558
33.	Hydrogen sulfide	H_2S	34.076	383.4	212.9	394.0	26.8	342.8	190.3	0.79[e]
34.	Air		28.966	142.4	79.0	—	—	—	—	0.856[h]
35.	Argon (ref. 2)	Ar	39.944	157.7	87.5	—	—	151.4	84[c]	—
36.	Helium (ref. 2)	He	4.003	8.0	4.3	—	—	< 2.04	< 1.0[c]	—
37.	Hydrogen	H_2	2.016	37.0	20.4	—	—	25.4	14.0	0.07[h]
38.	Oxygen	O_2	32.000	162.6	90.2	—	—	98.2	54.4	1.14[h]
39.	Nitrogen	N_2	28.016	139.6	77.4	—	—	114.0	63.2	0.808[h]
40.	Chlorine	Cl_2	70.914	430.7	239.1	158	10.8	310.2	172.2	1.414
41.	Fluorine (ref. 2)	F_2	38.00	155.4	86.2	—	—	90.6	50.2[c]	1.11[i]
42.	Ammonia	NH_3	17.032	431.8	239.8	212	14.4	352.1	195.5	0.6173
43.	Water	H_2O	18.016	672.0	373.2	0.9492	0.065	492.0	273.2	1.000

[a] Sublimation point. [b] At saturation pressure (Triple point). [c] From reference 3.
[d] Apparent value for methane at 60°F. [e] Saturation pressure and 60°F.
[f] Specific gravity 119°F/60°F (Sublimation point). [g] At 20°C/4°C from reference 3.
[h] Density of liquid g/ml at normal boiling point. [i] At —187°C from reference 3.
[j] Average value from octane numbers of more than one sample. °R = °F + 460. °K = °C + 273.2.

of Common Pure Substances

Specific Gravity Gas (air = 1.0)	Critical Properties						Riedel Factor, α_c (Eq. 2.86)	Acentricity Factor, ω (Eq. 2.10)	Refractive Index, n, at 68°F	Item
	Pressure, P_c		Temperature, T_c		Volume, V_c					
	psia	atm	°R	°K	ft^3/lb	cm^3/g mole				
Ref. 1	Ref. 1		Ref. 1		Ref. 1		Ref. 2	Ref. 2	Ref. 1	
0.5538	673.1	45.8	344.22	191.1	0.0991	99.2	5.86	0.013	—	1.
1.038	709.8	48.3	550.32	305.6	0.0788	147.9	6.28	0.105	—	2.
1.522	617.4	42.0	666.26	370.0	0.0728	200.4	6.54	0.152	—	3.
2.006	550.7	37.5	765.62	425.2	0.0702	254.7	6.77	0.201	1.3326[e]	4.
2.006	529.1	36.0	734.96	408.2	0.0724	262.7	6.71	0.192	—	5.
2.491	489.5	33.3	845.5	469.6	0.0690	310.8	7.03	0.252	1.35748	6.
2.491	483	32.9	829.0	460.4	0.0685	308.5	6.87	0.206	1.35373	7.
2.491	464.0	31.6	781.08	433.8	0.0674	303.6	6.78	0.195	1.342[e]	8.
2.975	440.0	29.9	914.1	507.7	0.0685	368.5	7.27	0.290	1.37486	9.
3.459	396.8	27.0	972.62	540.2	0.0682	426.6	7.27	0.352	1.38764	10.
3.943	362.1	24.6	1023.7	568.6	0.0682	486.3	7.76	0.408	1.39743	11.
4.428	332	22.6	1070.5	594.6	0.0679	543.7	7.94	0.441	1.40542	12.
4.912	304	20.7	1111.9	617.6	0.0679	603.1	8.18	0.586	1.41189	13.
0.9684	742.1	50.5	509.82	283.1	0.0706	123.6	6.18	0.073	—	14.
1.453	667	45.4	657.4	365.1	0.0689	181.0	6.40	0.143	—	15.
1.937	583	39.7	755.6	419.6	0.0689	241.3	6.74	0.203	—	16.
1.937	600	40.8	784.3	435.6	0.0503	176.2	6.74	0.273	—	17.
1.937	600	40.8	771.9	428.7	0.0503	176.2	6.99	0.234	—	18.
1.937	579.8	39.5	752.5	417.9	0.0513	179.7	6.76	0.201	—	19.
2.421	586	39.9	836.9	464.8	0.0672	294.2	7.16	0.238	1.37148	20.
0.8988	905	61.6	557.4	309.5	0.0695	113.0	6.94	0.186	—	21.
2.697	714	48.6	1012.0	562.1	0.0535	260.9	6.83	0.215	1.50112	22.
3.181	590	40.1	1065.5	591.8	0.0564	324.4	7.12	0.279	1.49693	23.
3.665	540	36.7	1111.2	617.2	0.0556	368.5	7.35	0.322	1.49588	24.
1.106	1174.2	79.9	922.97	512.6	0.0589	117.8	8.48	0.556	1.3288	25.
1.590	925.3	63.0	929.58	516.3	0.0580	166.8	8.98	0.635	1.3614	26.
2.0745	749.5	51.0	966.3	536.7	0.0582	218.2	8.85	0.600	—	27.
2.5589	640.7	43.6	1013.6	563.0	0.0593	274.6	8.17	0.596	—	28.
0.9670	507	34.5	240	133.2	0.0532	93.0	6.04	0.041	—	29.
1.519	1071	72.9	547.9	304.3	0.0342	94.0	6.92	0.420	—	30.
2.212	1145	77.9	775.5	430.7	0.0306	122.4	8.55	0.273	—	31.
1.259	1198	81.5	584.5	324.6	0.0208	47.3	6.39	0.133	—	32.
1.176	1306	88.9	672.7	373.6	0.0459	97.6	6.25	0.100	—	33.
1.000	547	37.2	238.7	132.5	0.0517	93.5	—	—	—	34.
1.3790	705.4	48.0	272	151	0.0302	75.2	5.76	−0.002	—	35.
0.1368	33.21	2.26	9.8	5.3	0.2313	57.8	4.74	0	—	36.
0.0696	188.1	12.8	60.2	33.3	0.5167	65.0	4.74	0	—	37.
1.105	736.9	50.1	278.9	154.8	0.0382	76.3	5.92	0.021	—	38.
0.9672	493.0	33.5	227.6	126.3	0.0514	89.9	5.98	0.040	—	39.
2.448	1118.4	76.1	751	417.1	0.0281	124.4	6.18	0.074	—	40.
1.3119	808	55	259.4	144	—	—	6.33	0.115	—	41.
0.5880	1636	111.3	730.3	405.6	0.0681	72.4	7.00	0.250	—	42.
0.6220	3208	218.3	1165.6	647.4	0.0500	56.2	7.39	0.348	1.3330	43.

REFERENCES

1. Natural Gas Processors Suppliers Association *Engineering Data Book*, 8th Edition, 1966.
2. R. C. Reid, and T. K. Sherwood, *The Properties of Gases and Liquids*, McGraw-Hill Book Co., New York, 1966.
3. J. H. Perry, *Chemical Engineers' Handbook*, McGraw-Hill Book Co., New York, 1963.

TABLE A-2

Values[a] of $Z^{(0)}$ for Compressibility Factor Calculation (See Table A-3 for Additional Data in the Region Enclosed by Dotted lines)

T_r	P_r 0.2	0.4	0.6	0.8	1.0	1.2	1.4	1.6	1.8	2.0	2.2	2.4	2.6	2.8	3.0
0.80	0.851	0.066	0.100	0.133	0.164	0.192	0.225	0.258	0.287	0.318	0.347	0.376	0.405	0.433	0.461
0.85	0.882	0.067	0.101	0.134	0.165	0.194	0.226	0.258	0.287	0.316	0.345	0.374	0.403	0.431	0.459
0.90	0.904	0.778	0.102	0.135	0.167	0.198	0.229	0.258	0.288	0.316	0.345	0.373	0.402	0.430	0.458
0.95	0.920	0.819	0.697	0.145	0.176	0.205	0.235	0.262	0.292	0.321	0.347	0.375	0.403	0.430	0.457
1.00	0.932	0.849	0.756	0.638	0.291	0.231	0.250	0.278	0.304	0.331	0.356	0.381	0.407	0.433	0.458
1.05	0.942	0.874	0.800	0.714	0.609	0.470	0.341	0.320	0.332	0.350	0.372	0.393	0.417	0.441	0.466
1.10	0.950	0.893	0.833	0.767	0.691	0.607	0.512	0.442	0.408	0.402	0.410	0.421	0.440	0.462	0.484
1.15	0.958	0.908	0.858	0.805	0.746	0.684	0.620	0.562	0.514	0.484	0.477	0.478	0.485	0.498	0.513
1.20	0.963	0.921	0.879	0.835	0.788	0.737	0.690	0.640	0.598	0.568	0.553	0.545	0.544	0.548	0.554
1.25	0.968	0.930	0.896	0.858	0.820	0.778	0.740	0.702	0.664	0.636	0.618	0.606	0.599	0.597	0.598
1.30	0.971	0.940	0.909	0.878	0.846	0.811	0.780	0.749	0.718	0.691	0.671	0.657	0.619	0.644	0.642
1.40	0.977	0.952	0.929	0.908	0.883	0.859	0.838	0.817	0.795	0.777	0.759	0.745	0.734	0.725	0.720
1.50	0.982	0.963	0.945	0.927	0.909	0.892	0.875	0.859	0.844	0.831	0.819	0.808	0.800	0.794	0.790
1.60	0.985	0.971	0.957	0.944	0.930	0.917	0.904	0.893	0.882	0.872	0.863	0.855	0.848	0.843	0.840
1.70	0.988	0.977	0.966	0.956	0.946	0.936	0.926	0.919	0.911	0.903	0.896	0.889	0.883	0.879	0.875
1.80	0.991	0.982	0.974	0.966	0.958	0.950	0.944	0.937	0.931	0.926	0.921	0.916	0.913	0.910	0.908
1.90	0.993	0.986	0.980	0.974	0.968	0.962	0.958	0.952	0.948	0.944	0.940	0.936	0.933	0.931	0.930
2.00	0.995	0.989	0.984	0.979	0.975	0.971	0.968	0.964	0.961	0.959	0.956	0.954	0.953	0.953	0.952
2.50	1.000	0.999	0.999	0.998	0.998	0.998	0.998	0.997	0.999	1.000	1.001	1.001	1.002	1.004	1.006
3.00	1.001	1.002	1.003	1.004	1.005	1.007	1.008	1.010	1.012	1.014	1.016	1.019	1.022	1.025	1.028
3.50	1.002	1.004	1.006	1.008	1.011	1.013	1.015	1.018	1.020	1.022	1.024	1.027	1.030	1.033	1.036
4.00	1.003	1.005	1.008	1.010	1.013	1.015	1.017	1.020	1.022	1.024	1.026	1.029	1.032	1.035	1.038

[a] Values of $Z^{(0)}$ up to $P_r = 9.0$ taken from K. S. Pitzer, D. Z. Lippmann, R. F. Curl, Jr., C. M. Huggins, and E. D. Peterson, *J. Amer. Chem. Soc.*, **77**, 3433 (1955).

Values for $P_r = 10$ to 14 taken from A. Satter and J. M. Campbell, *Soc. Petrol Engrs. J.* **3**, 333 (1963).

TABLE A-2 (*continued*)

T_r	P_r 3.2	3.4	3.6	3.8	4.0	4.5	5.0	6.0	7.0	8.0	9.0	10	11	12	13	14
0.80	0.490	0.519	0.547	0.576	0.605	0.675	0.746	0.883	1.017	1.15		1.417	1.551	1.686	1.819	1.952
0.85	0.487	0.515	0.542	0.569	0.597	0.663	0.730	0.861	0.990	1.115		1.367	1.493	1.619	1.745	1.871
0.90	0.485	0.512	0.538	0.565	0.591	0.655	0.718	0.842	0.960	1.089	1.21	1.330	1.451	1.573	1.694	1.815
0.95	0.484	0.510	0.536	0.561	0.587	0.647	0.709	0.828	0.947	1.066	1.185	1.303	1.421	1.539	1.657	1.775
1.00	0.484	0.509	0.534	0.557	0.582	0.642	0.702	0.819	0.932	1.048	1.166	1.281	1.398	1.514	1.631	1.748
1.05	0.489	0.512	0.535	0.557	0.580	0.639	0.700	0.814	0.923	1.032	1.147	1.249	1.358	1.467	1.576	1.686
1.10	0.504	0.525	0.547	0.567	0.589	0.643	0.699	0.810	0.916	1.019	1.129	1.224	1.327	1.430	1.533	1.636
1.15	0.529	0.546	0.563	0.581	0.600	0.651	0.705	0.809	0.911	1.008	1.113	1.201	1.299	1.395	1.492	1.589
1.20	0.563	0.574	0.587	0.601	0.618	0.664	0.714	0.810	0.907	1.000	1.100	1.183	1.275	1.365	1.458	1.549
1.25	0.602	0.609	0.618	0.629	0.643	0.682	0.726	0.816	0.907	0.994	1.088	1.172	1.261	1.348	1.435	1.524
1.30	0.642	0.645	0.651	0.659	0.668	0.701	0.740	0.824	0.910	0.992	1.078	1.162	1.248	1.332	1.418	1.504
1.40	0.718	0.718	0.722	0.727	0.734	0.754	0.781	0.844	0.921	0.994	1.071	1.150	1.230	1.311	1.390	1.471
1.50	0.785	0.784	0.784	0.786	0.790	0.805	0.826	0.877	0.934	1.000	1.070	1.141	1.212	1.283	1.354	1.425
1.60	0.836	0.834	0.833	0.834	0.835	0.844	0.860	0.904	0.953	1.010	1.075	1.137	1.200	1.262	1.325	1.388
1.70	0.873	0.872	0.872	0.873	0.874	0.882	0.895	0.930	0.972	1.023	1.082	1.137	1.194	1.251	1.307	1.364
1.80	0.907	0.906	0.906	0.907	0.908	0.914	0.925	0.955	0.993	1.030	1.091	1.141	1.181	1.244	1.293	1.345
1.90	0.929	0.929	0.930	0.932	0.934	0.941	0.950	0.976	1.010	1.051	1.097	1.144	1.190	1.238	1.284	1.330
2.00	0.952	0.953	0.954	0.954	0.956	0.962	0.972	0.996	1.027	1.064	1.106	1.149	1.191	1.234	1.276	1.318
2.50	1.008	1.009	1.012	1.014	1.018	1.026	1.035	1.055	1.079	1.105	1.136	1.167	1.198	1.230	1.261	1.291
3.00	1.030	1.033	1.036	1.038	1.041	1.049	1.058	1.077	1.10	1.124	1.150	1.173	1.198	1.222	1.247	1.271
3.50	1.039	1.042	1.045	1.048	1.051	1.058	1.067	1.086	1.105	1.126	1.148	1.168	1.189	1.210	1.230	1.250
4.00	1.041	1.044	1.047	1.050	1.053	1.060	1.068	1.086	1.104	1.124	1.143	1.160	1.179	1.197	1.216	1.235

TABLE A-3

Values[a] of $Z^{(0)}$ in the Critical Region

T_r	P_r 1.0	1.1	1.2	1.3	1.4	1.5	1.6	1.7	1.8	1.9	2.0
0.98	0.193	0.204	0.217	0.230	0.244	0.257	0.270	0.284	0.299	0.313	0.326
0.99	0.205	0.210	0.223	0.235	0.247	0.260	0.273	0.287	0.301	0.315	0.328
1.00	0.291	0.220	0.231	0.241	0.250	0.265	0.278	0.290	0.304	0.317	0.331
1.01	0.476	0.283	0.243	0.248	0.259	0.271	0.283	0.294	0.307	0.319	0.331
1.02	0.525	0.402	0.273	0.260	0.270	0.278	0.291	0.300	0.311	0.323	0.334
1.03	0.558	0.466	0.34	0.29	0.283	0.288	0.297	0.306	0.316	0.328	0.339
1.04	0.586	0.509	0.41	0.33	0.307	0.302	0.307	0.314	0.324	0.334	0.343
1.05	0.609	0.543	0.470	0.375	0.341	0.324	0.320	0.323	0.332	0.341	0.350
1.06	0.628	0.572	0.505	0.423	0.370	0.349	0.336	0.333	0.343	0.348	0.358
1.07	0.645	0.597	0.534	0.468	0.408	0.379	0.358	0.349	0.356	0.358	0.367
1.08	0.663	0.618	0.562	0.504	0.445	0.412	0.385	0.373	0.370	0.369	0.375
1.09	0.677	0.636	0.587	0.535	0.480	0.443	0.412	0.396	0.387	0.383	0.387
1.10	0.691	0.652	0.607	0.564	0.512	0.473	0.442	0.422	0.408	0.400	0.402
1.11	0.703	0.667	0.625	0.584	0.538	0.502	0.469	0.448	0.428	0.418	0.417
1.13	0.726	0.693	0.658	0.621	0.584	0.549	0.520	0.494	0.472	0.456	0.450
1.15	0.746	0.715	0.684	0.652	0.620	0.589	0.562	0.536	0.514	0.495	0.484

Values of $Z^{(0)}$ Near the Two Phase Region

T_r	P_r 0.4	0.5	0.6	0.7	0.8	0.9	1.0
0.90	0.778	0.701	0.102	0.118	0.135	0.151	0.167
0.91	0.787	0.715	0.104	0.120	0.136	0.152	0.168
0.92	0.796	0.728	0.650	0.122	0.138	0.153	0.169
0.93	0.805	0.740	0.666	0.124	0.140	0.155	0.170
0.94	0.812	0.751	0.681	0.125	0.142	0.157	0.173
0.95	0.819	0.762	0.697	0.612	0.145	0.160	0.176
0.96	0.826	0.772	0.711	0.632	0.149	0.164	0.180
0.97	0.832	0.782	0.724	0.652	0.56	0.170	0.186
0.98	0.838	0.791	0.735	0.669	0.591	0.177	0.193
0.99	0.844	0.800	0.746	0.685	0.616	0.514	0.205
1.00	0.849	0.807	0.757	0.699	0.638	0.554	0.291
1.01	0.854	0.813	0.767	0.713	0.654	0.583	0.476
1.02	0.860	0.820	0.776	0.726	0 672	0.608	0.525
1.03	0.865	0.826	0.784	0.737	0.687	0.630	0.558
1.04	0.870	0.833	0.793	0.748	0.701	0.648	0.586
1.05	0.874	0.838	0.800	0.758	0.714	0.665	0.609

[a] From Pitzer *et al.*, *J. Amer. Chem. Soc.*, **77**, 3433 (1955).

TABLE A–4

Values[a] of $Z^{(1)}$ for Compressibility Factor Calculation
(See Table A-5 for Additional Data in the Region Enclosed by Dotted Lines)

T_r	P_r 0.2	0.4	0.6	0.8	1.0	1.2	1.4	1.6	1.8	2.0	2.2	2.4	2.6
0.80	−0.095	−0.028	−0.044	−0.058	−0.07	−0.08	−0.10	−0.11	−0.12	−0.13	−0.14	−0.15	−0.16
0.85	−0.067	−0.031	−0.049	−0.064	−0.08	−0.09	−0.11	−0.12	−0.13	−0.14	−0.15	−0.16	−0.17
0.90	−0.042	−0.09	−0.053	−0.068	−0.085	−0.10	−0.11	−0.12	−0.13	−0.14	−0.15	−0.16	−0.17
0.95	−0.025	−0.050	−0.10	−0.072	−0.091	−0.10	−0.11	−0.12	−0.12	−0.13	−0.14	−0.15	−0.15
1.00	−0.012	−0.016	−0.020	−0.05	−0.080	−0.090	−0.099	−0.108	−0.115	−0.123	−0.13	−0.13	−0.14
1.05	0.000	+0.001	+0.005	+0.015	+0.02	+0.01	−0.01	−0.04	−0.06	−0.07	−0.08	−0.09	−0.10
1.10	+0.002	0.008	0.016	0.030	0.055	0.082	+0.11	+0.082	+0.035	0.000	−0.02	−0.03	−0.05
1.15	0.004	0.012	0.021	0.040	0.064	0.093	1.12	0.140	0.136	+0.100	+0.07	+0.04	+0.02
1.20	0.006	0.014	0.025	0.043	0.07	0.10	0.13	0.16	0.17	0.17	0.16	0.14	0.12
1.25	0.007	0.016	0.028	0.046	0.07	0.10	0.13	0.16	0.18	0.19	0.19	0.18	0.16
1.30	0.008	0.018	0.031	0.05	0.07	0.10	0.13	0.16	0.18	0.20	0.20	0.20	0.20
1.40	0.012	0.025	0.039	0.05	0.07	0.10	0.13	0.16	0.18	0.19	0.20	0.21	0.21
1.50	0.016	0.032	0.051	0.06	0.08	0 10	0.13	0.15	0.17	0.18	0.20	0.20	0.21
1.60	0.018	0.035	0.051	0.07	0.08	0.10	0.12	0.14	0.16	0.17	0.18	0.19	0.20
1.70	0.018	0.035	0.051	0.07	0.08	0.10	0.11	0.13	0.15	0.16	0.17	0.18	0.19
1.80	0.018	0.035	0.051	0.07	0.08	0.10	0.11	0.13	0.14	0.15	0.17	0.18	0.19
1.90	0.018	0.035	0.051	0.07	0.08	0.10	0.11	0.13	0.14	0.15	0.17	0.18	0.19
2.00	0.018	0.035	0.051	0.07	0.08	0.10	0.11	0.13	0.14	0.15	0.17	0.18	0.19
2.50	0.018	0.035	0.051	0.07	0.08	0.10	0.11	0.13	0.14	0.15	0.17	0.18	0.19
3.00	0.018	0.035	0.051	0.07	0.08	0.10	0.11	0.13	0.14	0.15	0.17	0.18	0.19
3.50	0.018	0.035	0.051	0.07	0.08	0.10	0.11	0.13	0.14	0.15	0.17	0.18	0.19
4.00	0.018	0.035	0.051	0.07	0.08	0.10	0.11	0.13	0.14	0.15	0.17	0.18	0.19

[a] Values of $Z^{(1)}$ up to $P_r = 9.0$ taken from K. S. Pitzer, D. Z. Lippmann, R. F. Curl, Jr., C. M. Huggins, and D. E. Peterson, *J. Amer. Chem. Soc.*, **77**, 3433 (1955).
Values for $P_r = 10$ to 14 taken from A. Satter and J. M. Campbell, *Soc. Petrol. Engrs. J.*, **3**, 333 (1963).

TABLE A-4 (*continued*)

	P_r												
T_r	2.8	3.0	4.0	5.0	6.0	7.0	8.0	9.0	10	11	12	13	14
0.80	−0.17	−0.18	−0.23	−0.26	−0.29	−0.32	−0.35	−0.37	−0.400	−0.430	−0.460	−0.490	−0.520
0.85	−0.18	−0.18	−0.22	−0.25	−0.28	−0.31	−0.34	−0.36	−0.390	−0.420	−0.450	−0.480	−0.510
0.90	−0.17	−0.18	−0.21	−0.24	−0.27	−0.30	−0.32	−0.35	−0.380	−0.410	−0.440	−0.470	−0.500
0.95	−0.16	−0.17	−0.20	−0.22	−0.25	−0.28	−0.31	−0.34	−0.370	−0.400	−0.430	−0.460	−0.490
1.00	−0.14	−0.15	−0.17	−0.20	−0.23	−0.26	−0.30	−0.33	−0.360	−0.390	−0.418	−0.445	−0.470
1.05	−0.10	−0.11	−0.14	−0.17	−0.20	−0.24	−0.28	−0.31	−0.336	−0.361	−0.384	−0.404	−0.424
1.10	0.06	−0.07	−0.10	−0.13	−0.16	−0.21	−0.25	−0.28	−0.307	−0.330	−0.352	−0.371	−0.391
1.15	0.00	−0.01	−0.04	−0.08	−0.12	−0.16	−0.20	−0.24	−0.262	−0.290	−0.318	−0.343	−0.368
1.20	+0.09	+0.07	0.00	−0.04	−0.08	−0.12	−0.16	−0.19	−0.220	−0.248	−0.274	−0.298	−0.320
1.25	0.14	0.12	+0.05	0.00	−0.03	−0.07	−0.11	−0.13	−0.164	−0.190	−0.215	−0.240	−0.261
1.30	0.19	0.18	0.10	+0.04	0.00	−0.04	−0.07	−0.09	−0.108	−0.125	−0.140	−0.157	−0.175
1.40	0.21	0.20	0.15	0.11	+0.07	+0.04	+0.01	−0.01	−0.030	−0.049	−0.065	−0.081	−0.095
1.50	0.21	0.21	0.20	0.17	0.14	0.11	0.09	+0.07	+0.054	+0.036	+0.019	+0.001	+0.017
1.60	0.20	0.21	0.22	0.21	0.19	0.17	0.15	0.14	0.125	0.109	0.094	0.080	0.064
1.70	0.20	0.21	0.24	0.25	0.26	0.25	0.24	0.22	0.201	0.183	0.163	0.145	0.125
1.80	0.20	0.21	0.26	0.29	0.31	0.32	0.32	0.30	0.290	0.270	0.247	0.223	0.199
1.90	0.20	0.21	0.26	0.30	0.35	0.38	0.40	0.40	0.385	0.368	0.346	0.321	0.295
2.00	0.20	0.21	0.26	0.30	0.35	0.40	0.43	0.45	0.454	0.448	0.435	0.416	0.395
2.50	0.20	0.21	0.26	0.30	0.35	0.40	0.45	0.50	0.530	0.537	0.527	0.510	0.487
3.00	0.20	0.21	0.26	0.30	0.35	0.40	0.45	0.50	0.545	0.590	0.601	0.596	0.580
3.50	0.20	0.21	0.26	0.30	0.35	0.40	0.45	0.50	0.545	0.590	0.630	0.670	0.665
4.00	0.20	0.21	0.26	0.30	0.35	0.40	0.45	0.50	0.545	0.590	0.630	0.670	0.709

TABLE A-5

Values[a] of $Z^{(1)}$ in the Critical Region

T_r	P_r 1.0	1.2	1.4	1.6	1.8	2.0
0.98	−0.090	−0.099	−0.109	−0.118	−0.125	−0.130
0.99	−0.087	−0.095	−0.104	−0.114	−0.121	−0.127
1.00	−0.080	−0.090	−0.099	−0.108	−0.115	−0.123
1.01	−0.02	−0.080	−0.091	−0.102	−0.10	−0.100
1.02	−0.01	−0.065	−0.082	−0.095	−0.09	−0.09
1.03	0.00	−0.047	−0.068	−0.085	−0.08	−0.09
1.04	+0.01	−0.025	−0.050	−0.073	−0.07	−0.08
1.05	+0.02	+0.01	−0.01	−0.04	−0.06	−0.07
1.06	+0.03	+0.06	+0.07	−0.02	−0.05	−0.073
1.07	+0.04	+0.08	+0.09	0.000	−0.038	−0.059
1.08	+0.047	+0.08	+0.10	+0.030	−0.015	−0.041
1.09	+0.050	+0.08	+0.11	+0.056	+0.012	−0.022
1.10	+0.055	+0.082	+0.11	+0.082	+0.035	0·000
1.11	+0.057	+0.085	+0.12	+0.099	+0.062	+0.020
1.13	+0.062	+0.089	+0.12	+0.123	+0.105	+0.060
1.15	+0.064	+0.093	+0.122	+0.140	+0.136	+0.100

Values of $Z^{(1)}$ Near the Two-Phase Region

T_r	P_r 0.4	0.6	0.8	1.0
0.90	−0.09	−0.053	−0.068	−0.085
0.91	−0.08	−0.053	−0.069	−0.087
0.92	−0.072	−0.18	−0.070	−0.089
0.93	−0.066	−0.15	−0.071	−0.090
0.94	−0.058	−0.12	−0.072	−0.091
0.95	−0.050	−0.10	−0.072	−0.091
0.96	−0.042	−0.08	−0.072	−0.091
0.97	−0.035	−0.065	−0.14	−0.091
0.98	−0.027	−0.050	−0.11	−0.090
0.99	−0.021	−0.033	−0.08	−0.087
1.00	−0.016	−0.020	−0.05	−0.080
1.01	−0.012	−0.012	−0.02	−0.02
1.02	−0.008	−0.006	0.00	−0.01
1.03	−0.005	−0.001	+0.005	0.00
1.04	−0.002	+0.002	+0.010	+0.01
1.05	+0.001	+0.005	+0.015	+0.02

[a] From Pitzer *et al.*, *J. Amer. Chem. Soc.*, **77**, 3433 (1955).

TABLE A-6

Constants[a] for the Alani and Kennedy Equation

Substance	K	n	$m \times 10^4$	C
Methane 70°–300°F	9,160.6413	61.893223	3.3162472	0.50874303
Methane 301°–460°F	147.47333	3,247.4533	−14.072637	1.8326695
Ethane 100°–249°F	46,709.573	−404.48844	5.1520981	0.52239654
Ethane 250°–460°F	17,495.343	34.163551	2.8201736	0.62309877
Propane	20,247.757	190.24420	2.1586448	0.90832519
i-Butane	32,204.420	131.63171	3.3862284	1.1013834
n-Butane	33,016.212	146.15445	2.902157	1.1168144
n-Pentane	37,046.234	299.62630	2.1954785	1.4364289
n-Hexane	52,093.006	254.56097	3.6961858	1.5929406
n-Heptane	82,295.457	64.380112	5.2577968	1.7299902
n-Octane	89,185.432	149.39026	5.9897530	1.9310993
n-Nonane	124,062.650	37.917238	6.7299934	2.1519973
Decane	146,643.830	26.524103	7.8561789	2.3329874

Note: The constants for methane and ethane refer to the gaseous state. They should not be used to calculate the densities of the pure liquids but are applicable to the calculation of the density of mixtures containing methane and ethane.

Units: P = psia; T = °R; v_m = molal volume, ft³/lb mole;

$$R = 10.7335 \ \frac{\text{lb ft}^3}{\text{in.}^2 \ ^\circ\text{R lb mole}}.$$

[a] G. H. Alani and H. T. Kennedy, *Petrol. Trans., A.I.M.E.*, **219**, 288 (1960).

TABLE A-7

Reduced Liquid Volumes for the Lyckman *et al.* Equation[a]

T_r	$V_r^{(0)}$	$V_r^{(1)}$	$V_r^{(2)}$
0.560	0.3814	−0.1372	0.1321
0.570	0.3828	−0.1298	0.1133
0.580	0.3844	−0.1226	0.0949
0.590	0.3861	−0.1157	0.0771
0.600	0.3879	−0.1090	−0.0598
0.610	0.3899	−0.1025	0.0430
0.620	0.3920	−0.0962	0.0266
0.630	0.3943	−0.0901	0.0108
0.640	0.3967	−0.0842	−0.0046
0.650	0.3992	−0.0784	−0.0196
0.660	0.4019	−0.0729	−0.0340
0.670	0.4047	−0.0676	−0.0480
0.680	0.4076	−0.0625	−0.0616
0.690	0.4107	−0.0575	−0.0746
0.700	0.4140	−0.0528	−0.0872
0.710	0.4174	−0.0482	−0.0994
0.720	0.4209	−0.0439	−0.1110
0.730	0.4247	−0.0397	−0.1221
0.740	0.4285	−0.0358	−0.1327
0.750	0.4326	−0.0320	−0.1428
0.760	0.4368	−0.0285	−0.1523
0.770	0.4412	−0.0252	−0.1612
0.780	0.4459	−0.0222	−0.1695
0.790	0.4507	−0.0194	−0.1772
0.800	0.4558	−0.0169	−0.1841
0.810	0.4611	−0.0146	−0.1903
0.820	0.4667	−0.0127	−0.1957
0.830	0.4726	−0.0111	−0.2003
0.840	0.4788	−0.0098	−0.2038
0.850	0.4854	−0.0090	−0.2063
0.860	0.4924	−0.0086	−0.2076
0.870	0.4998	−0.0087	0.2076
0.880	0.5078	−0.0094	−0.2060
0.890	0.5164	−0.0107	−0.2027
0.900	0.5257	−0.0128	−0.1972
0.910	0.5359	−0.0159	−0.1893
0.920	0.5472	−0.0200	−0.1782
0.930	0.5598	−0.0256	−0.1633
0.940	0.5742	−0.0331	−0.1433
0.950	0.5910	−0.0431	−0.1165
0.960	0.6113	−0.0566	−0.0797
0.970	0.6370	−0.0759	−0.0274
0.980	0.6727	−0.1056	0.0533
0.990	0.7327	−0.1605	0.2032
1.000	1.0000	0	0

[a] E. W. Lyckman, C. A. Eckert, and J. M. Prausnitz, *Chem. Eng. Sci.*, **20**, 703 (1965).

TABLE A-8

Values of the Constant *n* in the Goldhammer Equation (Eq. 2.26)[a]

Alcohol and water	0.25
Hydrocarbons and ethers	0.29
All other organic compounds	0.31
All inorganic compounds except water	0.33

[a] S. H. Fishtine, *Ind. Eng. Chem. Fundam.*, **2**, 149 (1963).

TABLE A-9

Typical Liquid Compressibilities[a]

Substance	Temperature, °C	Pressure, atm	Compressibility $\times 10^6$, atm^{-1}
Methyl alcohol	15	22.7	104.357
	20	197.4	96.251
	20	394.8	81.054
	20	493.5	65.856
	20	987.0	54.711
	20	11,844.0	8.105
Ethyl alcohol	14	22.7	101.317
	20	493.5	63.830
	20	987.0	54.711
	20	11,844.0	8.105
Propyl alcohol (*n*)	20	197.4	78.014
	20	394.8	67.882
	20	493.5	65.856
	20	987.0	47.619
	20	11,844.0	7.092
Butyl alcohol (iso)	18	7.896	98.278
	20	197.4	82.067
	20	394.8	64.843
	20	493.5	56.738
	20	987.0	46.606
	20	11,844.0	8.105
Hexane	20	197.4	118.541
	20	394.8	92.199
Benzene	17	4.935	90.172
	20	197.4	78.014
	20	394.8	67.882
Toluene	20	197.4	74.975
	20	394.8	64.843
Xylene (meta)	20	197.4	69.909
	20	394.8	60.790
Kerosene	20	493.5	55.724
	20	987.0	45.593
	20	11,844.0	8.105
Mercury	20	296.1	4.002
	22	493.5	4.022
	22	987.0	3.961
	22	11,844.0	2.401
Water	20	12.831	49.645
	20	197.4	43.566
	20	394.8	41.540
	20	493.5	39.514
	40	493.5	38.500
	40	987.0	33.435
	40	11,844.0	9.119

[a] Modified after Table 106, *Smithonian Physical Tables*, Smithonian Institution, Washington, D.C., 1933.

TABLE A-10

Gas-Viscosity Temperature Function for the Bromley and Wilke Equation[a]

$(1.33T_r)$	$f(1.33T_r)$	$(1.33T_r)$	$f(1.33T_r)$	$(1.33T_r)$	$f(1.33T_r)$
0.30	0.1969	1.65	1.0174	4.0	2.0719
0.35	0.2252	1.70	1.0453	4.1	2.1090
0.40	0.2540	1.75	1.0729	4.2	2.1457
0.45	0.2834	1.80	1.0999	4.3	2.1820
0.50	0.3134	1.85	1.1264	4.4	2.2180
0.55	0.3440	1.90	1.1529	4.5	2.2536
0.60	0.3751	1.95	1.1790	4.6	2.2888
0.65	0.4066	2.00	1.2048	4.7	2.3237
0.70	0.4384	2.1	1.2558	4.8	2.3583
0.75	0.4704	2.2	1.3057	4.9	2.3926
0.80	0.5025	2.3	1.3547	5.0	2.4264
0.85	0.5346	2.4	1.4028	6.0	2.751
0.90	0.5666	2.5	1.4501	7.0	3.053
0.95	0.5985	2.6	1.4962	8.0	3.337
1.00	0.6302	2.7	1.5417	9.0	3.607
1.05	0.6616	2.8	1.5861	10	3.866
1.10	0.6928	2.9	1.6298	20	6.063
1.15	0.7237	3.0	1.6728	30	7.880
1.20	0.7544	3.1	1.7154	40	9.488
1.25	0.7849	3.2	1.7573	50	10.958
1.30	0.8151	3.3	1.7983	60	12.324
1.35	0.8449	3.4	1.8388	70	13.615
1.40	0.8744	3.5	1.8789	80	14.839
1.45	0.9036	3.6	1.9186	90	16.010
1.50	0.9325	3.7	1.9576	100	17.137
1.55	0.9611	3.8	1.9962	200	26.80
1.60	0.9894	3.9	2.0343	400	41.90

[a] J. H. Perry, *Chemical Engineers' Handbook*, McGraw-Hill, New York, 1963. Used with permission of McGraw-Hill Book Co.

TABLE A-11

Liquid Molal Volume at the Normal Boiling Point[a]

Substance	v_b, cm^3/g mole
Methane	37.7
Propane	74.5
Heptane	162
Cyclohexane	117
Ethylene	49.4
Acetylene	42.0
Benzene	96.5
Fluorobenzene	102
Bromobenzene	120
Chlorobenzene	115
Iodobenzene	130
Methanol	42.5
n-Propyl alcohol	81.8
Dimethyl ether	63.8
Ethylpropyl ether	129
Acetone	77.5
Acetic acid	64.1
Isobutyric acid	109
Methyl formate	62.8
Ethyl acetate	106
Diethylamine	109
Acetonitrile	57.4
Methyl chloride	50.6
Carbon tetrachloride	102
Dichlorodifluoromethane	80.7
Ethyl mercaptan	75.5
Diethyl sulfide	118
Phosgene	69.5
Ammonia	25.0
Chlorine	45.5
Water	18.7
Hydrochloric acid	30.6
Hydriodic acid	45.6
Sulfur dioxide	43.8

[a] From R. C. Reid and T. K. Sherwood, *The Properties of Gases and Liquids*, 2nd Ed., McGraw-Hill, New York, 1966. Used with permission of McGraw-Hill Book Co.

TABLE A-12

Structural Contributions to Calculate the Constant *B* in Equation 2.54[a]

C	−0.462
H	+0.249
O	+0.054
Cl	+0.340
Br	+0.326
I	+0.335
Double bond	+0.478
C_6H_5	+0.385
S	+0.043
CO	+0.105 (ketones and esters)
CN	+0.381 (nitriles)

[a] L. H. Thomas, *J. Chem. Soc.*, Part **2,** 573 (1946).

TABLE A-13

Group Contributions to the Parachor, [*P*][a]

CH_2 in—$(CH_2)_n$:		
$n < 12$		40.0
$n > 12$		40.3[b]
C		9.0
H		15.5
in OH		10.0
in HN		12.5
O		19.8
O_2 in esters		54.8
N		17.5
S		49.1
P		40.5
F		26.1
Cl		55.2
Br		68.0
I		90.3
Se		63
Si		31
Al		55
Sn		64.5
As		54
Ethylenic bond:		
Terminal		19.1
2,3 position		17.7
3,4 position		16.3
Triple bond		40.6
Carbonyl bond in ketones:		
Total no. of C atoms =	3	22.3
	4	20.0
	5	18.5
	6	17.3
	7	17.3
Total no. of C atoms =	8	15.1
	9	14.1
	10	13.0
	11	12.6
Single bond		0.0
Semipolar bond		0.0
Singlet linkage		−9.5
Hydrogen bridge		−14.4[c]
Chain branching, per branch		−3.7[d]
Secondary-secondary adjacency		−1.6
Secondary-tertiary adjacency		−2.0
Tertiary-tertiary adjacency		−4.5
Alkyl groups:[e]		
1-Methylethyl		133.3
1-Methylpropyl		171.9
1-Methylbutyl		211.7
2-Methylpropyl		173.3
1-Ethylpropyl		209.5
1,1-Dimethylethyl		170.4
1,1-Dimethylpropyl		207.5
1,2-Dimethylpropyl		207.9
1,1,2-Trimethylpropyl		243.5
Position differences in benzene:		
Ortho-meta		1.8–3.4
Meta-para		0.2–0.5
Ortho-para		2.0–3.8
Ring closure:		
3-membered ring		12.5
4-membered ring		6.0
5-membered ring		3.0
6-membered ring		0.8
7-membered ring		4.0

[a] O. R. Quayle, *Chemical Reviews*, **53**, 439 (1953), copyright (1953) by the American Chemical Society. Reprinted by permission of the copyright owner.

[b] Somewhat greater beyond 20—CH_2—.

[c] Includes any ortho effect.

[d] Varies from −2.1 to −6.5.

[e] Complete group, including branching.

TABLE A-14

Group Contributions to the Molar Refraction, $[R_D]$[a]

C	2.418
H	1.100
$-CH_2-$	4.618
O (hydroxyl)	1.525
O (ether)	1.643
O (carboxyl)	2.211[b]
COO—(ester)	6.154
Cl	5.967
Br	8.865
I	13.90
N (primary amine)	2.322
N (secondary amine)	2.502
N (tertiary amine)	2.840
N (nitrile)	5.516
3-membered ring	0
4-membered ring	0
5-membered ring	0
6-membered ring	0
7-membered ring	0
Double bond	1.733
Semipolar double bond	1.733
Triple bond	2.398

[a] From R. C. Reid and T. K. Sherwood, *The Properties of Gases and Liquids*, 2nd Ed., McGraw-Hill, New York, 1966. Used by permission of McGraw-Hill Book Co.

[b] Includes allowance for double bond.

TABLE A-15

Constant[a] for the Eykman Equation

Substance	C, cm^3/g
Benzene	0.75000
Toluene	0.75447
Ethylbenzene	0.75292
Isopropylbenzene	0.75101
Styrene	0.78910
p-Bromoethylbenzene	0.53105
p-Isopropylchlorobenzene	0.65852
o-Chlorostyrene	0.66972
n-Hexane	0.76093
n-Heptane	0.75738
p-Ethyl phenyl ethyl alcohol	0.69962
Water	0.44907
Ethyl alcohol	0.61376

[a] R. R. Dreisbach, *Ind. Eng. Chem.*, **40**, 2269 (1948).

TABLE A-16

Constants[a] for the Pelofsky Equation

Substance	*A*, g/cm	*B*, g/cm sec
Ethane	−0.1391	36.61
Propane	−0.1696	34.64
n-Butane	−0.1734	34.06
n-Pentane	−0.1865	35.45
n-Hexane	−0.1851	33.23
Benzene	−0.1916	38.87
Ethylbenzene	−0.1937	38.65
n-Propylbenzene	−0.2092	36.98
Aniline	−0.2505	45.46
Toluene	−0.1913	39.40
Phenol	−0.3119	41.63
Nitrobenzene	−0.2705	50.32
Iodobenzene	−0.2787	46.67
Bromobenzene	−0.2502	45.22
Chlorobenzene	−0.2216	43.80
o-Xylene	−0.1975	38.15
m-Xylene	−0.1936	39.17
p-Xylene	−0.1809	37.16
Acetone	−0.1871	42.19
Methyl propyl ketone	−0.1867	36.52
Dimethyl ketone	−0.1816	36.70
Water	−0.08366	79.10
Methyl alcohol	−0.1502	29.29
Ethyl alcohol	−0.1955	26.49
n-Propyl alcohol	−0.2691	26.93
n-Butyl alcohol	−0.2215	26.65

[a] A. H. Pelofsky, *J. Chem. Eng. Data*, **11**, 394 (1966).

TABLE A-17

Values[a] of the Constant *a* in the Meissner and Michaels Equation

Compound	Carbon Atoms in Molecule	$a \times 10^4$
Propionic acid	3	26
n-Propyl alcohol	3	26
Isopropyl alcohol	3	26
Methyl acetate	3	26
n-Propyl amine	3	19
Methyl ethyl ketone	4	19
n-Butyric acid	4	7.0
Isobutyric acid	4	7.0
n-Butyl alcohol	4	7.0
Isobutyl alcohol	4	7.0
Propyl formate	4	8.5
Ethyl acetate	4	8.5
Methyl propionate	4	8.5
Diethyl ketone	5	8.5
Ethyl propionate	5	3.1
Propyl acetate	5	3.1
n-Valeric acid	5	1.7
Isovaleric acid	5	1.7
n-Amyl alcohol	5	1.7
Isoamyl alcohol	5	1.7
Propyl propionate	6	1.0
n-Caproic acid	6	0.75
n-Heptanoic acid	7	0.17
n-Octanoic acid	8	0.034
n-Decanoic acid	10	0.0025

[a] H. P. Meissner and A. S. Michaels, *Ind. Eng. Chem.*, **41**, 2782 (1949).

Index

www.ingramcontent.com/pod-product-compliance
Ingram Content Group UK Ltd.
Pitfield, Milton Keynes, MK11 3LW, UK
UKHW020700041026
15590UKWH00001B/374

9 781555 631390